Brian L Bayne.

April 1981.

Analysis of Marine Ecosystems

Analysis of Marine Ecosystems

A. R. LONGHURST

Bedford Institute of Oceanography
Dartmouth, Nova Scotia

1981

ACADEMIC PRESS

A Subsidiary of Harcourt Brace Jovanovich, Publishers

London   New York   Toronto   Sydney   San Francisco

ACADEMIC PRESS INC. (LONDON) LTD
24/28 Oval Road
London NW1 7DX

United States Edition published by

ACADEMIC PRESS INC.
111 Fifth Avenue
New York, New York 10003

British Library Cataloguing in Publication Data

Analysis of marine ecosystems.
1. Marine ecology
I. Longhurst, A R
574.5′2636   QH541.5.S3   80-41229
ISBN 0-12-455560-8

Filmset in 'Monophoto' Times New Roman by
Eta Services (Typesetters) Ltd., Beccles, Suffolk
Printed in Great Britain by Galliard Printers Ltd., Norfolk

List of Contributors

Richard T. Barber Duke University Marine Laboratory, Beaufort, North Carolina 28516, USA

Brian L. Bayne Natural Environment Research Council, Institute for Marine Environmental Research, Plymouth PL1 3DH, England

Maurice Blackburn 741 Washington Way, Friday Harbour, Washington 98250, USA

Carl M. Boyd Department of Oceanography, Dalhousie University, Halifax, Nova Scotia, Canada B3H 4J1

Robert J. Conover Marine Ecology Laboratory, Bedford Institute of Oceanography, Dartmouth, Nova Scotia, Canada B2Y 4A2

David H. Cushing Fisheries Laboratory, Ministry of Agriculture, Fisheries and Food, Lowestoft NR33 0HT, Suffolk, England

Richard W. Eppley Institute of Marine Resources, Scripps Institution of Oceanography, University of California, San Diego, La Jolla, California 92037, USA

Felix Favorite Northwest and Alaska Fisheries Center, National Marine Fisheries Service, Seattle, Washington 98112, USA

Glen Harrison Marine Ecology Laboratory, Bedford Institute of Oceanography, Dartmouth, Nova Scotia, Canada B2Y 4A2

Alex R. Hiby Sea Mammal Research Unit, British Antarctic Survey, Cambridge CB3 0ET, England

Ian R. Joint Natural Environment Research Council, Institute for Marine Environmental Research, Plymouth PL1 3DH, England

Taivo Laevastu Northwest and Alaska Fisheries Center, National Marine Fisheries Service, Seattle, Washington 98112, USA

John B. Lewis Redpath Museum, McGill University, Montreal, Canada H3A 2K6

Denby S. Lloyd Institute of Marine Science, University of Alaska, Fairbanks, Alaska 99701, USA

Alan R. Longhurst Bedford Institute of Oceanography, Department of Fisheries and Oceans, Dartmouth, Nova Scotia, Canada B2Y 4A2

C. Peter McRoy Institute of Marine Science, University of Alaska, Fairbanks, Alaska 99701, USA

Michael N. Moore Natural Environment Research Council, Institute for Marine Environmental Research, Plymouth PL1 3DH, England

Takahisa Nemoto Ocean Research Institute, University of Tokyo, Nakano-ku, Tokyo 164, Japan

Robert W. Owen Southwest Fisheries Center, National Marine Fisheries Service, La Jolla, California 92037, USA

Philip J. Radford Natural Environment Research Council, Institute for Marine Environmental Research, Plymouth PL1 3DH, England

G. Carleton Ray Department of Environmental Sciences, University of Virginia, Clark Hall, Charlottesville, Virginia 22903, USA

Gilbert T. Rowe Oceanographic Sciences Division, Brookhaven National Laboratory, Upton, New York 11973, USA

John R. Sargent Natural Environment Research Council, Institute of Marine Biochemistry, Aberdeen AB1 3RA, Scotland

William L. Silvert Marine Ecology Laboratory, Bedford Institute of Oceanography, Dartmouth, Nova Scotia, Canada B2Y 4A2

Robert L. Smith Oregon State University, Corvallis, Oregon 97365, USA

Yu. I. Sorokin Southern Department, Institute of Oceanology, Academy of Sciences of USSR, Gelendzhik-7, Krasnodar 383470, USSR

Grover C. Stephens Department of Developmental and Cellular Biology, University of California, Irvine, California 92664, USA

Mikhail E. Vinogradov P. P. Shirshov Institute of Oceanology, Academy of Sciences of the USSR, Moscow 117218, USSR

John J. Walsh Oceanographic Sciences Division, Brookhaven National Laboratory, Upton, New York 11973, USA

Kevin J. Whittle Torry Research Station, Ministry of Agriculture, Fisheries and Food, Aberdeen AB1 3RA, Scotland

Bernt Zeitzschel Institute für Meereskunde an der Universität Kiel, Düsternbrooker Weg 20, 23 Kiel, Federal German Republic

Preface

During the last ten years the new legislative procedures for resource management and environmental protection have had to be increasingly supported by evidence from marine ecology; what is required in drafting or responding to an environmental impact statement goes far beyond the established practices of applied ecologists in their management of fisheries and wildlife. The extent to which marine ecology is sufficiently evolved as a predictive science to respond satisfactorily to the new demands placed upon it is far from clear; at times it has seemed that the questions posed during an assessment of ecological consequences have been irrelevant, in the sense that there has seemed to be no way in which they could be answered objectively and quantitatively at the present time.

It is certainly hard to accept that our present level of understanding of the dynamics of ecosystem processes and their response to stress is sufficient to serve us well in the long term; it is important to remember not only that ecology as a quantitative science is still young, but that the formal application of ecology to environmental impact assessment is less than ten years old. We have to assume that in the future not only will our present day ecological concepts be replaced, but also that our application of them in the whole field of applied ecology will seem naïve and primitive.

To measure our progress from the observational and qualitative ecology of the relatively recent past towards the predictive, quantitative ecology which is already demanded but not yet available, it is occasionally good to sit back for a synoptic view of the current status of marine ecology. Only by measuring our rate of progress in some such manner can we hope to retain credibility for the view (which must be shared by the majority of marine ecologists) that continued support should be given to the long-term aspects of ecological research in the coming decades so as to avoid a future in environmental management constrained by scientific undercapitalisation.

This volume is intended to provide such a synoptic view of the status of marine ecology at the turn of the decade, and more especially those aspects of the subject that are concerned with the analysis of ecosystem function; though we have avoided extensive direct discussion of the application of ecology to the problems of the day, we hope that our volume will be useful to applied ecologists and resource managers, by making clearer to them our present level of uncertainty about ecological processes, and our current rate of progress towards deeper understanding.

In fact, this book had two origins. The first was an argument during a rainy walk on Dartmoor several years ago with an oceanographer friend from Woods Hole who tried to convince me that marine ecology had made disappointingly slow progress over the couple of decades that he and I had been involved with it. The second was an invitation from Academic Press to put together a collection of invited chapters to form a modern text on marine ecology. I welcomed the opportunity to develop at some length my conviction that, in fact, the progress of marine ecology has been consistent and rapid over the last two decades.

In soliciting contributions to this volume an attempt was made to review most aspects of marine ecology as it is practised currently, though without the intention of achieving a comprehensive coverage. Though the main emphasis of the volume is on the analysis of marine ecosystems by numerical methods, some attention has been given to reviewing our current concepts of how a series of exemplary ecosystems function as units; I have taken this approach, and placed these chapters prominently in the first section of the book, because I think that one of the principal failures of marine ecologists in recent years has been to become so intellectually involved in the pursuit of processes, and the measurement of rates as to neglect synthesis of their findings into quantitative descriptions of ecosystems. This is perhaps an inevitable consequence of the fact that graduate students commonly receive little or no training in the intellectual processes required for integration and synthesis: if fledgling ecologists were taught how to think like engineers, as well as how to exercise their critical faculties, we might make faster progress in putting our analyses all together.

This book, then, is divided into three parts. In Part 1 we examine current concepts of how some selected ecosystems function, our selection being biassed in two ways: the examples are mostly concerned with the pelagic realm and with ecosystems of the oceans. There is a deliberate under-representation of the benthic realm and of littoral ecosystems. This reflects an opinion that not only are our quantitative studies of benthic ecosystems less complete than those of plankton but also that in benthic and littoral communities it may be harder to draw conclusions capable of such wide extrapolation as in pelagic systems: local topographic effects are notoriously difficult to filter out.

In Part 2 we examine the present status of our understanding of a series of selected ecological processes, with emphasis placed on those which seem to have attracted most attention from marine ecologists during the last decade, and on those which seem likely to attract attention in the future because new concepts, or challenges to classical concepts, promise to emerge. We have covered a very wide range of processes that regulate the interrelations between organisms and control their numbers or locations; in so doing we have covered not only population ecology but also such questions as predator strategies, and biochemical processes internal to marine invertebrates. Again, we have not sought for comprehensiveness, but rather to illustrate the nature of experimental marine ecology at the start of a new decade.

In Part 3, some of the field, microcosm and numerical techniques currently used in the analysis of marine ecosystems are examined and evaluated; again, we do not seek to be comprehensive and the reader will find little reference to instrumental measuring techniques or to such matters as the use of optical and acoustic remote sensing for the assessment of populations. This bias stems from an opinion that the greatest lack of understanding in ecology is the measurement of rate functions and that the manipulation of the biota to be measured contains much greater uncertainties than the design and calibration of probes and instruments.

Obviously, there are dangers in presenting a text which is as wide as ours but is not truly comprehensive; not everybody will agree with my bias in seeking contributions, and other arrangements would certainly have been possible. Perhaps my own greatest concern is that our choice of a series of ecosystems in Part 1 has inevitably resulted in emphasis on places in the ocean characterised by dynamic processes; in retrospect, I think our book, and marine ecology as a whole, has neglected to place enough emphasis on the more stable and less 'interesting' areas which, in fact, make up the greater part of the oceans.

Finally, there is a sense in which, after editing the book, I have to admit that my argumentative and sceptical friend was correct; I am once again struck by the way in which the work of some of the early ecologists has a continuing sense of freshness, and continuity with current concepts. If you re-read the work of Petersen, Hensen, Kofoid, Harvey, Russell or Hardy, and add the jargon of the present day to their simple language, you may wonder if we have come so very far conceptually since their work. However, much progress has in fact been made in recent decades, despite the chronic inability of marine ecologists to state their larger objectives so clearly or with such general agreement as can other marine scientists. This is especially true in comparison with marine geophysics whose exploitation of the break-through into plate tectonic concepts that occurred in the early sixties has no parallel in marine ecology; indeed, it has been one of our mistakes in recent years when planning very long-term programmes such as the International Decade of Ocean Exploration to seek such a parallel driving force in marine ecology, and to believe that its lack denoted that something was seriously wrong with our subject. It has been my principal intention, in inviting authors to contribute to this book, to illustrate the contrary case: that marine ecology is well and healthy, and proceeding much faster over a very broad front than is sometimes apparent.

It will also be clear to our readers that this progress has not occurred without dissent. I have not attempted to conceal this fact in this volume, and there are several instances where disagreement between authors is evident. In some such cases – where the difference is not fundamental – I have tried to achieve a compromise by bringing the authors together in correspondence. In other cases, where the disagreement appears to me to be fundamental, and to challenge established concepts, I have rather encouraged it than tried to find a compromise.

It is a very great pleasure to acknowledge the readiness of our large team of authors to conform so willingly to the general pattern that I set out in my original chapter-synopses for the book. I apologise to them for what I have done to their texts in the cause of conformity, and I am grateful for their readiness to accede to my very Draconian deadlines. I think that the up-to-date text was worth all the rush that I imposed. I am also grateful to Sylvia Smith for her patience in making sense out of my disorganised editorial procedures, and to Olive Ross for processing my almost wholly unreadable corrections to manuscripts. I am also grateful to my employer, the Canadian Department of Fisheries and Oceans, for permission to undertake the work of editing the book.

November, 1980   A. R. LONGHURST

Contents

Part 1   Current Concepts of Marine Ecosystems

This section comprises reviews of current understanding of the trophic and energetic relationships within a range of exemplary marine ecosystems. The general community composition of the biota in each ecosystem, and its area of occurrence, is taken as understood by the reader, or else restricted to a very short introductory section.

6 Shelf-Sea Ecosystems   159

JOHN J. WALSH

Part 2   Functions within Ecosystems

This section comprises reviews of some important processes which are common to many marine ecosystems and have been widely studied in recent years.

13 The Role of Large Organisms   **397**

G. CARLETON RAY

14 Significance of Spatial Variability   **415**

ALAN R. LONGHURST

15 Temporal Variability in Production Systems   **443**

DAVID H. CUSHING

16 Comparative Function and Stability of Macrophyte-based Ecosystems 473

C. PETER McROY and DENBY S. LLOYD

17 Lipids and Hydrocarbons in the Marine Food Web 491

JOHN R. SARGENT and KEVIN J. WHITTLE

18 Elemental Accumulation in Organisms and Food Chains   535

MICHAEL N. MOORE

Part 3 Simulation and Experimental Studies of Marine Ecosystems

These chapters are devoted to studies of marine ecosystems by the use of numerical simulation, by the use of microcosms, and by the manipulation of natural ecosystems.

Part 1

Current Concepts of Marine Ecosystems

This section comprises reviews of current understanding of the trophic and energetic relationships within a range of exemplary marine ecosystems. The general community composition of the biota in each ecosystem, and its area of occurrence, is taken as understood by the reader, or else restricted to a very short introductory section.

1 Low Latitude Gyral Regions

MAURICE BLACKBURN

1 Introduction

In the top 500 m of the Atlantic and Pacific, a huge anticyclonic gyre or group of gyres occupies most of the area between the equatorial divergence and the eastward mid-latitude currents, in each hemisphere. Each gyre is a few thousand km in its largest dimension. The gyres become smaller in area below 500 m. A homologous gyre exists at similar latitudes in the South Indian Ocean. These features are probably ecosystem habitats or contain such habitats, and most of this chapter is devoted to them. Other low-latitude gyres in open ocean waters are probably not ecosystem habitats and are considered only briefly. They are the cyclonic and anticyclonic gyres of the eastern tropical Pacific, Atlantic and Indian Oceans, and the seasonally reversing monsoon gyre of the Indian Ocean. Within this system of large and moderate sized gyres there is a profusion of temporary smaller ones, which are generally called eddies or rings. Their importance in the anticyclonic gyre systems is briefly discussed. The great anticyclones are often called subtropical or central gyres, and those terms are sometimes used here, although 'subtropical' is not strictly correct.

2 The major anticyclonic gyres

PHYSICAL BACKGROUND

Until quite recently, most authors recognised only one great anticyclone in each ocean, i.e. in the North and South Atlantic, North and South Pacific and South Indian Ocean. Each gyre was assumed to be bounded by a westward equatorial current, a poleward western boundary current, an eastward mid-latitude current, and an eastern boundary current flowing towards the Equator. These five gyres appear on many charts as rather smooth features, but it now appears that they are much more complicated. Figure 1 shows the mean annual dynamic topography of the sea surface in the Pacific, from Wyrtki (1974, 1975). In addition to the expected features, this chart reveals a zonal ridge in the topography between 20° and 25°N in the western Pacific, indicating an eastward current. This current was first recognised by Yoshida and Kidokoro (1967). Another zonal ridge, not

3

shown in Fig. 1, was found by Hasunuma and Yoshida (1978) at about 17°N in the western Pacific. Those authors believed that the area to the east of the Kuroshio, previously considered as one anticyclonic gyre, actually has three gyres. There is a similar eastward current at about 10°S in the western Pacific, first described by Reid (1959, 1961). It appears to split the South Pacific anticyclone, since it has been traced right across the ocean. This current is recognisable only at 9°S, 170°E in Fig. 1, but it appears more clearly in similar charts for certain months (Wyrtki, 1974, 1975). Yoshida and Kidokoro (1978) thought that similar features exist in some of the major anticyclones of other oceans.

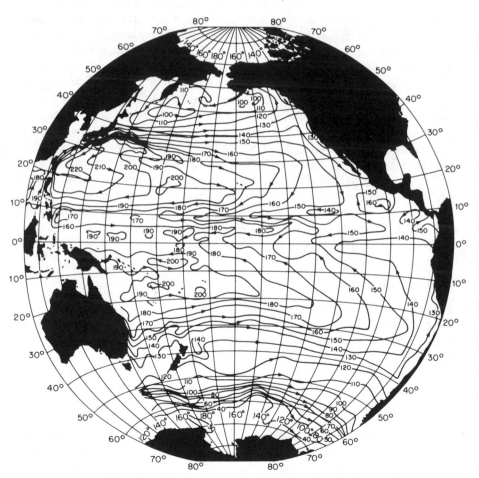

FIG. 1 Mean annual dynamic topography (dyn. cm) of the sea surface relative to 1000 db in the Pacific. (From Wyrtki, 1975)

The North Pacific anticyclone may also be split meridionally, as indicated by Sverdrup et al. (1942). Many later investigators did not support that view, but some recent observations are consistent with it (Shulenberger, 1978). Worthington (1976) distinguished two large anticyclones in the North Atlantic, the one usually recognised and another between 40° and 50°N. The South Pacific gyre has no true western boundary current, but a series of anticyclonic eddies (Hamon, 1965; Nilsson et al., 1977).

These observations, and the rapidly growing literature on mesoscale oceanic eddies, show that the anticyclone regions are physically more complex than previously thought. Some present conceptions about the biology of the regions may change as additional physical and biological studies are made.

The general physico-chemical features of the major anticyclones were summarised by Burkov and Monin (1973) and Reid et al. (1978). The gyres are warm, clear, and convergent. Surface current velocities are up to 250 cm s^{-1} in western boundary currents and much lower elsewhere, down to 5 to 10 cm s^{-1} in the gyre centers. The upper mixed layer is generally from 15 to 120 m thick, depending upon area and season. The thermocline is about 100 to 200 m thick. Waters in the upper 500 m generally have high salinity, fairly high dissolved oxygen content, and low nutrient concentrations. Property measurements change only slightly with distance in the horizontal plane, except near the gyre edges. However, temperature, salinity, and dissolved oxygen are higher, and nutrient concentrations lower, on the western sides of anticyclone areas than on the eastern, at depths such as 100 m (Muromtsev, 1958; Reid et al., 1978). Seasonal changes in properties are slight to moderate. The gyres persist to depths over 1 km (Wyrtki, 1974, 1975; Worthington, 1976; Reid et al., 1977).

PRODUCTION AND BIOMASS LEVELS

The subtropical gyral regions have long been considered the biologically poorest parts of the ocean, except for some arctic waters. Table I summarises the approximate ranges of several biological and related measurements from the gyres, and compares them with data from equatorial divergences and subpolar cyclonic gyres. The subtropical gyres are represented by fewer data per unit of area. Their upper waters are better lighted but much poorer in nitrate at a given depth, than the other two regions. Nitrate is often unmeasurable in the upper parts of the euphotic zone, and the deep permanent pycnocline restricts its replenishment from below. Nitrogen or phosphorus is generally considered to limit the rate of production of phytoplankton in ocean waters where light is not limiting (Dugdale, 1967), and most studies indicate nitrogen rather than phosphorus (Eppley et al., 1973; Thomas, 1977). Nitrate is a major source of nitrogen for phytoplankton, but ammonia and urea, excreted by animals, are also very important (Dugdale and Goering, 1967; McCarthy, 1972).

The rate of primary production estimated from ^{14}C uptake is lower in the subtropical gyres than in the other two regions. Some authors would consider the rates in Table I to be too low, perhaps by 50% (Banse, 1974) or much more (Sorokin, 1978). The standing stock of phytoplankton, measured as chlorophyll, is also lowest in the subtropics, but it tends to be higher in subpolar than equatorial waters, despite the similar range of production rates in those two regions. The production/biomass ratio (P/B; in this case $mgCmg^{-1}$ chlorophyll h^{-1} or day^{-1}, often called assimilation number) is lower for phytoplankton in cold seas than in warm (Eppley, 1972). Thus the cold water biomass must be higher if production is the same. Many workers including Eppley believe that the assimilation number, and the related growth measurement in cell doublings day^{-1}, varies more or less directly with nitrogen concentration in warm, well-lighted waters. Others have suggested that the relation is more complex, or that one does not exist (Vedernikov, 1976; Walsh, 1976; Goldman et al., 1979). In any

TABLE I   Typical approximate values for biological and related measurements in three kinds of ocean environments

Measurement[a]	Subpolar gyres	Equatorial divergences	Subtropical gyres	Sources[b]
Mixed layer depth, summer (m)	15–30	25–65	15–50	1
Mixed layer depth, winter (m)	>120		65–120	1
Euphotic layer depth (m)	20–50	45–85	75–150	2, 3, 4, 5
NO_3-N, 0 m (µg atoms l^{-1})	5–25	5–10	0–1	4, 5, 6, 7
NO_3-N, 100 m (µg atoms l^{-1})	10–25	10–25	0–5	6, 7
Primary production ($gCm^{-2}day^{-1}$)	0.1–0.5	0.1–0.5	⩽0.1	8
Chlorophyll (mg m^{-2})	15–150	15–30	5–25	3, 4, 9
Herbivore mean weight (µgC)	40–400	4–40	4–40	3
Zooplankton, 0–200 m (g m^{-2})	30–50	7–25	<7	10, 11
Zooplankton, 0–1000 m (g m^{-2})	150	8–13	9	12
Micronekton, 0–200 m (g m^{-2})	0.6	0.6–0.8	0.3	13
Micronekton, 0–1000 m (g m^2)	2	1–3	1	12, 14
Benthos (g m^{-2})	0.1–1.0	0.1–1.0	<0.05	15

[a] Primary production and chlorophyll integrated for the euphotic layer. For zooplankton and micronekton, 1 ml displacement volume assumed to equal 1 g

[b] 1 Robinson (1976), 2 Ryther (1963), 3 Taniguchi (1973), 4 Krey and Babenerd (1976), 5 Thomas (1979), 6 Walsh (1976), 7 Anderson et al. (1969), 8 Koblentz-Mishke et al. (1970), 9 Blackburn et al. (1970), 10 Bogorov et al. (1968), 11 Reid et al. (1978), 12 Vinogradov (1968), 13 Blackburn (1977), 14 Legand et al. (1972), 15 Belyayev et al. (1973)

case the total daily production in a water column would still be low if the biomass remained small. The steady-state biomass must itself be limited by some chemical concentration. Biomass in nitrogen-poor water can be increased by adding nitrogen (Thomas, 1970).

Zooplankton biomass in a given part of the water column is also lower in the central gyres than elsewhere. It is lower in equatorial regions than subpolar. The smaller zooplankters of the warm waters have a shorter generation length and thus their P/B ratio might be 5 or 10 times higher (Greze, 1978), whereas production itself is probably at least as high in the subpolar as in the warmer seas. The biomass of micronekton (active pelagic animals from 1 to 10 cm) is lower in the subtropics than in the other two regions, in which it may be similar, although there are few data. Zooplankton and micronekton measurements in Table I must both be too low, because of animals that pass through net meshes (microzooplankton) or avoid nets altogether. The biomass of microzooplankton may be about 25% of that of net-caught zooplankton (Beers and Stewart, 1969, 1971). Measurements of acoustic scattering in ocean waters, which are associated with some kinds of zooplankton and micronekton, show minima in the subtropics (Farquhar, 1977). Similar information about biomass of large nekton is very scarce. Commercial fish catches are not easily expressed as biomass and many nektonic animals are not fished. Catches of fish and fish larvae were used to estimate an average biomass of $0.03 \, g \, m^{-2}$ of epipelagic fish in a large area from $0°$ to $20°N$ in the central Pacific (Blackburn, 1976). Thorne et al. (1977) estimated the corresponding biomass in a small coastal upwelling area as $60 \, g \, m^{-2}$, by acoustic methods. Neither estimate included micronekton. The subtropical gyres are fished commercially for various tunas, as are most waters between $45°N$ and S.

The benthos biomass is also generally lower in the anticyclone regions than elsewhere in oceanic waters. Rowe (1971) considered it a function of primary production and depth, except on some oxygen-deficient bottoms. This implies that the concentration of particulate organic matter (POM) varies spatially like the primary production, not only near the surface but also in deep waters, where it is scarcer and mostly detritus. Menzel (1967, 1974) denied or doubted such a relationship in deep waters, but Riley (1970) and Gordon (1977) thought it probable. Gordon's observations between Halifax and Bermuda showed a large decrease in POM towards the subtropical gyre, both in surface and deep waters. If the concentrations vary with area in the same way at all depths, the material must descend before deep currents can redistribute it. Vertically migrating animals may facilitate this descent (Vinogradov, 1968). Bottom sediments below subtropical gyres are principally red clays, and contain very little organic matter (Lisitzin, 1970).

Sorokin (1978) reviewed data, very largely from his earlier papers, on the distribution of biomass and production of planktonic bacteria. Biomass in the euphotic zone, where it is highest, is 10 to 200 mgCm^{-3} in the equatorial divergence and 1 to 20 mgCm^{-3} in other tropical ocean waters. Corresponding estimates of daily production are respectively 10 to 40 and 2 to 12 mgCm^{-3}, rather close to values of phytoplankton production cited by Sorokin. These production estimates have been criticised by Banse (1974), who concluded that not enough dissolved organic carbon (DOC) could be present to support so much production. Sorokin assumed the DOC is advected from other regions, although little is known about geographical variation in DOC (Wangersky, 1978).

BIOLOGICAL VARIABILITY WITH AREA

Small standing stocks should show little variation in a physically uniform environment. Thus biological conditions within a subtropical gyre might be rather monotonous in the horizontal plane. Venrick (1972) found that only 8 out of 20 diatom species exhibited contagious distributions over 20 km in the North Pacific central gyre. On a similar scale, chlorophyll and zooplankton biomasses showed much less spatial variation in that gyre than in the California current (McGowan, 1977). Skipjack tuna are ubiquitous in subtropical Pacific waters but rarely found in surface aggregations, except near islands (Matsumoto, 1975; Waldron, 1964).

Two zonal traverses across the subtropical South Indian Ocean showed that standing stock of chlorophyll was very uniform in the top 100 m (Krey and Babenerd, 1976). Phosphate concentrations, and primary production where measured, were also uniform. More variation was seen in a traverse along 28°N in the Pacific (Shulenberger, 1978). There were two areas in which isopleths of temperature, salinity, density, nitrate, nitrite, and chlorophyll domed towards the sea surface. Zooplankton biomass was fairly uniform. A seasonally repeated meridional traverse in the eastern Indian Ocean showed little change in nitrate, chlorophyll, and primary production in the subtropical portion at a given time of year, but zooplankton biomass was more variable because of advection (Tranter, 1973; Tranter and Kerr, 1977).

In the Sargasso Sea, which is part of the central North Atlantic gyre, areal variability of biota is enhanced by 'cold-core rings'. These rings, really cyclonic eddies, are a few hundred km in diameter and about 1 km deep. They originate as pinched-off meanders of the Gulf Stream and contain continental slope water (Fuglister, 1971; Richardson et al., 1973; Lai and Richardson, 1977). They transfer some of the relatively abundant nutrients, phytoplankton and zooplankton of the slope waters into the oligotrophic Sargasso, in discrete moving parcels of water which remain biologically rich for several months (Wiebe *et al.*, 1976; Ortner *et al.*, 1978). Such rings may cover about 10% of the area of the northern Sargasso Sea at any one time.

BIOLOGICAL VARIABILITY WITH TIME

Seasonal changes in biota should be small to moderate in subtropical gyres, since temperature, light, and mixed layer depth do not change greatly. The same should be true of tropical waters with the same characteristics, and even of equatorial divergences if the upwelling lasts all year.

Pertinent observations have been made in the North Atlantic anticyclone, at stations near Bermuda (Menzel and Ryther, 1960, 1961; Deevey, 1971; Deevey and Brooks, 1971) and Barbados (Steven, 1971; Moore and Sander, 1977). Primary production rate at Bermuda varied by a factor of 16 (maximum/minimum), with the highest values in late winter and spring and the lowest in summer. This cycle was related to fairly deep mixing in late winter and spring and formation of a shallow summer thermocline. The deep mixing increased the concentration of inorganic nitrogen in the euphotic zone. Menzel and Ryther pointed out that such seasonal change in mixing does not occur in the more southern parts of the Sargasso Sea. Zooplankton biomass in the 0 to 500 m layer varied in almost the same way as primary production, but some of the inflections occurred a month or two later. The same zooplankton cycle was recognisable in the 500 to 1000 m layer, but not deeper. At Barbados the primary production and zoo-plankton biomass varied by a factor about 20, but there was no regular pattern that could be called seasonal. Individual zooplankton species showed no seasonal changes in abundance or reproduction.

Statistically significant differences in various biological measurements were observed over a year in the eastern tropical Pacific (Owen and Zeitzschel, 1970; Blackburn *et al.*, 1970). They were of smaller amplitude than those found at Bermuda. The sampled layer was 0 to 200 m. The area was 16°N to 3°S, 98° to 119°W. It included the equatorial divergence and reached the edge of the North Pacific gyre. Nitrate levels were high in the divergence, and elsewhere almost as low as in central gyre waters (Thomas, 1977, 1979). Sampling was done every two months.

Primary production varied with season and longitude, but not latitude. The seasonal curves were similar on the four sampled meridians, with two maxima and minima in the year differing by a factor of 4 or less (Fig. 2). These changes have not been explained, and Thomas (personal communication) does not believe they can be explained by the small temporal changes in observed nutrient concentrations. The corresponding curves of chlorophyll standing stock showed only slight indications of the two maxima of the production curves, and on only two of the meridians. In general they showed a rather flat maximum from April–May to August–

September, followed by a minimum in October–November. Thus they agreed with the production curves for only half the year, and the start of their maximum was two months behind the first maximum of production. Amplitudes were like those in Fig. 2.

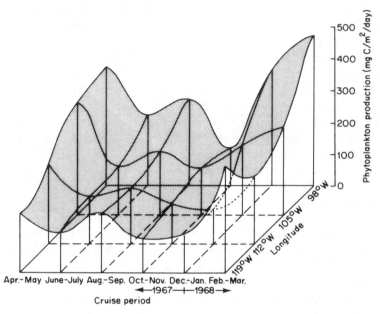

FIG. 2   Variation in primary production with period and longitude in the eastern tropical Pacific. (From Owen and Zeitzschel, 1970)

Zooplankton biomass varied in synchrony with the chlorophyll, in day samples. Night samples showed no significant change with time. Logarithms of primary production, chlorophyll, and day and night zooplankton were positively correlated with each other in each season and area (Blackburn, 1973). Biomass of fish-cephalopod micronekton varied like the other stocks but with a 4-month lag. The temporal changes in the stocks are reasonable, except for the above-mentioned lag of chlorophyll on production during part of the year. A simulation model of time change in various biomasses in tropical waters showed that a phytoplankton peak may be followed 20 days later by a herbivore peak, with a peak of carnivorous zooplankton after a further 20 days (Vinogradov et al., 1973). Since the primary production curve has not been explained, and the observations have not been repeated in another year, it cannot be assumed that there is any regular seasonal cycle. The main point of interest is that all the observed time changes were small, and that those for phytoplankton and zooplankton biomass were in phase.

Tranter (1973) made seasonally repetitive measurements of biological properties along 110°E in the South Indian Ocean. Seasonal variation, of amplitude less than in the Sargasso Sea but more than in the eastern tropical Pacific, was evident in primary production and stocks of chlorophyll and zooplankton. Nitrate varied less. Changes were generally synchronous between the nitrate-rich tropical and nitrate-poor subtropical waters, as in the eastern tropical Pacific. They were less synchronous than expected between biological properties in the same area, and advective effects were suspected. Shomura and Nakamura (1969) described changes in zooplankton biomass over a 19-month period at Hawaii, in the North Pacific central gyre. The variation, again small, was not seasonally regular. It seemed to result entirely from advection.

BIOLOGICAL VARIABILITY WITH DEPTH

A subsurface maximum of chlorophyll occurs at all seasons in tropical and subtropical waters. It lies generally in the upper part of the thermocline, and is therefore deeper in anticyclone waters than equatorial (Vinogradov et al., 1970; Venrick et al., 1973; Longhurst, 1976a). This maximum has been attributed to a decrease in the sinking rate of plant cells at the depth of the maximum (Steele and Yentsch, 1960; Venrick et al., 1973). Alternatively it may result from a depth-differential in grazing (Lorenzen, 1967; Longhurst, 1976a). Longhurst concluded that the biomass of epi-zooplankton in the eastern tropical Pacific was highest near the depth of maximum primary production, which lies above the chlorophyll maximum.

Vinogradov (1968) compared bathymetric profiles of zooplankton concentration to over 4 km in different parts of the Pacific. The sharpest decline with depth was seen in the warmer parts of the northern and southern anticyclone areas. In the cooler parts of those areas between 20° and 40°, and in equatorial and subarctic waters, the concentrations declined more gradually. Vinogradov made a separate study of the bathymetric distribution of macroplankton (included in micronekton in Table I), which showed very striking regional differences. Equatorial waters had maxima near the surface and at 1000 to 2000 m; the concentration at the lower maximum exceeded that of zooplankton. In subtropical and subarctic waters the macroplankton maximum occurred at 500 to 1000 m. The bathymetric limits of macroplankton were about 2000, 3000 and 4000 m in the subtropical, equatorial, and subarctic regions respectively.

Diel vertical migrations of some kinds of zooplankton and micronekton are well known in warm ocean waters. They would be expected to have a relatively large range in areas of deep light penetration. Median daytime depths of sound scattering layers in the Pacific are respectively 375, 335 and 285 m in the northern anticyclone, the equatorial region, and the subarctic (Tont, 1977). Vinogradov (1968) postulated a 'ladder of migrations' in which different groups of animals made large vertical movements in different parts of the entire water column, diurnally in warm waters and seasonally in cold. In this way organic material would be transferred to progressively greater depths as live animals and faeces, providing food for the animals living there. Longhurst (1976b) stated, however, that there is yet no evidence of migration below 1700 m.

An analysis of samples of zooplankton and micronekton from the upper 1200 m of the tropical Pacific was made (Legand et al., 1972). The areas of study were at the Equator and 10°S. Three groups of animals were recognised: those living entirely above and below 450 m, and those living below 450 m by day and above that level at night. The last group, called interzonal fauna, included most species of mesopelagic fish, euphausiids, and decapod shrimps. Roger and Grandperrin (1976) did similar work in other tropical Pacific areas, including central gyre waters at 22°S, with similar results. Figure 3 shows their schematic diagram. Angel and Fasham (1973) distinguished five depth zones of different zooplankton composition in the upper 1000 m of the subtropical North Atlantic.

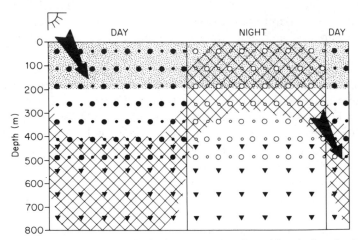

FIG. 3   Diagram showing the principal components of trophic relations in the upper pelagic tropical Pacific. Strippling means phytoplankton. Large solid dots and large open circles mean tunas and epipelagic micronekton, feeding by day (dots) and not by night (circles). Small solid dots and small open circles mean zooplankton, feeding by day (dots) and not by night (circles). Cross-hatching means vertically migrating fauna. Triangles mean non-migrating deep fauna. (From Roger and Grandperrin, 1976)

Concentration of total POM declines below the surface layer to some depth about 200 to 800 m, in all regions. Some authors detected a further decline (in carbon) with depth (Wangersky, 1976; Gordon, 1977), and others did not (Riley, 1970; Menzel, 1974). DOC declines with depth to about 200 m, and its concentration seems to remain uniform at greater depths in all regions (Wangersky, 1978). The vertical distribution of biomass of bacterioplankton has been studied in warm ocean waters by Sorokin (1978). There can be four maxima in the first few hundred m, before the values decline to uniform low levels at greater depths.

TAXONOMIC DIVERSITY

It is well known that the plant and animal plankton of warm seas contains more species than that of cold seas. This diversity may be highest in the subtropics. In the Pacific, the central gyres and their western boundary currents have more species of euphausiids, pteropods, and chaetognaths than other regions, except for chaetognaths in the eastern tropical Pacific (Reid et al., 1978). The copepod fauna is more diverse in the North Pacific gyre, and the rank order of species abundance more uniform with area, than in the California Current (McGowan, 1977).

It is often argued that diversity in an ecosystem promotes stability, although this view is not unanimous (Odum, 1969). If diversity has this effect, it is probably by ensuring that some kind of prey will be available for each kind of animal at any time (Parsons and Boom, 1974).

DISCRETENESS OF GYRE COMMUNITIES

As noted, the gross biomass differences between the subtropical gyres and adjacent waters persist right down the water column. Hence, the individual gyres are possibly ecosystem habitats. Several workers have pointed out consistent differences in the species-composition of zooplankton and nekton taxa between these gyres and other areas, suggesting that the gyre communities may indeed be discrete. For instance *Euphausia brevis* occurs in all five anticyclonic gyres of the world ocean but not elsewhere, except in the Caribbean Sea, and some other species have similar distributions (Reid et al., 1978). Other species of zooplankton are similarly characteristic of equatorial, subpolar, and transition areas, and some are cosmopolitan. Transition areas lie between subtropical and subpolar waters.

From such observations on many species, McGowan (1974) recognised the following eight areas of the Pacific as 'biotic provinces' and probable ecosystem habitats: the two anticyclonic gyres, the two subpolar areas, the two transition areas, equatorial waters, and the eastern tropical Pacific (Fig. 4). Similar studies on the distribution of mesopelagic fish in the North Atlantic indicated a difference between tropical and subtropical faunas

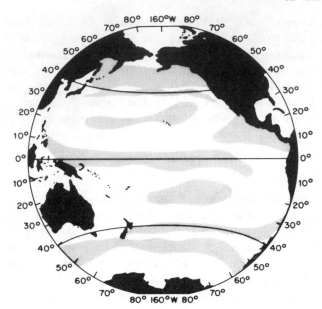

FIG. 4   Core regions of the basic biotic provinces of the oceanic Pacific. (From McGowan, 1974)

(Backus *et al.*, 1970; Backus and Craddock, 1977; Badcock and Merrett, 1977). Barnett (1975) recognised two distinct mesopelagic fish faunas, not one, in the North Pacific gyral region. He connected this finding to the recognition of two gyres in that region by Sverdrup *et al.* (1942). Species of phytoplankton tend to be more cosmopolitan than those of zooplankton and small nekton (Reid *et al.*, 1978). The tuna species that occur in subtropical gyres are found also in equatorial or transition zones or both (Blackburn, 1965).

The fact that some species of plants and animals occur in more than one biotic province does not deny the possibility that the province floras and faunas are discrete, because species can be composed of quasi-independent populations. Complete separation of province faunas is very unlikely, however, because of advection. Geynrikh (1976) considered that many species in the peripheral areas of the North and South Pacific gyres were expatriates from equatorial, neritic, and transition zone waters. Similarly there were gyre and neritic species in equatorial waters. A well-documented study of expatriation refers to the euphausiid *Nematoscelis megalops*, which is transported from western Atlantic slope waters into the Sargasso Sea within cold-core rings (Wiebe and Boyd, 1978; Boyd *et al.*, 1978). In this case the expatriated populations do not persist.

FUNCTIONING OF THE ECOSYSTEMS

In view of the relative discreteness of their physical habitats and biological communities, the core areas of the anticyclonic gyres (Fig. 4) may be ecosystems. There is some consensus of opinion as to how such ecosystems could work, at least in the top few hundred m, and there are many unresolved questions.

The upper mixed layer of an anticyclonic gyre is thinner in summer, and generally no thicker in winter, than the euphotic layer (Table I). Thus plant cells will not become light-limited by being mixed out of the euphotic zone. They will not be entirely nutrient-limited either, since they take up ammonia and urea excreted by animals, whether or not they have other nitrogen sources (Dugdale, 1967, 1976; Dugdale and Goering, 1967; McCarthy, 1972; Eppley et al., 1973). So net primary production will continue all year. Probably secondary production is also continuous, at least above the thermocline. As noted there are many functionally herbivorous species, mostly small and reproducing continuously. Because of their size and the high water temperatures they tend to have short generations, and high daily rates of metabolism and population growth (Ikeda, 1970; Taniguchi, 1973; Sheldon et al., 1973; Fenchel, 1974; Greze, 1978).

Therefore, herbivore production in the upper part of the water column *could* be adjusted to the level of plant production without much time lag, so that the herbivores consumed approximately each day's production of algae and perhaps other things, and the herbivores and other pelagic animals supplied enough nitrogen for a similar increment of primary production next day, for a considerable time. A similar balance is required between herbivores and their predators, so that the former do not excessively graze the steady-state phytoplankton biomass as well as the production. Temporal changes in plant production, resulting from slight or moderate changes in light, temperature, or nutrient conditions, could be paralleled by the herbivore production. Some of those changes might occur at the same time each year and others not. Plant and herbivore biomasses might sometimes go out of balance, for instance by changes in the grazing population caused by horizontal advection. Any plants not eaten, together with faecal pellets, moults, and remains of dead animals, would become sinking recyclable detritus.

The foregoing inductive hypothesis has been put forward or endorsed in one form or another by most workers in the field. Its principal architects include Cushing (1959), Menzel and Ryther (1960, 1961), Heinrich (1962), and Dugdale and Goering (1967). It has not been verified adequately. Time-series data on secondary production are not available from gyres, or most other waters. There are such data on total net-caught zooplankton biomass but the herbivore fraction is not stated, and microzooplankton are omitted. Published P/B ratios for various zooplankters (Mullin, 1969; Greze, 1978) suggest that production may be related linearly to biomass for each type and size of animal in a warm sea. Then if the composition of subtropical zooplankton were uniform by taxa and sizes, it might be assumed that secondary production, and also production at each higher trophic level, varied like the total biomass with time. Zooplankton biomass has been compared with primary production over a year or more as noted, at Bermuda and Barbados. The two curves were roughly parallel in each case, although small to moderate inflections, not always synchronous, were seen. These limited results are consistent with the hypothesis stated. Wroblewski (1977) made a numerical model to simulate plant and herbivore biomasses and nutrient concentration in the North Pacific central gyre over a few weeks. The three curves soon became flat. However, in the corresponding model for the North Pacific subarctic gyre, they exhibited the expected large variations in amplitude and phase. Eppley *et al.* (1973) considered that their observations in the North Pacific central gyre indicated close coupling between primary production and the grazing and excretion rates of the zooplankton. If there is close coupling, patches of plants or animals are unlikely to be large or persistent.

The hypothesis could apply equally well to most other warm ocean waters, including equatorial. Mixed layer depths are again no greater than euphotic depths, and herbivores are again small (Taniguchi, 1973; Thomas, 1979; Table I). Primary production, chlorophyll, and zooplankton biomass varied slightly with time in about the same way as each other over the offshore eastern tropical Pacific, including equatorial divergence waters, as mentioned. So the equatorial system is approximately balanced between plants and herbivores like the subtropical one, notwithstanding the much higher levels of nitrate, primary production, and both biomasses at the Equator. Ammonia concentrations are also higher in equatorial than in gyre waters (Eppley *et al.*, 1973; Thomas, 1977).

It seems likely that plant production or biomass or both are kept to low levels in central gyres by the meagre nitrogen concentrations in the euphotic zone. That opinion does not ignore the question about effects of nutrient concentrations on plant P/B ratios mentioned earlier. An equatorial divergence and an adjacent subtropical gyre might be likened to two lakes, one enriched with nutrient and the other not. The possibility exists that the equatorial biomass is depressed by grazing (Walsh, 1976; Thomas, 1979).

Ammonia and urea excreted by animals account for a substantial fraction of the nitrogen required by plants in gyre waters (Dugdale and Goering, 1967; Eppley *et al.*, 1973; Ikeda and Motoda, 1978). Bacteria may also regenerate some plant-utilisable nitrogen in the euphotic zone, from detritus. However the total nitrogen regenerated in that zone cannot be enough to maintain the production at the same general level indefinitely, because some of the POM sinks or is transported to greater depths (McGowan and Hayward, 1978). There has been much discussion about the sources of the rest of the nitrogen in a subtropical gyre. Dissolved organic nitrogen is, at best, a minor source for plants (Thomas et al., 1971; Wangersky, 1978). Some phytoplankters may obtain nitrogen by making diel vertical movements into the nutricline (Eppley et al., 1968). A few can fix molecular nitrogen (Stewart, 1971; Mague et al., 1974). Rain may be a very minor source of ammonia and nitrate. Carpenter and McCarthy (1975) considered nitrogen fixation and rain to be negligible sources of nitrogen in the Sargasso Sea.

It therefore appears necessary to consider nitrate from the nutricline as a source, but the ways in which it could reach the euphotic zone of a gyre are not clear (McGowan and Hayward, 1978). The water column is well stratified and vertical diffusion is probably low. The mixed layer seldom goes as deep as the top of the nutricline. It must therefore generally contain little nitrate, as is observed. An exception occurred in the northern Sargasso Sea by deeper mixing in late winter and spring, as noted; primary production rose at that time. McGowan and Hayward (1978) observed higher primary production in the North Pacific gyre in the summer of 1969 than in five other summers. The upper mixed layer was not much deeper than usual, but it was hypothesised that mixing occurred from below, possibly by internal wave action, so as to bring nutricline water into the euphotic zone. The supposedly increased nutrient concentration was not detected in the euphotic zone, and was assumed to be taken up too fast to be measured. Other physical mechanisms for bringing up nutrients may exist, such as 'frontal upwelling' (Woods, 1977) and unusually deep wind-mixing in storms (Iverson, 1977). Such events may be relatively rare, like the one suggested by McGowan and Hayward.

Horizontal advection, of water made nitrogen-rich in the euphotic zone outside the gyres, is probably important. Thomas (1972) found a layer of nitrate-rich water overlying poor water in the north-east part of the South Pacific anticyclone area, which must have come from an upwelling region. Such effects could occur in other eastern parts of subtropical gyres. The cold core rings which transfer continental slope water into the Sargasso Sea are another example. Nitrate concentrations in the ring waters are intermediate between those of the slope and Sargasso (Wiebe et al., 1976). Similar cyclonic eddies are probably formed in the Kuroshio and travel into the North Pacific gyre (Wyrtki, 1975). Other mesoscale eddies originate elsewhere, for instance near islands and seamounts (Barkley, 1972; Royer, 1978), and those that are cyclonic are likely to elevate the nutricline within the eddies. Then some of the nutrient-richer water would eventually be mixed into surrounding areas. Eddy energy is higher in the western parts of subtropical gyres than in the central and eastern parts (Dantzler, 1976; Wyrtki et al., 1976; Bernstein and White, 1977). Nevertheless there are substantial amounts of this energy in the gyre centers (Wyrtki et al., 1976), except perhaps in the north-eastern North Pacific (Bernstein and White). If two major anticyclones exist in the North Pacific there should be some cyclonic flow between them. Observations by Shulenberger (1978) are consistent with such flow.

Another question that frequently arises is whether the phytoplankton is quantitatively sufficient to meet all the food needs of herbivores, and, if not, what else is eaten. Similar questions are asked about transfers between other trophic levels. Ikeda and Motoda (1978) estimated the mean daily herbivore ration as 18 to 72% of daily primary production in the most oligotrophic parts of the Kuroshio region. Taniguchi (1973) indicated that the ration and production were approximately equal in both cold and warm seas. Both authors estimated the herbivore demand from zooplankton biomass measured only in the upper 150 m, and without microzooplankton. Both used primary production rates comparable with those in Table I. It was estimated by Riley (1970) that consumption of POM in the entire water column of the Sargasso Sea would approximately balance a primary production rate of about $0.35\,gCm^{-2}day^{-1}$, which is considerably higher than the estimate for such waters given in Table I. Finenko and Zaika (1970) calculated that zooplankton in the upper 500 m of oligotrophic tropical Atlantic waters require, for respiration only, 1.5 times the amount of food that the primary production could supply. Moreover they used a primary production rate about $0.25\,gCm^{-2}day^{-1}$, again higher than in Table I. They considered detritus and bacteria to be additional food sources.

Sorokin (1978) referred to other discrepant findings of the same kind, and ascribed them in part to the use of low primary production estimates. In addition he considered that most of the phytoplankton of the world ocean is not grazed alive. Instead it becomes detritus and dissolved organic matter, which re-enters the living production cycle as bacteria. These bacteria supply about half the food needs of 'herbivores' including microzooplankton, the other half being live plants and detritus, according to Sorokin. Petipa (1979) indicated that the principal zooplankters of non-upwelling tropical Pacific waters eat all of the following: animals, plants, bacteria, and detritus. Gametes are another food source for zooplankton (Isaacs, 1976).

It has been said that the zooplankton biomass of an oligotrophic warm ocean area is 45% or more carnivorous, in the upper layers (Grice and Hart, 1962; Greze et al., 1969; Timonin, 1973). Petipa (1979) listed data on biomass and food ration of the principal groups of zooplankton of such an area, including microzooplankton. The total listed biomass could be called 29% carnivorous and 47% detritus-feeding, including all groups in which animals and detritus accounted for 80% or more of the ration. On the same basis the remaining 24% of biomass could be called omnivorous, since no group obtained more than 26% of its ration from either living plants or bacteria. The composition of the salp ration was not given.

In the light of the foregoing data it would now be very difficult, and perhaps meaningless, to characterise central gyre waters by number of trophic levels. Ryther (1969) suggested five levels between phytoplankton and man, although he was aware of some of the complexities. If one assumes tunas are apex predators (although some are eaten by billfishes), there is a well-defined food chain some distance down the trophic pyramid as follows: tunas ← epipelagic micronekton ← certain copepods, amphipods, and euphausiids, some of which eat smaller animals (Legand et al., 1972; Roger and Grandperrin, 1976). There must be at least four levels before plants in this particular chain. The tunas and epipelagic micronekton are shown in Fig. 3 as day-feeders. The prey of the micronekton are also assumed to be day-feeders, although this has not been shown for all of them (Roger and Grandperrin). Figure 3 also shows the interzonal fauna of mesopelagic fishes, euphausiids, and decapod shrimps, which rise at night. These are only minor constituents of the diet of tunas in most areas, despite occasional suggestions to the contrary in literature.

Trophic relations of the interzonal and deeper faunas were discussed by Vinogradov (1968), Legand et al. (1972), Omori (1974), Harding (1974), and Hopkins and Baird (1977). They are not considered further here, except to emphasise that the principal predators of mid-ocean mesopelagic fishes are still not known (Clarke, 1973). Giant squids eat such fishes in waters off Peru and Chile (Nesis, 1970).

For the large area of the eastern tropical Pacific mentioned earlier, Blackburn (1973) found significant positive regressions of chlorophyll and zooplankton biomasses on primary production, zooplankton on chlorophyll, and biomass of fish-cephalopod micronekton (measured four months later) on zooplankton, all in logarithms. Each dependent variable varied with the independent variable to a power < 1, in all seasons and areas, as in the following examples:

$$\text{zooplankton} \propto \text{primary production}^{0.2}$$
$$\text{micronekton} \propto \text{zooplankton}^{0.2}.$$

A constant relation was assumed between production and biomass in each of the two animal groups. The results then suggested that the transfer of material from phytoplankton to zooplankton, and zooplankton to micronekton, is more efficient in oligotrophic tropical seas than in eutrophic. Cushing (1973) confirmed this by plotting estimated production of zooplankton against measured primary production for 5° rectangles in the tropical and subtropical Indian Ocean. The coefficients of transfer fell from 20% in areas of low plant production to 5% in areas of high production. It does not matter here whether these transfers involved one trophic link or more. The described trend in ecological efficiency has not been explained. Herbivore assimilation does not appear to decline as food concentration increases (Conover, 1978).

Taniguchi (1973) regressed mean biomass of herbivorous zooplankton on mean primary production for several ocean areas from the Equator to the Bering Sea. The equation was:

$$\text{herbivores} \propto \text{primary production}^{1.47}$$

which is very different from the corresponding relation given above. The two relationships do not appear incompatible, since Blackburn dealt only with tropical waters. Herbivore biomass is expected to be higher in relation to herbivore production in cold seas than in warm, as mentioned earlier; and probably also higher in relation to primary production, although with a smaller difference between cold and warm, from the transfer coefficients above.

Since production and consumption appear to be approximately balanced at most times in both equatorial and subtropical waters, those systems are probably efficient in the way the communities utilise organic material. There is no conspicuous waste, such as phosphatic or guano deposits (Cushing, 1971), in either system. The subtropical gyre communities seem however to be more efficient than those of the more productive equatorial divergences, since they transfer a greater fraction of the primary production to higher trophic levels, however they do it.

3   Other low-latitude gyral regions

At the terminus of the equatorial countercurrent in the eastern tropical Pacific, part of the water turns to the left and part to the right. These flows eventually join the north and south equatorial currents respectively (Wyrtki, 1966). They create a cyclonic gyre, called the Costa Rica dome, and an anticyclonic gyre, each with a diameter of a few hundred km, which are recognisable in surface current charts. The dome is seen in Fig. 1 at approximately 10°N, 90°W. It persists all year, and has been well described (Wyrtki, 1964). At least two other cyclonic flows (domes) occur in the eastern tropical Pacific. One is in the Gulf of Tehuantepec (15°N, 95°W) and the other is in the Panama Bight (5°N, 80°W). They were described by Blackburn (1962), and Wooster (1959) and Forsbergh (1969), respectively. Being seasonal (strongest in the northern winter), they are not apparent in Fig. 1. These domes are geostrophic upwellings, like the Costa Rica dome, although wind-induced coastal upwelling also occurs at Panama, and may occur at Tehuantepec. Some anticyclonic flow occurs to the west of the Tehuantepec dome. Physically, chemically, and biologically, the domes are like equatorial divergence waters (Owen and Zeitzschel, 1970; Blackburn *et al.*, 1970; Longhurst, 1976a; Thomas, 1979; references above). The eastern tropical Pacific in general has a much shallower and thicker oxygen minimum layer than equatorial waters, however (Wyrtki, 1966). Neither the abovementioned gyres nor the equatorial currents persist much below 500 m (Wyrtki, 1974, 1975).

The eastern Atlantic homologue of the Costa Rica dome is the Guinea dome, which is probably seasonal (Mazeika, 1967). Primary production, zooplankton biomass, and other biological measurements are fairly high in that area (Corcoran and Mahnken, 1969; Zeitzschel, 1969). A similar feature, again seasonal, exists south of Java in the Indian Ocean (Wyrtki, 1962). Associated nutrient and biological conditions resemble those of other tropical divergences (Tranter, 1973).

McGowan (1974) considered the eastern tropical Pacific as one of his biotic provinces (Fig. 4), since it has a relatively distinct fauna. It seems very unlikely that this region or the corresponding regions in the other oceans represent ecosystems, since so much water goes horizontally in and out of the upper 500 m. McGowan admitted moderate 'leakage' from the eastern tropical Pacific.

The monsoon gyre of the Indian Ocean is very variable. In the north-east monsoon season it extends meridionally from about 5°N to 8°S and flows counter-clockwise; in the south-west monsoon season it extends from about 2°N to 10°S and flows clockwise (Wyrtki, 1973). The gyre is present across almost the entire width of the ocean in each season. In general the monsoon gyre waters are richer in nutrients and biota than those of the subtropical anticyclone to the south. This difference, however, is less obvious during the north-east monsoon than it is at the south-west monsoon season, when there is more vertical mixing and upwelling (Tranter, 1973; Rao, 1973; Krey and Babenerd, 1976). It seems highly improbable that the monsoon gyre is an ecosystem.

Acknowledgements

I thank the Friday Harbor Laboratory of the University of Washington for giving me access to literature in its library, and in other libraries of the University. The following investigators gave me valuable help with literature and illustrations: K. Banse, B. V. Hamon, A. R. Longhurst, J. A. McGowan, J. L. Reid, W. H. Thomas, J. S. Wroblewski, K. Wyrtki, and B. Zeitzschel. Hiroko Yoshida sent me works of her late husband Kozo Yoshida, some of which were used in this review.

References

Anderson, G. C., Parsons, T. R. and Stephens, K. (1969). Nitrate distribution in the subarctic Northeast Pacific Ocean. *Deep-Sea Res.* **16**, 329–334

Angel, M. V. and Fasham, M. J. R. (1973). SOND Cruise 1965: factor and cluster analysis of the plankton results, a general summary. *J. Mar. Biol. Ass. U.K.* **53**, 185–231

Backus, R. H., Craddock, J. E., Haedrich, R. L. and Shores, D. L. (1970). The distribution of mesopelagic fishes in the Equatorial and Western North Atlantic Ocean. *J. Mar. Res.* **28**, 179–201

Backus, R. H. and Craddock, J. E. (1977). Pelagic faunal provinces and sound-scattering levels in the Atlantic Ocean. *In* "Oceanic Sound Scattering Prediction" (N. R. Andersen and B. J. Zahuranec, eds) pp. 529–547. Plenum Press, New York

Badcock, J. and Merrett, N. R. (1977). On the distribution of midwater fishes in the eastern North Atlantic. *In* "Oceanic Sound Scattering Prediction" (N. R. Andersen and B. J. Zahuranec, eds) pp. 249–282. Plenum Press, New York

Banse, K. (1974). On the role of bacterioplankton in the tropical ocean. *Mar. Biol.* **24**, 1–5

Barkley, R. A. (1972). Johnston Atoll's wake. *J. Mar. Res.* **30**, 201–216

Barnett, M. A. (1975). Studies on the patterns of distribution of mesopelagic fish faunal assemblages in the central Pacific and their temporal persistence in the gyre. Ph.D. dissertation, Scripps Institution of Oceanography, University of California, San Diego

Beers, J. R. and Stewart, G. L. (1969). Micro-zooplankton and its abundance relative to the larger zooplankton and other seston components. *Mar. Biol.* **4**, 182–189

Beers, J. R. and Stewart, G. L. (1971). Micro-zooplankters in the plankton communities of the upper waters of the eastern tropical Pacific. *Deep-Sea Res.* **18**, 861–883

Belyayev, G. M., Vinogradova, N. G., Levenshyteyn, R. Y., Pasternak, F. A., Sokolova, M. N. and Filatova, Z. A. (1973). Distribution patterns of deep-water bottom fauna related to the idea of the biological structure of the ocean. *Oceanol.* **13**, 114–120. (Eng. transl.)

Bernstein, R. L. and White, W. B. (1977). Zonal variability in the distribution of eddy energy in the mid-latitude North Pacific Ocean. *J. Phys. Oceanogr.* **7**, 123–126

Blackburn, M. (1962). An oceanographic study of the Gulf of Tehuantepec. *U.S. Fish. Wildl. Serv., Spec. Sci. Rep. Fish.* 404

Blackburn, M. (1965). Oceanography and the ecology of tunas. *Oceanogr. Mar. Biol. Ann. Rev.* **3**, 299–322

Blackburn, M. (1973). Regressions between biological oceanographic measurements in the eastern tropical Pacific and their significance to ecological efficiency. *Limnol. Oceanogr.* **18**, 552–563

Blackburn, M. (1976). Review of existing information on fishes in the Deep Ocean Mining Environmental Study (DOMES) area of the tropical Pacific. Inst. Mar. Res., Univ. Cal., IMR Ref. 76–1, 77 pp.

Blackburn, M. (1977). Studies on pelagic animal biomasses. *In* "Oceanic Sound Scattering Prediction" (N. R. Andersen and B. J. Zahuranec, eds) pp. 283–298. Plenum Press, New York

Blackburn, M., Laurs, R. M., Owen, R. W. and Zeitzschel, B. (1970). Seasonal and areal changes in standing stocks of phytoplankton, zooplankton and micronekton in the eastern tropical Pacific. *Mar. Biol.* **7**, 14–31

Bogorov, V. G., Vinogradov, M. D., Voronina, N. M., Kanaeva, I. P. and Suetova, I. A. (1968). Distribution of zooplankton biomass within the superficial layer of the world ocean. *Dokl. Akad. Nauk SSSR* **182**, 1205–1207. (In Russian)

Boyd, S. H., Wiebe, P. H. and Cox, J. L. (1978). Limits of *Nematoscelis megalops* in the Northwestern Atlantic in relation to Gulf Stream cold core rings. II. Physiological and biochemical effects of expatriation. *J. Mar. Res.* **36**, 143–159

Burkov, B. A. and Monin, A. S. (1973). Global features of water circulation in the world ocean. *Oceanol.* **13**, 37–48. (Eng. transl.)

Carpenter, E. J. and McCarthy, J. J. (1975). Nitrogen fixation and uptake of combined nitrogenous nutrients by *Oscillatoria* (*Trichodesmium*) *thiebautii* in the western Sargasso Sea. *Limnol. Oceanogr.* **20**, 389–401

Clarke, T. A. (1973). Some aspects of the ecology of lanternfishes (Myctophidae) in the Pacific Ocean near Hawaii. *Fish. Bull. U.S.* **71**, 401–434

Conover, R. J. (1978). Transformation of organic matter. *In* "Marine Ecology" (O. Kinne, ed.) Vol. IV, pp. 221–499. Wiley, Chichester

Corcoran, E. F. and Mahnken, C. V. W. (1969). Productivity of the tropical Atlantic Ocean. *In* "Proceedings of the Symposium on the Oceanography and Fisheries Resources of the Tropical Atlantic (Abidjan, Ivory Coast, 1966). Review Papers and Communications", pp. 57–67. UNESCO, Paris

Cushing, D. H. (1959). The seasonal variation in oceanic production as a problem in population dynamics. *J. Cons. Perm. Internat. Explor. Mer* **24**, 455–464

Cushing, D. H. (1971). Upwelling and the production of fish. *Adv. Mar. Biol.* **9**, 255–334

Cushing, D. H. (1973). Production in the Indian Ocean and the transfer from the primary to the secondary level. *In* "The Biology of the Indian Ocean" (B. Zeitzschel, ed.) pp. 475–486. Springer-Verlag, New York

Dantzler, H. L. (1976). Geographic variations in intensity of the North Atlantic and North Pacific oceanic eddy fields. *Deep-Sea Res.* **23**, 783–794

Deevey, G. B. (1971). The annual cycle in quantity and composition of the zooplankton of the Sargasso Sea off Bermuda. I. The upper 500 m. *Limnol. Oceanogr.* **16**, 219–240

Deevey, G. B. and Brooks, A. L. (1971). The annual cycle in quantity and composition of the zooplankton of the Sargasso Sea off Bermuda. II. The surface to 2000 m. *Limnol. Oceanogr.* **16**, 927–943

Dugdale, R. C. (1967). Nutrient limitation in the sea: dynamics, identification and significance. *Limnol. Oceanogr.* **12**, 685–695

Dugdale, R. C. (1976). Nutrient cycles. *In* "The Ecology of the Seas" (D. H. Cushing and J. J. Walsh, eds) pp. 141–172. Blackwell, Oxford

Dugdale, R. C. and Goering, J. J. (1967). Uptake of new and regenerated forms of nitrogen in primary productivity. *Limnol. Oceanogr.* **12**, 196–206

Eppley, R. W. (1972). Temperature and phytoplankton growth in the sea. *Fish. Bull. U.S.* **70**, 1063–1085

Eppley, R. W., Holm-Hansen, O. and Strickland, J. D. H. (1968). Some observations on the vertical migration of dinoflagellates. *J. Phycol.* **4**, 333–340

Eppley, R. W., Renger, E. H., Venrick, E. L. and Mullin, M. M. (1973). A study of plankton dynamics and nutrient cycling in the central gyre of the North Pacific Ocean. *Limnol. Oceanogr.* **18**, 534–551

Farquhar, G. B. (1977). Biological sound scattering in the oceans: a review. *In* "Oceanic Sound Scattering Prediction" (N. R. Andersen and B. J. Zahuranec, eds) pp. 493–527. Plenum Press, New York

Fenchel, T. (1974). Intrinsic rate of natural increase: the relationship with body size. *Oecologia* **14**, 317–326

Finenko, Z. Z. and Zaika, V. E. (1970). Particulate organic matter and its role in the productivity of the sea. *In* "Marine Food Chains" (J. H. Steele, ed.) pp. 32–44. Oliver and Boyd, Edinburgh

Forsbergh, E. D. (1969). On the climatology, oceanography and fisheries of the Panama Bight. *Inter-Amer. Trop. Tuna Commn., Bull.* **14**, 49–385. (In English and Spanish)

Fuglister, F. C. (1971). Cyclonic rings formed by the Gulf Stream 1965–66. *In* "Studies in Physical Oceanography" (A. Gordon, ed.) pp. 137–168. Gordon and Breach, New York

Geynrikh, A. K. (1976). Role of displaced species in the structure of the Pacific tropical plankton communities. *Oceanol.* **15**, 492–495. (Eng. transl.)

Goldman, J. C., McCarthy, J. J. and Peavey, D. G. (1979). Growth rate influence on the chemical composition of phytoplankton in oceanic waters. *Nature* **279**, 210–215

Gordon, D. C. (1977). Variability of particulate organic carbon and nitrogen along the Halifax-Bermuda section. *Deep-Sea Res.* **24**, 257–270

Greze, V. N. (1978). Production in animal populations. *In* "Marine Ecology" (O. Kinne, ed.) Vol. IV, pp. 89–114. Wiley, Chichester

Greze, V. N., Gordejava, K. T. and Shmeleva, A. A. (1969). Distribution of zooplankton and biological structure in the tropical Atlantic. *In* "Proceedings of the Symposium on the Oceanography and Fisheries Resources of the Tropical Atlantic (Abidjan, Ivory Coast, 1966). Review Papers and Communications", pp. 85–90. UNESCO, Paris

Grice, G. D. and Hart, A. D. (1962). The abundance, seasonal occurrence and distribution of the epizooplankton between New York and Bermuda. *Ecol. Monogr.* **32**, 287–309

Hamon, B. V. (1965). The East Australian Current, 1960–1964. *Deep-Sea Res.* **12**, 899–921

Harding, G. C. H. (1974). The food of deep-sea copepods. *J. Mar. Biol. Ass. U.K.* **54**, 141–155

Hasunuma, K. and Yoshida, K. (1978). Splitting of the subtropical gyre in the western North Pacific. *J. Oceanogr. Soc. Japan* **34**, 160–172

Heinrich, A. K. (1962). The life histories of plankton animals and seasonal cycles of plankton communities in the oceans. *J. Cons. Perm. Internat. Explor. Mer* **27**, 15–24

Hopkins, T. L. and Baird, R. C. (1977). Aspects of the feeding ecology of oceanic midwater fishes. *In* "Oceanic Sound Scattering Prediction" (N. R. Andersen and B. J. Zahuranec, eds). Plenum Press, New York

Ikeda, T. (1970). Relationship between respiration rate and body size in marine plankton animals as a function of the temperature of habitat. *Bull. Fac. Fish. Hokkaido Univ.* **21**, 91–112

Ikeda, T. and Motoda, S. (1978). Estimated zooplankton production and their ammonia excretion in the Kuroshio and adjacent seas. *Fish. Bull. U.S.* **76**, 357–368

Isaacs, J. D. (1976). Reproductive products in marine food webs. *Bull. S. Cal. Acad. Sci.* **75**, 220–223

Iverson, R. L. (1977). Mesoscale oceanic phytoplankton patchiness caused by hurricane effects on nutrient distribution in the Gulf of Mexico. *In* "Oceanic Sound Scattering Prediction" (N. R. Andersen and B. J. Zahuranec, eds) pp. 767–778. Plenum Press, New York

Koblentz-Mishke, O. J., Volkovinsky, V. V. and Kabanova, J. G. (1970). Plankton primary production of the world ocean. *In* "Scientific Exploration of the South Pacific" (W. S. Wooster, ed.) pp.183–193. National Academy of Sciences, Washington

Krey, J. and Babenerd, B. (1976). "Phytoplankton Production. Atlas of the International Indian Ocean Expedition". Institut für Meereskunde, Kiel

Lai, D. Y. and Richardson, P. L. R. (1977). Distribution and movement of Gulf Stream rings. *J. Phys. Oceanogr.* **7**, 670–683

Legand, M., Bourret, P., Fourmanoir, P., Grandperrin, R., Gueredrat, J. A., Michel, A., Rancurel, P., Repelin, R. and Roger, C. (1972). Relations trophiques et distributions verticales en milieu pélagique dans l'Océan Pacifique intertropical. *Cah. O.R.S.T.O.M.*, *sér. Océanogr.* **10**, 303–393

Lisitzin, A. P. (1970). Sedimentation and geochemical considerations. *In* "Scientific Exploration of the South Pacific" (W. S. Wooster, ed.) pp. 89–132. National Academy of Sciences, Washington

Longhurst, A. R. (1976a). Interactions between zooplankton and phytoplankton profiles in the eastern tropical Pacific Ocean. *Deep-Sea Res.* **23**, 729–754

Longhurst, A. R. (1976b). Vertical migration. *In* "The Ecology of the Seas" (D. H. Cushing and J. J. Walsh, eds) pp. 116–137. Blackwell, Oxford

Lorenzen, C. J. (1967). Vertical distribution of chlorophyll and phaeo-pigments: Baja California. *Deep-Sea Res.* **14**, 735–746

McCarthy, J. J. (1972). The uptake of urea by natural populations of marine phytoplankton. *Limnol. Oceanogr.* **17**, 738–748

McGowan, J. A. (1974). The nature of oceanic ecosystems. *In* "The Biology of the Oceanic Pacific" (C. B. Miller, ed.) pp. 9–28. Oregon State University Press, Corvallis

McGowan, J. A. (1977). What regulates pelagic community structure in the Pacific? *In* "Oceanic Sound Scattering Prediction" (N. R. Andersen and B. J. Zahuranec, eds) pp. 423–443. Plenum Press, New York

McGowan, J. A. and Hayward, T. L. (1978). Mixing and oceanic productivity. *Deep-Sea Res.* **25**, 771–793

Mague, T. H., Weare, N. M. and Holm-Hansen, O. (1974). Nitrogen fixation in the North Pacific Ocean. *Mar. Biol.* **24**, 109–119

Matsumoto, W. (1975). Distribution, relative abundance, and movement of skipjack tuna, *Katsuwonus pelamis*, in the Pacific Ocean based on Japanese tuna longline catches. 1964–67. *NOAA Tech. Rep. NMFS SSRF* 695

Mazeika, P. A. (1967). Thermal domes in the eastern tropical Atlantic Ocean. *Limnol. Oceanogr.* **12**, 537–539

Menzel, D. W. (1967). Particulate organic carbon in the deep sea. *Deep-Sea Res.* **14**, 229–238

Menzel, D. W. (1974). Primary productivity, dissolved and particulate organic matter and the sites of oxidation of organic matter. *In* "The Sea" (E. D. Goldberg, ed.) Vol. V, pp. 659–678. Wiley, New York

Menzel, D. W. and Ryther, J. H. (1960). The annual cycle of primary production in the Sargasso Sea off Bermuda. *Deep-Sea Res.* **6**, 351–367

Menzel, D. W. and Ryther, J. H. (1961). Zooplankton in the Sargasso Sea off Bermuda and its relation to organic production. *J. Cons. Perm. Internat. Explor. Mer* **26**, 250–258

Moore, E. and Sander, F. (1977). A study of the offshore zooplankton of the tropical western Atlantic near Barbados. *Ophelia* **16**, 77–96

Mullin, M. M. (1969). Production of zooplankton in the ocean: the present status and problems. *Oceanogr. Mar. Biol. Ann. Rev.* **7**, 293–314

Muromtsev, A. M. (1958). "The Principal Hydrological Features of the Pacific Ocean". Gimiz, Leningrad, in Russian; Eng. transl. 1963. Israel Programme for Scientific Translations, Jerusalem

Nesis, K. N. (1970). The biology of the giant squid of Peru and Chile, *Dosidicus gigas*. *Oceanol.* **10**, 108–118. (Eng. transl.)

Nilsson, C. S., Andrews, J. C. and Scully-Power, P. (1977). Observations of eddy formation off east Australia. *J. Phys. Oceanogr.* **7**, 659–669

Odum, E. P. (1969). The strategy of ecosystem development. *Science* **164**, 262–270

Omori, M. (1974). The biology of pelagic shrimps in the ocean. *Adv. Mar. Biol.* **12**, 233–324

Ortner, P. B., Wiebe, P. H., Haury, L. and Boyd, S. (1978). Variability in zooplankton biomass distribution in the northern Sargasso Sea: the contribution of Gulf Stream cold core rings. *Fish. Bull. U.S.* **76**, 323–334

Owen, R. W. and Zeitzschel, B. (1970). Phytoplankton production: seasonal changes in the oceanic eastern tropical Pacific. *Mar. Biol.* **7**, 32–36

Parsons, T. R. and Boom, B. R. L. (1974). The control of ecosystem processes in the sea. *In* "The Biology of the Oceanic Pacific" (C. B. Miller, ed.) pp. 29–58. Oregon State University Press, Corvallis

Petipa, T. S. (1979). Studies in trophic relationships in pelagic communities of the southern seas of the USSR and in the tropical Pacific. *In* "Marine Population Mechanisms" (M. J. Dunbar, ed.) pp. 233–250. Cambridge University Press, London

Rao, T. S. S. (1973). Zooplankton studies in the Indian Ocean. *In* "The Biology of the Indian Ocean" (B. Zeitzschel, ed.) pp. 243–255. Springer-Verlag, New York

Reid, J. L. (1959). Evidence of a South Equatorial Countercurrent in the Pacific Ocean. *Nature* **184**, 209–210

Reid, J. L. (1961). On the geostrophic flow at the surface of the Pacific Ocean with respect to the 1,000-decibar surface. *Tellus* **13**, 489–502

Reid, J. L., Nowlin, W. D. and Patzert, W. C. (1977). On the characteristics and circulation of the southwestern Atlantic Ocean. *J. Phys. Oceanogr.* **7**, 62–91

Reid, J. L., Brinton, E., Fleminger, A., Venrick, E. L. and McGowan, J. A. (1978). Ocean circulation and marine life. *In* "Advances in Oceanography" (H. Charnock and G. Deacon, eds) pp. 65–130. Plenum Press, New York

Richardson, P. L., Strong, A. E. and Knauss, J. A. (1973). Gulf Stream eddies: recent observations in the western Sargasso Sea. *J. Phys. Oceanogr.* **3**, 297–301

Riley, G. A. (1970). Particulate organic matter in sea water. *Adv. Mar. Biol.* **8**, 1–118

Robinson, M. K. (1976). Atlas of North Pacific Ocean monthly mean temperatures and mean salinities of the surface layer. *Naval Oceanogr. Office, Ref. Publ.* 2, Washington

Roger, C. and Grandperrin, R. (1976). Pelagic food webs in the tropical Pacific. *Limnol. Oceanogr.* **21**, 731–735

Rowe, G. T. (1971). Benthic biomass and surface productivity. *In* "Fertility of the Sea" (J. D. Costlow, ed.) Vol. II, pp. 441–454. Gordon and Breach, New York

Royer, T. C. (1978). Ocean eddies generated by seamounts in the North Pacific. *Science* **199**, 1063–1064

Ryther, J. H. (1963). Geographic variations in productivity. *In* "The Sea" (M. N. Hill, ed.) Vol. II, pp. 347–380. Interscience, New York

Ryther, J. H. (1969). Photosynthesis and fish production in the sea. *Science* **166**, 72–76

Sheldon, R. W., Sutcliffe, W. H. and Prakash, A. (1973). The production of particles in the surface waters of the ocean with particular reference to the Sargasso Sea. *Limnol. Oceanogr.* **18**, 719–733

Shomura, R. S. and Nakamura, E. L. (1969). Variations in marine zooplankton from a single locality in Hawaiian waters. *Fish. Bull. U.S.* **68**, 87–100

Shulenberger, E. (1978). The deep chlorophyll maximum and mesoscale environmental heterogeneity in the western half of the North Pacific central gyre. *Deep-Sea Res.* **25**, 1193–1208

Sorokin, Y. I. (1978). Decomposition of organic matter and nutrient regeneration. *In* "Marine Ecology" (O. Kinne, ed.) Vol. IV, pp. 501–616. Wiley, Chichester

Steele, J. H. and Yentsch, C. S. (1960). The vertical distribution of the chlorophyll. *J. Mar. Biol. Ass. U.K.* **39**, 217–226

Steven, D. M. (1971). Primary productivity of the tropical western Atlantic Ocean near Barbados. *Mar. Biol.* **10**, 261–264

Stewart, W. D. P. (1971). Nitrogen fixation in the sea. *In* "Fertility of the Sea" (J. D. Costlow, ed.) Vol. II, pp. 537–564. Gordon and Breach, New York

Sverdrup, H. U., Johnson, M. W. and Fleming, R. H. (1942). "The Oceans: their Physics, Chemistry and General Biology". Prentice-Hall, New York

Taniguchi, A. (1973). Phytoplankton-zooplankton relationships in the Western Pacific Ocean and adjacent seas. *Mar. Biol.* **21**, 115–121

Thomas, W. H. (1970). Effect of ammonium and nitrate concentration on chlorophyll increases in natural tropical Pacific phytoplankton populations. *Limnol. Oceanogr.* **15**, 386–394

Thomas, W. H. (1972). Nutrient inversions in the southeastern tropical Pacific Ocean. *Fish. Bull. U.S.* **70**, 929–932

Thomas, W. H. (1977). Nutrient-phytoplankton interrelationships in the eastern tropical Pacific Ocean. *Inter-Amer. Trop. Tuna Commn., Bull.* **17**, 173–212. (In English and Spanish)

Thomas, W. H. (1979). Anomalous nutrient-chlorophyll interrelationships in the offshore eastern tropical Pacific Ocean. *J. Mar. Res.* **37**, 327–335

Thomas, W. H., Renger, E. H. and Dodson, A. N. (1971). Near-surface organic nitrogen in the eastern tropical Pacific Ocean. *Deep-Sea Res.* **18**, 65–71

Thorne, R. E., Mathisen, O. A., Trumble, R. J. and Blackburn, M. (1977). Distribution and abundance of pelagic fish off Spanish Sahara during CUEA Expedition JOINT-1. *Deep-Sea Res.* **24**, 75–82

Timonin, A. G. (1973). Structure of pelagic communities. Trophic structure of zooplankton communities in the northern part of the Indian Ocean. *Oceanol.* **13**, 85–93. (Eng. transl.)

Tont, S. A. (1977). Daytime depths of sound scattering layers in the major biogeographic regions of the Pacific Ocean. *In* "Oceanic Sound Scattering Prediction" (N. R. Andersen and B. J. Zahuranec, eds) pp. 811–815. Plenum Press, New York

Tranter, D. J. (1973). Seasonal studies of a pelagic ecosystem (Meridian 110°E). In "The Biology of the Indian Ocean" (B. Zeitzschel, ed.) pp. 487–520. Springer-Verlag, New York

Tranter, D. J. and Kerr, J. D. (1977). Further studies of plankton ecosystems in the eastern Indian Ocean. III. Numerical abundance and biomass. *Aust. J. Mar. Freshwat. Res.* **28**, 557–583

Vedernikov, V. I. (1976). Dependence of the assimilation number and concentration of chlorophyll-a on water productivity in different temperature regions of the World Ocean. *Oceanol.* **15**, 482–485. (Eng. transl.)

Venrick, E. L. (1972). Small-scale distributions of oceanic diatoms. *Fish. Bull. U.S.* **70**, 363–372

Venrick, E. L., McGowan, J. A. and Mantyla, A. W. (1973). Deep maxima of photosynthetic chlorophyll in the Pacific Ocean. *Fish. Bull. U.S.* **71**, 41–52

Vinogradov, M. E. (1968). "Vertical Distribution of the Oceanic Zooplankton". Nauka, Moscow, in Russian; Eng. transl. 1970. Israel Programme for Scientific Translations, Jerusalem

Vinogradov, M. E., Gitelzon, I. I. and Sorokin, Y. I. (1970). The vertical structure of a pelagic community in the tropical ocean. *Mar. Biol.* **6**, 187–194

Vinogradov, M. E., Krapivin, V. F., Menshutkin, V. V., Fleyshman, B. S. and Shushkina, E. A. (1973). Mathematical model of the functions of the pelagical ecosystem in tropical regions. *Oceanol.* **13**, 704–717. (Eng. transl.)

Waldron, K. D. (1964). Fish schools and bird flocks in the central Pacific Ocean, 1950–1961. *U.S. Fish. Wildl. Serv., Spec. Sci. Rep. Fish.* 464

Walsh, J. J. (1976). Herbivory as a factor in patterns of nutrient utilization in the sea. *Limnol. Oceanogr.* **21**, 1–13

Wangersky, P. J. (1976). Particulate organic carbon in the Atlantic and Pacific Oceans. *Deep-Sea Res.* **23**, 457–465

Wangersky, P. J. (1978). Production of dissolved organic matter. In "Marine Ecology" (O. Kinne, ed.), Vol. IV, pp. 115–220. Wiley, Chichester

Wiebe, P. H., Hulbert, E. M., Carpenter, E. J., Jahn, A. E., Knapp, G. P., Boyd, S. H., Ortner, P. B. and Cox, J. L. (1976). Gulf Stream cold core rings: large-scale interaction sites for open ocean plankton communities. *Deep-Sea Res.* **23**, 695–710

Wiebe, P. H. and Boyd, S. H. (1978). Limits of *Nematoscelis megalops* in the Northwestern Atlantic in relation to Gulf Stream cold core rings. I. Horizontal and vertical distributions. *J. Mar. Res.* **36**, 119–142

Woods, J. D. (1977). Turbulence as a factor in sound scattering in the upper ocean. In "Oceanic Sound Scattering Prediction" (N. R. Andersen and B. J. Zahuranec, eds) pp. 129–145. Plenum Press, New York

Wooster, W. S. (1959). Oceanographic observations in the Panama Bight, "Askoy" Expedition 1941. *Amer. Mus. Nat. Hist., Bull.* **118**, 115–151

Worthington, L. V. (1976). "On the North Atlantic Circulation". Johns Hopkins University Press, Baltimore

Wroblewski, J. S. (1977). Vertically migrating herbivorous plankton – their possible role in the creation of small scale phytoplankton patchiness in the ocean. In "Oceanic Sound Scattering Prediction" (N. R. Andersen and B. J. Zahuranec, eds) pp. 817–847. Plenum Press, New York

Wyrtki, K. (1962). The upwelling in the region between Java and Australia during the south-east monsoon. *Aust. J. Mar. Freshwat. Res.* **13**, 217–225

Wyrtki, K. (1964). Upwelling in the Costa Rica Dome. *Fish. Bull. U.S.* **63**, 355–372

Wyrtki, K. (1966). Oceanography of the eastern equatorial Pacific Ocean. *Oceanogr. Mar. Biol. Ann. Rev.* **4**, 33–68

Wyrtki, K. (1973). Physical oceanography of the Indian Ocean. *In* "The Biology of the Indian Ocean" (B. Zeitzschel, ed.) pp. 18–36. Springer-Verlag, New York

Wyrtki, K. (1974). The dynamic topography of the Pacific Ocean and its fluctuations. *Hawaii Inst. Geophys., Univ. Hawaii*, HIG-74-5

Wyrtki, K. (1975). Fluctuations of the dynamic topography in the Pacific Ocean. *J. Phys. Oceanogr.* **5**, 450 459

Wyrtki, K., Magaard, L. and Hagen, J. (1976). Eddy energy in the oceans. *J. Geophys. Res.* **81**, 2641–2646

Yoshida, K. and Kidokoro, T. (1967). A subtropical countercurrent. II. A prediction of eastward flows at lower subtropical latitudes. *J. Oceanogr. Soc. Japan* **23**, 231–246

Zeitzschel, B. (1969). Productivity and microbiomass in the tropical Atlantic in relation to the hydrographical conditions (with emphasis on the eastern area). *In* "Proceedings of the Symposium on the Oceanography and Fisheries Resources of the Tropical Atlantic (Abidjan, Ivory Coast, 1966). Review Papers and Communications", pp. 69–84. UNESCO, Paris

2 Coastal Upwelling Ecosystems

RICHARD T. BARBER and ROBERT L. SMITH

1 Introduction

Does coastal upwelling merit consideration as an ecosystem that can be clearly distinguished from other coastal or continental shelf ecosystems? An ecosystem is a functional unit of physical and biological organisation with characteristic trophic structure and material cycles, some degree of internal homogeneity, and recognisable boundaries (Odum, 1977). Researchers who have worked in upwelling regions, such as Margalef (1978a,b), Vinogradov and Shushkina (1978), Walsh (1977), and Boje and Tomczak (1978), agree that upwelling ecosystems are distinct from other classes of marine ecosystems, but they disagree as to what the nature of the coastal upwelling ecosystem is. In exception to the general agreement that coastal upwelling ecosystems are a distinct class of ecosystem, Cushing (1971, 1978) has argued that coastal upwelling ecosystems are not distinct ecosystems and that more understanding will accrue if it is recognised that 'production cycles in upwelling areas and temperate waters are essentially the same' (Cushing, 1971). Cushing's theses that upwelling and temperate waters form a continuum and that upwelling food webs are less efficient than others controvert Ryther's (1969) synthesis of primary production and fish yield information for the entire world ocean in which it was argued that 'two sets of variables – primary production and the associated food chain dynamics – may act additively to produce differences in fish production' that set upwelling ecosystems apart from other shallow or open-ocean marine ecosystems. Ryther's (1969) statement of the uniqueness of upwelling ecosystems and his assertion that 50% of the world's fish catch comes from the 0.1% of the ocean's area where coastal upwelling occurs have been the central tenet of upwelling research of the last decade, while Cushing's (1971) disagreement has been the major heresy regarding upwelling ecosystems. Neither view can be comfortably accepted or rejected *in toto* because the necessary comparison experiments between types of ecosystems have not been done. The conceptual climate for an experimental test of the Ryther (1969) and Cushing (1971) ideas seems not to exist yet, perhaps because of the uncertainty concerning the specific nature of the upwelling ecosystem: if it is not possible to specify what the nature is, is it possible to design an experiment comparing coastal, open-ocean and upwelling areas? Another factor prompting caution in the design of a large, between-ecosystem comparative experiment is recent awareness of pronounced large-scale interannual variations in ecosystem structure and function. Experience gained in 1976 from the east coast of the USA (Sharp, 1976), from the coast of Peru (Dugdale et al., 1977), and from the eastern equatorial Pacific (Cowles et al., 1977) emphasises that temporal variation in the quantitative and qualitative character of coastal, upwelling, and open-ocean ecosystems can be greater than the between-ecosystem differences. Until the sources of temporal variations within each type of ecosystem are resolved it will be difficult to test the between-ecosystem hypothesis formalised by Ryther (1969).

Among workers supporting the idea that coastal upwelling ecosystems are a distinct class (e.g. Walsh, 1977; Margalef, 1978a,b; Vinogradov and Shushkina, 1978; and Boje and Tomczak, 1978), each author has given a different description of the coastal upwelling ecosystem. It is not clear whether these differences arise from studying the same entity from a different perspective or whether the geographic and temporal variations in coastal upwelling ecosystems are so great that separate ecosystem studies will each produce a unique holistic description. Furthermore, it is unknown whether there is a single coastal upwelling ecosystem demonstrating a continuum of variability as a function of latitude, shelf width, atmospheric forcing, etc., or whether there are a set of coastal upwelling ecosystems that are alternative steady-states (Sutherland, 1974). Is coastal upwelling an ecosystem or a class of ecosystems?

2 What is coastal upwelling?

Coastal upwelling is the process in which subsurface water is brought to the surface and moved away from the area of vertical transport by horizontal surface flow. Because horizontal divergence of the surface waters from the coast induces ascending vertical transport to replace the water moved off, it is necessary to understand the forces that initiate the surface divergent flow and the subsurface flow that converges with the coast to understand how the upwelling circulation functions as a direct couple between the atmosphere and the biota. To begin, however, it is useful to establish why coastal upwelling ecosystems differ from other regions of the ocean. The reason simply stated is that in coastal upwelling optimal conditions of nutrient supply are provided by vertical transport of subsurface water into the euphotic layer, and optimal light conditions for phytoplankton are maintained in the stabilised horizontal divergent flow of the surface layer. Optimal nutrient and light conditions are maintained for long periods of time each year, so the annual amount and pattern of biological productivity are distinct from other regions of the ocean.

The fundamental generalisation describing regulation of productivity in ocean waters was made by Redfield (1934), Riley (1947), and Sverdrup (1953) who established that the supply of inorganic nutrients and amount of available light determine the overall geographic pattern and temporal sequence of ocean productivity. In low and mid-latitude regions where and when there is a net positive heat flux to the ocean, the surface waters warm, producing a density barrier (pycnocline) that prevents mixing of the surface and subsurface waters. After dissolved inorganic nutrients are taken out of solution and converted to the particulate phase by phytoplankton the warmed and stabilised surface waters remain depleted of dissolved inorganic nutrients. In mid and high latitudes when and where there is a net negative heat flux to the ocean, the surface waters cool, become unstable and mix with the underlying waters. Under conditions of heat loss, strong winds, and low light input, the concentration of inorganic nutrients will be high due to deep mixing, but the mixing will carry phytoplankton below the euphotic zone where respiration will exceed photosynthesis. In accordance with the critical depth formulation of Sverdrup (1953) the strength of the mixing will directly regulate growth of the phytoplankton population and will prevent net population increases when the mixing exceeds a critical value. The synthesis of new organic matter by phytoplankton proceeds in temperate and polar regions because vertical mixing and stratification occur in a seasonal time sequence in the same spatial domain.

Coastal upwelling is a circulation pattern that overrides both the nutrient limitation of stratified waters and the light limitation of well mixed waters that develops when the wind transports water offshore in the surface layer, resulting in an equal amount of water welling up near the coast to replace this water. The surface layer directly affected by the wind is the order of tens of meters deep and is called the Ekman layer after V. W. Ekman who provided the theoretical basis for understanding the effect of the wind on ocean currents (Ekman, 1905). Because of the Coriolis force, the direction of net transport is 90° to the right (left) of the wind in the northern (southern) hemisphere. Ekman transport is defined as the mass of water moved, in a direction perpendicular to the wind, through a strip of unit width extending from the surface to the bottom of the Ekman layer. Nutrient rich subsurface water is thus brought to the surface and, once at the surface, it flows horizontally away from the site of upwelling in a coherent surface flow. The injection of new nutrients to the surface and the formation of a shallow wind-driven surface Ekman layer that is separated from the subsurface onshore flow by a density gradient are analogous to the seasonal mixing and stratification cycle occurring in temperate and polar waters. What is different in coastal upwelling is that a spatial sequence is substituted for a seasonal sequence; therefore, the limits on duration of the productivity event are removed but spatial restraints are imposed. The total annual flux of new organic material to the ecosystem will be much greater, but the size and location are limited.

In seasonally mixed ecosystems there is one major productivity event per year, while in coastal upwelling systems the duration of conditions favorable for maximum productivity is significantly longer. On the coast of Peru upwelling is continuous throughout the year between 5°S and 25°S, although there is a clear seasonal cycle in the strength of upwelling (Wooster, 1970). Off Oregon at 45°N latitude, during the six-month upwelling season (Bakun et al., 1974), there are usually 4 or 5 significant upwelling events (Huyer, 1976), each of which has the quantitative impact of an annual spring bloom, in addition to many of lesser intensity. In this context the difference between coastal upwelling ecosystems and other coastal and ocean ecosystems is a matter of quantity: the annual flux of new inorganic nutrients, as defined by Dugdale and Goering (1967), to the euphotic zone is much higher. Thus the quantity of organic material that may be exported from the system as a commercial catch, as a loss to the sediments or to the adjacent mid-depth water column, is 10 to 100-fold higher than the material export rates that other systems can support. Cushing's (1971) argument that there is no fundamental difference in the structure or function of seasonally mixed temperate zone waters and coastal upwelling but that they are both points on a quantitative continuum prompts two observations. First, the quantitative difference alone gives coastal upwelling regions their biological, chemical, geological and economic character; the saying that quantity has a quality of its own describes why coastal upwelling ecosystems produce mid-water oxygen minimum layers, high organic sediments, anoxia, denitrification and hydrogen sulfide events in subsurface waters, large bird populations, and guano deposits. The biotic, geologic and chemical consequences of increasing the intensity and duration of productivity in a small spatial domain are nonlinear to the degree that coastal upwelling ecosystems have qualitative features that are not present in seasonally mixed coastal and continental shelf ecosystems. Quantity does appear to have a quality of its own. A more serious flaw in Cushing's (1971) hypothesis is the overlooking of the effect of circulation in coastal upwelling ecosystems in providing structure and energy to the biota. In the upwelling circulation the diverging alongshore surface circulation and subsurface countercurrent provide vertical and horizontal sensory cues to the biota that enable them to maintain themselves in the ecosystem and these flows provide a subsidy of auxiliary energy that enhances productivity (Margalef, 1978a and Fig. 11). Observational and model studies have demonstrated how the coastal upwelling circulation is exploited by zoo-plankton (Peterson et al., 1979; Brockmann, 1979) and fish (Mathisen, 1980) to maintain themselves in the ecosystem.

The simultaneous occurrence of three processes in a narrow coastal band makes coastal upwelling ecosystems different in a modal sense from other marine ecosystems. The processes are: 1. vertical transport of inorganic nutrients to the euphotic zone, 2. formation of a divergent surface Ekman layer that is distinct from the convergent subsurface flow and 3. the presence of a two-layered alongshore flow with a poleward undercurrent beneath the equatorward flow. The development of these three processes varies greatly in time even during the upwelling season and varies greatly in space down to a few kilometers in an upwelling region; but although varying and intermittent in space and time, these processes have enough regular structure that organisms have evolved mechanisms to exploit the processes. The physical processes of coastal upwelling have enough time-space stability, to use Sanders' (1969) terminology, to permit the evolution of coastal upwelling ecosystems; but one of the 'stable' characteristics is the regular occurrence of strong variations in the physical processes. The variability versus constancy question in coastal upwelling ecosystems has led to some semantic confusion. In response to variation in the local winds the upwelling process will start and stop (Figs 4 and 5); but the pattern of response to the atmospheric driving is characteristic for each upwelling region, so it is possible to describe a characteristic pattern of processes for each system.

3 The physical setting

The major coastal upwelling regions are located along the eastern boundaries of the oceans where predominant equatorward winds are part of the more or less stationary mid-ocean atmospheric high pressure systems. Figure 1 shows the four high-pressure systems that are responsible for the large-scale wind patterns that drive the four major regions of coastal upwelling off North America, South America, Northwest Africa and Southwest Africa. A fifth region, the Somali and Arabian coast, exhibits strong upwelling when the southwest monsoon directly transports surface water away from the coast. Wherever the wind is parallel to the coast, with the coast on the left (right) relative to the wind direction in the northern (southern) hemisphere, one may expect some coastal upwelling but it is along the eastern ocean boundaries (and the Somali and Arabian coasts) that the wind is persistently, in a climatological sense, favorable for coastal upwelling for enough time and over a sufficiently large length of coastline that a distinct coastal upwelling ecosystem can develop.

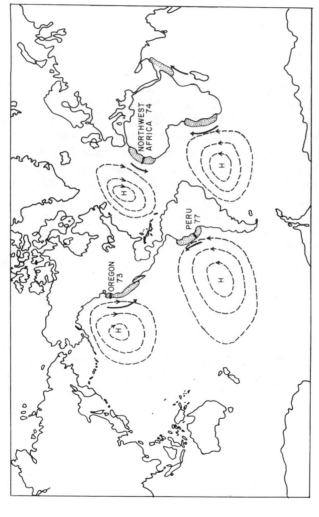

FIG. 1   Major coastal upwelling regions of the world and the sea-level atmospheric pressure systems (anticyclones) that influence them. The dashed circles represent mean idealised positions of isobars during the season of maximum upwelling in a given region. Major areas of upwelling are shown by the stippled areas; heavy lines within show the areas of study discussed in this chapter. Arrows indicate the location of the (a) California current off USA, (b) Peru current off Peru, (c) Canary current off Northwest Africa, (d) Benguela current off Southwest Africa, and (e) the Somali current in the Indian Ocean east of Africa. From Thompson (1977)

In the last decade a number of multidisciplinary research groups from several countries have worked in all of the major coastal upwelling regions; from this work a consensus about the physical dynamics of coastal upwelling ecosystems has evolved. A major idea is that coastal upwelling ecosystems are dependent on the large-scale atmospheric and oceanic processes controlling the winds and the vertical distribution of density and nutrients, but the process of upwelling *per se* is inherently mesoscale (5 to 50 km and 1 to 10 days) so that the density, nutrient and productivity distributions are dependent on both mesoscale and large-scale processes. This idea is supported by observations and modelling (O'Brien *et al.*, 1977).

The actual upwelling, that is, the vertical motion into the euphotic zone, occurs in an inshore region that has an intrinsic scale (perpendicular to the coast) which is much smaller than the spatial scale of the wind stress. Theoretical considerations (Charney, 1955; Yoshida, 1967) show this radius of deformation of the density field to be on the order of $H\bar{N}/f$ where f is the Coriolis parameter, $\bar{N}$ the mean Brunt-Väisälä (or stability) frequency and H the depth of the water. For the deep ocean at mid-latitudes the radius of deformation is on the order of 80 km. For the coastal regions the water is shallower although the average stratification may be greater, so that the radius of deformation is less, about 10–20 km.

Coastal upwelling is a mesoscale ocean response to large-scale wind driving. Implicit in this concept is the notion that the water upwelling to the surface layer is coming from just below the pycnocline or thermocline. The pycnocline along the eastern boundaries of the ocean basin is shallower than in mid ocean or along the western boundaries consistent with the equatorward flowing eastern boundary currents. Thus the large-scale wind system provides the driving force and the large-scale circulation pattern determines the water properties (density and nutrient content) for upwelling.

The wind strength varies, not only seasonally, but on shorter time scales. The energetic fluctuations occur on the several-day fluctuations (the weather) and, in many coastal regions, on a diel period (the land-sea breeze). The time for the upwelling circulation to respond to the wind is of the order of f^{-1}, the inertial period; the diel fluctuations in wind are usually averaged out for low and mid latitudes because the inertial period is from 17 h at 45° to 5 days at 5°, but the diel sea breeze may have biological effects due to the changes in the depth of the mixed layer it induces. In the temporal domain there is a mixture of scales, as there is in space. The seasonal or longer term large-scale wind systems provide the mean winds favorable to upwelling but high frequency weather events may limit productivity (Huntsman and Barber, 1977) or the survival of organisms with critical life cycle stages (Lasker, 1978; Walsh, 1978).

To describe the physical environment of coastal upwelling ecosystems we will describe the wind (or wind stress), stratification, bathymetry and flow pattern (the currents) using observations recently made in three major upwelling regions. In Oregon in 1973, Northwest Africa in 1974, and Peru in 1977, a continuous two-month long series of wind and current measurements were made at a mid-shelf location near, but offshore from, the region of maximal vertical transport; and detailed hydrographic and productivity measurements were made on a section through the current meter moorings and perpendicular to the coast (Fig. 2). These observations were made by the Coastal Upwelling Ecosystems Analysis program during an interdisciplinary investigation of coastal upwelling. For comparison, we have chosen equivalent data sets from three expeditions: JOINT-I off Northwest Africa near 21°40'N during March and April 1974, CUE-II off Oregon at 45°N during July and August 1973, and JOINT-II off Peru at 15°S during April and May 1977. The reduced data from these expeditions are available in data reports, listed in Barber (1977, 1979), from Duke University, Beaufort, North Carolina 28516, USA.

In comparing coastal upwelling regions, one must confront the problem of scales: the physical processes important in determining the productivity and character of a coastal upwelling ecosystem occur on different time and space scales. Arbitrarily we define three: the scale of fronts and plumes (0.1 to 50 km and hours to a few days); the scale of wind events (10 to 1000 km and days to a few weeks); the scale of seasonal and interannual phenomena, that is, El Nino (500 to 5000 km and months to years). We will not address the first scale. Some significant advances in our understanding of fronts and plumes will probably occur in the next few years, but our present ideas are mostly speculative. The largest scale is mentioned to introduce differences among the three regions. The winds are favorable for coastal upwelling throughout the year off both Northwest Africa (but strongest from March through July at 22°N) and Peru (but strongest in June and August at 15°S). Off Oregon at 45°N the wind is favorable for upwelling from April through

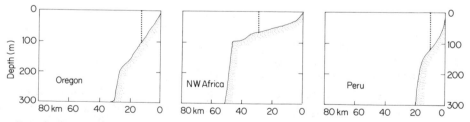

FIG. 2   Cross-shelf bathymetry and location of the mid-shelf current meter array and meteorology buoys in the three upwelling ecosystems described here; the section off Oregon is at 45°16'N, off Northwest Africa at 21°40'N and off Peru at 15°S

September, strongest in July, and unfavorable the other six months. Discussions of the seasonal cycle in the three regions can be found in Wooster *et al.* (1976), Zuta and Guillen (1970) and Bakun *et al.* (1974). Of the three regions, only Peru seems to suffer the occasional catastrophic warming called El Nino. Although the effects of El Nino are similar to what one might expect from a cessation of coastal upwelling, coastal winds do not diminish (Wyrtki, 1975) and the coastal upwelling process continues – but with warmer, and unproductive, water. In general, the degree and mode of coupling of the large-scale circulation to that in the coastal upwelling region are not yet well understood, but connections between the large-scale distribution of properties (density, nutrients) are clear.

Table I shows that in all three regions the mean alongshore component of wind stress is favorable to upwelling during the observation periods but Oregon has the greatest range and most temporal variability relative to the mean and Peru has the least. Upwelling off the coast of Oregon is intermittent during the May to September upwelling season (Huyer, 1976) but is continuous, although varying in strength, off the coast of Peru (Wooster and Sievers, 1970). Northwest Africa in the region between 20° to 25°N latitude has year-round upwelling (Wooster *et al.*, 1976) but with well defined and intermittent upwelling events (Barton *et al.*, 1977) because of the variations and reversals in the coastal winds (Table I).

The cross-shelf profiles of density or temperature from the three regions show the onset of upwelling in response to wind. The source water for upwelling comes from a depth of between 100 and 200 m off Oregon and

TABLE I Mean alongshore component of wind stress of dynes·cm^{-2} ± one standard deviation in three coastal upwelling regions during the upwelling season. Positive indicates equatorward (northward) wind stress in the southern hemisphere and negative indicates equatorward (southward) wind stress in the northern hemisphere. Wind stress is the force per unit area exerted by the wind on the sea surface and has the same direction as the wind but is proportional to the square of the wind speed. The alongshore component of the wind stress divided by the Coriolis parameter is an index of the coastal upwelling (Bakun *et al.*, 1974)

Region	$\bar{X}$	S.D.	Range
Oregon 5 July to 28 August 1973	−0.50	0.7	+1.10 to −3.48
Northwest Africa 10 March to 6 April 1974	−1.55	1.0	+0.52 to −3.34
Peru 5 March to 13 May 1977	+0.79	0.4	0.00 to +1.74

Northwest Africa (Fig. 3), but off Peru (Fig. 4) the source water comes from a shallower depth, 50 to 75 m. Water at these depths off Oregon, Northwest Africa, and Peru contains high concentrations of inorganic nutrients (nitrate, silicate, phosphate), so the upwelling process transports significant quantities of nutrients to the surface layer. The shallowness of the pycnocline in the eastern boundary regions of the ocean basins places high nutrient water within reach of the vertical upwelling circulation. In western boundary regions this condition is not met; Fahrbach and Meincke (1979) have shown that water upwells from 150 m in the Cabo Frio upwelling region on the Atlantic coast of Brazil but source water at that depth has relatively low nutrient content. The physical process of upwelling at Cabo Frio transports small quantities of nutrients to the euphotic zone, so there are relatively minor biological consequences.

Differences in stratification and mixing in the three regions are marked. Off Oregon (Fig. 3) the stratification across the narrow shelf is strong and the depth of the mixed layer is limited to less than 20 m. In contrast, off Northwest Africa the mixed layer frequently reaches to the bottom across the inner shelf; this deep mixing limited the overall productivity of the region (Fig. 10 and Table III) during the period when these physical and biological observations were made (Huntsman and Barber, 1977). The strong sea breeze on the coast of Peru drives a diel cycle of mixing but the depth was usually restricted to the upper 10 m and seldom exceeded 20 m (Brink *et al.*, 1980). The mean profile off Peru in 1977 at the mid-shelf location had continuous stratification in the upper 30 m (Fig. 10), indicating that deep mixing does not limit primary productivity in Peru as it does in Northwest Africa.

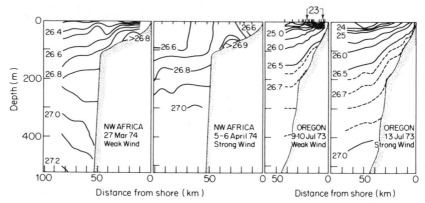

FIG. 3   Cross-shelf density profiles from Oregon and Northwest Africa that show the rapid transition from no upwelling to strong upwelling during strong winds. From Huyer (1976)

Detailed vertical profiles of mean cross-shelf and alongshore currents were made with heavily instrumented mooring in the three regions at mid shelf just offshore from the zone of maximum vertical transport; the mid-shelf location was selected for intensive study on the assumption that it is the site of maximum productivity. The vertical current profiles (Fig. 5) show classical cross-shelf upwelling current profiles with wind-driven offshore flow at the surface. The thickness of the time-averaged offshore flow was about 33 m off Africa, 19 m off Oregon, and about 24 m off Peru in 1977. The compensatory flow pattern, that is, the onshore flow beneath the wind-

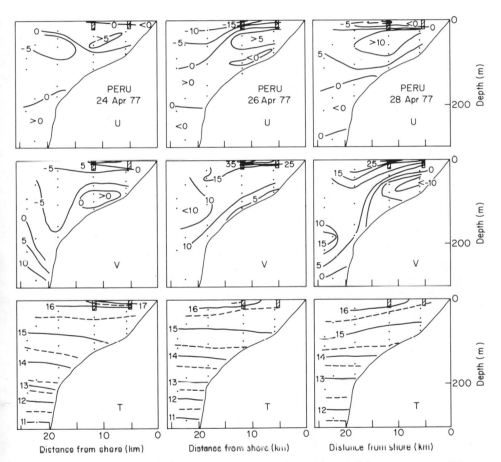

FIG. 4 Cross-shelf profiles of the changes in onshore/offshore currents (U), alongshore currents (V) and temperature (T) in response to wind changes; on 23 April the wind stress was zero, on 26 April 1.5 dynes·cm^{-2}, and on 28 April 1.0 dynes·cm^{-2}. From Brink et al. (1980)

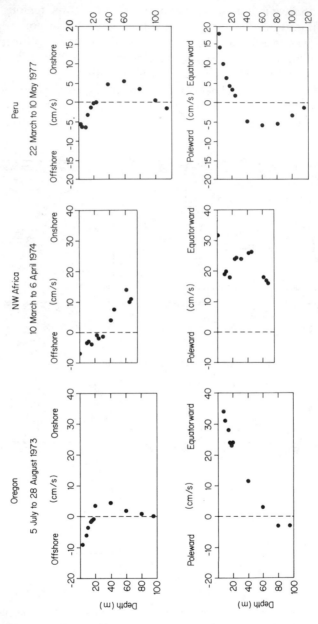

FIG. 5   Mean current profiles from the mid-shelf locations drawn in Fig. 2 for the three coastal upwelling ecosystems

driven upper layer, differs in the three regions. Off Northwest Africa the compensatory flow is strongest at the deepest current meter at mid shelf and seems to occur in a bottom Ekman layer; off Oregon, and off Peru, the mean flow at mid depth on the shelf is onshore. One feature of the flow pattern in Fig. 5 that immediately captures the attention of ecologists is that the thickness of the surface offshore flow is less than the thickness of the euphotic layer off Peru and Oregon, so phytoplankton in the subsurface onshore flow in those two regions will be exposed to low but significant light levels for several days while being advected onshore across the shelf to the inshore site of upwelling. Such preconditioning to light may be a regular feature off Oregon and Peru; Barber *et al.* (1971) described rapid initial growth rates and absence of a lag phase in newly upwelled and subsurface water as being characteristic off Peru in 1969.

The stratification and flow pattern interact in a manner that affects the distribution of both phytoplankton and zooplankton. Off Peru and Oregon the depth of the surface mixed layer is restricted to a depth that is less than the thickness of the offshore flow. Plankton and particle exchanges between the onshore and offshore flow take place by sinking or swimming; off Northwest Africa the mixing (Fig. 3) that accompanies each wind and upwelling event thoroughly mixes the assemblages in the two flows at intervals of a few days.

The vertical pattern of currents demonstrates why the sedimentary environments in these three regions differ. Strong onshore and alongshore (10 to 15 cm s^{-1}) currents are present next to the bottom off Northwest Africa; these currents prevent the accumulation of fine-grain, highly organic material and maintain the sediments in a coarse and well oxidised condition (Rowe *et al.*, 1977). The sediments where the Peru vertical current profile was measured were highly organic and reducing (Rowe, 1971) and bottom currents were weak or nonexistent (Fig. 5). The current profiles and density distributions in response to wind (Figs 3 and 4) account for why organic debris does not accumulate on the shelf in Northwest Africa but does off Peru.

Another feature of the three systems, and probably of all major coastal upwelling systems, is the poleward undercurrent (Smith, 1968). Off Northwest Africa and Oregon the mean flow is in the direction of the wind and the flow expected in eastern boundary currents is present throughout much of the water column. But in both regions a persistent counterflow, or undercurrent, exists. Off Peru a strong and shallower manifestation of the poleward undercurrent dominates flow in the upwelling region. The evidence is that the poleward undercurrent is a slope phenomenon, never occurring on the shelf off Africa, but frequently extending onto the shelf off Oregon, and dominating the shelf circulation off Peru.

4 Chemical environment

Coastal upwelling ecosystems exist where upwelling transports inorganic
nutrients from below the pycnocline to the euphotic layer; the high
concentrations of 'new' nutrients (Dugdale and Goering, 1967) are the
salient and long recognised (Gunther, 1936) property of coastal upwelling
ecosystems. In other marine ecosystems new nutrients are transported or
mixed into the euphotic zone only once or in a few episodes annually and
primary production proceeds through the rest of the year using nutrients
regenerated or recycled within the ecosystem (Dugdale, 1967; Eppley et al.,
1977, 1979b; McCarthy and Goldman, 1979). In coastal upwelling the
concentrations of nutrients in the euphotic zone may vary under the control
of the wind-driven variations in vertical transport (Figs 3 and 6) in
Northwest Africa and Oregon, or the concentrations may be more or less
continuously high over the shelf (Fig. 8) as seen in Peru in 1977. The high
mean nutrient conditions in major coastal upwelling ecosystems are well
documented (Gardner, 1977; Dugdale, 1972) and the sources of variability
have been described (Codispoti and Friederich, 1978). Over the shelf
nutrients are present episodically (Fig. 6) or continuously (Fig. 8) in
concentrations higher than the half saturation concentration (K_s) for
maximal nutrient uptake by phytoplankton (Dugdale, 1976). The presence
of saturating concentrations of nutrients results in the biological expression
of growth regulating processes unexpressed in ecosystems where nutrients
are in low and regulating concentrations (Dugdale, 1976). Since nutrients
are in saturating concentrations, the short-term or instantaneous regulation
of phytoplankton production is not controlled by nutrient concentrations
per se but by other factors and combinations of factors; off Northwest
Africa in 1974 and Peru in 1977 the nutrient conditions were different (Figs
9, 10 and Table III) but the mean primary production, biomass and specific
productivity were similar. The rich nutrient condition will support high
growth rates if the light supply is optimal; since the light supply to
phytoplankton is controlled by the advective and turbulent regimes, the
optimal chemical environment sets the stage for physical processes regulating primary production. In addition to regulating the rate of photosynthesis
through light, the transport and mixing processes will determine which
plankton, and especially which phytoplankton, will be present or able to
remain in the restricted space domain of coastal upwelling. This transport
relationship may be the major one determining what plankton will occur in
coastal upwelling ecosystems. Margalef (1978a) elegantly expressed this
idea:

Faced with the overwhelming importance of circulation and turbulence, a careful
consideration of the niceties of light distribution and of kinetics of absorption of

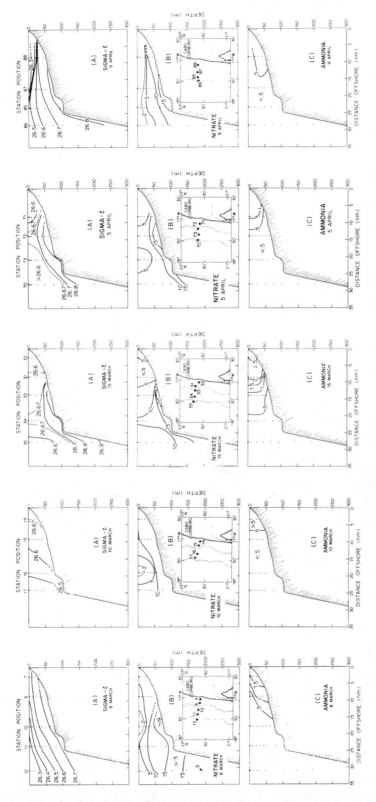

FIG. 6 Cross-shelf profiles of sigma-t, nitrate and ammonia from Northwest Africa showing the wind-driven and regeneration effects. From Codispoti and Friederich (1978)

nutrients may appear irrelevant. The ecosystem, in fact, is driven or controlled by external forces and any modest simulation of the orders of magnitude involved in the phenomena shows, at least, how immaterial is the choice among the different models of dependence of production on light and nutrients. Primary production appears simply as a function of the external energy supplied to the system and degraded in it.

We do not argue with the essence of Margalef's (1978a) suggestion, but we are not bold enough to omit further discussion of the chemical processes in coastal upwelling. In particular, it is necessary to consider connections between circulation, the chemical environment and microbial and macro-faunal benthic communities.

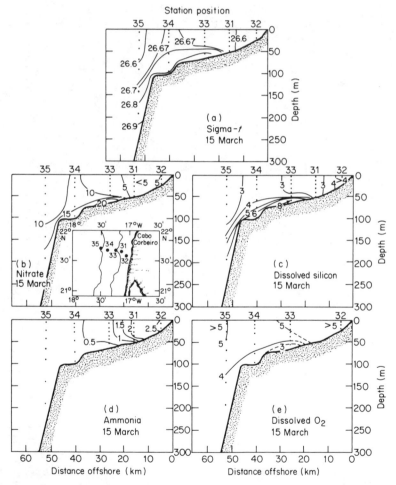

FIG. 7   Cross-shelf profiles of density, nutrients and oxygen from Northwest Africa during a period of strong winds

The central role played by the advection of new nutrients has over-shadowed the role of nutrient regeneration by the biota (including microorganisms) and nutrient trapping by the circulation pattern. Advection brings new nutrients into the ecosystem but regeneration and partitioning in the two-layered circulation maintains the concentrations of biogenic elements (N, P, Si and perhaps micronutrient metals) at high levels. Redfield *et al.* (1963) observed that concentrations of phosphate-phosphorus increase along any isohaline or isopycnal as one approaches a coastal upwelling area from offshore. This phenomenon is attributed to differential advection of water at different depths and the process at work in estuarine nutrient and sediment trapping (Redfield, 1955), in which sinking particulate matter (especially fecal pellets), living zooplankton, and fish transport biogenic elements in particulate phase out of the surface layer that flows out of the system and into the subsurface layer flowing into the system. The vertical transport of C, N, P, and Si by organisms and fecal pellets moves the elements in particulate phase between the differential flows and releases them in dissolved phase in the incoming subsurface flow. The particulate (living and dead) vertical transport between the differential flows and the dissolved horizontal transport within the flows results in counter-current concentration.

In Northwest Africa and Peru nutrient regeneration is quantitatively important (Whitledge, 1978 and Table II) and the sediment is a major site where dissolved nutrient elements are released after being transported out of the surface layer in organic particulates (Rowe, 1980; Rowe *et al.*, 1977).

TABLE II   The percent of nitrogen uptake by phytoplankton that is supplied by regenerated forms of nitrogen and the percent originating from different sources. From Whitledge (1980)

Upwelling area	Regenerated N Total N (%)	Zoo-plankton (%)	Micro-nekton (%)	Pelagic fish (%)	Demersal fish (%)	Benthos and sediments (%)
Northwest Africa						
Cap Blanc Shelf	72	33	—	15	—	24
Cap Blanc Outer Shelf	61	24	—	29	—	8
Peru						
Shelf	56	19	—	18		19
Offshore	42	23	—	5	1	13
Baja California						
Shelf – March	47	3	32	1	—	11
Shelf – April	52	16	23	1	—	12

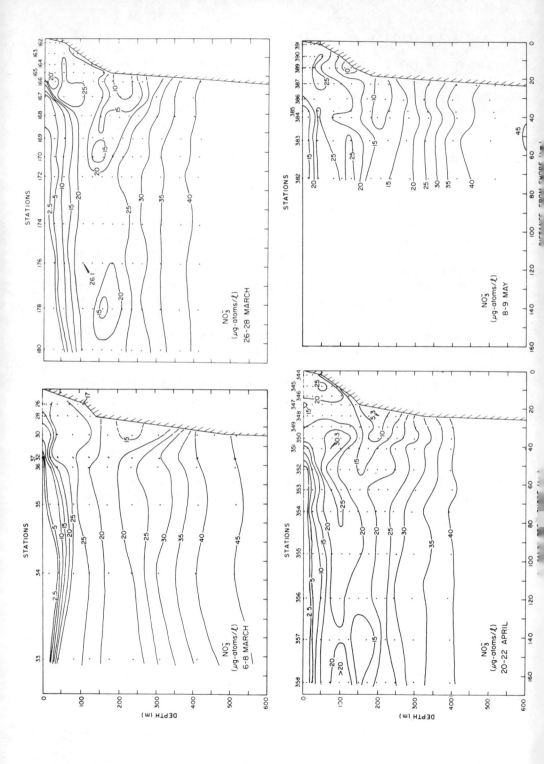

The results of Whitledge (1980) and Rowe (1980) showing high absolute rates of nutrient recycling within the ecosystem follow from the relationship defined by Eppley *et al.* (1979a):

the injection of new nitrogen into the euphotic zone results not only in increased phytoplankton production but also, as will be shown, in accelerated activities of the heterotrophic organisms responsible for remineralization.

... the nature of the local pelagic food web is such that inputs of new nitrogen ... elevate primary production and at the same time the flux of regenerated nitrogen, via animals and bacteria, returning to the phytoplankton ... Thus an input of nitrate can be expected to increase nitrogen, carbon, and energy flux throughout the food web, at least to the degree that the elements of the food web are present and responsive.

The potential importance of vertical transport of organic material in particulate phase has been recognised but proved to be difficult to demonstrate; sampling with conventional water bottles, Menzel (1967) found no 'raining down' of phytoplankton-sized particles under highly productive upwelling plumes off Peru. Observations with drifting and moored sediment traps made a decade later in the same locations (Staresinic, 1978; Rowe, 1980) indicated that 10 to 20 % of the surface primary productivity was rapidly transported to the bottom. The nutrient consequence of elemental trapping by the 'estuarine partitioning' (Redfield, 1955) is maintenance of elevated concentrations of phosphorus, nitrogen, and silicon in the bottom waters of the shelf and slope (Figs 6 and 7). Trapping of nutrient elements by the circulation and short-term storage of these elements in the sediments provide a flywheel mechanism to the nutrient supply process that damps out variations that would occur if wind-driven vertical transport were the only process providing nutrients to the euphotic zone.

An additional consequence of two-layered advective partitioning by the upwelling circulation is creation of zones of intense oxygen depletion or anoxia. Oxygen depletion is usually associated with stagnation of a lower layer of water relative to an upper layer that is supplying organic material. Redfield *et al.* (1963) demonstrate that stagnation is not a necessary condition: it is the ratio of the velocities in two layers that determines the magnitude of the accumulation of biogenic elements (C, N, P, and Si) or depletion of oxygen in the lower layer (Equations 7a, b and c in Redfield *et al.*, 1963). Coastal upwelling ecosystems have strong two-layered flows (Fig. 5), high surface productivity (Figs 10 and 15), and transport of material in

---

FIG. 8 Cross-shelf profiles of nitrate from sections at 15°S off Peru in March/April/May 1977 showing nitrate always higher than 20 µg atoms/l in shelf waters and the nitrate deficit due to partial denitrification in the shelf and slope bottom waters. From Hafferty *et al.* (1979)

particulate phase from the surface to subsurface layer (Rowe, 1980), so oxygen depletion will necessarily develop in the subsurface counterflow. Furthermore, according to Redfield *et al.* (1963), the depletion increases in magnitude when the subsurface flow accelerates, increasing the velocity ratio. This relationship accounts for why anoxia, denitrification and hydrogen sulfide generation were found off Peru (Dugdale *et al.*, 1977) in March 1976 when the poleward undercurrent was most strongly developed.

Denitrification relates oxygen, nitrogen, productivity and differential advection and may set upper bounds on the intensity of productivity in an upwelling region. Increased differential advection will enhance productivity by more efficient nutrient trapping in the upwelling source water as the ratio of the velocities in the two layers increases; but simultaneously oxygen depletion in the subsurface poleward flow will increase. When oxygen has been exhausted, microbial processes fueled by the organic material being partitioned into the subsurface layer will use nitrate and nitrite as a terminal electron acceptor producing N_2 in the process known as denitrification (Goering, 1978). By removing or reducing the nitrate in the upwelling source water denitrification will sharply limit the productivity of the surface layer, thereby interrupting the cycle that generated the denitrification (and oxygen depletion). This speculation supports Piper and Codispoti's (1975) suggestion that denitrification is a feedback process that limits the intensity of productivity in the ocean.

Complete denitrification in ocean waters is probably a rare event, but it was documented in Peru (Dugdale *et al.*, 1977); however, partial denitrification (Fig. 8) and oxygen depletion are always present off Peru and

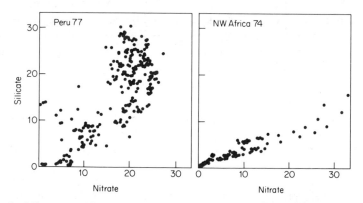

FIG. 9 Nitrate vs silicate in μg atoms/l in the euphotic zone off Peru in 1977 and Northwest Africa in 1974; these data are from stations used in the calculation of mean profiles in Fig. 10 and the integrated values in Table III

Southwest Africa. Off Northwest Africa and Oregon these processes are much less pronounced (Fig. 7); the difference may follow simply from the absence of a poleward undercurrent on the shelf in Northwest Africa (Fig. 5), which prevents countercurrent partitioning and steady-state maintenance of high organic oxygen demand in the subsurface waters of the shelf and upper slope.

Off Northwest Africa the strong equatorward alongshore currents and onshore cross-shelf currents that sweep the bottom prevent sedimentary accumulation (Diester-Haass, 1978) and advect the organic load of phytoplankton, detritus and fecal pellets (that is, the oxygen demanding materials) through the ecosystem along with a supply of moderately well oxygenated water (Fig. 7). Accumulation of an organic load is prevented both in the water column and on the bottom but the flux of food to the macrofauna and microorganisms is high. The advective pattern off Northwest Africa produces a benthic community with relatively high biomass and diversity (Thiel, 1978). The accumulation of organic oxygen demand by the two-layered circulation and the weak alongshore and cross-shelf currents that are present at the bottom off Peru (Fig. 5) account for the anoxic and reducing character of the sediments on the shelf and slope (Rowe, 1980).

It is not surprising that variations in the position and strength of the alongshore bottom and mid-depth currents predominate in determining the

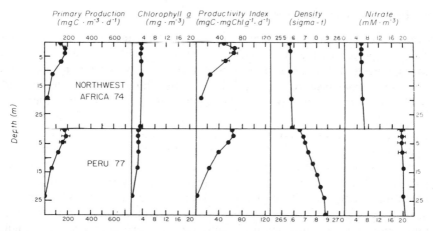

FIG. 10   Vertical profiles of properties in the surface layer off Northwest Africa and Peru. Primary production and productivity index are given to the bottom of the euphotic zone or the 1% light depth. The density profile for 1977 is a composite of hydrographic and current meter observations; other observations are from conventional hydrographic casts

biological character of the benthos and sediment by determining whether the chemical environment will be oxidising or reducing. It is surprising that advective variations in the poleward undercurrent can, and do, affect the productivity of the surface layer positively by nutrient trapping and negatively by the sequence: organic trapping → intensification of oxygen depletion → denitrification. Since Smith (1978) has shown that off Peru the dominant perturbations in the subsurface alongshore flow originate remotely and propagate along the coast, we hypothesise that a portion of the variability in productivity in coastal upwelling ecosystems is controlled by physical processes outside the coastal region. Margalef (1978b) has said, 'Upwelling can be understood only in the frame of models covering a large space, perhaps all the Ocean, where the upwelling centers appear as nodal points of stress. The definition of boundaries is an academic question.' Results from Peru and Northwest Africa show that the circulation pattern of the ecosystem determines the chemical and biological character. In the sense that this advective character is controlled by ocean basin-wide processes we agree with Margalef (1978b) that the dynamic boundaries of coastal upwelling include the whole ocean basin. This observation does not mean that future coastal upwelling studies have to be bigger and more expensive; Namias (1973), Wyrtki (1977) and Quinn et al. (1978) have shown that investigations of large-scale phenomena can be done at low cost from islands and coastal stations.

5 Biological structure

Coastal upwelling is time varying in that vertical transport occurs after the wind has been blowing for a certain period of time (Figs 3 and 4) and space varying in that the vertical transport occurs at a specific place in the ecosystem either inshore (Fig. 4) or moving as a locus from inshore to mid-shelf (Fig. 6). The nature of coastal upwelling makes it possible to identify a time or place where temperature is the lowest, nutrients are the highest and biomass is the lowest and to call that time or place the beginning; events that follow in time and space can be interpreted in the conceptual framework of ecological succession. Odum (1969) defined succession as a process of community development that 1. is orderly, directional and, therefore, predictable; 2. results from modification of the physical environment by organisms; and 3. culminates in a stabilised ecosystem in which maximum biomass, information content or biotic interactions are maintained per unit of energy flux through the ecosystem. For heuristic purposes we shall use the concept of succession in coastal upwelling from the viewpoint of Margalef (1978a,b) and of Vinogradov and his colleagues (Vinogradov and Menshutkin, 1977; Vinogradov and Shushkina, 1978;

TABLE III Integrations of vertical profiles (given in Fig. 9) of properties in two upwelling ecosystems. The data are from stations in the vicinity of the sections and moorings shown in Figs 2–7

Region	Primary production ($mgCm^{-2}d^{-1}$)	Chlorophyll a ($mg\,m^{-2}$)	Productivity index ($mgC\,mgChla^{-1}day^{-1}$)	Stability (N^2) ($cycles^2h^{-2}$)	Nitrate ($mM\,m^{-2}$)	Light ($Ly\,day^{-1}$)
Northwest Africa $n = 39$ March/April 74	1964	64	34	2	142	588
Peru $n = 24$ March/April 77	1894	43	47	28	498	523

Vinogradov *et al*., 1980) and contrast both viewpoints with observations made by a number of workers in the waters off Oregon, Northwest Africa and Peru. (We apologise to Margalef and Vinogradov for liberties we take in simplifying their complex and multifaceted ideas; readers should examine the original papers and reach their own conclusions.) The thesis of Margalef (1978a,b) is that an enhanced supply of energy to the coastal upwelling ecosystem makes that ecosystem unique (Fig. 11). Energy in the form of wind drives the circulation, and wind and sunlight control the turbulence, with sunlight (heat) suppressing turbulence. Circulation and turbulence in Margalef's view determine the biological qualitative and quantitative character of the ecosystem. This idea is expressed in the quotation

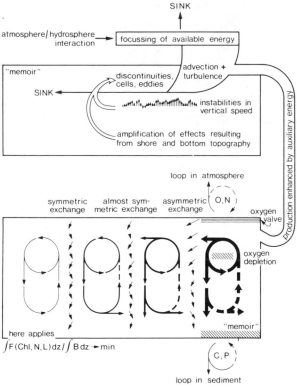

FIG. 11 A rough sketch of an upwelling ecosystem, on an *x*, *z* plane. The upwelling, at right, is the result of local availability of auxiliary energy that, among other things, enhances production. Some physical distributions (*above*) are reflected on the biological structure (*below*). The interfaces (air-water, sediment) act as a 'memoir' of past events. The coastline is at the extreme right; the open ocean at the left. From Margalef (1978b)

previously given from Margalef (1978a) and shown graphically in Fig. 11. Corollaries to the major thesis are the ideas that input of energy prevents succession from occurring and holds coastal upwelling ecosystems in an immature state, that spatially coastal upwelling ecosystems are 'local deformation of the ecological fields', and that there is a 'simple dependence of primary production on the auxiliary energy made available' (Margalef, 1978b and Fig. 11).

Vinogradov and Shushkina (1978) paint a different picture; they argue that in a pelagic ecosystem food relationships are paramount and that, therefore, the trophic exchanges of energy and matter are the most important processes determining the nature of the coastal upwelling ecosystem. The trophic structure (Fig. 12) and the trophic sequence (Fig. 13) over a section from the Peru coast to 240 km offshore are given to describe the ecosystem. The sequence of development of communities (succession) is divided into production and destruction periods; these periods are broken down into a series of stages based on trophic relationships. If we interpret Vinogradov and Shushkina's (1978) description of the trophic sequence to mean that ecological succession is occurring and is the major process structuring the ecosystem, then, using the critieria of Odum (1969), the observations should indicate 1. an orderly and predictable sequence of community developments in the newly upwelled water as it moves offshore; 2. modification of the environment by organisms; and 3. production of a stabilised high biomass or high diversity terminal community downstream and offshore. How well do observations from Oregon, Northwest Africa and Peru agree with the holistic descriptions of the coastal upwelling ecosystems that are given in Figs 11, 12 and 13?

The keystone of Margalef's description is the determining role played by the input of auxiliary kinetic energy in the form of wind. As described earlier (Figs 5, 7 and 8), many features of the chemical environment, especially those regarding nutrients, oxygen, denitrification and organic distributions and the benthic and sedimentary character, can be accounted for in Northwest Africa and Peru by differences in the circulation. A difference is that Margalef (1978a,b) describes the relationship as a quantitative one in which there is a simple dependence on the amount of external energy that enters the ecosystem. Observations reported in this chapter suggest it is the specific pattern or structure of the physical processes that is important. When a great deal of auxiliary energy enters the Northwest Africa ecosystem the primary production is reduced in the short term by the mixing of water resulting from the very weak density structure (Fig. 10 and Table III). The alternation of strong winds and periods of heating and stratification at an optimal frequency will enhance productivity off Northwest Africa; the relationship seems to be more with the pattern of external energy than with

TABLE IV Characteristics of elements of plankton communities in Pacific upwellings that are shown in Fig. 12. From Vinogradov and Shushkina (1978)

Trophic levels and groups	Size (length), L	Predominant weight, w (mcal)	Cal mg wet weight^{-1}	Metabolism $R = aw^b$		Assimilability U^{-1}
				a	b	
Phytoplankton, p						
Nanno	4–7 μm		1.0			
Small, p_1	8–20 μm		0.7			
Medium, p_2	21–100 μm		0.4			
Large, p_3	>100 μm		0.2			
Bacteria, b	1–5 μm		1.0			1.0
Unicellular heterotrophs, a						
Flagellates, a_1	3–5 μm	4×10^{-5}	0.8			0.6
Ciliates, a_2	10–100 μm	5×10^{-3}	0.8	0.26	0.76	0.6
Non-carnivorous metazoan plankton						
Fine filterers, m						
Meroplankton, m_1	0.1–3.5 mm	2–30	0.7	0.59	0.76	0.6
Appendicularians, m_1	0.1–2.5 mm	1–10	0.1	5.27	0.66	0.6
Doliolids, m_1	1–2.5 mm	4–10	0.01	5.27	0.66	0.6
Small calanoids, m_2	<1 mm	1–5	0.7	0.80	0.73	0.6
Coarse filterers, f						
Medium-sized calanoids, f_1	>1 mm	30–100	0.7	0.80	0.73	0.6
Juvenile euphausiids, f_2	<10 mm	120–2000	0.7	0.42	0.90	0.6
Predatory zooplankton						
Cyclopoids, s_1	0.2–1.5 mm	3–6	0.7	0.44	0.60	0.7
Calanoids, s_2	1.0–4 mm	500–2000	0.7	0.80	0.73	0.7
Small tomopterids, s_2	<3 mm	30–80	0.7	0.59	0.76	0.7
Small coelenterates, s_2	<5 mm	10–30	0.03	6.38	0.58	0.7
Chaetognaths, v	<20 mm	500–1200	0.7	5.30	0.52	0.7
Polychaetes, v	>3 mm	1300–8400	0.7	0.59	0.76	0.7

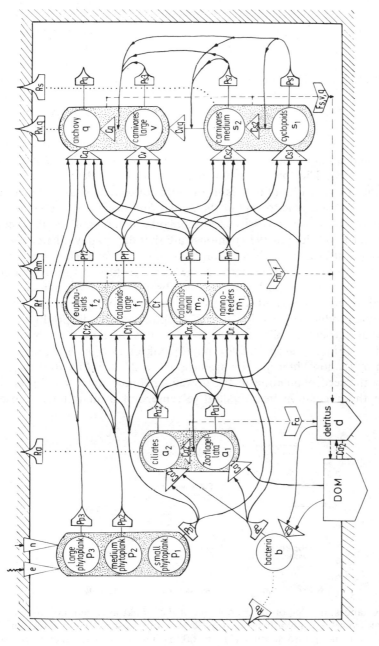

FIG. 12 Scheme of energy transfer through a plankton community in the zone of upwelling. e: solar radiation; n: nutrients. For explanation of other terms, please refer to Table IV. From Vinogradov and Shushkina (1978)

its amount. Similar arguments apply to the benthic and sedimentary components of the ecosystem. The Northwest Africa pattern of one-layered alongshore flow on the shelf, strong bottom currents and episodic mixing of euphotic layer phytoplankton to depth determine that an abundant and diverse benthic assemblage of macrofauna (Thiel, 1978) and demersal fish (Mathisen et al., 1978) will exist off Northwest Africa. The two-layered alongshore flow, weak bottom currents and complete absence of mixing of the entire water column off Peru result in efficient nutrient trapping, oxygen depletion, high organic deposition in the shelf sediments, and the development of abundant microbial populations, microfaunal (nematode) communities, and unique benthic mats of the filamentous microbe, *Thioploca* (Gallardo, 1977). The total biomass in the benthos off Peru is probably much higher (Rowe, 1980) but it is in the form of sulfur-oxidising bacteria and other organisms able to exploit the anoxic and organic-rich condition.

Analyses by Blasco et al. (1980) on phytoplankton and Smith et al. (1980) on zooplankton off Peru in 1976 demonstrate that day to day differences in species composition at a single location or on a section are determined to a large degree by advective transport. In the mean profiles in Fig. 10, based mostly on data from mid-shelf stations, the absolute primary production in 1977 was low when compared with observations from other years (Fig. 15). The absolute productivity was low in 1977 despite high specific productivity, and optimal nutrient, stability and light conditions, because the subsurface onshore flow (Fig. 5) transported few phytoplankton into the upwelling site as shown by the low chlorophyll concentration at 23 m in Fig. 10. It is not clear why the seed population in the upwelling source water was lower in 1977, but the circulation in a qualitative sense, not a quantitative sense,

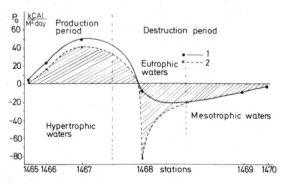

FIG. 13   Changes in value of net production of the plankton community in a 130-mile section from Cape Pakasmaio into the ocean. Figure shows net production of community not taking into account (1) and taking into account (2) anchovies and large euphausiids. From Vinogradov and Shushkina (1978)

seemed responsible for ecologically significant differences in primary production (Fig. 15) in Peru shelf waters in 1977 relative to other years.

Our attempt to fit observations into Margalef's conceptual scheme
supports the essence of his argument for the dominance of circulation and
turbulence, but we disagree with the causality that he evokes. Rather than
the amount of energy that is put in, it is the pattern (or information content)
of the physical processes that determines the character and richness of the
coastal upwelling ecosystem. Over the range of energy inputs (wind stress)
observed (Table I) there is always a complex relationship with primary
productivity in which low input causes low productivity because of a lack of
vertical transport of nutrients and high input causes low productivity by
turbulent mixing. Margalef's thesis that auxiliary (wind) energy makes
coastal upwelling ecosystems unique seems correct; but the determinant is
how topography, large-scale stratification, circulation and latitudinal effects
structure the auxiliary energy, not simply how much auxiliary energy enters
the ecosystem.

A second area of disagreement is with the idea that the wind/vertical
transport link in the ecosystem sequence is the beginning or initial step in

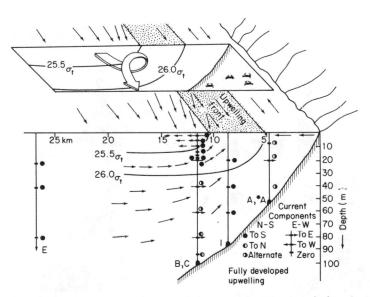

FIG. 14   A conceptual model of cross-shelf and alongshore circulation during fully
developed upwelling off the coast of Oregon. Positions of the current meter
moorings of the CUE-II study, identified as Oregon 1973 in this chapter, are
indicated as vertical lines. The vertical current profile given in Fig. 5 is from the B, C
mid-shelf mooring at the 100 m isobath. From Peterson et al. (1979)

the sequence of trophic steps. The inshore wind-driven vertical transport is where the ecosystem extracts energy from the local atmosphere, but conceptually it is more the mid point of the ecological sequence (Fig. 14) than the initial, or time zero, step (Fig. 11 or Fig. 13). Circulation on three space scales, large (>200 km), middle (20 to 200 km) and small or plume (<20 km), determines what will be in the onshore, subsurface flow that is upstream of the vertical transport. In specific incidences when phytoplankton growth rate (Barber *et al.*, 1971), nutrient uptake (MacIsaac and Dugdale, 1969) or primary production (Figs 10 and 15) was anomalously low the result seemed to follow from the changes in the upstream processes that determine what is present in the newly upwelled water. The residence

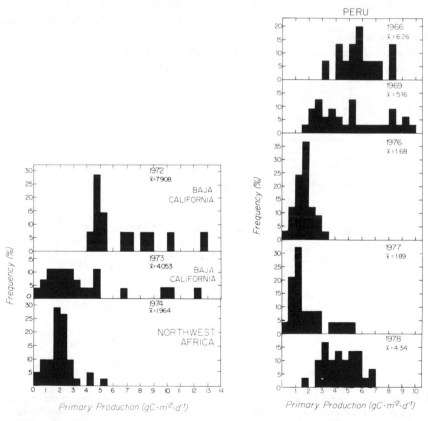

FIG. 15 Frequency vs intensity graphs of primary production from three upwelling ecosystems during different years; Northwest Africa 1974 and Peru 1977 are compared in detail in Figs 3–10

time of the surface layer in any one upwelling locus or plume is so short that productive character depends on what is in the water when it reaches the surface, which emphasises that the inshore vertical transport is a mid point in the sequence where the phytoplankton consequences are preordained.

The determination of upstream biological character in the upwelling source water would be easier to explain if there were downwelling and 'two-celled' closed circulation patterns. Such patterns have been evoked by biologists (Walsh, 1977; Peterson et al., 1979) and physical oceanographers (Mooers et al., 1976) to explain the observed distribution of properties, but two-celled cross-shelf circulation with down-welling offshore of the upwelling locus has not been observed by mid-shelf moored current meter arrays off Oregon, Northwest Africa or Peru. The Fig. 5 mean vertical current profiles show no evidence of a mean two-celled circulation; more importantly, an analysis of the records on the basis of 6-h intervals shows that a two-celled pattern does not appear off Oregon (Halpern, 1976), Northwest Africa (Halpern et al., 1976), or Peru (Brink et al., 1980) in the data filtered to remove tides and higher frequency fluctuations. Short duration profiling current meter records have suggested the transitory existence of a two-celled circulation pattern (Johnson and Johnson, 1979), but the ephemeral nature of the observed events seems to preclude a major role in the transport processes. The thorough mixing of the entire shelf water column that accompanies strong winds off Northwest Africa provides the physical link between the downstream surface community and the upwelling source water (Fig. 3); cross-shelf variations in the turbulent vertical transport of phytoplankton and organic matter probably have produced distributions that have been interpreted as evidence of downwelling.

Off Oregon Peterson et al. (1979) argue that a combination of behaviour and circulation can explain the observed copepod distributions; two species of copepod that are abundant in the Oregon upwelling, Acartia longiremis and Calanus marshallae, actively swim down to the subsurface onshore flow and are returned to the inshore surface region of dense phytoplankton. This behavioral mechanism together with the short duration of Oregon upwelling events, relaxation and reversal of the surface flow, and the outer boundary given by the Columbia River plume make it possible for these two species to maintain large populations in the upwelling system. Further analysis of the type done by Peterson et al. (1979) is required for other key species at all trophic levels. The problem is difficult because behavior is usually a species-specific property. Mathisen (1980) suggests that vertical migration between onshore and offshore flowing layers is important in keeping the Peruvian anchoveta in its acceptable inshore habitat, and Brockmann (1979) modeled the transport consequences using diel migration of Euphausia krohnii and a

Northwest Africa circulation model. Of the zooplankton and nekton that characterise upwelling it is clear that euphausids are especially adapted to coastal upwelling ecosystems. At or just beyond the shelf break in all major coastal upwelling regions, large populations of euphausids are present (Thiriot, 1978); the trophic sequence described by Vinogradov and Shushkina (1978) and shown in Fig. 13 indicates the important quantitative role euphausids play off Peru. Euphausids consume as much of the productivity as anchoveta (Fig. 13) and because of their shelf-break habitat they are the major link through which shelf productivity is transferred to fish such as jack mackeral and hake. (See Figure 2 in Cushing, 1978.)

The idea that a particular succession of trophic relationships determines the biological character of coastal upwelling ecosystems is central to Vinogradov and Shushkina's (1978) description and modelling of the ecosystem. While recognising that external energy inputs set limits on the possible range of trophic relationships, these authors emphasise the degree to which the structure of the ecosystem is internally determined by the biology of the species making up the community. It seems unlikely that Vinogradov and Shushkina would agree in toto with Margalef's (1978a) comment on the 'overwhelming importance of circulation and turbulence'. A debate on the subject would probably be quite lively.

The sequence of trophic balance from inshore to 240 km offshore (Fig. 13) implies that succession, as defined by Odum (1969), occurs in the zonal dimension. In this sequence a bolus of water is upwelled, isolated from the upwelling source or site and moved offshore maintaining continuity as a patch of upwelled water. The specific biological character that will be observed depends on when in its life cycle the upwelled patch is sampled. This physical pattern of upwelling has not been detected by the CUEA research group in Peru (or elsewhere) despite large areal surveys by ships and aircraft; but it has been observed by others off the coast of Peru. Strickland et al. (1969) and Beers et al. (1971) described patches of upwelling and the changes that occurred as the patches were monitored. Strickland et al. (1969) observed that one patch of high nutrient, low chlorophyll water was followed for five days without detectable changes in nutrients or chlorophyll. Strickland et al. (1969) believed that grazing prevented phytoplankton growth; we have frequently seen the high nutrient/ no growth situation in waters of the eastern tropical Pacific (Barber and Ryther, 1969; Barber et al., 1971; Cowles et al., 1977) and experimental removal of zooplankters did not increase the phytoplankton growth. Our interpretation of this phenomenon is that the upwelled water did not contain sufficient, or the right kind of, phytoplankton to exploit the optimal nutrient (and light) conditions provided by the stable, high nutrient patch of water. In this view the productivity of the upwelled patch depends on the phytoplankton present in the upwelling source water, so the trophic sequence is externally regulated by processes that regulate where the upwelled water will be drawn from and what kind of water will be present at that depth. If this is true, then the successional sequence is predictable only if advective variations in the subsurface, alongshore flow are known. The well documented trophic structure (Fig. 12) and sequence (Fig. 13) described by Vinogradov and Shushkina (1978) are, then, one predictable successional sequence. Strickland *et al.* (1969) and Beers *et al.* (1971) describe other sequences including the sequence in which the nutrients are never taken up, but are contributed to the anomalous, high nutrient surface waters occasionally present off Peru.

A second characteristic of succession is that the process driving it is modification of the abiotic portion of the ecosystem by the biotic components. The organisms in the coastal upwelling ecosystem cannot alter the circulation but modification of the chemical environment is a major process characteristic of upwelling ecosystems. The uptake of nutrients and the synthesis of organic matter drives the transition from autotrophic to heterotrophic; the transition may not necessarily require continuity of water *per se* from the inshore site of vertical transport to the offshore terminus. In Oregon, Northwest Africa and Peru the relatively fast surface alongshore flow is restricted to the shelf; the process of euphausids feeding in the outside edge of that flow and being preyed on by fish just offshore will transport organic matter offshore even though there is no flow continuity.

The last question deals with production of a stable, high biomass, high density 'climax' community at the end of the succession. In including this characteristic in a definition of succession Odum (1969) is drawing from classical terrestrial examples; as Margalef (1978a) specifies, gravity partitions N, P, and Si vertically out of ocean ecosystems particularly in the stable, open ocean situation. Thus the high biomass analog of a rain forest or temperate zone hardwood forest does not exist at the terminus of the upwelling sequence. The accumulation of information and particularly the information content per unit of energy flux does increase downstream as the upwelling ecosystem matures into open ocean systems (Longhurst, 1967). The same selective forces that enhance the 'banking' of information in terrestrial ecosystems as they age seem to be at work in coastal upwelling.

This chapter has drawn heavily on the results obtained in the mesoscale CUEA studies of three specific upwelling areas, so it is prudent to provide a caveat about how representative these systems are. The Oregon, Northwest Africa and Peru studies were not intended to represent the entire range of structure and function of coastal upwelling ecosystems nor are they necessarily 'typical' examples, but it seems to us that they are valid examples. If the studies had been done somewhere else using the same methods, a different picture would emerge. For example, a few hundred kilometers south of the Cap Blanc region of Northwest Africa where much American, French, Spanish and German work was concentrated in 1974 is the Cap Timiris upwelling; here an intensely productive, plume-like ecosystem develops in response to upwelling being focused by a submarine canyon. Since this ecosystem is well described by repeated, interdisciplinary studies (Herbland *et al.*, 1973; Herbland, 1978) it is known that the character of the Cap Timiris ecosystem has more in common with the ecosystem at 15°S off the coast of Peru than it does with its neighbor at Cap Blanc. As emphasised earlier in the chapter, upwelling is inherently a mesoscale process; thus the results of a particular study characterise conditions only within a mesoscale range of the study. The proximity of the dissimilar Cap Blanc and Cap Timiris ecosystems demonstrates the diversity of ecosystem structure that can occur in a geographic region, but the similarity of Cap Timiris and the Peru ecosystem at 15°S indicates that generalizations about coastal upwelling ecosystems can be made with a modest degree of confidence.

Acknowledgements

The authors thank Dr David Halpern for providing the surface layer current meter data used in Fig. 5. CUEA is supported by the IDOE office of the US National Science Foundation.

References

Bakun, A., McLain, D. R. and Mayo, P. V. (1974). The mean annual cycle of coastal upwelling off western North America as observed from surface measurements. *Fish. Bull.* **72**, 843–844

Barber, R. T. (1977). The JOINT-I expedition of the Coastal Upwelling Ecosystem Analysis program. *Deep-Sea Res.* **24**, 1–6

Barber, R. T. (1979). List of CUEA data reports. *CUEA Newsletter* **8**, 33–41

Barber, R. T. and Ryther, J. H. (1969). Organic chelators: factors affecting primary production in the Cromwell Current upwelling. *J. Exp. Mar. Biol. Ecol.* **3**, 191–199

Barber, R. T., Dugdale, R. C., MacIsaac, J. J. and Smith, R. L. (1971). Variations in phytoplankton growth associated with the source and conditioning of upwelling water. *Inv. Pesq.* **35**, 171–193

Barton, E. D., Huyer, A. and Smith, R. L. (1977). Temporal variation observed in the hydrographic regime near Cabo Corbeiro in the north-west African upwelling region, February to April 1974. *Deep-Sea Res.* **24**, 7–24

Beers, J. R., Stevenson, M. R., Eppley, R. W. and Brooks, E. R. (1971). Plankton populations and upwelling off the coast of Peru June 1969. *Fish. Bull.* **69**, 859–876

Blasco, D., Estrada, M. and Jones, B. (1980). Phytoplankton distribution and composition in the Northwest Africa upwelling region near Cabo Corbeiro. *Deep-Sea Res.* In press

Boje, R. and Tomczak, M. (1978). Ecosystem analysis and the definition of boundaries in upwelling regions. In "Upwelling Ecosystems" (R. Boje and M. Tomczak, eds) pp. 3–11. Springer-Verlag, New York

Brink, K. H., Halpern, D., Huyer, A. and Smith, R. L. (1980). Nearshore circulation near 15°S: the physical environment of the Peruvian upwelling system. In "Productivity of Upwelling Ecosystems" (R. T. Barber and M. E. Vinogradov, eds). Elsevier, Amsterdam

Brink, K. H., Halpern, D. and Smith, R. L. (1980). Circulation in the Peru upwelling system near 15°S. J. Geophys. Res. 85, 4036–4048

Brockmann, C. (1979). A numerical upwelling model and its application to a biological problem. Meeresforsch. 27, 137–146

Charney, J. G. (1955). The generation of oceanic currents by wind. J. Mar. Res. 14, 477–498

Codispoti, L. A. and Friederich, G. E. (1978). Local and mesoscale influences on nutrient variability in the Northwest African upwelling region near Cabo Corbeiro. Deep-Sea Res. 25, 751–770

Cowles, T. J., Barber, R. T. and Guillen, O. (1977). Biological consequences of the 1975 El Nino. Science 195, 285–287

Cushing, D. H. (1971). A comparison of production in temperate seas and the upwelling areas. Trans. Roy. Soc. South Africa 40, 17–33

Cushing, D. H. (1978). Upper trophic levels in upwelling areas. In "Upwelling Ecosystems" (R. Boje and M. Tomczak, eds) pp. 101–110. Springer-Verlag, New York

Diester-Haass, L. (1978). Sediments as indicators of upwelling. In "Upwelling Ecosystems" (R. Boje and M. Tomczak, eds) pp. 261–281. Springer-Verlag, New York

Dugdale, R. C. (1967). Nutrient limitation in the sea: dynamics, identification and significance. Limnol. Oceanogr. 12, 685–695

Dugdale, R. C. (1972). Chemical oceanography and primary production in upwelling regions. Geoforum 11, 47–61

Dugdale, R. C. (1976). Nutrient cycles. In "Ecology of the Seas" (D. H. Cushing and J. J. Walsh, eds) pp. 141–172. W. B. Saunders Co., Philadelphia

Dugdale, R. C. and Goering, J. J. (1967). Uptake of new and regenerated forms of nitrogen in primary productivity. Limnol. Oceanogr. 12, 196–206

Dugdale, R. C., Goering, J. J., Barber, R. T., Smith, R. L. and Packard, T. T. (1977). Denitrification and hydrogen sulfide in the Peru upwelling region during 1976. Deep-Sea Res. 24, 601–608

Ekman, V. W. (1905). On the influence of the earth's rotations on ocean-currents. Arkiv för matematik, astronomi, och fysik 2, 1–53

Eppley, R. W., Renger, E. H. and Harrison, W. G. (1979a). Nitrate and phytoplankton production in southern California coastal waters. Limnol. Oceanogr. 24, 483–494

Eppley, R. W., Renger, E. H., Harrison, W. G. and Cullen, J. J. (1979b). Ammonium distribution in southern California coastal waters and its role in the growth of phytoplankton. Limnol. Oceanogr. 24, 495–509

Eppley, R. W., Sharp, J. H., Renger, E. H., Perry, M. J. and Harrison, W. G. (1977). Nitrogen assimilation by phytoplankton and other microorganisms in the surface waters of the central North Pacific Ocean. Mar. Biol. 39, 111–120

Fahrbach, E. and Meincke, J. (1979). Some observations on the variability of the Cabo Frio upwelling. CUEA Newsletter 8, 13–18

Gallardo, V. A. (1977). Large benthic microbial communities in sulphide biota under the Peru-Chile Subsurface Countercurrent. Nature 268, 331–332

Gardner, D. (1977). Nutrients as tracers of water mass structure in the coastal upwelling off northwest Africa. *In* "A Voyage of Discovery" (M. Angel, ed.) pp. 305–326. Pergamon Press, New York

Goering, J. J. (1978). Denitrification in marine systems. *In* "Microbiology 1978" (D. Schelessinger, ed.) pp. 357–361. Am. Soc. Microbiol. Publ., Washington, D.C.

Gunther, E. R. (1936). A report on oceanographic investigations in the Peru coastal current. *Discovery Rep.* **13**, 107–276

Hafferty, A. J., Lowman, D. and Codispoti, L. A. (1979). JOINT-II, *Melville* and *Iselin* bottle data sections March–May 1977. CUEA Technical Report 38. Duke University Marine Lab., Beaufort, N.C.

Halpern, D. (1976). Structure of a coastal upwelling event observed off Oregon during July 1973. *Deep-Sea Res.* **23**, 495–508

Halpern, D., Smith, R. L. and Mittelstaedt, E. (1976). Cross-shelf circulation on the continental shelf off Northwest Africa during upwelling. *J. Mar. Res.* **35**, 787–796

Herbland, A. (1978). Heterotrophic activity in the Mauritanian upwelling. *In* "Upwelling Ecosystems" (R. Boje and M. Tomczak, eds) pp. 155–166. Springer-Verlag, New York

Herbland, A., LeBorgne, R. and Voituriez, B. (1973). Production primaire, secondaire et regeneration des sels nutritifs dans l'upwelling de Mauritanie. *Doc. Scient. Centre Rech. Oceanog. Abidjan* **4**, 1–75

Huntsman, S. A. and Barber, R. T. (1977). Primary production off northwest Africa: the relationship to wind and nutrient conditions. *Deep-Sea Res.* **24**, 25–34

Huyer, A. (1976). A comparison of upwelling events in two locations: Oregon and Northwest Africa. *J. Mar. Res.* **34**, 531–546

Johnson, D. R. and Johnson, W. R. (1979). Vertical and cross-shelf flow in the coastal upwelling region off Oregon. *Deep-Sea Res.* **26**, 399–408

Lasker, R. (1978). The relation between oceanographic conditions and larval anchovy food in the California Current: identification of factors contributing to recruitment failure. *Rapp. P.-V. Reun. Cons. Int. Explor. Mer* **173**, 212–230

Longhurst, A. (1967). Diversity and trophic structure of zooplankton communities in the California Current. *Deep-Sea Res.* **14**, 393–408

MacIsaac, J. J. and Dugdale, R. C. (1969). The kinetics of nitrate and ammonia uptake by natural populations of marine phytoplankton. *Deep-Sea Res.* **16**, 45–57

Margalef, R. (1978a). Life-forms of phytoplankton as survival alternatives in an unstable environment. *Oceanologica Acta* **1**, 493–509

Margalef, R. (1978b). What is an upwelling ecosystem? *In* "Upwelling Ecosystems" (R. Boje and M. Tomczak, eds) pp. 12–14. Springer-Verlag, New York

Mathisen, O. A. (1980). Adaptation of the anchoveta (*Engraulis ringens* J.) to the Peruvian upwelling system. *In* "Productivity of Upwelling Ecosystems" (R. T. Barber and M. E. Vinogradov, eds). Elsevier, Amsterdam

Mathisen, O. A., Thorne, R. E., Trumble, R. J. and Blackburn, M. (1978). Food consumption of pelagic fish in an upwelling area. *In* "Upwelling Ecosystems" (R. Boje and M. Tomczak, eds) pp. 111–123. Springer-Verlag, New York

McCarthy, J. J. and Goldman, J. C. (1979). Nitrogenous nutrition of marine phytoplankton in nutrient-depleted waters. *Science* **203**, 670–672

Menzel, D. W. (1967). Particulate organic carbon in the deep sea. *Deep-Sea Res.* **14**, 229–238

Mooers, C. N. K., Collins, C. A. and Smith, R. L. (1976). The dynamic structure of the frontal zone in the coastal upwelling region off Oregon. *J. Phys. Oceanogr.* **6**, 3–21

Namias, J. (1973). Response of the equatorial countercurrent to the subtropical atmosphere. *Science* **181**, 1244–1245

O'Brien, J. J., Clancy, R. M., Clarke, A. J., Crepon, M., Elsberry, R., Gammelsrod, T., MacVean, M., Roed, L. P. and Thompson, J. D. (1977). Upwelling in the ocean: Two and three dimensional models of upper ocean dynamics and variability. *In* "Modelling and Prediction of the Upper Layers of the Ocean" (E. B. Kraus, ed.) pp. 178–228. Pergamon Press, New York

Odum, E. (1969). The strategy of ecosystem development. *Science* **164**, 262–270

Odum, E. P. (1977). The emergence of ecology as a new integrative discipline. *Science* **195**, 1289–1293

Peterson, W. T., Miller, C. B. and Hutchinson, A. (1979). Zonation and maintenance of copepod populations in the Oregon upwelling zone. *Deep-Sea Res.* **26A**, 467–494

Piper, D. Z. and Codispoti, L. A. (1975). Marine phosphorite deposits and the nitrogen cycle. *Science* **188**, 15–18

Quinn, W. H., Zopf, D. O., Short, K. S. and Kuo Yang, T. W. (1978). Historical trends and statistics of the southern oscillation, El Nino, and Indonesian droughts. *Fish. Bull.* **76**, 663–678

Redfield, A. C. (1934). On the proportions of organic derivatives in sea water and their relation to the composition of plankton. James Johnstone Memorial Volume, pp. 176–192. Liverpool

Redfield, A. C. (1955). The hydrograph of the Gulf of Venezuela. *Papers Mar. Biol. Oceanogr., Deep-Sea Res.* Suppl. to Vol. 3, 115–133

Redfield, A. C., Ketchum, B. H. and Richards, F. A. (1963). The influence of organisms on the composition of seawater. *In* "The Sea" (M. N. Hill, ed.) Vol. 2, pp. 26–77. Interscience, New York

Riley, G. A. (1947). Factors controlling phytoplankton populations on Georges Bank. *J. Mar. Res.* **6**, 54–73

Rowe, G. T. (1971). Benthic biomass in the Pisco, Peru upwelling. *Inv. Pesq.* **35**, 127–135

Rowe, G. T. (1980). Benthic production and processes off Baja California, Northwest Africa and Peru. *In* "Productivity of Upwelling Ecosystems" (R. T. Barber and M. E. Vinogradov, eds). Elsevier, Amsterdam

Rowe, G. T., Clifford, C. H. and Smith, K. L. Jr. (1977). Nutrient regeneration in sediments off Cap Blanc, Spanish Sahara. *Deep-Sea Res.* **24**, 57–63

Ryther, J. H. (1969). Photosynthesis and fish production in the sea. *Science* **166**, 72–76

Sanders, H. L. (1969). Benthic marine diversity and the stability-time hypothesis. *Brookhaven Symp. Biol.* **22**, 71–80

Sharp, J. H. (1976). "Anoxia on the Middle Atlantic Shelf During the Summer of 1976". National Science Foundation, Washington, D.C.

Smith, R. L. (1968). Upwelling. *Oceanogr. Marine Biol. Ann. Rev.* **6**, 11–46

Smith, R. L. (1978). Poleward propagating perturbations in currents and sea levels along the Peru coast. *J. Geophys. Res.* **83**, 6083–6092

Smith, S. L., Brink, K. H., Santander, H., Cowles, T. J. and Huyer, A. (1980). The effect of advection on variations in zooplankton at a single location near Cabo Nazca, Peru. *In* "Coastal Upwelling Research, 1980" (F. A. Richards, ed.) American Geophysical Union, Washington, D.C. In press

Staresinic, N. (1978). The vertical flux of particulate organic matter in the Peru coastal upwelling as measured with a free-drifting sediment trap. Ph.D. Thesis. Woods Hole Oceanographic Institution and Mass. Inst. Tech.

Strickland, J. D. H., Eppley, R. W. and Rojas de Mendiola, B. (1969). Phytoplankton populations, nutrients and photosynthesis in Peruvian coastal waters. *Bol. Inst. Mar. Peru* **2**, 1–45

Sutherland, J. P. (1974). Multiple stable points in natural communities. *Am. Nat.* **108**, 859–873

Sverdrup, H. U. (1953). On conditions for the vernal blooming of phytoplankton. *J. Conseil Exp. Mer.* **18**, 287–295

Thiel, H. (1978). Benthos in upwelling regions. *In* "Upwelling Ecosystems" (R. Boje and M. Tomczak, eds) pp. 124–138. Springer-Verlag, New York

Thiriot, A. (1978). Zooplankton communities in the West African upwelling area. *In* "Upwelling Ecosystems" (R. Boje and M. Tomczak, eds) pp. 32–61. Springer-Verlag, New York

Thompson, J. D. (1977). Ocean deserts and ocean oases. *In* "Desertification" (M. N. Glantz, ed.). Westview Press, Boulder

Vinogradov, M. E. and Menshutkin, V. V. (1977). The modeling of open-sea ecosystems. *In* "The Sea" (E. D. Goldberg, I. N. McCave, J. J. O'Brien and J. H. Steele, eds) Vol. 6, pp. 891–921. Interscience, New York

Vinogradov, M. E. and Shushkina, E. A. (1978). Some development patterns of plankton communities in the upwelling areas of the Pacific Ocean. *Mar. Biol.* **48**, 357–366.

Vinogradov, M. E., Shushkina, E. A. and Lebedeva, L. P. (1980). Production characteristics of plankton communities in coastal waters of Peru. *In* "Productivity of Upwelling Ecosystems" (R. T. Barber and M. E. Vinogradov, eds). Elsevier, Amsterdam

Walsh, J. J. (1977). A biological sketchbook for an eastern boundary current. *In* "The Sea" (E. D. Goldberg, I. N. McCave, J. J. O'Brien and J. H. Steele, eds) Vol. 6, pp. 923–968. Interscience, New York

Walsh, J. J. (1978). The biological consequences of interaction of the climatic, El Nino, and event scales of variability in the eastern tropical Pacific. *Rapp. P.-V. Reun. Cons. Int. Explor. Mer* **173**, 182–192

Whitledge, T. E. (1978). Regeneration of nitrogen by zooplankton and fish in the Northwest Africa and Peru upwelling ecosystems. *In* "Upwelling Ecosystems" (R. Boje and M. Tomczak, eds) pp. 90–100. Springer-Verlag, New York

Whitledge, T. E. (1980). The role of nutrient recycling in upwelling ecosystems. *In* "Productivity of Upwelling Ecosystems" (R. T. Barber and M. E. Vinogradov, eds). Elsevier, Amsterdam

Wooster, W. S. (1970). Eastern boundary currents in the South Pacific. *In* "Scientific Exploration of the South Pacific" (W. S. Wooster, ed.) pp. 60–68. National Academy of Sciences, Washington, D.C.

Wooster, W. S., Bakun, A. and McLain, D. R. (1976). The seasonal upwelling cycles along the eastern boundary of the North Atlantic. *J. Mar. Res.* **34**, 131–141

Wooster, W. S. and Sievers, H. A. (1970). Seasonal variations of temperature, drift and heat exchange in surface waters off the west coast of South America. *Limnol. Oceanogr.* **15**, 595–605

Wyrtki, K. (1975). El Nino – The dynamic response of the equatorial Pacific Ocean to atmospheric forcing. *J. Phys. Oceanogr.* **5**, 572–584

Wyrtki, K. (1977). Sea level during the 1972 El Nino. *J. Phys. Oceanogr.* **7**, 779–787

Yoshida, K. (1967). Circulation in the eastern tropical oceans with special reference to upwelling and undercurrents. *Jap. J. Geophys.* **4**, 1–75

Zuta, S. and Guillen, O. (1970). Oceanografia de las aguas costeras del Peru. *Bol. Inst. Mar. Peru-Callao* **2**, 157–324

3 Ecosystems of Equatorial Upwellings

MIKHAIL E. VINOGRADOV

1 Introduction

Within the immense low productivity zone of the tropical ocean are distinct regions of semi-permanent upwelling of intermediate waters where nutrient salts rise into the euphotic zone and assure a high plankton productivity. Tropical upwelling depends on one of three basic causes: 1. Local cyclonic gyres which induce dome-like features with upwelling from great depths and the approach of the thermocline near to the surface. Such domes occur in many regions of the ocean. 2. Offshore wind drift and wind induced divergence of coastal currents causing longshore compensatory upwelling of intermediate waters; this is the well-known phenomenon of coastal upwelling discussed in chapter 2. 3. Wind-induced transverse circulation within currents produced by the slope of isopycnal surfaces. It is this which gives rise to the phenomenon of equatorial upwelling that we are concerned with in this chapter. Upwelling in the equatorial zone is to a considerable degree connected with the system of powerful trade winds, and its intensity changes with the spatial (longitudinal) and temporal (seasonal, year-to-year) variability in trade wind stress.

The region of equatorial upwelling *sensu lato*, relatively narrow and undeveloped in the western part of the oceans, widens to the east where the phenomenon is strongly developed (Wyrtki, 1966). In the Atlantic Ocean, upwelling is permanent east of 20°–30°W, from 6–7°N to 7–8°S, but in some seasons merges with the African coastal upwelling from which it is not completely distinct. In the eastern part of the Pacific, equatorial upwelling is more constant and may be distinctly traced east of 160°–180°E. It gradually widens eastwards until it extends from 8–12°N to 6–8°S. In the Indian Ocean, owing to the monsoon circulation, coastal upwellings are well developed, but strong upwelling on the Equator is also sometimes observed in the eastern part of the ocean.

The intensity of upwelling directly at the Equator, the equatorial upwelling *sensu stricta* or equatorial divergence, likewise increases from west to east reaching its maximum development at 93–100°W in the Pacific and 5–

10°W in the Atlantic Ocean. Simultaneously the thermocline shoals towards the east, thus contributing to a still greater nutrient enrichment of the surface layers. The upper limit of the thermocline in the Pacific Ocean (Fig. 1) lies at a depth of about 120 m at 150°W–160°E, and at 10–15 m at 90°W where, in some seasons, it may rise to the surface (Austin and Rinkel, 1958; Austin, 1960; Blackburn et al., 1970; Jones, 1973; Fedorov et al., 1975).

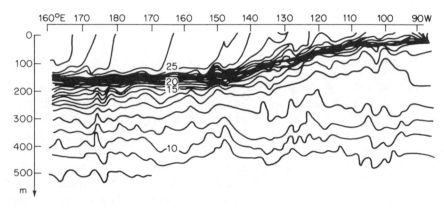

FIG. 1   Change in the depth of the thermocline along the Equator in the eastern Pacific. The position of the upper boundary of Cromwell current is indicated by a heavy line (Gueredrat, 1971)

2   The origin of equatorial upwelling

The equatorial regions of the oceans are occupied by a system of strong currents. The general westward flow induced by trade winds is, as it were, pierced by the equatorial countercurrents and undercurrents (Lomonosov in the Atlantic, Tareev in the Indian, and Cromwell in the Pacific Oceans), which perform significant eastward mass transport.

In the western Atlantic this system of eastward countercurrents consists of three branches. The northern branch (the equatorial countercurrent proper) extends between 3° and 8°N and has a velocity of 40–50 cm s^{-1}. The central branch (the Lomonosov undercurrent) runs almost exactly along the Equator at a rate of 100–110 cm s^{-1}. The southern branch, between 3°30' and 5°S, has a velocity of 40–60 cm s^{-1} (Khanaytchenko et al., 1965; Metcalf et al., 1962; Khanaitchenko, 1969, 1973). It is noteworthy that the fastest of these currents practically coincides with the Equator. It also has the most constant flow in respect to time. This pattern of zonal currents has its analogue in the Pacific Ocean.

The general character of vertical water movement in this complicated system of currents may be demonstrated by the example of a section drawn along 25°W in the Atlantic Ocean (Fig. 2). Vertical transport is most intense at the boundaries of opposing zonal flows. The intense transverse circulation which exists in the Lomonosov current causes strong upward transport in the region of the Equator, i.e. along the middle part of the current. From this rise, due to the change of sign of the Coriolis parameter

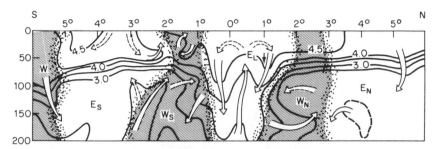

FIG. 2   Circulation in the surface layer of the equatorial Atlantic Ocean in a meridional section along 20–25°W. E_N, northern branch of the equatorial countercurrent; W_N, northern branch of the southern equatorial current; E_L, Lomonosov current; W_S, southern branch of the south equatorial current; E_S, southern branch of the equatorial countercurrent; W, south equatorial current. Currents flowing westwards are indicated by shading. Isolines represent change in oxygen concentration (Khanaychenko, 1973)

between north and south hemispheres, drift currents in the Ekman layer diverge both northwards and southwards. Downwelling along the margins of the current takes place in the region of 1°N and 1°S. This is demonstrated by the distribution of such conservative elements as strontium-90 and caesium-137 (Fig. 3).

An analogous picture occurs in the Pacific Ocean, as seen in a meridional temperature section along 120°W (Fig. 4), where separate zones of descending isotherms in convergences at the boundaries of zonal flows may be easily traced against a background of a general rise of isotherms in the equatorial region.

Let us now consider the water masses involved in the equatorial upwelling and, first of all, on the Equator itself.

THE ORIGIN OF UPWELLING WATER

The Cromwell and Lomonosov currents extend from the lower boundary of the quasihomogeneous surface layer, or some intermediate level within this layer, to the depth of the maximum density gradient. It is this water that

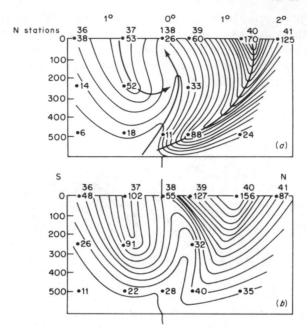

FIG. 3   Distribution of (*a*) strontium-90 and (*b*) caesium-137 in a section along 35°W (Khanaychenko, 1971)

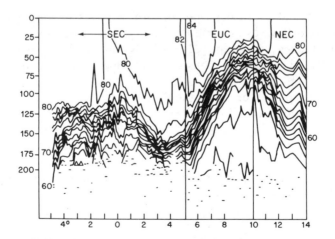

FIG. 4   Distribution of temperature (°F) in a meridional section along 150°W in the eastern Pacific. NEC, north equatorial current; EUC, equatorial countercurrent; SEC, south equatorial current; E, equatorial divergence (Austin and Rinkel, 1958)

rises towards the surface at the equator. According to Rotschi (1970) the Cromwell current is formed from the southern tropical and subtropical waters of the Pacific Ocean (upper part of the current) and by the waters of the equatorial countercurrent and the Coral Sea (core and lower part of the current). It has been assumed that southern tropical water is transported within the core of the current from the regions of its origin towards the east (Tsuchiya, 1968; Bubnov and Egorikhin, 1975). But the Cromwell current entrains surrounding water intensively both from the south and the north, including water from the tropical gyres (Knauss, 1966; Rotschi, 1970). Recently obtained data (Fedorov et al., 1975) also point to a constant peripheral entrainment of surrounding water into the flow of the current. These data suggest that the residence time of a parcel of water in the Cromwell current varies between only 5 and 50 days. Approximately the same time (18 days) is suggested by Knauss (1966). During this time a parcel will be carried eastwards by the current for no more than 200–2000 km. Therefore, it seems hardly possible for the Cromwell current to transport water from the Coral Sea to the East Pacific. The same holds for the Lomonosov current in the Atlantic Ocean.

The great velocity of transport in subsurface countercurrents and the relatively rapid water exchange are conducive to a high degree of turbulence. The coefficients of momentum exchange in the Cromwell current exceed 10^2 cm s^{-1} (Bubnov, 1975) which seems to be the highest velocity of dissipation of kinetic turbulent energy anywhere in the ocean (Gibson and Williams, 1972; Fedorov, 1975). Nevertheless, the flow of the Cromwell current is stratified, this stratification being caused by the different origin of its components (Montgomery and Stroup, 1962; Tsuchiya, 1968; Rotschi, 1970).

The brevity of a parcels residence within the Cromwell current accounts for the banded pattern of spatial distribution and high variability of temperature and salinity. This is the only explanation which makes it possible to reconcile such a high degree of turbulence (Gibson and Williams, 1972) with the presence of numerous heterogeneous and poorly mixed parcels.

INTENSITY OF UPWELLING AT THE EQUATOR

The equatorial upwelling entrains the water of the Lomonosov and Cromwell currents that lie above their cores (Knauss, 1966; Rotschi, 1973). In the Pacific Ocean under non-stationary conditions the water at the equator may ascend at a rate of 10^{-2} and even 10^{-1} cm s^{-1} (Rotschi and Jarrige, 1968; Chekotillo, 1969). Analogous values (10^{-2} cm s^{-1}) have been obtained for the Atlantic Ocean (Polosin, 1967).

Theoretical models (Hidaka, 1960; Yoshida, 1967) predict values of

10^{-2}–10^{-3} cm s^{-1} stationary conditions. According to the calculations made by Fedorov (1975), based on direct measurements of the zonal components of the Cromwell current in January on the Equator, in the eastern part of the Pacific Ocean, the rate of upwelling at the boundary between the upper drift current and the Cromwell current, reaches 10^{-2}–10^{-3} cm s^{-1}, tending to increase from east to west with the deepening of the surface of the boundary. At the same time, the intensity of the rise of deep water to the surface layers tends to increase eastwards as the Cromwell current draws nearer to the surface, and its effect on the processes in the euphotic zone increases. Therefore, it may be assumed that the phenomenon of equatorial upwelling intensifies from west to east.

Temporal and spatial variability of vertical transport within the upwelling zone results in the formation of lenses of water rich in nutrients and in a 'patchy' plankton distribution. According to data from the bioluminesce field, such patches may have a horizontal extent of the order of several hundred metres (Levin et al., 1975).

3   The temporal variability of equatorial upwellings

SEASONAL VARIABILITY

In the eastern part of the Pacific Ocean equatorial upwelling is most intense at the end of the northern winter when strong easterly (or southeasterly) winds are blowing on the Equator. Lowest wind strengths and, consequently, the least active upwelling and highest surface temperature are usually observed in March. Seasonal variability has been studied in greater detail in the Atlantic Ocean, but nearly all the available information concerns the regions near Africa, west of 10°W, where seasonal variability depends on a great extent on the local monsoon.

In the open part of the Gulf of Guinea the most intensive vertical transport occurs during the period of strengthening southeastern trade winds and of the southwestern monsoon in June–September. The intensity of vertical transport depends first of all on wind eddying and divergence which reach their highest values during the cold season (Table I).

Thus, the change of seasons on the Equator in the region of the Gulf of Guinea takes the following course: the warm season (March–May) is characterised by a slackening of southeast trades, weaker vertical water transport, highest temperature and stability of the upper water layers, and lower salinity. The pycnocline is well developed both in the upper layers and in depths of 300–400 m. In June–July, southeast trades of force 4–5 prevail in the equatorial zone. The pycnocline shoals in some areas and decays in others. Surface temperature falls abruptly (Fig. 5). The upwelling water enriches the photic layer with phosphates. Further upwelling in August–

TABLE I   Seasonal wind stress and divergence

	March–May	June–August	September–November	December–February
Ω^a	−34.9	−80.2	−65.3	−58.6
D^a	12.2	19.2	18.4	14.3

[a] Mean values for the vertical component of the rotor of the tangential wind stress ($\Omega \times 10^{10}$ dynes·cm⁻³) and of the surface current divergence ($D \times 10^{10}$ dynes·cm⁻³) in the equatorial zone in the Atlantic Ocean, 0–5°S, 0–15°W (according to Sedykh and Lutoshkina, 1971)

September is accompanied by the decay of the pycnocline. Waters from increasingly greater depths become involved in the vertical re-transport, whose velocity may be one order higher than during the warm season. In August–September the upwelling water is drawn from depths greater than 500–600 m, as compared with no more than 200–300 m during the warm season. This period is characterised by lowest surface temperatures.

Towards the end of September and in October, the southeast trades slacken, turbulent exchange decreases, the pycnocline is restored, the temperature of the surface layer rises, and although by now an immense water column to depths of 900 m is entrained in the vertical movement its intensity in the upper layers gradually subsides. The process culminates in January–February with single eddies strongly developed in the vertical plane. By March, vertical stratification is fully restored, temperature gradually reaches its highest value, and salinity is at a minimum (Sedykh and Lutoshkina, 1971).

The position of the zones of upwelling and downwelling is also subjected to considerable seasonal changes. However, between 1°N and 3°S there lies

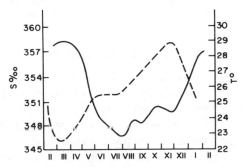

FIG. 5   Seasonal cycle of temperature (——) and salinity (– – – –) in the equatorial zone (1°N–3°S) at 5°W (Sedykh and Lutoshkina, 1971)

a zone of upwelling throughout the year, with only slight shifting of its boundaries from month to month. This zone is characterised by the greatest velocities of upward vertical transport in the Atlantic Ocean of up to 0.46 $\times 10^{-2}$ cm s^{-1} (Palyi, 1971).

YEAR-TO-YEAR VARIABILITY
The variability of equatorial upwelling from one year to another is determined by the variability of global atmospheric circulation. Thus, in 1957, in the Pacific Ocean, exceptionally weak winds caused a cessation of upwelling which, naturally, was followed by a substantial reorganisation of the equatorial currents, which must have profoundly affected the functioning of equatorial communities. It is obvious that the causes of these anomalies are to be sought in the peculiarities of atmospheric circulation in the southern hemisphere (Bjerknes, 1969).

Today some information about the variability of equatorial upwellings may be derived from satellite observations, but adequate data are still lacking on the parallel changes in the marine biota.

Besides the seasonal and year-to-year variability of currents, short-term non-stationary variability associated with large-scale turbulence is often observed, expressed as small latitudinal shifts of frontal zones, splitting of the main flow of the current into a number of separate streams and the inclusion in it of relatively small parcels of water moving in opposite directions.

All this shows that in the equatorial zone we have to deal with a system of circulation which is variable in time, and spatially complex, comprising zonal flows moving with great velocity in opposite directions and divided by zones of more or less intense ascents and sinkings. The meridional water transport proceeds from upwelling zones to downwelling zones, but its velocities are an order lower than the velocities of zonal flow. It is this intricate circulation that determines the structure and functional peculiarities of pelagic communities in the equatorial zone.

4   The formation of plankton communities

The composition, distribution and productivity of the communities of the equatorial zone are determined by upwelling into the euphotic zone of water rich in nutrients, the transport of developing communities by zonal flows and the transverse meridional transport of communities from zones upward to zones of downward vertical transport. The narrow zones of upwelling along the equator, alternating with as narrow zones of downwelling, impart a banded pattern to the distribution of plankton biomass. Thus in the East Pacific upward movement is most intense on the equator and at the divergence at the northern boundary of the equatorial countercurrent at about 10°N. Along these zones extend well-defined maxima of plankton biomass (Fig. 6) observed by many authors (Brandhorst, 1958; King and Hida, 1957; King and Iversen, 1962; Vinogradov and Voronina, 1963). During the observations carried aboard the *Vityaz* by the end of 1961, two less important fronts of upwelling were recorded at 3°N and 5°N. They too were characterised by high plankton biomasses.

A similar banded pattern of biomass distribution in the Atlantic zone of equatorial upwelling is observed (Gruzov, 1971a), as well as in the Indian Oceans (Vinogradov and Voronina, 1962a,b).

THE EFFECT OF MERIDIONAL TRANSPORT

Already by 1937, Steemann Neilsen had shown that newly upwelled deep waters are poor in phyto- and zooplankton. During the period required for the development of algae the upwelled water is displaced so that the growth of zooplankton following the phytoplankton bloom occurs at a considerable distance from the upwelling site: thus, upwelling sites and plankton maxima prove to be spatially separated. This time-space sequence in the development of pelagic communities enables us to investigate some concepts regarding the spatial-dynamic aspects of plankton communities (Margalef,

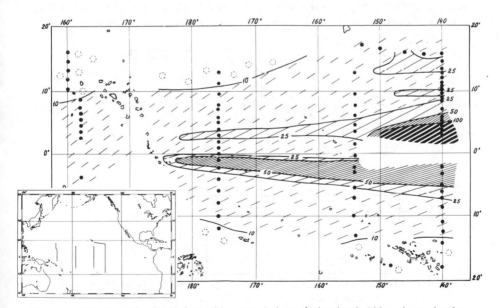

FIG. 6   Distribution of plankton biomass (ml m^{-3}) in the 0–100 m layer in the equatorial region of the Pacific (Vinogradov and Voronina, 1962a)

1967, 1968; Vinogradov *et al.*, 1971; Vinogradov, 1977). According to Sette (1955) the development of all the trophic levels within a community takes from 50 to 150 days, while the data of Blackburn *et al.* (1970) show that the maximum of small fish and cephalopods in the equatorial region lags the maximum of chlorophyll by about four months. Analogous values (70–100 days) were obtained from model calculations carried out by Vinogradov *et al.* (1973). In the Pacific Ocean a community may be carried a distance of 1800–2500 km by the zonal component of transport during this time, and be shifted by the meridional component 250–450 km laterally from the Equator. The lateral shift will be greater the more prolonged the development of the species, or the farther up the food chain is the given organism from the producers. Therefore the aggregations of macroplankton and large fish feeding on it will as a rule be displaced far downstream along the zonal flow or sideways from the divergence zone (Fig. 7).

It was found in the Pacific Ocean that the maximal number of adult individuals of the abundant copepod species *Undinula darwini* and *Rhincalanus cornutus*, actively feeding on phytoplankton, are confined to two narrow strips within the eutrophic region of the equatorial divergence, situated at a distance of 0°30′–1° from the Equator: *Neocalanus robustior*, *Eucalanus attenuatus* and other herbivores and omnivores have a similar distribution. However, the maximum numbers of adult individuals of the predatory surface copepods *Euchaeta marina* and *E. wolfendeni* occur still farther from the Equator (Fig. 8), though the juveniles of these species reach their

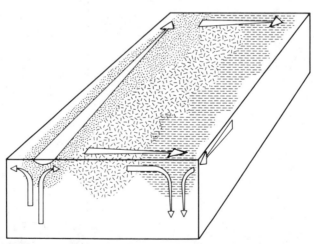

FIG. 7   Block-diagram of the distribution of the trophic links of plankton in the zone of equatorial divergence. ▦ , region of maximum development of phytoplankton; ▨ , of herbivorous zooplankton; ▤ , of predatory macroplankton and concentrations of large fish. Arrows indicate water movement

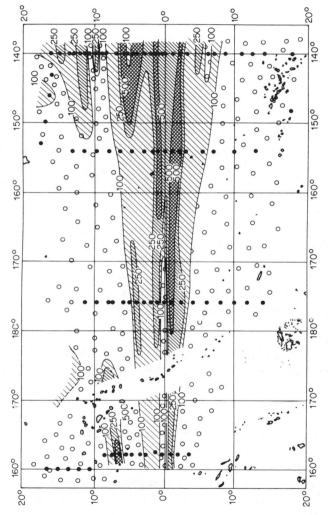

FIG. 8 Distribution of adult *Euchaeta marina* (indiv. m^{-2}) in the 0–500 m layer) in the equatorial zone in September–November, 1961 (after Vinogradov and Voronina, 1963)

maximum abundance directly on the equatorial divergence (Vinogradov and Voronina, 1963). An analogous situation occurs in the Indian Ocean (Vinogradov et al., 1961; Vinogradov and Voronina, 1962a,b).

As the development of macroplankton animals is still more prolonged, their epiplanktonic species are carried still farther by the meridional component of the current and, usually, concentrate at the nearest convergence at a lateral distance of 250–350 km from the zone of upwelling, and it is but natural that active fishes such as tuna, and other predators feeding on macroplankton, should themselves concentrate near these convergences. Indeed, in the eastern equatorial part of the Pacific Ocean both macroplankton and tunas are observed to concentrate at the convergence (between the south equatorial current and the equatorial countercurrent) to the north of the zone of abundance of small zooplankton (King, 1958; Murphy and Shomura, 1958; Sette, 1955; Parin, 1962).

Many of the macroplankton forming the DSL perform diel vertical migrations of such amplitude that at night they occur in surface water layers with a meridional current component directed away from the Equator, and in the daytime sink to depths in which the water is moving toward the Equator. In the region of the equator these animals rise through the thermocline at dark (Vinogradov, 1974) into the surface layers which move westwards, while at dawn they sink through the Cromwell current, below its core, into waters which also are moving westwards. As a result of the complicated interaction of these currents moving in opposite directions at different depths, the diel migrants of the macroplankton are drawn west along the Equator without being subjected to significant meridional transport (Voronina, 1964).

THE VARIABILITY OF COMMUNITIES ALONG THE EQUATOR

Let us consider the changes in the structural characteristics of a community with decreasing intensity of upwelling from east to west, using the example of the equatorial communities in the eastern Pacific.

The communities of the euphotic zone of the upper mixed layer formed in the zone of upwelling are carried westwards along the equator by the surface waters of the southern equatorial countercurrent. On the way they undergo clear successional changes. The biomasses of all their elements, including zooplankton, macroplankton and fishes, associated with changes in the intensity of upwelling (King and Demond, 1953; Blackburn, 1966, 1968; Blackburn et al., 1970; Vinogradov and Voronina, 1964) (Table II).

We can see the quantitative changes occurring in the components of equatorial communities in greater detail in the data obtained during the cruise of the Akademik Kurchatov in January 1974 on the Equator between 97°–155°W.

TABLE II   Plankton community biomass[a]

Group of organisms	155°W	140°W	122°W	97°W
Phytoplankton	4.8[a]	4.5	5.1	46.5
Bacteria	2.9	5.2	2.8	16.5
Flagellates	0.51	0.91	0.83	1.7
Ciliates	2.1	0.54	1.7	0.24
Other protozoans (radiolarians, foraminifera)	0.05	0.12	0.25	0.04
Total protozoans	2.6	1.6	2.8	2.0
Fine filterers, metazoans (Appendicularia,				
Doliolidae, small calanoids)	1.7	2.0	1.1	2.7
Copepod nauplii	0.5	0.3	0.4	1.6
Coarse filterers (calanoids, juv. euphausiids)	1.1	1.1	0.87	3.3
Total non-carnivorous metazoans	3.3	3.4	2.37	7.6
Cyclopoids	0.39	0.91	1.4	2.3
Predatory calanoids	1.4	0.49	0.38	0.83
Other predators (Chaetognatha, Polychaeta,				
Amphipoda, etc.)	0.69	0.52	0.38	1.5
Total, mainly carnivorous metazoans	2.5	1.9	2.1	4.6
Total zooplankton (including protozoans)	8.4	6.9	7.27	14.2
Total plankton	16.1	16.6	15.17	77.2

[a] Biomass (wet weight gm^{-2}) of the components of plankton communities along the Equator in the Pacific Ocean in the 0–120 m layer in January, 1974; data from Tumantseva and Sorokin (1975) and Vinogradov and Shushkina (1976) with correction

Phytoplankton. In the region of greatest upwelling intensity at 97°W the upper boundary of the thermocline was located at a depth of 10–13 m. Below the thermocline the biomass of phytoplankton was 1.5 gm^{-3}, the number of cells $10.4 \times 10^6 \ m^{-3}$. The quantity of phytoplankton decreased sharply from east to west with increasing depth of the mixed layer, down to 110 m at 155°W, so that at 140–150°W its biomass was reduced by a factor of 12–13, and the number of cells 40 times less than at 97°W (Table III). Analogous values were recorded in this region by the *Alize* expedition (Desrosiers, 1969).

The diurnal value of primary production at 97°W was found to vary substantially during several days of observation. The fluctuations of production observed in 5–8 serial measurements were probably caused by patchiness in the phytoplankton community which the drifting ship passed through (Sorokin *et al.*, 1975a). These conditions are characteristic of upwelling regions in general (Beers *et al.*, 1971).

The recorded values of daily primary production (5–15 $mgCm^{-3}$) must be recognised as being low, and the very fact of the permanent presence of high

TABLE III   Eastern Pacific equatorial phytoplankton[a]

Lat. W	Depth of mixed layer (m)	Depth of euphotic zone (m)	Number of cells in 0–100 m	Biomass (wet weight g/m²)	Nanophyto-plankton in total biomass (%)	Average daily production (gCm⁻²)	Range of daily values
97°	10	40	1742[b]	49.5	27.8	3.15	0.5–6.0
122°	20	80	38	4.5	29.9	0.59	0.34–1.09
140°	50	100	25	3.1	47.2	0.50	0.20–0.81
155°	80	100	26	3.0	47.7	0.48	0.41–0.67

[a] Data for eastern Pacific, January, 1974 (Sorokin et al., 1975a)
[b] Thousands per ml

concentrations of nutrients (0.6 mcg atoms P/1 and more) to the surface (Sorokin et al., 1975a) creates a situation unusual in tropical regions of the open ocean. Even in cold waters (for example, in the Sea of Japan) such nutrient concentrations ensure a production of 50 mg cm^{-3} and a biomass of phytoplankton of 10–15 g m^{-3}, so that the potential productivity of the region of equatorial upwelling must be higher by at least one order than is actually observed. It has been suggested that the main cause limiting the development of phytoplankton in young waters is a deficit of organic iron complexes, vitamins and other growth-promoting metabolites which appear in the water with its ripening upon the development of microflora (Barber and Ryther, 1969; Sorokin et al., 1975a; Vedernikov et al., 1975). However, a mathematical simulation of phytoplankton development in the upwelling region (Menshutkin and Finenko, 1975) showed that in fact the limiting factor is the intense turbulent mixing which carries the cells beyond the compensation depth, thus preventing the development of a phytoplankton bloom. Under such conditions the role of grazing by zooplankton seems to be negligible compared with the effect of mixing.

Bacterioplankton. The number and biomass of bacteria in the region of intense upwelling at 97°W are commensurate with those recorded in mesotrophic and eutrophic lakes and reservoirs (Sorokin et al., 1975b). In the upper 50 m these values reach more than 2.5×10^6 cells ml^{-1}, and about 300 mg m^{-3}. In the most productive region (97°W) bacterioplankton is about 30% of the phytoplankton biomass, and may reach a still higher value in areas of less intense upward vertical transport (Table II).

The diet of so-called herbivorous copepods may consist of up to 30% of bacteria and to 20–50% of phytoplankton, while the diet of microzooplankton (ciliates and flagellates) almost entirely of bacterioplankton. Bacterial metabolism is 30–50% of the figure for primary phytoplankton production at 97–105°W and exceeds it farther west. It may be assumed that there the bacterioplankton subsists partly on allochthonous organic matter brought to the surface in the process of upwelling (Sorokin et al., 1975b).

Microzooplankton is the least studied element of marine plankton communities. In the eastern region of intense upwelling the nauplii of copepods are important in its composition, while protozoans become dominant farther west. The numerical abundance of microzooplankton decreases by a factor of three in this region of less intense upwelling but its biomass does not change considerably (Table II). Microzooplankton is a minimal part of total plankton at 97°W and increases towards the west. In the zone of maximum upwelling microzooplankton consumes about 10% of the diurnal production of bacterio- and phytoplankton; downstream this value increases to 36%.

Mesozooplankton. The biomass of mesozooplankton, like the biomasses of all other plankton groups, is greatest in the zone of intense upwelling, then shows a continuous decrease to 170–160°E and even farther to the west. Concomitant changes occur in the species, size composition and the trophic relationships of the mesozooplankton. The concept of succession in a plankton community implies that in the early stage of its development herbivorous forms should prevail while during the later stages predatory forms gain in importance. Indeed, the observations carried out along the Equator (86°W–151°E) by the *Alize* confirmed the dominance of herbivores in the young communities of intense upwelling. *Eucalanus subtenuis* alone accounted for 45 % of all copepods. In the more mature communities of western waters the predatory species *Euchaeta marina* and *Pleuromamma abdominalis* were dominant (Gueredrat, 1971). Maximum concentrations of predatory forms, and above all of migratory macroplankton, are formed west of 160°W (Voronina, 1964). The Shannon index of species diversity also points to a westward trend in the process of maturation of plankton communities (Gueredrat, 1971).

But there is a peculiar feature of the equatorial upwelling ecosystem which has to be taken into account: here, the upwelled water carries not only the nutrients that assure the development of phytoplankton, and of all the successive links of the food chain, but also the animal population of the subsurface undercurrents, belonging to the relatively mature communities of the waters entrained by the upwelling. Hence, the whole system acquires features characteristic both of juvenile and mature communities within the area of its formation. As a result, a heterogeneous community is created within the south equatorial current and subsurface undercurrents. Since in the eastern equatorial upwelling the waters of the south equatorial current move to the west in a relatively thin layer (0–20 m) and practically the entire water column, down to more than 200 m, is occupied by waters of the undercurrent moving eastwards, carnivorous forms (usually chaetognaths), characteristic of older serial stages may dominate, permanently or periodically, in the upwelling communities.

In January, in the upper 200 m layer, carnivores accounted for 62 % of the total biomass of mesozooplankton at 97°W, 59 % at 122°W, 68 % at 140°W and for 54 % at 155°W (Vinogradov and Semenova, 1975; Flint, 1975). Such a pattern must affect the production and other functional characteristics of the communities, as will be shown later.

The changes along the equator of abiotic and biotic variables of their environment are accompanied by changes in the abundance of certain individual species. Some are most numerous in the highly eutrophic eastern part of the equatorial region, but disappear entirely or almost entirely in the oligotrophic waters of the west (*Eucalanus subtenuis, Euchaeta concinna,*

Pleuromamma quadrungulata) while others, on the contrary, are most numerous in the poorer western equatorial regions and nearly, or entirely absent in the east (*Undeuchaeta plumosa, Rhincalanus cornutus, Eucalanus attenuatus, Candacia aethiopica, C. bispinosa*, etc.) (Gueredrat, 1971; Vinogradov and Voronina, 1962a, 1964).

5 Seasonal changes in plankton composition and abundance

Seasonal changes in the plankton communities of the tropics are determined by the wind regime and water circulation in this region. In the zone of upwelling, seasonal changes depend mainly on the intensity of upwelling and the location of zones of divergence which in their turn depend on the seasonal strengthening or weakening of the trade winds.

Seasonal differences in plankton abundance in the equatorial zone of the eastern Pacific were first reported by King and Demond (1953) and King (1958), though these later proved to be statistically not significant (King and Hida, 1957).

Special investigations of seasonal variability in the vast equatorial zone (100°W–121°W and 16°N–10°S) were carried out by the EASTROPAC expedition, which disclosed substantial seasonal fluctuations of diurnal primary production (127–318 mgCm^{-2}). A major peak of production was observed in February–May, and a second smaller one in August–September: these were separated by two minima, in June–July and in October–November (Owen and Zeitzschel, 1970).

The seasonal change in chlorophyll concentration presented a different picture: generally high concentrations from April to September with slight indication of two maxima, one in June–July and a second, minor peak, in December–January; minimal values occurred in October–November (Blackburn *et al.*, 1970). The biomass of zooplankton in daytime hauls varied in phase with the chlorophyll, but over a very narrow range, while in the night hauls neither zooplankton nor crustacean micronekton showed a statistically significant seasonal variation. However, at the Equator itself the biomass of crustacean micronekton increased in June and July by 15 times as compared with the preceeding two-month period. A similar increase was recorded in February and March, 1968, but not in the same period in 1967, so that there is no certainty that these peaks of micronekton were not accidental. Finally, significant seasonal fluctuations were discovered in some small fish and cephalopods; these fluctuations were of a very small amplitude and lagged four months behind the variations in chlorophyll concentration.

The small changes mentioned were statistically significant. However, they have not been explained or reported in another year. Therefore it remains uncertain that they are really seasonal and occur regularly each year.

In the Atlantic Ocean the seasonal changes in plankton communities along the Equator in the Gulf of Guinea have been investigated in more detail. In the eastern equatorial Atlantic two distinct periods may be recognised in the annual cycle of plankton communities (Gruzov, 1971a,b; Vinogradova, 1971). The first, in February to May, is a period of low biomass characterised by a relatively stable stratification of the water column and a progressive warming of the surface layers. At this time, and particularly toward the end of the period, the equatorial zone from 1°N to 5°S differs only slightly in plankton abundance from the oligotrophic trophical regions, a slight increase in biomass taking place only directly under the Equator (Fig. 9). Blue-green algae are important in the im-

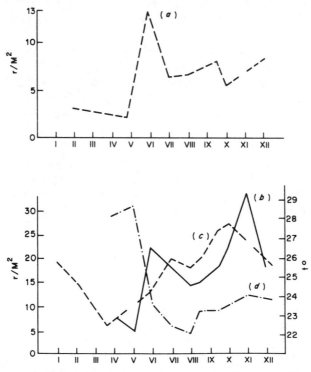

FIG. 9 Seasonal changes in plankton abundance in the equatorial region of the eastern Atlantic (1°N–5°S); (a) total stock of phytoplankton (wet weight); (b) biomass of zooplankton (wet weight in g m^{-2} in the 0–100 m layer) at 5°W; (d) temperature in the 0–30 m layer at 5°W (after Gruzov, 1971a)

poverished phytoplankton and the average body size of herbivorous copepods reaches its annual minimum.

The second period, from June to December is characterised by intensified upwelling. The early part of this period is marked by a rapid development of phytoplankton, first of all of diatoms and naked flagellates, the optimal diet for herbivores, and both growth and biomass of algae reach their annual maximum. These conditions favour a rapid growth of herbivorous copepods and a rejuvenation of their populations which, in its turn, results in intensive renewal of phytoplankton further enhanced by the development of large concentrations of herbivorous pyrosomes (their colonies in July may number $100-200 \text{ m}^{-5}$). This period of rich food supply is characterised by a high fecundity of herbivores and the appearance of abundant generations consisting of very large individuals. As the biomass of zooplankton increases rapidly, the biomass of algae stabilises or slightly decreases while peridineans and coccoliths gain in relative importance. Toward the middle of the second period (August) the growth of phytoplankton slows down, its species composition undergoes a radical change so that peridineans and coccolithophorids dominate.

In October, with the renewal of upwelling, diatoms again become important, among them large *Rhizosolenia* and *Thalassiothrix*. A mass development of tunicates (*Thalia democratica*) occurs, followed by a wave of filter-feeding crustaceans.

The biomass of zooplankton reaches its yearly maximum during the period of reduced upwelling; in 1965 this occurred in November, in 1963 in September. At this time reduced phytoplankton production can no longer satisfy the food requirements of the mass of herbivorous zooplankton. This period of food shortage leads gradually back to the low biomass situation of the first half of the year (Gruzov, 1971b).

6 Functional characteristics of the equatorial plankton community

As we are discussing the equatorial upwelling ecosystem as an entity, we are interested not only in the variability of the components of this system but also in their relationships, and in the functional characteristics of the community. These characteristics, like the structural ones, undergo substantial changes from east to west along the Equator.

A functional analysis of the plankton community of the equatorial upwelling in the eastern part of the Pacific Ocean permits us to draw some conclusions about its development as it is transported from the highly productive regions of the east to the poorer western regions (Vinogradov *et al.*, 1976; Vinogradov and Shushkina, 1978; Vinogradov, 1978).

The decreased biomass of phyto- and bacterioplankton caused by the westward diminishing intensity of upwelling reflects upon the production of plankton animals. The degree to which food requirements are satisfied $(\delta = C/C_{max})$ decreases 1.5–2.0 times in nearly all the trophic groups, and remains almost unchanged only for predators (Table IV). Grazing pressure on different groups increases considerably, while efficiency of food assimilation $(\eta = A/R = (P + R)/R$, where A–assimilated food, R–respiration, P– production) decreases. At the same time as greater stress in trophic relations is observed, a more stable balance of production and consumption between some trophic levels $(\varepsilon_j = P_j/\sum_i T_{ij})$, where T–particular ration) and a higher efficiency of energy transfer through the system $(\omega_{ij} = \sum_i T_{ij}/C_j$, where C– full ration) or food 'consumption' occurs. Thus, ecological efficiency in flagellates was 3% at 97°W and 17% at 155°W, in infusorians 14% and 21– 26% and in fine filter-feeders 5% and 11% respectively. Due to the relatively high concentration of predators in the whole zone of equatorial upwelling (44–55% of zooplankton biomass), from 50 to 80% of the production of mesozooplankton is consumed internally, and in the meso-trophic waters of reduced upwelling intensity the production of mesozoo-plankton shows a significant decrease.

Calculations of energy flow in the equatorial upwelling community show the prevalent importance of the detritus food chain, comprising the utilization of dead organisms by bacteria and of these by microzooplankton. Bacterioplankton and microplankton act as intermediate links between the primary producers and the mesozooplankton consumers. The presence of these intermediate links lengthens the food chain, dissipating a considerable part of the energy of primary production. Therefore, to assure the food supply for mesozooplankton through the detritus food chain, a several-fold

TABLE IV   Nutrition of equatorial zooplankton[a]

Group of organisms	$\delta = C/C_{max}$			
	97°	122°	140°	155°
Bacteria	0.73	0.46	0.62	0.40
Protozoans	0.78	0.54	0.66	0.52
Filter-feeders (metazoan)	0.90	0.67	0.70	0.63
Predators (metazoan)	0.91	0.77	0.82	0.84

[a] Variation in apparent degree of satisfaction of nutritive requirements within the equatorial ecosystem at various longitudes in the Pacific Ocean. C = observed ration, C_{max} = ration required for maximal growth rate

greater amount of energy is required to enter the system as compared with the direct plant herbivore chain.

Allochthonous dissolved organic matter may be introduced into the food web through the detritus food chain. According to Sorokin *et al.* (1975), in the region of intense equatorial upwelling at 97°W up to 80% of the energy from primary production enters the community through the heterotrophic organisms, but even in poorer waters (155°W) no less than 50–70% enters the system in this way. This clearly demonstrates the immense importance of the bacterioplankton and its principal consumers, the naked flagellates and infusorians, in the general energy balance of the community.

The production of all trophic groups of the plankton community is biggest within the zone of most intense upwelling. Here too the positive net production of the community as a whole is of significance although its absolute value is not high (Table V). The average daily net production of the community is equivalent to 12% at 97°W of the biomass. With the westward transport of the community of the surface layers by the southern equatorial current net production becomes negative: the organic matter produced by phytoplankton does not suffice for the energy expenditure of the bacterial flora and animals.

This means that the ecosystem of this region subsists not only on the organic matter that it produces, but also on the energy accumulated earlier, or on alochthonous organic matter introduced into the community, for instance, by the ascent of Antarctic Intermediate Water. The energy deficit

TABLE V   Production rates within plankton ecosystem[a]

Longitude (W)	Phyto-plankton (P_p)	Bacteria (P_b)	Protozoans (P_a)	Metazoans Herbi-vorous (P_f)	Carni-vorous (P_s)	Entire community P_o	$c = P_o/B_o$
97°	24.1	6.6	1.9	2.8	1.0	5.9	0.12
122°	5.3	2.9	0.94	0.61	0.25	−3.8	−0.36
140°	4.5	2.3	0.89	1.0	0.33	−3.2	−0.25
155°	5.7	5.0	0.82	0.83	0.31	−7.8	−0.66

[a] Production rates (kcal m^{-2}day^{-1}) of functional groups within the equatorial plankton ecosystem in the eastern Pacific at various longitudes. Net production of community $P_o = P_p - \sum_{i=b}^{s} \times R_i$ where R_i = metabolic rate, B_o = community biomass, i = groups of organisms, (p = phytoplankton, b = bacteria, a = protozoa, f = herbivorous zooplankton, s = predatory zooplankton). Some values differ somewhat from earlier published but similar data (Vinogradov *et al.*, 1976; Vinogradov, 1978) in the light of more recent investigations

increases from east to west and, concurrently, a decrease takes place in the efficiency of primary organic production ($K_{3_p} = P_p/\sum_{i=b}^{s} R_i = P_p/D$) which may be assessed by the ratio of primary production to total heterotrophic activity within the community. The value of K_{3_r} decreases from 1.3 at $97°W$ to 0.4 in the central Pacific. The K_{3_p} value, and the value of primary production may be used to characterise the waters of the region of intense upwelling as eutrophic and of the remaining part of the upwelling region as mesotrophic (Vinogradov and Shushkina, 1978).

When we take into account the fact that part of the mesoplankton is consumed by animals not included in our calculations since they belong to higher trophic levels (macroplankton and fishes) which are most abundant on the Equator westwards of the zone of intense upwelling, a basic property of the developing communities becomes clearly apparent: accumulation of biomass and energy ($K_{3_p} > 1$) in the earlier stages of development and expenditure of energy in more mature communities west of $120°W$. This process of accumulation and subsequent expenditure of energy suggests the existence of a general succession of plankton communities along the equator.

During equatorial upwelling this process is modified by the addition of nutrients to the euphotic zone, by the lateral shift of developing communities away from the equator by the meridional current component, and, finally, by the effect of the waters of subsurface undercurrents bringing eastwards the highly mature communities of western waters. Notwithstanding these disturbances, the changes in the basic functional characteristics of the communities follow a rather regular course in conformity with the westward attenuation of upwelling.

References

Austin, T. S. (1960). Oceanography of the east central equatorial Pacific as observed during expedition Eastropic. *Fish. Bull.* **168**. *Fish and Wildl. Serv.* **60**, 257–282

Austin, T. S. and Rinkel, M. O. (1958). Variation in upwelling in the equatorial Pacific. *Proc. 9th Pacific Sci. Congr.* **16**, 67–71

Barber, R. I. and Ryther, J. H. (1969). Organic chelators as factor effecting primary production in Cromwell Current upwelling. *J. Exp. Mar. Biol. Ecol.* **3** (2), 191–199

Beers, J. R., Stevenson, M. R., Eppley, R. W. and Brooks, E. R. (1971). Plankton populations and upwelling off the coast of Peru. June 1969. *Fish. Bull. U.S. Department of Commerce* **69** (4), 859–876

Bjerknes, I. (1969). Large-scale ocean-atmosphere interaction. Morning review lectures of the second international oceanographic Congress, UNESCO, 11–20

Blackburn, M. (1966). Relationships between standing crops at three successive trophic levels in the eastern trophic Pacific. *Pacif. Sci.* **20**, 36–59

Blackburn, M. (1968). Micronekton of the eastern tropical Pacific Ocean: family composition, distribution, abundance, and relations to tuna. *Fishery Bull. Fish. Wildl. Serv. U.S.* **67**, 71–115

Blackburn, M., Laurus, R. M., Owen, R. W. and Zeitzschel, B. (1970). Seasonal and areal changes in standing stocks of phytoplankton, zooplankton and micronekton in the eastern tropical Pacific. *Mar. Biol.* **7** (1), 14–31

Brandhorst, W. (1958). Thermocline topography, zooplankton standing crop, and mechanisms of fertilization in the eastern tropical Pacific. *J. Cons. perm. Int. Explor. Mer.* **24**, 16–31

Bubnov, V. A. (1975). Vertical turbulent exchange in the ocean near the Equator. *Trans. Inst. Oceanol.* **102**, 47–50

Bubnov, V. A. and Egorikhin, V. D. (1975). Structure of currents at the Equator in the eastern part of the Pacific Ocean. *Trans. Inst. Oceanol.* **102**, 34–41

Desrosières, R. (1969). Surface macrophytoplankton of the Pacific Ocean along the Equator. *Limnol. and Oceanogr.* **14**, 626–632

Fedorov, K. N. (1975). Estimation of vertical velocity in Equatorial upwelling. *Trans. Inst. Oceanol.* **102**, 41–46

Fedorov, K. N., Prokhorov, V. I., Bubnov, V. A. (1975). Thermohaline mesostructure of the equatorial current system in the eastern half of the Pacific. *Trans. Inst. Oceanol.* **102**, 24–33

Flint, M. V. (1975). Trophic structure and vertical distribution of trophic groups of mesoplankton of the Equator (97°W). *Trans. Inst. Oceanol.* **102**, 238–244

Gibson, C. H. and Williams, R. B. (1972). Measurements of turbulence and turbulent mixing in the Pacific Equatorial Undercurrent. Oceanography of the South Pacific. *N.Z. Nat. Commn.* UNESCO

Gruzov, L. N. (1971a). The formation of zooplankton accumulations in the pelagic parts of the Gulf of Guinea. *Proceedings of the Atlantic Division of the All-Union Research Institute of Marine Fisheries and Oceanography (VNIRO)*, **37**, 406–428

Gruzov, L. N. (1971b). On the balance between the reproduction and the consumption processes in a plankton community of the Equatorial Atlantic. *Proceedings of the AtlantNIRO* **37**, 429–449

Gueredrat, I. A. (1971). Evolution d'une population de copépodes dans le systéme des courants equatoriaux de l'Ocean Pacifique. Zoogéographic, écologie et diversité spécifique. *Mar. Biol.* **9** (4), 300–314

Hidaka, K. (1960). On the equatorial upwelling. *Mem. Kobe Mar. Observ.* **14**, 32–34

Jones, J. H. (1973). Vertical mixing in the Equatorial Undercurrent. *J. Phys. Oceanogr.* **3** (3), 286–296

Khanaychenko, N. K. (1969). Confirmation of the existence of the southern branch of the equatorial countercurrent. *Doklady AN SSSR* **187**, No. 6

Khanaychenko, N. K. (1973). Peculiarities of water circulation in the Tropical Atlantic determining the character of the biological productivity of the ocean. *In* "The Tropical Zone of the World Ocean" pp. 141–148. Nauka, Moscow

Khanaychenko, N. K., Khlystov, N. Z. and Zhidkov, V. G. (1965). On the system of equatorial countercurrent in the Atlantic Ocean. *Okeanologia* **5** (2), 222–229

King, J. E. (1958). Variation in abundance of zooplankton and forage organisms in the central Pacific in respect to the equatorial upwelling. *Proc. 9th Pacific Sci. Congr.* **16**, 98–107

King, J. E. and Demond, J. (1953). Zooplankton abundance in the central Pacific. *Fishery Bull. Fish. Wildl. Serv. U.S.* **54**, (82), 111–144

King, J. E. and Hida, T. S. (1957). Zooplankton abundance in the central Pacific. Part 2. *Fishery Bull. Fish. Wildl. Serv. U.S.* **57** (118), 365–395

King, J. E. and Iversen, R. T. B. (1962). Midwater trawling for forage organisms in the central Pacific, 1951–1956. *Fishery Bull. Fish. Wildl. Serv. U.S.* **62**, 271–321

Knauss, J. A. (1966). Further measurements and observations on the Cromwell Current. *J. Mar. Res.* **24** (2), 205–210

Levin, L. A., Utyushev, R. N. and Artemkin, A. S. (1975). Intensity of the field of bioluminescence in the equatorial part of the Pacific Ocean. *Trans. Inst. Oceanol.* **102**, 94–102

Margalef, R. (1967). Some concepts relative to the organization of plankton. *Oceanogr. Mar. Biol. Ann. Rev.* **5**, 257–289

Margalef, R. (1968). "Perspectives in Ecological Theory". University of Chicago Press

Menshutkin, V. V. and Finenko, Z. Z. (1975). Mathematical simulation of the process of phytoplankton development under conditions of oceanic upwellings. *Trans. Inst. Oceanol.* **102**, 175–184

Metcalf, W. G., Voorhis, A. O. and Staclup, M. C. (1962). The Atlantic Equatorial undercurrent. *J. Geophys. Res.* **67** (6)

Montgomery, R. B. and Stroup, E. D. (1962). Equatorial waters and currents at 150°W in July–August 1952. The Johns Hopkins Oceanogr. Studies No. 1. Johns Hopkins Press, Baltimore

Murphy, G. J. and Shomura, R. (1958). Variations in yellowfin abundance in the central equatorial Pacific. *Proc. 9th Pacific Sci. Congr.* **16**

Owen, R. W. and Zeitzschel, B. (1970). Phytoplankton production seasonal change in the oceanic eastern tropical Pacific. *Mar. Biol.* **7** (1), 32–36

Paliy, N. F. (1971). On the vertical circulation in the western part of the Gulf of Guinea. *Proceedings of the AtlantNIRO* **37**, 97–111

Parin, N. V. (1962). Some peculiar features involved with the distribution of mass pelagic fishes in the zone of equatorial currents in the Pacific Ocean (based on the materials of the 34 cruise of R/V *Vityaz*. *Okeanologia* **2** (6), 1075–1082

Polosin, A. S. (1967). On the zero surface in the equatorial zone of the Atlantic Ocean. *Okeanologia* **7** (1), 89–97

Rotschi, H. (1970). "Variations of Equatorial Currents. Scientific Exploration of the South Pacific". 75–83 *Nat. Acad. Sci.*, Washington, D.C.

Rotschi, H. (1973). Hydrology at 170°E in the South Pacific. Scientific exploration of the South Pacific. *Nat. Acad. Sci.*, Washington, D.C.

Rotschi, H. and Jarrige, F. (1968). Sur le renforcement d'un upwelling equatorial. *Cah. O.R.S.T.O.M., Sér. Océanogr.* **34**, 87–90

Sedykh, K. A. and Lutoshkina, B. N. (1971). Hydrological aspects of the formation of the equatorial productive zone in the Gulf of Guinea. *Proceedings of the AtlantNIRO* **37**, 31–80

Sette, O. E. (1955). Consideration of midocean fish production as related to oceanic circulatory systems. *J. Mar. Res.* **14** (4), 398–414

Sorokin, Yu. I., Sukhanova, I. N., Konovalova, G. V. and Pavelyeva, E. B. (1975a). Primary production and phytoplankton in the area of equatorial divergence in the Equatorial Pacific. *Trans. Inst. Oceanol.* **102**, 108–122

Sorokin, Yu. I., Pavelyeva, E. B., Vasilyeva, M. I. (1975b). Productivity and trophic role of bacterioplankton in the area of equatorial divergence. *Trans. Inst. Oceanol.* **102**, 184–198

Tsuchiya, M. (1968). "Upper waters of the Intertropical Pacific Ocean". Johns Hopkins Oceanogr. Studies, No. 4

Tumantseva, N. I. and Sorokin, Yu. I. (1975). Microzooplankton of the area of equatorial divergence in the eastern part of the Pacific Ocean. *Trans. Inst. Oceanol.* **102**, 200–212

Vedernikov, V. I., Koblentz-Mishke, O. J., Sukhanova, I. N., Karabashev, G. S. and Fisher, J. (1975). A comparison of vertical changes in the quantities of particulate matter, chlorophyll, phytoplankton, and the intensity of pigment luminescence in the Equatorial and Peruvian regions of the Eastern Pacific. *Trans. Inst. Oceanol.* **102**, 165–175

Vinogradov, M. E. (1974). On the depth of the night-time upward migration of the deep-scattering layers in the central Pacific. *Okeanologia* **14** (6), 1082–1086

Vinogradov, M. W. (1977). A spatial-dynamic aspect of the existence of the pelagic communities. *In* "Okeanologia, Biology of the Ocean", Vol. 2, pp. 14–23

Vinogradov, M. E. (1978). Some physical and biological features of equatorial upwellings. *Byull. Mosk. Obshch. Ispyt. Prir. (ser. Biol.)* **83** (1), 5–16

Vinogradov, M. E. and Semenova, T. N. (1975). A trophic characterization of pelagic communities in the equatorial upwelling. *Trans. Inst. Oceanol.* **102**, 232–238

Vinogradov, M. E. and Shushkina, E. A. (1976). Some peculiarities of the vertical structure of a planktonic community in the equatorial Pacific upwelling. *Okeanologia* **16** (4), 667–684

Vinogradov, M. E. and Shushkina, E. A. (1978). Some development patterns of plankton communities in the upwelling areas of the Pacific Ocean. *Mar. Biol.* **48** (4), 357–366

Vinogradov, M. E. and Voronina, N. M. (1962a). The distribution of different groups of plankton in accordance with their trophic level in the Indian equatorial current area. *Rapp. P.-V. Réun. Cons. perm. int. Explor. Mer.* **153**, 200–204

Vinogradov, M. E. and Voronina, N. M. (1962b). Some data on the distribution of zooplankton in the northern Indian Ocean. *Trans. Inst. Oceanol.* **58**, 80–113

Vinogradov, M. E. and Voronina, N. M. (1963). Quantitative distribution of plankton in the upper layers of the Pacific equatorial currents. 1. *Trans. Inst. Oceanol.* **71**, 22–59

Vinogradov, M. E. and Voronina, N. M. (1964). Some peculiarities of the plankton distribution in the Pacific and Indian Ocean's Equatorial current area. *Okeanol. Issled., 10 Sect. ICY Programme* **13**, 128–136

Vinogradov, M. E., Voronina, N. M. and Sukhanova, I. N. (1961). The horizontal distribution of the tropical plankton and its relation to some peculiarities of the structure of water in the open sea areas. *Okeanologia* **1**, 283–293

Vinogradov, M. E., Gitelzon, I. I. and Sorokin, Yu. I. (1971). The vertical structure of a pelagic community in the tropical ocean. *Mar. Biol.* **6** (3), 187–194

Vinogradov, M. E., Krapivin, V. F., Menshutkin, V. V., Fleishman, B. S. and Shushkina, E. A. (1973). Mathematical simulation of functioning of the pelagic ecosystem in the tropical ocean (based on the materials of the 50th cruise of the R/V *Vityaz*). *Oceanology* **13** (5), 852–866

Vinogradov, M. E., Shushkina, E. A. and Kukina, I. N. (1976). Functional characteristics of the planktonic community in the equatorial upwelling. *Okeanologia* **16** (1), 122–138

Vinogradova, L. A. (1971). Seasonal development of phytoplankton in the Gulf of Guinea. *Proceedings of the AtlantNIRO* **37**, 117–159

Voronina, N. M. (1964). The distribution of macroplankton in the waters of equatorial currents of the Pacific Ocean. *Okeanologia* **4** (5), 884–895

Wyrtki, K. (1966). Oceanography of the eastern equatorial Pacific Ocean. *Oceanogr. Mar. Biol.* **4**, 33–68

Yoshida, K. (1967). Circulation in the Eastern Tropical Oceans with special references to upwelling and undercurrents. *Jap. J. Geophys.* **4**, 1–75.

4 High Latitude Ecosystems

TAKAHISA NEMOTO and GLEN HARRISON

1 Introduction

Ecological research on high latitude marine ecosystems has been the subject
of a number of recent reviews (Holdgate, 1967; Knox, 1970; Everson, 1977;
Dunbar, 1968, 1977; Llano, 1978). Because of present interests in marine
renewable and non-renewable resources, research in both polar regions has
been stimulated resulting in a considerable body of new scientific information, especially from the Antarctic.

High latitude oceans are characterised by their extremes in environment
conditions: pronounced seasonal oscillations in solar radiation, low
sea water temperatures and prevalence of sea ice, especially in winter. These
factors impose limits on productivity, growth rate, and size of marine
organisms, resulting in generally lower species diversity, shorter food chains,
and higher abundance of species present.

This chapter deals with some special aspects of the biology of organisms
in polar waters, mainly in the Antarctic. Emphasis will be placed on
comparative phytoplankton distribution and production, zooplankton
swarms and their relation to phytoplankton and the predation of larger
organisms upon zooplankton and micronekton. An example of a food chain
in the Antarctic will conclude the chapter.

2 Characteristics of the marine environment

PHYSICAL AND CHEMICAL

Solar radiation

The unique light environment at high latitudes results from extreme
seasonal oscillations in intensity and low angles of incidence. Seasonal
differences in day length (and daily solar energy flux) increase with latitude.
In the tropics, daily solar radiation varies by no more than 10% seasonally.
At mid-latitudes, the variation increases to fourfold and in regions greater
than 65° latitude, prolonged periods (months) of total light or darkness are
characteristic: for example, 24 h of daylight occurs at 65° in December and
only 4 h in June, and at 80 to 85° latitude, there is total darkness for 5

months of the year. The extreme seasonal amplitudes in day length at high latitudes may, however, result in total daily radiation values in summer equal to or greater than that of the tropics. Still, annual total radiant energy at high latitudes (60°) may be only half on the average that received in tropical latitudes (see Table 25 in Sverdrup *et al.*, 1946).

Solar altitude also varies with latitude. At higher latitudes, solar angles are lower and consequently the amount of incident radiation penetrating the sea surface is less as a result of higher surface reflection. At a solar altitude of 5°, as much as 40 % of the incoming radiation can be reflected (Sverdrup *et al.*, 1946). El-Sayed (1966) measured 50 % in Marguerite Bay (68°S lat.) in February.

Finally, the presence of sea ice and snow affect the amount of solar energy penetrating the sea surface. Bunt and Lee (1970) normally found less than 1 % of surface light penetrating one metre sea ice with snow cover. Clear ice penetration was greater than 10 % for one metre thickness.

Temperature

Despite strong seasonal variations in solar radiation, the seasonal range in high latitude mixed layer temperatures is very small relative to that at mid-latitudes. Surface temperatures in the Antarctic (south of the convergence) range by no more than 3 to 5°C annually and close to the continent (e.g. McMurdo Sound) −1.7 to −1.9°C (Bunt, 1964; Knox, 1970). Arctic temperatures under the ice are normally less than −1.5°C year-round (Kawamura, 1967; English, 1961) while in sub-Arctic waters (Dunbar, 1953) temperatures may range 6°C annually near the surface (Grainger, 1975) and exceed those of the sub-Antarctic around the Antarctic convergence.

Sea ice

The extensive ice cover and its annual oscillations in both polar regions are a consequence of the high latitude solar radiation and temperature conditions. The source of sea ice is primarily from sea water freezing *in situ* but can have significant input from glaciers. Seasonal variations in heat exchange in the polar regions result in the marked annual geographic excursion of sea ice. In the Antarctic, the areal extent of sea ice may vary tenfold between late winter and late summer, while in the Arctic coverage varies less: 80 % in winter to 60 % in summer (Zenkevitch, 1963). Because the Arctic Ocean is almost entirely at latitudes higher than the Antarctic and because it is relatively confined by land, in contrast to the open boundaries of the Antarctic, Arctic sea ice is more likely to recirculate, accounting for its smaller seasonal excursion and the prevalence of multiyear ice (Foster, 1978).

Inorganic nutrients

It is well documented that surface waters at high latitudes are enriched with inorganic nutrients relative to mid- and tropical latitudes (see, for example, Figs 45, 49, 52 in Sverdrup *et al.*, 1946) and that they play a particularly important role in the high biological production in the Antarctic (e.g. Walsh, 1971). Nitrates, phosphates and silicates are generally in high concentrations throughout the year (Hori, 1966; El-Sayed, 1970; etc.) and are presumed high enough to prevent any significant limitation to primary production (e.g. Holdgate, 1967). Nitrates, silicates, and phosphates range from 15 to 30, 40 to 90, and 1 to 2 mg atoms m^{-3}, respectively, in the euphotic zone (Holm-Hansen *et al.*, 1977). In the Arctic, values are comparably high in winter (e.g. Kawamura, 1967; Grainger, 1975, 1977) but may drop to low levels (particularly nitrate) near the surface after the spring phytoplankton bloom (Grainger, 1975; McRoy *et al.*, 1972; Alexander, 1974; Jones and Coote, unpublished data). The persistent influx of nutrients to south polar region results from upwelling along the Antarctic divergence. Dunbar (1968) has suggested that water column stability and near surface nutrient exhaustion are responsible for the relatively lower production in the Arctic. Although sea ice formation tends to concentrate nutrients near the ice-sea-water interface (e.g. Meguro *et al.*, 1967), this appears to be an insignificant source to the water column during spring melt because of the large build-up of nutrients in the water over winter (Grainger, 1977).

3 Phytoplankton and primary production

The abundance and productivity of phytoplankton in polar seas are strongly driven by the seasonal oscillations in their physical/chemical environment. As a consequence, high latitude growing seasons are characterised by a single intense peak in phytoplankton production (Fig. 1). The magnitude of the production peak also increases with latitude, thought to be primarily a result of the progressive latitudinal lag in zooplankton grazing interactions (Cushing, 1975).

ANTARCTIC

Phytoplankton and primary production studies in the Antarctic have been extensive and have been summarised in a number of recent reviews (Knox, 1970; El-Sayed, 1970, 1978; El-Sayed and Turner, 1977; Holm-Hansen *et al.*, 1977; Fogg, 1977). Since the early descriptive work of Hart (1934), the south polar sea has been recognised as an extremely productive and relatively diverse region. At least 100 species of diatoms, 60 dinoflagellates, and several silicoflagellates have been identified in Antarctic waters (Hasle, 1969; El-Sayed, 1978). One explanation for this high production is the already-mentioned continuous upwelling of nutrient-rich water associated with the Antarctic divergence zone (Walsh, 1971). Primary production in the Antarctic does, however, differ from that in tropical/subtropical upwelling regions in that it has a short (generally less than 4 months, El-Sayed and Turner, 1977) and unimodal productive period during Austral summer.

Geographical variations in biomass and production of the Antarctic (south of the Antarctic convergence) and sub-Antarctic waters (north of the Antarctic convergence) have been described by El-Sayed (1970). Of the three oceans bounding the Antarctic continent, production in the Atlantic sector is highest (Table I). Surface values of chlorophyll and productivity show

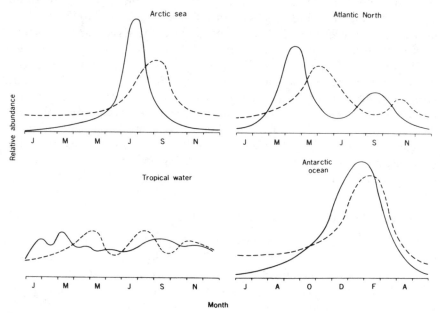

FIG. 1 Seasonal cycles in plankton biomass (modified from Parsons and Takahashi, 1973)

similar variations and are on the average four times higher than in the Pacific. Indian Ocean values are probably intermediate (El-Sayed and Turner, 1977). These general longitudinal differences have been attributed to a 'land mass' effect: Atlantic stations were primarily coastal, Indian Ocean and particularly Pacific stations were oceanic and typically more remote from land masses. Within each sector, biomass and productivity exhibit similar and high degrees of variability.

Conspicuous latitudinal variations have also been noted (El-Sayed, 1970; El-Sayed and Turner, 1977; Holm-Hansen et al., 1977). Phytoplankton surface biomass and productivity are notably lowest in the area of the Antarctic convergence (50 to 60°S lat.). Peak production and standing crops are found between 70 and 75°S. North of the polar front (sub-Antarctic waters), dinoflagellates (*Ceratium, Peridinium* species) are often important components of the phytoplankton communities while south of the convergence, diatoms (*Thalassiothrix, Rhizosolenia, Corethron, Thalassiosira*, etc.) usually dominate. The significance of nanoplankton (less than 10 µm) has also been found to vary latitudinally, decreasing from 90% of the total biomass and production in sub-Antarctic waters to 60% in Antarctic waters near the continent (El-Sayed and Turner, 1977).

TABLE I   Standing crop of phytoplankton (Chl) and primary production (PP) from Antarctic and sub-Antarctic waters (taken from El-Sayed, 1970; El-Sayed and Turner, 1977)

		Minimum	Maximum	Mean	Obs.
Atlantic sector					
Surface	Chl	0.01	118.35	0.89	518
	PP	0.02	97.44	5.25	458
Integral	Chl	0.72	81.38	15.94	79
	PP	3.43	337.49	50.70	87
Pacific sector					
Surface	Chl	0.01	5.80	0.26	723
	PP	0.03	22.50	1.22	656
Integral	Chl	0.23	41.32	12.62	217
	PP	3.54	194.73	32.01	213
Overall					
Surface	Chl			0.29	
	PP			3.93 (day^{-1})	
Integral	Chl			17.40	
	PP			134.00 (day^{-1})	

Surface: Chl = mg m^{-3}, PP = mgCm^{-3}h^{-1}
Integral: Chl = mg m^{-2}, PP = mgCm^{-2}h^{-1} in euphotic zone

The vertical distribution of phytoplankton biomass in Antarctic waters is generally uniform within the euphotic zone although subsurface maxima in chlorophyll are not uncommon (El-Sayed, 1970, 1978). Significant amounts of chlorophyll can be found at depths in excess of 200 m and the total amount below the level of light penetration may equal or surpass that in the euphotic zone. Photosynthesis, on the other hand, is usually maximal at the surface. On very bright days (100 Ly/half light day), however, maximum photosynthesis has been observed below the surface; photoinhibition at the shallow depths has been implicated (Holm-Hansen *et al.*, 1977).

Data on seasonal variations in biomass and productivity in the Drake Passage (Atlantic sector) showed peak surface chlorophyll values (greater than 0.50 mg m^{-3}) coinciding with the productivity maximum (greater than 3 mgCm^{-3}h^{-1}) during Austral spring and summer (December–March) (El-Sayed, 1970). Maximum chlorophyll and productivity values for the Pacific sector were greater than 0.4 mg m^{-3} and greater than 2.5 mgC fixed m^{-3}h^{-1}, respectively. Converted to daily integral production rates, maximum photosynthetic rate (3 to 4 gCm^{-2}day^{-1}) are comparable to values from some of the world's most productive oceanic waters (Platt and Subba Rao, 1975).

The magnitude and variations in Antarctic phytoplankton production have been attributed to environmental conditions of temperature, light, nutrients and to biological control by grazing pressure. El-Sayed (1978) has argued that temperature and nutrients are of minor importance. Physiological studies by Bunt et al. (1966) and Bunt (1968) suggest that Antarctic phytoplankton are obligate psychrophiles with the ability for significant growth at extremely low temperatures. El-Sayed (1971) observed phytoplankton bloom conditions in the southwestern Weddell Sea when surface temperatures were only −1.7 °C. High nutrient concentrations have also been observed during and after such blooms, suggesting continuous upwelling nutrient inputs always in excess of demand.

The availability of phytoplankton for primary herbivores depends on its particle size composition. In the study of environmental factors for phytoplankton growth, Parsons (1976) suggested that appreciable growth of large-cell or chain-forming phytoplankton would occur in the Antarctic upwelling areas during the summer, using a theoretical equation (Parsons and Takahashi, 1973). These chain-forming and large phytoplankton may be utilised directly by the large filter-feeding plankton such as krill *Euphausia superba* and salps, the filtering mechanisms of which are effective for filtering them. Diatom species in water where feeding conditions for baleen whales is poor are far larger (*Chaetoceros* spp., *Corethron criophilum*, *Rhizosolenia* spp., *Dactyliosolen antarctica*, *Thalassiothrix antarctica*) which are themselves unsuitable food for krill (Barkely, 1940).

Light has been considered the most likely limiting factor for south polar primary production. Holm-Hansen et al. (1977), for example, have noted a strong correlation between euphotic zone primary production and incident radiation. Although net primary production may occur at 0.1 or even 0.01 % surface intensities in Antarctic species, this low light level adaptation may make them particularly sensitive to high light intensities (photoinhibition) in near surface waters.

Water column stability has also been suggested as important for the productivity of Antarctic waters (El-Sayed, 1978) although the direct mechanism would be through the effect of light. Water column instability and deep mixing of phytoplankton in the convergence zone, for example, have been suggested as reason for the low production there (Hasle, 1956; Walsh, 1971). El-Sayed (1978) has considered the most significant factor determining large seasonal variations in Antarctic production to be the sea ice conditions and their effect on available light for primary production.

Overall, primary production in the Antarctic has been considered high. Ryther (1963) estimated $100\ gCm^{-2}yr^{-1}$, while Currie (1964) calculated $43\ gCm^{-2}yr^{-1}$. Horne *et al.* (1969) estimated $130\ gCm^{-2}yr^{-1}$ in inshore waters and Whitaker (cited in Fogg, 1977) estimated 25 and $80\ gCm^{-2}yr^{-1}$ for two other years in the same general area. El-Sayed (1978) and Holm-Hansen *et al.* (1977) have recalculated the overall Antarctic yearly production using a more comprehensive data base and concluded that the earlier estimates were too high, primarily because data for earlier estimates came mostly from the richer inshore areas. The most recent estimate is about $16\ gCm^{-2}yr^{-1}$, suggesting that Antarctic production is significantly lower than previously believed.

ARCTIC

Relatively little information exists on phytoplankton and primary production in Arctic seas compared with the Antarctic. Most published studies have been concentrated in the sub-Arctic regions of the Pacific (including the Gulf of Alaska, Bering, Chukchi, and Beaufort Seas). Sparse data exists for the Atlantic (Eastern Canadian Arctic, Norwegian and Barents Seas) and studies from both oceans have been principally confined to coastal waters in summer. Although possessing many common environmental characteristics of the Antarctic, the Arctic as a whole is considered relatively unproductive (e.g. Dunbar, 1968).

Phytoplankton and primary production in the Bering Sea and associated waters have been most extensively studied (Koblenz-Mishke, 1965; Taniguchi, 1969; Larrance, 1971; Taguchi, 1972; McRoy *et al.*, 1972; McRoy and Goering, 1974, 1976; Motoda and Minoda, 1974). Marked regional variations in phytoplankton crops and production have been observed. Taniguchi (1969), for example, noted high surface crops (greater than $0.5\ mgChlm^{-3}$) and production (max. $5.04\ mgCm^{-3}h^{-1}$) along the Alaskan coast and north of the Aleutians and relatively lower values in the Bering Sea gyre ($0.15\ mgChlm^{-3}$, $1.6\ mgCm^{-3}h^{-1}$). McRoy *et al.* (1972) observed extremely high values in the Bering Strait ($7\ mgChlm^{-3}$, $18\ mgCm^{-3}h^{-1}$), and noted that chlorophyll and production were significantly higher ($4\times$) at the ice edge than in the adjacent water. Taguchi (1972) estimated average summer inshore euphotic zone production to be about $460\ mgCm^{-2}day^{-1}$ and oceanic production about $330\ mgCm^{-2}day^{-1}$

while McRoy and Goering (1976) estimate that shelf production accounted for 52% of the total for the Bering Sea on an annual basis. Levels of phytoplankton standing stocks and production for coastal areas in the Atlantic are generally comparable. Summer euphotic zone production values off Greenland averaged 300 to 500 $mgCm^{-2}day^{-1}$ (Steeman-Nielsen, 1958). In the Barents Sea, summer chlorophyll values at the surface averaged 0.8 mg m^{-3} while production averaged 15.7 $mgCm^{-3}h^{-1}$ (Vedernikov and Solov'yeva, 1972). Recent unpublished data from the eastern Canadian Arctic (greater than 65°N lat.) showed average surface chlorophyll and primary production values of 0.7 $mgChlm^{-3}$ and less than 1 $mgCm^{-3}h^{-1}$, respectively. Although these production estimates appear low, integrated (m^{-2}) daily production within the euphotic zone averaged 223 $mgCm^{-2}day^{-1}$. English (1961) observed extremely low *in situ* production rates in the high Arctic (80° to 85°N lat.). In July and August 1957, daily production ranged from 0.1 to 0.5 $mgCm^{-3}$ in the euphotic zone where chlorophyll ranged from 0.3 to 1.2 mg m^{-3}. During the following year the highest production recorded was 1.4 mg m$^{-3}h^{-1}$; chlorophyll was slightly greater than 1 mg m^{-3}.

The vertical structure of production and phytoplankton is not as well documented in the Arctic as in the Antarctic. In summer, subsurface maxima in chlorophyll and phytoplankton numbers have been associated with development of the seasonal thermocline (e.g. Taniguchi, 1969) and may also be associated with the strong halocline resulting from melting sea ice (McRoy and Goering, 1974). Significant amounts of viable chlorophyll and numbers of phytoplankton have been observed below the euphotic zone and pycnocline in the Arctic (Taniguchi, 1969; McRoy et al., 1972; MacLaren-Marex, Inc., 1979) as has been found in the Antarctic. Production maxima are generally found at or near the surface (McRoy et al., 1972) although this is not always the case (Taniguchi, 1969). Surface photoinhibition of primary production in Arctic waters has not been described.

Because of relatively smaller geographical differences in winter/summer ice conditions in the Arctic, the open water productive season is probably more restricted than in the Antarctic. McRoy et al. (1972) and McRoy and Goering (1974), for example, indicate that the Bering Sea is largely ice covered from November to May with a maximum of 75% of the sea surface covered (1 to 2 m thickness) during March and April. Significant production begins in late February with the development of the ice community which reaches its maximum just before ice melt. The first plankton bloom develops in the wake of the receding ice and the major open water plankton bloom follows in summer. McRoy et al. (1972) have measured winter (February) production under the ice and found it to be quite low (15 to 20

$mgCm^{-2}day^{-1}$); chlorophyll averaged 8.8 mg m^{-2}. Summer (June) maximum production rates recorded were 870 $mgCm^{-2}day^{-1}$; chlorophyll averaged 56 mg m^{-2}. McRoy and Goering (1976) have estimated that 54%, 19%, 25%, and 2% of the annual production occurs during spring, summer, fall, and winter, respectively (including sea ice and river production).

In the Atlantic production began in March off the west coast of Greenland, reached a maximum (greater than 1000 $mgCm^{-2}day^{-1}$) in May and June, and effectively ceased by October (Steeman-Nielsen, 1958). Grainger (1975) observed phytoplankton biomass (chlorophyll) increasing in June, following ice melt in Frobisher Bay, reaching a maximum of 200 mg m^{-2} by the middle of August, and essentially disappearing by early December with the advance of ice. Peak planktonic production in the Davis Strait (60°N lat.) was over 2 $gCm^{-2}day^{-1}$ (160 $mgChlm^{-2}$) in May. Late summer production and biomass averaged 20 $mgCm^{-2}day^{-1}$ and 10 mg m^{-2}, respectively (MacLaren-Marex Inc., 1979). In the high Arctic (greater than 80°N lat.) the production season may be extremely short and open water is very restricted. English (1961) showed that plankton chlorophyll peaks (1 to 2 mg m^{-3}) under the ice coincided with brief periods in early summer (May, June) when snow cover was absent (Fig. 2).

Dunbar (1970) has reviewed the important environment properties which might affect the level of primary production in the Arctic Ocean. As in the Antarctic, temperature has been considered of minor importance since phytoplankton production is adapted to the relatively low and uniform conditions. Light and nutrient limitation are probably most important, although Larrance (1971) has suggested that grazing pressure could be important in the sub-Arctic Pacific. Dunbar has considered the strong vertical stability of the water column during the growing season as the factor most responsible for the low Arctic production when compared with the Antarctic. Although early in the productive period a stable shallow mixed layer (induced primarily by surface ice melt) enhances Arctic production (Marshall, 1957; Taniguchi, 1969; McRoy and Goering, 1974; Grainger, 1975) the concomitant reduction of essential nutrient salts in the euphotic zone (e.g. Alexander, 1974; Grainger, 1975) and hence restricted input from nutrient-rich waters below the pycnocline cause lowered production overall during most of the summer. Light conditions are also important in determing the level of production. Larrance (1971), for example, observed a strong correlation, as Holm-Hansen et al. (1977) in the Antarctic, between primary production and incident radiation. McRoy and Goering (1974) and Grainger (1975) have explained year-to-year variations in production by variations in average light conditions as reflected through the extent of sea ice. Although nutrient concentrations may be relatively low in summer, in the high Arctic where ice and snow conditions prevail year-round, light

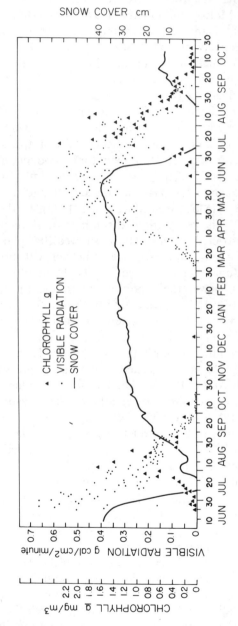

Fig. 2   Relationships between visible radiation, snow cover and concentration of chlorophyll a, Drift Station Alpha, 1957–1958. Chlorophyll a from 5 to 6 m depth (Redrawn from English, 1961)

conditions must be of the greatest importance in determining the productive cycle (see Fig. 2). In general, ice and light conditions may play a significantly more important role in Arctic production than in the Antarctic.

In comparing annual primary production estimates of the north and south polar and adjacent regions, it is apparent from the available data that the sub-Arctic regions are probably more productive than the sub-Antarctic (Table II). Marked latitudinal variations in Arctic production are, however, obvious. Annual estimates at latitudes below 70°N latitude are around 50 to 100 $gCm^{-2}yr^{-1}$ and above 70°N are 10 to 25 $gCm^{-2}yr^{-1}$; under the polar ice cap, 1 to 5 $gCm^{-2}yr^{-1}$ or lower. Comparison is not entirely valid since the sub-Arctic studies have been mostly at coastal stations and sub-Antarctic almost entirely in oceanic waters. The land mass effect is apparent.

Overall, northern high latitude production is probably considerably lower than the estimates from most of the studies between 50° and 70°N latitude since the greatest portion of the north waters (Arctic Ocean) is under ice year-round. Nonetheless, considering the areas whose waters are influenced by ice at some time during the year, northern high latitude regions may be more productive and southern high latitude regions less productive than earlier believed.

TABLE II Annual primary production from selected areas in the Arctic and sub-Arctic Oceans

Location	Latitude (°N)	Primary production ($gCm^{-2}yr^{-1}$)	Source
Arctic Ocean	80–87	1–5	English (1961)
Pacific			
Beaufort Sea	71	10–15	Alexander (1974)
Bering Seas (shelf)	55–65	250	McRoy and Goering (1976)
Bering Seas (oceanic)		75	McRoy and Goering (1976)
Mid-sub-Arctic Pacific	45–55	80–100	Larrance (1971)
		55–91	Koblentz-Mishke (1965)
Gulf of Alaska	50	48	McAllister (1969)
South sub-Arctic Pacific	45	35–55	Koblentz-Mishke (1965)
Atlantic			
Barents Sea	70	25	Vedernikov and Solov'yeva (1972)
West Greenland	69	90	Anderson (1977)
West Greenland	64	95–98	Steemann-Nielsen (1958)
Baffin Island	63	40–70	Grainger (1975)
South Greenland	60	29	Steemann-Nielsen (1958)
Davis Strait	60	50	MacLaren-Marex, Inc. (1979)

ICE BIOTA

One of the striking features at high latitudes is the presence of ice biota associated with pack and fast ice. The brown or greenish-brown colouration is caused by microalgae and has been observed in both north and south polar waters (Apollonio, 1961; Meguro, 1962; Bunt, 1964; Burkholder and Mandelli, 1965; etc.). Horner (1976) and Hoshiai (1979) have recently reviewed work on sea ice organisms and their ecology. Two distinctly coloured layers are sometimes observed in summer in the Antarctic: one on the surface of the ice which develops under snow cover (Meguro, 1962; Burkholder and Mandelli, 1965) and the other on the bottom of the ice, called the epontic community. Ice algae which have been completely frozen within the ice over winter lose their pigmentation and metabolic activity but can resume growth the following spring. Recently, Hoshiai (1977) and Ackley *et al.* (1979) reported another type of algal community in the ice of the Antarctic Sea. High chlorophyll biomass was observed at an intermediate depth within the ice. The intermediate ice algae may be residual from a previous growing season bloom in the ice or water column while the epontic community is usually present season *in situ* growth (Hoshiai, 1979).

Ice algal communities are composed of a diversity of species although pennate diatoms are most abundant (Horner and Alexander, 1972; Hoshiai, 1979). *Nitzschia, Fragilariopsis, Pleurosigma,* and *Amphiprora* are common genera in the Antarctic (El-Sayed, 1971; Hoshiai, 1979) while *Nitzschia, Navicula* and *Amphiprora* dominate in the Arctic. The algal biomass may be considerable. In the Antarctic, Burkholder and Mandelli (1965) measured greater than $400 \, mgChlm^{-2}$ which was 30 times higher than in the water column. Bunt (1963), Whitaker (1977), and Hoshiai (1979) recorded maximum concentrations in the epontic community of 617, 800, and $829 \, mgChlm^{-3}$, respectively. In the Arctic, Apollonio (1961) found $90 \, mgChlm^{-3}$ while Meguro *et al.* (1967) recorded $120 \, mg \, m^{-3}$; 100 times more concentrated than in the water column. Photosynthetic rates can also be high. Burkholder and Mandelli (1965), for example, found over $1 \, g \, C$ fixed $m^{-2}h^{-1}$ in the 'snow community' of the Antarctic although rates are usually much lower because of significant light limitation, particularly for the epontic communities (Bunt and Lee, 1970). Bunt (1968) gives a peak production rate of about $4 \, mgCm^{-3}h^{-1}$. Nonetheless, ice algae can provide a significant amount of the yearly primary production in polar seas. Alexander (1974) estimated ice algae could contribute as much as 25 to 30% of the annual production in shallow coastal areas although it has been found of less significance in other studies (e.g. McRoy and Goering, 1974; Anderson, 1977; MacLaren-Marex Inc., 1979).

Andriashev (1968, 1970) and Gruzov *et al.* (1967) have stressed the significance of the cryobiology of the whole peculiar community below the ice, composed not only of the rich epontic microalgal flora but also of animal species associated with the lower part of the fast ice. The microalgae provide a source of food for the epontic fauna and for other organisms which come to the under surface of the ice to feed on it. The fauna is divided into two groups: a strictly ice and a sub-ice fauna (Andriashev, 1968). The latter includes animals which do not actually enter the loose ice but nevertheless have trophic connection with the ice algae. Polychaetes (*Harmothoe* sp., *Pionosyllis* sp.), cyclopoid copepods (*Oithona* sp.), harpacticoid copepods (*Tisbe* sp., *Harpacticus* sp.), and *Bovallia walkeri* are reported in the ice fauna (Andriashev, 1968). Numbers of individuals and biomass of some of the ice fauna are high and several hundred cyclopoid and harpacticoid copepods per square metre sometimes occur on the lower ice surface (Andriashev, 1968). Numbers of amphipods (*Orchomenopsis* sp.) may reach 3040 individuals with a biomass of 86 g m^{-2}. The sub-ice fauna includes larger adult fishes and fish fry (*T. borchgrevinki*, *Dissostichus mawsoni*) (Andriashev, 1968). Krill, mainly *E. crystallorophias* but partly *E. superba*, are also important components of this group.

Thus, three important contributions of ice-biota to polar ecosystems may be listed. The ice-biota, based on the production of the epontic algae themselves, form a unique ice community; ice-algae produce considerable amounts of organic material which serve in part, at least, to support the pelagic and benthic organisms of the water column below. Dunbar (1977) considered that the Arctic ice-biota served to reduce the effects of seasonality in polar environments. A sea ice cover may reduce water column photosynthesis, but microalgae in and on the ice contribute to the primary production during the growing season in high latitudes, damping the strong seasonal oscillation, and contributing to a more stable system (Dunbar, 1977) because the effective production season can be significantly extended by ice community growth (McRoy and Goering, 1974; Alexander, 1974). The probable seeding of ice-associated algae with the melting of ice to the pelagic and benthic communities is also considered important (Ackley et al., 1979) but is as yet unquantified.

4   Some zooplankton problems

The dominant characteristic of the ecology of high latitude zooplankton is their response to the extremely strong seasonal pulse of phytoplankton production which occurs over a very restricted time period. Though a single peak of short duration may occur in zooplankton biomass in response to seasonal high phytoplankton biomass, there may not be sufficient time within a single season for a complete zooplankton generation to occur, and generation time may extend over several seasons. These effects dominate zooplankton ecology in polar oceans.

Biomass distributions along meridional sections in the southern ocean show that high biomass occurs especially near the Antarctic convergence and decreases towards low latitudes (Voronina, 1966; Foxton, 1956). Abundant herbivorous salps may also be related to upwelling at the Antarctic divergence as in other upwelling areas as suggested by Longhurst (1967). The principal zooplankton species in the Atlantic sector of the Antarctic and the Drake Passage are given in Table III. Zooplankton caught by 1.0 m net in the upper 100 m are dominated by *Calanoides acutus*, *Rhincalanus gigas*, and *Euphausia superba*. On the other hand, zooplankton species collected by smaller nets of 70 cm hauled vertically mainly comprise smaller species of copepods and chaetognaths. *Euphausia superba* and *Thysanoessa* spp. especially seem not to be caught effectively by vertical tows, probably because they are often distributed in dense swarms in the sea. Mackintosh (1934) noted that *Euphausia superba*, and sometimes *Limacina belea* and *Salpa fusiformis* also form heavy shoals. Later works on *E. superba* (Marr, 1962, etc.) and salps (Foxton, 1966) confirm this suggestion.

The trophic importance of zooplankton in high latitude waters is enhanced by their habit of forming heavy swarms. Rather many animals high in the food chain, such as baleen whales, feed principally on swarming krill: *E. crystallorophias*, *E. superba*, and *Thysanoessa macrura* in the Antarctic, and *E. vallentini* in the sub-Antarctic waters. Although also distributed in these areas, *E. triacantha* is not found in the stomachs of baleen whales as it does not form dense swarms (Baker, 1959) and is therefore not feasible for the whales to harvest effectively. Marr (1962) illustrated the shape and size of krill swarms in the Antarctic, based on

TABLE III   The seven relatively most abundant zoo-plankters in the Antarctic Ocean, as indicated by two different sampling methods (Mackintosh, 1934, 1937)

N. 100 oblique sample	N. 70 vertical sample
Calanus acutus	*Rhincalanus gigas*
Rhincalanus gigas	*Eukrohnia hamata*
Euphausia superba	*Calanoides acutus*
Limacina balea	*Eucalanus* spp.
Thysanoessa spp.	*Pleuromamma robusta*
Calanus propinquus	*Sagitta maxima*
Chaetognatha spp.	*Pareuchaeta* spp.

earlier observations by Gunther, and Mackintosh (1966) showed that although shape and size of swarms varied considerably, their edges were usually sharply defined (Mackintosh, 1966). Swarms are more or less rounded, but sometimes irregular, and appear to gather just after development to late furcilia larval stages.

Marr (1962), and recent Japanese krill fisheries experience in the Antarctic, clearly indicate that dense swarms of *E. superba* in the upper mixed layer are usually about 30 m in diameter or less, but sometimes extend up to some hundreds or more. Acoustic surveys of krill also prove the presence of deeper swarms (down to 100 m in the summer time), sometimes extending horizontally over several kilometres, with average density of swarms amounting to 100 km^{-2}. Since their orientation of individuals is to a uniform direction in the daytime, it is likely that swarms are maintained by visual cues, as are the swarms of the anomuran crab *Munida gregaria* in the southern ocean.

The biomass concentration of *E. superba* within its swarms is important to quantify in order to understand the mechanisms of swarm formation and predation by larger organisms. The range of values obtained so far is rather large, from a few to 33 kg m^{-3}, based on data compiled by Everson (1977). Recent analysis of Japanese data estimated much lower figures, with around 200 g m^{-3} as the most probable mean value.

Heavy swarms of calanoid copepods are also common in sub-Antarctic waters. Kawamura (1974) described such swarms of *Calanus tonsus* and *Calanus simillimus* in the sub-Antarctic zone, where many sei and right whales feed on them. These copepod swarms or patches maintain their shape close beneath the surface even under rough conditions with 8.0 to 10.7 m s^{-1} wind velocities and a sea state of 5 (Kawamura, 1974). The swarms extend to up to some hundred metres in dimension, but their edge is somewhat irregular in shape, compared with those of *E. superba*. Density of *Calanus tonsus* swarms ranges from 330 to 23 680 individuals m^{-3} against a background density of only 0.1 individuals m^{-3}. The biomass of these copepods reaches 34 g m^{-3} (Kawamura, 1974) which is a sufficient concentration for skimming feeding of sei and right whales. Although *Calanoides acutus* is frequently caught by plankton nets in Antarctic waters surprisingly few large animals are known to prey on this species (Andrews, 1966). This anomaly is, like the case of *Euphausia triacantha* due to the lack of swarm formation in *C. acutus*.

As elsewhere (Longhurst, 1976), zooplankton species may maintain their general geographical coordinates in the Antarctic Ocean by directed seasonal vertical migration related to differential water mass transport with increasing depth. Ontogenetic and seasonal vertical migration patterns in zooplankton organisms is reported in both the Arctic and the Antarctic

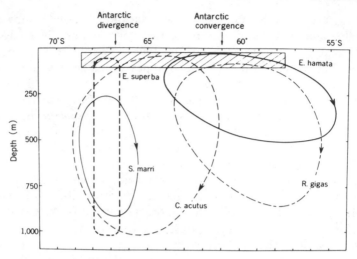

FIG. 3   Seasonal vertical migration patterns of zooplankton in the Antarctic based
on the data by Mackintosh (1937), and Marr (1962). Summer feeding zone is shaded

Ocean. Three abundant zooplankters (*Rhincalanus gigas*, *Eukrohnia hamata*, and *Calanoides acutus*) concentrate near the surface during the Antarctic summer (Mackintosh, 1937; Andrews, 1966) and make up the bulk of the whole Antarctic macroplankton, but descend into deep water in winter. By the action of different current directions at different depths within the Antarctic Ocean, these species are maintained within the limits of their normal species distribution, as shown in Fig. 3. The seasonal vertical movement of these species in the Antarctic from 400 to 600 m thus results in horizontal displacements of some hundreds of miles (Mackintosh, 1937). *E. superba*, on the other hand, releases its ova in the upper 200 m, and these subsequently sink to as much as 1000 m, then again progressively approach the surface during development. This vertical ontogenetic migration is carried out within the Antarctic zone, but the diel and seasonal vertical migration of adults and late furcilia are still unclear, especially in winter.

Dawson (1978) summarised seasonal vertical migration by *Calanus hyperboreus* in the Arctic Ocean; ova are released in the winter or early spring rather deep, and progeny come to the surface as copepodite Stage III from the spring to summer after 300 days development. They feed actively and again descend to the deeper levels as Stage IV copepodites to overwinter. Stage V *C. hyperboreus* come up again to the surface to feed in the following second summer, and then yet again descend deeper levels in autumn for the adult moult. This high Arctic type of ontogenetic and seasonal vertical migration is an extreme form of the pattern observed in other calanoid copepods in high latitudes in the North Pacific. Here, *Calanus cristatus* and *C. plumchrus* concentrate in the upper 200 m in summer, but descend into very deep water in winter, to moult to the adult stage and to spawn. Their generation time is thus much shorter than *C. hyperboreus*. (Heinrich, 1962).

Arctic herbivores have life histories matched to the strong seasonal oscillation of primary production. McLaren (1964) pointed out that the most demanding stress to which Arctic organisms are exposed is that of the very short productive season and the very long winter when food is in short supply. In the Arctic Ocean and in ice-edge waters of the Antarctic, zooplankton reproduce just after the phytoplankton bloom, and each generation of herbivores may share more than one annual phytoplankton bloom. Dunbar (1968) observed that life cycles of secondary producers are usually relatively long especially in polar seas for this reason. Digby (1954) described a seasonal change of nutrition in the Arctic Ocean copepods so that herbivorous or 'mixed-diet' species may become strictly carnivorous during the phytoplankton-deficient dark Arctic winter. Davis (1977) confirmed that the omnivorous copepod *Acartia longiremis* principally utilises organisms such as juvenile *Sagitta* sp. during early winter. Detritus, too, is a possible source of nutrition in the winter, as Hopkins (1969) discussed for the Arctic Ocean. Perhaps non-living particulate organic matter also plays some role in the nutrition of Antarctic herbivores although there are few data available.

Observation of the development of the oil sacs of copepods and the storage of oil by euphausiids in high latitudes suggests that lipids may be used as a food reserve during the winter (Littlepage, 1964). Lee (1974) examined the lipid composition of oil sacs of the copepod *Calanus hyperboreus* in the Arctic, and found differences in the lipid composition corresponding to seasonal feeding changes. Summer-feeding *C. hyperboreus* utilising phytoplankton have large oil sacs in which wax ester forms more than 80% of total lipids. On the other hand, *C. hyperboreus* collected in winter in deeper waters contain higher levels of polyunsaturated fatty acids probably due to their winter diet of microzooplankton (Lee, 1974). The general tendency that wax ester should be relatively high in the lipids of pelagic organisms living towards high latitudes and deep in the water column is also stressed by Benson and Lee (1975), although apparent anomalies remain to be explained: triglyceride-rich *E. superba*, and wax ester-rich *E. crystallorophias* are distributed very closely in the Antarctic.

Pelagic organisms in cold water tend to be larger than related species in warmer water (Wolf, 1962). Mauchline (1972) and McLaren (1966) considered that relative size results from the interaction of several factors of which differences in growth rate and longevity are the most important.

Among euphausiids, the Antarctic *E. superba* was thought to show the fastest growth rate (Bargman, 1945; Ruud, 1932; Nemoto, 1959; etc.) although the growth of *E. superba* is now considered much slower, living more than 2 years. The north boreal *Thysanoessa inermis* in the Arctic Ocean apparently lives three years (Einarsson, 1945) similar to the generation time of *Calanus hyperboreus* in the central Arctic Ocean (above).

5 Distribution and feeding of baleen whales

The Antarctic Ocean and the Arctic Sea are the most important feeding areas for the larger marine mammals in summer, though the former appear to carry the largest stocks of feeding baleen whales and seals. The Antarctic Ocean contains some 56 % of all individuals and 79 % of biomass of seals in the world oceans (Laws, 1977), and there are presently about four times as many whales in the Antarctic as in the northern hemisphere (Mackintosh, 1965). The main attraction of polar seas as feeding areas for large mammals is the presence there of swarm-forming zooplankton and micronekton which grow rapidly in the short summer season. The availability of krill *Euphausia superba* as standing stock biomass in the Antarctic increases by over 100-fold during the summer months (Laws, 1977). The principal difference between the Arctic and the Antarctic Oceans as an environment and as feeding grounds for whales and seals is the average sea depth. Laws (1977) pointed out that in Antarctica the area of continental shelf uncovered by permanent ice is very limited compared with the Arctic continental shelves. The grey whale (*Eschrichtius robustus*), the walrus (*Odobenus rosmarus*), and the northern bearded seal (*Erignathus barbatus*) apparently feed largely on benthic organisms in the Arctic continental shelf waters, but no bottom-feeding mammal exists in the Antarctic, probably since the Antarctic lacks comparable broad shelf areas.

The feeding patterns of baleen whales have been discussed by Nemoto (1959), Kulmov (1961), and Nemoto (1970). Three kinds of mechanism occur in baleen whales, each being associated with morphological characteristics of the feeding apparatus: the form and fineness of the filtration apparatus of the baleen whales; the extent to which the throat is grooved; the form of the tongue; and, finally, the general shape of the head.

Two principal feeding methods are used. All finner whales (blue, fin, Brydes, humpback, and the minkes, all of the family Balaenopteridae with the possible exception of the sei whale) feed in rather the same manner as does a pelican: that is, they engulf huge volumes of water, not necessarily near the surface, this volume reaching as much as 60 m³ (Pivorunas, 1979) which is momentarily held in the enormously expanded throat before filtration through the baleen plates. This engulfment requires that the lower jaw be so articulated as to permit opening downwards to 90° from the body axis.

In the second main feeding technique, practiced by the three right whales (family Balaenidae) the baleen plates are used in the same manner as a plankton net: with the mouth open and the baleen plates suspended below the arched anterior part of the skull, the whale skims dense concentrations of small organisms either at the surface or in the subsurface layers discussed in chapter 4. Finally, the third mechanism is that of the grey whale (*Eschrichtius*) which obtains its benthic food by a lateral biting at, and filtration of, the surface deposits in benthic-rich areas. Variations and combinations of the three basic engulfing, skimming, and biting types occur: the sei whale, for instance, is known both to engulf ('swallow' of Nemoto, 1970) and to skim. As has frequently been pointed out in the past (refs. in Nemoto, 1970) and more recently by Brodie *et al.* (1978), for the feeding of such large animals to be energy-effective, their tiny prey-organisms must be very highly concentrated, and many observations indicate that surface and subsurface layers of copepods are located, swarms or layers of euphausiids are utilised, and aggregations of nektonic fish (clupeids, myctophids) and crustacea (*Pleuroncodes, Munida*) are engulfed.

As for the density of food organisms required by baleen whales, Nemoto and Kawamura (1977) postulated a relationship between density in swarms of food expressed as number of individuals per unit volume, and biomass of food (expressed as weight of food organisms) of baleen whales. Baleen whales particularly select certain species of plankton as their dominant food, and this selection depends on the size–density relationship of each food species, as is shown diagramatically in Fig. 4. The larger the size of food

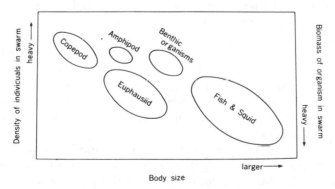

FIG. 4   Relationship between body size of organisms, their density of individuals, and population biomass in swarms of food of baleen whales

organisms, the fewer individuals but not necessarily the smaller biomass per unit volume of sea water (Nemoto and Kawamura, 1977). Among such plankton organisms as copepods, amphipods and euphausiids, the larger species with less numerous individuals dominate the food of baleen whales in high latitudes which feed by 'engulfing'. *Euphausia superba* forms very heavy swarms, the biomass of which is very large through the numbers of individuals are relatively small. Conversely, the number of *Calanus* co-pepods in a swarm is large but their total biomass is small. The small size and biomass of copepods do not suit the feeding of engulfing whales such as the blue and humpback.

Nemoto and Kawamura (1977) described the ranking of food items of larger baleen whales in the Antarctic and indicate the relative selection of different food organisms. The food types of baleen whales in the Antarctic appears to be less diverse than those of the North Pacific including the Bering Sea. Euphausiids form more than 97% of the total food on Antarctic blue, fin, and humpback whales, but only blue whales take so many euphausiids in the North Pacific. North Pacific fin and humpback whales take 64% and 77% krill respectively, the rest of their diet comprising copepods, fish and other items (Nemoto and Kawamura, 1977). As for the sei whale, however, the diversity of food items are not so different between the northern and southern oceans.

Thus, considered as feeding grounds for baleen whales, the Antarctic Ocean and the northern part of the North Pacific appear quite different. Where presented as Eltonian pyramids (Nemoto and Kawamura, 1977), not only is the base far larger in the Antarctic, but the primary level of organic production is also larger there. Fewer trophic levels lead to baleen whales, such as sei and humpback, in the Antarctic, when we compare food selection in both hemispheres. A comparison of the trophic levels below baleen whales in the North Pacific and in the Antarctic is given in Table IV, which indicates the greater complexity of the diet in fin and humpback whales. However, for right whales, there is little regional difference in diet.

An ecological distinction between the feeding patterns of different whales occurs in the Antarctic as well as in the North Pacific. Larger whales tend to arrive earlier in the Antarctic (Laws, 1977), and Mackintosh (1965) suggested that differential southerly penetration by blue, fin, humpback, and sei whales is due to an intraspecific competition related to body size. According to Nemoto (1959), blue and fin whales occupy different feeding areas even in close proximity in the pack ice.

Baleen whales which feed by engulfing and require heavy swarms of food organisms migrate to the high latitude waters where such swarms are present, and consequently undertake long migrations. On the other hand whales which feed by skimming (e.g. right whales) do not penetrate into such high latitudes in the Antarctic. Right whales generally feed along the Antarctic convergence where swarms of *Calanus* species are concentrated. In the Arctic Sea, the skimming Greenland whales feed on less aggregated plankton, but they never descend to mid-latitudes of the North Pacific. Grey whales (*Eschrichtius robustus*), migrate into the Arctic and the adjacent seas in the summertime, where they mainly feed on benthic organisms in shelf water; zooplankton standing stocks in their feeding areas are comparatively low but they can obtain sufficient food solely from benthic animals (Rice and Wolman, 1971). They spend the winter off Baja California after a long coastal migration, there to calve in coastal lagoons and sometimes may feed on the local pelagic red crab (or lobster krill) *Pleuroncodes planipes*, which occurs off that coast in exceedingly dense pelagic concentrations (Rice and Wolman, 1971).

TABLE IV Trophic levels of food of baleen whales according to preference. A, Antarctic; P, North Pacific, x, species absent, diet ranked from heavy to negligible on scale 3, 2, 1 and —

Whales	Trophic levels[a]							
	1		2		3		4	
	A	P	A	P	A	P	A	P
Blue	3	3	—	—	—	—	—	—
Pygmy blue	3	x	—	x	—	x	—	x
Fin	3	3	—	3	—	1	—	1
Bryde's	3	3	3	3	—	2	—	—
Sei	3	3	3	2	—	1	—	1
Humpback	3	3	1	3	—	1	—	—
Minke	3	3	—	3	—	—	—	—
Right	3	3	—	—	—	—	—	—
Grey	x	—	x	3	x	—	x	—

[a] 1, Herbivorous euphausiids, *Calanus* copepods, schooling fish, e.g. anchovy; 2, Amphipods, zooplankton-feeding schooling fish, saury, mackerel, omnivorus euphausiids; 3, Schooling fish that feed on smaller fish, e.g. mackerel and squids; 4, Squids

Parsons (1976) postulated that a pathway to stability in a predator-prey situation might be for the predator to migrate vertically into colder water where it could hibernate (or aestivate) until its particular prey had again increased in quantities sufficient for efficient consumption. He noted that in such a case, temperature and migration would be the chief components in stabilising the community. As discussed here, the high latitude marine environments are very productive of heavy swarming zooplankton or micronekton useful as forage for baleen whales which can adjust to cold as well as warm water temperatures, and their horizontal migrations are supported by the heavy swarming organisms in summer in high latitudes. The energy stored by baleen whales during the polar summer is enough to spend a relatively foodless wintertime in the low latitudes until the swarming zooplankton again increase in sufficient quantities in the high latitudes. Thus, this is another pathway to stabilise prey-predator relationship by horizontal long migration, which also takes advantage of the metabolic gain of wintering in warm water to reduce heat loss and demands on stored food reserves.

6 Sea birds

Sea birds also play an important role in the Antarctic food chain. A biomass of about 500 000 tons of penguins is considered to occur in the Antarctic, and 50 000 tons of other birds, including Diomedeidae and Procellariidae (Everson, 1977). The penguins include 7 species, of which adélie are the most important and comprise about 90 % of the biomass of all penguins in the pack ice region. Although the weight of the individual birds is small, their food consumption annually is more than 30 million tons of krill, squid and fish. As the feeding ranges of parent penguins from the ice edge and island shores are rather restricted, the localised consumption of krill by sea birds must be very intense (Everson, 1977).

Feeding methods of sea birds are discussed by Ainley (1977) who suggests that birds in polar and subpolar waters take food mainly by diving for it, but a few species use the plunging method. The prevalence of the diving method coincides remarkably well with the distribution of the abundance of zooplankton volumes and krill in the Antarctic. The variety of food items taken by penguins is described and discussed by Emison (1968), Watson (1971) and Watson et al. (1975). Adélies in the Ross Sea select organisms over 15 mm in length (Emison, 1968), so that shoaling silver-fish (*Pleuragramma antarcticum*) and swarm-forming krill are the main food items taken there.

The status as scavengers of giant petrels (*Macronectes giganteus*), in taking carrion at sea and on land in certain seasons and places is stressed by Johnston (1977). Two albatrosses (*D. chrysostoma* and *D. melanophris*) generally feed on fish, squid and krill, but sometimes tunicate salps are clearly important (Tickell, 1962). As there are few recorded instances of any organisms feeding on salps, this predation on herbivorous organisms by sea birds sometimes forms a very special link in Antarctic food chains.

7 Antarctic food chains

The Antarctic food web has been considered to be simple, with only a few organisms as keystone species. A food chain model for the Antarctic is given in Fig. 5, combining those of Hart (1942), El-Sayed (1971), Nemoto (1968), and

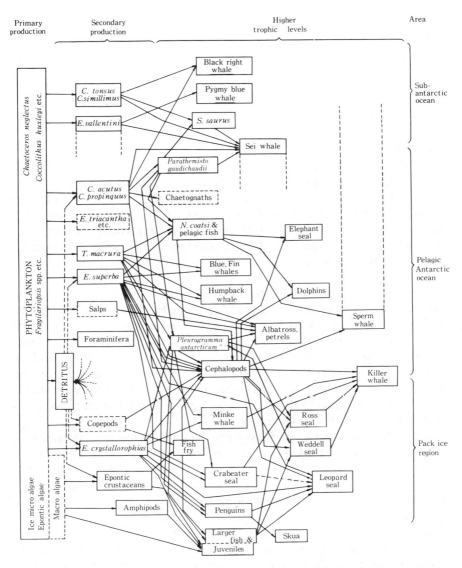

FIG. 5 Main food chains in the Antarctic Ocean compiled from Hart (1942), Nemoto (1968), El-Sayed (1971) and Everson (1977). Non-swarm-forming plankton are shown in the chain boxes

Everson (1977). Primary production is utilised mainly by krill and copepods though specific distributions differ.

The most important keystone zooplankter in the pack ice region and pelagic Antarctic waters, *Euphauia superba*, is well adapted to be a herbivore (Hart, 1942; Barkley, 1940; Nemoto, 1968). Not only the external, but also the internal, feeding structures such as the shape of the stomach and the clusters of setae on its walls are well adapted to capture and digest phytoplankton, even those with hard skeletons. *E. superba* generally feeds on chain-forming diatoms and other moderate sized phytoplankters, including *Fragilariopsis* spp. Two main food chains which lead to baleen whales are described by Nemoto (1968): (*i*) chain-forming diatoms – krill – blue, fin and humpback whales; and (*ii*) phytoplankton – copepods – amphipods – sei whales. The former dominates the high Antarctic waters around the Antarctic divergence region, the second in the sub-Antarctic, where the krill *E. superba* is replaced by *E. vallentini*, the main food of pygmy blue whales *B. musculus brevicauda. Calanus tonsus* and *C. simillimus* also occur in the low Antarctic and sub-Antarctic zones where they are fed by skimming type southern black right whales. *E. vallentini* in summer remains in the upper layers in daytime, and is fed on there by baleen whales.

Two pelagic fishes also contribute greatly to the Antarctic food chain. The relatively abundant pelagic silver fish (*Pleuragramma antarcticum*) also feed on krill and copepods (DeWitt and Hopkins, 1977), especially on *E. superba* at the outer edge of the pack ice, and on *E. crystallorophias* well inside in the Ross Sea. *Pleuragramma* is the main food item for many larger predators, such as penguins and seals. Another pelagic fish, *Notolepis coasti* also depends on krill very heavily (Kawamura, personal communication). This species is commonly found in the stomachs of baleen whales, along with *E. superba*, and also occurs in commercial trawls set for krill.

Food chains in the pack and fast ice zone along the Antarctic continent or around the offshore islands are more complex. Primary producers include not only phytoplankton, but also epontic and ice-microflora and macroalgae. The ice itself also provides seals and birds with resting places which expands their feeding ranges out to sea. During the austral summer, Antarctic seals are drifted with the pack ice (Gilbert and Erickson, 1977), and in the pack ice their main food is *E. superba* and *E. crystallorophias* (Öristland, 1977). As with whales, there are important food differences between seal species. Crab-eater seals take 94 % krill and 3 % fish, while Weddell seals take 53 % fish and 11 % cephalopods, those fish in turn feed on fish fry, krill, benthic crustaceans, algae fragments and ctenophores. Ross seals take 22 % fish and 64 % cephalopods. Leopard seals possibly stand higher in the food web, for they take 37 % krill, 11 % cephalopods, 13 % fish, but 39 % birds, seals and vertebrate carrion (Öristland, 1977). These results suggest that food selection in seal species is, like baleen whales, directly connected with their physical characteristics and behaviour. The clearest example of this is the beautiful adaptation of crab-eater seals which have filtering teeth analogous to baleen in order to take krill.

Some authors have attempted a quantitative analysis of the Antarctic ecosystem. Holdgate (1967) compiled values for production and standing crops of organisms for the Antarctic, and Everson (1977) describes the quantitative interactions in the southern ocean. Calculations by Marr (1962) on the biomass of krill in the Antarctic and by Foxton (1956) on other zooplankton suggest that standing stocks of krill and zooplankton are of nearly the same magnitude of biomass: about 50 mg m^{-3} krill and 55 mg m^{-3} zooplankton are assumed by Holdgate (1967). These authors suggest that about 105 mg m^{-3} zooplankton is taken by zooplankton consumers having a biomass of 14.5 mg m^{-3}. As these consumers do not include fish and cephalopods which also actively feed on krill and zooplankton, the consumption of zooplankton including krill must actually be greater, and Everson (1977) addresses himself to this problem. The estimates of squid and fish biomass consumed by predatory sperm whales, seals, and birds are enormous, and their demand for krill and zooplankton through these organisms must be considerable. This suggests that the biomass of krill may be far larger than that estimated from assumed predation by whales at their pre-whaling stock, namely about 200 million tons. At least 200 million tons of krill must be consumed annually by predators other than whales.

Holm-Hansen et al. (1977), and El-Sayed (1978), also calculated relationships within some Antarctic food webs. Using Öristland's figure of the food consumption by seals of 8.5 × 10^6 tons, as carbon, per year (90% of which is krill) Holm-Hansen et al. obtained a consumption of about 1.3% of the total annual primary production, which they considered to be reasonable. El-Sayed (1978) also calculated annual potential yields using a three-trophic level assumption, and different conversion factors, and the range of annual production reviewed by Everson (1977). El-Sayed's estimate using a production assumption of 16 gCm^{-2}yr^{-1} and 10% conversion factor gives 61 × 10^6 tons of herbivores and 6 × 10^6 tons of primary carnivores in terms of carbon. If the ecological efficiency is as high as it is in an upwelling region (Ryther, 1969), primary carnivore carbon biomass may be as high as 2 × 10^7 tons. Perhaps the solution of this problem must await a direct assessment of the abundance of krill, as is now in progress in the international BIOMASS Project in the Antarctic.

Great changes in the Antarctic ecosystem have apparently occurred following the decrease in the number of baleen whales during this century. Krill, mainly of E. superba, consumed by blue, fin, and humpback whales in the early 1900s must have amounted to 190 million tons with a small addition by sei and minke whales. This consumption in the 1970s must have decreased to about 43 million tons (Laws, 1977). About 150 million tons of krill annually are apparently thus surplus for other krill-feeding organisms.

Although not based on quantitative data, some change in biological characteristics and a population increase of the predators of krill is suggested by Sladen (1964). Laws (1977) also summarised the consequences of increases of krill for predators including baleen whales and showed that there are at least two current trends in the Antarctic. One is the increase in the population size of animals such as the chinstrap, adélie, and gentoo penguins, and of fur seals at South Georgia. Another is the behaviour and the growth and reproduction rate of marine mammals. Nemoto (1959) and Gambell (1968) suggested that sei whales now penetrate into high Antarctic latitudes more than before in the 1960s, at about the time that other large baleen whales, the blue, fin and humpback, were depleted. Sei whales, generally occurring near the Antarctic convergence (Nemoto and Yoo, 1970), only entered the Antarctic in early February before the 1950s; but recently they have been found to go south earlier, during January (Laws, 1977) where they feed on *E. superba*, thus considerably extending their feeding range and diet.

The pregnancy rates of blue, fin, and sei whales have also increased after their population decrease. The mean age at sexual maturity of baleen whales and crab-eater seals became increasingly younger from 1945 to 1956, this earlier maturity possibly resulting from increased food intake (Laws, 1977). The effects of a surplus of krill becoming available for other fishes and birds may also be important, but there is no clear evidence for an increase in their population or for any physiological changes.

References

Ackely, S. F., Buck, K. R. and Taguchi, S. (1979). Standing crop of algae in the ice of the Weddell Sea region. *Deep-Sea Res.* **26**, 269–281

Ainley, D. G. (1977). Feeding methods in sea birds: A comparison of polar and tropical nesting communities in the eastern Pacific Ocean. *In* "Adaptations within the Antarctic Ecosystems" (G. A. Llano, ed.) pp. 669–685. Gulf Pub. Co., Houston

Alexander, V. (1974). Primary productivity regimes of the nearshore Beaufort Sea, with reference to potential roles of ice biota. *In* "The Coast and Shelf of the Beaufort Sea (J. C. Reed and J. E. Sater, eds) pp. 609–632. Arctic Institute of North America

Anderson, J. S. (1977). Primary productivity associated with sea ice at Godhaven, Disko, West Greenland. *Ophelia* **16**, 205–220

Andrews, K. J. H. (1966). The distribution and life-history of *Calanoides acutus* (Giesbrecht). *Discovery Rep.* **34**, 117–162

Andriashev, A. P. (1968). The problems of the life community associated with the Antarctic fast ice. *In* "Symposium on Antarctic Oceanography" pp. 147–155. Scott Polar Research Institute, Cambridge

Andriashev, A. P. (1970). Cryopelagic fishes of the Arctic and Antarctic and their significance in polar ecosystems. *In* "Antarctic Ecology" (M. W. Holdgate, ed.) pp. 297–304, Academic Press, London

Apollonio, S. (1961). The chlorophyll content of Arctic sea ice. *Arctic* **14**, 197–199

Baker, A. de C. (1959). The distribution and life history of *Euphausia triacantha* Holt and Tattersall. *Discovery Rep.* **29**, 309–340

Bargman, H. E. (1945). The development and life history of adolescent and adult krill *Euphausia superba*. *Discovery Rep.* **23**, 103–176

Barkley, E. (1940). Nahrung und Filterapparat des Walkrebschens *Euphausia superba* Dana. *Z. Fisch.* **2**, 65–156

Benson, A. A. and Lee, R. F. (1975). The role of wax in oceanic food chains. *Sci. Amer.* **232**, 77–86

Brodie, P. F., Sameoto, D. D. and Sheldon, R. W. (1978). Population densities of euphausiids off Nova Scotia as indicated by net samples, whale stomach contents, and sonar. *Limnol. Oceanogr.* **23**, 1264–1267

Bunt, J. S. (1963). Diatoms of Antarctic sea ice as agents of primary production. *Nature* **199**, 1255–1257

Bunt, J. S. (1964). Primary productivity under sea ice in Antarctic waters. *Antarct. Res. Series* **1**, 13–26

Bunt, J. S. (1968). Some characteristics of microalgae isolated from Antarctic sea ice. *Antarct. Res. Series* **11**, (Biology of the Antarctic Seas, III), pp. 1–14

Bunt, J. S. and Lee, C. C. (1970). Seasonal primary production in Antarctic sea ice at McMurdo Sound in 1967. *J. Mar. Res.* **23**, 304–320

Bunt, J. S., Owens, O. van H. and Hoch, G. (1966). Exploratory studies on the physiology and ecology of a psychrophilic marine diatom. *J. Phycol.* **2**, 96–100

Burkholder, D. R. and Mandelli, E. F. (1965). Productivity of microalgae in Antarctic sea ice. *Science* **149**, 872–874

Currie, R. I. (1964). Environmental features in the ecology of Antarctic seas. *In* "Biologie Antarctique" (R. Carrick *et al.*, eds) pp. 87–94. Herman, Paris

Cushing, D. H. (1975). "Marine Ecology and Fisheries". Cambridge University Press

Davis, C. C. (1977). *Sagitta* as food for *Acartia*. *Astarte* **10**, 1–3

Dawson, J. K. (1978). Vertical distribution of *Calanus hyperboreus* in the central Arctic Ocean. *Limnol. Oceanogr.* **23**, 950–957

DeWitt, H. H. and Hopkins, T. L. (1977). Aspects of the diet of the Antarctic silverfish *Pleurogramma antarcticum*. *In* "Adaptations within the Antarctic Ecosystems" (G. A. Llano, ed.) pp. 557–567. Gulf Pub. Co., Houston

Digby, P. S. B. (1954). The biology of marine planktonic copepods of Scoresby Sound, East Greenland. *J. Anim. Ecol.* **23**, 298–338

Dunbar, M. J. (1953). Arctic and subarctic marine ecology; Immediate problems. *Arctic* **6**, 75–90

Dunbar, M. J. (1968). "Ecological Development in Polar Regions". Prentice-Hall, Englewood Cliffs, New Jersey

Dunbar, M. J. (1970). On the fishery potential of the sea waters of the Canadian north. *Arctic* **23**, 150–174

Dunbar, M. J. (1977). The evolution of polar ecosystems. *In* "Adaptations within the Antarctic Ecosystems" (G. A. Llano, ed.) pp. 1063–1076. Gulf Pub. Co., Houston

Einarsson, H. (1945). Euphausiacea I. North Atlantic species. *Dana Rep.* **27**, 1–185

El-Sayed, S. Z. (1966). Prospects of primary productivity studies in Antarctic waters. *In* "Symposium on Antarctic Oceanography", pp. 227–239. Scott Polar Research Institute, Cambridge

El-Sayed, S. Z. (1970). On the productivity of the southern ocean (Atlantic and Pacific sectors). *In* "Antarctic Ecology" (M. W. Holdgate, ed.) pp. 119–135. Academic Press, London and New York

El-Sayed, S. Z. (1971). Biological aspects of the pack ice ecosystem. *In* "Symposium on Antarctic Ice and Water Masses" (G. Deacon, ed.) pp. 25–54. SCAR

El-Sayed, S. Z. (1978). Primary productivity and estimates of potential yields of the southern ocean. *In* "Polar Research" (M. A. McWhinnie, ed.) pp. 141–160. AAAS Selected Symposium 7. Westview Press, Boulder

El-Sayed, S. Z. and Turner, J. T. (1977). Productivity of the Antarctic and subtropical regions: A comparative study. *In* "Polar Oceans" (M. Dunbar, ed.) pp. 463–504. Proceedings of SCOR/SCAR Polar Oceans Conference, Montreal, Canada, May 1974

Emison, W. B. (1968). Feeding preferences of the Adélie penguin in Cape Crozier, Ross Island. *In* "Antarctic Bird Studies" (O. L. Austin, ed.), *Antarct. Res. Series* **12**, 191–212

English, T. S. (1961). Some biological oceanographic observations in the central north Polar Sea, Drift Station Alpha, 1957–1958. Arctic Institute of North America, Research Paper 13, viii–80

Everson, I. (1977). "The Living Resources of the Southern Ocean. Southern Ocean Fisheries Survey Programme", GLO/SO/77/1. FAO, Rome.

Fogg, G. E. (1977). Aquatic primary production in the Antarctic. *Phil. Trans. R. Soc. Lond. B.* **279**, 27–38

Foster, T. D. (1978). Polar Oceans: Similarities and differences in their physical oceanography. *In* "Polar Research" (M. A. McWhinnie, ed.) pp. 117–140. AAAS Selected Symposium 7. Westview Press, Boulder

Foxton, P. (1956). The distribution of the standing crop of zooplankton in the southern ocean. *Discovery Rep.* **28**, 191–236

Foxton, P. (1966). The distribution and life-history of *Salpa thompsoni* Foxton with observation on a related species, *Salpa gerlacki* Foxton. *Discovery Rep.* **34**, 1–116

Gambell, R. (1968). Seasonal cycles and reproduction in sei whales of the southern hemisphere. *Discovery Rep.* **35**, 31–134

Gilbert, J. R. and Erickson, A. W. (1977). Distribution and abundance of seals in the pack ice of the Pacific sector of the southern ocean. *In* "Adaptations within the Antarctic Ecosystems" (G. A. Llano, ed.) pp. 703–740. Gulf Pub. Co., Houston

Grainger, E. H. (1975). A marine ecology study in Frobisher Bay, Arctic Canada. *In* "Energy Flow – Its Biological Dimensions. A Summary of the IBP in Canada, 1964–1974" (L. W. Billingsly and T. W. M. Cameron, eds). Canadian Committee for the IBP Roy. Soc. Can.

Grainger, E. H. (1977). The animal nutrient cycle in sea-ice. *In* "Polar Oceans" (M. Dunbar, ed.) pp. 285–299. Proceedings of SCOR/SCAR Polar Oceans Conference, Montreal, Canada, May 1974

Gruzov, Y. N., Propp, M. V. and Pushkin, A. F. (1967). Biological associations of coastal areas of the Davis Sea (based on the observations of divers). *Soviet Antarctic Exped. Inform. Bull.* **6**, 523–533

Hart, T. J. (1934). On the phytoplankton of the South-west Atlantic and the Bellingshausen Sea, 1929–31. *Discovery Rep.* **8**, 1–268

Hart, H. J. (1942). Phytoplankton periodicity in Antarctic surface waters. *Discovery Rep.* **21**, 261–356

Hasle, G. R. (1956). Phytoplankton and hydrography of the Pacific part of the Antarctic Ocean. *Nature* **177**, 616–617

Hasle, G. R. (1969). Analysis of the phytoplankton of the Pacific southern ocean. *Hvalradets Skrifter* **52**, 1–168

Heinrich, A. K. (1962). The life histories of plankton animals and seasonal cycles of plankton communities in the oceans. *J. Cons. Int. Explor. Mer* **27**, 15–24

Holdgate, M. W. (1967). The Antarctic ecosystem. *Phil. Trans. R. Soc. London Ser B* **252**, 363–389

Holm-Hansen, O., El-Sayed, S. Z., Franceschini, G. A. and Cuhel, R. L. (1977). Primary production and the factors controlling phytoplankton growth in the southern ocean. *In* "Adaptations within the Antarctic Ecosystems" (G. A. Llano, ed.) pp. 11–50. Gulf Pub. Co., Houston

Hopkins, T. L. (1969). Zooplankton standing crop in the Arctic Basin. *Limnol. Oceanogr.* **14**, 80–85

Hori, S. (1966). On the water masses of the Southern Ocean in the section between Queen Maud Land and South Africa. *Ant. Res.* **27**, 66–77

Horne, A. T., Fogg, G. E. and Eagle, D. J. (1969). Studies *in situ* of the primary production of an area of inshore Antarctic sea. *J. Mar. Biol. Assoc. U.K.* **49**, 393–405

Horner, R. A. (1976). Sea ice organisms. *In* "Oceanogr. Mar. Biol. 14" (H. Barnes, ed.) pp. 167–182. Aberdeen Univ. Press, Aberdeen

Horner, R. A. and Alexander, V. (1972). Algal population in Arctic sea ice: An investigation of heterotrophy. *Limnol. Oceanogr.* **22**, 454–458

Hoshiai, T. (1977). Seasonal change of ice communities in the sea ice near Syowa Station Antarctica. *In* "Polar Oceans" (M. J. Dunbar, ed.) pp. 307–318. Proceedings of SCOR/SCAR Polar Oceans Conference, Montreal, Canada, May 1974

Hoshiai, T. (1979). Notes on the ecology of ice biota. *Bull. Coast. Oceanogr.* **17**, 25–32

Johnston, G. W. (1977). Comparative feeding ecology of the giant petrels *Macronectes giganteus* (Gmelin) and *M. halli* (Mathews). *In* "Adaptations within Antarctic Ecosystems" (G. A. Llano, ed.) pp. 647–668. Gulf Pub. Co., Houston

Kawamura, A. (1967). Observation of phytoplankton in the Arctic Ocean in 1964. *Inf. Bull. Planktology Japan, Comm. No. Dr. Y. Matsue's 60th Birthday* 151–171

Kawamura, A. (1974). Food and feeding ecology in the southern sei whale *Sci. Rep. Whales Res. Inst.* **26**, 25–143

Knox, G. A. (1970). Antarctic marine ecosystems. *In* "Antarctic Ecology" (M. W. Holdgate, ed.) pp. 69–96. Academic Press, London and New York

Koblenz-Mishke, O. J. (1965). Primary production in the Pacific. *Oceanol.* **5**, 104–116

Kulmov, S. K. (1961). Plankton and the feeding of the whalebone whales (Mystacoceti). *Trudy Inst. Okeanol.* **51**, 142–156

Larrance, T. D. (1971). Primary production in the mid-subarctic Pacific region, 1966–1968. *Fish. Bull. U.S.* **69**, 595

Laws, R. M. (1977). Seals and whales of the Southern Ocean. *Phil. Trans. R. Soc. London Ser. B* **279**, 81–96

Lee, R. F. (1974). Lipid composition of the copepod *Calanus hyperboreus* from the Arctic Ocean with depth and season. *Mar. Biol.* **26**, 313–318

Littlepage, J. L. (1964). Seasonal variation in lipid content of two Antarctic marine Crustacea. *In* "Biologie Antarctique" (R. Carrick *et al.*, eds) pp. 463–470. Herman, Paris

Llano, G. H. (1978). Polar Research: A synthesis with special reference to biology. *In* "Polar Research" (M. A. McWhinnie, ed.) pp. 27–62. AAAS Selected Symposium 7. Westview Press, Boulder

Longhurst, A. R. (1967). Vertical distribution of zooplankton in relation to the eastern Pacific oxygen minimum. *Deep-Sea Res.* **14**, 51–63

Longhurst, A. R. (1976). Vertical migration. *In* "The Ecology of the Seas" (D. H. Cushing and J. J. Walsh, eds) pp. 116–137. Blackwell Sci. Pub., Oxford

Mackintosh, N. A. (1934). Distribution of the macroplankton in the Atlantic sector of the Antarctic. *Discovery Rep.* **9**, 65–160

Mackintosh, N. A. (1937). The seasonal circulation of the Antarctic macroplankton. *Discovery Rep.* **16**, 365–412

Mackintosh, N. A. (1965). "The Stocks of Whales". Fishing News Ltd., London

Mackintosh, N. A. (1966). The swarming of krill and problems of estimating the standing stock. *Norsk Hvalfangst Tid.* **11**, 113–116

MacLaren-Marex, Inc. (1979). Primary productivity studies in the water column and pack ice of the Davis Strait. April, May and August 1978. Report, Indian and Northern Affairs, Canada

Marr, J. S. W. (1962). The natural history and geography of the Antarctic krill (*Euphausia superba*). *Discovery Rep.* **32**, 33–464

Marshall, P. T. (1957). Primary production in the Arctic. *J. Conseil.* **23**, 173–177

Mauchline, J. (1972). The biology of bathypelagic organisms, especially Crustacea. *Deep-Sea Res.* **19**, 753–780

McAllister, C. D. (1969). Aspects of estimating zooplankton production from phytoplankton production. *J. Fish. Res. Board Can.* **26**, 199–220

McLaren, I. A. (1964). Marine life in Arctic water. *In* "The Unbelievable Land" (I. N. Smith, ed.) pp. 93–97. Queen's Printer, Ottawa

McLaren, I. A. (1966). Adaptive significance of large size and long life of the chaetognath *Sagitta elegans* in the Arctic. *Ecology* **47**, 852–855

McRoy, C. P. and Goering, J. T. (1974). The influence of ice on the primary productivity of the Bering Sea. *In* "The Oceanography of the Bering Sea" (D. Hood and E. Kelley, eds) pp. 403–421. Univ. Alaska Inst. of Marine Sciences

McRoy, C. P. and Goering, J. T. (1976). Annual budget of primary production in the Bering Sea. *Mar. Sci. Comm.* **2**, 255–267

McRoy, C. P., Goering, T. J. and Shiels, W. S. (1972). Studies of primary productivity in the eastern Bering Sea. *In* "Biological Oceanography of the Northern North Pacific Ocean" (A. Y. Takenouti, ed.) pp. 199–216. Idemitsu Shoten, Tokyo

Meguro, H. (1962). Plankton ice in the Antarctic Ocean. *Ant. Res.* **14**, 72–79

Meguro, H., Ito, K. and Fukushima, H. (1967). Ice flora (bottom type): A mechanism of primary production in Polar seas and the growth of diatoms in sea ice. *Arctic* **20**, 114–133

Motoda, S. and Minoda, T. (1974). Plankton of the Bering Sea. *In* "The Oceanography of the Bering Sea" (D. Hood and E. Kelley, eds) pp. 207–241. Univ. Alaska Inst. Marine Science

Nemoto, T. (1959). Food of baleen whales with reference to whale movements. *Sci. Rep. Whales Res. Inst.* **14**, 149–290

Nemoto, T. (1968). Feeding of baleen whales and krill and the value of krill as a marine resource in the Antarctic. *In* "Symposium on Antarctic Oceanography", pp. 240–255. Scott Polar Research Institute, Cambridge

Nemoto, T. (1970). Feeding pattern of baleen whales in the ocean. *In* "Marine Food Chains" (J. H. Steele, ed.) pp. 241–252. Oliver and Boyd, Edinburgh

Nemoto, T. and Kawamura, A. (1977). Characteristics of food habits and distribution of baleen whales with special reference to the abundance of North Pacific sei and Bryde's whales. *Rep. Int. Whales Comm.* (Special Issue 1) 1977, 80–87

Nemoto, T. and Yoo, K. I. (1970). An amphipod *Parathemisto gaudichaudii* as a food of the Antarctic sei whales. *Sci. Rep. Whales Res. Inst.* 22, 153–158

Öristland, T. (1977). Food consumption of seals in the Antarctic pack ice. *In* "Adaptations within the Antarctic Ecosystems" (G. A. Llano, ed.) pp. 749–768. Gulf Pub. Co., Houston

Parsons, T. R. (1976). The structures of life in the sea. *In* "The Ecology of Seas" (D. H. Cushing and J. J. Walsh, eds) pp. 81–97. Blackwell Sci. Pub., Oxford

Parsons, T. R. and Takahashi, M. (1973). "Biological Oceanographic Processes". Pergamon Press, Oxford

Pivorunas, A. (1979). 'The feeding mechanisms; Baleen Whales' *Am. Scientist* 67, 432–440

Platt, T. and Subba Rao, D.v. (1975). Primary production of marine microphytes, pp. 249–280. *In* "Photosynthesis and Productivity in Different Environments" (J. P. Cooper, ed.). Cambridge Univ. Press

Rice, D. W. and Wolman, A. L. (1971). "The Life History and Ecology of the Grey Whale (*Eschrichtius robustus*)". The American Soc. Mammal, Special Pub. No. 3

Ruud, J. T. (1932). On the biology of the southern Euphausiidae. *Hvalradets Skrifter* 2, 1–105

Ryther, J. H. (1963). Geographical variations in productivity. *In* "The Sea 2" (M. N. Hill, ed.) pp. 347–380. John Wiley and Sons, New York

Ryther, J. H. (1969). Photosynthesis and fish production in the sea. *Science* 166, 72–76

Sladen, W. J. L. (1964). The distribution of the Adélie and Chinstrap penguin. *In* "Biologie Antarctique" (R. Carrick *et al.*, eds) pp. 359–365. Herman, Paris

Steemann-Nielsen, E. (1958). A survey of recent Danish measurements of the organic productivity in the sea. *Rapp. Cons. Explor Mer* 144, 92–95

Sverdrup, H. U., Johnson, M. W. and Fleming, R. H. (1946). "The Oceans, their Physics, Chemistry and General Biology". Prentice-Hall, New York

Taguchi, S. (1972). Mathematical analysis of primary production in the Bering Sea in summer. *In* "Biological Oceanography of the Northern North Pacific Ocean" (A. Y. Takenouti, ed.) pp. 253–262. Idemitsu Shoten, Tokyo

Taniguchi, A. (1969). Regional variations of surface primary productivity in the Bering Sea in summer and the vertical stability of water affecting the production. *Bull. Fac. Fish. Hokkaido Univ.* 20, 169–179

Tickell, W. L. N. (1962). Feeding preferences of the Albatrosses *Diomedea melanophris* and *D. chrysostoma* at South Georgia. *In* "Biologie Antarctique" (R. Carrick *et al.*, eds) pp. 383–387. Hermann, Paris

Vedernikov, V. I. and Solov'yeva, A. A. (1972). Primary production and chlorophyll in the coastal waters of the Barents Sea. *Oceanol.* 12, 559–565

Voronina, N. M. (1966). The zooplankton of the Southern Ocean; some study results. *Oceanology* 6, 557–563

Walsh, J. J. (1971). Relative importance of habitat variables in predicting the distribution of phytoplankton at the ecotone of the Antarctic upwelling system. *Ecol. Monogr.* 41, 291–309

Watson, G. E. (1971). Birds of the Antarctic and sub-Antarctic. *Antarct. Map. Folio Ser.* 1971 (14), 1–18

Watson, G. E., Angle, J. P. and Harper, P. C. (1975). Birds of the Antarctic and sub-Antarctic. *Antarct. Res. Series* 1975

Whitaker, T. M. (1977). Sea ice habitats of Signy Island (South Orkneys) and their primary productivity. *In* "Adaptations within the Antarctic Ecosystems" (G. A. Llano, ed.) pp. 75–82. Gulf Pub. Co., Houston

Wolf, T. (1962). The systematics and biology of bathyal and abyssal Isopoda Asellota. *Galathea Rep.* **6**, 1–320

Zenkevitch, L. (1963). "Biology of the Seas of the U.S.S.R.". George Allen and Unwin Ltd., London

5 Coral Reef Ecosystems

JOHN B. LEWIS

1 Introduction

Coral reefs are shallow-water, tropical marine ecosystems, characterised by a tremendous variety of plants and animals and by high rates of production in nutrient-poor and plankton-impoverished oceans. In attempting to identify trophic relationships of organisms on a coral reef and to define pathways of energy flow, one is faced with an exceedingly complex and diverse ecosystem with innumerable biotic associations in which the success of the symbiotic way of life has blurred relationships. The dominant organisms, the corals themselves, have combined autotrophic and hetero-trophic modes of production. This results in recycling and conservative mechanisms which are largely responsible for the high rates of production in the ecosystems.

The abundance and exuberance of life on coral reefs has excited biologists since the time of Darwin. A sketch of a Caribbean reef, shown in Fig. 1, does little to convey its real beauty and interest. Reefs are constructional features, composed of the skeletal components of corals and other limestone-secreting organisms. The primary framework, consisting mainly of herma-typic (reef) corals, is bound together by calcareous algae or other sessile organisms and consolidated by biogenic activity and sediment fill. Sediments are derived from the organic and physical degradation of the reef builders and may have a bulk of many times the frame itself. Thus a reef is composed of organisms that produce and bind the hard structure as well as those which erode and destroy it. Reef-building corals have their greatest diversity in the Indo-Pacific region where there are about 80 genera and 700 species. The Atlantic reef coral fauna is limited to only 26 genera and some 50 species (Goreau and Wells, 1967).

Three basic types of reefs are recognised: fringing reefs growing along shallow coastlines, barrier reefs occurring on offshore banks at some distance from land, and atoll reefs formed by the submergence of oceanic volcanic peaks. Temperature and light are critical factors for reef forma-tion. The best developed reefs flourish where mean annual water tempera-tures lie between 23 and 25 °C and no significant reef development occurs where the mean annual temperature falls below about 18 °C. Light

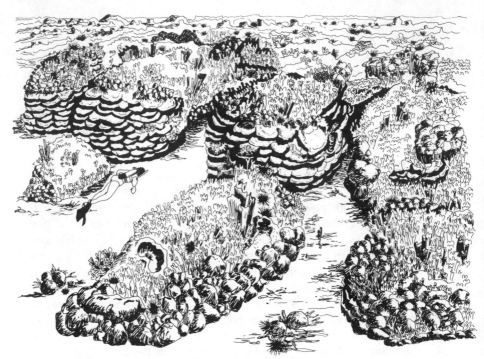

FIG. 1   Sketch of a Barbados fringing reef. (Redrawn from Stearn *et al.*, 1977)

determines the depth to which photosynthesis can occur in the zo-oxanthellae of coral tissue and regulates calcification rates. Coral reefs are also sensitive to the other fundamental properties of the marine environment, to salinity, oxygen and nutrient supply as well as to water turbidity, circulation and sedimentation.

In response to environmental gradients, reefs are conspicuously zoned as bands roughly parallel to the shore. The seaward edge, exposed to the prevailing wave direction, is the most actively growing portion of the reef and is characterised by a steep profile and by a spur and groove development. The inner region, often exposed at low tide, is composed of partly consolidated coral debris and calcareous sediments. The flow of water across the reef is important in supplying water onto the reef and in the transport of nutrients and organic matter.

2   Reef primary production

Estimates of gross production developed in waters over reefs fall between about $300\text{--}5000 \text{ gCm}^{-2}\text{yr}^{-1}$. A comparison of reported rates from a number of different reefs is shown in Table I. The highest value for gross production reported (Gordon and Kelly, 1962) was 11 680 gCm^{-2}yr^{-1} from a reef in Hawaii. In general, rates of production (P) to community respiration (R) are greater than one; that is, most reefs appear to produce more organic matter than is utilised within the system. From all accounts, these high levels of production occur in spite of low nutrient values in surrounding waters. When rates of primary production on coral reefs are compared with those of other marine ecosystems, it is apparent that they are as high as in the most fertile waters (Westlake, 1963).

How this production is achieved was first considered by Sargent and Austin (1949) at Rongelap Atoll in the Pacific. Similar studies by other authors listed in Table I have confirmed the general conclusions reached by Sargent and Austin (1954) that the production developed in the water flowing over reefs greatly exceeds production in oceanic waters in the vicinity. Gross production rates of phytoplankton are 10 to 100 times lower than the production measured over reefs. Doty and Oguri (1956) recorded gross production rates of 21–37 gCm^{-2}yr^{-1} in oceanic waters in the vicinity of Hawaii; Steemann-Nielson (1954) recorded rates of 28 gCm^{-2}yr^{-1} at Rongelap in the Pacific, while Beers *et al.* (1963) found production rates of 139 gCm^{-2}yr^{-1} at Barbados in the Caribbean. Sournia and Ricard (1976) found values of 37–153 gCm^{-2}yr^{-1} in French Polynesia in a deep lagoon subject to fertilisation from land. Thus it appears that the sources of autochthonous production are much greater than the allochthonous inputs from the plankton system. The source of this internal production is the large standing crop of benthic algae.

SOURCES OF PRIMARY PRODUCTION

The predominance of benthic primary producers over phytoplankton is a characteristic feature of reef ecosystems. Benthic producers are highly diversified and specialised in habit. The various categories of benthic algae include: 1. fleshy macrophytes, 2. filamentous endolithic algae, 3. filamentous epilithic and sand-dwelling algae, 4. encrusting coralline algae, and 5. the symbiotic zooxanthellae. Also important are 6. marine grasses.

Fleshy macrophytes are generally an inconspicuous element as benthic producers on coral reefs (Johnston, 1969; Dahl, 1974) but may be abundant in back-reef shallow habitats and make an important contribution to total reef production locally. Brown and green algae are the commonest types; *Sargassum* sp. is a prominent brown algae while many of the green algae such as *Halimeda* and *Penicillus* are carbonate-secreting.

Boring blue-green filamentous algae form a band or layer in the surface skeletons of living hermatypic corals and inhabit the surface of dead reef rock. Odum and Odum (1955) found that they comprised the bulk of the

TABLE I Comparison of estimates of primary production of coral reefs made by different authors[a]

Locality	Gross production (gCm^{-2}yr^{-1})	Community respiration (gCm^{-2}yr^{-1})	P/R	Ref.
Hawaiian coral reef, Coconut Island	7300	12 370	0.59	Gordon and Kelly (1962)
Fringing coral reef, North Kapaa, Hawaii	2427	2200	1.1	Kohn and Helfrich (1957)
Eniwetok Atoll, Marshall Islands	4200	4200	1.0	Odum and Odum (1955)
El Mario reefs, Puerto Rico	4450	4100	1.1	Odum et al. (1959)
Eastern Reef, Rongelap Atoll, Marshall Is.	1250	1090	1.1	Sargent and Austin (1954)
Kavaratti lagoon, Laccadives	4715	3482	1.3	Qasim et al. (1972)
Kavaratti reef, Laccadives (not corrected for diffusion)	2250	880	2.5	Qasim et al. (1972)
Eniwetok Atoll, algal flat	4234	2190	1.9	Smith and Marsh (1973)
One Tree Island, lagoon reef	1387	1314	1.1	Kinsey (1972)
Eniwetok Atoll, coral algal flat	2190	2190	1.0	Smith and Marsh (1973)
Guam, reef flat	6900	2600	2.6	Marsh (1974)
Eniwetok Atoll, windward reef flat	3285	2190	1.5	Smith (1974)
Eniwetok Atoll, algal flat	5329	2190	2.4	Smith (1974)
Moorea, French Polynesia, fringing reef	2628	3052	0.86	Sournia (1976)

[a] Estimates of production and of biomass are given in gCm^{-2}yr^{-1} wherever possible, throughout this article. Data given by authors in other units have been converted by factors contained in the International Biological Programme Report by Winberg (1971)

plant material (94%) contained in coral skeletons. It appears, however, that they do not make an important contribution to reef primary production because only a small percentage (0.001%) of the incident light is able to penetrate the filamentous layer (Halldal, 1968). Nevertheless, Sournia (1977) regards the importance of this group of primary producers as being underestimated.

Filamentous epilithic algae, including both green and blue-green, form dense algal mats on reef rock surfaces and bind sediment particles. On patches of sand at the base of reef structures dense mats of Cyanophytes such as *Oscillatoria* develop in back-reef habitats.

Coralline red algae are abundant on rocky substrates of reefs and intertidal areas in the tropics and are responsible for the algal ridge, a prominent constructional feature of Pacific reefs. They thrive in high wave energy environments.

Symbiotic zooxanthellae are found in the tissue of all hermatypic corals. The functional relationship between hermatypes and their zooxanthellae is still a controversial one but the consensus of opinion is that the latter interact with their hosts by supplying organic carbon, affect the recycling of nutrients, and regulate the calcium deposition of corals. According to Odum and Odum (1955) they form a small proportion (10%) of the corallum tissue biomass.

The marine grasses such as *Thalassia* and *Cymodocea*, are often associated with back-reef and lagoonal habitats of coral reefs. They are important in stabilizing sediments (Ginsburg and Lowenstam, 1958), create habitats for other benthic organisms (Fenchel, 1970), and are also important nitrogen fixers (Patriquin, 1972).

The quantitative production by the various benthic producers obviously varies with location. Rates of gross and net production of each type are listed in Table II and are expressed in gCm^{-2} plant cover. Most net production rates vary between about 200 and 1000 $gCm^{-2}yr^{-1}$ although some reported rates are considerably higher. The P/R ratios are generally higher than for the reef as a whole.

REASONS FOR HIGH PRIMARY PRODUCTION

A fundamental reason for the high rates of primary production is, of course, the abundant light energy available for benthic producers in warm, shallow, well-lit waters. High levels of productivity of shallow-water emergent and submerged vegetation are characteristic of salt marshes and other intertidal and littoral areas (Odum, 1971; Parsons *et al.*, 1977).

A second important reason concerns the fixation of atmospheric nitrogen. At Eniwetok Atoll, Johannes *et al.* (1972) and Wiebe *et al.* (1975) found that nutrient-poor water flowing over the reef became enriched with various

forms of nitrogen. The source of at least part of this nitrogen was found
by Wiebe *et al.* (1975) to be due to nitrogen fixation by the blue-green algae
Calothrix crustacea. There is apparently a broad spectrum of nitrogen fixers
on reefs and in reef sediment. Patriquin (1972), and Patriquin and Knowles
(1972) detected nitrogen fixation by anaerobic bacteria in the reducing
substrate of the *Thalassia* root layer. Goering and Parker (1972) found
nitrogen fixation by epiphytes growing on *Thalassia*, and Capone (1977)
observed fixation by epiphytes on macroalgae and coral rubble. Both Di

TABLE II Comparison of estimates of primary production of benthic algae,
Thalassia and corals

Organisms and locality	Gross production $(gCm^{-2}yr^{-1})$	Net production $(gCm^{-2}yr^{-1})$	P/R	Ref.
Corallinaceae, Eniwetok and Hawaii	547	241	1.8	Marsh (1970
Corallinaceae, Hawaii	2555	2080	5.4	Littler (1973)
Corallinaceae, Curacao	890	370	1.7	Wanders (1976a)
Halimeda, Jamaica	1460	839	2.3	Hillis-Colinvaux (1974)
Oscillatoria, Moorea	416	226	2.2	Sournia (1976)
Fleshy and filamentous algae, Curacao	712	452	2.7	Wanders (1976a)
Sargassum platycarpum, Curacao	3840	2550	2.9	Wanders (1976b)
Chlorophyceae, Laccadives	1402	693	1.9	Qasim *et al.* (1972)
Fleshy and filamentous algae, Virgin Islands	10 950	6278	2.3	Connor and Adey (1977)
Thalassia, Jamaica	—	704	—	Greenway (1974)
Thalassia, Cuba	—	585	—	Buesa (1974)
Thalassia, Barbados	—	368	—	Patriquin (1973)
Thalassia, Florida	—	534–712	—	Zieman (1968)
Thalassia, Puerto Rico	—	800	—	Burkholder *et al.* (1959)
Gorgonacea, Fla.	2080	533	3.9	Kanwisher and Wainwright (1967)
Scleractinia, Fla.	2385	769	3.1	Kanwisher and Wainwright (1967)

Salvo (1973) and Sorokin (1974) have shown how important reef bacteria are for nitrogen fixation.

It appears that nitrogen is not, after all, a limiting nutrient but is fixed in substantial quantities and is readily available to primary producers. Wiebe *et al.* (1975) found fixation rates of 65.7 $gNm^{-2}yr^{-1}$ by the algae *Calothrix*, while Patriquin and Knowles (1972) estimated rates of 10–50 $gNm^{-2}yr^{-1}$ in a stand of *Thalassia*. The fixation of nitrogen by epiphytes of *Thalassia* was found by Goering and Parker (1972) to be of the order of 105 $gNm^{-2}yr^{-1}$. Odum and Odum (1955) considered that 263 $gNm^{-2}yr^{-1}$ was required by primary producers on a reef at Eniwetok and about 219 g were available from inorganic nitrate. The balance was presumably made up from fixed nitrogen.

A third mechanism involves the retention and recycling of nutrients within the reef system. In some circumstances nutrient retention may be simply a function of the hydrographic circulation. For example, at Fanning Atoll in the Pacific, Gallagher *et al.* (1971) found that it took about one month for complete exchange of water in the lagoon to be affected. As a consequence, only a small proportion of the nutrients and of the daily production of the lagoon was lost to the surrounding water. Kinsey and Domm (1974) obtained evidence for nutrient retention in artificially fertilised, isolated reef pools on the Great Barrier Reef. As a result of fertilisation and nutrient retention the test areas were maintained as autotrophic systems.

On the grounds that there was no decrease in nutrient concentrations in water flowing over a reef at Eniwetok Atoll in the Pacific, Johannes *et al.* (1972) concluded that a recycling mechanism was present. Similarly, Pilson and Betzer (1973) at the same locale found that although levels of phosphates were consistently low, there was no difference in concentrations in water at upstream and downstream stations. Gilmartin (1960) found phosphate values to be consistently higher in bottom water than in surface water over the reef, suggesting that nutrient regeneration takes place at the water/substrate interface. Pomeroy *et al.* (1974) also found that there was a continuous turnover of phosphorus within reef components with active uptake during the daytime and release to the water column at night.

Recycling of nutrients also takes place within reef corals themselves. There is some evidence that reef corals can take up nitrate from the water (Franzisket, 1974; D'Elia and Webb, 1977). It has also been found that dissolved nitrogen in the form of ammonium is excreted by coral tissue and this ammonium is taken up by the symbiotic zooxanthellae (Kawaguti, 1953; Sloterdijk, 1975; D'Elia and Webb, 1977; Muscatine and D'Elia, 1978). Uptake and retention of ammonium are enhanced by light and the daylight uptake is sufficient to sustain net ammonium retention during the night. The zooxanthellae then appear to have the ability to retain virtually all of the animal-excreted ammonium and to scavenge ammonium and nitrate from the environment.

Finally, Di Salvo (1974) emphasised the role that bacteria play in retention and recycling of nutrients in reef spaces and sediments, and Patriquin (1972) concluded that sediments in stands of the sea grass *Thalassia* acted as reservoirs and sources of organic phosphates.

3 Reef secondary producers

BENTHIC INVERTEBRATES

While there exists a good deal of descriptive information on the benthic fauna of coral reefs there are very few quantitative data on which to base estimates of production. We have much more informative data on reef primary production than on secondary production.

If one groups the benthic fauna on reefs in a conventional manner, that is, epifauna and infauna, then the epifauna will include the corals and other sessile organisms such as alcyonarians and massive sponges. The infauna will include those animals which burrow into the hard rock substrate, for example worms and bivalves, and the meiofauna and other soft-bottom forms which burrow in the adjacent reef sediments. Although great variety or diversity of both components is characteristic of reefs, it is evident that corals are the dominant organisms.

With few exceptions, corals are essentially hard substrate colonisers. Encrusting, branching and massive colonies form a mosaic of species in response to environmental and biotic factors. Estimates of bottom coverage range between 10 and 90%. If one categorises reef corals as sessile suspension feeders (Lewis, 1976; Muscatine and Porter, 1977), then they fall within one of the most successful ecological groups of shallow-water invertebrates. Crisp (1975) regards populations of suspension feeders as having characteristically high biomass and production and as being intrinsically efficient converters of food. The fact that the most prolific coral growth occurs on the seaward edge of reefs exposed to strong currents, suggests that this feeding strategy is fundamental to their success.

Furthermore, coral colonies provide a substrate and shelter for a host of both permanent and temporary residents. A distinction may be made between infaunal forms boring into the coral skeleton and cryptic forms which live externally in the shelter provided by branching colonies. Boring organisms cause destruction of corals not only by carbonate degradation but also by exposure of fresh skeleton to chemical dissolution and structural weakening. Sponges, sipunculids, polychaete worms, molluscs and crustacea are all able to bore into coral rock by chemical or mechanical means.

Estimates of amounts of coral skeletons removed and reworked by borers may be as high as 60% of the whole corallum (MacGeachy and Stearn, 1976). Among the organisms that utilise branching corals, holes and crevices in the reef structures for protection include echinoderms, molluscs, polychaete worms and crustacea. The number of species from a single coral head is remarkable – Gibbs (1971) obtained 2000 individuals belonging to 220 species, McCloskey (1970) obtained 1517 individuals of 37 species from several colonies, and Grassle (1973) counted 1441 polychaetes of 103 species in one colony.

Patton (1976) has reviewed the fauna associated with corals and, in addition to describing the many kinds of interactions between corals and their inhabitants, he has emphasised the frequency of symbiotic or obligatory relationships. In many of these symbionts the trophic position is obscure either because the food source is unknown or because there are several food sources.

Attached or burrowing symbionts are often highly modified in structure for life in coral colonies and induce changes in skeleton growth of the host. Some striking examples include the Caribbean sponge *Mycale* which grows on the undersurface causing it to form upturned peripheral folds above the sponge ossicles, numerous crustaceans such as barnacles of the genus *Pyrgoma* and hapalocarcinid crabs which are covered by coral skeleton or form galls, and various vermetid gastropods whose tubes are surrounded by coral skeleton and tissue.

Examples of coral associates with more intimate nutritional connections with the host include sponges of the genus *Clione* which have a network of branching filaments throughout the skeleton of living colonies only of *Pocillopora*, commensal shrimp and crabs in galls or cysts which have been observed to feed upon mucus extruded by the coral, and the mytilid bivalve *Fungiacava* which lies within a cavity of the host skeleton and extends the siphon into the stomodeum of the coral polyp to feed on material in the coelenteron (Goreau *et al.*, 1970).

While the coral associates may be the most diverse and numerically abundant group, they are chiefly filter feeders and suspension feeders and do not utilise the standing crop of benthic primary producers. The benthic algae are consumed principally by a few groups, the sea urchins, crustacea and gastropod molluscs. The most conspicuous and abundant of the herbivores are the long-spined urchins of the genus *Diadema* with closely related species in the Atlantic, Pacific and Indian Oceans. *Diadema* feeds upon a wide variety of algal types and will resort to carnivory. The stout spined Cidarids may also be abundant and both groups have an important role in energy transformation and reef erosion (Glynn *et al.*, 1979; Hawkins, 1979). Indeed, Stearn and Scoffin (1977) have shown that *Diadema* may destroy a reef more rapidly than the carbonate structure can be replaced. Other regular urchins feed upon sea grasses and may occur in large enough numbers to graze down large areas of vegetation (Greenway, 1974). Herbivorus gastropod molluscs and crustaceans are probably less important ecologically than the echinoids. With some exceptions, most are small forms, and their low biomass on reefs probably has relatively little effect upon total primary production.

In considering the third trophic level invertebrates, the most striking example of a carnivore is the notorious starfish, *Acanthaster*. By all accounts (Chesher, 1970; Endean, 1973) this coral predator can lay waste vast areas of reef and they have been reported in densities as high as $14\,000\,\mathrm{km}^{-2}$. Various views have been put forth to account for the sudden outbreaks of large populations of *Acanthaster*. Some workers have proposed that the infestations are a natural phenomenon, others that the outbreaks are unique, or that they have been caused by human interference by collection of predators or by marine pollution. At this point perhaps the most that can be said is that such outbreaks of a top carnivore were unexpected and the cause is as yet unknown.

Other benthic invertebrate predators are less dramatic in their effect. They include mainly molluscs and crustaceans. The gastropod genus *Conus* which has been intensively studied by Kohn (1959, 1968) is a widespread predator on worms and molluscs. *Polynices* is a common infaunal predator on burrowing bivalves. Perhaps the most interesting group are the coral predators. These have been reviewed by Robertson (1970) who found a surprising variety of predatory and parasitic forms. Most are facultative predators but some crustaceans and molluscs consistently live within and feed upon coral tissue. The gastropods are among the most numerous and five families of coral feeders are known. Ott and Lewis (1972) attempted to quantify the importance of the gastropod *Coralliophila* as a coral predator and concluded it was of minor importance.

Although supporting data are sparse, one has the impression that, with the exception of the herbivorous echinoids and the starfish *Acanthaster*, epifaunal herbivores and carnivores do not constitute a significant proportion of the biomass of secondary producers on reefs. Most of the biomass appears to be made up of filter feeders and suspension feeders living as infaunal forms or closely associated with branching colonies. Hutchings (1974) regarded the polychaetes as the dominant group of invertebrates living in coral reef habitats while Thomassin (1974) emphasised the importance of the carcinological fauna. However, the bulk of information on biomass of these groups is based on numbers of animals only. There is a great need to quantify biomass of invertebrates on reefs and for information on their feeding ecology. The food resource for this apparently substantial group must come from processed benthic primary production as well as from allocthonous sources.

CORAL REEF FISH

Fish play an important part in the economy of coral reefs, as herbivores and predators and also in processes of carbonate erosion and sediment production. There is a considerable recent literature on the ecology of reef fish which has been reviewed by Goldman and Talbot (1976) and by Ehrlich (1975). Most studies have emphasised the diversity of species of reef fish – a recurring theme in this account. The maximum number of species (about 3000) is to be found in the Indo-West Pacific region. While a large number of species may be regarded as permanent residents on reefs, others may simply seek shelter there or prey upon the permanent residents.

Feeding relationships of coral reef fish have been reviewed by Hiatt and Strasburg (1960), by Randall (1967), and by Goldman and Talbot (1976). The most important herbivores are the Scaridae (parrot fish) and the Acanthuridae (surgeon fish). These herbivores are mong the most specialised and advanced groups. They consume principally the algal turfs of filamentous and other low-growing algae and upon surface diatoms although at least one species of Scarid feeds on *Thalassia*. Bakus (1967) found that Scaridae consumed about 5% of the total production of Cyanophytes on a reef flat at Eniwetok Atoll. In spite of their specialisation for feeding Choat (1969) has shown that there is considerable overlapping in resource utilisation. Bakus (1966) and Ogden and Lobel (1978) considered that herbivory by reef fish strongly influenced community structure of algae and of secondary producers.

There are more species of carnivorous fish on reefs than of other feeding types and most are opportunistic in their feeding habits (Talbot, 1965). Many carnivorous fish utilise different resources at different growth stages. Amongst the more prominent of the carnivores are the Chaetodontidae (angel fish and butterfly fish) which feed upon sessile organisms, the Serranidae and Lutjanidae (groupers and snappers) which are generalised carnivores, the Sphyraenidae and Muraenidae (barracudas and moray eels) which are fish feeders, and the zooplankton feeders, Clupeidae (herring), Chromidae (damselfish) and Apogonidae (cardinal fish). There are a number of specialised predators such as the cleaner fish (*Labroides*) and the trumpet fish (Aulostomidae) with their elongate snouts for selecting individual prey organisms.

Of special interest are those reef fish which feed upon coral polyps. On Caribbean reefs, Randall (1967) found that the commonest coral feeders were the Balistidae (trigger fish), Monocanthidae (file fish), Tetraodontidae (puffers) and Chaetodontidae. On the whole, the incidence of fish feeding on polyps was low. On the other hand, Hiatt and Strasburg (1960) concluded that coral feeding fish were common on Pacific reefs and the Scaridae were the most important coral predators. Talbot (1965) found that 20% by weight of the fish on a reef in East Africa were coral feeders.

4   Basis for secondary production

Given that high rates of primary production are characteristic of coral reefs we must next enquire as to how this is mobilised to support the higher trophic levels. On almost every reef studied (Table I) gross production by the primary producers exceeds community respiration, but how is this energy channeled into reef building? In addition to direct utilisation of primary production by herbivores, several other possible pathways exist for the support of secondary production. These include translocation of organic matter within the corals themselves, decomposition of detritus from primary producers, and utilisation of suspended particulate matter from autochthonous and allochthonous sources.

PHYTOPLANKTON

With the exception of some isolated atoll lagoons the gross production of phytoplankton in reef waters is too low to support any substantial standing crops of secondary production. Exceptions noted by Sournia and Ricard (1975) showed production rates of 37–154 $gCm^{-2}yr^{-1}$ in the Pacific and by Jones (1963) of 307 $gCm^{-2}yr^{-1}$ in Florida. However, these values are still far below those for production of benthic algae. The generally low values for phytoplankton production appears to result in small populations of planktonic herbivores, for Qasim (1970) found that of the average yearly net primary production of 124 gCm^{-2} only about 30 were required by the plantonic herbivores in the estuary.

ZOOPLANKTON

The supply of zooplankton in oceanic waters flowing over coral reefs has been considered to be inadequate to supply the nutritional needs of corals and other benthic invertebrates. Johannes and Tepley (1974) found that the zooplankton biomass present could supply only 20% of the respiratory requirements of *Porites lobata*, while Porter (1974a) observed that the Caribbean coral *Montastrea annularis* could only obtain a small percentage of its maintenance requirements from the plankton. While other studies (Tranter and George, 1972; Glynn, 1973a; Johannes et al., 1970) have supported this view, the conclusion is based upon the low plankton biomass from net capture estimates. Feeding rates of corals on zooplankton in the field have not been measured except for Porter's study (1974a). It has been shown in the laboratory (Coles, 1969) that corals can capture enough zooplankton to cover their respiratory requirements. Oceanic plankton is undoubtedly captured by corals for depletion of plankton biomass of between 23 and 60% has been accounted for by Tranter and George (1972) and by Glynn (1973a).

More recent work has indicated that there is an additional source of zooplankton which has not previously been accounted for. It is now generally accepted that reefs harbour a resident population, resident taxa which are components of the reef itself and not of the ocean in which the reef is situated (Emery, 1968; Porter, 1974a; Sale et al., 1976; Porter et al., 1977). This reef fauna is composed of benthic components as well as epibenthic and holoplanktonic forms (Alldredge and King, 1977; Sale et al., 1976; Hamner and Carleton, 1979). Benthic and epibenthic forms presumably remain associated with the bottom during the daytime and migrate surfaceward at night. Holoplanktonic forms require other behavioural adaptions (Sale et al., 1976), and Steven (1963) and Waterman (1961) have described the response to directional light which would make retention of holoplanktonic forms on the reef possible. While the contribution in terms of biomass to the plankton on a reef by the resident taxa has not been quantified or its contribution to secondary production estimated, it does appear that zooplankton standing crops are much higher than was previously thought. Hamner and Carleton (1979) have recorded extraordinarily high densities (500 000–1 500 000 individuals m^{-3}) of copepods swarming on reefs in Australia and Palau.

ORGANIC PARTICULATE MATERIAL

A number of studies have shown that water, on passing over the reef, becomes enriched with particulate matter (Qasim and Sankaranarayanan, 1970; Johannes and Gerber, 1974). This particulate matter may take the form of algal fragments, fecal pellets, coral mucus and miscellaneous aggregated organic matter (Gerber and Marshall, 1974). The flux of this organic matter in its various forms has been estimated at 20–40 g dwt.m^{-2}day^{-1} by Glynn (1973a), as a net import of 0.27 gCm^{-2}day^{-1} by Johannes and Gerber (1974), while Qasim and Sankaranarayanan (1970) found concentrations to be equivalent to 20% of the gross production of the reef. This autochthonous material is a source of energy for secondary production and is obviously derived from primary production and from material being recycled by the second trophic level. Sournia (1977) observed that most of the suspended organic compounds in reef areas are detrital in nature. On reefs where herbivores are abundant a large proportion of the detrital matter is derived from fecal matter. Hawkins (1979) estimated that 20% of the available primary production was consumed by Diadema and that 18% of food consumed was released as fecal pellets. This amounted to about 26 kcal m^{-2}yr^{-1}.

This material will certainly be of use to filter feeders and suspension feeders on the reef. Recently, Lewis and Price (1975, 1976) have shown that Atlantic reef corals can act as suspension feeders. By means of nets and strands of mucus they are able to capture a wide range of particulate material as well as entrap zooplankton with their tentacles. Species with large active tentacles were found to capture food with their tentacles as well as by mucus filaments, while short-tentacled species (Agaracidae) appear to be restricted to a suspension feeding strategy. Corals have been shown to clear the surrounding water of particulate material (Lewis, 1976) at rates adequate to cover their energy costs in respiration (Lewis, 1977a).

Considerable interest has been shown in the mucus, which is produced by corals for feeding and cleaning, as a basis for production. Johannes (1967) observed that mucus released by corals subsequently formed sticky aggregates, while Marshall (1972) observed quantities of zooxanthellae entangled in mucus. It has been suggested that these mucus flocs could serve as nutritional sources for a variety of suspension feeders on the reef. The mucus with entrapped particles was found by Coles and Strathmann (1973) to have higher organic C and N content than other suspended particles in the water column, and Benson and Muscatine (1974) found it contained energy-rich wax esters. The latter authors observed mucus being fed upon by reef fish.

Richman *et al.* (1975) have estimated that mucus was produced at a mean rate of 51 mg organic matter m^{-3}day^{-1} on a reef at Eilat. This was about 2% of the total concentration of organic material in the water.

MICROBIAL PRODUCTION

Coral reef microorganisms have only recently come under detailed study, although they are obviously of basic importance. One cannot conceive of such systems operating without microorganisms functioning as decomposers, nitrogen fixers and in biogeochemical processes. Di Salvo (1973) observed that bacteria have a regenerative function within coral reef ecosystems and provide active mechanisms for nutrient retention. The bulk of our information on trophodynamics of bacteria on reefs comes from the work of Sorokin (1971a,b; 1973a,b,c; 1974; 1978).

Microbial populations exist within the water flowing over the reefs, within the sediments, in internal reef spaces and upon suspended detritus. Sorokin (1974) has shown that the stock of microbial biomass is high and corresponds to the level of an eutrophic neritic ecosystem. Estimates of microbial biomass and production in various reef habitats in water flowing over reefs are shown in Table III. Comparisons of microbial production with production of microalgae in the sediments and of phytoplankton are presented in the same table. From Table III it is evident that production by bacteria is of the same order as photosynthesis and that the highest bacterial production (and biomass) was found on sediment washed from dead corals. Production by bacteria in waters flowing over reefs was as much as 70 times higher than photosynthetic production in surrounding waters. If these figures are recalculated on a m^2 basis (Sorokin, 1974), from comparison with other production data in Tables I and II we find that production by phytobenthos was about $300 \, gCm^{-2}yr^{-1}$, while production by phytoplankton was only about $11 \, gCm^{-2}yr^{-1}$. These data are similar to the rates found by Sournia (1976) who measured production of about $400 \, gCm^{-2}yr^{-1}$ in sands in Takapoto Lagoon in the Pacific. Somewhat lower values were obtained from bacteria on the surface of dead corals in Hawaii by Sorokin (1978). Production ranged from 4.4 to $40 \, gCm^{-2}yr^{-1}$ and biomass ranged from 0.02 to $0.17 \, gCm^{-2}$.

Bacterial production rates are thus only slightly lower than rates of primary production by some benthic macroalgae noted in Table II. Of significance in bacterial production is the high rate of turnover, a mechanism which will allow retention of organic matter in reef waters and in reef spaces. Sorokin (1974) has calculated that bacterial populations in reef sands are renewed every two days.

This production is important not only in the flux of organic matter from bacterial decomposition but the bacterial biomass in itself is available as a source of food for benthic filter feeders and suspension feeders. In experiments using ^{14}C techniques, Sorokin (1978) has shown that benthic and planktonic reef fauna can feed on bacteria at concentrations usually found in reef environments. Sponges, oysters and colonial tunicates had filtration rates such that their daily ration was equal to 2–5% of their body weight and of polychaetes 8–10%. Corals were also able to consume labelled bacteria and four species obtained 1.7–4.7% of their body weight in the daily ration. Sorokin concluded that the normal concentrations of bacteria in reef water and in the sediments were sufficient to supply the daily maintenance needs of benthic consumers including corals. Thus the bacteria are a potential source of food for filter feeders and suspension feeders and have an important function in the internal metabolism of the reef by mineralisation and nutrient regeneration.

5 Biomass and rates of secondary production on reefc

Rates of secondary production on coral reefs are hard to come by and even estimates of biomass are rare. Glynn (1973b) has compiled a list of numbers of individuals and of biomass on a reef in the Caribbean (Table IV). The biomass of corals, including the skeletons, is many times higher than the total of all other invertebrates. However, in terms of tissue biomass, corals comprise about 75% of the total on a m² basis. Coral biomass was 285.5 g protein m⁻² or about 144 gCm⁻², while total biomass of all invertebrates was 376.7 g protein m⁻² or about 196 gCm⁻².

Odum and Odum (1955) also determined biomass of invertebrates and fish on a reef in the Pacific and reported a mean value of 143 g dry wt.m⁻² excluding skeletal material. If one allows for 20% organic carbon per g of dry weight, then organic biomass was about 30 gCm⁻². Of this amount, between 10–20 gCm⁻² was coral tissue.

Biomass of polychaetes in coral clumps has been determined by Vittor and Johnson (1977) and by Clausade (1970) in terms of colony volumes.

TABLE III   Biomass and production of bacteria in (a) coral sediments and (b) reef waters and in the open ocean. (From Sorokin, 1974)

(a) Biomass and production of bacteria in coral sediments

Position of sampling and character of sediment	Total number of bacteria ($10^9 \mathrm{gm}^{-1}$)	Biomass of bacteria (mgCgm^{-1})	Production of bacteria ($\mathrm{mgCgm}^{-1}\mathrm{day}^{-1}$)	Photosynthetic production ($\mathrm{mgCgm}^{-1}\mathrm{day}^{-1}$)
Kaneohe Bay, Hawaii internal reef, sand among dead corals	2.6	66	30	18.5
Kaneohe Bay, external reef, coarse sand	0.3	4	0.8	3.2
Kaneohe Bay, detrital sediment washed from dead corals	9.7	910	436	550
Majuro Atoll, internal reef, fine sand among corals	3.5	88	35	74
Majuro Atoll, scraped material from dead corals	0.32	17	11	18
Fanning Atoll, coral sand	3.1	91	23	38
Great Barrier Reef (Australia), Heron Island, coral sand	1.68	52	28	61

The data of Vittor and Johnson from the Bahamas can be recalculated on a m^2 basis to give a value of $0.19 \, gCm^{-2}$ for total polychaetes. Kohn and Lloyd (1973) determined biomass values of polychaetes in the Indian Ocean of between 2 and 6 g dry wt.m^{-2} (0.4–$1.2 \, gCm^{-2}$ approximately). Webb *et al.* (1977) determined a value of about $3.6 \, gCm^{-2}$ for *Holothuria atra* in the Pacific.

Rates of production are known for the giant clam *Tridacna* in the Pacific and for the echinoid *Diadema* in the Caribbean. Richard (1977) determined production rates of $0.08 \, gCm^{-2}yr^{-1}$ with a P/B ratio of 0.23 for *Tridacna*, while Hawkins (1979) calculated rates of $61.2 \, gCm^{-2}yr^{-1}$ for *Diadema* with a P/B ratio of 2.

TABLE III—*cont*.

(*b*) Biomass and production of bacteria in reef waters and in the open ocean

Place of sampling	Total number of bacteria $(10^3 ml^{-1})$	Biomass of bacteria $(mgCm^{-3})$	Production of bacteria $(mgCm^{-3}day^{-1})$	Phytosynthesis of phytoplankton $(mgCm^{-3}day^{-1})$
Surface tropical waters, north trade-wind current, Pacific Ocean	68	1.7	2.6	0.5
Kaneohe Bay, Hawaii, water over the reef	1070	43	28.0	36
Majuro Atoll, water over the reef	490	19	7.4	4
Butaritari Atoll, water over the reef	2830	170	41.5	37
Same, water in lagoon	1970	79	24.0	17
Ninigo Atoll, water over the reef, low tide	620	32	21.0	9
Same, high tide	290	13	7.2	4
Saint Andrew reefs (Admiralty Islands) water over the reef, low tide	740	21	14.1	7
Same, high tide	370	11	7.4	3

TABLE IV   Numbers and biomass of invertebrates per m^2 on a *Porites furcata* reef. (From Glynn, 1973b)

Group	Number of individuals	Biomass (g dry wt. incl. skeletons)	Biomass (g protein)
Scleractinia	several colonies	12 002.3	285.5
Zooanthidae	2 colonies	1.1	0.4
Polychaeta	23 160	6.7	3.2
Sipunculida	20	0.9	0.7
Crustacea	2668	40.3	14.9
Mollusca	137	59.4	10.0
Echinodermata	2002	263.3	52.6
Pisces	15	6.1	3.6

There are a number of estimates of fish biomass on reefs but only Bardach (1959) has estimated production rates. He has quoted a figure of $22 \, \text{kcal m}^{-2}\text{yr}^{-1}$ as energy production on a Bermuda reef. Biomass estimates of fish on a number of reefs are shown in Table V. The highest value was estimated at $35 \, \text{gCm}^{-2}$ by Randall (1963) on an artificial reef, while Goldman and Talbot (1976) have estimated that about $40 \, \text{gCm}^{-2}$ is the maximum which could be supported by the majority of reefs.

TABLE V   Estimates of abundance of fish on reefs. (From Stevenson and Marshall, 1974)

Location	Wet weight (gm^{-2})	gCm^{-2}	Ref.
Offshore reef, Great Barrier Reef	209	4.2	Talbot and Goldman (1972)
Fringing reef, Hawaii	62	1.2	Brock (1954)
Reef flat, Eniwetok Atoll	9	0.9	Odum and Odum (1955)
Patch reef, Bermuda	49	1.0	Bardach (1959)
Fringing reef, Virgin Islands	160	3.2	Randall (1963)
Fringing reef, Virgin Islands	38	0.1	Dammann (1969)
Artificial reef, Virgin Islands	1750	35.0	Randall (1963)

Any consideration of coral reef production must take into account the development of the physical structure of the reef, the rate of growth of corals, the degradation of the reef rock by biological action, and its subsequent infilling and cementation. Thus rates of growth of reefs are linked with both primary and secondary trophic levels.

Chave et al. (1972) have attempted to quantify carbonate production by coral reefs and have divided it into three categories: potential, gross and net production. Potential production, or the amount of calcium carbonate produced by limestone secreting oganisms per unit area of reef organism, was estimated at 1×10^4 to 1×10^5 gCaCO$_3$m^{-2}yr^{-1}. Gross production, or the total amount of calcium carbonate produced by the reef community per unit area of bottom, ranged from 4×10^2 to 6×10^4 gCaCO$_3$m^{-2}yr^{-1}. The net production is the amount of carbonate retained by the reef and was estimated at about 10×10^3 gCaCO$_3$m^{-2}yr^{-1}.

Potential and gross production are primarily biological considerations of how benthic organisms grow and calcify. Net production is of particular interest to geologists but it is also a measure of the stability of the physical structure that is spatially the reef ecosystem. The spatial structure may increase or decrease depending upon rates of net production of carbonate. This is determined by rates of organic production of reef organisms at the primary and secondary levels. The growth and erosion of coral reefs may be regarded as a special ecosystem feature in which the spatial structure is produced and modified by the organisms themselves.

The connection between reef rock and calcium carbonate production and the second trophic level has been illustrated by Stearn and Scoffin (1977). On a fringing reef in Barbados it was found that corals and algae fixed about 163.3×10^6 g yr^{-1} of calcium carbonate over the whole reef. Of this amount, 25×10^6 g CaCO$_3$ was removed by boring organisms such as sponges, bivalves, sipunculids and polychaetes. A small amount of erosion by the parrot fish, *Sparisoma viridis*, was also calculated at 1×10^6 gCaCO$_3$yr^{-1} but the most important bioerosive agent was the sea urchin *Diadema antillarum*. Erosion rates calculated for *Diadema* amounted to 163×10^6 gCaCO$_3$yr^{-1} over the whole reef. These figures indicate a deficit of 26×10^6 gCaCO$_3$yr^{-1} and suggest that the reef is being destroyed faster than it is growing.

6 Trophic relationships

From the numerous studies of plankton input into the reef ecosystem, it is apparent that the 'normal' phytoplankton/zooplankton basis for benthic production is of minor importance. Rather the system is a self-contained one in which secondary production is dependent upon the dominant role played by benthic primary producers. There is much to suggest the similarity to estuarine communities.

The importance of this view is indicated in the early work of Odum and Odum (1955) who illustrated biomass relationships of reef components at Eniwetok Atoll in the Pacific (Fig. 2). Ten different kinds of primary producers were enumerated with an average total biomass of 703 g dwt.m^{-2}. The standing crop of herbivores (including corals) was 132 g dwt.m^{-2}, or 18.9% of the biomass of primary producers. Carnivores comprised 8.3% of the biomass of herbivores. While this quantitative trophic structure conforms to general ecological theory, decomposer components are not included and corals should not be included as a single herbivore level.

Odum and Odum (1955) have also defined reefs as stable ecosystems and climax communities in the sense that there is little net increase in living biomass, in spite of the enormous accumulation of calcium carbonate. The ecosystem is an open, steady-state system since production is just about balanced by community respiration. This notion of the reef as a steady-state system has also been considered by Wangersky (1978) who compared an atoll reef to a chromatographic column. Inorganic nutrients and organic matter are brought onto the reef by incoming currents of water, processed and recycled by primary and secondary producers and exported off the reef at the same rate they are imported. Any mechanism which would increase the hold-up time of nutrients and organics would be advantageous – symbiotic relationships, for example. However, the analogy to a chromatographic column may not be entirely apt when we consider that the particulate matter content is considerably augmented in water passing over the reef (Johannes, 1967; Qasim and Sankaranarayanan, 1970; Coles and Strathmann, 1973).

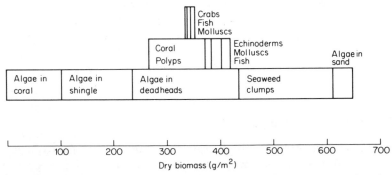

FIG. 2   Biomass pyramid of weights of living organisms from a coral reef, Eniwetok Atoll (from Odum and Odum, 1955)

Retention of organic material on the reef is best illustrated by corals. Hermatypic corals occupy a key position in the trophodynamics of the reef ecosystem. They not only form the main structural component of most modern reefs but also provide a substrate, food and shelter for a host of associated epifaunal and infaunal organisms. But perhaps most important is the symbiosis of reef corals and their endosymbiotic algae. The ecological consequence of this symbiosis is that it conveys a potential for autotrophy and provides a conservative mechanism adapted to a nutrient-poor environment.

In a provocative paper, Muscatine and Porter (1977) have described the polytrophic feeding capacity of reef corals. They not only capture zooplankton with their tentacles but are also suspension feeders on a wide variety of suspended particulate material (Lewis, 1976). Sorokin (1973) and Di Salvo (1972) have demonstrated that corals can feed upon bacteria, and Goreau et al. (1971) have described deposit feeding on substrate detritus by means of mesenterial filaments. Finally, Stephens (1962) has shown how the coral *Fungia* may act as a saprotroph taking up dissolved organic substances from the water. Each of these feeding methods can potentially contribute to the daily energy needs of corals and provides them with a very broad range of food resources.

The potential for autotrophy is realised through the symbiotic association with zooxanthellae, and corals may be considered as 'holding captive' a significant proportion of reef primary production.

The movement of photosynthetically-fixed carbon in the zooxanthellae has been considered in a review by Muscatine (1973) who marshalled evidence to show that there is translocation of algal products as proteins, carbohydrates and lipids to the coral tissue. Von Holt and Von Holt (1968) found that 40% of the ^{14}C fixed by photosynthesis was recovered from tissue of the coral *Scolymia*, while Muscatine and Cernichiari (1969) recovered 36–50% of the translocated carbon. Although a significant proportion of photosynthetically-fixed carbon is transferred to the coral host, it is not known whether this amount is significant in terms of the nutritional requirements of the corals. Muscatine and Porter (1977) estimated that 86.6 to 137.6% of the reduced organic carbon required by the coral tissue was translocated from the algae. However, their estimates were based on a number of broad assumptions and Yonge (1963) and Goreau et al. (1971) considered that the zooxanthellae could only supply a fraction of the coral's daily energy requirements. Details of the energy budgets of corals are required before this question can be answered.

Undoubtedly the zooxanthellae play a major role in coral growth and reef building by causing accelerated calcification in light (Goreau, 1959; Goreau and Goreau, 1959, 1960a,b). Photosynthesis appears to produce substances which are translocated to calcification sites and stimulate deposition (Pearse and Muscatine, 1971; Vandermeulen et al., 1972; Barnes and Taylor, 1973; Vandermeulen and Muscatine, 1974). Goreau (1959) proposed that zooxanthellae enhanced precipitation of calcium carbonate by removing carbon dioxide from the calcifying milieu during photosynthesis. Simkiss (1964a,b) suggested that zooxanthellae can take up organic and inorganic phosphates which act as crystal poisons and interfere with aragonite formation. Pearse and Muscatine (1971) and Vandermeulen and Muscatine (1974) suggest that the zooxanthellae supply material needed for the organic matrix of the skeleton. Crossland and Barnes (1974) have proposed that excreted ammonia neutralises protons during carbonate precipitation and thus stimulates calcification. Thus, while several hypotheses have been put forth, the biological mechanism of calcium carbonate movement in corals is still unknown, as is the quantitative utilisation of the photosynthetic products of the zooxanthellae in skeleton formation. It would appear from the work of Chalker (1976) that calcium transport in corals is an energy-consuming process.

It has also been proposed that zooxanthellae influence the metabolism of reef corals by taking up metabolic wastes and thus consuming and recycling dissolved organic matter (Muscatine, 1973). There is evidence that corals can take up nitrate from the ambient water (Franzisket, 1974; D'Elia and Webb, 1977), but more important, they can also take up ammonium. Uptake and retention of ammonium are enhanced by light, mediated by the zooxanthellae (Kawaguti, 1953; Sloterdijk, 1975; Muscatine and D'Elia, 1978). The inference is that ammonia from the coral excretion is taken up by the zooxanthellae and they use it as a source of nitrogen.

Another ecological consequence of the symbiotic relationship between corals and their zooxanthellae lies in competition for space on a reef. Because of the need to obtain light energy for photosynthesis many corals have developed a branching morphology. The shape of branches may be partly a result of growth in response to light conditions as in terrestrial plants, but also involved is the tendency to increase the plankton-capturing surface area. In any event, Porter (1974b) has shown that branching growth leads to an 'overtopping strategy' in which one coral overgrows another species in competition for space. One would expect such competitive interactions to occur on reefs where up to 90% of the benthic cover is made up of corals. Indeed, Lang (1973) has shown that when corals come into actual contact with each other, competitively superior corals can cause disintegration of tissue of its neighbour by means of mesenterial filaments and overgrow its dead skeleton.

A number of schemes illustrating trophic relationships on reefs has been attempted. Sorokin (1973) has emphasised the importance of bacteria in trophic relationships based on his findings of the importance of those microorganisms for food for suspension feeders and as producers. The role of detritus has also been noted (Lewis, 1977b) on the assumption that a large proportion of the primary production is transformed to detritus before becoming available to higher trophic levels.

As noted previously, the difficulty in illustrating straightforward trophic relationships in a reef ecosystem lies in the many symbiotic relationships within reefs, especially in the corals. In fact, as several authors have mentioned, reef corals may be considered as an ecosystem within themselves. In Fig. 3 an attempt has been made to sketch trophic relationships by separating the corals from the rest of the community. Benthic algae are regarded as the organic base for the food web. Organic matter from benthic algae is consumed by herbivores such as *Diadema* as well as being acted upon by bacteria. Decomposition results in detritus production which is directed to the coral trophic levels and to sessile suspension feeders. Detritus is augmented by particulate material from the phytoplankton/zooplankton levels and from fecal material and mucus from the second and third trophic levels. There is, in fact, an important decomposition channel through which organic material passes. Production by herbivores and suspension feeders

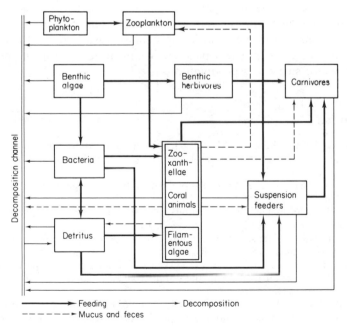

FIG. 3   Trophic relationships and energy transformation in coral reef ecosystems

and from the several coral trophic levels passes to the carnivore level. There are thus numerous pathways for energy flow to consumers and many feedback loops.

7 Summary

(i) Rates of primary production on coral reefs vary between about 300 and 5000 $gCm^{-2}yr^{-1}$ and are comparable to the most fertile marine waters.

(ii) The sources of this production are the benthic algae, including the zooxanthellae of reef corals.

(iii) Nutrient concentrations in water flowing over reefs is consistently low but nutrients are retained within the system by conservative, regenerative and recycling mechanisms.

(iv) There are few quantitative data on secondary production on reefs but opinion has it that reefs support large standing crops of benthic secondary producers.

(v) The support for secondary production comes from autochthonous rather than allochthonous sources, from the breakdown and decomposition of the primary producers. There are numerous pathways of energy transformation within the system.

(vi) Reef corals are seen as multitrophic organisms with the capacity for autotrophy and hetertrophy, and as key organisms in the building and sustaining of the reef framework. Their symbiosis with zooxanthellae is an adaption to living in a nutrient-poor environment and one of the principal mechanisms for sustaining production on reefs.

References

Alldredge, A. L. and King, J. M. (1977). Distribution, abundance and substrate preferences of demersal reef zooplankton at Lizard Island Lagoon, Great Barrier Reef. *Mar. Biol.* **41**, 317–333

Bakus, G. J. (1966). Some relationships of fishes to benthic organisms on coral reefs. *Nature* **210**, 280–284

Bakus, G. J. (1967). The feeding habits of fishes and primary production at Eniwetok, Marshall Islands. *Micronesica* **3**, 135–149

Bardach, J. E. (1959). The summer standing crop of fish on a shallow Bermuda reef. *Limnol. and Oceanogr.* **18**, 673–678

Barnes, D. J. and Taylor, D. L. (1973). In situ studies of calcification and photosynthetic carbon fixation in the coral *Montastrea annularis*. *Helgo. wissen. Meeres.* **24**, 284–291

Beers, J. R., Steven, D. M. and Lewis, J. B. (1963). Primary production in the Caribbean Sea off Jamaica and the tropical North Atlantic off Barbados. *Bull. mar. Sci.* **18**, 86–104

Benson, A. A. and Muscatine, L. (1974). Wax in coral mucus: Energy transfer from corals to reef fishes. *Limnol. and Oceanogr.* **19**, 810–814

Brock, V. E. (1954). A preliminary report on a method of estimating reef fish populations. *J. of Wildlife Management* **18**, 297–308

Buesa, R. J. (1974). Population and biological data on turtle grass 'Thalassia testudinum König, 1805' on the northwestern Cuban shelf. *Aquaculture* **4**, 207–226

Burkholder, P. R., Burkholder, L. M. and Rivero, J. A. (1959). Some chemical constituents of turtle grass, *Thalassia testudinum*. *Bull. Torrey Bot. Club* **86**, 88–93

Capone, D. G. (1977). $N_2(C_2H_2)$ fixation by macroalgal epiphytes. *Proc. Third Int. Coral Reef Symp.* 1 (D. L. Taylor, ed.) pp. 337–342. Rosenstiel School of Marine and Atmospheric Science, Miami

Chalker, B. E. (1976). Calcium transport during skeletogenesis in hermatypic corals. *Comp. Bioch. Physiol.* **54A**, 455–459

Chave, K. E., Smith, S. V. and Roy, K. J. (1972). Carbonate production by coral reefs. *Mar. Geol.* **12**, 123–140

Chesher, R. H. (1970). *Acanthaster planci:* impact on Pacific coral reefs. *United States Department of the Interior Publication 187631*

Choat, J. H. (1969). Studies on labroid fishes (Labridae and Scaridae) at Heron Island, Great Barrier Reef. Ph.D. Thesis, University of Queensland, Australia

Clausade, M. (1970). Répartition qualitative et quantitative des Polychetes vivant dans les alvéoles des constructions organogènes épirécifales de la portion septentrionale du Grand Récife de Tuléar. Rec. Trav. Stat. mar. Endoume, fasc. hors. série suppl. **10**: *Trav. Stat. mar. Tuléar* 107–109

Coles, S. L. (1969). Quantitative estimates of feeding and respiration for three scleractinian corals. *Limnol. and Oceanogr.* **14**, 949–953

Coles, S. L. and Strathmann, R. (1973). Observations on coral mucus 'flocs' and their potential trophic significance. *Limnol. and Oceanogr.* **18**, 673–678

Connor, J. L. and Adey, W. H. (1977). The benthic algal composition, standing crop and productivity of a Caribbean algal ridge. *Atoll Res. Bull.* **211**, 1–40

Crisp, D. J. (1975). Secondary productivity in the sea. *Proc. Symp. Productivity of World Ecosystems*, Seattle, Washington, 31 August–1 September 1972 (D. E. Reichle, J. E. Franklin and D. W. Goodall, eds) pp. 71–89. National Academy of Sciences, Washington

Crossland, C. J. and Barnes, D. J. (1974). The role of metabolic nitrogen in coral calcification. *Mar. Biol.* **28**, 325–332

Dahl, A. L. (1974). The structure and dynamics of benthic algae in the coral reef ecosystem. *Proc. Second Int. Symp. Coral Reefs* 1, Australia, 22 June–2 July 1973 (A. M. Cameron, B. M. Campbell, A. B. Cribb, R. Endean, J. S. Jell, O. A. Jones, P. Mather and F. H. Talbot, eds) pp. 21–25. The Great Barrier Reef Committee, Brisbane

Dammann, A. E. (1969). Study of the fisheries potential of the Virgin Islands. *Virgin Islands Ecological Research Station Contributions I*, 1–197

D'Elia, C. F. and Webb, K. L. (1977). The dissolved nitrogen flux of coral reefs. *Proc. Third Int. Coral Reef Symp.* 1 (D. L. Taylor, ed.) pp. 325–330. Rosenstiel School of Marine and Atmospheric Science, Miami

Di Salvo, L. H. (1972). Some aspects of the regenerative function and microbial ecology of coral reefs. *Proc. Symp. Corals and Coral Reefs*, Mandapam Camp, India, 12–16 January 1969 (C. Mukundan and C. S. P. Pillai, eds) pp. 67–69. Marine Biological Association of India, Cochin

Di Salvo, L. H. (1973). Microbial ecology. *In* "Biology and Geology of Coral Reefs" (O. A. Jones and R. Endean, eds) Vol. II, Biology 1, pp. 1–15. Academic Press, New York and London

Di Salvo, L. H. (1974). Soluble phosphorus and amino nitrogen released to sea water during recoveries of coral reef regenerative sediments. *Proc. Second Int. Symp. Coral Reefs* 1, Australia, 22 June–2 July 1973 (A. M. Cameron, B. M. Campbell, A. B. Cribb, R. Endean, J. S. Jell, O. A. Jones, P. Mather and F. H. Talbot, eds), pp. 11–19. The Great Barrier Reef Committee, Brisbane

Di Salvo, L. H. Gundersen, K. (1971). Regenerative functions and microbial ecology of coral reefs I. Assays for microbial population. *Can. J. Microb.* 17, 1081–1089

Doty, M. S. and Oguri, M. (1956). The island mass effect. *J. du Cons. Int. pour l'Explor. de la Mer* 22, 33–37

Ehrlich, P. R. (1975). The population ecology of coral reef fishes. *Ann. Rev. Ecology Systematics* 6, 211–247

Emery, A. R. (1968). Preliminary observations on coral reef plankton. *Limnol. and Oceanogr.* 13, 293–303

Endean, R. (1973). Population explosions of *Acanthaster planci* and associated destruction of hermatypic corals in the Indo-West Pacific region. *In* "Biology and Geology of Coral Reefs" (O. A. Jones and R. Endean, eds) Vol. II, Biology 1, pp. 389–438. Academic Press, New York and London

Fenchel, T. (1970). Studies on the decomposition of organic detritus from the turtle grass, *Thalassia testudinum. Limnol. and Oceanogr.* 15, 14–20

Franzisket, L. (1974). Nitrate uptake by reef corals. *Int. Rev. ges. Hydrobiol.* 59, 1–7

Gallagher, B. S., Shimada, K. M., Gonzales, F. I. Jr. and Stroup, E. D. (1971). Tides and currents in Fanning Atoll lagoon. *Pac. Sci.* 25, 191–205

Gerber, R. P. and Marshall, N. (1974). Ingestion of detritus by the lagoon pelagic community at Eniwetok Atoll. *Limnol. and Oceanogr.* 19, 815–824

Gibbs, P. E. (1971). The polychaete fauna of the Solomon Islands. *Bull. Brit. Mus. (Nat. Hist.) Zool.* 21, 101–211

Gilmartin, M. (1960). The ecological distribution of the deep water algae of Eniwetok Atoll. *Ecology* 41, 209–221

Ginsburg, R. N. and Lowenstam, H. A. (1958). The influence of marine bottom communities on the depositional environment of sediments. *J. Geol.* 66, 310–318

Glynn, P. W. (1973a). Ecology of a Caribbean coral reef. The *Porites* reef-flat biotope: Part II. Plankton community with evidence for depletion. *Mar. Biol.* 22, 1–21

Glynn, P. W. (1973b). Aspects of the ecology of coral reefs in the western Atlantic region. *In* "Biology and Geology of Coral Reefs" (O. A. Jones and R. Endean, eds) Vol. II, Biology 1, pp. 271–324. Academic Press, New York and London

Glynn, P. W., Wellington, G. M. and Birkland, C. (1979). Coral reef growth in the Galapagos; limitation by sea urchins. *Science* 203, 47–48

Goering, J. J. and Parker, P. L. (1972). Nitrogen fixation by epiphytes on sea grasses. *Limnol. and Oceanogr.* 17, 320–323

Goldman, B. and Talbot, L. H. (1976). Aspects of the ecology of coral reef fishes. *In* "Biology and Geology of Coral Reefs" (O. A. Jones and R. Endean, eds) Vol. III, Biology 2, pp. 125–154. Academic Press, New York and London

Gordon, D. C. Jr. (1971). Organic carbon budget of Fanning Island lagoon. *Pac. Sci.* 25, 222–227

Gordon, M. S. and Kelly, H. M. (1962). Primary productivity of an Hawaiian coral reef: A critique of flow respirometry in turbulent water. *Ecology* 43, 473–480

Goreau, T. F. (1959). The physiology of skeleton formation in corals. I. A method for measuring the rate of calcium deposition by corals under different conditions. *Biol. Bull.* **116**, 59–75

Goreau, T. F. and Goreau, N. I. (1959). The physiology of skeleton formation in corals. II. Calcium deposition by hermatypic corals under various conditions in the reef. *Biol. Bull.* **117**, 239–250

Goreau, T. F. and Goreau, N. I. (1960a). The physiology of skeleton formation in corals. III. Calcification rate as a function of colony weight and total nitrogen content in the reef coral *Manicina areolata* (Linnaeus). *Biol. Bull.* **118**, 419–429

Goreau, T. F. and Goreau, N. I. (1960b). The physiology of skeleton formation in corals. IV. On isotonic equilibrium exchanges of calcium between corallum and environment in living and dead reef corals. *Biol. Bull.* **119**, 416–427

Goreau, T. F. and Wells, J. W. (1967). The shallow water Scleractinia of Jamaica: Revised list of species and their vertical distribution range. *Bull. mar. Sci.* **17**, 442–453

Goreau, T. F., Goreau, N. I., Yonge, C. M. and Neumann, Y. (1970). On feeding and nutrition in *Fungiacava eilatensis* (Bivalvia, Mytilidae), a commensal living in fungiid corals. *J. Zool. Lond.* **160**, 159–172

Goreau, T. F., Goreau, N. I. and Yonge, C. M. (1971). Reef corals: autotrophs or heterotrophs? *Biol. Bull.* **141**, 247–260

Grassle, J. F. (1973). Variety in coral reef communities. *In* "Biology and Geology of Coral Reefs" (O. A. Jones and R. Endean, eds) Vol. II, Biology 1, pp. 247–270. Academic Press, New York and London

Greenway, M. (1974). The effects of cropping on the growth of *Thalassia testudinum* (König) in Jamaica. *Aquaculture* **4**, 199–206

Halldal, P. (1968). Photosynthetic capacities and photosynthetic action spectra of endozoic algae of the massive coral *Favia*. *Biol. Bull.* **134**, 411–424

Hamner, W. M. and Carleton, J. H. (1979). Copepod swarms: Attributes and role in coral reef ecosystems. *Limnol. and Oceanogr.* **24**, 1–14

Hawkins, C. M. (1979). Ecological energetics of the tropical sea urchin, *Diadema antillarum Philippi* in Barbados, West Indies. Ph.D. Thesis, McGill University

Hiatt, R. W. and Strasburg, D. W. (1960). Ecological relationships of the fish fauna on coral reefs of the Marshall Islands. *Ecol. Monog.* **30**, 65–127

Hillis-Colinvaux, L. (1974). Productivity of the coral reef alga *Halimeda* (Order Siphonales). *Proc. Second Int. Symp. Coral Reefs* 1, Australia, 22 June–2 July 1973 (A. M. Cameron, B. M. Campbell, A. B. Cribb, R. Endean, J. S. Jell, O. A. Jones, P. Mather and F. H. Talbot, eds) pp. 35–42. The Great Barrier Reef Committee, Brisbane

Hutchings, P. A. (1974). A preliminary report on the density and distribution of invertebrates living on coral reefs. *Proc. Second Int. Symp. Coral Reefs* 1, Australia, 22 June–2 July 1973 (A. M. Cameron, B. M. Campbell, A. B. Cribb, R. Endean, J. S. Jell, O. A. Jones, P. Mather and F. H. Talbot, eds) pp. 285–296. The Great Barrier Reef Committee, Brisbane

Johannes, R. E. (1967). Ecology of organic aggregates in the vicinity of a coral reef. *Limnol. and Oceanogr.* **12**, 189–195

Johannes, R. E., Alberts, J., D'Elia, C., Kinzie, R. A., Pomeroy, L. R., Sottile, L. R., Wiebe, W., Marsh, J. A., Jr., Helfrich, P., Maragos, J., Meyer, J., Smith, S., Crabtree, D., Roty, A., McCloskey, L. R., Betzer, S., Marshall, N., Pilson, M. E. Q., Telek, G., Clutter, R. I., DuPaul, W. D., Webb, K. L. and Wells, J. M. Jr. (1972). The metabolism of some coral reef communities: a team study of nutrient and energy flux at Eniwetok. *Bioscience* **22**, 541–543

Johannes, R. E., Coles, S. L. and Kuenzel, N. T. (1970). The role of zooplankton in the nutrition of some scleractinian corals. *Limnol. and Oceanogr.* **15**, 579–586

Johannes, R. E. and Gerber, R. (1974). Import and export of net plankton by an Eniwetok coral reef community. *Proc. Second Int. Symp. Coral Reefs* 1, Australia, 22 June–2 July 1973 (A. M. Cameron, B. M. Campbell, A. B. Cribb, R. Endean, J. S. Jell, O. A. Jones, P. Mather and F. H. Talbot, eds) pp. 97–104. The Great Barrier Reef Committee, Brisbane

Johannes, R. E. and Tepley, L. (1974). Examination of feeding of the reef coral *Porites lobata* in situ using time lapse photography. *Proc. Second Int. Symp. Coral Reefs* 1, Australia, 22 June–2 July 1973 (A. M. Cameron, B. Campbell, A. B. Cribb, R. Endean, J. S. Jell, O. A. Jones, P. Mather and F. H. Talbot, eds) pp. 127–131. The Great Barrier Reef Committee, Brisbane

Johnston, C. S. (1969). The ecological distribution and primary production of macrophytic marine algae in the Eastern Canaries. *Int. Rev. ges. Hydrobiol.* **54**, 473–490

Jones, J. A. (1963). Ecological studies of southeastern Florida patch reefs. Part I. Diurnal and seasonal changes in the environment. *Bull. mar. Sci.* **13**, 282–307

Kanwisher, J. W. and Wainwright, S. A. (1967). Oxygen balance in some reef corals. *Biol. Bull.* **33**, 378–390

Kawaguti, S. (1953). Ammonium metabolism of the reef corals. *Biol. J. Okayama Univ.* **I**, 171–176

Kinsey, D. W. (1972). Preliminary observations on community metabolism and primary productivity of the pseudo-atoll reef at One Tree Island. *Proc. Symp. Corals and Coral Reefs*, Mandapam Camp, India 12–16 January 1969 (C. Mukudan and C. S. P. Pillai, eds) pp. 13–32. Marine Biological Association of India, Cochin

Kinsey, D. W. and Domm, A. (1974). Effects of fertilization on a coral reef environment – primary production studies. *Proc. Second Int. Symp. Coral Reefs* 1, Australia 22 June–2 July 1973 (A. M. Cameron, B. M. Campbell, A. B. Cribb, R. Endean, J. S. Jell, O. A. Jones, P. Mather and F. H. Talbot, eds) pp. 49–66. The Great Barrier Reef Committee, Brisbane

Kohn, A. J. (1959). The ecology of *Conus* in Hawaii. *Ecol. Monogr.* **29**, 47–90

Kohn, A. J. (1968). Microhabitats, abundance and food of *Conus* on atoll reefs in the Maldive and Chagos Islands. *Ecology* **49**, 1046–1062

Kohn, A. J. and Helfrich, P. (1957). Primary organic productivity of a Hawaiian coral reef. *Limnol. and Oceanogr.* **2**, 241–251

Kohn, A. J. and Lloyd, M. C. (1973). Polychaetes of truncated reef limestone substrates on Eastern Indian Ocean coral reefs: Diversity, abundance and taxonomy. *Int. Rev. ges. Hydrobiol.* **58**, 369–399

Lang, J. C. (1973). Interspecific aggression by scleractinian corals. 2. Why the race is not only to the swift. *Bull. mar. Sci.* **23**, 260–279

Lewis, J. B. (1976). Experimental tests of suspension feeding in Atlantic reef corals. *Mar. Biol.* **36**, 147–150

Lewis, J. B. (1977a). Suspension feeding in Atlantic reef corals and the importance of suspended particulate material as a food source. *Proc. Third Int. Coral Reef Symp.* 1 (D. L. Taylor, ed.) pp. 405–408. Rosenstiel School of Marine and Atmospheric Science, Miami

Lewis, J. B. (1977b). Processes of organic production on coral reefs. *Biol. Rev.* **52**, 305–347

Lewis, J. B. and Price, W. S. (1975). Feeding mechanisms and feeding strategies of Atlantic reef corals. *J. Zool. London* **176**, 527–544

Lewis, J. B. and Price, W. S. (1976). Patterns of ciliary currents in Atlantic reef corals and their functional significance. *J. Zool. London* **178**, 77–89

Littler, M. M. (1973). The productivity of Hawaiian fringing-reef Corallinaceae and an experimental evaluation of production methodology. *Limnol. and Oceanogr.* **18**, 946–952

MacGeachy, J. K. and Stearn, C. W. (1976). Boring by macro-organisms in the coral *Montastrea annularis* on Barbados reefs. *Int. Revue ges. Hydrobiol.* **61**, 715–745

Marsh, J. A. (1970). Primary productivity of reef-building calcareous red algae. *Ecology* **51**, 255–263

Marsh, J. A. (1974). Preliminary observations on the productivity of a Guam reef flat community. *Proc. Second Int. Symp. Coral Reefs* 1, Australia, 22 June–2 July 1973 (A. M. Cameron, B. M. Campbell, A. B. Cribb, R. Endean, J. S. Jell, O. A. Jones, P. Mather and F. H. Talbot, eds) pp. 139–145. The Great Barrier Reef Committee, Brisbane

Marshall, N. (1972). Notes on mucus and zooxanthellae discharged from reef corals. *Proc. Symp. Corals and Coral Reefs*, Mandapam Camp, India, 12–16 January 1969 (C. Mukudan and C. S. P. Pillai, eds) pp. 59–65. Marine Biological Association of India, Cochin

McCloskey, L. R. (1970). The dynamics of the community associated with a marine scleractinian coral. *Int. Revue ges. Hydrobiol.* **55**, 13–81

Muscatine, L. (1973). Nutrition of corals. *In* "Biology and Geology of Coral Reefs" (O. A. Jones and R. Endean, eds) Vol. II, Biology 1, pp. 77–115. Academic Press, New York and London

Muscatine, L. and Cernichiari, E. (1969). Assimilation of photosynthetic products of zooxanthellae by a reef coral. *Biol. Bull.* **137**, 506–523

Muscatine, L. and Porter, J. W. (1977). Reef corals: Mutualistic symbioses adapted to nutrient-poor environments. *Bioscience* **27**, 454–460

Muscatine, L. and D'Elia, C. F. (1978). The uptake, retention and release of ammonium by reef corals. *Limnol. and Oceanogr.* **23**, 725–734

Odum, E. P. (1971). "Fundamentals of Ecology". W. B. Saunders, Philadelphia

Odum, H. T. and Odum, E. P. (1955). Trophic structure and productivity of a windward coral reef community on Eniwetok Atoll. *Ecol. Monog.* **25**, 291–320

Odum, H. T., Burkholder, P. R. and Rivero, J. (1959). Measurements of productivity of turtle grass flats, reefs and the Bahai Fosforescente of southern Puerto Rico. *Publ. Inst. mar. Sci.* **6**, 159–170

Ogden, J. C. and Lobel, P. S. (1978). The role of herbivorous fishes and urchins in coral reef communities. *Env. Biol. Fish.* **3**, 49–63

Ott, B. S. and Lewis, J. B. (1972). The importance of the gastropod *Coralliophila abbreviata* (Lamarck) and the polychaete *Hermodice carunculata* (Pallas) as coral reef predators. *Can. J. Zool.* **50**, 1651–1656

Parsons, T. R., Takahashi, M. and Hargrave, B. (1977). "Biological Oceanographic Processes", 2nd edn. Pergamon Press, Oxford

Patriquin, D. G. (1972). The origin of nitrogen and phosphorus for growth of the marine angiosperm *Thalassia testudinum*. *Mar. Biol.* **15**, 35–46

Patriquin, D. G. (1973). Estimation of growth rate, production and age of the marine angiosperm, *Thalassia testudinum* König. *Carib. J. Sci.* **13**, 111–123

Patriquin, D. and Knowles, R. (1972). Nitrogen fixation in the rhizosphere of marine angiosperms. *Mar. Biol.* **16**, 49–58

Patton, W. K. (1976). Animal associates of living reef corals. *In* "Biology and Geology of Coral Reefs" (O. A. Jones and R. Endean, eds) Vol. III, Biology 2, pp. 1–36. Academic Press, New York and London

Pearse, V. B. and Muscatine, L. (1971). Role of symbiotic algae (zooxanthellae) in coral calcification. *Biol. Bull.* **141**, 350–363

Pilson, M. E. Q. and Betzer, S. B. (1973). Phosphorus flux across a coral reef. *Ecology* **54**, 581–588

Pomeroy, L. R., Pilson, M. E. Q. and Wiebe, W. J. (1974). Tracer studies of the exchange of phosphorus between reef water and organisms on the windward reef of Eniwetok Atoll. *Proc. Second Int. Symp. Coral Reefs* 1, Australia, 22 June–2 July 1973 (A. M. Cameron, B. M. Campbell, A. B. Cribb, R. Endean, J. S. Jell, O. A. Jones, P. Mather and F. H. Talbot, eds) pp. 87–96. The Great Barrier Reef Committee, Brisbane

Porter, J. W. (1974a). Zooplankton feeding by the Caribbean reef-building coral *Montastrea cavernosa*. *Proc. Second Int. Symp. Coral Reefs* 1, Australia, 22 June–2 July 1973 (A. M. Cameron, B. M. Campbell, A. B. Cribb, R. Endean, J. S. Jell, O. A. Jones, P. Mather and F. H. Talbot, eds) pp. 111–125. The Great Barrier Reef Committee, Brisbane

Porter, J. W. (1974b). Community structure of coral reefs on opposite sides of the isthmus of Panama. *Science* **186**, 543–545

Porter, J. W., Porter, K. G. and Batac-Catalan, Z. (1977). Quantitative sampling of Indo-Pacific demersal reef plankton. *Proc. Third Int. Coral Reef Symp.* 1 (D. L. Taylor, ed.) pp. 105–112. Rosenstiel School of Marine and Atmospheric Science, Miami

Qasim, S. Z. (1970). Some problems related to the food chain in a tropical estuary. *In* "Marine Food Chains" (J. H. Steele, ed.) pp. 45–51. Oliver and Boyd, Edinburgh

Qasim, S. Z., Bhattathiri, P. M. A. and Reddy, C. V. G. (1972). Primary production of an atoll in the Laccadives. *Int. Revue ges Hydrobiol.* **57**, 207–225

Qasim, S. Z. and Sankaranarayanan, V. N. (1970). Production of particulate organic matter by the reef on Kavaratti Atoll (Laccadives). *Limnol. and Oceanogr.* **15**, 574–578

Randall, J. E. (1963). An analysis of the fish populations of artificial and natural reefs in the Virgin Islands. *Carib. J. Sci.* **3**, 31–48

Randall, J. E. (1967). Food habits of the reef fishes of the West Indies. *Studies in Tropical Oceanography*, Institute of Marine Science, University of Miami 5, 665–847

Richard, G. (1977). Quantitative balance and production of *Tridacna maxima* in the Takapoto Lagoon (French Polynesia). *Proc. Third Int. Coral Reef Symp.* 1 (D. L. Taylor, ed.) pp. 599–605. Rosenstiel School of Marine and Atmospheric Science, Miami

Richman, S., Loya, Y. and Slobodkin, L. B. (1975). The rate of mucus production by corals and its assimilation by the coral reef copepod *Acartia negligens*. *Limnol. and Oceanogr.* **20**, 918–923

Robertson, R. (1970). Review of the predators and parasites of stony corals, with special reference to symbiotic prosobranch gastropods. *Pac. Sci.* **24**, 43–54

Roffman, B. (1968). Patterns of oxygen exchange in some Pacific corals. *Comp. Biochem. Physiol.* **27**, 405–418

Sale, P. F., McWilliam, P. S. and Anderson, D. T. (1976). Composition of the near-reef zooplankton at Heron Reef, Great Barrier Reef. *Mar. Biol.* **34**, 59–66

Sargent, M. C. and Austin, T. S. (1949). Organic productivity of an atoll. *Trans. Amer. Geophys. Union* **30**, 245–249

Sargent, M. C. and Austin, T. S. (1954). Biologic economy of coral reefs. *United States Geological Survey Professional Papers 260-E*, 293–300

Simkiss, K. (1964a). Possible effects of zooxanthellae on coral growth. *Experientia* **20**, 140

Simkiss, K. (1964b). Phosphates as crystal poisons of calcification. *Biol. Rev.* **39**, 487–505

Sloterdijk, H. H. (1975). The importance of zooxanthellae for the nitrogenous excretion of some hermatypic corals. M.Sc. Thesis, McGill University

Smith, S. V. (1974). Coral reef carbon dioxide flux. *Proc. Second Int. Symp. on Coral Reefs* 1, Australia, 22 June–2 July 1973 (A. M. Cameron, B. M. Campbell, A. B. Cribb, R. Endean, J. S. Jell, O. A. Jones, P. Mather and F. H. Talbot, eds) pp. 77–85. The Great Barrier Reef Committee, Brisbane

Smith, S. V. and Marsh, J. A. (1973). Organic carbon production and consumption on the windward reef flat at Eniwetok Atoll. *Limnol. and Oceanogr.* **18**, 953–961

Sorokin, Y. I. (1971a). On the role of bacteria in the productivity of tropical ocean waters. *Int. Revue ges Hydrobiol.* **56**, 1–48

Sorokin, Y. I. (1971b). Bacterial populations as components of oceanic ecosystems. *Mar. Biol.* **11**, 101–105

Sorokin, Y. I. (1973a). On the feeding of some scleractinian corals with bacteria and dissolved organic matter. *Limnol. and Oceanogr.* **18**, 380–385

Sorokin, Y. I. (1973b). Trophical role of bacteria in the ecosystem of the coral reef. *Nature, London* **242**, 415–417

Sorokin, Y. I. (1973c). Microbial aspects of the productivity of coral reefs. In "Biology and Geology of Coral Reefs" (O. A. Jones and R. Endean, eds) Vol. II, Biology 1, pp. 17–45. Academic Press, New York and London

Sorokin, Y. I. (1974). Bacteria as a component of the coral reef community. *Proc. Second Int. Symp. Coral Reefs* 1, Australia, 22 June–2 July 1973 (A. M. Cameron, B. M. Campbell, A. B. Cribb, R. Endean, J. S. Jell, O. A. Jones, P. Mather and F. H. Talbot, eds) pp. 3–10. The Great Barrier Reef Committee, Brisbane

Sorokin, Y. I. (1978). Microbial production in the coral-reef community. *Arch. Hydrobiol.* **83**, 281–323

Sournia, A. (1976). Oxygen metabolism of a fringing reef in French Polynesia. *Helgo. wissen. Meeres.* **28**, 401–410

Sournia, A. (1977). Analyse et bilan de la production primaire dans les récifs coralliens. *Ann. l'Inst. Océanogr., Paris* **53**, 47–74

Sournia, A. and Ricard, M. (1975). Phytoplankton and primary productivity in Takapato Atoll, Tuamato Islands. *Micronesica* **11**, 159–166

Sournia, A. and Ricard, M. (1976). Phytoplankton and its contribution to primary productivity in two coral reef areas of French Polynesia. *J. Exp. mar. Biol. Ecol.* **21**, 129–140

Stearn, C. W. and Scoffin, T. P. (1977). Carbonate budget of a fringing reef, Barbados. *Proc. Third Int. Coral Reef Symp.* 2 (D. L. Taylor, ed.) pp. 471–476. Rosenstiel School of Marine and Atmospheric Science, Miami

Stearn, C. W., Scoffin, T. P. and Martindale, W. (1977). Calcium carbonate budget of a fringing reef on the west coast of Barbados. Part I. Zonation and productivity. *Bull. mar. Sci.* **27**, 479–510

Steemann-Nielsen, E. (1954). On organic production in the ocean. *J. du Cons. Int. pour l'Explor. de la Mer* **39**, 309–328

Stephens, G. C. (1962). Uptake of organic material by aquatic invertebrates. I. Uptake of glucose by the solitary coral, *Fungia scutaria. Biol. Bull.* **123**, 648–659

Steven, D. M. (1963). The dermal light sense. *Biol. Rev.* **38**, 204–240

Stevenson, D. K. and Marshall, N. (1974). Generalizations on the fisheries potential of coral reefs and adjacent shallow-water environments. *Proc. Second Int. Symp. Coral Reefs* 1, Australia, 22 June–2 July 1973 (A. M. Cameron, B. M. Campbell, A. B. Cribb, R. Endean, J. S. Jell, O. A. Jones, P. Mather and F. H. Talbot, eds) pp. 147–156. The Great Barrier Reef Committee, Brisbane

Talbot, F. H. (1965). A description of the coral structure of Tutia Reef (Tanganyika Territory, East Africa), and its fish fauna. *Proc. Zool. Soc. Lond.* **145**, 431–470

Talbot, F. H. and Goldman, B. (1972). A preliminary report on the diversity and feeding relationships of reef fishes of One Tree Island, Great Barrier Reef System. *Proc. Symp. Corals and Coral Reefs*, Mandapam Camp, India, 12–16 January 1969 (C. Mukudan and C. S. P. Pillai, eds) pp. 425–443. Marine Biological Association of India, Cochin

Thomassin, B. A. (1974). Soft bottom carcinological fauna *sensu lato* on Tuléar coral reef complexes (S. W. Madagascar): Distribution importance, roles played in trophic food-chains and in bottom deposits. *Proc. Second Int. Symp. Coral Reefs* 1, Australia, 22 June–2 July 1973 (A. M. Cameron, B. M. Campbell, A. B. Cribb, R. Endean, J. S. Jell, O. A. Jones, P. Mather and F. H. Talbot, eds) pp. 297–320. The Great Barrier Reef Committee, Brisbane

Tranter, D. J. and George, J. (1972). Zooplankton abundance at Kavaratti and Kalpeni Atolls in the Laccadives. *Proc. Symp. Corals and Coral Reefs*, Mandapam Camp, India, 12–16 January 1969 (C. Mukudan and C. S. P. Pillai, eds) pp. 239–256. Marine Biological Association of India, Cochin

Vandermeulen, J. H., Davis, N. and Muscatine, L. (1972). The effects of inhibitors of photosynthesis on zooxanthellae in corals and other marine invertebrates. *Mar. Biol.* **16**, 185–191

Vandermeulen, J. H. and Muscatine, L. (1974). Influence of symbiotic algae on calcification in reef corals: Critique and Progress Report, 1–19. *In* "Symbiosis in the Sea" (W. B. Vernberg, ed.). University of South Carolina Press, Columbia, South Carolina

von Holt, C. and von Holt, M. (1968). Transfer of photosynthetic products from zooxanthellae to coelenterate hosts. *Comp. Biochem. Physiol.* **24**, 73–81

Vittor, B. A. and Johnson, P. G. (1977). Polychaete abundance, diversity and trophic role in coral reef communities at Grand Bahama Island and the Florida Middle Ground. *Proc. Third Coral Reef Symp.* 1 (D. L. Taylor, ed.) pp. 163–168. Rosenstiel School of Marine and Atmospheric Science, Miami

Wanders, J. B. W. (1976a). The role of benthic algae in the shallow reef of Curaçao (Netherlands Antilles) I. Primary productivity in the coral reef. *Aqua. Bot.* **2**, 235–270

Wanders, J. B. W. (1976b). The role of benthic algae in the shallow reef of Curaçao (Netherlands Antilles) II. Primary productivity of the *Sargassum* beds on the north-east coast submarine plateau. *Aqua. Bot.* **2**, 327–335

Wangersky, P. J. (1978). Production of dissolved organic matter. *In* "Marine Ecology" (O. Kine, ed.) Vol. IV, Dynamics, Chapt. 4, pp. 115–220. John Wiley and Sons, New York.

Waterman, T. H. (1961). Light sensitivity and vision. *In* "The Physiology of Crustacea" (T. H. Waterman, ed.) Vol. II, pp. 1–64. Academic Press, New York

Webb, K. L., Du Paul, W. D. and D'Elia, C. F. (1977). Biomass and nutrient flux measurements on *Holothuria atra* populations on windward reef flats at Eniwetok, Marshall Islands. *Proc. Third Int. Coral Reef Symp.* 1 (D. L. Taylor, ed.) pp. 409–415. Rosenstiel School of Marine and Atmospheric Science, Miami

Westlake, D. F. (1963). Comparisons of plant productivity. *Biol. Rev.* **38**, 385–425

Wiebe, W. J., Johannes, R. E. and Webb, K. L. (1975). Nitrogen fixation in a coral reef community. *Science* **188**, 257–259

Winberg, G. G. (ed.) (1971). Symbols, units and conversion factors in studies of fresh water productivity. *Publ. Int. Biol. Prog.* Central Office, London, England

Yonge, C. M. (1963). The biology of coral reefs. *Adv. Mar. Biol.* **1**, 209–260

Zieman, J. C. (1968). A study of the growth and decomposition of the sea-grass, *Thalassia testudinum*. M.Sc. Thesis, University of Miami

6 Shelf-Sea Ecosystems

JOHN J. WALSH

1 Introduction

Several years ago I climbed overboard into the clear waters of Haiti, and after a copper helmet had been lowered over my head and shoulders I slid slowly down a rope two, four, eight, ten fathoms and finally at sixty-three feet my canvas shoes settled into the soft ooze near a coral reef. I made my way to a steep precipice, balanced on the brink, and looked down, down into the green depths where illumination like moonlight showed waving sea-fans and milling fish far beyond the length of my hose. As I peered down I realized I was looking toward a world of life almost as unknown as that of Mars or Venus ... Modern oceanographic knowledge ... is comparable to the information of a student of African animals, who has trapped a small collection of rats and mice but is still wholly unaware of antelope, elephants, lions, and rhinos. The hundreds of nets I have drawn through the depths of the sea, from one-half to two miles down, have yielded a harvest which has served only to increase my desire actually to descend into this no-man's zone.

Beebe, 1934

Fifty years ago, Beebe's direct observations of tropical continental shelves and Bigelow's more traditional shipboard analyses of temperate shelves marked the beginning of modern studies of this ecosystem (Allee, 1934). Today these regions, which comprise 10% of the area of the world's ocean, yield 99% of the global fish catch. The physical extent of the present shelf ecosystem is about 75 km wide reaching 130 m depth at the shelf-break (Shepard, 1963), and has a most recent origin of about 15 000 years (Milliman and Emery, 1968; Edwards and Merrill, 1977). Glacial retreat and rising sea level are the major factors in the submergence of modern shelves (Shepard, 1936). During each of the four recent ice ages of the Pleistocene period (1.8×10^6 yr B.P.) the continental slope each time may have adjoined the coastline, without much of an intervening shelf which was then dry land (Hay and Southam, 1977).

The probable absence of most continental shelves during the 1×10^5 yr of the last Wisconsin (Würm) glaciation thus raises doubts about the evolutionary uniqueness or greater intrinsic ecological efficiency of the species of the food web of this rich coastal ecosystem. Because marine speciation takes a minimum of about 5×10^5 yr (Day, 1963), some epipelagic and benthic species of the Pleistocene coastal zone, i.e. the upper 200 m of slope waters,

159

probably just radiated inshore during the last Holocene reappearance of the shelves. The high yield of present shelves is instead attributable to their shallow depths, which concentrate fish for economic harvesting and allow greater nutrient recycling with the overlying water column. Seasonal wind mixing is able to reach the decomposing organic matter on the shelf bottom, with consequent rapid return of nutrients to the euphotic zone. In contrast, the widely dispersed fish of the open ocean, except for tuna, have yet to be commercially exploited and the deep permanent thermocline inhibits rapid return of nutrients to the surface waters, which only occurs during winter overturn. As a result, both the nutrient content and daily primary production of shelf waters are usually an order of magnitude higher than surface waters of the open ocean (Table I), e.g. the Sargasso Sea (Menzel and Ryther, 1960). In some oceanic areas such as the North Pacific (Anderson and Munson, 1972) and the equatorial divergence (Walsh, 1976) where nutrients are high, the phytoplankton stock is always low because of constant grazing stress, and so no spring bloom occurs.

The yield of individual continental shelves (Table I) varies as a function of: their width; the duration of wind induced nutrient input and available light (Walsh, 1974); the seasonal temperature cycle which may affect the ability of planktonic herbivores to crop the primary production; the subsequent input of food to the benthos; and the utilisation of these pelagic and demersal resources by the fish, through a different number of steps in each shelf food web. Before overfishing, for example, the Peru upwelling ecosystem yielded about 20% of the world's annual fish catch as a single species yield – anchoveta (Walsh et al., 1980). At Peruvian latitudes, light and temperature are conducive to high rates of primary production throughout the year. Upwelling adds nutrients throughout the year to the euphotic zone off Peru to maintain a level of carbon fixation sufficient to support annual pelagic fish landings that used to be two orders of magnitude larger, per unit area, than the present Bering Sea demersal fishery.

In contrast to Peru, the Bering Sea has less annual primary production (McRoy and Goering, 1976) because ice reduces incident radiation by covering most of the shelf until March each year; when the spring bloom of phytoplankton begins, water temperatures are 10 °C lower, and surface nutrients become exhausted by the end of summer as wind mixing declines. The yield of Alaska pollock (Pruter, 1973) from the Bering Sea is about 3% of the world fish catch, and the adult fish are restricted to the outer shelf (150–200 m). If the annual pollock harvest is derived only from this area of the total shelf (10–20%), however, then the fish yield km^{-2} of the Bering Sea is similar to that of other mid-latitude shelves, e.g. New York, Oregon, or the North Sea.

TABLE I A comparison of habitat variability and food chain productivity within shelf ecosystems at ~20°, 40°, and 60° latitude

	Peru (10–15°S)	Oregon (42–47°N)	New York (37–42°N)	Bering Sea (55–60°N)
Temperature Δt of the inshore mixed layer	5°C yr^{-1}	5°C yr^{-1}	20°C yr^{-1}	10°C yr^{-1}
Cumulative nitrate within the euphotic zone (30 m)	7.0 g atom m^{-3}yr^{-1}[a]	2.5 g atom m^{-3}yr^{-1}	1.0 g atom m^{-3}yr^{-1}	2.5 g atom m^{-3}yr^{-1}[b]
Primary production of continental shelf	>1000 gCm^{-2}yr^{-1}	200–250 gCm^{-2}yr^{-1}	200–300 gCm^{-2}yr^{-1}	200–250 gCm^{-2}yr^{-1}
Organic carbon content of shelf sediments	>2.0%	<0.5%	<0.5% to 1%	<0.5% to 1.5%
Fish yield of the shelf ecosystem	~100 tons km^{-2}yr^{-1}	~10 tons km^{-2}yr^{-1}	~10 tons km^{-2}yr^{-1}	~1–2 tons km^{-2}yr^{-1}

[a] Sargasso Sea cumulative nitrate = 0.1 g atom m^{-3}yr^{-1}
[b] Station 'P' cumulative nitrate = 3.5 g atom m^{-3}yr^{-1}

The annual primary production and the carbon content of the sediments (Sharma, 1974; Gross, 1968; Milliman, 1973) do not vary between these mid-latitude shelves (Table I). Similar carbon inputs and outputs (as fish and sediment carbon) suggest that the greater width of the Bering Sea shelf (~ 500 km), in contrast to New York (~ 100 km) and Oregon (~ 25 km), may allow segregation of the benthic and pelagic food webs between the inner and outer parts of the high latitude shelf. With low incident radiation, the inshore (< 100 m) benthic secondary production appears to be transferred to long-lived mammals and large invertebrates in the Bering Sea rather than to demersal fish as at lower latitudes. A detailed comparison of two shelf ecosystems at about the same latitude, Oregon and New York–Georges Bank, may thus provide insight into how wind forcing and shelf width can structure the physical habitat, leading to different coupling between similar species in continental shelf systems having similar light regimes.

2   Environmental variables

WIND FORCING

Wind events are an important source of habitat variability on the continental shelf in contrast to the open ocean (Walsh, 1976; Beardsley et al., 1976) and are responsible for both the generation of currents and for vertical mixing. Because of the alignment of the North American coasts, a southerly wind tends to favor offshore surface flow as a result of Coriolis force on the east coast of the United States (i.e. to the right) in contrast to a northerly wind on the west coast. Nutrient rich, cold subsurface water moves onshore and upwells at the coast to replace the warmer, nutrient impoverished surface water transported offshore by winds favorable to upwelling (Walsh, 1975). Downwelling (onshore flow of surface water and sinking at the coast), occurs when winds are from the respective opposite directions off New York and Oregon. Coastal upwelling is a boundary process and most of the water is upwelled within a zone only 10–20 km from the coast, with offshore secondary cross-shelf flows set up as a function of the shelf width (Walsh, 1977). Finally, under conditions of weak stratification and strong winds, vertical mixing of the water column also occurs in addition to the upwelling/downwelling cross-shelf circulation pattern (Walsh et al., 1978).

A progressive vector diagram (Fig. 1) shows wind forcing (in km) for a year at the Ambrose Light Tower off New York, and at the Newport jetty off Oregon. This analysis was compiled by plotting, head to tail, the successive vectors of wind speed and direction, every 3 h throughout April 1974–March 1975. The average offshore wind speed between Cape Hatteras and Rhode Island rises from 5 m s^{-1} in July to 8 m s^{-1} in January, though

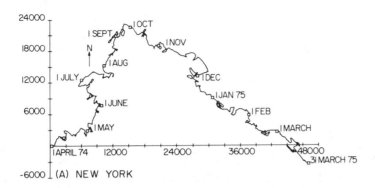

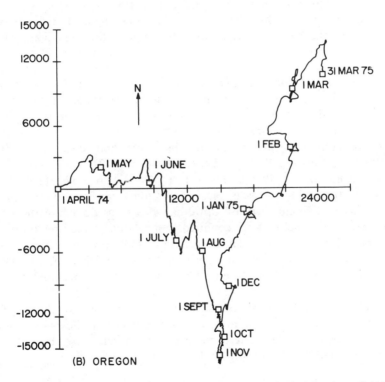

FIG. 1   Progressive vector diagram (km) of the seasonal wind forcing off (A) New York and (B) Oregon

winds of > 20 m s^{-1} have been recorded during each month over the last 50 years at New York City (Lettau *et al.*, 1976). Spectral analysis (O'Brien and Pillsbury, 1974) of these two sets of wind data indicates that the dominant frequencies of variability of the seasonal wind forcing within the two ecosystems are similar. Wind events, or storms, are more frequent in the winter, for example, and decline within the New York Bight from 5–6 in January to half that number by September (Walsh *et al.*, 1978). The cross-shelf advective input of nutrients (Table I) also involves about the same period of 4–5 months of favorable spring-summer upwelling winds off both New York and Oregon (Fig. 1).

At a level of ecological complexity corresponding to trophic levels, the two shelves appear to have the same biological structure and productivity despite their differences in width. Their annual primary production is of the same order of magnitude, with a 5 year mean of 193 gCm^{-2}yr^{-1} at the edge of the shelf off Oregon (Small *et al.*, 1972), in contrast to a 5 year estimate of 200–300 gCm^{-2}yr^{-1} within the whole New York Bight. The diversity of nearshore zooplankton species is also the same within 18–25 km of the coast; using a similar size of net mesh, 29 and 26 species of copepods respectively were found to be abundant in the New York Bight (Judkins *et al.*, 1980) and off Oregon (Peterson and Miller, 1975). The abundance of estuarine-dependent fish is similar (McHugh, 1976), constituting 44.1% by weight of the US commercial catch in the mid-Atlantic Bight, in contrast to 45.2% off Oregon-Washington, both in 1970. Finally, the number of common fish species on the slope and shelf is the same (~ 200) off Oregon (Pearcy, 1972; Alton, 1972) and New York (Hennemeuth, 1976), while the annual fish yield is also similar (Table I).

However, the timing of species succession and energy flow off Oregon and New York are different, because upwelling processes dominate the whole narrow west coast shelf but are confined to a relatively small region on the broad shelf off the east coast. The upwelling and downwelling responses are mainly confined to waters < 30 m depth on the New York shelf, thus within 10–20 km of the coast (Scott and Csanady, 1976). A residual flow of 5 cm s^{-1} to the south is present at most times of year deeper on the New York shelf due to a longshore pressure gradient and thermohaline forcing. Similarly, off Oregon, the equatorward California current flows throughout the year in deeper waters but is replaced on the shelf by the poleward Davidson current during the winter downwelling situation.

TEMPERATURE

Consequently, the annual temperature cycle at the 60 m isobath, 60 km off New York (Fig. 2A) is that of a typical boreal shelf; 10 km off Oregon, at the 60 m isobath, it is that of a summer upwelling regime (Fig. 2B). The

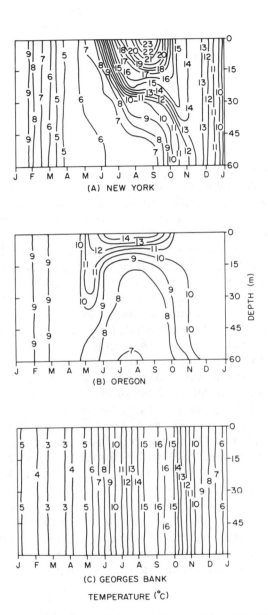

FIG. 2   Annual cycle of temperature (°C) at the 60 m isobath off (A) New York, (B) Oregon, and (C) Georges Bank

vernal heating cycle, river runoff, and a decline in wind events lead to stratification of surface waters by August in the New York Bight with an annual temperature range of $\sim 20°C$ (Table I) in the mid-shelf region. Because of the persistent westerly wind component in both ecosystems, temperature of the well-mixed water on the shelf in winter is lowered by wind off the land on the east coast and raised by wind off the ocean on the west coast. Higher winter temperatures, and the summer upwelling over most of the Oregon shelf together lead to an annual temperature range at the mid-shelf of only $\sim 5\,°C$ (Table I).

TIDAL MIXING

On some shelves, tidal motion is an important factor in vertical mixing in addition to wind. Large tidal velocities of 55 cm s^{-1} maximum amplitude, for example, can lead to roughly the same vertical mixing as a 13 m s^{-1} wind (Pingree et al., 1978); tidal mixing energy is applied from below as opposed to the wind energy from above. Maximal tidal velocities at > 30 m depth are ~ 15–25 cm s^{-1} in the New York Bight and ~ 55–110 cm s^{-1} on Georges Bank. The latter velocities are similar to those of 1 m s^{-1} around the British Isles (Simpson and Pingree, 1978), where an index of the tidal mixing (Simpson and Hunter, 1974) has been formulated as the ratio of depth (h) to the cube of the amplitude (u) of the tidal stream, i.e. hu^{-3}. The reciprocal of this ratio and a constant drag coefficient, $Ch^{-1}u^3$, is the mean tidal energy dissipation rate per unit mass (Pingree et al., 1978). Because of the large range of values in both parameters, a log scale is used to estimate areas of stratification ($\log hu^{-3} \geqslant 2$ or $\log h^{-1}u^3 \leqslant -2$), transitional areas or fronts (1.5 or -1.5), and tidally well-mixed areas ($\leqslant 1$ or $\geqslant -1$). Larger tidal velocities and shallower depths favor increased tidal mixing.

At depths of 40–60 m in the New York Bight the tidal velocity of 0.25 m s^{-1} suggest values of 3.41–3.59 for $\log hu^{-3}$, and stratification of the water column. A tidal velocity of 1.10 m s^{-1} at similar depths on Georges Bank, however, suggest values of 1.48–1.65 and an area of tidal mixing. The seasonal temperature cycle at 60 m on Georges Bank does, in fact, reflect tidal mixing with an isothermal vertical structure at all times (Fig. 2C). Because of tidal and wind mixing on Georges Bank, the winter temperature minimum is about the same as that of the New York shelf but the summer temperature maximum is similar to that off Oregon.

LIGHT

Because light penetration decays exponentially with depth in the ocean, phytoplankton which are mixed over the whole water column on Georges Bank experience less light than those in surface layers of stratified deeper waters (Riley, 1942). Thus, less chlorophyll is found in the highly mixed shallow regions of Georges Bank despite high nutrient concentrations at the beginning of the spring bloom. Similarly on the European shelf, higher chlorophyll concentrations occur on the stratified side of a shelf front rather than on the tidally mixed side (Pingree et al., 1978), though higher concentrations occur at the front between two such regions.

The relation between vertical mixing, light intensity, and phytoplankton growth was quantified as a critical depth concept (Sverdrup, 1953), below which the 24 h respiration of the water column exceeds the integrated daily photosynthesis. This critical depth is $h_c \simeq 0.2\ I_o\ (kI_c)^{-1}$, where I_o is the incident radiation, k is the extinction coefficient, and I_c is the compensation light intensity at which algal photosynthesis equals respiration (~ 0.3 ly h^{-1}). Sverdrup's concept is that if $h_c < h$, the depth to which the phytoplankton are mixed as a result of wind and/or tidal stirring, no bloom will occur even in the presence of high nutrient content.

In turbid, seasonally well-mixed shelf waters with chlorophyll concentrations of ~ 2.0 µgchl l^{-1} (Fig. 3), Secchi disk depth is ~ 11 m at the 60 m isobath in the New York Bight. Empirically, k is related to either the Secchi depth, h_s, by $k = 1.44\ h_s^{-1}$ (Holmes, 1970), i.e. $1.44\ (11)^{-1} = 0.131$ m^{-1}, or to chlorophyll content by $k = 0.04 + 0.0088\ chl + 0.054\ chl^{2/3}$ (Riley, 1956), i.e. 0.143 m^{-1}. If the average illumination of a mixed water column $\bar{I}_h = I_o$ $(kh)^{-1}$, increases to an intensity at which gross photoplankton production is greater than loss rates from respiration, grazing, and sinking, then a bloom will ensue. Riley (1967) and Hitchcock and Smayda (1977) found that above an in situ threshold of $\bar{I}_h = 40$ gcal cm^{-2}day^{-1} (ly day^{-1}) phytoplankton blooms occurred in Long Island Sound and Narragansett Bay.

At the beginning of March off New York and Oregon at the 60 m isobath, the incident radiation is <250 gcal cm^{-2}day^{-1} and $\bar{I}_h$ is thus 32 gcal cm^{-2}day^{-1} within the well mixed 60 m water column; the critical depth is also too shallow at 53 m. However, incident illumination increases to over 300 gcal cm^{-2}day^{-1} by the end of March off New York (Walsh et al., 1978) and Oregon (Small et al., 1972), with associated increases in critical depth and average in situ light intensity. Similar changes in spring light intensity and critical depth have been calculated for the Irish Sea, with mean I_o and h_c increasing from 250 gcal cm^{-2}day^{-1} and 60 m in mid-March to 430 gcal cm^{-2}day^{-1} and 100 m in mid-April (Pingree et al., 1976).

3 Pelagic food web

PHYTOPLANKTON

Early in the year, the isothermal March water column contains more than 4–5 µg atoms NO$_3$l^{-1} off both New York (Walsh et al., 1978) and Oregon

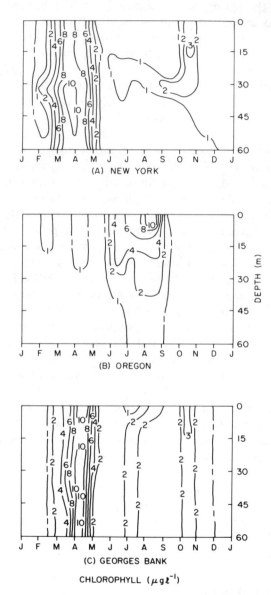

FIG. 3   Annual cycle of chlorophyll (μg l^{-1}) at the 60 m isobath off (A) New York, (B) Oregon, and (C) Georges Bank

(Stefansson and Richards, 1964), as well as around the British Isles (Pingree *et al.*, 1976). Having an appropriate combination of light and nutrients, the spring bloom thus begins at the 60 m isobath in the New York Bight and on Georges Bank (Fig. 3) before the onset of seasonal stratification in the former, and despite tidal mixing in the latter (Fig. 2). Bloom conditions also occur after a surface light intensity of 300 gcal cm^{-2}day^{-1} in the Sargasso Sea (Steele and Menzel, 1962). Sporadic nearshore blooms of phytoplankton occur at the 18 m isobath, within 2 km of the Oregon coast, from March to June (Petersen and Miller, 1977); up to 6 µgchl l^{-1} are found even earlier at the 30 m isobath off New York in December, but these populations are relicts of the fall bloom. In contrast, mid-shelf chlorophyll concentrations, similar to the east coast spring bloom, are not found at the 60 m isobath off Oregon until July–August during the middle of the upwelling season (Fig. 3).

During summer stratified conditions in the New York Bight, nitrate is undetectable in the euphotic zone, while only 0.1–0.2 µg atoms NO$_3$l^{-1} occur in tidally mixed areas of Georges Bank and the English Channel. High chlorophyll concentrations (4–5 µgchl l^{-1}) are then found only in subsurface maxima within the 10 km nearshore upwelling zone of the New York Bight, and on the surface, at the edge of fronts, on Georges Bank and in the English Channel. After the fall overturn in October, there is more than 2.5 µg atoms NO$_3$l^{-1} throughout the water column off New York, Georges Bank, the English Channel, and Oregon. This nitrogen input is partially consumed by a small fall bloom in New York, the English Channel, and Georges Bank, but again not off Oregon. By October, incident light is less than 300 gcal cm^{-2}day^{-1} in all four areas, with the critical depths < 50 m: the decline in both chlorophyll and light then continues towards their winter minima.

Downwelling begins off Oregon in October and ends in May (Fig. 1), with an intervening upwelling season of varying inter-annual intensity (Petersen and Miller, 1975). Despite sufficient light and nutrients off Oregon in March and October, a bloom is not observed perhaps because the sinking water entrains phytoplankton onshore and below the euphotic zone. For example, the zooplankton species at these times suggests that the water on the shelf has come from the south as part of the downwelling circulation (Petersen and Miller, 1977). Similarly during the upwelling season off Oregon, cross-shelf distributions of phytoplankton and suspended material (Small and Ramberg, 1971) also suggest sinking within a convergence front at mid-shelf (Mooers *et al.*, 1976). Downwelling has been observed during 'northeasters' (March, July and October storms) in the New York Bight, but phytoplankton loss is restricted to the inner part of this wider shelf. At mid-shelf depth, chlorophyll appears to be resuspended from the sediments particularly during strong southwesterly winter-spring winds in the New York Bight (Walsh et al., 1978).

NITROGEN DEMAND

Unlike Oregon, however, nutrients limit phytoplankton population growth over most of the mid-Atlantic Bight during summer, when nitrogen demand must be supplied by recycling by zooplankton, bacterioplankton, and benthos (Walsh et al., 1978). A phytoplankton C:N ratio of 5:1 and an annual primary production of 300 $gCm^{-2}yr^{-1}$ suggests an annual nitrogen demand of 60 $gNm^{-2}yr^{-1}$, or ~ 71 µg atoms $Nl^{-1}yr^{-1}$ at the 60 m isobath, part of which is supplied by biological regeneration of nitrogen as ammonium and part by physical input of nitrate.

In a revised carbon/nitrogen budget for the shelf between Cape Hatteras and Georges Bank (the mid-Atlantic Bight), the estimated sources of nitrogen are now 11.0 µg atoms $Nl^{-1}yr^{-1}$ from zooplankton excretion, 7.0 from bacterioplankton, 11.0 from benthos, 3.0 from nitrification, 0.5 from rainfall, 3.0 from estuarine input, 9.0 from the upstream boundary across the shelf to the north, and 26.0 from the longshore boundary at the edge of the shelf. Approximately 46% of the nitrogen demand of the annual primary production of the New York Bight may be supplied by recycling, which suggest that the pelagic food web of this shelf ecosystem may be more rightly coupled in a biological sense to the demersal food web than those of the upwelling system off Oregon.

COPEPODS

The seasonal pulses of copepod biomass off New York, Georges Bank, and Oregon (Fig. 4) reflect regional differences in both temperature cycles and the timing of phytoplankton blooms. For example, *Pseudocalanus minutus* and *Centropages typicus* are the dominant copepods off New York and Georges Bank (Judkins et al., 1980, Sherman, 1978; Grice and Hart, 1962; Clark, 1940; Sears and Clarke, 1940; Bigelow and Sears, 1939) with the former found across the shelf in the winter-spring and the latter inshore during summer-fall; *Calanus finmarchicus* is also abundant offshore after the spring bloom. The New York offshore peak of these zooplankters (> 50 m) occurs during May–June, as offshore waters warm about one month after the offshore phytoplankton bloom (> 50 m) in March–April. Inshore zooplankton (< 50 m) becomes abundant during the even warmer period of July–August, following by one month the inshore phytoplankton growth period (< 50 m) of May–June.

The peaks of zooplankton biomass on Georges Bank and in the Gulf of Maine (Fig. 4C) have similar spatial and seasonal patterns (Bigelow, 1926; Redfield, 1941; Riley, 1947) to the respective inshore and offshore copepod

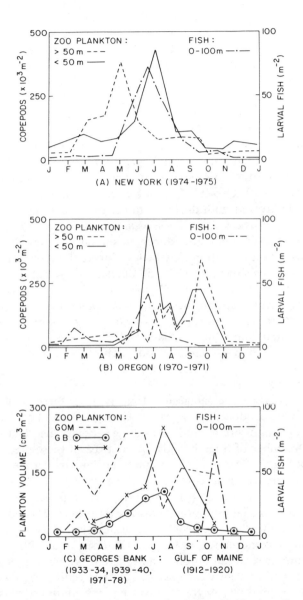

Fig. 4 Annual cycle of copepods on the inner (<50 m) and outer (>50 m) continental shelf and of larval fish over the upper 100 m off (A) New York, (B) Oregon and (C) Georges Bank: Gulf of Maine

communities within the New York Bight (Fig. 4A). The cross-shelf distribution of the same genera of copepods is also similar off Oregon: *Pseudocalanus* sp. is found across the shelf, *Calanus marshallae* offshore and *Centropages abdominalis* inshore (Fig. 4B) during the summer upwelling period (Petersen *et al.*, 1979). However, only *Pseudocalanus* sp. dominates both the single offshore (> 50 m) summer peak of zooplankton in July–August, that coincides with the mid-shelf phytoplankton bloom, and the smaller cross-shelf peaks of copepods that occur in the absence of high chlorophyll during October (Petersen and Miller, 1975, 1977).

   Pseudocalanus is a cold water form (Corkett and McLaren, 1978), which grows well at low temperatures (Vidal, 1978) and intermittent food supply (Dagg, 1977), in contrast to *Centropages*. The summer Δt between Oregon and New York is $\sim 10\,°C$, a full Q_{10} range, and *Pseudocalanus* would thus be metabolically favored both during summer upwelling off Oregon and in the colder spring off New York. *Pseudocalanus* females also lay fewer eggs than *Centropages*, yet adult abundance is similar off New York (Dagg, 1979), suggesting that predation may be less during the growth phase of the cold water form. The increase (Fig. 5) of inshore (< 50 m) chaetognath biomass after the spring peak of *Pseudocalanus minutus* may allow the slower growing *Centropages* to succeed *Pseudocalanus* off New York during the summer. Moreover, *Centropages* is also an omnivore, in contrast to *Pseudocalanus*, and may thus partially subsist on its own or other invertebrate nauplii within the summer food web off New York.

INVERTEBRATE PREDATORS

Clarke *et al.* (1943) have suggested that two to three generations of *Sagitta elegans* are produced each year on Georges Bank. There are, in fact, indications of maturation in May–June, September–October, and November–December of three broods, or cohorts, of these predators in all three systems (Fig. 5). The main chaetognath peak is smaller and later in the year off Oregon, however, in contrast to those off New York and Georges Bank (Redfield and Beale, 1940). Similarly, major pulses of ctenophore predators occur in the fall off both California (Hirota, 1974) and New York (Malone, 1977), after the peak of *Centropages typicus*, compared to the summer bloom of ctenophores off Oregon (Petersen and Miller, 1976). The low temperature, lack of fall bloom, and different timing of predator stress may thus all act to prevent *Centropages* from succeeding *Pseudocalanus* as the dominant copepod off Oregon. Conversely, the continued existence of a herbivore off Oregon during summer and fall may not allow a fall bloom to develop in October, just as the lack of a spring bloom in the North Pacific is due to constant grazing pressure of *Calanus cristatus* (McAllister *et al.*, 1960).

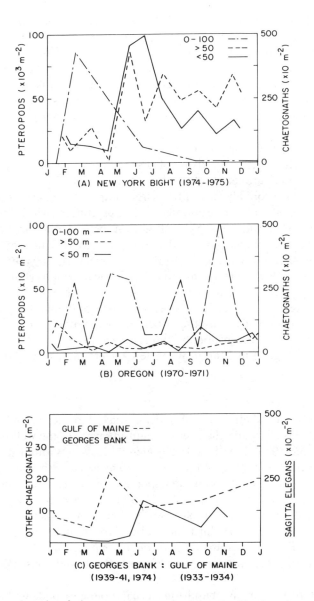

FIG. 5 Annual cycle of chaetognaths on the inner (<50 m) and outer (>50 m) continental shelf and of pteropods over the upper 100 m off (A) New York, (B) Oregon and (C) Georges Bank: Gulf of Maine

ICHTHYOPLANKTON

Seasonal cycles of larval fish (Fig. 4) tend to follow those of the invertebrate herbivores and carnivores. The May–June peak of larval fish in the New York Bight comprises mainly yellowtail flounder and mackerel, while the smaller winter peak on Georges Bank is of cod, haddock, and sand lance following the herring maximum: data were not available for the summer period on Georges Bank but yellowtail and mackerel spawn in this area as well (Colton et al., 1979). Similarly off Oregon, the winter peak of larval fish represents sand lance, butter sole, and english sole, while the summer maximum comprises capelin (Richardson and Pearcy, 1977).

Although the larval survival of fish may be the critical factor in determining fluctuations of adult abundance (Lasker, 1975; Walsh, 1978), icthyoplankton probably have little direct impact on energy flow through lower trophic levels. For example, daily respiration within the summer water column of the New York Bight of peak larval fish populations represents a maximum oxygen demand of only $1.6 \, \mathrm{mlO_2 m^{-2} day^{-1}}$ in contrast to $48 \, \mathrm{mlO_2 m^{-2} day^{-1}}$ for the chaetognaths, $360 \, \mathrm{mlO_2 m^{-2} day^{-1}}$ for the benthos, and $500 \, \mathrm{mlO_2 m^{-2} day^{-1}}$ for the copepods (Falkowski et al., 1980). Furthermore, a nitrogen budget for the New York shelf (Walsh et al., 1978) suggests that 53% of the nearshore phytoplankton biomass in August is consumed daily by the copepods, when predators are abundant, whereas only 6% of the inshore bloom is grazed in May, and 7% of the mid-shelf bloom (Fig. 3) in March, when few chaetognaths and larval fish are present.

PTEROPODS

With an incident radiation of $150 \, \mathrm{gcalcm^{-2} day^{-1}}$ at the end of January, the critical depth is 32 m and the average light intensity of the well-mixed water column at the 30 m isobath is $38.5 \, \mathrm{gcalcm^{-1} day^{-1}}$. Without downwelling, an inshore phytoplankton bloom at the 30 m isobath should occur in the New York Bight after the February increase of incident radiation. In fact, phytoplankton blooms ($\sim 10 \, \mathrm{\mu gchl \, l^{-1}}$) are then observed nearshore (Evans et al., 1979) despite the presence of herbivorous pteropods (mainly Limacina retroversa). These are the third most abundant zooplankters, with a peak occurrence (Fig. 5) before Pseudocalanus minutus in February–March (Judkins et al., 1980) within 60 km of the Long Island coast (Walsh et al., 1978).

These pteropods are as ubiquitous as calanoid copepods within the Mid-Atlantic Bight. Redfield (1939) described the growth of a population of Limacina retroversa during a 7 month drift from its possible winter spawning area on the Nova Scotian shelf until it was entrained southward within the gyre around the Gulf of Maine, and pteropods have also been used to delimit long-shore shelf habitat changes at Cape Hatteras (Myers, 1967). Although abundance within the copepod maxima off Oregon and New York are of the same order (Fig. 4), the Oregon pteropod maximum of *Limacina helicina* is only 2% of that off New York and occurs during October (Fig. 5) at the same time as the *Pseudocalanus* sp. fall peak. Thus downwelling appears to prevent a winter bloom of phytoplankton and zooplankton off Oregon, whereas off New York the February inshore primary production is cropped before mid-March by both pteropods and tintinnids. This is indicated by diatom chain length, fo:fa ratios, and PN:chl ratios (Walsh *et al.*, 1976).

4 Demersal food web

BENTHOS

As a result of different shelf width and similar wind forcing, an inshore zooplankton species succession of *Limacina retroversa*, *Pseudocalanus minutus*, and *Centropages typicus* occurs in the New York Bight, while *Pseudocalanus* sp. remains the dominant shelf herbivore off Oregon. The nearshore winter and summer primary production ($\sim 100 \, \mathrm{gCm^{-2}yr^{-1}}$) is thus transferred to protozoa and zooplankton off New York, while the production of the March–May spring bloom at mid-shelf ($\sim 200 \, \mathrm{gCm^{-2}yr^{-1}}$) is apparently transferred instead to the bottom (Fig. 6). A regression of benthic macrofaunal biomass on average chlorophyll of the March–May euphotic zone, in fact, yields an r^2 of 0.80 for other areas in the Canadian coastal zone, the North Sea, and Long Island Sound (Mann, 1976). In contrast, the lack of a spring bloom on most of the Oregon shelf, suggests that little transfer of carbon to the bottom occurs at this time of year.

In the preliminary carbon budget for Cape Hatteras-Georges Bank already discussed, the utilisation of the primary production by the herbivorous zooplankton (growth efficiency 20% and assimilation efficiency 60%) leads to an annual secondary production of $20 \, \mathrm{gCm^{-2}yr^{-1}}$ available for consumption by the pelagic food web and a fecal pellet flux of $40 \, \mathrm{gCm^{-2}yr^{-1}}$ to the demersal food web (Fig. 6). Flux of fecal pellets and phytodetritus to the bottom ($\sim 240 \, \mathrm{gCm^{-2}yr^{-1}}$) can be partitioned into the larger macrobenthic invertebrates (detritovores, herbivores and omnivores) and the smaller more abundant heterotrophic organisms, the meiofauna and microbiota (Fig. 6). Finally, the food requirements of the pelagic and demersal fish and the yield to man from all commercial living resources, are independent estimates (Edwards and Bowman, 1979; Sherman *et al.*, 1978) at the output side of this carbon budget.

The number of meiobenthic organisms in some areas, south of Martha's Vineyard ($0.1–1.0 \times 10^6$ animals $\mathrm{m^{-2}}$; Wigley and McIntyre, 1964) for example, far outnumber the macrobenthos (1.4×10^3 animals m^{-2}; Wigley and Theroux, 1979). They appear to be responsible, along with the bacteria and protozoa, for remineralisation of material on the bottom rather than serving as a pathway to higher trophic levels (Fenchel, 1969; McIntyre *et al.*, 1970; Coull, 1973; Tenore, 1977). The detritus flux is thus assumed to be either buried, exported, or mineralised by both the meiobenthos and the microbenthos in this carbon budget. It is possible that meiobenthos production is consumed by brittle stars and polychaetes, as well as by other deposit feeders. An inverse spatial distribution has been observed between the meiofauna and macrofauna biomass in the New York Bight, with perhaps as much as 50% of the meiobenthic production consumed by the macrobenthos (J. Tietjen, personal communication). A similar budget of benthic interactions has been constructed for the Baltic Sea (Ankar and Elmgren, 1976).

Some phytodetritus and fecal matter would be used as an energy source by all three size categories of benthos, and several studies have estimated total community metabolism off New York; for example, Thomas *et al.*

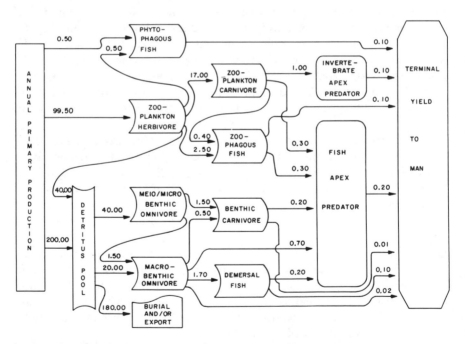

FIG. 6   A carbon budget of both yield to man and a biological sink for atmospheric CO_2 on the north-east continental shelf

(1979) measured nearshore bottom oxygen utilisation by incubating cores aboard ship at *in situ* temperatures. Their results of 360 $mlO_2m^{-2}day^{-1}$ (or 54 $gCm^{-2}yr^{-1}$ with an R.Q. of 0.75) were similar to seasonal oxygen demand of the bottom biota (Smith *et al.*, 1974), within *in situ* bell-jar incubations. Recent core incubations over a broader area of the shelf of the mid-Atlantic Bight (G. T. Rowe, personal communication) suggest levels of infaunal utilisation of organic carbon in different parts of the shelf to be:

Nearshore (< 30 m depth)	70 $gCm^{-2}yr^{-1}$
Central shelf depths	43 $gCm^{-2}yr^{-1}$
Georges' Bank	27 $gCm^{-2}yr^{-1}$
Outershelf margin	20 $gCm^{-2}yr^{-1}$

If an average value for inner shelf respiration measurements of $\sim 50 gCm^{-2}yr^{-1}$ is assumed to be representative of all the mid-Atlantic Bight, only about 25% of the total direct loss (about 240 $gCm^{-2}yr^{-1}$) from the water column can be attributed to utilisation by microfauna, meiofauna, and macrofauna. If this is the case, either the fecal pellets and phytoplankton carbon produced during the winter-spring bloom is buried and exported, or it must be utilised by organisms not considered in these assumptions. Another approach to this enigmatic carbon loss is independently to estimate carbon utilisation from known macrofaunal biomass (Wigley and Theroux, 1979).

Large areas of the mid-Atlantic shelf have a wet weight benthic biomass of more than 100 g m^{-2} ($\sim 6 gCm^{-2}$), most of which is composed of the long-lived macrobenthos with low rates of secondary production (Warwick and Price, 1975; Buchanan and Warwick, 1974). If the average macrofaunal biomass on the shelf is on the order of 100 g m^{-2} and a P/B of 0.5 is assumed, then their production would be about 50 $gm^{-2}yr^{-1}$, or 3 $gCm^{-2}yr^{-1}$. With a growth effiency of 15%, total carbon consumption by the macrofauna would be about 20 $gCm^{-2}yr^{-1}$ (Fig. 6). This is reasonable, because attempts to partition carbon flow through benthic communities in the past have shown that only a small portion is used by larger organisms (Smith *et al.*, 1973; Smith, 1973). If the benthic fluxes of the carbon budget (40 $gCm^{-2}yr^{-1}$ to small organisms and 20 $gCm^{-2}yr^{-1}$ to macrofauna, see Fig. 6) are in fact reasonable estimates, then the benthic community on the shelf of the mid-Atlantic Bight does not consume much of the annual carbon flux to the bottom, and there must be burial or carbon export.

SEDIMENTS

Most of the sediment on the mid-Atlantic shelf is relict sand with < 0.5% carbon (Emery and Uchupi, 1972), however, deposited during or soon after the last Wisconsin glaciation, when the shelf was dry land 15 000 years ago.

Seaward of the 60 m isobath, the sediments contain more silt (Freeland and Swift, 1978), and, beyond the shelf-break, the organic content finally increases to 1–2% carbon, but still lower in organic content than the >4% carbon muds (Table I) off the Peru coast (Rowe, 1971). The Hudson Canyon, areas of Georges Bank (Hathaway et al., 1979), and the 'mud hole,' a region of 4×10^3 km^2 southwest of Martha's Vineyard between the 50–100 m isobaths, are the exceptions where 1–2% carbon muds are found on the mid-Atlantic shelf. In these areas, ^{14}C and ^{210}Pb dating suggest modern deposits with a sedimentation rate of at least 50–100 cm 1000 yr^{-1} (Drake et al., 1978; Bothner et al., 1980) in contrast to about 15 cm 1000 yr^{-1} on the slope (MacIlvaine, 1973) and 3 cm 1000 yr^{-1} on the rise (Milliman, 1973).

The C:N ratio (Fig. 7) of mid-Atlantic shelf sediments (Milliman, 1973) can be used as an index of areas of modern carbon deposition and possibly also of the fate of particles during transport in this ecosystem (Fig. 6). The average C:N ratio of marine bacteria, phytoplankton, copepods, polychaetes, other zooplankton and benthic organisms, as well as fish are all less than 6, while land plants and aquatic vascular plants, e.g. Zostera and Thalassia, are greater than 15 (Müller, 1977). Sediments with C:N content <6 may indicate a marine origin if also rich in carbon (Müller, 1977) and those >10 would indicate a terrestrial or marine vascular plant source. The carbon content of the shelf sediments with a C:N ratio <6 in the Middle Atlantic Bight (Fig. 7) is, in fact, twice those with a >10 C:N ratio, while δ^{13}C measurements (Hunt, 1966) suggest the surface sediments with a C:N <6 south of Martha's Vineyard (Fig. 7) are of marine origin. Similar δ^{13}C measurements for the >10 C:N sediments off New York and New Jersey (Turekian and Benoit, 1980) as well as the N-alkane hydrocarbon fraction of the sediment carbon pool in this region both suggest that this material instead had terrestrial and/or petrochemical origin (Farrington and Tripp, 1977). In contrast off South Carolina, the shelf sediments have a C:N ratio <6 (Emery and Uchupi, 1972) and an N-alkane hydrocarbon fraction that suggests a marine origin (Hathaway et al., 1979).

The C:N ratio of the shelf edge sediments (Fig. 7) thus suggests terrestrial or vascular plant carbon deposition, presumably when relict sands were deposited during both the last Wisconsin and previous glaciations. The depth of Pleistocene sediments on the upper slope is ~ 300 m in contrast to only ~ 100 m on the shelf (Hathaway et al., 1978), i.e. average accumulation rates of respectively 15 cm 1000 yr^{-1} and 5 cm 1000 yr^{-1}. As the river mouths, marshes, and sea grass beds retreated towards the present shore line during the Holocene transgression, sediments of >10 C:N content were continually laid down and are still observed across the shelf in some areas,

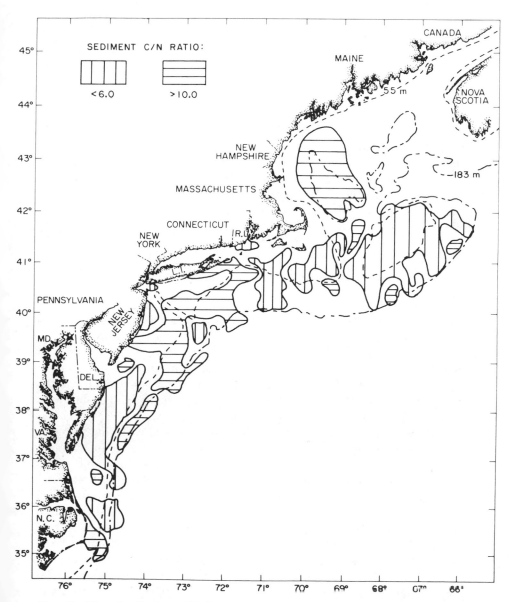

FIG. 7   The distribution of relict (C/N > 10) and modern (C/N < 6) sediments between Cape Hatteras and Georges Bank (modified from Milliman, 1973)

e.g. south of Montauk Point and east of Atlantic City (Fig. 7). In other areas, e.g. Georges Bank (Hathaway et al., 1979), Nantucket Shoals, the 'mud hole', Hudson Cañyon, and at the mouth of estuaries, a surface layer of modern (Holocene) biogenic <6 C:N sediments may now overlay the relict sediments with their >10 C:N ratios. For example, after a sea-level standstill of ~1500 yr at −40 to −45 m of the present coastline (Jansen, 1976), an 'allochthonous' sediment of 15–17 C:N ratio was deposited 8700 B.P. in the northern North Sea (Jansen et al., 1979). This terrestrial or vascular plant material is overlain by a surface layer 1.5 m thick with a C:N ratio between 4 to 10, suggesting a biogenic deposition rate of 17 cm 1000 yr^{-1} in a system where most of the primary production is consumed by the zooplankton (Steele, 1974). In contrast, recent dating of cores on northern Georges Bank (Hathaway et al., 1979) suggested that the deposition rate of diatoms and organic matter in this region may be 363 cm 1000 yr^{-1}.

5 Pelagic-demersal coupling

INPUT

A longshore drift of 5 cm s^{-1} occurs in the Mid-Atlantic Bight, and regions of <6 C:N content in the surface sediments may either be downstream of the areas where primary production is not consumed, or be the sites of deposition near estuaries, as a result of entrainment within subsurface onshore flow of water. During an April spring bloom of ~4 μgchl l^{-1} at the 65 m isobath near 55°N in the Baltic Sea, long chains of diatoms were observed (Smetacek et al., 1978), despite the presence of Pseudocalanus elongatus with few carnivores, thus with low grazing loss as during spring in the New York Bight. The C:N content of particulate matter in sediment traps within this water column varied from ~5 to 10, with lower values mainly observed during the first 18 days of active sedimentation. Because of plankton patchiness and a current of 15 cm s^{-1} within the euphotic zone, these sediment traps may have underestimated the local detrital flux, but an average downward carbon input to the shelf of 0.2 gCm^{-2}day^{-1}, or 15% of the mean primary production, was found above sediments containing 3% carbon. During upwelling at 700 m on the upper slope off California, a similar carbon flux of 0.1 gCm^{-2}day^{-1} was also estimated from sediment traps (Knauer et al., 1979), but a flux of only 0.01 gCm^{-2}day^{-1} occurs in the deep ocean (Rowe and Gardner, 1979). If only 15% of the annual primary production off New York also sinks directly on the shelf, i.e. 45 gCm^{-2}yr^{-1}, this is close to the benthic flux of 60 gCm^{-2}yr^{-1} obtained in the preliminary carbon budget (Fig. 6).

LONGSHORE STRUCTURE

Assuming sinking velocities of $1-10 \text{ m day}^{-1}$ (Smayda, 1970) for live phytoplankton, and a longshore current of 5 cm s^{-1} ($\sim 4 \text{ km day}^{-1}$), a bloom could travel for 6–60 days, over 25–250 km along the 60 m isobath, before sinking to the bottom in the mid-Atlantic Bight. The latter is the downstream distance of the 'mud hole' from Georges Bank (Fig. 8), while the former is close to the separation distance between the 'mud hole' and Nantucket Shoals. Both upstream areas have high chlorophyll in the water column at all times of year in addition to low C:N ratios in the sediments (Fig. 7). Longshore studies in May at the 60 m isobath show that vertical isopleths of chlorophyll do, in fact, slope downstream from the surface towards the bottom, to intersect the sediments at both the Hudson Shelf Valley and the 'mud hole'.

Before the fall overturn in October, however, little chlorophyll remains above the bottom of the water column in the New York Bight. Subsurface nitrate concentrations along a transect at the 65 m isobath instead increase near the 'mud hole' (Fig. 8). The longshore difference in integrated chlorophyll between the eastern tip of Georges Bank and New York Bight (Fig. 8) was $\sim 70 \text{ mgchlm}^{-2}$ in October 1978, or about 40 mg atoms Nm^{-2} (assuming a C:chl of 35:1 and a C:N of 5:1) in contrast to the dissolved nitrate change of 130 mg atoms NO_3m^{-2}. If the difference in particulate nitrogen represented by the phytoplankton in the water column (40 mg atoms Nm^{-2}) is subtracted from that of the dissolved stocks, an estimate of 90 mg atoms NO_3m^{-2} is obtained for nitrate increase by nitrification during the 150 days drift from Georges Bank to New York. In contrast, ammonium concentration at each end of the 65 m isobath transect was 192 mg atoms NH_3m^{-2} in the north and 105 mg atoms NH_3m^{-2} in the south. This ammonium decrease independently suggests that 87 mg atoms NH_3m^{-2} may have been converted to NO_3 during the southwesterly drift along the shelf. This longshore nitrogen budget thus implies that near bottom nitrifying bacteria might have consumed $8.4 \text{ mg NH}_3\text{m}^{-2}\text{day}^{-1}$ before the fall overturn.

Direct observations of a bottom flux of $21 \text{ mg NH}_3\text{m}^{-2}\text{day}^{-1}$ (Rowe et al., 1975), the above estimate of nitrification, and an oxygen consumption of $148 \text{ mgCm}^{-2}\text{day}^{-1}$ (Thomas et al., 1979), suggest a C:N ratio of 5 for summer carbon respiration:nitrogen regeneration in the New York sediments. The areas of sediment C:N content > 10 may thus indicate regions where marine biogenic deposition has not overlain relict sediments on the shelf, that is to say the southern and offshore boundaries of present carbon deposition and decomposition (Fig. 7). Presumably high C:N values do not indicate areas of the shelf where differential nitrogen recycling on the bottom might occur with respect to carbon mineralisation. The implication

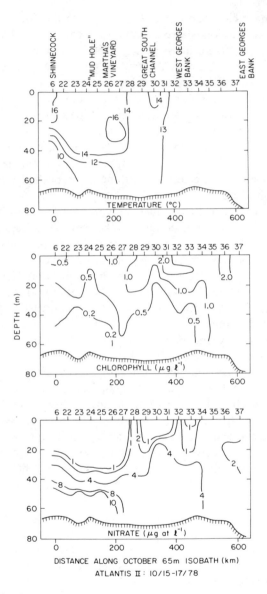

Fɪɢ. 8   The October longshore distribution of temperature (°C), chlorophyll (μg l[-1]), and nitrate (μg atoms l[-1]), at the 65 m isobath between New York and Georges Bank

is that during longshore transport that carbon which sinks to the bottom is assimilated by the benthos with little accumulation in the sediments; in fact, there is a suggestion of a declining downstream gradient in sediment carbon and macrobenthos populations from Georges Bank to Cape Hatteras (Wigley and Theroux, 1979). What then is the fate of the seasonal spring pulses of phytoplankton input to the bottom (Fig. 6), if the 'mud hole' and northern Georges Bank are the only major downstream (offshore) consequences of continuous detrital input from the tidally mixed parts of this shelf ecosystem?

CROSS-SHELF EXCHANGE

Away from the 'mud hole', a composite transect of C:N sediment content from land to the deep ocean basin of the North Atlantic (Fig. 9) suggests two areas of low C:N ratios. The nearshore region is located off the estuaries (Fig. 7) and reflects sinking and onshore entrainment of phytoplankton beneath surface offshore river plumes (Malone and Chervin, 1979). The second region of low C:N values is at 800–900 m, beyond the high C:N zone at the shelf-break (i.e. the presumed area of relict deposition of terrigenous and/or vascular plant material) and represents offshore transport of biogenic material from the shelf to the slope (Schubel and Okubo, 1972; Hay and Southam, 1977). Off North Carolina, half as many meiofauna are found at 400 m as at 800 m (Coull et al., 1977), at which depth the offshore transition from sandy to muddy sediments occurs (Tietjen, 1971). Off Nova Scotia at 800 m, the sediment chlorophyll is 30 times that either on the shelf or on the lower slope (Mills and Fournier, 1979). This organic matter at 800–900 m had either bypassed the natural sedimentary traps at the mouths of estuaries or has originated at mid-shelf; we recall that estimates of particle flux to the bottom of the shelf and the upper slope are similar (Smetacek et al., 1978; Knauer et al., 1979), but accumulation of Pleistocene sediments is greater on the upper slope.

The vertical distribution of particulate and dissolved organic carbon in the sea is, in fact, homogeneous below 200–300 m in the Atlantic and Pacific Oceans (Menzel and Ryther, 1970). The entire biochemical cycle of organic matter, including production, decomposition and solution, thus appears to occur mainly above these depths: beyond the shelves there may be little amount of direct input of carbon to the bottom from the water column above. For example, little change occurs in C:N ratio of the sediments from 1500 to 5000 m (Fig. 9), and this intermediate C:N ratio of the deep sediment reflects the long residence time of open ocean source material and thus slow reworking of nitrogen compounds in the overlying waters (Müller, 1977; Knauer et al., 1979). Any lowering of the C:N ratio of sediments on the upper slope (Fig. 9) would thus require a modern biogenic shelf source and some mechanism for relatively fast lateral seaward transport past the shelf-

As a result of upwelling and low salinity water influx at the coast (Ketchum and Keen, 1955), there is a continual transport of surface water seaward in the New York Bight. Seabed drifters (Bumpus, 1974) and a simple physical model (Csanady, 1976), also suggest a bottom layer flow seabed of the 60 m isobath. Hence, there may be offshore flow at the bottom as well as at the surface on the outer part of the shelf, with a return onshore flow in the middle of the water column to maintain the salt balance. Organic

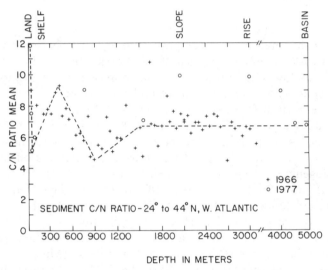

FIG. 9   A composite of sediment C/N ratios on (1) the north-east continental shelf away from the 'mud hole' and (2) from the continental slope to the ocean basin between 24° to 44°N in the western Atlantic

matter tends to be co-transported with fine grain sediment fractions (Gross, 1968). A simple advection–diffusion model of the transport of fine grain inorganic matter across the north-east continental shelf (Schubel and Okubo, 1972) thus provides insight into the origin of low C:N values on the upper slope (Fig. 9). For example, with an offshore advective and diffusion 'velocity' of 2 cm s^{-1} and a fine grain sinking velocity of ~ 1 m day^{-1}, their calculations suggest that any fine grain sediment which had escaped the estuaries would bypass the shelf and be deposited on the upper slope, within 100 km of the shelf edge.

The mid-shelf bloom during March off Long Island occurs between the 60 m isobath, ~ 60 km from shore, and the shelf-break (Walsh *et al.*, 1978). Assuming an offshore flow of 2 cm s^{-1}, an average depth of 250 m and a sinking velocity of 1–10 m day^{-1}, algal cells within the bloom would travel seaward 40–400 km, or 0–360 km past the shelf-break before reaching the bottom. With weaker offshore flow, of 0.5 cm s^{-1}, which is 1/10 of the longshore flow, and a sinking velocity of 2 m day^{-1}, a phytoplankter could sink to the bottom at only 10 km seaward of the shelf-break. In this area, the bottom topography decreases from 200 to 2000 m within ~40 km of the New York shelf, and, in fact, the lowest C:N ratios are found at 800–900 m (Fig. 9), while 2–3% carbon sediments are observed at the head of the Hudson Canyon in waters of 200–300 m depth. Off Oregon, the shelf width is ~1/4 that of New York, and deposits with 2–3% organic carbon occur at 1000–2500 m (Gross, 1968), roughly the same distance (125 km) as off the coast at New York.

In contrast, the Bering Sea shelf is about five times wider than that off New York with a similar weak flow regime (~1–5 cm s^{-1}) which is dominated by tidal mixing (Coachman and Charnell, 1979). The Pleistocene sediments are generally less than 100 m thick on the Bering Sea shelf, with little Quarternary (Pleistocene + Holocene) deposits found nearshore (Nelson et al., 1974). After a sea-level standstill at −40 to −45 m of the present coastline ~10–12 000 B.P. in the Chukchi (Creager and McManus, 1965) and Laptev (Holmes and Creager, 1974) Seas, the Holocene sedimentation rate of nearshore Bering Sea areas, e.g. Norton Sound, has been less than 20 cm 1000 yr^{-1} (Nelson et al., 1974). The C:N content of the surface sediments in the tidally mixed nearshore areas (<50 m) is >10 in contrast to a C:N ratio of <6 for the middle and outer shelf areas (Lisitzin, 1966; R. Iversen, personal communication). Within a 200 km wide area of the middle shelf where Pseudocalanus is the major herbivore, a similar carbon budget for the southeast Bering Sea suggests that 50% of the annual primary production is buried within or exported from this region. Less than 0.5% carbon is found within surface sediments of the middle shelf (Lisitzin, 1966; Sharma, 1974), however, whereas >1% carbon is found ~125 km seaward in the <6 C:N sediments of the outer shelf (Gershanovitch, 1962; R. Iverson, personal communication). Because of a greater shelf width, the carbon export of the Bering Sea appears to be partly deposited on the outer shelf rather than on the upper slope off New York and on the lower slope off Oregon.

6 Consequences

These examples suggest that the ocean's continental shelves, with an average width of 75 km, may represent a major carbon sink from the water column, and source for the export of phytodetritus and fecal pellets to the present adjacent slopes. An important aspect of carbon export to the slope from the

New York Bight, however, is that a mass balance of the elements must presumably be maintained, in an annual steady-state. If CO_2 does not limit plant production in the ocean, the same amount of nitrogen (Ryther and Dunstan, 1971) must be returned each year to the shelf as is lost in the proposed detrital flux, i.e. $\sim 180 \, gCm^{-2}yr^{-1}$ (Fig. 6), or $\sim 36 \, gNm^{-2}yr^{-1}$ assuming a C:N ratio of 5. The fact that our independent nitrogen budget (p. 170) indicated that about $33 \, gNm^{-2}yr^{-1}$ must be supplied from the upstream, offshore, and coast boundaries, suggests that a mass balance of nitrogen is apparently maintained on the New York shelf.

The nitrate-rich pool of upper slope water at the shelf-break is large and seasonally constant at $\sim 9500 \, mg$ atoms Nm^{-2} down to $500 \, m$, the maximum depth of the winter mixed layer (Leetma, 1977). In comparison, the whole shelf water column has a maximum winter nitrate content of $400 \, mg$ atoms Nm^{-2} at the $60 \, m$ isobath, over an order of magnitude less than the offshore source water. Nitrate may thus be supplied to the shelf from intermediate offshore depths, while particulate nitrogen is lost to the slope bottom, with a significant time delay (years to decades; Broecker *et al.*, 1979) between nutrient remineralisation on the bottom of the slope (1000–2000 m) and replenishment of nitrate stocks to the offshore surface layer by vertical mixing or diffusion through the deep thermocline. In this mass balance, carbon could also be removed from the shelf in the form of fixed CO_2, and stored deep on the slope as detrital carbon, forming a CO_2 sink in the ocean without nitrogen limitation (Walsh, 1981).

The flux of CO_2 from past fossil fuel burning, cement production, and deforestation, its present storage pools (in the atmosphere, on land, or in the sea), and the impact of our future use of fossil carbon are poorly understood processes. At the present rate of increase in fuel consumption of $4.3 \% \, yr^{-1}$, a doubling of the atmospheric CO_2 and a 2–3 °C increase in temperature from the 'greenhouse' effect (Chamberlin, 1899) could occur by 2035; present ocean temperatures are, in fact, only ~ 2.3 °C warmer than during the last Wisconsin glaciation (CLIMAP, 1976). Up to 1950, most of the CO_2 emitted from fossil fuel burning and from the use of limestone for cement was thought to have been absorbed in the ocean with little accumulation in the atmosphere (Revelle and Suess, 1957). Although the amount of CO_2 in the atmosphere has steadily increased since the industrial revolution from many sources, the cumulative addition of 70–80×10^9 tons C to this pool is only equivalent to about half of the CO_2 produced by burning fossil fuel, a great deal of which has occurred in just the last two decades (Bolin, 1977). Finally, though land biota were at one time thought to be a sink for CO_2, the terrestrial ecosystem now instead may be a source of CO_2 as a result of deforestation. The rate of carbon flux from the land to the atmosphere or ocean is a subject of some controversey (Woodwell *et al.*, 1978; Ralston, 1979; Broecker *et al.*, 1979), comparable to earlier disagreements about the amount of harvestable carbon from the sea (Ryther, 1969; Alverson et al., 1970).

Without consideration of marine photosynthesis as a net loss term, a global budget for the inorganic storage of CO_2 in the ocean from 1850 to 1950 (Stuvier, 1978) overestimated the initial amount of carbon in the pre-industrial atmosphere by $\sim 50 \times 10^9$ tons (Keeling, 1978), suggesting that at least 0.5×10^9 tons C yr^{-1} might have been lost to a biotic sink. Models of abiotic storage of CO_2 in the ocean by vertical mixing over just the last few decades can account for only 37% of the CO_2 emitted from fossil fuel burning and similarly ignored biological fixation of carbon in the sea; they also cannot account for the missing CO_2 presumably emitted from deforestation (Siegenthaler and Oeschger, 1978). Furthermore, chemical considerations (Bacastow and Keeling, 1973) of the buffer capacity of surface sea water suggest that with an increased CO_2 content leading to more acidic conditions, the upper mixed layer might become an increasingly effective barrier to CO_2 absorption by the deep ocean. Specific considerations of CO_2 emitted from just wood burning (Wong, 1978) suggests, in fact, that almost all of this CO_2 source, i.e. 1.2×10^9 tons C yr^{-1} could be lost in the form of detrial carbon deposition on the continental shelves.

Of the three major inorganic nutrients utilised by marine phytoplankton, nitrogen is most likely to be limiting. Despite its low concentrations in sea water, phosphate recycles very rapidly in the marine environment, and does not appear to limit significantly the productivity of the world's oceans. Silicate recycles slowly, but is not required by all marine phytoplankton; when it is limiting, the primary effect is probably on phytoplankton species composition rather than total primary production.

Current theory suggests that it is the rate of supply of nitrogen to the euphotic zone which limits primary production, with the input rate of nitrate as an estimate of the 'new' production of the system (Chapter 11). Since CO_2 presumably does not limit marine primary production, although it has been observed to stimulate photosynthesis in lakes (Schindler et al., 1972), the extent to which marine food chains may act as a sink for atmospheric CO_2 thus depends upon the rates of nitrogen recycling and the removal of carbon which is *not* associated with respiratory costs of the nitrogen recycling.

From an early CO_2 model of the atmosphere and an ocean without continental shelves (Keeling and Bolin, 1967), approximately 2.7 $\times 10^9$ tons C yr^{-1}, or about 10% of the global marine primary production ($\sim 25 \times 10^9$ tons C yr^{-1}; Woodwell et al., 1978) was estimated to sink to the sediments. Based on a more recent nitrogen budget for 'new' production of offshore waters (Eppley and Petersen, 1980) and again ignoring the coastal zone, the same flux of carbon was assumed to occur in the open ocean. Most of this carbon apparently does not leave the upper 300 m of the sea (Menzel and Ryther, 1970), however, and below 300 m, sediment trap studies of the oceanic detrital carbon flux suggest a loss of 0.005–0.010 gCm^{-2}day^{-1} (Wiebe et al., 1976; Rowe and Gardner, 1979; Knauer et al., 1979); over the whole 2.6×10^8 km^2 of open ocean, this is equivalent to a yearly carbon sink of only 0.5–1.0×10^9 tons C yr^{-1}.

The major organic carbon sink in the global CO$_2$ budget (Keeling and Bolin, 1967) probably occurs instead on the continental shelves. Recent analysis of the Peru upwelling ecosystem (Walsh, 1980) suggests that about 50% of the > 1000 gCm^{-2}yr^{-1} primary production of this shelf is exported to the adjacent continental slope. With a minimum marine primary production of 200 gCm^{-2}yr^{-1} (Table I), a total shelf area of 3×10^{-7} km^2, and a 50% ratio of carbon export to production (Fig. 6), about 6×10^9 tons C yr^{-1} would be fixed and 3×10^9 tons C yr^{-1} lost to the slope regions. Current CO$_2$ budgets require a presently unknown sink, in fact, of $\sim 3 \times 10^9$ tons C yr^{-1} during the last decade (Bolin, 1977) to account for the missing CO$_2$, which must have been released by increased fossil fuel burning and deforestation, but not observed in the atmosphere or apparently mixed to the deep sea in such a short time.

The sedimentation rate at the head of the Hudson Canyon has doubled over the last few thousand years (Drake et al., 1978), and nitrogen content of rain has doubled in recent decades, largely by the combustion of fossil fuel (Hall et al., 1978). Rainfall contributed less than 1% of the annual nitrogen demand of primary production in the New York Bight, however; more importantly the amount of organic carbon to be exported from the shelf is fixed during the spring bloom when nutrients are not limiting. The total yield to man as commercial fish from the New York continental shelf over the last 10 years has been only ~ 0.63 gCm^{-2}yr^{-1}, while the carbon export may be at least 180 gCm^{-2}yr^{-1} (Fig. 6); the former provides food, but the latter may, by reducing the 'greenhouse' effect, actually be of greater importance to us. This chapter presents a preliminary, but a reasonably consistent proposal for steady-state carbon fluxes which could both support the 99% of the global fish catch which is produced by the shelf ecosystem and, at the same time, serve as a major sink in the global CO$_2$ budget.

Major alteration of carbon flow through any one of the shelf food web components (Fig. 6) can be attributed to interannual changes of species survival patterns as a function of external perturbations. For example, because of their proximity, the continental shelves have tended to become the refuse pit of industrialised maritime nations. We are coming to realise, however, that dilution of waste material in the sea can no longer be considered a permanent removal process in either the open ocean or nearshore waters. The increasing utilisation of the continental shelf for oil drilling and transport, cooling of nuclear power plants, planned and inadvertent waste disposal, as well as for food and recreation, require careful management of man's future activities in this ecosystem. Certainly as we continue to learn more about the organisms and control functions of the shelf-sea ecosystem, man will be able to more rationally manage the coastal zone by avoiding overfishing and carefully selecting sites of industrial activity. The assimilatory capacity of this ecosystem to store toxicants must now be considered, however, with respect to its present resiliency (Holling, 1973) to withstand perturbations at much shorter time scales than either glacial or evolutionary events.

Acknowledgements

This analysis was mainly sponsored by Contract No. EY-76-C-02-0016 from the Department of Energy as part of the Atlantic Coastal Ecosystem (ACE) Program, with additional support by grants from the International Decade of Ocean Exploration, NSF, as part of the Coastal Upwelling Ecosystem Analysis (CUEA) Program, and by grants from the Division of Polar Programs, NSF, as part of the Processes and Resources of the Bering Sea (PROBES) Program. I wish to thank my colleagues in the ACE, CUEA, and PROBES Programs for many discussions of the ideas and data presented in this chapter.

References

Allee, W. C. (1934). Concerning the organization of marine coastal communities. *Ecol. Monogr.* **4**, 541–554

Alton, M. S. (1972). Characteristics of the demersal fauna inhabiting the outer continental shelf and slope off the northern Oregon Coast. *In* "The Columbia River Estuary and Adjacent Waters" A. T. Pruter and D. L. Alversen, eds) pp. 583–636. University of Washington Press, Seattle

Alversen, D. L., Longhurst, A. R. and Gulland, J. A. (1970). How much food from the sea? *Science* **168**, 503–505

Anderson, G. C. and Munson, R. E. (1972). Primary productivity studies using merchant vessels in the North Pacific Ocean. *In* "Biological Oceanography of the Northern North Pacific Ocean" (A. Y. Takenouti, ed.) pp. 245–252. Idemitsu Shoten, Tokyo

Ankar, S. and Elmgren, R. (1976). The benthic macro- and meiofauna of the Askö-Landsort area (Northern Baltic Proper). *Cont. Askö Lab.* **11**, 1–115

Bacastow, R. and Keeling, C. D. (1973). Atmospheric carbon dioxide and radiocarbon in the natural carbon cycle. II. Changes from A D 1700 to 2070 as deduced from a geochemical model. *In* "Carbon and the Biosphere" (G. M. Woodwell and E. V. Pecan, eds) AEC Symp. Ser. 30, 86–135, Springfield

Beardsley, R. C., Biocourt, W. C. and Hansen, D. V. (1976). Physical oceanography of the middle Atlantic Bight. ASLO Spec. Symp. **2**, 20–34

Beebe, W. (1934). "Half Mile Down", pp. 1–344. Harcourt, Brace and Co., New York

Bigelow, H. B. (1926). Plankton of the offshore waters of the Gulf of Maine. *Bull. Bur. Fish., Wash.* **40** (Part II), 1–509

Bigelow, H. B. and Sears, M. (1939). Studies of the waters of the continental shelf, Cape Cod to Chesapeake Bay. III. A volumetric study of the zooplankton. *Mem. Comp. Zoo.* **54**, 183–378

Bolin, B. (1977). Changes of land biota and their importance for the carbon cycle. *Science* **196**, 613–615

Bothner, M. H., Spiker, E. and Johnson, P. P. (1980). ^{14}C and ^{210}Pb profiles in the "mud patch" on the continental shelf off Southern New England: evidence for modern accumulation. *Geol. Soc. Amer.* Abstract

Broecker, W. S., Takahashi, T., Simpson, H. J. and Peng, T. H. (1979). Fate of fossil fuel carbon dioxide and the global carbon budget. *Science* **206**, 409–418

Buchanan, J. B. and Warwick, R. M. (1974). An estimate of benthic macrofaunal production in the offshore mud of the Northumberland coast. *J. Mar. Biol. Assn U.K.* **54**, 197–222

Bumpus, D. F. (1974). A description of the circulation on the continental shelf of the east coast of the United States. *Prog. Oceanogr.* **6**, 111–157

Chamberlin, T. C. (1899). An attempt to frame a working hypothesis of the cause of glacial periods on an atmospheric basis. *J. Geol.* **7**, 575, 667, 751

Clarke, G. L. (1940). Comparative richness of zooplankton in coastal and offshore areas of the Atlantic. *Biol. Bull.* **78**, 226–255

Clarke, G. L., Pierce, E. L. and Bumpus, D. G. (1943). The distribution and reproduction of *Sagitta elegans* on Georges Bank in relation to hydrographic conditions. *Biol. Bull.* 85, 201–226

CLIMAP Project members (1976). The surface of the ice-age earth. *Science* **191**, 1131–1137

Coachman, L. K. and Charnell, R. L. (1979). On lateral water mass interaction – a case study, Bristol Bay, Alaska. *J. Phys. Oceanogr.* **9**, 278–297

Colton, J. B., Smith, W. G., Kendall, W. A., Berrien, P. L. and Fahay, M. P. (1979). Principal spawning areas and times of marine fishes, Cape Sable to Cape Hatteras. *Fish Bull.* **76**, 911–915

Corkett, C. J. and McLaren, I. A. (1978). The biology of *Pseudocalanus*. *Adv. Mar. Biol.* **15**, 1–231

Coull, B. C. (1973). Estuarine meiofauna: a review: relationships and microbial interaction. *In* "Estuarine Microbial Ecology" (L. H. Stevenson and R. R. Colwell, eds) pp. 499–512. University of South Carolina Press, Columbia

Coull, B. C., Ellison, R. L., Fleeger, J. W., Higgins, R. P., Hope, W. D., Hummon, W. D., Rieger, R. M., Sterrer, W. E., Thiel, H. and Tietjen, J. H. (1977). Quantitative estimates of the meiofauna from the deep sea off North Carolina, U.S.A. *Mar. Biol.* **39**, 233–240

Creager, J. S. and McManus, D. A. (1965). Pleistocene drainage patterns on the floor of the Chukchi Sea. *Mar. Geol.* **3**, 279–290

Csanady, G. T. (1976). Mean circulation in shallow seas. *J. Geophys. Res.* **81**, 5389–5399

Dagg, M. (1977). Some effects of patchy food environments on copepods. *Limnol. Oceanogr.* **22**, 99–107

Dagg, M. J. (1978). Estimated, *in situ*, rates of egg production for the copepod *Centropages typicus* (Krøyer) in the New York Bight. *J. exp. mar. Biol. Ecol.* **34**, 183–196

Day, J. H. (1963). The complexity of the biotic environment. *In* "Speciation in the Sea", **5**, 31–49. Syst. Assoc. Publ.

Drake, D. E., Hatcher, P. and Keller, G. (1978) Suspended particulate matter and mud deposition in upper Hudson Submarine Canyon. *In* "Sedimentation in Submarine Canyons, Fans and Trenches" (D. J. Stanley and G. Kelling, eds) pp. 33–41. Dowden, Hutchinson and Ross, Stroudsburg

Edwards, R. L. and Merrill, A. S. (1977). A reconstruction of the continental shelf area of eastern North America for the times 9500 B.P. and 12 500 B.P. *Archeology of Eastern North America* **5**, 1–44

Edwards, R. L. and Bowman, R. F. (1979). Food consumed by continental shelf fishes. *In* "Predator–Prey Systems in Fish Communities and their Role in Fisheries Management", pp. 387–406. Sport Fish. Instit., Washington

Emery, K. O. and Uchupi, E. (1972). Western North Atlantic Ocean topography. Rocks, structure, water, life and sediments. *Amer. Assoc. Petrol. Geol. Mem.* **17**, 1–532

Eppley, R. W. and Petersen, B. J. (1980). Particulate organic matter flux and planktonic new production in the deep ocean. *Nature* **282**, 677–680

Evans, C. A., O'Reilly, J. E. and Thomas, J. P. (1979). Report on chlorophyll measurements made on MARMAP surveys between October 1977–December 1978. N.E.F.C. Sandy Hook Lab. Rept. No. SHL 79-10, 1–247.

Falkowski, P. G., Hopkins, T. S. and Walsh, J. J. (1980). An analysis of factors affecting oxygen depletion in the New York Bight. *J. Mar. Res.* **38**, 479–506

Farrington, J. W. and Tripp, B. W. (1977). Hydrocarbons in western North Atlantic surface sediments. *Geochim. Cosmochim. Acta* **41**, 1627–1641

Fenchel, T. (1969). The ecology of the marine microbenthos. IV. Structure and function of the benthic ecosystem, its chemical and physical factors and the microfauna communities with special reference to the ciliated protozoa. *Ophelia* **6**, 1–182

Freeland, G. L. and Swift, D. J. (1978). Surficial sediments. *MESA New York Bight Atlas Monogr.* **10**, 1–93

Gershanovitch, D. E. (1962). New data on recent sediments of the Bering Sea. *In* "Issledovanye po Progrome Mezhdunarodenovo Geofizicheskovo Goda" (L. G. Vinogradova and M. V. Fedosova, eds) pp. 128–164. Pishchem Promizdat, Moscow

Grice, G. D. and Hart, A. D. (1962). The abundance, seasonal occurrence, and distribution of the epizooplankton between New York and Bermuda. *Ecol. Monogr.* **32**, 287–309

Gross, M. G. (1968). Organic carbon in surface sediment from the northeast Pacific Ocean. *Int. Oceanol. Limnol.* **1**, 46–54

Hall, C. S., Rowe, G. T., Ryther, J. H. and Woodwell, G. M. (1978). Acid rain, zooplankton fecal pellets and the global carbon budget. *Biol. Bull.* **153**, 427–428

Hathaway, J. C., Poag, C. W., Valentine, P. C., Miller, R. E., Schultz, D. M., Manheim, F. T., Kohout, F. A., Bothner, M. H. and Sangrey, D. A. (1979). U.S. Geological Survey core drilling on the Atlantic shelf. *Science* **206**, 515–527

Hay, W. W. and Southam, J. R. (1977). Modulation of marine sedimentation by the continental shelves. *In* "The Fate of Fossil Fuel CO_2 in the Oceans" (N. R. Anderson and A. Malahoff, eds) pp. 569–604. Plenum Press, New York

Hennemuth, R. C. (1976). Fisheries and renewable resources of the Northwest Atlantic Shelf. *In* "Effects of Energy-Related Activities on the Atlantic Continental Shelf" (B. Manowitz, ed.) BNL 50484, 146–166, Brookhaven National Laboratory

Hirota, J. (1974). Quantitative natural history of *Pleurobrachia bachei* in La Jolla Bight. *Fish. Bull.* **72**, 295–352

Hitchcock, G. L. and Smayda, T. J. (1977). The importance of light in the initiation of the 1972–1973 winter–spring diatom bloom in Narragansett Bay. *Limnol. Oceanogr.* **22**, 126–131

Holling, C. S. (1973). Resilience and stability of ecological systems. *Annl. Rev. Ecol.* **4**, 1–23

Holmes, M. L. and Craeger, J. S. (1974). Holocene history of the Laptev Sea continental shelf. *In* "Marine Geology and Oceanography of the Arctic Seas" (Y. Herman, ed.) pp. 211–229. Springer-Verlag, New York

Holmes, R. W. (1970). The Secchi disk in turbid coastal waters. *Limnol. Oceanogr.* **15**, 688–694

Hunt, J. M. (1966). The significance of carbon isotope variations in marine sediments. *In* "Advances in Organic Geochemistry" (S. D. Hobson and G. C. Speers, eds) pp. 27–35. Pergamon Press, Oxford

Jansen, J. H. F. (1976). Late pleistocene and holocene history of the northern North Sea, based on acoustic reflection records. *Neth. J. Sea Res.* **10**, 1–43

Jansen, J. H. F., Doppert, J. W. C., Hoogendoorn-Toering, K., De Jong, J. and Spaink, G. (1979). Late pleistocene and Holocence deposits in the Witch and Fladen Ground area, northern North Sea. *Neth. J. Sea Res.* **13**, 1–39

Judkins, D. C., Wirick, C. D. and Esaias, W. E. (1980). Composition, abundance, and distribution of zooplankton in the New York Bight, September 1974– September 1975. *Fish. Bull.* **77**, 669–683.

Keeling, C. D. (1978). Atmospheric carbon dioxide in the 19th century. *Science* **202**, 1109

Keeling, C. D. and Bolin, B. (1967). The simultaneous use of chemical tracers in oceanic studies. I. General theory of reservoir models. *Tellus* **19**, 566–581

Ketchum, B. H. and Keen, D. J. (1955). The accumulation of river water over the continental shelf between Cape Cod and Chesapeake Bay. *Deep-Sea Res.* **3** (Suppl.), 346–357

Knauer, G. A., Martin, J. H. and Bruland, K. W. (1979). Fluxes of particulate carbon, nitrogen, and phosphorous in the upper water column of the northeast Pacific. *Deep-Sea Res.* **26**, 97–108

Lasker, R. (1975). Field criteria for survival of anchovy larvae: the relation between inshore chlorophyll maximum layer and successful first feeding. *Fish. Bull.* **73**, 453–462

Leetma, A. (1977). Effects of the winter of 1976–1977 on the northwestern Sargasso Sea. *Science* **198**, 188–189

Lettau, B., Brower, W. A. and Quayle, R. G. (1976). Marine climatology. *MESA New York Bight Monogr.* **7**, 1–239

Lisitzin, A. P. (1966). "Recent Sedimentary Processes in the Bering Sea", pp. 1–300. Nauka, Moscow

MacIlvaine, J. C. (1973). Sedimentary processes on the continental slope off New England. Ph.D. dissertation, MIT

Malone, T. C. (1977). Plankton systematics and distribution. *MESA New York Bight Monogr.* **13**, 1–45

Malone, T. C. and Chervin, M. B. (1979). The production and fate of phytoplankton size fractions in the plume of the Hudson River, New York Bight. *Limnol. Oceanogr.* **24**, 683–696

Mann, K. H. (1976). Production on the bottom of the sea. *In* "Ecology of the Seas" (D. H. Cushing and J. J. Walsh, eds) pp. 225–250. Blackwell, Oxford

McAllister, C. D., Parsons, T. R. and Strickland, J. D. H. (1960). Primary productivity and fertility at station "p" in the northeast Pacific Ocean. *J. Cons. Explor. Mer* **25**, 240–259

McHugh, J. L. (1976). Estuarine fisheries: are they doomed? *In* "Estuarine Processes" (M. Wiley, ed.) Vol. I, pp. 15–27, Academic Press, New York

McIntyre, A. D., Muro, A. L. S. and Steele, J. H. (1970). Energy flow in a sand ecosystem. *In* "Marine Food Chains" (J. H. Steele, ed.) pp. 19–31. Oliver and Boyd, London

McRoy, C. P. and Goering, J. J. (1976). Annual budget of primary production in the Bering Sea. *Mar. Sci. Comm.* **2**, 255–267

Menzel, D. W. and Ryther, J. H. (1960). The annual cycle of primary production in the Sargasso Sea off Bermuda. *Deep-Sea Res.* **6**, 351–367

Menzel, D. W. and Ryther, J. H. (1970). Distribution and cycling of organic matter in the oceans. *In* "Organic matter in Natural Waters" (D. W. Hood, ed.) **1**, 31–54. Inst. Mar. Sci. Occ. Pub., Fairbanks

Milliman, J. D. (1973). Marine Geology. *In* "Coastal and Offshore Environmental Inventory, Cape Hatteras to Nantucket Shoals" (S. P. Saila, ed.) pp. 10–92. Mar. Pub. Ser. No. 3, Univ. Rhode Island, Kingston

Milliman, J. D. and Emery, K. O. (1968). Sea levels during the past 35 000 years. *Science* **162**, 1121–1123

Mills, E. L. and Fournier, R. O. (1979). Fish production and the marine ecosystems of the Scotian shelf, Eastern Canada. *Mar. Biol.* **54**, 101–108

Mooers, C. N. K., Collins, C. A. and Smith, R. L. (1976). The dynamic structure of the frontal zone in the coastal upwelling region off Oregon. *J. Phys. Oceanogr.* **6**, 3–21

Müller, P. J. (1977). C/N ratios in Pacific deep-sea sediments: effect of inorganic ammonium and organic nitrogen compounds sorbed by clays. *Geochim. Cosmochim. Acta* **41**, 765–776

Myers, T. D. (1967). Horizontal and vertical distribution of thecosomatous pteropods off Cape Hatteras. Ph.D. dissertation, Duke University

Nelson, C. H., Hopkins, D. M. and Scholl, D. W. (1974). Cenozoic sedimentary and tectonic history of the Bering Sea. *In* "Oceanography of the Bering Sea with Emphasis on Renewable Resources" (D. W. Hood and E. J. Kelley, eds) **2**, 485–516. Inst. Mar. Sci. Occ. Pub., Fairbanks

O'Brien, J. J. and Pillsbury, R. D. (1974). Rotary wind spectra in a sea breeze regime. *J. Apl. Meteor.* **13**, 820–825

Pearcy, W. G. (1972). Distribution and ecology of oceanic animals off Oregon. *In* "The Columbia River Estuary and Adjacent Waters" (A. T. Pruter and D. L. Alversen, eds) pp. 351–377. University of Washington Press, Seattle

Petersen, W. T. and Miller, C. B. (1975). Year-to-year variations in the planktology of the Oregon upwelling zone. *Fish. Bull.* **73**, 642–653

Petersen, W. and Miller, C. B. (1976). Zooplankton along the continental shelf off Newport, Oregon 1969–1972: distribution, abundance, seasonal cycle, and year-to-year variations. ORESU-T-76-002, pp. 1–111. OSU Sea Grant Program. Pub., Corvallis

Petersen, W. T. and Miller, C. B. (1977). Seasonal cycle of zooplankton abundance and species composition along the central Oregon coast. *Fish. Bull.* **75**, 717–724

Petersen, W. T., Miller, C. B. and Hutchinson, A. (1979). Zonation and maintenance of copepod populations in the Oregon upwelling zone. *Deep-Sea Res.* **26**, 467–494

Pingree, R. D., Holligan, P. M. and Mardell, G. T. (1978). The effects of vertical stability on phytoplankton distribution in the summer on the northwest European shelf. *Deep-Sea Res.* **25**, 1011–1028

Pingree, R. D., Holligan, P. M., Mardell, G. T. and Head, R. N. (1976). The influence of physical stability on spring, summer, and autumn phytoplankton blooms in the Celtic Sea. *J. Mar. Biol. Ass. U.K.* **56**, 845–873

Pruter, A. T. (1973). Development and present status of bottom fish resources in the Bering Sea. *J. Fish. Res. Bd. Canada* **30**, 2373–2385

Ralston, C. W. (1979). Where has all the carbon gone? *Science* **204**, 1345–1346

Redfield, A. C. (1939). The history of a population of *Limacina retroversa* during its drift across the Gulf of Maine. *Biol. Bull.* **76**, 26–47

Redfield, A. C. (1941). The effect of the circulation of water on the distribution of the calanoid community in the Gulf of Maine. *Biol. Bull.* **80**, 86–110

Redfield, A. C. and Beale, A. (1940). Factors determining the distribution of populations of chaetognaths in the Gulf of Maine. *Biol. Bull.* **79**, 459–487

Revelle, R. and Suess, H. E. (1957). Carbon dioxide exchange between atmosphere and ocean and the question of an increase of atmospheric CO_2 during the past decades. *Tellus* **9**, 18–27

Richardson, S. L. and Pearcy, W. G. (1977). Coastal and oceanic fish larvae in an area of upwelling off Yaquina Bay, Oregon. *Fish. Bull.* **75**, 125–145

Riley, G. A. (1942). The relationship of vertical turbulence and spring diatom flowerings. *J. Mar. Res.* **5**, 67–87

Riley, G. A. (1947). A theoretical analysis of the zooplankton population of Georges Bank. *J. Mar. Res.* **6**, 104–113

Riley, G. A. (1956). Oceanography of Long Island Sound. II. Physical oceanography. *Bull. Bingham Oceanogr. Coll.* **15**, 15–46

Riley, G. A. (1967). The plankton of estuaries. *In* "Estuaries" (G. H. Lauff, ed.) **83**, pp. 316–326. Publ. Am. Assoc. Adv. Sci., Washington

Rowe, G. T. (1971). Benthic biomass in the Pisco, Peru upwelling. *Invest. Pesq.* **35**, 127–135

Rowe, G. T., Clifford, C. H., Smith, K. L. and Hamilton, P. L. (1975). Benthic nutrient regeneration and its coupling to primary productivity in coastal waters. *Nature* **225**, 215–217

Rowe, G. T. and Gardner, W. D. (1979). Sedimentation rates in the slope water of the northwest Atlantic Ocean measured directly with sediment traps. *J. Mar. Res.* **37**, 581–600

Ryther, J. H. (1969). Photosynthesis and fish production in the sea. *Science* **166**, 72–76

Ryther, J. H. and Dunstan, W. M. (1971). Nitrogen, phosphorus, and eutrophication in the coastal marine environment. *Science* **171**, 1008–1013

Schindler, D. W., Brunskill, G. J., Emerson, S., Broecker, W. S. and Peng, T. H. (1972). Atmospheric carbon dioxide: its role in maintaining phytoplankton standing crops. *Science* **177**, 1192–1194

Schubel, J. and Okubo, A. (1972). Comments on the dispersal of suspended sediment across continental shelves. *In* "Shelf Sediment Transport: Process and Pattern" (D. J. Swift, D. B. Duane and O. H. Pilkey, eds) pp. 333–346. Dowden, Hutchinson and Ross, Stroudsburg

Scott, J. T. and Csanady, G. T. (1976). Nearshore currents off Long Island. *J. Geophys. Res.* **81**, 5401–5409

Sears, M. and Clarke, G. L. (1940). Annual fluctuations in the abundance of marine zooplankton. *Biol. Bull.* **79**, 321–328

Sharma, G. D. (1974). Contemporary sedimentary regimes of the eastern Bering Sea. *In* "Oceanography of the Bering Sea with Emphasis on Renewable Resources" (D. W. Hood and E. J. Kelley, eds) **2**, 517–540, Inst. Mar. Sci. Occ. Pub., Fairbanks

Shepard, F. S. (1963). "Submarine Geology", pp. 206–278. Harper and Row, New York

Sherman, K. (1978). MARMAP, a fisheries ecosystem study in the NW Atlantic: fluctuations in ichthyoplankton-zooplankton components and their potential for impact on the system. *Proc. Workshop on Advanced Concepts in Ocean Measurements*, 24–28 October, 1978, pp. 1–38. Belle W. Baruch Institute for Marine Biology and Coastal Research, University of South Carolina, Columbia

Sherman, K., Cohen, E., Sissenwine, M., Grosslein, M., Langton, R. and Green, J. (1978). Food requirements of fish stocks in the Gulf of Maine, Georges Bank, and adjacent waters. Unpublished ICES ms., C.M. 1978/Gen: **8**, pp. 1–12

Siegenthaler, U. and Oeschger, H. (1978). Predicting future atmospheric carbon dioxide levels. *Science* **199**, 288–295

Simpson, J. H. and Hunter, J. R. (1974). Fronts in the Irish Sea. *Nature* **250**, 404–406

Simpson, J. H. and Pingree, R. D. (1978). Shallow sea fronts produced by tidal stirring. *In* "Oceanic Fronts in Coastal Processes" (M. J. Bowman and W. E. Esaias, eds) pp. 29–42. Springer-Verlag, New York

Small, L. F. and Ramberg, D. A. (1971). Chlorophyll a, carbon and nitrogen in particles from a unique coastal environment. *In* "Fertility of the Sea" (J. Costlow, ed.) Vol. II, pp. 475–492. Gordon and Breach, New York

Small, L. F., Curl, H. and Glooschenko, W. A. (1972). Estimates of primary production off Oregon using an improved chlorophyll light technique. *J. Fish. Res. Bd. Canada* **29**, 1261–1267

Smayda, T. J. (1970). The suspension and sinking of phytoplankton in the sea. *Oceanogr. Mar. Biol. Ann. Rev.* **8**, 353–414

Smetacek, V., von Brockel, K., Zietzschel, B. and Zenk, W. (1978). Sedimentation of particulate matter during phytoplankton spring bloom in relation to the hydrographical regime. *Mar. Biol.* **47**, 211–226

Smith, K. L. (1973). Respiration of a sublittoral community. *Ecology* **54**, 1065–1075

Smith, K. L., Rowe, G. T. and Nichols, J. A. (1973). Benthic community respiration near the Woods Hole sewage outfall. *Estuar. Coast. Mar. Sci.* **1**, 65–70

Smith, K. L., Rowe, G. T. and Clifford, C. H. (1974). Sediment oxygen demand in an outwelling and upwelling area. *Tethys* **6**, 223–230

Steele, J. H. (1974). "The Structure of Marine Ecosystems", pp. 1–178. Harvard University Press, Cambridge

Steele, J. H. and Menzel, D. W. (1962). Conditions for maximum primary production in the mixed layer. *Deep-Sea Res.* **9**, 39–49

Stefansson, U. and Richards, F. A. (1964). Distributions of dissolved oxygen, density, and nutrients off the Washington and Oregon coasts. *Deep-Sea Res.* **11**, 355–380

Stuiver, M. (1978). Atmospheric carbon dioxide and carbon reservoir changes. *Science* **199**, 253–258

Sverdrup, H. U. (1953). On conditions for the vernal blooming of phytoplankton. *J. Cons. Explor. mer* **18**, 287–295

Tenore, K. R. (1977). Food chain pathways in detrital feeding benthic communities, a review, with new observations on sediment resuspension and detrital recycling. *In* "Ecology of Marine Benthos" (B. C. Coull, ed.) pp. 37–54. University of South Carolina Press, Columbia

Thomas, J. C., O'Reilly, J. E., Draxler, A., Babinchak, J. A., Robertson, C. N., Phoel, W. C., Waldhauer, R., Evans, C. A., Matte, A., Cohn, M., Nitkowski, M. and Dudley, S. (1979). Biological processes: productivity and respiration. *In* "Oxygen Depletion and Associated Benthic Mortalities in the New York Bight, 1976" (C. J. Sindermann and R. L. Swanson, eds) NOAA Prof. Pap. **11**, 231–262

Tietjen, J. H. (1971). Ecology and distribution of deep-sea meiobenthos off North
    Carolina. *Deep-Sea Res.* **18**, 941–957
Turekian, K. K. and Benoit, G. J. (1980). Radiocarbon in New York Bight
    sediments and the use of carbon isotopes in delineating carbon sources. *Environ.*
    Sci. Tech. (In press.)
Vidal, J. (1978). Effects of phytoplankton concentration, temperature, and body size
    on rates of physiological processes and production efficiency of the marine
    planktonic copepods, *Calanus pacificus* Brodsky and *Pseudocalanus* sp. Ph.D.
    dissertation, University of Washington
Walsh, J. J. (1974). Primary production in the sea. *Proc. First Int. Congr. Ecology,*
    PUDOC, Wageningen, pp. 150–154
Walsh, J. J. (1975). A spatial simulation model of the Peru upwelling ecosystem.
    Deep-Sea Res. **22**, 201–236
Walsh, J. J. (1976). Herbivory as a factor in patterns of nutrient utilization in the sea.
    Limnol. Oceanogr. **21**, 1–13
Walsh, J. J. (1977). A biological sketchbook for an eastern boundary current. *In*
    "The Sea" (J. H. Steele, J. J. O'Brien, E. O. Goldberg and I. N. McCave, eds)
    Vol. 6, pp. 923–968. Wiley Interscience, New York
Walsh, J. J. (1978). The biological consequences of interaction of the climatic, El
    Niño, and event scales of variability in the Eastern Tropical Pacific. *Rapp. P.-V.*
    Reun. Cons. int. Explor. Mer **173**, 182–192
Walsh, J. J. (1980). A carbon budget for overfishing off Peru. *Nature* (in press)
Walsh, J. J. (1981). Concluding remarks: marine photosynthesis and the global CO_2
    cycle. *In* "Primary Production in the Sea" (P. G. Falkowski, ed.) pp. 497–506.
    Academic Press, New York
Walsh, J. J., Whitledge, T. E., Howe, S. O., Wirick, C. D., Castiglione, L. J. and
    Codispoti, L. A. (1976). Transient forcings of the lower trophic levels during the
    spring bloom within the New York Bight. *ASLO Spec. Symp.* **2**, 273–274
Walsh, J. J., Whitledge, T. E., Barvenik, F. W., Wirick, C. D., Howe, S. O., Esaias,
    W. E. and Scott, J. T. (1978). Wind events and food chain dynamics within the
    New York Bight. *Limnol. Oceanogr.* **23**, 659–683
Walsh, J. J., Whitledge, T. E., Esaias, W. E., Smith, R. L., Huntsman, S. A.,
    Santander, H. and DeMendiola, B. R. (1980). The spawning habitat of the
    Peruvian anchovy, *Engraulis ringens. Deep-Sea Res.* **27**, 1–27
Warwick, R. M. and Price, R. (1975). Macrofauna production in an estuarine mud-
    flat. *J. Mar. Biol. Assn. U.K.* **55**, 1–18
Wiebe, P. H., Boyd, S. H. and Winget, C. (1976). Particulate matter sinking to the
    deep-sea floor at 2000 m in the tongue of the ocean, Bahamas, with a description
    of a new sedimentation trap. *J. Mar. Res.* **34**, 341–354
Wigley, R. L. and McIntyre, A. D. (1964). Some quantitative comparisons of
    offshore meiobenthos and macrobenthos south of Martha's Vineyard. *Limnol.*
    Oceanogr. **9**, 485–493
Wigley, R. L. and Theroux, R. B. (1979). Macrobenthic invertebrate fauna of the
    middle Atlantic Bight region: Faunal composition and quantitative distribution.
    U.S. Geol. Survey Prof. Pap. **10**, 1–395
Wong, C. S. (1978). Atmospheric input of carbon dioxide from burning wood.
    Science **200**, 197–200
Woodwell, G. M., Whittaker, R. H., Reiners, W. A., Likens, G. E., Delwiche, C. C.
    and Botkin, D. B. (1978). The biota and the world carbon budget. *Science* **199**,
    141–146

7 Fronts and Eddies in the Sea: Mechanisms, Interactions and Biological Effects

ROBERT W. OWEN

1 Introduction

This chapter concerns fronts and eddies, two major classes of water motion that create and change patterns of biological distributions. Flow of water across topographic features in the open sea produces patterns of vertical circulation ('fronts') and eddy-like motions which can reorder life processes and distributions in a variety of ways. Such patterns may also arise independently of sea bottom topography from ocean current confluence, from wind stress applied unevenly in space and time, or from heat and water exchange across the sea surface. Frontal and eddy circulations can occur at any depth and affect populations of organisms at all trophic levels, including benthic forms from great depths to the littoral zone. Life forms as diverse as phytoplankton, protozoans, crustaceans, fish, sea snakes, marine mammals and birds are found to alter their distributions in the presence of such flow patterns.

The ocean is widely held to be turbulent upwards of the scale of a few centimeters and seconds, but also to contain quasi-ordered velocity fields. Recent experimental work suggests that such ordered fields may be more prevalent in turbulent fluids than previously believed (Laufer, 1975). Fronts and eddies are two such classes of quasi-ordered velocity fields. We are concerned here with flow patterns that locally alter biological distributions, but it is obvious that various processes causing fronts and eddies can occur at the same place and time. This frequently leads to a degree of complexity that approaches fully turbulent motion, i.e. chaos. To the degree that organisms interact with mixed regimes of physical processes, patchiness of organisms induced by behavioral mechanisms alone may be obliterated.

There is a considerable body of evidence that patterned patchiness frequently arises from patterned circulation. Environmental circumstances under which such patterns arise are often rather unexceptional, and thus may be more widespread than case studies imply. There is also some promise of prediction of patterning (or de-patterning) circulations, which would be of value in estimating yield of fish stocks affected by patterning of food and supply. Accounting for such patterning of organisms, which is recurrent and widespread in many instances, should also enhance the accuracy of biological production models constructed for organisms at any position in the food web.

Fronts and eddies are engendered by a variety of physical processes and have a variety of biological consequences. The terms 'front' and 'eddy' are used here to denote localised zones where quasi-ordered singularities of lateral flow induce or intensify vertical flow. A front is a line or linear zone that defines an axis of laterally convergent flow, below or above which vertical flow is induced (Fig. 1). A frontal system denotes sets of convergences. These usually produce alternating zones of downwelling and upwelling flow. An eddy denotes an area of closed horizontal streamlines where vertical motion is induced or sustained.

Fronts and eddies are recurrent or persistent singularities in the ocean's usually turbulent flow. They are found singly, in groups, and in combination in all seas and at all depths, although most frequently and to clearest ecological effect in the surface layers. Characteristic motions have been detected on scales that range from 1 to 10^5 m laterally and 1 to 10^3 m in depth, and persist from a few hours to virtual permanence. The magnitude of ecological effects, however, is not necessarily scale-dependent. Time and space scales are related: the smallest fronts and eddies are the most ephemeral and the largest are the most permanent. This relationship holds because driving forces are more ephemeral on small scales, because size of circulations tend to increase where the driving force is maintained, and because when the driving force relaxes, larger circulations persist longer.

A basic and attractive tenet implicit in much of the work in this area is that vertical displacements have greater ecological effects than similar horizontal displacements because environmental gradients, such as light, pressure, temperature, salinity, oxygen, nutrients and flow, are by far the steepest in the vertical (chapter 14). Vertical motion in fronts is often vigorous and highly localised, whereas in eddies it is slow and spread over much larger areas. For this reason, lateral gradients of temperature, salinity and food supply are correspondingly sharper across convergent flows than across eddies. Direct responses of organisms have not often been demonstrated, and causal effects are notoriously difficult to isolate.

Vertical motion in eddies occurs to compensate for lateral flow into or out of the circulation. Such convergent or divergent lateral flow occurs as a result of deflection due to the Coriolis parameter of relative motion on the rotating earth (Von Arx, 1962). Cyclonic eddies of sufficient size exhibit divergent flow due to the Coriolis deflection and compensatory upward movement occurs. Anticyclonic eddies, conversely, exhibit downward

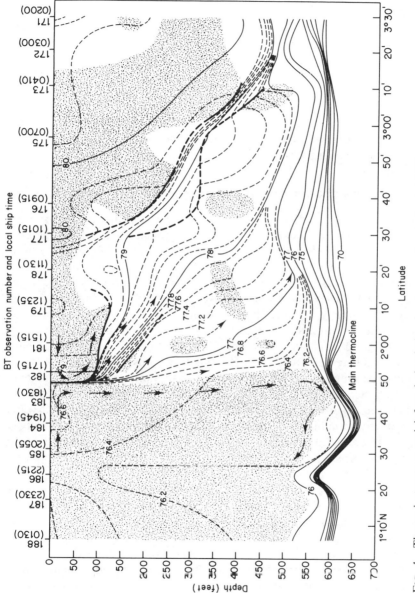

FIG 1 Thermal structure and inferred circulation in an ocean front along 172°W near the Equator. From Cromwell and Reid (1956)

motion in their interiors. Estimates of upward velocities in large cyclonic eddies (cf. Owen, 1980) are of the same order as sinking rates of some phytoplankton and detrital particles and can also transport substantial nutrient supplies upward (Broenkow, 1965). Whether cyclonic or not, eddies are zones of horizontal recirculation and thus locally maintain and transport their contained populations and substances.

Difficulty of detection and measurement at sea in part accounts for the fact that we can only guess at the full extent, nature and ecological impact of some of the interactions reviewed here. We currently look to advances in instrumentation and observation platforms (especially appropriately equipped and deployed spacecraft) to add needed perspective and we regard the study of such singularities to be in its early stages.

2 Interactive characteristics of organisms and substances

Most macroscopic particles, living or not, have properties that make possible their differential concentration in the presence of fronts or eddies of appropriate scale and intensity. These properties are: differential motion (sinking, floating and swimming) and proliferation (production and reproduction). Characteristics of these properties are more or less specific to the type of particle. They operate interactively with vertical circulations either directly or indirectly as rates and directions of sinking and swimming and on local rates of production. Rates and directions are in turn responsive mainly to food (or nutrient) supply and to environmental gradients, especially light, temperature and food concentrations.

It must be emphasised that a degree of randomness is typical of both water motion and organism characteristics. Sinking and swimming rates vary between organisms of the same species, size or population, and water motion is usually subject to random variations so that trapping efficiency of a particular circulation is accordingly diminished, as is the sharpness of the resulting patterns of organism concentration. It seems clear from case studies, however, that factors creating pattern frequently predominate over those that disrupt it. Vertical movement, for example, tends to be more consistent in rate and direction among individuals of the same types than among individuals of different types.

The same arguments apply to production rates: among plankton species represented at a site, several are likely to respond with similar division rates to a change in, say, nutrients supplied by a front or eddy. Such species become differentially concentrated as a set if their vertical motions are either similar or not important. Production rates tend to be more similar within than between organism types, ranging upward to 2 doublings/day among some phytoplankton populations.

Particle characteristics that determine sinking rates are size, form and density. Volume and density of a particle govern its buoyant forces whereas its surface area and form are the primary determinants of local fluid drag.

Detrital particles of filterable size sink in still water at rates depending primarily on their size and form. Detritus is predominantly microscopic but can exceed 1 m in largest dimension in shapes as diverse as "burst balloons", strings and sheets. These can have occluded bubbles and lumps that make estimates of density and sinking rate uncertain. The bulk of detritus, however, has little apparent potential for passive collection except in the slowest convergences and perhaps in cyclonic (upwelling) eddies. This is because most detrital particles sink slowly: they are either quite small or are in an aggregated form known as 'marine snow'. An important exception is fecal pellets of zooplankton, particularly those of copepods since they are so numerous. Smayda (1969) found still water sinking rates of naturally occurring fecal pellets to range from 1.5 to 15.6 m h^{-1}, mostly about 6 m h^{-1}. Large pellets sank faster than small ones, and shape effects also were discerned. Effects of diet on pellet buoyancy and shape are apparent from observations of Marshall and Orr (1955) and may be expected to influence sinking rate of the pellets. Disintegration and bacterial or fungal growth probably affect sinking rates of older pellets.

Sinking of diatoms, which have little or no other motility, was reviewed comprehensively by Smayda (1970). He cites sinking rates up to 1.26 m h^{-1} for a large centric diatom and average rates of about 1 m day^{-1}, determined on live cells sinking in still water. Smayda also cites studies of sinking rates derived from field observation of rates at which diatom layers changed depth. These ranged from 1–5 m day^{-1}, greater than still water sinking rates of the same or similar species.

Many diatoms apparently exercise a degree of control over their sinking rate through regulation of their density, size and shape. Changes in sinking rates favorable to the population have been noted to occur in apparent response to environmental factors, notably light and nutrient levels. although sinking predominates widely, diatoms may also become positively buoyant on occasion, changing buoyancy by ionic regulation of cell sap density or by accumulation of fats. Smayda (1970) cites several cases of centric diatom blooms at the sea surface that dramatically demonstrate floating. As would be expected if cell density were independent of its volume, the main influence on still water sinking rates of phytoplankton is cell size (or aggregate size, if colonial). Eppley et al. (1967) summarise this effect and also emphasise the physiological condition of cells as a determinant of sinking rate. Cells from senescent cultures sink about four times faster than cells from growth-phase cultures. Increased sinking rates would place nutrient-deficient or light-blitzed cells at water depths more favorable in either respect.

The determination of sinking rates in dinoflagellates is more complex because most are motile. A passive sinking rate of *Gonyaulax polyedra* in growth phase was reported by Eppley *et al.* (1967) at 2.8 m day^{-1}, whereas swimming at 1 to 2 m h^{-1} is indicated by studies of this and other dinoflagellate species (Hasle, 1950, 1954; Hand *et al.*, 1965). In any case, dinoflagellates move faster by swimming than by sinking.

Phototaxis is not restricted to dinoflagellates, but is widely represented among motile plankton and fish species. The resulting diel patterns of vertical migration thus occur in every phylum represented in the sea and for a wide variety of apparent advantages to the migrants. Longhurst (1976) provides a recent discussion of the mechanisms and advantages of vertical migrations as well as entry to the extensive literature on patterns of vertical movement of zooplankton. In circumstances where phototaxic migrants move into or through persistent vertical circulations, patterns due to their interaction must necessarily have diel periodicity. Zones of higher organism concentration in daylight would thus become rarified at night, as the migrators would instead collect, if at all, at convergences of the opposite direction. Diel variation of convective circulation in the surface layer (Woods, in press) also would affect such patterning.

Hardy and Bainbridge (1954) measured swimming speeds of various zooplankters by placing individual animals in a water-filled torus, which was then rotated on its horizontal axis at recorded rates to keep the swimmer at the same height. Sustained upward rates of nearly 30 m h^{-1} were achieved by the robust copepod, *Centropages*, and 90 m h^{-1} by the euphausiid, *Meganyctiphanes norvegica*. Downward speeds often exceeded upward speeds: e.g. *Calanus* sustained speeds of 47 m h^{-1} downward versus 15 m h^{-1} upward. Enright (1977), from open ocean catches of the 2–3 mm copepod *Metridia pacifica* in several sets of serial net hauls made vertically above and below the base of the thermocline, determined upward swimming speeds from 30 to 90 m h^{-1} for more than an hour. This was greater by 2 to 10 times than previous field estimates he noted. These values give some idea of the maximum depth range that could be traversed in one or two hours of vertical swimming in the absence of vertical flow.

Sinking rates of zooplankton are usually small compared to their vertical motility, but could be important during periods of inactivity. One such period is at night: after the organisms have completed their main upward migration, passive sinking may occur for several hours before downward migration is induced near dawn (Raymont, 1963). Sinking has been invoked to explain the 'midnight scattering' apparent among various vertical migrants that spend the night in the upper layer. This scattering could as well be due to random vertical swimming in the absence of light stimulus. Midnight scattering may be followed by upward swimming near dawn just before the migrants swim downward to seek their daytime levels.

Diel vertical migrations of fish usually follow the pattern of most zooplankton migrants; rising to the surface layers around sunset, dispersal during darkness, and returning to deeper layers by sunrise. Adults and larvae of sardine, herring, pilchard, coalfish and plaice have been noted to migrate roughly according to this pattern, and many myctophids and gonostomatids perform regular diel migrations to the sea surface layer from several hundred meters depth. These larger zooplankton and fishes do not necessarily interact directly with patterning circulations, but may find feeding advantages where their prey are so concentrated.

Organisms not directly affected by patterned vertical motions may nevertheless respond to substances or to organisms that are so affected. For example, nutrient salts brought up to the euphotic zone locally stimulate production of phytoplankton, whereas zooplankton patterned by vertical motion may locally graze down phytoplankton or attract predators, producing secondary patterning of either. Similarly, flotsam and surface active materials collect at surface-convergent fronts where they subsequently can attract or affect organisms. Some, if not most, organisms respond selectively to gradients that are created or modified by fronts and eddies.

Attraction to food commonly accounts for aggregation of animals at or near fronts larger than the Langmuir pattern. They may arrive by a variety of ways. Passive floating behavior by *Pelamis platurus*, the sea snake, causes large aggregations at fronts in the eastern tropical Pacific where they feed on small fish (Kropach, 1975). Small fish, in turn, aggregate in these fronts for plankton. Several species arrive with the floating wood and debris to which they are attracted (Hunter and Mitchell, 1967). The phenomenon is commonplace and commonly attracts larger and smaller animals of recurrent species, leading Dunson and Ehlert (1971) to suggest that the fronts create a community. Most often seen nearshore, such assemblages occur in offshore water to sometimes dramatic degrees. In the offshore eastern tropical Pacific, fronts occur that assemble plankton, fishes, birds, whales and porpoises, as well as flotsam with expatriated littoral and neritic animals and plants. Beebe's (1926) classical description of such a front lists several species in each of these categories. They arrived in numbers by every conceivable pathway

These observations have been since confirmed and extended, e.g. for bird aggregation by Ashmole and Ashmole (1967), and for mammals by Gaskin (1968 and 1976). Food aggregations were the attractants in each case.

3 Patterned flow types, scales and genesis

Where circulation characteristics are determined by the presence of coast-lines, islands, banks, seamounts, etc. the pattern is under topographic control. Circulations under dynamic control are relatively independent of topography, and include fronts and eddies arising from atmospheric forcing and from current confluence, meandering, or other flow instabilities. Fronts and eddies of both types occur at all sizes and frequencies important to organisms. Those of the dynamic type occur virtually anywhere in the sea, whereas the topographic type occur most frequently and noticeably along continental margins and in island wakes.

Due to their proximity to external driving forces, the surface layers of the sea tend to be the most energetic, both physically and biologically. Topographic control is thus most pronounced where the bottom is shoal enough to intercept and modify surface layer processes. Although active over far less ocean area than dynamic processes, topographic fronts and eddies operate in zones already high in biological activity. Fronts and frontal systems that affect measurable biological conditions operate on scales as small as meters and hours and as large as 10^3 km and years. On the other hand, eddies of less than a few km in radius and a week in duration have not been noticed to produce biological patterning. Although small eddies transport and perhaps conserve materials, their main effect is to broach or attenuate features of larger scale and thus act more to disrupt than to create pattern.

SMALL-SCALE FRONTAL SYSTEMS

Patterned convergences on the smallest scales at the sea surface are usually accounted for by cellular convections due to wind or to thermocline effects. Such convections take the form of paired, vortical cells of opposing rotation around their horizontal axes. Water downwells where surface flow con-verges between cells and upwells where subsurface flow converges. Thermohaline convection occurs when denser water forms rapidly at the sea surface and is responsible for the smallest patterns seen to collect organisms. Denser water forms by cooling due to evaporation, back radiation and heat conduction from the very surface. The density increase by cooling may be augmented by the salinity increase from evaporation. Formation of denser water at the sea surface creates vertical instability. At high rates of formation, this water is hydrodynamically most easily discharged by sinking of the dense water along preferred planes rather than by simple diffuse sinking as small filaments. This process has been observed to create patterns of small-scale convection cells which are elongated Bénard cells. Bénard cells may be commonplace under light wind conditions and probably augment convection at higher wind speeds. Defant (1961) reviewed criteria for such convections and gave evidence that Bénard circulations are frequent and widespread in equatorial and mid-latitudes, extending in depth to 25 m or more in the absence of overriding motions. Instability of surface layers, due mainly to evaporation, is cited to occur in the entire pelagic sector of the Atlantic from 20°N to 50°S and to be most pronounced between 15°S and 20°S.

The smallest coherent frontal patterns that have been noted to collect organisms in the open sea were attributable to the Bénard mechanism. Surface convergences penetrated to less than a meter in depth, extended visibly for more than 30 m in unbroken length, spaced at intervals of about 1.5 m. The patterns persisted in the absence of wind for at least two days (Owen, 1966). Taken singularly, such 'micro-fronts' would have little ecological impact. This particular frontal system, however, extended over several hundred km², so that even such fine-scale patterns can produce widespread alteration of concentration patterns and, presumably, concentration-dependent biological processes. In this instance, the surface convergences were marked by red bands less than 10 cm wide that consisted of upward swimming oikopleurans that had been highly concentrated in the upper 5 cm of the convergence planes. Their still water swimming speed of about $0.2 \, \text{cm s}^{-1}$ set an upper limit on downwelling water speed in the convergence planes.

Under freshening and sustained winds, convergence systems of the better known and (usually) larger scale Langmuir circulation either develop extant Bénard cells or create their own patterns, detectable as parallel windrows. Windrows lie along the wind axis and are often marked by assemblages of any floating (i.e. strongly buoyant) substances in the neighborhood. These are swept by convergent surface flow to the frontal zones. Trapping efficiency of buoyant materials is virtually 100% under steady wind direction. Surface convergences can become strong enough on occasion to submerge even highly buoyant *Sargassum* weed, requiring a downwelling circulation of $5-7 \, \text{cm s}^{-1}$ (Woodcock, 1950). Organic films on the sea surface also are rapidly collected along such convergences to give windrows their characteristic banded appearance. Under lower winds, collected oils dampen capillary waves to give the windrows the appearance of parallel slicks. At higher wind speeds, the organics are partly frothed and mark convergences by foam lines.

The size and circulation strength of these convection cells increase with time and wind speed. Spacing of convergence lines several meters to over 100 m is typical, and downwelling speeds may exceed $10 \, \text{cm s}^{-1}$. Depth of penetration is some fraction of the distance between convergences, usually half or somewhat less.

Extant theories that attempt to account for the Langmuir circulation include instability of vertical shear in the water, forcing by atmospheric vortices, wind-oriented thermal (Bénard) convection, wind profile modification by surface films, lateral radiation pressure on surface films, convergence of wave trains (reviewed by Scott et al., 1969), and vertical decay of surface wave oscillations (Faller, 1969). Failure of their data to support or rule out any single mechanism led Scott et al. (op. cit.) to conclude that combinations could occur of mechanisms they considered. A subsequent study (Harris and Lott, 1973) lends support to the possibility of development of Bénard cells into Langmuir cells: mean downwelling velocities ranging from 2 to 10 cm s^{-1} in Langmuir convergences correlated well with wind speed during net surface cooling but not as well during net surface heating. Despite the lack of agreement on mechanisms, Langmuir circulations are now widely held to be the most important mechanism for vertical transfer of heat and substances in surface waters of lakes as well as oceans.

The theoretical and mathematical description of particle trajectories in cellular convections was formulated by Stommel (1949) and subsequently elaborated to include swimming (e.g. Stavn, 1971). Direct observations confirm the applicability of the model to real particles and plankters in such circulations, despite the rather large difficulties of field measurement and the less-than-ideal behavior of both the fluid and plankters.

A number of field studies confirm the effectiveness of the Langmuir circulation in collecting and patterning organisms and substances. The most visible examples involve buoyant animals and substances such as *Physalia* (Woodcock, 1944), *Sargassum* weed (Faller and Woodcock, 1964; Woodcock, 1950) and surface films (Szekielda et al., 1972). Sutcliffe et al. (1963) demonstrated that whitecapping and perhaps other wind action converts dissolved organics to particulate form available to zooplankton, and that a Langmuir convection downwelling at 3–6 cm s^{-1} produced subsurface concentrated zones of these particulates. The soluble inorganic phosphate concentration was also shown to be higher in such particulates. Sutcliffe et al. (1971) showed, by detecting an increase in particle concentrations with increase in wind speed, that the process is highly effective. They found peak production to be particles of 6 μm diameter. Higher concentrations of particles 2–13 μm diameter were found below convergence zones.

Patterning of phytoplankton and microzooplankton by Langmuir circulations has been more widely reported than one would expect from the mismatch of their reported rates of vertical motion. Bainbridge (1957) reviews several observations of dinoflagellate concentrations in Langmuir-scale patterns. The mobile and prolific ciliate *Mesodinium rubrum*, which can occur in blooms ('red-water') concentrations, has been reported to gather in Langmuir convergences. Powers (1932) found them in windrows off Maine, and Bary (1953) found windrows of the same ciliate to orient along a changing wind axis in New Zealand. Packard *et al.* (1978) describe *Mesodinium* windrows spaced at 100 m intervals under 10–20 knot winds, and their dispersion through the water column at winds over 30 knots.

Patterning of zooplankton in Langmuir circulations is less apparent and notoriously difficult to sample. The best examples of Langmuir patterning of zooplankton thus are from lakes, where populations of cladocerans and copepods migrate vertically into Langmuir vortices in the epilimnion and form rows of higher concentration in convergence planes (George and Edwards, 1973).

INTERNAL WAVE PATTERNS

Internal waves generate banded patterns on the sea surface on a range of scales extending from that of the Langmuir circulation. Visible rows spaced 10–100 m apart are common on continental shelves, particularly when insolation produces a thin, light surface layer. Ewing (1950) first identified internal waves as responsible for light-wind band slicks and presented evidence that slicks progressed with and marked convergent displacements on the surface layer, moving at 1–100 cm s^{-1}. They are made visible from collection of surface films in compression zones (Garrett, 1967) or directly from the effect of billow currents of internal waves (Gargett and Hughes, 1972), either of which locally damps surface capillary waves. Rows may extend for several km and lie usually more or less parallel to shore, to isobaths and to one another. Row orientation is independent of wind direction, but rows are erased (or replaced by the Langmuir circulation) when wind speed increases about 3 m s^{-1} for 30 minutes or more.

Internal waves on the continental shelf are generated remotely by a variety of mechanisms and altered by changes in density stratification, currents and current shear, and proximity to the sea surface and bottom (see review by Garrett and Munk, 1979). Wind storms and tides are the likeliest common source of their generation. Unlike circulations discussed here, there is no net transfer of water or organisms with the passage of internal waves unless they break. Thus there is no actual circulation of substances and organisms are not differentially collected, except perhaps ephemerally at the very surface. Bacteria depending on the surface film for support are sometimes highly concentrated in internal wave slicks, but more often show no particular relation to them (ZoBell, 1946).

Internal waves, however, may interact with current boundaries and shelf-break fronts. Curtin and Mooers (1975) presented measurements that indicated generation of large amplitude, high frequency internal waves by breakdown of the semidiurnal internal tide at the shelf-break front off Oregon. These induced major perturbation of a sonic scattering layer of organisms inshore of the front. Earlier, Yasui (1961) presented a theoretical framework arguing for generation and propagation of tidal period internal waves along fronts due to current confluence, which are zones of high lateral shear as well as vertical circulation. Localised interaction between internal waves and eddies in mid-ocean also has been detected (Frankignoul, 1976) which may affect rate of vertical diffusion. As yet, these phenomena seem to be unexplored, but may be expected to modify effects of fronts on biota. Kamykowski (1974), for example, presented a reasonable mechanism by which phytoplankton can interact with the semidiurnal internal tide to produce patterns of species and biomass.

Internal waves are sensitive to current shear and to vertical density gradients, both of which are altered at fronts and eddies. It will not be surprising to find evidence of biological responses to these coactive physical processes if adequate sampling techniques can be devised.

4   Coastal fronts and eddies

Topography of the sea floor and obstacles to flow exert locally significant control of convergence systems above the scale of the Langmuir pheno-menon and spawn eddies large enough to affect biological processes.

TIDAL FRONTS

Simpson and Pingree (1978), Pingree *et al.* (1977) and Pingree *et al.* (1974) described cases of genesis and maintenance of shallow sea fronts by tidal mixing on the European continental shelf. Surface convergent fronts of considerable physical and biological activity occur in zones where water types of different mixing histories impinge. Where tidally-mixed water contacts stratified water, a convergent front occurs. Thermal images of the Celtic Sea from the NOAA-5 satellite confirm the broad occurrence of tidal fronts and show the genesis of cold eddies from frontal meanders (Fig. 2). These fronts respond to the neap and spring tidal cycle and its effect on phytoplankton chlorophyll concentrations in the stratified water beyond the front. The quoted studies suggest that blooms of a somewhat distant dinoflagellate population were caused by periodic release of nutrients from the destratified side of the front.

Admixture of water types at shelf fronts locally stimulates phytoplankton photosynthesis and thus sustains higher phytoplankton concentrations in the frontal zones. Savidge (1976) and Savidge and Foster (1978) described such a process from surface measurements of chlorophyll, temperature and photosynthesis along transects on the European shelf between Ireland and Wales.

Zooplankton concentrated in shelf fronts is also observed. Pingree *et al.*

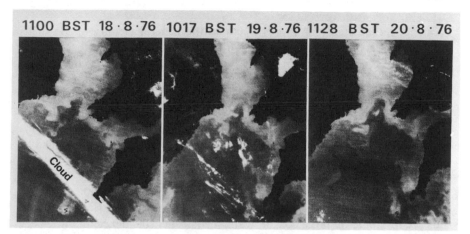

FIG. 2   Infrared images from NOAA-5 satellite of thermal patterns in the Celtic Sea from 18 to 20 August 1976. Lighter shades represent cooler surfaces. From Simpson and Pingree (1978)

(1974) described a vigorous tidal front of 15 km extent in the English Channel from temperature, salinity and current measurements and from direct underwater and surface observations of its effects on organisms and substances. Surface net hauls showed much greater frontal concentrations of copepods, crustacean meroplankton, the euphausiid *Nyctiphanes* sp. and (perhaps) young fishes.

Plankton concentrated by the front was confirmed by subsurface visual observation. Underwater observation of motions of plankton and of dye injected at several depths also confirmed the turbulent and convergent character of the fronts that was apparent from temperature and salinity structure and from drogue measurements. The list provided of animals associated with floating macrophytes concentrated at the front consisted of a variety of post-larval fishes and crustacean species. Puffins, shearwaters and terns were seen feeding along the front, which also collected debris and oil lumps. The authors remark on the similarity of this front and its biota to the large, deep-ocean fronts described by Beebe (1926) and Amos *et al.* (1972).

Marine mammals are documented to exploit the biota of tidal and other shelf fronts. Gaskin (1976), for example, cited feeding by minke whales on herring or capelin aggregated in coastal tide slicks of eastern Canada and feeding by fin whales on surface concentrations of the euphausiid *Meganyctiphanes* in a convergence in the Bay of Fundy. Mackerel schools had forced the euphausiids to the surface.

SHELF-BREAK AND UPWELLING FRONTS

Fronts of large extent commonly occur in the general vicinity of the shelf-break, i.e. where the gradient of the continental shelf steepens to become the continental slope, generally at about 100–300 m depth. Mooers *et al.* (1978) usefully distinguished between prograde fronts, those that separate more saline shelf water from offshore water, and retrograde fronts, those that separate less saline shelf water from offshore water (Fig. 3). Shelf waters of both types usually are colder than offshore.

Prograde fronts arise by upwelling on the continental shelf. Seasonal and episodic wind-driven upwelling is characteristic of eastern boundary current systems, but occurs inshore of western boundary currents as well, where upwelling occurs by topographic deflection or trapping of currents. Retrograde fronts are due to shelf-water dilution from terrestrial runoff and thus predominate off large coastal watersheds. Their incidence and intensity usually vary with season.

Prograde shelf-break fronts off upwelling zones bound plankton rich shelf waters. There is a dearth of studies that demonstrate higher plankton concentrations in these fronts than inshore, at least in part because field studies have been concerned more with the upwelling system as a whole and do not sufficiently sample the frontal interface. Packard *et al.* (1978) described the incidence and habitat of the pigmented ciliate, *Mesodinium rubrum*, in an upwelling zone off Baja, California. Characteristics of its environment in combination with its phototaxic behavior, rapid reproduction rate and ability to use both inorganics and organics for food, were cited to potentiate surface-layer concentrations of the ciliate at prograde fronts.

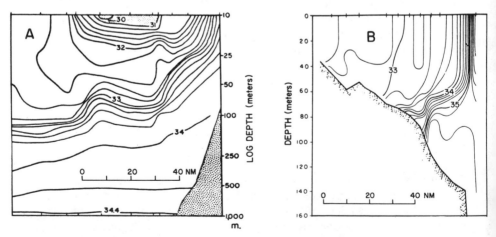

FIG. 3   Structure of prograde (upwelling) fronts and retrograde (estuarine) fronts. Panel A from Owen (1968). Panel B from Mooers *et al.* (1978)

Bang (1973) described measurements showing a strong upwelling front extended northwest off Cape Town in the eastern boundary current of South Africa. Mixing cells occurred and episodic overturn at the front produced subthermocline sheets of homogeneous water. Concentrations of large zooplankton or micronekton were detected at the front by fathometer. The area inshore of this front has begun to support a fishery on the lantern fish *Lampanyctodes hectoris* incidental to the extant anchovy and pilchard fishery of the region (Ahlstrom et al., 1976).

Albacore make annual trans-Pacific migrations to feed on relatively high concentrations of large plankton and small fishes in the California current, seaward of coastal upwelling zones. The offshore extent of these upwelling zones are frequently denoted by prograde (upwelling) fronts. Albacore caught, tagged with ultrasonic transmitters and released were tracked in the vicinity of such a front off Monterey Bay, California (Laurs et al., 1977). Their results showed that such temperature fronts influence local concentration of albacore, with albacore concentrating in the vicinity of the front, moving away with the degradation of the front as upwelling ceased. Albacore also were seen to move more slowly when crossing fronts and to spend little time on the colder side (t < 15° C) indicating a thermal barrier effect which would result in their collecting at fronts. The enriching effects of large cyclonic eddies that predominate outside upwelling zones, and eddy transport from upwelling zones described by Owen (1980) may also support higher forage levels beyond such fronts.

An important aspect of upwelling fronts is their periodic relaxation or breaching, and the subsequent release into offshore water of nutrients and plankton previously accumulated in the shelf water. Frontal relaxation occurs upon cessation of upwelling, which can be either seasonal or episodic depending on the wind system. Breaching occurs more locally than does relaxation. The causes of breaching have not been defined but likely involve meanders and eddies due to local instabilities of frontal currents, outside eddy impingement, and perhaps wave interactions with the fronts. When clear weather permits, breaching is clearly evident in thermal images of sea surface temperature fields off California transmitted from satellites. Bernstein et al. (1977) demonstrated the possibility of using such infrared images to follow the genesis, lysis and incidence of eddies, fronts and upwelling zones where they involve temperature, as in the California current system.

ESTUARINE FRONTS

Retrograde (estuarine) fronts exhibit salinity gradients, since freshwater sources dilute surface waters on one side of such fronts. Iisuka and Arie (1969) showed concentrations of *Trichodesmium* (flagellate) for 80 km along an estuarine front off Japan. LeFèvre and Grall (1970) showed similar convergent aggregations of *Noctiluca scintillans* off Brittany, and emphasised local competition for diatoms between *Noctiluca* and copepods.

Owen (1968) studied effects of the Columbia River plume, which albacore encounter in their early summer arrival off Oregon. Based on research vessel catch rates and oceanographic data, results showed that albacore moved through the pronounced outer plume boundary, defined by the $32.2°/_{oo}$ isohaline that demarcated the outer estuarine boundary of the salinity front of the plume-sea interface. Catches were usually lower in mid-plume than in the vicinity of its frontal limits, indicating an attraction to the fronts *per se* or possibly a combined attraction/aversion to the warmer temperatures/ lower salinities of the plume core. Catch rates may be seen to be much higher near the outer plume limit than near the inshore limit, where even stronger thermohaline fronts result from combined effects of the plume and coastal upwelling (cf. Fig. 3, panel A).

On smaller scales near shore, Tsujita (1957) showed seasonal on-shore/offshore movements of stocks of the Japanese sardine off western Japan. Spawning and hatching occurs at depth in oceanic water near a retrograde (estuarine) coastal front. Larvae are transported or migrate through the shelf-break front and seek nearshore feeding grounds, growing to 7–10 cm and diffusing seaward in summer in the upper layer of the coastal zone.

A large saline front found spanning the Equator 600 km off Sumatra and south of the Bay of Bengal by Amos *et al.* (1972) was formed at the confluence of the saline Equatorial Countercurrent with water from the north, diluted by effluent of the Ganges, Irrawaddy and Salween rivers. Hardly a shelf-break feature, the front still was of estuarine character and exhibited high biological activity: obviously associated with the front were small sharks and flying fish, and feeding sea birds. The front also collected a large quantity of debris and abundant plankton. Sea snakes and dolphin fish also were seen at the convergence. Under no immediate topographic control, the front was nevertheless of estuarine origin.

5 Deep-sea fronts and eddies

Beyond immediate topographic control, confluenced waters of different mixing histories produce a front or frontal system in the equatorial Pacific (Wyrtki, 1966). Pak and Zaneveld (1974) demonstrated increased particle concentrations associated with a front east of the Galapagos Islands (downstream with respect to the equatorial undercurrent) that was defined by large horizontal gradients of temperature, salinity, density and nitrate concentration. They attributed the front to the confluence of the Peru current with the south equatorial current, and cite earlier physical oceanographic investigations of the same feature that indicate its permanence and the seasonality of its location and intensity. It seems possible that this is the same front described by Beebe (1926) that dramatically assembled a large fauna.

Cromwell and Reid (1956) described such fronts (or the same front) from close-spaced temperature profiles along 120°W and 172°W and suggested cell-like circulation from their data (see Fig. 1). Knauss (1957) described a second front crossing at 120°W with even closer temperature profiles that defined the frontal edge. Knauss mentioned high concentrations at the front of squid, *Pyrosoma*, flying fish (2 species), sauries and lantern fish (2 species), and the absence of floating debris. Water characteristics on the cold sides of these fronts indicated the involvement of the east-flowing equatorial undercurrent in their creation and maintenance.

Murphy and Shomura (1972) documented increased availability of yellow-fin tuna near island groups of the central Pacific, particularly of younger fish in surface schools. Three assessment methods, sightings of surface schools, surface trolling and deep longlining clearly indicated their increased abundance within 100 km of islands. Concentration of tuna forage by topographic fronts and eddies was suggested to account for the effect.

Murphy and Shomura (1972) pointed out that 'the very existence of schools of carnivores would seem to require schools or aggregations of prey, for if prey were distributed at random, it could be most effectively harvested . . . by individual predators'. They showed a close correspondence between the incidence of thermal fronts and tuna school sightings on transects from 10°N to 5°S in the central Pacific beyond influence of islands. They concluded that concentrating mechanisms were more important than overall levels of forage in determining incidence of surface schools, but that the overall forage levels were of more influence on distribution of the much less aggregated deep-swimming tunas.

At higher latitudes, front-prone zones correspond with transition zones of the atmosphere between the easterly tradewinds and westerlies. The region between 22°N and 32°N in the Atlantic appears particularly active in late winter and spring (Voorhis and Hersey, 1964). Voorhis (1969) described shipboard physical measurements of a large, meandering thermal front extending east-west for nearly 1000 km at 27–30°N in the Sargasso Sea. A similar front was later found to persist for 3 months by repeated surface temperature surveys from low-flying aircraft equipped with an infrared detector.

Biological sampling across the Sargasso Sea at other times has shown north-south discontinuities at the front of small particles (Spilhaus, 1968), phytoplankton production (Ryther and Menzel, 1960), species composition and standing stock of phytoplankton (Hulbert, 1964), zooplankton (Colton *et al.*, 1975) and mesopelagic fishes (Backus *et al.*, 1969).

In the Pacific, the annual feeding migration pattern of albacore tuna from waters off Japan to the American west coast appears to be bounded by fronts defining the northern and southern limits of the transition zone separating Pacific central and subarctic water masses. Shomura and Otsu (1956) and Graham (1957) found that mid-Pacific albacore catches were characteristically associated with the subarctic front and transition zone. This zone is probably a consistent source of surface layer enrichment (McGary and Stroup, 1956) indicated by elevated phosphate levels and reduced water transparency due to plankton production. Laurs and Lynn (1977) studied albacore catch rates in relation to transition zone fronts, both under intensive coverage as the fish approached the west coast feeding grounds in late spring. Large catch rates did not persist in any one area for more than a few days, indicating movement of the schools; their degree of constraint between oceanographically-defined frontal zones is apparent in Fig. 4.

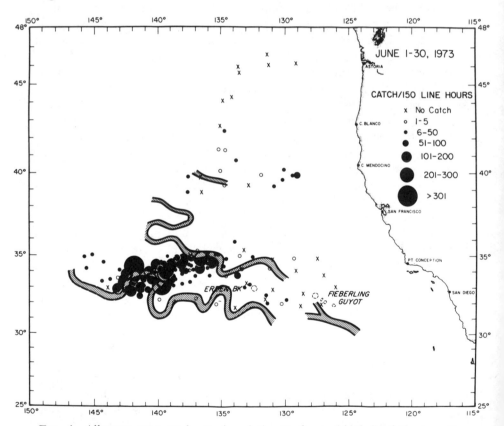

FIG. 4   Albacore tuna catch rate in relation to fronts (shaded) of the transition zone in June 1973. From Laurs and Lynn (1977)

Although not clear from the limited direct measurements available, this transition zone and its fronts probably furnish more food to migrators than waters beyond. Albacore have a fondness for saury and squid, which have been noted to concentrate at frontal structures elsewhere (Uda, 1938, 1952; Han and Gong, 1970). The subtlety of temperature and salinity gradients that define transition zone fronts make it seem unlikely that albacore respond directly to these gradients to determine their movements, although Neill *et al.* (1976) suggested a mechanism by which tunas may orient to temperature gradients as small as $0.1° km^{-1}$.

Seckel (1972) suggested that convergent flow along the large-scale saline front between North Pacific central and equatorial water masses may operate directly to concentrate skipjack tunas as well as their food. This front is demarcated by a horizontal salinity gradient that periodically reaches or passes the Hawaiian Islands; the abundance of skipjack, indicated by skipjack catch by local fishing, increases at these times. This suggests at least a bounding effect on the fish. The northward passage of the front is detected by the sharp decrease (here, in May) of salinity at a shore station on Oahu.

A growing body of evidence suggests that topographically independent baroclinic instabilities (wavelike motions in large-scale flows with vertical shear in which Coriolis and buoyancy forces are important) are significant in large-scale ocean circulations. Hart (1979) reviewed the theoretical basis of this large eddy-producing phenomenon and Hide and Mason (1975) reviewed annulus convection experiments in laboratory flows. In addition to the meanders and eddies of the Gulf Stream and Polar Front cited below, examples of such activity are apparent in the central North Pacific. Bernstein and White (1974), from four sets of temperature observations, verified the existence of a mosaic of baroclinic eddies well removed from islands, ocean boundaries and the Kuroshio.

TOPOGRAPHIC CONTROL OF DEEP-SEA EDDIES

It is equally apparent that continental margins dissipate energy through generation of large eddies. Even in the weak and diffused eastern boundary currents, evidence of boundary effects extend well past continental margins (shelf and slope areas). These are evidenced by thermal patterns from satellites (Bernstein *et al.*, 1977) (Fig. 5). Owen (1980) reviewed eddy genesis and incidence in the California current system and demonstrated the preponderance of cyclonicity among large eddies detected in geostrophic flow by extensive surveys of the California Cooperative Oceanic Fisheries Investigations (Wyllie, 1966). One particularly persistent feature is the southern California Eddy which appears to collect and recirculate biota from the upwelling zone north of Pt. Conception. Owen (1980) indicated

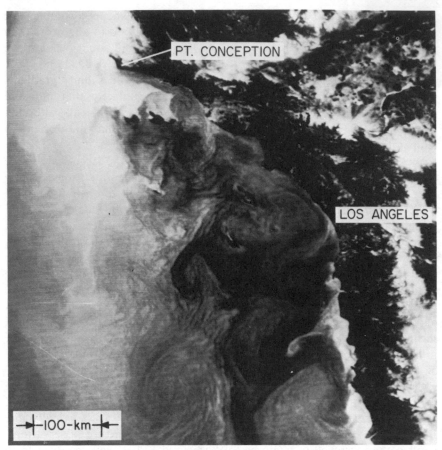

FIG. 5   Infrared image from NOAA-6 satellite of thermal patterns in the southern California Bight, September 1979. Point Conception and the Channel Islands are visible in the upper left quadrant. Lighter shades represent cooler surfaces. Photo courtesy of Remote Sensing Facility, Scripps Institution of Oceanography

the role of the Eddy in nutrient enrichment of the surface layer by upward flow and nutricline displacement. Washdown of diatom assemblages was indicated by Allen (1945) and higher chlorophyll concentrations in the Eddy were shown by Owen (1980). Brinton (1976) examined life history and local distribution of *Euphausia pacifica* which suggested the Eddy to be a reproductive refuge for the population. The correspondence is unmistakable between the Eddy and patterns of concentration of both sardine eggs (Sette and Ahlstrom, 1948) and diatom concentrations (Sargent and Walker, 1948) in spring of 1941.

Another large, stationary eddy is even more isolated; the Costa Rica Dome is produced by topographic deflection of the north equatorial countercurrent. The Dome, named for its upward bulging thermocline layer, centers in the bight of this deflection (Fig. 6). Upward displacement and active upwelling in the core region cause nutrient enrichment of the photic zone (Broenkow, 1965). This sustains detectably higher stocks of phytoplankton, zooplankton and small nekton according to authors cited by Owen (1980).

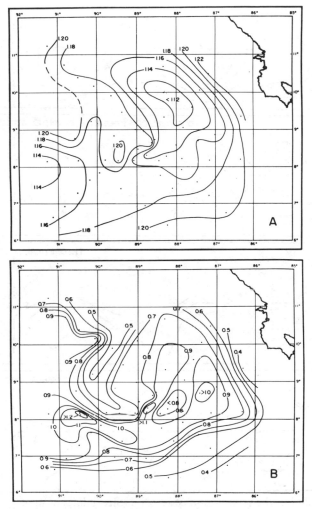

FIG. 6   (A) Flow represented by geopotential topography of the Costa Rica Dome at about 50 m depth relative to 1000 m. (B) Phosphate concentration (µg atoms l⁻¹) at 50 m depth. From Broenkow (1965)

Islands, seamounts and headlands act as obstacles to sea and air flow and induce instabilities that can develop into eddies. Eddies may be stationary, "attached" to the obstacle or, with increased flow or obstacle size, may be shed in series. Barkley (1972) found remarkable agreement between observed current patterns downstream of Johnson Atoll and an adaptation of

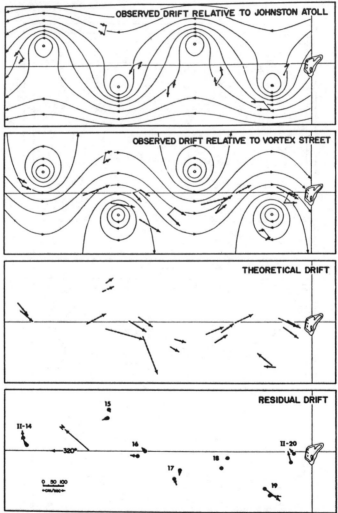

FIG. 7   Eddies in the wake of Johnston Atoll. Streamlines in upper panels derived from von Kármán model of vortex street, fitted to current measurements. Degree of fit is apparent from the lower panels; residual drift values are small compared with theoretical and observed values. From Barkley (1972)

Von Kármán's model of flow past cylinders (Fig. 7). Patzert (1970) described eddy systems in the lee of the Hawaiian group, where eddies appeared to form by local wind forcing. White (1973) detected eddy patterns east of the Galapagos, with respect to the equatorial undercurrent. Wind or current effects at headlands produce eddies, perhaps acting as half an island. Arthur (1965) identified the importance of flow vorticity at headlands of west coastlines. Where the thermocline and nutricline are shallow, the eddy formation produces cool, enriched water at the surface equatorward of such promontories (Reid et al., 1958).

Boden (1952) and Boden and Kampa (1953) determined summer and winter circulation patterns over the submerged platform of the seamount from which Bermuda emerges. Distribution of density in both seasons indicated a cyclonic eddy spanning the entire seamount, and convergent fronts at its margins. The planktonic larvae of benthic animals sampled in summer were considered to be returned to and retained on the platform by both eddy and frontal circulations, enhancing their eventual recruitment to the bottom fauna. Intense mixing zones close to the platform sides were indicated by local vertical instabilities as density inversions.

Hogg et al. (1978) detected eddies near the Bermuda Platform (Fig. 8). Fine-structure of temperature variation on intervals from 0.2 to 25 m indicated a large degree of vertical mixing in patches close to the island, a consequence they felt due to strong longshore flow between the eddies. Hogg et al. (1978) attribute one of the eddies to a meander in the Polar Front. Intense mixing zones occurred on small scales (0.2–1 m variances) next to the platform walls (note Boden and Kampa's result above) and on larger scales (5–25 m) at about 10 miles distance. Velocity profiles showed that the vertical scales of mixing increased with distance from the island. Internal waves were discounted as a source of the mixing. Osborn (1978), using measurements of velocity shear microvariation, found zones of high energy dissipation (by mixing) within 5 km of the island of Santa Maria in the Azores. Mixing zones were as much as 45 m thick, at the base of the upper mixed layer. At 80 km distance, maximum dissipation zones were below the thermocline at depths over 250 m.

Emery (1972) described evidence from drift bottle and current meter data favoring a persistent system of shed eddies in the wake downstream of Barbados Island (West Indies). Weekly evening plankton tows made 3 km off Barbados for over a year indicated that retention or return of meroplankton near the shelf of the island was enhanced by the eddies.

Jones (1962) cited evidence for effects of the Marquesas Islands in the equatorial Pacific. He found significantly higher zooplankton volumes in quantitative net tows approaching the islands from any direction in the zone from 10 to 180 miles. The higher plankton concentrations were not due to

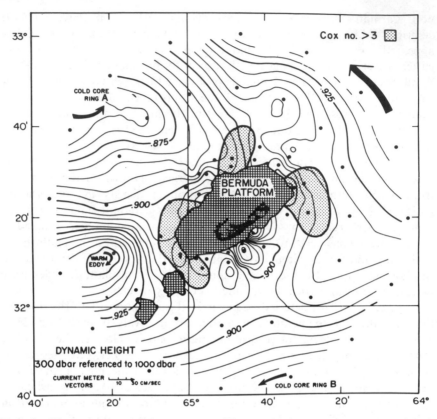

FIG. 8 Flow represented by geopotential topography and intense small-scale mixing zones at about 300 m depth near the Bermuda Platform, according to Hogg *et al.* (1978)

washout of island plankton as most were dominated by euphausiids and siphonophores. Neither group is considered land dependent. The decrease in water transparency approaching the island zone is consistent with higher phytoplankton stocks.

FRONT-EDDY INTERACTIONS

Large-scale associations of fronts and eddies are apparent from several examples. Where fronts are vigorous for their size or meet an obstacle, flow becomes unstable and frontal meanders may pinch off and become eddies as shown in Fig. 9 (Fuglister and Worthington, 1951). Western boundary currents and their seaward extensions produce large intense fronts and eddies. A well-documented example is spawning of eddies from the North

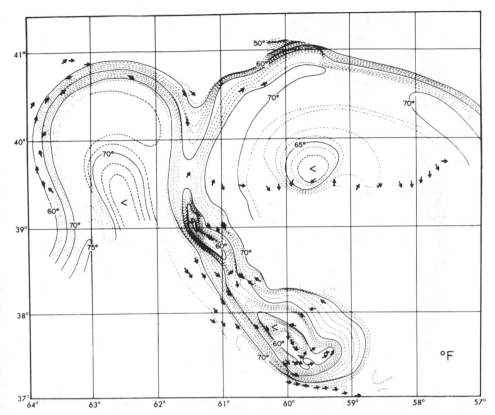

FIG. 9   Cyclonic eddy genesis from a Polar Front meander. Temperature pattern at 200 m depth and measured current vectors (→) delimit a cold core meander and eddy pattern. From Fuglister and Worthington (1951)

Atlantic Polar Front. The front is a product of the seaward extension of the Gulf Stream axis into mid-Atlantic and is defined by the large temperature gradient between cold shelf or subarctic water to the north and warm Sargasso Sea water to the south. Flow instability periodically produces a large meander of the Front that is subsequently shed as an eddy. If shed south or east of the front, the eddy contains colder water from the north side and is cyclonic, whereas warm-core, anticyclonic eddies spin off into colder water to the north and west (Fig. 10). Cyclonic eddies so form several times a year and are physically identifiable for as much as 2 years (Parker, 1971). Initially, they range to 300 km diameter laterally and extend from the sea surface to 3 km in depth. Such eddies do not always extend to the sea surface. Howe and Tait (1967) detected a sub-thermocline eddy generated

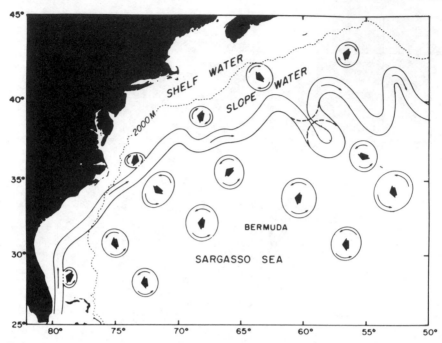

FIG. 10   A schematic representation of the path of the Gulf Stream and the distribution and movement of ring eddies by Richardson (1976)

by a meander of the Polar Front. This eddy centered at about 650 m depth and extended from 150 m to perhaps 1200 m. Circulation strength at the core depth was about 10 cm s^{-1} and the eddy extended laterally about 60 n.mi. Surface layer involvement may have been eradicated by local processes (e.g. wind-induced transport) prior to its detection and measurement. Biological effects peculiar to subsurface eddies are easily imagined but were not assessed.

Predominantly under dynamic control, both the Polar Front and eddy genesis appear to be under local topographic control where the front crosses the mid-Atlantic Ridge. From detailed oceanographic station patterns, Dietrich (1964) noted marked weakening at the Ridge of the temperature and salinity gradients that define the Polar Front at 200 m depth. A major dislocation of the front also occurred at its Ridge crossing: the dislocation took the form of a meander of the type that is a cold-core eddy precursor (Fig. 11, Panel A). Flow direction and speed of the sea surface relative to that at about 1000 m show this meander as well as the blocking or dissipative effect on the Front of the Ridge (weaker gradients of dynamic height east of the Ridge in Fig. 11, right). Topography of the sea floor is elsewhere unlikely to affect the Polar Front and its eddies, once clear of the American continental slope.

Eddies from the Gulf Stream and Polar.Front tend to conserve the biomass and community structure of plankton isolated when they form. Noting their potential for biological interaction, Wiebe *et al.* (1976) summarised biological measurements in several cold-core eddies from the Polar Front. Initially containing plankton stocks of cold-side origin, young cold-core eddies exhibited relatively higher plankton biomass and species assemblages that were distinguishable from surrounding Sargasso Sea waters. With increasing age, such eddies attenuate and assume the physical and biological character of the surrounding water, but may be biologically distinguishable for more than a year.

Phytoplankton biomass and species composition differences in the eddies usually attenuated in about 6 months, or less in one case when storm mixing of the phytoplankton habitat occurred. One young eddy may have generated its own phytoplankton assemblage. Phytoplankton composition in this eddy was shown to differ from that of the source waters as well as from that of Sargasso Sea waters.

Zooplankton biomass and species ensembles of cold-core eddies took considerably longer to attenuate. Consistently higher zooplankton stocks and distinctive euphausiid species were evident in eddies as old as a year. Diel migration patterns of several zooplankton species also appear to have been affected by the presence of the eddies. Daytime depths reached by warm-water euphausiid species that migrate was consistently and markedly deeper beyond than within a cold-core meander of the Polar Front.

Weibe and Boyd (1978) later examined distribution limits of one of the cold-water euphausiids, *Nematoscelis megalops*, involved in the Polar Front and eddy complex. The study illustrates a compund front-eddy effect. The Polar Front (like other western boundary currents) is shown to be an effective thermal barrier to some oceanic plankton species, including *Nematoscelis*. Cold-core eddies shed from the front, however, violate the barrier by transporting *Nematoscelis* far south of where they would otherwise occur. Reduced numbers and nutritional condition of individuals in older eddies indicated that expatriate populations attenuate in time by starvation.

Haedrich (1972) reported midwater trawl catches of myctophids and gonostomatid fishes in the Northwest Atlantic. Much reduced numbers of species and fish biomass were noted in catches at a station in a newborn warm-core eddy on the continental slope north of the Polar Front, compared with species diversity and biomass of catches made at four slope and six warm-water stations.

Dynamic generation of large-scale fronts and eddies is not confined to the

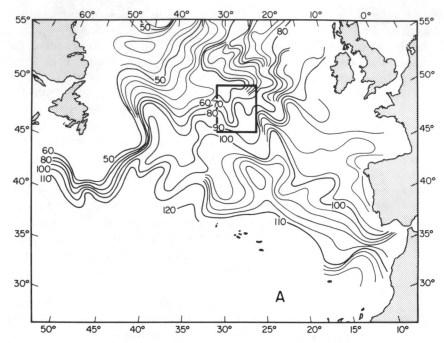

FIG. 11   (A) Flow represented by geopotential topography of the sea surface (relative to that at about 1000 m) in the northern North Atlantic in late summer 1958.

Gulf Stream extension but appears to be characteristic of extensions of other western boundary currents as well. The zone of departure from the South African coast of the Agulhas Current is characterised by complex patterns of vigorous fronts and eddies (Fig. 12). Bang (1970) described the cyclonic eddy 'K' of 50 n.mi. diameter circulating at as much as 6.6 kt at the surface and extending to more than 1600 m depth. Eddy induced vertical displacements of about 500 m were measured in the temperature field. Retroflexion of Agulhas Current water (temp. > 25 °C) is seen from the shaded part of the figure, as this water is redirected eastward by the West Wind Drift. The narrow strip of cold surface water (14–20 °C) above the 'north wall' of the Agulhas Current emphasises the locally divergent nature of what is characterised by Bang as a strongly convergent and eddy-prone frontal system.

Bass (1970) identified the role of the Agulhas retroflexion zone (Agulhas Bank) and the Natal coast as nursery areas for sharks. Four species use Agulhas Bank, 8 other species use the Natal coast. Nursery grounds are on fringes of adult distribution.

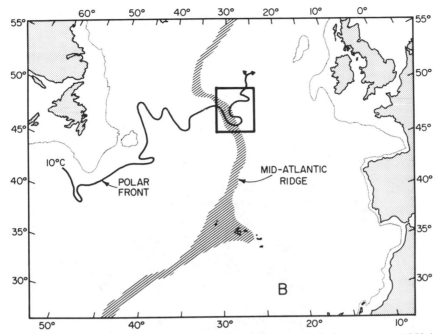

Fig. 11   (B) The oceanic Polar Front axis at 200 m depth in late summer 1958 in relation to the axis of the mid-Atlantic Ridge. Both panels from Dietrich (1964)

Western boundary currents of the Pacific also are well-known to produce fronts and eddies. Strong flow south and flow along the Australian east coast produce large and persistent eddy circulations (e.g. Nilsson *et al.*, 1977). Some of these may be recurrent, or quasi-stationary, suggesting topographic control. Anticyclonic eddies greater than 250 km diameter are particularly common southeast of Sydney. Sufficiently intense biological investigation of the anticyclonic eddies has only recently begun.

The Kuroshio Current off Japan's east coast and its confluence with the south-flowing Oyashio Current are also active generators of large pulsative fronts and eddies. These have long been recognised by Japanese fishermen to attract squids, fishes and mammals (Uda, 1938). Regularity of the eddy patterns along the Oyashio Front, which defines the axis of confluence, led Barkley (1968) to note that these patterns take the form of two vortex streets lying side by side. This mechanism suggests that the Front acts as a perturbing obstacle to the flow of both impinging currents. Because much of the energy of both currents is expended in eddy formation in the confluence region, the Polar and subtropical fronts extending eastward therefore are weaker than the topograpically deflected Gulf Stream.

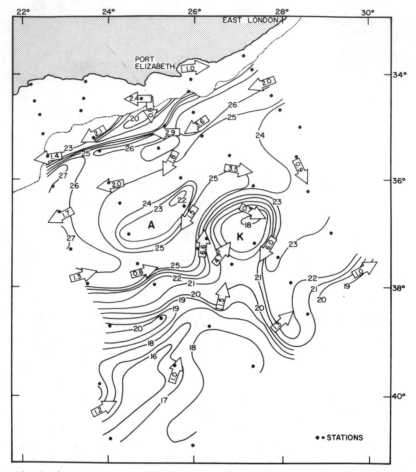

FIG. 12   Surface temperatures, navigation set vectors in knots, and station positions
in the Agulhas retroflexion area (southeast tip of Africa) in March 1969. 'A' and 'K'
designate adjacent cold-core eddies. From Bang (1970)

Eddy generation also appears to be mediated by flow of air from land
masses. Worthington (1972), noting great mixed layer depths and high
pressure zones associated with outbreaks of continental polar air over the
Pacific south of Japan and over the Atlantic south of New England and east
of Newfoundland, believes that convergent sinking of surface water occurs
as a result, generating anticyclonic eddies. Anderson *et al.* (1970) consider
that flow instabilities of the Agulhas Current off Durban are induced by
offshore wind components indicated by local pressure variations.

Since Uda's (1938) comprehensive study of fronts off Japan and their genesis of eddies, the relation of both to the regional Japanese fisheries has received wide attention. Particularly well documented is the effect of the Kurile Front on saury fishing. The Kurile Front is the northernmost of a series of fronts usually found in the Oyashio (Uda, 1938) and is probably analogous to the Polar Front since it forms the southern limit of subarctic water. The most productive saury fishing areas in the northwest Pacific occur in series of eddies along the Kurile Front in summer and fall. This front extends only to 50 m depth. Han and Gong (1970) confirmed this association, showing sensitivity of the saury fishery off the east coast of Korea to the local equivalent of the Polar Front.

Uda (1938, 1952, 1973) and Uda and Ishino (1958) described various associations with fronts and frontal eddies of albacore, yellowfin, skipjack and bluefin tunas, mackerel, salmon, whale species and squid. Feeding, or spawning and feeding provide the biological driving forces for their inter-action with fronts and eddies.

6 Summary and conclusion

Combination of physical characteristics of fronts and eddies with interactive characteristics of organisms leads to the following catalog of ecological effects. Many of these have been shown to occur or to account for observed phenomena. Through organisms' sinking, swimming and production, as dictated by food needs and environmental gradients, such quasi-ordered velocity fields have accounted for quasi-ordered patchiness of marine organisms on widely different time and space scales. Fronts and eddies can:

mechanically affect local concentrations of organisms,
juxtapose populations that would not otherwise interact,
create new 'communities',
conserve and translocate selected species ensembles and concentrations,
attract and sustain large motile animals,
serve as reproduction refuges,
mechanically limit dispersal of meroplankton or neritic populations,
selectivity pattern components of populations or communities (quasi-ordered patchiness),
induce/sustain higher local production of organisms,
modify migration patterns as diverse as annual, trans-oceanic fish move-ments and diel vertical migrations of motile phytoplankton,
collect surface active and particulate substances.

We see that on every scale of frontal and eddy activity in the open sea, there are examples of ecologically significant effects operating through food web stability, community structure, local population dynamics, phytoplankton production and standing stocks, etc. This creates the increasingly distinct view that life forms tend to tesselate the environment in horizontal patterns that are partly governed by vertical exchange due to a variety of physical processes. The scale of the phenomenon and its ecological effect depend on the type and scale of the driving mechanisms, but the resulting biological pattern is also controlled by the organism, its own physical characteristics, behavior, food requirements and response to environmental change.

References

Ahlstrom, E. H., Moser, H. G. and O'Toole, M. J. (1976). Development and distribution of larvae and early juveniles of the commercial lanternfish, *Lampanyctodes hectoris* (Gunther), off the west coast of southern Africa with a discussion of phytogenetic relationships of the genus. *Bull. So. Cal. Acad. Sci.* **75**, 138–152

Allen, W. E. (1945). Vernal distribution of marine plankton diatoms offshore in southern California in 1940. *Bull. Scripps Inst. Oceanogr.* **5**, 335–369

Amos, A. F., Langseth, M. G. Jr. and Markl, R. G. (1972). Visible oceanic saline fronts, p. 49–62. *In* "Studies in Physical Oceanography" (A. L. Gordan, ed.) Vol. 1. Gordon and Breach, New York

Anderson, F. P., Sharp, S. O. and Oliff, W. D. (1970). The reaction of coastal waters off Durban to changes in atmospheric pressure. *In* "Collected Proceedings of the SANCOR Symposium, Oceanography in South Africa 1970", Paper H-2

Arthur, R. S. (1965). On the calculation of vertical motion from determinations of horizontal motion. *J. Geophys. Res.* **70**, 2799–2803

Ashmole, N. P. and Ashmole, M. J. (1967). Comparative feeding ecology of sea birds of a tropical oceanic island. *Bull. Peabody Mus. Nat. Hist.* **24**, 1–131

Backus, R. H., Craddock, J. E., Haedrich, R. L. and Shores, D. L. (1969). Mesopelagic fishes and thermal fronts in the western Sargasso Sea. *Mar. Biol.* **3**, 87–106

Bainbridge, R. (1957). The size, shape and density of marine phytoplankton concentrations. *Biol. Rev.* **32**, 91–115

Bang, N. D. (1970). Major eddies and frontal structures in the Agulhas Current retroflexion area in March 1969. *In* "Collected Proceedings of the SANCOR Symposium, Oceanography in South Africa 1970", Paper B-2

Bang, N. D. (1973). Characteristics of an intense ocean frontal system in the upwelling region west of Cape Town. *Tellus* 25: 256–265

Barkley, R. A. (1968). The Kuroshio-Oyashio front as a compound vortex street. *J. Mar. Res.* **26**, 83–104

Barkley, R. A. (1972). Johnston Atoll's wake. *J. Mar. Res.* **30**, 201–216

Bary, B. M. (1953). Sea-water discoloration by living organisms. *N.Z.J. Sci. Technol.* **34**, 393–407

Bass, A. J. (1970). Shark distribution and movements along the eastern coast of South Africa. *In* "Collected Proceedings of the SANCOR Symposium, Oceanography in South Africa 1970", Paper G-5

Beebe, W. (1926). "The Arcturus Adventure". Putnam and Sons, New York

Bernstein, R. L. and White, W. B. (1974). Time and length scales of baroclinic eddies in the central North Pacific Ocean. *J. Phys. Oceanogr.* **4**, 613–624

Bernstein, R. L., Breaker, L. and Whritner, R. (1977). California Current eddy formation: ship, air and satellite results. *Science* **195**, 353–359

Boden, B. P. (1952). Natural conservation of insular plankton. *Nature* **169**, 697–699

Boden, P. B. and Kampa, E. M. (1953). Winter cascading from an oceanic island and its biological implications. *Nature* **171**, 426–427

Brinton, E. (1976). Population biology of *Euphausia pacifica* off southern California. *Fish. Bull. U.S.* **74**, 733–762

Broenkow, W. W. (1965). The distribution of nutrients in the Costa Rica Dome in the eastern tropical Pacific Ocean. *Limnol. Oceanogr.* **10**, 40–52

Colton, J. B. Jr., Smith, D. E. and Jossi, J. W. (1975). Further observations on a thermal front in the Sargasso Sea. *Deep-Sea Res.* **22**, 433–439

Cromwell, T. and Reid, J. L. Jr. (1956). A study of oceanic fronts. *Tellus* **8**, 94–101

Curtin, T. B. and Mooers, C. N. K. (1975). Observations and interpretation of a high-frequency internal wave packet and surface slick pattern. *J. Geophys. Res.* **80**, 882–894

Defant, A. (1961). "Physical Oceanography", Vol. 1, p. 199. Pergamon Press, New York

Dietrich, G. (1964). Oceanic polar front survey in the North Atlantic. *In* "Research in Geophysics" (H. Odishaw, ed.) Ch. 12. MIT Press, Cambridge, Mass.

Dunson, W. A. and Ehlert, G. W. (1971). Effects of temperature, salinity and surface water flow on distribution of the sea snake *Pelamis*. *Limnol. Oceanogr.* **16**, 845–853

Emery, A. R. (1972). Eddy formation from an oceanic island: Ecological effects. *Carrib. J. Sci.* **12**, 121–128

Enright, J. T. (1977). Copepods in a hurry: sustained high-speed upward migration. *Limnol. Oceanogr.* **22**, 118–125

Eppley, R. W., Holmes, R. W. and Strickland, J. D. H. (1967). Sinking rates of marine phytoplankton measured with a fluorometer. *J. exp. mar. Biol. Ecol.* **1**, 191–208

Eppley, R. W., Holm-Hansen, O. and Strickland, J. D. H. (1968). Some observations on the vertical migration of dinoflagellates. *J. Phycol.* **4**, 333–340

Ewing, G. C. (1950). Slicks, surface films and internal waves. *J. Mar. Res.* **9**, 161–187

Faller, A. J. (1969). The generation of Langmuir circulations by eddy pressure of surface waves. *Limnol. Oceanogr.* **14**, 504–513

Faller, A. J. and Woodcock, A. H. (1964). The spacing of windrows of *Sargassum* in the ocean. *J. Mar. Res.* **22**, 22–29

Frankignoul, C. (1976). Observed interaction between oceanic internal waves and mesoscale eddies. *Deep-Sea Res.* **23**, 805–820

Fuglister, F. C. and Worthington, L. V. (1951). Some results of a multiple ship survey of the Gulf Stream. *Tellus* **3**, 1–14

Gargett, A. E. and Hughes, B. A. (1972). On the interaction of surface and internal waves. *J. Fluid Mech.* **52**, 179–191

Garrett, W. D. (1967). The organic chemical composition of the ocean surface. *Deep-Sea Res.* **14**, 221–227

Garrett, C. and Munk, W. (1979). Internal waves in the ocean. *Ann. Rev. Fluid Mech.* **11**, 339–369

Gaskin, D. E. (1968). Distribution of *Delphinidae* (*Cetacea*) in relation to sea surface temperatures off eastern and southern New Zealand. *N.Z.J. mar. Sci. Freshw. Res.* **2**, 527–534

Gaskin, D. E. (1976). The evolution, zoogeography and ecology of *Cetacea*. *In* "Oceanogr. Mar. Biol. Ann. Rev." (H. Barnes, ed.) Vol. 14, pp. 247–346

George, D. G. and Edwards, R. W. (1973). *Daphnia* distribution within Langmuir circulations. *Limnol. Oceanogr.* **18**, 798–800

Graham, J. J. (1957). Central North Pacific albacore surveys, May to November 1955. *U.S. Fish. Wildl. Svc. SSR-Fisheries* No. 212

Haedrich, R. L. (1972). Midwater fishes from a warm-core eddy. *Deep-Sea Res.* **19**, 903–906

Han, Hi Soo and Gong, Yeong (1970). Relation between oceanographical conditions and catch of saury in the eastern sea of Korea. *In* "The Kuroshio – A Symposium on the Japan Current" (J. Marr, ed.) pp. 585–592. East-West Center Press, Honolulu

Hand, W. G., Collard, P. A. and Davenport, D. (1965). The effects of temperature and salinity change on the swimming rate in dinoflagellates, *Goniaulax* and *Gymnodinium. Biol. Bull.* **128**, 90–101

Hardy, A. C. and Bainbridge, R. (1954). Experimental observation of the vertical migrations of plankton animals. *J. Mar. Biol. Ass. U.K.* **33**, 409–448

Harris, G. P. and Lott, J. N. A. (1973). Observations of Langmuir circulations in Lake Ontario. *Limnol. Oceanogr.* **18**, 584–589

Hart, J. E. (1979). Finite amplitude baroclinic instability. *Ann. Rev. Fluid. Mech.* **11**, 147–172

Hasle, G. R. (1950). Phototactic vertical migration in marine dinoflagellates. *Oikos* **2**, 162–175

Hasle, G. R. (1954). More on phototactic diurnal migration in marine dinoflagellates. *Nytt. Mag. Biol.* **2**, 139–147

Hide, R. and Mason, P. J. (1975). Sloping convection in a rotating fluid. *Adv. Phys.* **24**, 57–100

Hogg, N. G., Katz, E. J. and Sanford, T. B. (1978). Eddies, islands and mixing. *J. Geophys. Res.* **83**, 2921–2938

Howe, M. R. and Tait, R. I. (1967). A subsurface cold-core cyclonic eddy. *Deep-Sea Res.* **14**, 373–378

Hulbert, E. M. (1964). Succession and diversity in the planktonic flora of the western North Atlantic. *Bull. Mar. Sci. Gulf and Caribbean* **14**, 33–44

Hunter, J. R. and Mitchell, C. T. (1967). Association of fishes with flotsam in the offshore waters of central America. *Fish. Bull. U.S.* **66**, 13–29

Iisuka, S. and Irie, H. (1969). Anoxic status of bottom waters and occurrences of *Gymnodinium* red water in Omura Bay. *Bull. Plankton Soc. Japan* **16**, 99–115

Jones, E. C. (1962). Evidence of an island effect upon the standing crop of zooplankton near the Marquesas Islands, central Pacific. *J. Cons. perm. explor. Mer.* **27**, 223–231

Kamykowski, D. (1974). Possible interactions between phytoplankton and semi-diurnal internal tides. *J. Mar. Res.* **32**, 67–89

Knauss, J. A. (1957). An observation of an oceanic front. *Tellus* **9**, 234–237

Kropach, C. (1975). The yellow-bellied sea snake, *Pelamis*, in the eastern Pacific. *In* "The Biology of Sea Snakes" (W. A. Dunson, ed.) pp. 185–213. University Park Press, Baltimore

Laufer, J. (1975). New trends in turbulence research. *Ann. Rev. Fluid Mech.* **7**, 307–326

Laurs, R. M., Yuen, H. S. H. and Johnson, J. H. (1977). Small-scale movements of albacore, *Thunnus alalunga*, in relation to ocean features as indicated by ultrasonic tracking and oceanographic sampling. *Fish. Bull. U.S.* **75**, 347–355

Laurs, R. M. and Lynn, R. J. (1977). Seasonal migration of north Pacific albacore,

Thunnus alalunga, into North American coastal waters: distribution, relative abundance, and association with transition zone waters. *Fish. Bull. U.S.* **75**, 795–822.

LeFèvre, J. and Grall, J. R. (1970). On the relationships of *Noctiluca* swarming off the western coast of Brittany with hydrological features and plankton characteristics of the environment. *J. exp. mar. Biol. Ecol.* **4**, 287–306

Longhurst, A. R. (1976). Vertical migration. *In* "The Ecology of the Seas" (D. H. Cushing and J. J. Walsh, eds) pp. 116–137. W. B. Saunders, Philadelphia

Marshall, S. M. and Orr, A. P. (1955). "The Biology of a Marine Copepod". Oliver and Boyd, London

McGary, J. W. and Stroup, E. D. (1956). Mid-Pacific oceanographic, mid-latitude waters, January–March 1954. *U.S. Fish. Wildl. Svc. SSR-Fisheries* No. 180

Mooers, C. N. K., Flagg, C. N. and Boicourt, W. C. (1978). Prograde and retrograde fronts. *In* "Oceanic Fronts in Coastal Processes" (M. J. Bowman and W. E. Esaias, eds). Springer-Verlag, Berlin

Murphy, G. I. and Shomura, R. S. (1972). Pre-exploitation abundance of tunas in the equatorial central Pacific. *Fish. Bull. U.S.* **72**, 875–913

Neill, W. H., Chang, R. K. C. and Dizon, A. E. (1976). Magnitude and ecological implications of thermal inertia in skipjack tuna, *Katsuwonus pelamis* (Linnaeus). *Env. Biol. Fish.* **1**, 61–80

Nilsson, C. S., Andrews, J. C. and Scully-Power, P. (1977). Observations of eddy formation off east Australia. *J. Phys. Oceanogr.* **7**, 659–669

Osborn, T. R. (1978). Measurements of energy dissipation adjacent to an island. *J. Geophys. Res.* **83**, 2939–2957

Owen, R. W. (1966). Horizontal vortices in the surface layer of the sea. *J. Mar. Res.* **24**, 56–66

Owen, R. W. (1968). Oceanographic conditions in the northeast Pacific and their relation to the albacore fishery. *Fish. Bull. U.S.* **66**, 503–526.

Owen, R. W. (1980). Eddies of the California Current System: physical and ecological characteristics. *In* "The California Islands" (D. Power, ed.) Mus. Nat. Hist., Santa Barbara, California

Packard, T. T., Blasco, D. and Barber, R. T. (1978). *Mesodinium rubrum* in the Baja California upwelling system. *In* "Upwelling Ecosystems" (R. Boje and M. Tomczak, eds). Springer-Verlag, Berlin

Pak, H. and Zaneveld, J. R. V. (1974). Equatorial front in the eastern Pacific Ocean. *J. Phys. Oceanogr.* **4**, 570–578

Parker, C. E. (1971). Gulf Stream rings in the Sargasso Sea. *Deep-Sea Res.* **18**, 981–993

Patzert, W. C. (1970). Eddies in Hawaiian waters. *Hawaii Inst. Geophys. Rep. 69–8*

Pingree, R. D., Forster, G. R. and Morrison, G. K. (1974). Turbulent convergent tidal fronts. *J. mar. biol. Assoc. U.K.* **54**, 469–479.

Pingree, R. D., Holligan, P. M. and Head, R. N. (1977). Survival of dinoflagellate blooms in the western English Channel. *Nature* **265**, 266–269

Powers, P. B. A. (1932). *Cyclotrichium meunieri* sp. nov. (protozoa, ciliata); cause of red water in the Gulf of Maine. *Biol. Bull.* **63**, 74–80

Raymont, J. E. G. (1963). "Plankton and Productivity in the Oceans". Pergamon Press, New York

Reid, J. L. Jr., Roden, G. I. and Wyllie, J. G. (1958). Studies of the California Current System. *Calif. Coop. Oceanic Fish. Invest. Rep.* 1 July 1956 to 1 Jan. 1958

Richardson, P. (1976). Gulf Stream rings. *Oceanus.* **19** (3), 65–68

Ryther, J. H. and Menzel, D. W. (1960). The seasonal and geographic range of primary production in the western Sargasso Sea. *Deep-Sea Res.* **6**, 235–238

Sargent, M. C. and Walker, T. J. (1948). Diatom populations associated with eddies off southern California in 1941. *J. Mar. Res.* **7**, 490–505

Savidge, G. (1976). A preliminary study of the distribution of chlorophyll *a* in the Celtic and Western Irish Seas. *Estuarine & Coastal Mar. Sci.* **4**, 617–625

Savidge, G. and Foster, P. (1978). Phytoplankton biology of a thermal front in the Celtic Sea. *Nature* **271**, 155–157

Scott, J. T., Myer, G. E., Stewart, R. and Walther, E. G. (1969). On the mechanism of Langmuir circulations and their role in epilinmion mixing. *Limnol. Oceanogr.* **14**, 493–503

Seckel, G. (1972). Hawaiian-caught skipjack tuna and their physical environment. *Fish. Bull. U.S.* **70**, 763–787

Sette, O. E. and Ahlstrom, E. H. (1948). Estimations of abundance of the eggs of the Pacific pilchard (*Sardinops caerulea*) off southern California during 1940 and 1941. *J. Mar. Res.* **7**, 511–542

Shomura, R. S. and Otsu, T. (1956). Central North Pacific albacore surveys, January 1954–February 1955. *U.S. Fish Wildl. Svc. SSR-Fisheries* No. 173

Simpson, J. H. and Pingree, R. D. (1978). Shallow sea fronts produced by tidal stirring. *In* "Oceanic Fronts in Coastal Processes" (M. J. Bowman and W. E. Esias, eds). Springer-Verlag, Berlin

Smayda, T. (1969). Some measurements of the sinking rate of fecal pellets. *Limnol. Oceanogr.* **14**, 621–625

Smayda, T. (1970). The suspension and sinking of phytoplankton in the sea. *In* "Oceanogr. Mar. Biol. Ann. Rev." (H. Barnes, ed.) Vol. 8, pp. 353–414

Spilhaus, A. F. Jr. (1968). Observations of light scattering in sea water. *Limnol. Oceanogr.* **13**, 418–422

Stavn, R. H. (1971). The horizontal-vertical distribution hypothesis: Langmuir circulations and *Daphnia* distributions. *Limnol. Oceanogr.* **16**, 453–466

Stommel, H. (1949). Trajectories of small bodies sinking slowly through convection cells. *J. Mar. Res.* **8**, 24–29

Sutcliffe, J. H. Jr., Baylor, E. R. and Menzel, D. W. (1963). Sea surface chemistry and Langmuir circulation. *Deep-Sea Res.* **10**, 233–243

Sutcliffe, W. H. Jr., Sheldon, R. W., Prakash, A. and Gordon, D. C. Jr. (1971). Relations between wind speeds, Langmuir circulation and particle concentration in the ocean. *Deep-Sea Res.* **18**, 639–643

Szekielda, K. H., Kupferman, S. L., Klemas, V. and Polis, D. F. (1972). Element enrichment in organic films and foam associated with aquatic frontal systems. *J. Geophys. Res.* **77**, 5278–5282

Tsujita, T. (1957). The fisheries oceanography of the East China Sea and Tsuchima Strait. 1. The structure and ecological character of the fishing grounds. *Bull. Seikai Reg. Fish. Res. Lab.* **13**, 1–47

Uda, M. (1938). Researches on "Siome" or current rip in the seas and oceans. *Geophys. Mag.* **11**, 307–372

Uda, M. (1952). On the relation between the variation of important fisheries conditions and the oceanographical conditions in the adjacent waters of Japan. 1. *J. Tokyo Univ. Fish.* **38**, 376–381

Uda, M. (1973). Pulsative fluctuation of oceanic fronts with tuna fishing grounds and fisheries. *J. Fac. Mar. Sci. Technol., Tokai Univ.* **7**, 245–265

Uda, M. and Ishino, M. (1958). Enrichment pattern resulting from eddy systems in relation to fishing grounds. *J. Tokyo Univ. Fish.* **44**, 105–129

Von Arx, W. S. (1962). "An Introduction to Physical Oceanography". Addison-Wesley, London

Voorhis, A. D. (1969). The horizontal extent and persistence of thermal fronts in the Sargasso Sea. *Deep-Sea Res.* **16** (suppl.), 331–337

Voorhis, A. D. and Hersey, J. B. (1964). Oceanic thermal fronts in the Sargasso Sea. *J. Geophys. Res.* **69**, 3809–3814

White, W. B. (1973). An oceanic wake in the equatorial undercurrent downstream from the Galapagos Archipelago. *J. Phys. Oceanogr.* **3**, 156–161

Wiebe, P. H. and Boyd, S. H. (1978). Limits of *Nematoscelis megalops* in the Northwestern Atlantic in relation to Gulf Stream cold core rings. I. Horizontal and vertical distributions. *J. Mar. Res.* **36**, 119–142

Wiebe, P. H., Hulbert, E. M., Carpenter, E. J., Jahn, A. E., Knapp III, G. P., Boyd, S. H., Ortner, P. B. and Cox, J. L. (1976). Gulf Stream cold core rings: large-scale interaction sites for open ocean plankton communities. *Deep-Sea Res.* **23**, 695–710

Woodcock, A. H. (1944). A theory of surface water motion deduced from the wind-induced motion of the *Physalia*. *J. Mar. Res.* **5**, 196–205

Woodcock, A. H. (1950). Subsurface pelagic *Sargassum*. *J. Mar. Res.* **9**, 77–92

Woods, J. D. Diurnal and seasonal variation of convection in the wind mixed layer of the ocean. *Q.J. Roy. Met. Soc.* (In press.)

Worthington, L. V. (1972). Anticyclogenesis in the oceans as a result of outbreaks of continental polar air. *In* "Studies in Physical Oceanography" (A. L. Gordan, ed.) Vol. 1, pp. 169–178. Gordon and Breach, New York

Wyllie, J. (1966). Geostrophic flow of the California Current at the surface and at 200 m. *Calif. Coop. Oceanic Fish. Invest. Atlas* No. 4

Wyrki, K. (1966). Oceanography of the eastern equatorial Pacific ocean. *Oceanogr. Mar. Biol. Ann. Rev.* **4**, 33–68

Yasui, M. (1961). Internal waves in the open sea. *Oceanogr. Mag.* **12**, 157–183

Zaneveld, J. R., Andrade, M. and Beardsley, G. F. Jr. (1969). Measurements of optical properties at an oceanic front observed near the Galapagos Islands. *J. Geophys. Res.* **74**, 5540–5541

ZoBell, C. E. (1946). "Marine Microbiology". Chronica Botanica, Waltham, Mass.

8 The Deep-Sea Ecosystem

GILBERT T. ROWE

1 The deep-sea environment

Having the greatest areal extent and volume of any subsystem of the global oceans, the physical boundaries of the deep-sea system appear to be easier to define than in other marine ecosystems (Fig. 1). The upper boundary is the permanent thermocline (Sanders *et al.*, 1965) or the bottom of the euphotic zone (Menzies *et al.*, 1973; Smith, 1978a). Located at various depths, depending on local conditions, it is more or less equivalent to the outer margin of the continental shelf, or 200 m. The euphotic zone down to the 1% light level is equivalent to an epipelagic zone. Below this is the mesopelagic or twilight zone down to about 1 km, the depth of light penetration in the clearest of oceans. Below that lie the bathypelagic (to 6 km) and the hadal pelagic zones (to 10 km). These three deep-sea pelagic subdivisions are equivalent to the archibenthic (bathyal), abyssal and hadal zones on the bottom.

Continental slope sediments are generally muddy, but tend to have more

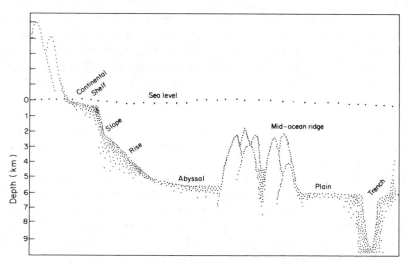

FIG. 1   Oceanic topographic realms

235

sand and silt than deeper muds on the rise and abyssal plains, which are often almost entirely silt and clay-sized particles. Sedimentary outcrops occur on the slope, but are rare deeper. The organic content of deep-sea sediments is generally low, except in areas of relatively high surface biological production or low bottom water oxygen concentrations (Rowe, 1971). Organic matter usually correlates inversely with grain size, rather than with depth or distance from land.

Local topography is not necessarily as monotonous as it is assumed to be. The upper continental slopes are often incised by steep submarine canyons that were cut during Pleistocene sea-level recessions. Some extend onto the continental rises, and a few reach the abyssal plains. Basin age decreases with depth, due to sea floor spreading; and this has important consequences for species evolution and the geologic evolution of community structure.

The regularity and pattern of supplies of food for deep-sea heterotrophs is poorly understood (Menzel, 1974; Rowe and Staresinic, 1979). While we cannot yet quantify this energy flux or its variation, most organic matter is certainly not only produced, but is remineralised in the upper 500 m. Little reaches great depths.

The deep-sea ecosystem, almost totally *heterotrophic*, is dominated by catabolism. Being dependent on the import of organic energy from outside the system, it is an *allochthonous* system. In addition to perennial darkness, most of the deep-sea is cold and seasonal variations in temperature are very small, if they exist at all.

The conservative physical and chemical characteristics of the deep water masses do not reflect local conditions, but rather those of high latitudes where the water mass was formed; isolated as it is from euphotic zone processes, each water mass tends to retain its own suite of characteristics. Many decades are required for deep water to spread across the thousands of kilometers of an ocean basin. The deep water of the Atlantic is probably 200–300 years old while that in the Pacific was formed about 1000 years ago. A number of chemical variables, particularly inorganic nutrients and oxygen, are non-conservative metabolic products or substrates in the euphotic zone, but in the deep sea are conservative. These reflect the origin of the water mass, and change very modestly as they travel and age from their sites of formation.

Hydrostatic pressure in the deep sea reaches about $1000 \, \mathrm{kg \, cm^{-2}}$ at 10 000 m, the depth of the deepest trenches. Pressure is the only physical variable that varies regularly in space, but has no appreciable variation in time. It has been demonstrated experimentally that pressure has effects at all taxonomic and biochemical levels (MacDonald, 1975; Marquis, 1976), but the relationship between pressure and ecosystem function has rarely been considered.

Physical and chemical variations are only significant on geologic time scales. Since the Oligocene, deep water temperatures in the North Atlantic have dropped from 10 °C to their present 2 °C (Lowenstam and Epstein, 1959), following glaciation. There is also evidence in deep-ocean cores that some basins became anoxic during the Cretaceous. Much more recently, high salinities during the Wisconsin glaciation (18 000 B.P.) are postulated to have caused deep stagnation (Worthington, 1968), but there is no evidence that the water became anoxic.

Until recently it has been assumed that current speeds in the deep sea were very modest, but it is now known that deep boundary currents, flowing generally equatorward below surface boundary currents on the western margins of basins, have predictable velocities up to about 25 cm s^{-1} (Wüst, 1958; Rowe and Menzies, 1968, and others). It has been proposed that such deep boundary currents strongly influence sediment redistribution (Heezen et al., 1966) and form strata in cores that can be identified as contourites. It has been suggested that these boundary currents contribute to the vertical zonation of the fauna (Rowe and Menzies, 1969).

Less predictable, but perhaps biologically more important, are the intermittent turbidity currents, or short-term density flows down slopes. The results of such flows are seen in cores as layers of turbidities, characterised by a gradual change from coarse to fine-grained particles (graded bedding). It has been suggested that such massive cascades of sediments, which probably were most common during periods of lowered sea-level, would both fertilise the bottom with allochthonous organic matter and bury the existing fauna (Heezen et al., 1955). Sampling of sediments and fauna in submarine canyons, where such events occur, does indicate that they transport organic-rich material down to great depths (Griggs et al., 1969; Cacchione et al., 1978; Vivier, 1978; Rowe et al., in press), but specific detrimental effects of turbidity currents remain unknown.

Many abyssal plains are not always flat and barren, but composed of pebble-sized concretions or encrustations of ferromanganese oxides. The nodules have been of great interest recently because of the possibility that they could be mined profitably. They are by weight 15 to 20 % Mn and 11 to 23 % Fe, but it may be the enrichment of Ni, Co, and Cu, all less than 1 % (Cronin, 1977), that could make them valuable eventually. Radiometric methods to age nodules, and thereby determine their rate of formation, suggest accretion occurs at a rate of millimeters per million years (mm/10^6yr). They occur in sediments with the lowest sediment accumulation rates, but these rates are on the order of mm/10^{-3}yr, or *3 orders of magnitude* faster than nodule growth. So how could nodules stay on the surface, if they really do not grow much faster (Ku, 1977)? Heathe (1979) found that the size distribution of nodules is consistent with a simple model of uniform growth of a few mm per million years, but what keeps them on the surface, spending about a million years there, is unknown.

Among the possible mechanisms that contribute to nodule formation and trace element uptake are a scavenging by clay particles or iron and manganese oxides, with original supplies of metals coming from submarine volcanism, sedimentation of particles and directly from sea water. Although biotic remains are associated with nodules (Greenslate, 1974; Dudley and Margolis, 1974), it seems unlikely to me that animals function in nodule formation. On the contrary, nodule concentration is highest where we find lowest benthic biomass, organic matter concentrations, and rates of metabolism. The function of microbes in the process remains unknown (Ehrlich, 1972).

2 The pelagic realm

QUANTITATIVE DISTRIBUTIONS
The quantity of pelagic organisms decreases exponentially though not regularly with depth. Each area of the open ocean can be characterised by a regression equation with different constants for initial abundance or biomass at the surface and the rate of their decrease with depth (Johnson, 1962; Banse, 1964; Vinogradov, 1968) (Fig. 2). The details and significance of variability within this decrease are discussed in chapter 14. Rowe et al. (1974) suggested that because benthos and zooplankton display the same exponential decrease with depth, both were supplied by the same sources of food (see Section 3).

MORPHOLOGY AND PHYSIOLOGY OF THE FAUNA
Bathypelagic crustaceans increase in size with depth (Mauchline, 1972) evidently increasing their abilities to forage. The cause of large size is thought to be increased longevity (Mauchline, 1972). Eggs are generally larger in bathypelagic than epipelagic forms, and in euphausiid crutaceans may be equal to about 10% of the body volume. Generally, deep-living forms produce fewer young, but these have a relatively large size when released (Mauchline, 1972).

There is a tendency for deep-living pelagos to reduce tissue density with increasing depth, either to increase relative metabolic efficiency or to conserve energy by increasing neutral buoyancy. Crustacea do this by increasing lipid content and decreasing protein and skeletal material, whereas fish with increasing depth increase their water content relative to carbohydrate, protein, skeletal ash, and lipid (Childress and Nygaard, 1973). Respiratory rates of bathypelagos decrease with increasing depth

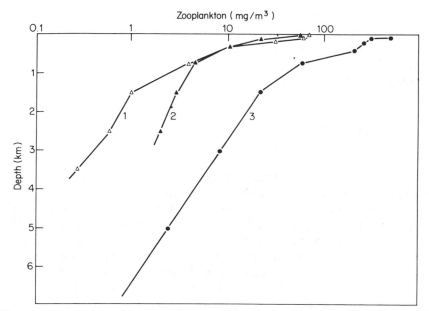

FIG. 2 Zooplankton biomass with depth in (1) the Sargasso Sea, (2) the Gulf Stream, and (3) the region of the Kurile-Kamchatka Trench (Vinogradov, 1968)

(Childress and Nygaard, 1973; Meek and Childress, 1973; Childress, 1971). This apparently is also true for deep demersal fish (Smith and Hessler, 1974) and may be an adaptation to a relatively low supply of food (Childress and Nygaard, 1973). Gut length in pelagic copepods is longer in deeper-living species (Vinogradov, 1968).

Small increases in pressure increase respiration, but this is usually counteracted by lowered temperature at depth, making respiration rate variations with depth relatively small (Napora, 1964; Teal and Carey, 1967; Teal, 1971; Meek and Childress, 1973). Vertical migrators might experience a 5 to 10 °C change in temperature, which would decrease metabolic rate as much as twofold. Among the salps, temperature change has less effect on swimming speed in species that migrate vertically than in those that do not (Campenot and Harbison, 1979).

Biochemical accommodation to pressure may resemble accommodation to high temperature, by modification of enzyme systems (Hochachka et al., 1970). Pressure inhibition of enzyme-substrate complex formation should be small when the volume change of activation, or V, is minimised over wide ranges of pressure, and a correlation has been found between a small V and small pressure effects on enzyme activity for an enzyme in a vertically migrating fish (Moon et al., 1971).

While vertical zonation of bathypelagic species is thought to result from competition along a gradient (Vinogradov, 1973), it has recently been demonstrated how pressure may also be important. Lactate dehydrogenases from two closely related scorpaenid fishes, although electrophoretically indistinguishable, exhibited different kinetic responses to pressure, with the greatest sensitivity in the species having the shallower depth range (Siebenaller and Somero, 1978).

Mesopelagic, bathypelagic and benthopelagic fishes are three categories into which deep-sea fishes can conveniently be grouped (Marshall, 1971). Each has a set of morphological and physiological characteristics that contribute to its survival. The mesopelagic or twilight zone fishes have big eyes that are very sensitive and they probably can see down-welled light to about 1 km in the clearest of oceans. Most have large mouths and teeth relative to their body size and sufficient 'red' muscle to maintain the continuous swimming required by their usual practice of diurnal vertical migration. Their eyes usually employ an aphakic (lensless) space or are tubular and capable of binocular vision. In some species the tubular eyes point up while in others they point horizontally. Numerous observations have been made of mesopelagic lantern fishes facing upward (Barham, 1971). Their jaws, especially in the hatchet fishes, are capable of opening to extreme widths, allowing them to either catch falling carcasses or to capture prey of a size equal almost to their own. The numerous photophores producing light on their ventral sides most likely disguise their silhouettes from predators looking up into the down-welling light (Marshall, 1971).

The bathypelagic fishes, living below any light (>1 km), exhibit the greatest morphological modifications and it is usually their representatives that most often catch the public eye. They too have huge jaws and teeth, and may also have long barbels, apparently to attract prey. The ceratioid or angler fishes dangle a light-emitting barbel from the tops of their heads while stomiatoid fishes have barbels with photophores hanging from their lower jaws. Eyes are small, usually less than 1 mm in diameter.

Survival in this large-volume habitat depends for the individual fish on finding food, but the species depend on a sufficient number of encounters between opposite sexes. Sexual dimorphism is common. Males, often very small, will have a well-developed olfactory system to help them find a female, whose olfactory system may be regressed. Females probably use an attractant or pheromone. In fishes over the continental slope there is evidence of a sonic sexual dimorphism. Males of some species have drumming muscles associated with a swim bladder while females do not, but both sexes have large otoliths, suggesting they hear well (Marshall, 1971).

Benthopelagic species, continues Marshall (1971), thought to live and feed near the bottom, are not as morphologically peculiar as many of the bathypelagic species. Rattails (macrourids), brotulids, rays, etc. may be capable of making noise, have normal-sized eyes, 'red' muscle, no long barbels, and no major sexual dimorphism. Associated with the bottom, they have been presumed to feed there because of the structure of their jaws. No doubt they do, but it has also been discovered recently that they swim way up into the water column where they scavenge a wide variety of material including man-made trash (Haedrich and Polloni, 1976; Haedrich and Henderson, 1974; Pearcy and Ambler, 1974).

If it is assumed that the most difficult environment to survive in is the one that demands the most radical morphological modifications, then it can be concluded that the bathypelagic zone is the most difficult. On the one hand it is physically the most stable, compared to the bottom boundary layer or the mesopelagic twilight zone. However, it is characterised by a decreasing supply of rare food particles, whether they be carcasses, fecal matter or living prey, that in general are larger at greater depths. In the vertical plane the importance of larger-sized particles may increase with depth even though total food biomass decreases with depth. Such a conclusion has been reached independently in the studies of fishes (Marshall, 1971) and the different sinking rates of particles (McCave, 1975). There is little doubt that the morphology of bathypelagic species is a manifestation of selection for the ability to capture rare prey particles with a minimum expenditure of energy.

3   The deep-sea benthic community

QUANTITATIVE DISTRIBUTION OF THE MACROFAUNA

Early studies of the quantity of life on the deep-sea bottom, based on widely dispersed samples (Spärck, 1935, 1936, 1956), discovered that most of the biomass, on the bottom, lines the continental margins (Zenkevitch, 1961). The abyssal plains under the sluggish central open ocean gyres have about 0.01 of the biomass of the continental shelves and slopes, their great reduction presumably somehow being a function of both increased depth and distance from shore. Unfortunately, few recent investigators have measured biomass. Differences in sieve size, faunal composition and other techniques make most studies difficult to compare, but six different sets of data from the north-west Atlantic, the Gulf of Mexico and off South America, were used (Rowe, 1971) to illustrate the statistical relationships between depth, euphotic zone primary production and benthic macrofaunal abundance (Fig. 3) and biomass. This study and another in the northwest Atlantic led Rowe et al. (1974) to suggest that benthic animal abundance or biomass follows the relationship:

$$Y = ae^{-bx}$$

where Y is abundance or biomass, x is depth and a is a constant directly related to surface primary productivity. Each continental margin and adjacent basin has a characteristic and statistically significant regression.

There are many exceptions to this generalisation that seem to reflect differences in the input of organic matter to the bottom. On the Cascadia abyssal plain off the west coast of the United States, for example, there was

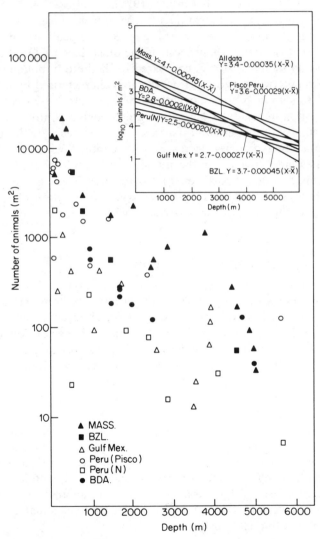

FIG. 3   Abundances of deep-sea macrobenthos in several ocean basins (Rowe, 1971)

more biomass within the canyon crossing the plain than beyond its channel (Griggs *et al.*, 1968). Rowe *et al.* (in press) on the US east coast found that only the section of Hudson canyon where sediment was presently accumulating had high biomass; in sections where there was geologic evidence that organic matter passes through the canyon to greater depths, biomass was not measurably different from outside the canyon. On a larger scale but for similar reasons, Dickinson and Carey (1978) were able to demonstrate that biomass at the base of the continental slope was greater than at equivalent depths (2800 m) about 100 km further offshore.

THE MEIOFAUNA AND THE MEGAFAUNA

Although most studies of deep-sea benthos have dealt with the 'macrofauna', some recent investigations have also included consideration of the meiofauna dominated by nematodes, harpacticoids, ostracods, turbellaria, kinorynchs, gastrotrichs, retained on a 40 μm or 63 μm mesh sieve, but which pass through sieves designed to take macrofauna; 0.25 mm up to 0.5 mm have usually been used in contemporary deep-sea studies, although shallow water studies often employ a mesh as big as 1 or 2 mm (Mare, 1942). This size fraction of benthos has been investigated on both sides of the deep North Atlantic (Dinet *et al.*, 1973; Tietjen, 1971; Thiel, 1975; Coull *et al.*, 1977) and in the Mediterranean (Vivier, 1978; Dinet and Vivier, 1979). Quantitatively, their abundance, as with the macrofauna, decreases with depth, but not at an exponential rate. The data of Coull *et al.* (1977) demonstrate this:

400 m depth	442 400 m^{-2}	(mean of 4 cores)
800	891 800	(mean of 2 cores)
4000	73 500	(mean of 2 cores)

as do those from Tietjen (1971) from the same area off North Carolina:

50–100 m	352–849 000 m^{-2}
250–500	138–1 174 000
600–750	110–164 000
800–2500	40–149 000

While the expected rate of change for the macrofauna would be a decrease of about 1 order of magnitude per one or two kilometers increase in depth, the meiofauna decreased only about 10 times over the whole depth range sampled (50–4000 m). This pattern led Thiel (1975) to suggest that meiofaunal importance increases in the deep sea relative to the macrofauna. Small size, he suggested, is selected for because of the limiting food supply.

Benthic organisms that are more than about 1 cm in diameter and are epibenthic can often be seen and identified to higher taxa in bottom photographs and can be caught in large trawls; however, they are infrequent enough to be rarely captured in quantitative infaunal samples (cores and grabs). Le Danois (1948) summarised early European information on such 'megafauna' for the eastern North Atlantic, while a number of studies have been conducted off the United States (Rowe and Menzies, 1969; Haedrich *et al.*, 1975, 1980; Musick and Markle, 1977; Grassle *et al.*, 1975, Carney and Carey, 1980). Sokolova (1976) has presented maps relating the feeding types of what can be considered megafauna to the trophic structure of the world ocean between depths of 3 and 6 km.

Haedrich and Rowe (1978) obtained photographic information and the weights for fishes and large invertebrates in trawl samples to derive biomass estimates for the megafauna. Comparing the depth-biomass regressions of the megafauna and the smaller macrofauna, they concluded that the megafauna comprises an increasingly larger fraction of the total biomass down to the lower continental rise. While the deep infauna is smaller than shallow infauna, they suggested that the ability to move about to find food is selected for, so that the fauna tends toward a bimodal distribution (Haedrich *et al.*, 1980; Polloni *et al.*, 1979). This reasoning is similar to that of Mauchline (1972) who suggested larger size is selected for in bathypelagic organisms compared to epipelagic ones. Khripounoff *et al.* (1980) found that the mean weight in some macrofaunal groups was higher in the Vema channel than in shallow comparative areas in the North Atlantic. How size-frequencies vary in the deep-sea benthos cannot yet be accurately quantified.

Big motile organisms undoubtedly exist in the deep sea. John Isaacs' baited deep-sea cameras have led to the rediscovery of abundant large scavenging lysianassid amphipod crustaceans attracted to bait near the bottom (Hessler *et al.*, 1972). Photographs show that macrourid fish aggregate around organic bait (Isaacs and Schwartzlose, 1975; Dayton and Hessler, 1972; Rowe and Staresinic, 1979) at abyssal depths (up to 6000 m), but not in trenches (6 km to 10 km) (Belyaev, 1966, Menzies *et al.*, 1973) where scavenging, at least in some trenches, is principally by amphipods (Hessler *et al.*, 1978). Rattail fishes, once thought to be exclusively bottom feeders, are now known to range far above the bottom to feed on mid-water organisms (Haedrich and Henderson, 1974; Haedrich and Polloni, 1975; Pearcy and Ambler, 1974), and undoubtedly link pelagic with benthic processes.

ZOOGEOGRAPHY AND AGE OF THE FAUNA

Deep-sea benthic genera, but not species, are cosmopolitan. As Vinogradova (1962) has shown, most species are endemic to particular deep basins, and they are zoned with depth, with certain depths being charac-

FIG. 4 Macrourids and brotulids consuming bait (shark and tuna) at 3650 m depth in the western North Atlantic

terised by 'species maxima' (Vinogradova, 1962). The shallower the basin, the less the endemism and at depths greater than 4000 m, almost no species occur in more than one basin. Between trenches, even those in close proximity, average endemism is 50–60% (Belyaev, 1966). Vinogradova (1962), Ekman (1935) and Menzies et al. (1973) have presented faunal province maps for the deep-sea benthos that reflect this isolation of deep basins. Gene flow between them must be very slight or absent because of sluggish deep circulation patterns. By default then, cosmopolitan species are confined to intermediate depths.

A hadal fauna living in trenches (6 to 10 km) has been described by Wolff (1956, 1970) and Belyaev (1966) as having its own endemic communities. The hadal species appear to be relatively young, dating from the Tertiary.

The antiquity of the deep-sea fauna is not agreed upon. Zenkevitch and Birstein (1960) and Zenkevitch (1966) maintain its great antiquity while Menzies and Imbrie (1958) contend that Paleozoic and Mesozoic types actually decrease with increasing depth. Only species with representatives in the Tertiary increase at depths greater than 200 m. Archaic species, too, as well as those that are cosmopolitan are found at intermediate depths (Madsen, 1961).

DIVERSITY

Sanders first recognised that in the marine benthos species diversity increases with the duration of environmental stability or predictability (Sanders, 1968). Hessler and Sanders (1967) found highest diversities in the deep-sea at intermediate (middle continental slope and upper continental rise) depths, not at the greatest depths sampled on the abyssal plain. Such a pattern, suggested by the younger geological age of the abyssal plain, would be expected as the greatest depths are continually populated by invasions from intermediate depth zones and higher latitudes. Though Menzies et al. (1973) have demonstrated temporal variation in the deep-sea environment over geological time, clearly Sanders is correct in saying that the deep-sea floor has been relatively unchanging for longer periods than other marine environments.

There may be additional causes of the high diversity in the deep-sea. Dayton and Hessler (1973), following Robert Paine's logic that community structure is controlled by top predators, suggested that large fishes non-selectively 'crop' smaller organisms, thereby preventing competitive exclusion and so allowing many species to coexist. However, Grassle and Sanders (1973) suggested, on the contrary, that high diversity is maintained by contemporaneous disequilibrium. Because of the way successional stages of assemblages or populations are distributed in a mosaic on the bottom, they can coexist with a minimum of competition. Following the reasoning of

Connell (1978), Rowe *et al.* (in press) suggest that high diversity in Hudson submarine canyon, with an unpredictable environment, compared with the Gulf of Maine, with homogeneous, apparently unvarying conditions, might result from the Gulf being closer to 'equilibrium' than the canyon.

Others (Jumars, 1975, 1976; Thistle, 1978) have sought to identify scales of patchiness in infaunal communities, using large subdivided boxcores, presuming that organisms provide the environmental heterogeneity that could serve to maintain high diversity. At this scale, interaction between species, especially the effects of predation, appear to be of great importance (Rex, 1977; Thistle, 1979).

ZONATION

Deep-sea benthic populations are zoned with depth even though the environment appears to vary only slightly across large depth gradients. Though no one contends that individual populations are depth zoned, the justification for dividing the deep ocean into large vertical subdivisions based on animal assemblages has been questioned (Sanders and Hessler, 1969). Although the latter authors could identify a faunal boundary only at the shelf-slope junction, others have found that a number of large species can be used to delimit additional faunal zones. Menzies *et al.* (1973) using isopods and Haedrich *et al.* (1975) using the megafauna have used the rate of faunal change between stations along the depth gradient to determine boundaries. Faunas have also been clustered, based on various measures of the similarity between pairs or groups of samples, to define depth zones (Rex, 1977; Haedrich *et al.*, 1975, 1980). Zones have now been identified for the continental shelf, the upper slope (down to about 1000 m to 1500 m), the lower slope (from the above down to about 2200 m), the continental rise (out to about 4000 m), the abyssal plain (about 4500 to > 5000 m), and the trenches (6000–10 000 m), in such diverse groups as gastropods (Rex, 1977), total megafauna (Rowe and Haedrich, 1977; Haedrich *et al.*, 1975, 1980), amphipods (Mills, 1972), isopods (Menzies *et al.*, 1973), and holothurians (Carney and Carey, 1977). Rowe *et al.* (in press) found the most abrupt faunal boundary at 1400 m in the Hudson submarine canyon rather than at the shelf-slope border.

The causes of zonation could be competition along a gradient of diminishing organic matter (Rex, 1977), bottom currents and sediments (Rowe and Menzies, 1969) and effects of pressure on enzyme systems (Siebenaller and Somero, 1978). These processes, coupled with the rates of successful invasion by geologically older continental margin populations, probably give rise to the patterns now observed. A nomenclature for the zones is given by Menzies *et al.* (1973).

4 Functioning of the deep-ocean ecosystem

ECOSYSTEM STRUCTURE

Studies of productivity in surface waters of the open ocean have demonstrated the relative conservative nature of dissolved and particulate organic carbon (POC) and oxygen in the deep sea (Menzel, 1967; Menzel and Ryther, 1969). While POC varies from about 50 to several thousand $\mu g \, l^{-1}$ in the upper few hundred meters, in deep water it is always between about 3 and $15 \mu g \, l^{-1}$. Because deep oxygen concentrations are almost invariable, and reflect the origin of the water mass rather than subsequent metabolism within it, these authors suggested that metabolism and the consumption of POC in deep water are uniform and almost negligible. Many have disagreed (e.g. Riley, 1971), and Menzel (1974) later suggested that the quantification of the cycling of organic matter in the deep sea is the major problem facing biological oceanography at the present time. The functioning of the deep-sea ecosystem as elucidated by the cycling of organic matter is the major theme of this chapter.

The advance from studies of community structure to studies of rates of processes derives from radical improvements in technology. From apparatus scarcely different from that used by Agassiz we have now advanced to the recent utilisation of research submersibles, acoustically-controlled, untethered vehicles, and remote or submersible-observed manipulators for placing and controlling sampling and experimental apparatus on the bottom and in mid-water.

The most useful tool at the Woods Hole Oceanographic Institution has been the research submersible *Alvin*. Supplemented by acoustically-controlled bottom releases and experimental packages, *Alvin* is being used by process-oriented biologists to quantify rates of biological and geochemical processes in the deep sea (Rowe, 1979).

Based on knowledge gathered over the last century, one could propose a deep ocean ecosystem with a structure similar to Fig. 6. The system's external forcing by organic matter, seen conceptually in Fig. 5, is poorly quantified. This includes the rain of particles (Fournier, 1972; Wiebe et al., 1976), dead carcasses, the zooplankton ladder (Vinogradov, 1968, Wiebe et al., 1979), detrital macrophytes (Schoener and Rowe, 1971; Menzies et al., 1967; Menzies et al., 1973; and Wolff, 1976), turbidity currents (Heezen et al., 1956) and chemolithotrophs, all discussed by Rowe and Staresinic (1979). Once into the system, these are consumed by bathypelagic organisms (Mauchline, 1972) or sink into the benthic boundary layer. Demersal or near-bottom-living fishes and crustaceans may migrate off the bottom to scavenge (C9, C8, C12) (Haedrich and Henderson, 1974; Haedrich and Polloni, 1976; Pearcy and Ambler, 1974); what cues such a foraging is

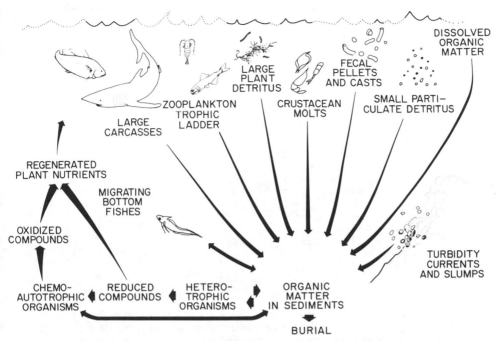

FIG. 5   Categories of possible sources of organic matter for deep-sea organisms

unknown. If the sinking detrital matter reaches bottom, it can be consumed by benthic invertebrates (C14, C13, C12) (Menzies, 1962) or heterotrophic bacteria (C15) (ZoBell, 1947). The remaining discussion in this section will present the preliminary results of studies on how the state variables of this box model are quantitatively related.

RATE OF VERTICAL FLUX OF POM

There have been attempts recently to quantify Agassiz's (1888) 'rain of particles' into great depths. In the box model (Fig. 6) these particles are represented by C1 (particulate organic matter) and C4 (fecal pellets and crustacean molts). Sedimentologists interested in silica and $CaCO_3$ dissolution rates have set large sedimentation traps moored near bottom to catch falling particles. These have successfully been deployed in both the Pacific (Berger and Soutar, 1967) and in the Atlantic, the latter at 5300 m depth (Honjo, 1978). Geochemists used Honjo's traps, set 300 m above the bottom, to assess the elemental composition of particles trapped over a three-month period (Spencer et al., 1978). Biologists have trapped the particles to measure the flux of organic matter. While the geochemists and geologists used large cones as traps, a design developed by Soutar, Wiebe et

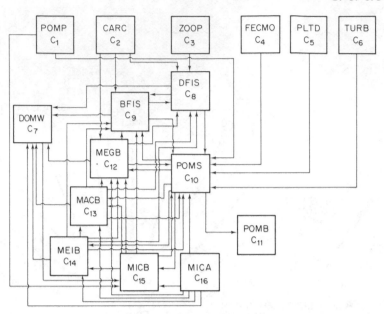

FIG. 6 Conceptual box model of the biological components of a deep benthic system and the sources and sinks of organic carbon in and to it. Respiration is implicit. C1, particulate organics; C2, carcasses of large animals; C3, zooplankton migrations; C4, fecal pellets and molts; C5, macrophyte detritus; C6, turbidity currents; C7, dissolved organics; C8, demersal foraging fishes; C9, benthic fishes, not foraging; C10, particulate organics in the sediments; C11, buried, refractory organics in the sediments; C12, megabenthos, crawling or swimming invertebrates; C13, macrobenthic infaunal invertebrates; C14, meiofaunal invertebrates; C15, heterotrophic microbiota; and, finally, autotrophic (chemolithotrophic) microbiota, C16

al. (1976) used a one meter square plexiglass box lined with large glass-fiber filters. Deployed in the Tongue of the Ocean in the Bahamas, that trap caught numerous fecal pellets but no molts. The organic carbon flux to that trap set 100 m above the bottom was 2.4 gm m^{-2}yr^{-1}. Rowe and Gardner (1979) deployed four arrays of cylindrical traps at 3 locations on the upper continental rise in the Atlantic. Organic carbon flux to the traps farthest above the bottom (500 m above bottom at 3600 m depth and 500 m above bottom at 2800 m depth) was 4.2 and 2.3 gm m^{-2}yr^{-1} (Fig. 7), much higher than found over the abyssal plain (0.36 gm m^{-2}yr^{-1}) by Honjo (1978).

Many deep ocean basins are characterised by a nepheloid or particle-rich layer (BNL) near the bottom. Because a major fraction of such layers is presumed to be resuspended and not of pelagic origin, the trapping experiments above have tried to exclude them by placing the traps several hundred meters above bottom. The variations in time and space and the origin of the BNL are not known, but these must be a function of contour current and turbidity current erosion. Downslope 'cascading' during the winter may contribute to the BNL by carrying particle-rich water down the continental slope (Emery and Uchupi, 1972). If the suggestion that continental shelves export large quantities of particulate organic detritus is valid (Walsh, chapter 6), the BNL and the shelf may be a source of organic carbon to deep, continental margin deposits.

BAROPHILIC MICROBIAL ASSEMBLAGES

In 1968, south of Massachusetts, Deep Submergence Research Vessel (DSRV) *Alvin* accidentally broke loose from the lifting cradle on its tender, the catamaran *Lulu*. Its occupants escaped after *Alvin* hit the water, but it sank to the bottom at 1540 m, where it remained for almost a year. Within the sphere, flooded during that period, was the lunch prepared for the pilot and scientists; when *Alvin* was recovered this food gave the first *in situ* data concerning the rate of microbial substrate utilisation in a deep-sea ecosys-

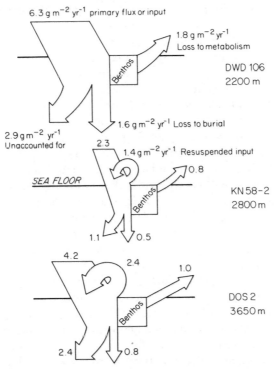

FIG. 7 Organic carbon budgets for the deep North Atlantic, from Rowe and Gardner (1979), in $gCm^{-2}yr^{-1}$

tem. In the conceptual model, Fig. 6, this rate is represented by the arrow linking POMS (the food in this case) to MICB (heterotrophic microbes). Though slightly dehydrated the sandwiches and apple were virtually undecomposed. That is, the flux between these two compartments was negligible (Jannasch et al., 1971)! This observation led to a series of in situ experiments at two permanent bottom stations in the North Atlantic. Using a sampler to recover microbes under pressure, Jannasch added labelled organic substrate without decompression to assess the microbes' responses to enrichment (Jannasch and Wirsen, 1977).

The work of Jannasch and his colleagues indicates that the generalities expressed by ZoBell and Morita (1957) may be subject to significant modification. A barophilic or barotolerant assemblage of bacteria did not occur in the water at these stations. Abyssal microbial populations may not react differently from shallow ones when subject to the extreme pressures and cold of deep water. Their turnover rate and substrate (organic matter) consumption seem to be decreased by several orders of magnitude, probably by inhibition of protein synthesis (Marquis, 1976).

ZoBell and Morita (1957) isolated slow-growing barophilic bacteria during the Galathea expedition. These are organisms that grow preferentially or solely at high pressures, but many other studies, such as that referred to above, have generally isolated only species that were barotolerant (Schwarz et al., 1976; Jannasch and Wirsen, 1977; Jannasch and Taylor, 1976). Recently, a barophilic spirillum bacterium has in fact been found that grows best at about 500 bars and 2° and 4 °C, even after decompression. Recovered from the decayed bodies of amphipod crustaceans that were maintained at high pressure for 5 months, the microorganism had an amazing generation time of only 3 to 4 days (Yayanos et al., 1979).

The early view by ZoBell and coworkers that barophils exist and the implication from Jannasch's work that they are not universal, leaves us with some important questions. If the deep sea is a major site of remineralisation, then microbes should be relatively common and active. However, observation of microbial populations suggests that this is not the case. Slow growth, a characteristic of some microbial assemblages, may have an advantage in the deep sea (Jannasch and Wirsen, 1977) and fast-growing barophils are more likely to be found in productive regions that are less food-limited. Perhaps the less productive the overlying water and the lower the organic matter in the sediments, the lower the probability that barophils will be found. This generalisation might explain the unsuccessful cultures of Jannasch and his co-workers in the Atlantic compared to ZoBell, Morita and Yayanos in the Pacific. Perhaps the only place in the deep-sea ecosystem where a fast growing assemblage of microbes always occurs is within the guts of larger organisms, as in the case of amphipods referred to above.

NITRIFICATION IN THE DEEP SEA

Phytoplankton in the euphotic zone often deplete the water of measurable nitrate, but the deep ocean, below the permanent thermocline and the euphotic zone, contains high concentrations (*ca.* 30 µg atoms l^{-1} in the Atlantic and *ca.* 40 µg atoms l^{-1} in the Pacific). The rate of production of nitrate by nitrifying bacteria in the deep sea therefore should equal total uptake in the surface ocean, assuming a steady-state and negligible input from other sources. It is difficult to account for the observed apparent production of NO_3 (Riley, 1967; Walsh *et al.*, 1975; Walsh, chapter 6, this volume) from attempts to culture nitrifiers (Carlucci and Strickland, 1968). Until recently it was presumed that deep-sea sediments contained little NO_3 (Lerman, 1977), but several recent studies suggest otherwise. Bender *et al.* (1977) found relatively high concentrations off Southwest Africa, with areas of inferred denitrification (pore water concentrations lower than those in the bottom water). Isaeva *et al.* (1978) encountered even higher values in surface sediments at depths of 5 km south-east of Australia. On the Hatteras abyssal plain data in Rowe and Clifford (1978) indicate there is a high positive concentration gradient in surface sediments (Fig. 8), suggesting that a large fraction of the nitrate in the deep sea is produced in the sediment.

Even though the nitrifiers are autotrophic (C16, MICA in Fig. 6), they probably produce only a small fraction ($<10\%$) of the organic matter in deep-sea sediments (C10). However, if they use free oxygen, this could account for a very large portion of the oxygen uptake Smith (1978) has measured and presumed to be respiration. If this be true, our calculations of

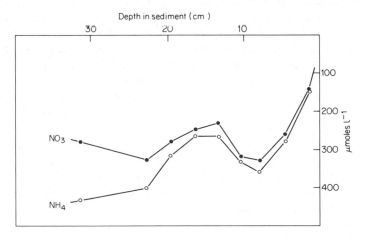

FIG. 8 Nitrate and ammonium ion concentrations in sediment pore water at a depth of 5 km on the Hatteras abyssal plain, western North Atlantic (Station 728 in Rowe and Clifford, 1978)

the amount of organic matter utilised by aerobic heterotrophs is significantly overestimated.

BENTHIC OXYGEN DEMAND

Bell jars equipped with oxygen electrodes have been used for the quantification of energy flow through various parts of shallow ecosystems (chapter 20), and K. L. Smith along with John Teal utilised them to measure deep bottom oxygen demand. At 1800 m south of Massachusetts (Smith and Teal, 1973) total oxygen consumption was $0.5 \, mlO_2 m^{-2} h^{-1}$, but in the San Diego trough at 1230 m it was much higher: $2.4 \, mlO_2 m^{-2} h^{-1}$ (Smith, 1974). The first bell jars were implanted by *Alvin*, but to reduce costs and weather limitations associated with submersibles, Smith and his colleagues developed acoustically controlled free vehicles that automatically placed bell jars or grab respirometers on the bottom (Smith et al., 1978). Using both *Alvin*-placed chambers and free vehicles on a transect across the US eastern continental margin out to the Hatteras abyssal plain, Smith (1978a) found that oxygen uptake fell as low as $0.02 \, mlO_2 m^{-2} h^{-1}$ at 5000 m. While some fraction of total uptake on shallow bottoms can be attributed to chemical oxygen demand (the oxidation of reduced products of anaerobic respiration), Smith could find no chemical uptake deeper than the continental shelf in the Atlantic. Later he found that in some places NO_3 appeared to be consumed, apparently due to denitrification, also a respiratory process (Smith et al., 1979).

These measurements have not been confined to sediments and benthos. Off California, benthopelagic fishes *Coryphaenoides acrolepis* and *Eptatretus deani* consumed 2.4 and $2.2 \, mlO_2 kg^{-1}$ (wet weight) h^{-1}, far lower than related fish at the same temperature but in shallow water (Smith and Hessler, 1974). More recently Smith (1978b) found that respiration in *Coryphaenoides armatus*, at 2.7 to $3.7 \, mlO_2 kg \, h^{-1}$, followed the relation $Y = 0.03 \, W^{0.65}$, where Y is $mlO_2 h^{-1}$ and W is wet weight.

In our box model, this quantifies a part of the consumption of C10 (POMs) by C8, C9, and C12 through C13, indirectly. A major fraction of the organic matter consumed by the heterotrophic organisms is used for metabolic energy; assuming a respiratory quotient of 0.8 (moles of CO_2 produced per O_2 consumed), 0.4 mg of organic carbon is consumed for every ml O_2. Such a calculation has led Smith (1978a) to suggest that only 1 to 2% of the surface primary production of the North Atlantic is consumed by the benthos in deep water.

Combining quantitative information about ecosystem structure with metabolic rate measurements on individuals, one can derive an estimate of the major flux through a state variable or its turnover rate, less losses and gains due to mortality, growth or reproduction. This is my principal goal in current work on the deep-ocean ecosystem, but there are few areas where we can combine these data sets yet. If both are available, one can combine respiration rates (Smith and Hessler, 1974; Smith, 1978b) with biomass data (Haedrich and Rowe, 1978) to estimate the total areal respiration for C8, C9, and C12 (Fig. 6). Such an estimate for total megabenthos would be 0.032 ml m^{-2}h^{-1} at 2000–4000 m. *Coryphaenoides armatus* comprises up to 80% of the total trawl biomass at sites studied in the Atlantic, but a calculation such as the above allows only a rough estimate of the respiration because we know so little about the other megafaunal taxa present.

This estimate of megafaunal respiration is about 6% of the bell-jar-estimated sediment oxygen demand at 1800 m in the Atlantic (Smith and Teal, 1973) or about 1% of the rate in the San Diego Trough in the Pacific (Smith and Hessler, 1974). Even if the Atlantic and the Pacific (San Diego Trough) oxygen demands are both highly underestimated, we can still conclude from this analysis that respiration of near-bottom fishes and other megabenthos is small compared to the total infaunal respiration and nitrification of the sediments (macrofauna, meiofauna and microbes).

By using the average respiration per unit biomass, the total biomass-depth regression can be used now to estimate oxygen utilisation out onto the abyssal plain:

$$0.0032 \text{ mlO}_2\text{g}^{-1}\text{h}^{-1} \{(\log_{10} \text{ g m}^{-2}) = 1.44 - 0.41 \times (\text{depth in km})\},$$

A better analysis would utilise the trawl-catch size distribution with Smith's weight-specific regression to obtain more exact estimates. And of course, similar regressions need to be developed for the big invertebrates. The contribution of the larger organisms, however, to total respiration, does appear to be minor, as in shallow-water assemblages (Banse et al., 1971).

A PRELIMINARY CARBON BUDGET: NORTH-WEST ATLANTIC

The data of Honjo (1978), Rowe and Gardner (1979) and Smith (1978) can now be used to construct a budget for organic carbon along the deep extension of the continental margin of the eastern United States. Dependent variables such as burial (C11) and respiration, as represented in Fig. 6, can be plotted against the input or forcing function, the rate of sedimenting organic carbon (C1 and C4), considered the independent variable (Fig. 9).

Several features of these relationships are worth noting. First, at each point the sum of burial and respiration is not quite equal to input, using an RQ of 1, but not including megafaunal respiration, or denitrification. Honjo's data exhibit the same relationships as those of Rowe and Gardner, falling between zero (no input) and the larger inputs on the continental rise, suggesting very different kinds of sediment traps (cones and cylinders) function rather similarly.

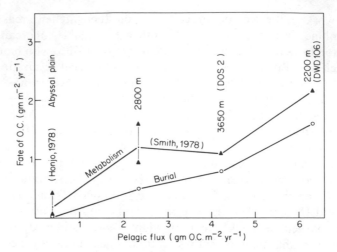

FIG. 9 Relationship between fates of organic carbon [respiration, from Smith (1978) and burial of organic carbon from Rowe and Gardner (1979)], and sedimenting POC from Rowe and Gardner (1979) and Honjo (1978)

Most structural features of the animal assemblages (biomass, species similarities, diversity, abundance, etc.) have historically been plotted against depth, utilising this as an independent variable, with the assumption that depth or something closely correlated with it (food supply, stability or predictability, pressure, etc.) controlled the dependent variables. As the depth of each station and carbon input are well correlated it is not surprising that the linear fit for carbon fate is about as good using depth as input of carbon.

Finally, these relationships can be used to define, in the form of simple, linear equations, the dynamic response of state variables to changes in the forcing function:

Burial rate = 0.25 (small particle sedimentation rate)
Oxygen demand (RQ of 0.8, disregarding nitrification) = 0.33 (small particle sedimentation rate)

An implication of this model is that only the rain of particles and the Vinogradov zooplankton ladder of migration are important in the mass budget of carbon in the deep sea. That is, large carcasses, large plant detritus and wood, chemolithotrophy and turbidity currents contribute only slightly in supplying organic carbon to the deep-sea biota. Intuitively, I feel this cannot be so. These other sources must be of more significance than the simple model implies because they contain large fractions of highly labile, or biologically available, organic matter (protein, carbohydrate, lipid), as opposed to the small particles, which are often considered highly refractory. The relationships presented are only correlations, may be spurious and may not be a real 'cause and effect'. The equations only suggest hypotheses to be tested.

GROWTH AND REPRODUCTIVE RATES

The above set of problems can be approached independently through the population dynamics of the component species in the deep-sea ecosystem, but such work indicates great variability in these functions. A small but common bivalve mollusc (*Tindaria callistiformis*) has been aged using a radiometric method (Thorium-228/Radium-228) and found to reach sexual maturity at about 50 years and maximum size in about 100 years, at only a few mms in length (Turekian *et al.*, 1975). On the other hand, xylophagous molluscs can grow to maturity in a few months (Turner, 1973). Galatheid decapods were observed to reach several centimeters in length in a few months when held in a plastic bag with borer-infested wood, perhaps by feeding on fecal matter produced by the molluscs (Turner, 1973, and 1977). Brachiopods also appear to have much faster growth rates than the bivalves dated by Turekian *et al.* (Zezina, 1975). The macrourid (rattail) fish *Nezumia* is believed to grow at about 25 mm per year (Rannou, 1976). Grassle (1977) has found that sterile mud boxes placed open on the bottom at 1800 m south of Massachusetts were colonised at slow rates that reflect the generally slow rates of succession and growth in the deep sea but Desbruyères *et al.* (1980) found growth rates to be surprisingly high in a similar experiment in the Bay of Biscay. Growth rates of all biota are implicit components of our model (Fig. 6), though it is apparent that such data will be some of the most difficult and time-consuming to obtain in future work.

In shallow water, reproduction is seasonal in many species, but as the deep sea exhibits no physical seasonality, the occurrence and causes of periodic spawning have been of great interest. Likewise they could be important for inferring taxonomic relationships and subtle cyclical inputs of energy. Size classes have been identified in deep-sea brittle stars in both the north-west (Schoener, 1968) and north-east Atlantic (Tyler and Gage, 1980) and it has been concluded they probably spawn seasonally. George and Menzies (1968) found a suggestion of a similar pattern in a species of isopod crustacean. Rokop (1974, 1977), considering a number of distantly related taxa at intermediate depths off California, concluded that in general at any given time a small percentage of each population will always be found to be spawning, and that periodic spawning is uncommon, if it occurs at all. Periodicity has been found in the North Atlantic, characterised by marked seasonal changes, whereas Rokop studied animals in the North Pacific, with less radical changes. The question remains open. Too little data are available to allow identification of seasonality in bathypelagic forms (Mauchline, 1972).

5 The hydrothermal vent community

The greatest surprise in contemporary deep-sea ecology has been the discovery of the hydrothermal vents between the Galapagos Islands and mainland Ecuador, South America (Lonsdale, 1978). In explorations by *Alvin* a group of earth scientists discovered that a peculiar invertebrate fauna of exceptionally high biomass lives on the rocks in water of about 10–15 °C flowing out of vents in the fracture zones between tectonic plates (Corliss *et al.*, 1979). The fauna is composed of galatheid and brachyuran decapod crustaceans, mussel-like bivalve molluscs, large pogonophoran worms, actinians and includes numbers of unidentified species and other taxa (Fig. 10).

This anomalous community obviously is contrary to the low-energy-flow generalities described above for the typical deep-sea ecosystem, and it is rational to presume that it is fueled from a different energy source. The hydrothermal effluent contains no oxygen. but is rich in hydrogen sulfide and numerous dissolved ions in a reduced state. The metabolic ions and temperature follow a linear dilution pattern as the vent water mixes with the surrounding cold, deep ocean (Corliss *et al.*, 1979). It is believed the sulfide and perhaps other reduced ions are used as an energy source by anaerobic bacteria apparently with the same metabolic processes as chemolithotrophic bacteria in shallow water (Karl *et al.*, 1980). The free energy liberated on the oxidation of sulfide and other reduced ions is used to synthesise cells and perform their metabolism (Rau and Hedges, 1979; Jannasch and Wirsen, 1979). A simplified model of the process (Fig. 11) illustrates the energy transfer that is assumed to occur deep down in the vent fissures. A milky blue effluent is known to carry with it fragmented strands of the bacteria that then are available to be fed on by the numerous higher organisms that line the outer fissure perimeters. These biota – crustaceans, molluscs, pogonophorans and coelenterates – appear to be dominated by mucophagous and filter-feeding forms. The composition of the fauna changes radically with increasing distance away from a vent and each vent so far discovered has a somewhat different assemblage. A clam 22 cm long appears to be less than 10 years old, using Th/Ra dating (Galapagos Biology Expedition Participants 1979). Although mussels and crabs live in warm water, most organisms live at ambient temperatures (2 °C) some distance from the vents (Galapagos Biology Expedition Participants 1979).

Another active spreading center has been found off the west coast of Mexico in what is called the East Pacific Rise. There the temperatures may

Fig. 10 Invertebrates clustered around hydrothermal vents in the Galapagos Rift area.
Courtesy Robert Ballard, Woods Hole Oceanographic Institute

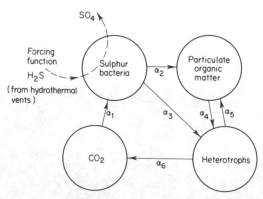

FIG. 11 Conceptual model of the synthesis of organic matter by the utilisation of energy in hydrogen sulfide by chemolithotrophic sulfur bacteria

exceed 300 °C and deposits of metals carpet areas around volcanic structures. Their biological features suggest they are similar to the Galapagos (*Anon.*, 1979).

These totally unexpected communities with a unique energy source for the deep sea present intriguing problems to physiologists, but also to zoogeographers and systematists. Pressure and temperature are of little importance to at least some of these forms, and we might presume that many are new species that have evolved only around such vents. The anaerobic or microaerophilic bacteria must have a very active metabolism, in spite of the apparent pressure restrictions on microbial activity (Jannasch and Wirsen, 1977), in order to supply the required organic matter for the larger populations. Research on these vents has only just begun.

Acknowledgement

This research was performed under the auspices of the United States Department of Energy under contract No. EY-76-C-02-0016.

References

Agassiz, A. (1888). "Three Cruises of the *Blake*," Vol. I. Houghton, Mifflin and Co., Boston

Anonymous (1979). *Oceanus* **22**, 10

Banse, K. (1964). On the vertical distribution of zooplankton in the sea. *Prog. Oceanogr.* **2**, 53–125

Banse, K., Nichols, F. H. and May, D. R. (1971). Oxygen consumption of the seabed. III. On the role of macrofauna at three stations. *Vie et Milieu* Suppl. No. 22, 31–52

Barham, E. G. (1971). Deep-sea fishes: lethargy and vertical orientation. *In* "Proc. Int. Symp. Biological Sound Scattering in the Ocean" (G. B. Farquhar, ed.) Vol. 5, pp. 100–118. Scient. Rep. Maury Center Ocean

Belyaev, G. M. (1972). "Hadal Bottom Fauna of the World Ocean". Institute of Oceanology, U.S.S.R. Academy of Sciences, Moscow, U.S.S.R.

Bender, M. L., Fanning, K. A., Froelich, P. N., Heath, G. R. and Maynard, V. (1977). Interstitial nitrate profiles and oxidation of sedimentary organic matter in the eastern equatorial Atlantic. *Science* **189**, 605–609

Berger, W. H. and Soutar, A. (1967). Planktonic foraminifera: field experiment on production rate. *Science* **1561**, 1495–1497

Bertelsen, E. (1951). The ceratioid fishes. *Dana Rept.* **39**, 1–276

Brunn, A. F. (1956). "The Galathea Deep-Sea Expedition". MacMillan, New York

Cacchione, D. A., Rowe, G. T. and Malahoff, A. (1978). Submersible investigation of outer Hudson Submarine Canyon. *In* "Sedimentation in Submarine Canyons, Fans, and Trenches" (D. J. Stanley and G. Kelling, eds) pp. 42–50. Dowden, Hutchinson and Ross, Inc., Stroudsburg, Pa.

Carlucci, A. F. and Strickland, J. D. H. (1968). The isolation, purification and some kinetic studies of marine nitrifying bacteria. *J. exp. mar. Biol. Ecol.* **2**, 156–166

Carney, R. S. and Carey, A. G. Jr. (1977). Distribution pattern of holothurians on the northeastern Pacific (Oregon, U.S.A.) continental shelf, slope and abyssal plain. Thallassia, Yugoslavia

Childress, J. J. (1971). Respiratory rate and depth of occurrence of midwater animals. *Limnol. Oceanogr.* **16**, 104–106

Childress, J. J. and Nygaard, M. H. (1973). The chemical composition of midwater fishes as a function of depth of occurrence off Southern California. *Deep-Sea Res.* **20**, 1093–1109

Connell, J. H. (1978). Diversity in tropical rain forests and coral reefs. *Science* **199**, 1302–1310

Corliss, J., Dymond, J., Gordon, L., Edmond, J., von Herzen, R., Ballard, R., Green, K., Williams, D., Bainbridge, A., Crane, K. and van Andel, T. (1979). Submarine thermal springs on the Galapagos Rift. *Science* **203**, 1073–1083

Coull, B. C., Ellison, R. L., Fleeger, J. W., Higgins, R. P., Hope, W. D., Hummon, W. D., Ruger, R. M., Sterrer, W. E., Thiel, H. and Tietjen, J. (1977). Quantitative estimates of the meiofauna from the deep sea off North Carolina. *Mar. Biol.* **39**, 233–240

Cronan, D. S. (1972). Deep-sea nodules: distribution and geochemistry. *In* "Marine Manganese Deposits" (G. P. Glasby, ed.) Ch. 2, pp. 11–44. Elsevier, Amsterdam

Dayton, P. K. and Hessler, R. R. (1972). Role of biological disturbance in maintaining diversity in the deep-sea. *Deep-Sea Res.* **19**, 199–208

Desbruyères, D., Bervas, J. and Khripounoff, A. (1980). Un cas de colonisation rapide d'un sediment profond. *Oceanologica Acta* **3**, 285–291

Dickinson, J. J. and Carey, A. G. (1978). Distribution of gammarid amphipoda (crustacea) in Cascadia Abyssal Plain (Oregon). *Deep-Sea Res.* **25**, 97–106

Dinet, A., Laubier, L., Soyer, J. and Vitiello, P. (1973). Resultats biologiques de la campagne Polymede. II. Le meiobenthos abyssal. *Rapp. P. v. Reun. Comm. int. Explor. scient. Mer Mediterr.* **21**, 701–704

Dinet, A. (1979). A quantitative survey of meiobenthos in the deep Norwegian Sea. *Ambio* Special Report **6**, 75–77

Dinet, A. and Vivier, M. H. (1976). Le Meiobenthos abyssal du Golfe de Gascogne. I – Considerations sur les données quantitatives. *Cah. Biol. Marine* **18**, 85–97

Dinet, Alain and Vivier, Marie-Hélène (1979). Le Meiobenthos abyssal du Golfe de Gascogne. II. Les peuplements de nematodes et leur diversité specifique. *Cahiers de Biologie Marine* 109–123

Dudley, W. C. and Margolis, S. V. (1974). Iron and trace element concentration in marine manganese nodules by benthic agglutinated Foraminifera. Abstr. *Progr. Ann. Mtg. Geol. Soc. Amer.* 6 (7), 716 (Abstract)

Ehrlich, H. L. (1972). The role of microbes in manganese nodule genesis and degradation. In "Ferromanganese Deposits on the Ocean Floor" (D. R. Horn, ed.) pp. 63–69. National Science Foundation, Washington, D.C.

Ekman, S. (1935). "Tiergeographie des Meeres". Akad. Verlagsges. Leipzig

Emery, K. O. and Uchupi, E. (1972). Western North Atlantic Ocean Topography, Rocks, Structure, Water, Life and Sediments. *Amer. Assoc. Petrol. Geologists Memoir* 17, 1–532

Filatova, Z. A. and Zenkevich, L. A. (1969). Present distribution of ancient primative Monoplacophoran mollusks in the Pacific Ocean and the fossilized Pogonophora in deposits of Cambrian Seas. *Oceanology* 9 (1), 126–136

Fournier, R. O. (1972). The transport of organic carbon to organisms living in the deep oceans. *Proc. Royal Society of Edinburgh* 73, 203–211

Galapagos Biology Expedition Participants (1979). Initial findings of a deep-sea biological quest. *Oceanus* 22, 2–10

George, R. Y. and Menzies, R. J. (1968). Further evidence for seasonal breeding cycles in the deep sea. *Nature* 220, 80–81

Grassle, J. F. and Sanders, H. L. (1973). Life histories and the role of disturbance. *Deep-Sea Res.* 20, 643–659

Grassle, J. F. (1977). Slow recolonization of deep-sea sediment. *Nature* 265, 618–619

Grassle, J. F., Sanders, H., Hessler, R., Rowe, G. and McClellan, T. (1975). Pattern and zonation: a study of the bathyal megafauna using the research submersible *Alvin. Deep-Sea Res.* 22, 457–481

Greenslate, J. (1974a). Microorganisms participate in the construction of manganese nodules. *Nature* 249, 181–183

Greenslate, J. (1974b). Manganese and biotic debris associations in some deep-sea sediments. *Science* 186, 529–531

Griggs, G. B., Carey, A. G. Jr. and Kulm, L. D. (1969). Deep-sea sedimentation and sediment fauna interaction in Cascadia Channel and on Cascadia Abyssal Plain. *Deep-Sea Res.* 16, 157–170

Haedrich, R. L. and Henderson, N. (1974). Pelagic food of *Coryphaenoides armatus*, a deep benthic rattail. *Deep-Sea Res.* 21, 739–744

Haedrich, R. L. and Polloni, P. T. (1976). A contribution to the life history of a small rattail fish, *Coryphaenoides carapinus. Deep-Sea Res.* 75, 203–211

Haedrich, R. L. and Rowe, G. T. (1978). Megafaunal biomass in the deep-sea. *Nature* 269, 141–142

Haedrich, R. L., Rowe, G. T. and Polloni, P. T. (1975). Zonation and faunal composition of epibenthic populations on the continental slope south of New England. *J. Mar. Sci.* 33, 191–212

Haedrich, R. L., Rowe, G. T. and Polloni, P. T. (1980). The megabenthic fauna in the deep-sea south of New England. *Mar. Biol.* 57, 165–179

Harbison, G. R. and Campenot, R. B. (1979). Effects of temperature on the swimming of salps (*Tunicata, Thaliacea*): Implications for vertical migration. *Limnol. Oceanogr.* 24, 1081–1091

Heath, G. R. (1979). Burial rates, growth rates, and size distribution of deep-sea manganese nodules. *Science* **205**, 903–904

Heezen, B., Ewing, M. and Menzies, R. (1955). The influence of submarine turbidity currents on abyssal productivity. *Oikos* **6**, 170–182

Heezen, B. C., Hollister, C. D. and Ruddiman, W. F. (1966). Shaping of the continental rise by deep geostrophic contour currents. *Science* **152** (3721), 502–508

Hessler, R. R. and Sanders, H. L. (1967). Faunal diversity in the deep-sea. *Deep-Sea Res.* **14**, 65–78

Hessler, R. R., Isaacs, J. D. and Mills, E. L. (1972). Giant amphipod from the abyssal Pacific Ocean. *Science* **175**, 636–637

Hessler, R. R., Ingram, C. L., Aristides Yayanos, A. and Burnett, B. (1978). Scavenging amphipods from the floor of the Philippine Trench. *Deep-Sea Res.* **25**, 1029–1047

Hochachka, P. W., Schneider, D. E. and Kuznetsov, A. (1970). Interacting pressure and temperature effects on enzymes of marine poikilotherms: catalytic and regulatory properties of FPPase from deep and shallow-water fishes. *Mar. Biol.* **1**, 285–293

Honjo, S. (1978). Sedimentation of materials in the Sargasso Sea at a 5367 m deep station. *J. Mar. Res.* **36**, 469–492

Issacs, J. D. and Schwartzlose, R. A. (1975). Active animals of the deep-sea floor. *Sci. Amer.* **233** (4), 85–91

Isaeva, A. B., Sapozhnikov, V. V. and Fedikov, N. F. (1978). Chemical characteristics of bottom and interstitial waters of Australian-New Zealand sector of Antarctic region. *Trans. Shirshov Inst. Oceanol.* **112**, 45–53

Jannasch, H. W., Eimhjellen, K., Wirsen, C. O. and Farmonfarmaian, A. (1971). Microbial degradation of organic matter in the deep sea. *Science* **171**, 672–675

Jannasch, H. W. and Wirsen, C. O. (1977). Microbial life in the deep-sea. *Sci. Amer.* **236**, 42–52

Jannasch, H. W. and Wirsen, C. O. (1979). Chemosynthetic primary production at east Pacific seafloor spreading centers. *BioScience* **29**, 592–598

Johnston, R. (1962). An equation for the depth distribution of deep-sea zooplankton and fishes. *Rapp. P.-v. Reun. Cons. perm. int. Explor. Mer.* **153**, 217–219

Jumars, P. A. (1975). Environmental grain and polychaete species diversity in a bathyal benthic community. *Mar. Biol.* **30**, 253–266

Jumars, P. A. (1976). Deep-sea species diversity: does it have a characteristic scale? *J. Mar. Res.* **34**, 217–246

Karl, D. M., Wirsen, C. O. and Jannasch, H. W. (1980). Deep-sea primary production at the Galapagos hydrothermal vent. *Science* **207**, 1345–1347

Khripounoff, A., Desbruyères, D. and Chardy, P. (1980). Les peuplements benthiques de la faille VEMA: données quantitatives et bilan d'énergie en milieu abyssal. *Oceanologica Acta* **3**, 187–198

Ku, T. L. (1977). Rates of accretion. *In* "Marine Manganese Deposits" (G. P. Glasby, ed.) Ch. 8, pp. 249–367. Elsevier, Amsterdam

Lerman, A. (1977). Migrational processes and chemical reactions in interstitial waters. *In* "The Sea" (E. Goldberg, I. McCave, J. O'Brien and I. Steele, eds) Vol. 6, Marine Modeling, pp. 695–738. Wiley, New York

LeDanois, E. (1948). "Les Profondeurs de la Mer". Payot, Paris

Lonsdale, P. (1977). Clustering of suspension-feeding macrobenthos near abyssal hydrothermal vents at oceanic spreading centers. *Deep-Sea Res.* **24**, 857–863

Lowenstam, H. A. and Epstein, S. (1959). Cretaceous paleotemperatures as determined by the oxygen isotope method, their relations to and the nature of rudisid reefs. *Proc. 20th Sess. Congr. Geol. Int., Mexico City*, 67–78

McCave, I. N. (1975). Vertical flux of particles in the ocean. *Deep-Sea Res.* **22**, 491–502

MacDonald, A. G. (1975). "Physiological Aspects of Deep-Sea Biology". Cambridge University Press, Cambridge

Madsen, J. F. (1961). On the zoogeography and origin of the abyssal fauna. *Galathea Rept.* **4**, 177–217

Mare, M. F. (1942). A study of a marine benthic community with special references to the micro-organisms. *J. Mar. Biol. Assoc., U.K.* **25**, 517–554

Marquis, R. E. (1976). High-pressure microbial physiology. *Adv. Micr. Physiol.* **14**, 159–241

Marshall, N. B. (1960). Swimbladder structure of deep-sea fishes in relation to their systematics and biology. *Discovery Rept.* **31**, 1–122

Marshall, N. B. (1971). "Explorations in the Life of Fishes". Harvard University Press, Cambridge

Mauchline, J. (1972). The biology of bathypelagic organisms, especially Crustacea. *Deep-Sea Res.* **19**, 753–780

Meek, R. P. and Childress, J. J. (1973). Respiration and the effect of pressure in the mesopelagic fish *Anoplogaster cornuta* (Beryciformes). *Deep-Sea Res.* **20**, 1111–1118

Menzel, D. W. (1967). Particulate organic carbon in the deep-sea. *Deep-Sea Res.* **14**, 229–238

Menzel, D. W. (1974). Primary productivity, dissolved and particulate organic matter and the sites of oxidation of organic matter. *In* "The Sea" (E. Goldberg, ed.) Vol. 5, pp. 659–678. Wiley, New York

Menzel, D. W. and Ryther, J. H. (1968). Organic carbon and the oxygen minimum in the South Atlantic Ocean. *Deep-Sea Res.* **15**, 327–337

Menzies, R. J. (1962). On the food and feeding habits of abyssal organisms as exemplified by the Isopoda. *Int. Rev. Ges. Hydrobiol.* **47**, 339–358

Menzies, R. J. and Imbrie, J. (1958). On the antiquity of the deep-sea bottom fauna. *Oikos* **9** (2), 192–201

Menzies, R. J., Imbrie, J. and Heezen, B. C. (1961). Further considerations regarding the antiquity of the abyssal fauna with evidence for a changing abyssal environment. *Deep-Sea Res.* **8**, 79–94

Menzies, R. J., Zeneveld, J. S. and Pratt, R. M. (1967). Transported turtle grass as a source of organic enrichment of abyssal sediments off North Carolina. *Deep-Sea Res.* **14**, 111–112

Menzies, R. J., George, R. Y. and Rowe, G. T. (1973). "Abyssal Environment and Ecology of the World Ocean". Wiley, New York

Mills, E. L. (1972). T. R. R. Stebbing, the *Challenger* and knowledge of deep-sea Amphipoda. *Proc. Roy. Soc. Edinburgh, Section B (Biology)* **72**, 67–87

Moon, T. W., Mustafa, T. and Hochachka, P. W. (1971). The adaptation of enzymes to pressure. II. By comparison of muscle pyravate kinases from surface and midwater fishes with the homologous enzyme from an offshore benthic species. *Am. Zool.* **11**, 491–502

Napora, T. A. (1964). The effect of hydrostatic pressure on the prawn *Systellaspis debilis*. *Proc. Symp. Exp. Mar. Ecol.* No. 2, 92–94

Pearcy, W. G. and Ambler, J. W. (1974). Food habits of deep-sea macrourid fishes off the Oregon coast. *Deep-Sea Res.* **21**, 745–759

Polloni, P. T., Haedrich, R. L., Rowe, G. T. and Clifford, C. H. (1979). The size-depth relationship in deep ocean animals. *Int. Rev. Ges. Hydrobiol.* **64**, 39–46

Rannou, M. (1976). Age et croissance d'un poisson bathyal *Nezumia sclerorhynchus* (Macrouride, Gadiformes) en Mer d'Alboran. *Cah. Biol. Mar.* **17**, 413–421

Ray, G. H. and Hedges, J. (1979). Carbon-13 depletion in a hydrothermal vent mussel: suggestion of a chemosynthetic food source. *Science* **203**, 648–649

Rex, M. A. (1977). Zonation in deep-sea gastropods: the importance of biological interactions to rates of zonation. *In* "Biology of Benthic Organisms" (B. Keegan, P. Ceidigh and P. Boaden, eds) pp. 521–530. Pergamon Press, New York

Rex, M. A. Maintenance and evolution of species diversity in the deep-sea benthos. *In* "The Sea" (G. Rowe, ed.) Vol. 8, Deep-Sea Biology. Wiley, New York. (In press)

Riley, G. (1970). Particulate and organic matter in sea water. *Adv. Mar. Biol.* **8**, 1–118

Rokop, F. J. (1974). Reproductive patterns in the deep-sea benthos. *Science* **186**, 743–745

Rokop, F. J. (1977). Patterns of reproduction in the deep-sea benthic crustaceans: a re-evaluation. *Deep-Sea Res.* **24**, 683–691

Rowe, G. T. (1971). Benthic biomass and surface productivity. *In* "Fertility of the Sea" (J. Costlow, ed.) pp. 441–454. Gordon and Breach, New York

Rowe, G. T. (1979). Monitoring with deep submersibles. *In* "Monitoring the Marine Environment" (D. Nichols, ed.) pp. 75–85. Biol. Soc. London, England

Rowe, G. T. and Clifford, C. H. (1978). Sediment data from short cores taken in the northwest Atlantic Ocean. Woods Hole Oceanographic Institution Tech. Rept. WHOI-78-46

Rowe, G. T. and Gardner, W. (1979). Sedimentation rates in the slope water of the northwest Atlantic Ocean measured directly with sediment traps. *J. Mar. Res.* **37**, 581–600

Rowe, G. T. and Haedrich, R. L. (1979). The biota and biological processes of the continental slope. *S.E.P.M. Spec. Publ.* (L. Doyle and O. Pilkey, eds) No. 27, pp. 49–59

Rowe, G. T. and Staresinic, N. (1979). Sources of organic matter to the deep-sea benthos. *Ambio Special Report* **6**, 19–27

Rowe, G. T. and Menzies, R. J. (1968). Deep bottom currents off the coast of N. Carolina. *Deep-Sea Res.* **15**, 711–719

Rowe, G. T. and Menzies, R. J. (1969). Zonation of large benthic invertebrates in the deep-sea off the Carolinas. *Deep-Sea Res.* **16**, 531–581

Rowe, G. T., Polloni, P. T. and Hornor, S. (1974). Benthic biomass estimates from the northwestern Atlantic Ocean and the northern Gulf of Mexico. *Deep-Sea Res.* **21**, 641–650

Rowe, G. T., Polloni, P. and Haedrich, R. L. Infaunal benthic community structure in the Hudson Submarine Canyon system and adjacent continental margin. *Deep-Sea Res.* (In press)

Sanders, H., Hessler, R. R. and Hampson, G. R. (1965). An introduction to the study of deep-sea benthic faunal assemblages along the Gay Head Bermuda transect. *Deep-Sea Res.* **12**, 845–867

Sanders, H. L. (1968). Marine benthic diversity: a comparative study. *Amer. Nat.* **102**, 243–282

Sanders, H. L. and Hessler, R. R. (1969). Ecology of the deep-sea benthos. *Science* **163**, 1419–1424

Schoener, A. and Rowe, G. T. (1970). Pelagic *Sargassum* and its presence among the deep-sea benthos. *Deep-Sea Res.* **17**, 923–925

Schoener, A. (1968). Evidence for reproductive periodicity in the deep-sea. *Ecology* **49**, 81–87

Schwarz, J. R., Yayanos, A. A. and Colwell, R. R. (1976). Metabolic activities of the intestinal microflora of a deep-sea invertebrate. *Appl. Environ. Microbiol.* **31**, 46–48

Siebenaller, J. and Somero, G. N. (1978). Pressure-adaptive differences in lactate dehydrogenases of congeneric fishes at different depths. *Science* **201**, 255–257

Smith, K. L. Jr. (1974). Oxygen demand of San Diego Trough sediments: an *in situ* study *Limnol. Oceanogr.* **19**, 939–944

Smith, K. L. Jr. (1978a). Benthic community respiration in the Northwest Atlantic: *In situ* measurements from 40 to 5200 meters. *Mar. Biol.* **47**, 337–347

Smith, K. L. Jr. (1978b). Metabolism of the abyssopelagic rattail *Coryphaenoides armatus* measured *in situ*. *Nature* **274**, 362–364

Smith, K. L. Jr. and Teal, J. M. (1973). Deep-Sea benthic community respiration: An *in situ* study at 1850 meters. *Science* **179**, 282–283

Smith, K. L. Jr. and Hessler, R. R. (1974). Respiration of benthopelagic fishes: *In situ* measurements at 1230 meters. *Science* **184**, 72–73

Smith, K. L. Jr., White, G. A. and Laver, M. B. (1979). Oxygen uptake and nutrient exchange of sediments measured *in situ* using a free vehicle grab respirometer. *Deep-Sea Res.* **26A**, 337–346

Spärck, R. (1935). On the importance of quantitative investigations of the bottom fauna in marine biology. *Cons. Perm. Int. Explor. Mer. Journ. du Conseil* **19** (1), 3–19

Spärck, R. (1936). Review of the Danish investigations on the quantitative composition of the bottom fauna in Iceland and Greenland waters. *Rapp. P.-v. Reun. Cons. perm. int. Exlor. Mer* **99**, 1–5

Spärck, R. (1956). The density of animals on the ocean floor. *In* "The Galathea Deep-Sea Expedition 1950–1952", pp. 196–201. Allen and Unwin, London

Spencer, D., Brewer, P., Fleer, A., Honjo, S., Krishnaswami, S. and Nozaki, Y. (1978). Chemical flux from a sediment trap experiment in the deep Sargasso Sea. *J. Mar. Res.* **36**, 493–523

Teal, J. M. and Carey, F. G. (1967). Effects of pressure and temperature on the respiration of euphaussiids. *Deep-Sea Res.* **14**, 725–733

Teal, J. M. (1971). Pressure effects on the respiration of vertical migrating decapod Crustacea. *Am. Zool.* **11**, 571–576

Thiel, H. (1975). The size structure of the deep-sea benthos. *Int. Rev. Ges Hydrobiol.* **60** (5), 575–606

Thiel, J. and Rumohr, H. (1979). Photostudio am Meeresboden. *Umschau 79* (1979), **15**, 469–472

Thistle, D. (1978). Harpacticoid dispersion patterns: implications for deep-sea diversity maintenance. *J. Mar. Res.* **36** (2), 377–397

Thistle, D. (1979). Deep-sea harpacticoid copepod diversity maintenance: The role of polychaetes. *Mar. Biol.* **52**, 371–376

Tietjen, J. H. (1971). Ecology and distribution of deep-sea meiobenthos off North Carolina. *Deep-Sea Res.* **18**, 941–957

Turekian, K. K., Cochran, J. K., Kharkar, D. P., Cerrato, R., Vaisnys, J., Sanders, H. L., Grassle, J. K. and Allen, J. (1975). The slow growth rate of a deep-sea clam determined by 228 Ra chronology. *Proc. Nat. Acad. Sci* **72**, 2829–2832

Turner, R. D. (1977). Wood, mollusks, and deep-sea food chains. *Bull. Amer. Malacol. Union, Inc.* 1977, 13–19

Turner, R. D. (1973). Wood-boring bivalves, opportunistic species in the deep sea. *Science* **180**, 1377–1379

Tyler, P. A. and Gage, J. D. (1980). Reproduction and growth of the deep-sea brittlestar *Ophiura ljungmani* (Lyman). *Oceanologica Acta* **3**, 177–185

Vinogradov, M. E. (1968). "Vertical Distribution of the Oceanic Zooplankton". Academy of Sciences of the U.S.S.R., Institute of Oceanography, Moscow

Vinogradov, M. E. (1972). Vertical stratification of zooplankton in the Kurile-Kamchatka trench. *In* "Biological Oceanography of the Northern North Pacific Ocean" (A. Y. Takenouti, ed.) pp. 333–340. Idemitsu Shoten, Japan

Vinogradova, N. (1962). Vertical zonation in the distribution of deep-sea benthic fauna in the ocean. *Deep-Sea Res.* **8**, 245–250

Vivier, M. (1978). Consequences d'un diversement de boue rouge d'alumine sur le meiobenthos profound (Canyon de Cassidaigne, Mediterranee). *Tethys* **8** (e) 1976 (1978), 249–262

Walsh, J. J. (1981). Shelf-sea ecosystems. Chapter 6, this volume

Walters, V. (1961). A contribution to the biology of the Giganturidae, with description of a new genus and species. *Bull. Mus. Comp. Zool.* **125**, 10

Wiebe, P., Boyd, S. and Winget, C. (1976). Particulate matter sinking to the deep-sea floor at 2000 m in the Tongue of the Ocean, Bahamas with a description of a new sedimentation trap. J. Mar. Res. **34**, 341–354

Wiebe, P. H., Madin, L., Haury, L., Harbison, G. R. and Philbin, L. (1979). Diel vertical migration by *Salpa aspera* and its potential for large-scale particulate organic matter transport to the deep sea. *Mar. Biol.* **53**, 249–255

Wiebe, P., Remsen, C. and Vaccaro, R. (1974). *Halosphaera viridis* in the Mediterranean Sea: size range, vertical distribution, and potential energy source for deep-sea benthos. *Deep-Sea Res.* **21**, 657–667

Wolff, T. (1976). Utilization of seagrass in the deep sea. *Aquatic Botany* **2**, 161–174

Worthington, L. V. (1968). Genesis and evolution of water masses. *Meteorological Monographs* **8**, 63–67

Wüst, G. (1958). Stromgeschwindigkeiten und strommegen in den Tiefen des Atlantischen Ozeans. Deutsche Atlantische Exped. Meteor. 1925–1927. *Wiss. Erg.* **6**, 35–420

Yayanos, A. A., Dietz, A. and van Boxtel, R. (1979). Isolation of a deep-sea barophilic bacterium and some of its growth characteristics. *Science* **205**, 808–810

Zenkevitch, L. A. and Birstein, J. A. (1960). On the problem of the antiquity of the deep-sea fauna. *Deep-Sea Res.* **7**, 10–23

Zenkevitch, L. (1961). Certain quantitative characteristics of the pelagic and bottom life of the ocean. *In* "Oceanography" (M. Sears, ed.) pp. 323–336. Amer. Assoc. Adv. Sci. 67

Zezina, O. N. (1975). On some deep-sea brachiopods from the Gay Head – Bermuda Transect. *Deep-Sea Res.* **22**, 903–912

ZoBell, C. (1946). "Marine Microbiology". Chronica Botanica, Waltham, Massachusetts

ZoBell, C. and Morita, R. (1957). Barophilic bacteria in some deep-sea sediments. *J. Bact.* **73**, 563–568

Part 2

Functions within Ecosystems

This section comprises reviews of some important processes which are common to many marine ecosystems and have been widely studied in recent years.

9 The Trophic Role of Dissolved Organic Material

GROVER C. STEPHENS

1 Introduction

Sea water is a rather concentrated solution of inorganic salts; its concentration is variable but usually contains about 35 grams of inorganic salts per kilogram. It is also a very dilute solution of organic material containing an average of approximately 3 milligrams organic material per litre. There is considerable variability with respect to total organic content and qualitative composition of organic materials. The analytical problems posed by a dilute solution of organic materials in the presence of high concentrations of salt have been overcome with respect to some constituents of the organic content, but we are far from having a complete inventory and characterisation of all organic compounds present in sea water. The general term, dissolved organic material (DOM), is used to refer to this complex and variable feature of the oceans.

The total amount of DOM in the oceans is very large. Table I lists the weight of organic material in several compartments of the biosphere for purposes of comparison, mostly taken from Woodwell *et al.* (1978). It will be immediately apparent that the marine DOM vastly outweighs total plant and animal biomasses. This simple fact requires emphasis since the DOM is imperceptible without sophisticated analytical tools. The preponderance of DOM is partly the result of its rather constant distribution with depth while organisms tend to be concentrated in the photic zone, in layers, and on the

TABLE I Organic carbon in the biosphere in gigatons (10^5g)

Dissolved in sea water[a]	1596
Plant biomass (total)	828
Annual primary productivity (total)	78
Marine plant biomass	1.7

[a] 5% of marine organic carbon is assumed to be particulate; this estimate does not include interstitial water of sediments

271

bottom. However, it should be stressed that the quantity of DOM exceeds that of organic matter in particulate form approximately by a factor of 10 even in productive areas that are relatively rich in plankton.

This complex solution of inorganic salts and organic material is the natural medium of marine organisms. Throughout their lives, they regulate exchanges of inorganic and organic solutes with their aqueous medium. Much of our information concerning the comparative physiology and physiological ecology of marine organisms is related to the physiological mechanisms and functional significance of these exchanges. Regulated exchange of inorganic ions has received most attention, but there are many instances in which organic compounds provide a source of information for orientation and behavior, act as triggers for key developmental events, or are produced and liberated to serve a variety of functional requirements.

The hypothesis that DOM may play a significant and direct role in the nutrition of aquatic organisms was introduced by Pütter (1909), and is usually associated with his name. It is an attractive possibility. Given the very large DOM reservoir, physiological and behavioral adaptations that might permit direct utilisation of this source of reduced carbon would be clearly advantageous. Often it is difficult to identify the particulate food resources on which particular animals subsist. In some cases (e.g. pogonophorans), animals have no digestive tract or a tract which is absent at some life stages or so reduced that it is difficult to understand how they might sustain themselves by particulate feeding methods. Reasons of this sort led Pütter to postulate his hypothesis and to numerous efforts to test it. The early literature is brilliantly reviewed by Krogh (1931) who concludes that no convincing experimental evidence had been advanced to support Pütter's hypothesis at that time. In the past two decades, interest in the possibility of a nutritional role of DOM has been rekindled, largely because of major advances in analytical procedures which now permit the simple and reliable determination of organic substrates at extremely low concentrations.

Despite the apparent simplicity of Pütter's hypothesis it is extremely difficult to devise a direct and convincing test of the effectiveness of DOM as a food resource. Natural concentrations of DOM are extremely low (in fact, lower than earlier investigators believed) so the relevance of enhanced growth or survival in concentrated solutions of organic substrates is dubious. Axenic techniques for rearing marine metazoans are virtually unavailable; thus the results of any direct approach are subject to the major criticism that the organisms under study may be supported indirectly by microorganisms rather than directly using the dissolved substrates that are provided. This problem can be avoided in studies employing marine phytoplankters in axenic culture. However, the problems of studying very dilute suspensions of cells in sea water containing the very low substrate levels characteristic of their environment are formidable and this approach has not been used.

Evidence concerning the contribution which dissolved resources can make to the nutrition of any organism is derived indirectly and must be based on the following information. First, its ability to transport external solutes and utilize them must be established quantitatively. Second, some reasonable estimate of its metabolic needs must be obtained. Third, levels of the relevant organic substrates normally present in its environment must be established. Given this information, one can then ask whether dissolved resources can be expected to provide an input which is a significant fraction of total needs.

We proceed to discuss briefly each of these categories of information. Selected examples for particular organisms are then summarised to illustrate the approach. Discussion is limited to consideration of free amino acid (FAA) though it should be pointed out that other fractions of the DOM pool may also provide significant inputs to marine organisms. What follows is intended as a summary and not a comprehensive review; readers interested in further details are referred to Stephens (1972), C. B. Jørgensen (1976), Sepers (1977), West *et al.* (1977), Wright and Stephens (in press).

2   FAA transport in marine organisms

There are many reports of studies of the uptake and assimilation of amino acids in invertebrates and algae employing radioactive tracers. Table II lists phyla and algal divisions for which this has been done. In the case of the invertebrate phyla, all marine representatives that have been examined show varying rates of uptake and assimilation of at least some amino acids. No fresh water metazoan (Stephens, 1964) and no vertebrate with the single exception of the hagfish (Stephens, 1972) has been shown unambiguously to have this capability; arthropods are the only major exception to the general rule that marine invertebrates accumulate amino acids from the ambient medium (Anderson and Stephens, 1969). Distribution of transport systems for FAA is less broad among algae. The divisions listed all contain some representatives for which FAA transport has been demonstrated, but not all members show this ability.

Transport of FAA in marine invertebrates is transepidermal. Solute is taken directly from the medium across the general body surface or enters by way of specialised organs such as the ctenidia of molluscs. The gut makes no significant contribution to the disappearance of labelled substrate or the appearance of labelled atoms in the organism (e.g. Stephens, 1963; Chien *et al.*, 1972).

Where it has been sought, there is evidence that labelled FAA taken from

TABLE II   List of animal Phyla and algal divisions in which one or more representatives have been examined for uptake of radio-labelled amino acids

Animal		Algal
Phylum	Class	division
Porifera	Demospongiae[a]	Chrysophyta
Coelenterata	Anthozoa[a]	Pyrrophyta
	Scyphozoa	
	Hydrozoa	Chlorophyta[a]
Platyhelminthes	Turbellaria	Phaeophyta
Nemertea	Anopla[a]	Rhodophyta
Brachiopoda	Articulata	
Ectoprocta	Gymnolaemata[a]	
Sipunculoidea[a]		
Annelida	Polychaeta[a]	
	Oligochaeta	
Mollusca	Bivalvia[a]	
	Gastropoda[a]	
Hemichordata	Enteropneusta[a]	
Pogonophora		
Echinodermata	Echinoidea[a]	
	Asteroidea[a]	
	Ophiuroidea[a]	
	Holothuroidea[a]	
Chordata	Ascidiacea[a]	

[a] Taxa demonstrated to have the capacity for a net accumulation of amino acid

the ambient medium enters catabolic and anabolic pathways. When ^{14}C-labelled substrate is provided, $^{14}CO_2$ appears in the medium after a short delay (N. L. G. Jørgensen, 1979). Labelled compounds have been found in all the major organic constituents of organisms when they are examined after pulse labelling (Stephens, 1972). The use of labelled substrates and subsequent autoradiographic examination has been much used to describe the distribiton of assimilated compounds. Since any radioactivity detected must be in a form which survives extraction with standard histological reagents (i.e. alcohols and xylene), this implies that the soluble substrates must have participated in anabolic synthetic pathways.

However, radiochemical techniques alone do not permit addressing the quantification of net exchanges between organisms and environment. Strictly speaking, studies which examine the disappearance of labelled substrate and the appearance of labelled atoms in an organism provide information only about influx. The uptake of labelled substrate can, in theory, be accompanied by losses of unlabelled substrate which are less than, equal to, or greater than the gains (Johannes, et al., 1969). As these workers pointed out, the well-known process of exchange diffusion in which substrate molecules are associated with a membrane carrier so that movement across the membrane is facilitated but no net transport occurs provides a tenable explanation of the observed disappearance of radioactive substrate. One need only postulate that labelled substrate molecules are exchanged one-for-one with unlabelled substrate molecules present in the internal FAA pool. It is possible to design experiments using radiochemical techniques to provide evidence that influx is large compared to efflux for particular substrates such as nonmetabolisable amino acid analogues (Stephens, 1972). However, this is clearly an indirect approach to the problem.

A much preferable approach is to measure simultaneously the net change in externally supplied labelled substrate using chemical techniques and compare the results with measurements of the same substrate using radiochemical techniques. Workers in the field have realised that for some time, but until recently the available techniques for determination of FAA in very dilute solution were relatively insensitive and application to sea water solutions at the requisite concentrations very time consuming. In 1975, fluorometric techniques for determination of FAA were introduced into this field making possible the simple and sensitive determination of these compounds at extremely low levels (North, 1975; Stephens, 1975).

In Table II, groups marked with an *a* have been shown to be capable of net influx by direct chemical techniques as well as FAA influx by radio-chemical techniques.

The data presented in Fig. 1 provide further insight into the relation between influx and efflux of FAA. In this case, a brittle star, *Ophionereis annulata*, was rinsed repeatedly in filtered sea water and placed in artificial sea water contining less than 0.1 μM total FAA. The medium was sampled periodically and a gradual increase in FAA was observed reaching a level of approximately 0.4 μM at the end of 3 hours of incubation. At that time, sufficient ^{14}C-gly was added to the medium to produce a glycine concentration of 6 μM. Periodic sampling and analysis of the medium showed that the added substrate was completely removed at the end of 40 minutes. Simultaneous radiochemical analysis of the medium indicated a concomitant decrease in radioactivity (measurements symbolised by open circles in Fig. 1). Notice that the gradual increase observed in the first 3 hours of sampling continued until the observations were terminated at the end of 5 hours. The dotted line is a regression line through the chemical measure-

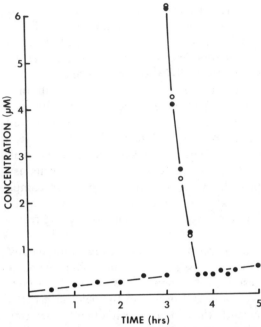

FIG. 1 Exchanges of primary amines between the brittle star, *Ophionereis*, and its environment. At zero time, an animal weighing 2.0 g was placed in 100 ml of artificial sea water. See text for details and interpretation

ments, excluding those made during the period of rapid entry of the added glycine substrate. Efflux rate of FAA estimated from this regression line is 50 nmoles $g^{-1}h^{-1}$. Entry rate of glycine (initial rate after addition of added substrate) is 820 nmoles $g^{-1}h^{-1}$ based on chemical determination data and 850 nmoles $g^{-1}h^{-1}$ based on radiochemical measurements. Net influx just after addition of the glycine is then 770–800 nmoles $g^{-1}h^{-1}$. In other experiments, it has been shown that this slow loss of FAA does not continue indefinitely but comes to a steady-state when ambient levels of 2–5 µM are achieved (Wright and Stephens, 1977; Stephens *et al.*, 1978) the rate of this efflux also appears independent of the level of exogenous substrate supplied in experiments of the kind described here (Davidsen, personal communication).

It is necessary for our purpose to describe FAA transport quantitatively. In general, observations relating influx to external substrate concentrations can be adequately described by the following relation:

$$Ji = \frac{J^i_{max}(S)}{K_t + (S)} \tag{1}$$

where J^i is influx, (S) is substrate concentration, J^i_{max} is the maximum rate of influx, and K_t is the substrate concentration at which $J^i = J^i_{max}/2$. This will be recognised as the familiar Michaelis-Menten equation describing enzyme catalysis of a substrate; flux replaces reaction velocity and K_t is analagous to the Michaelis constant, K_m. Qualitatively, one observes an increase in influx with increasing ambient substrate at low concentrations. At high concentrations, the system becomes saturated and influx is constant and independent of substrate concentrations (Fig. 2).

Assumptions are required for the theoretical justification of the use of this equation to describe transport (Schultz and Curran, 1970) but it fits experimental observations well. In some cases, data take the form illustrated by the dotted line in Fig. 2; this has been interpreted as a passive or diffusional component of transport superimposed on the typical saturation kinetics. Where this has been examined carefully, it has been possible to explain the apparent passive component as a result of systematic error in measuring true influx into the organism (Wright and Stephens, 1977). For energetic reasons it is very unlikely that marine organisms show any significant passive permeability to amino acids. Particularly in marine invertebrates, internal FAA pools are very high and normally exceed 10^{-1} M. Normal ambient concentrations are always low and never greater

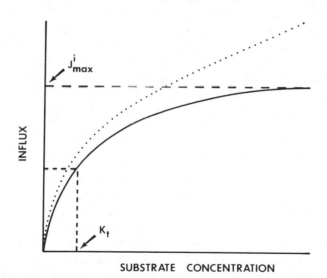

SUBSTRATE   CONCENTRATION

FIG. 2   Theoretical relation between substrate concentration and influx for a system which follows Michaelis-Menten kinetics. The constants J^i_{max} and K_t are illustrated. Actual data may fall along the dotted line but this should not in general lead to the conclusion that carrier-mediated transport is accompanied by a passive or diffusional component (see text)

than 10^{-4} M. Since diffusion is directly proportional to the concentration gradient, any passive influx from dilute experimental solutions would be accompanied by large losses down this externally directed gradient which would be energetically unacceptable as a continuing drain of reduced carbon and amino nitrogen.

The energetic requirements for transport of solute, even against the very large gradients present in marine invertebrates, are modest. The thermodynamic work required for transport of a mole of solute against a gradient is given by:

$$W = RT \ln (C_i/C_0) \tag{2}$$

where R is the gas constant, T is absolute temperature and C_i and C_0 the internal and external solution concentrations. Thus transport of a solute from a solution at 10^{-6} M to an internal pool at 10^{-1} M requires about 7 kcal mole^{-1}. This expression applies to a non-charged solute and an electrical term must be added to describe the energetics of amino acid transport. However, this term is small and does not modify the fact that thermodynamic transport costs are very low compared to the oxidative energy obtained by the resulting acquisition of a reduced carbon substrate.

The values of J^i_{max} and K_t in eqn (1) deserve comment. J^i_{max} is a measure of the capacity of a transport system. The capacity of transepidermal FAA transport systems expressed in this way is small compared to that for analagous systems such as intestinal FAA transport systems. The K_t is an expression of the affinity of the system for substrate (though not in the chemical sense of a dissociation constant for a carrier-substrate complex). In general K_t's for transepidermal FAA transport are low. These are low-capacity, high-affinity systems.

3   Metabolic requirements

In aerobic metabolism, there is a quantitative relation between food requirements, oxygen consumption, carbon dioxide production and heat production. This was established in the mid-18th century by Lavoisier and LaPlace (1780). It is this relation which is often used in estimating metabolic requirements for organisms using the indirect calorimetry method. Oxygen consumption is measured and related to food requirements by calculating the weight of food which would be fully oxidised per unit time at the measured Q_{O_2}. This provides a minimum estimate of food requirements since any deficit of food intake compared to oxygen consumption must reflect oxidation of food reserves or tissues and a condition of starvation.

This minimum estimate of food requirements will always be lower than actual requirements for several reasons: not all suitable substrates which are consumed are oxidised; there are losses of organic material (e.g. nitrogenous wastes) or investments of organic material in growth which are not reflected in this calculation. Finally, many organisms behave as partial or facultative anaerobes, deriving energy without complete oxidation of substrate, so that organic material (e.g. organic acids) is lost or sequestered. Oxygen consumption is not a measure of energy metabolism for such organisms. For all of these reasons, it would be preferable to estimate metabolic requirements on the basis of heat production (direct calorimetry). Such measurement of heat production has been accomplished for some marine invertebrates but it is a rather demanding technique, and few such data are available. Thus, despite its disadvantages, oxygen consumption continues to be used as a measure of metabolic rate, with the results treated as minimal figures. It provides a useful benchmark against which the rate of food intake by any feeding mechanism of interest can be laid. If the mechanism provides food at a rate which is broadly comparable to metabolic needs estimated from oxygen consumption, then it is a reasonable assumption that it provides a significant portion of those metabolic needs. It is not necessary that all food be promptly oxidised, since it can also enter anabolic pathways and may contribute to growth or repair or be converted to storage compounds.

Estimation of the metabolic requirements of algae is conceptually different. The forms which have been studied are obligate phototrophs which acquire reduced carbon substrates by photosynthesis. However, many of them are capable of using organic substrates for other purposes, in particular as sources of nitrogen. Thus the evaluation of FAA inputs has taken the form of estimating the fraction of nitrogen needs which can be supplied by this pathway. The information required includes the growth rate or doubling time of the alga, nitrogen content per cell, and influx rate of FAA per unit time. The first two measures give the amount of nitrogen which must be acquired for cell doubling, which allows estimation of the percentage of this nitrogen requirement that FAA can supply. We must also be concerned about losses of cell nitrogen by such pathways as extrusion of nitrogen-containing compounds, but these can be measured.

4   Availability of FAA in marine environments

Although DOM is an immense pool, its composition is poorly known; less than 10% of its components have been identified: roughly 1–2% of the total is FAA. Presumably the replacement time of DOM in the oceans is very slow. For example, ^{14}C-dating gives an estimated age of 3400 years for a sample of DOM taken at 3500 m in the eastern Pacific (Williams et al., 1969). However, turnover rates for the FAA fraction are much more rapid (e.g. Lee and Bada, 1977). Thus we are dealing with a small fraction of the total DOM resources but one which is apparently relatively very active.

Levels of FAA in sea water vary greatly depending on the specific source of the samples examined. For our purposes it is convenient to consider FAA levels in just two broad categories of habitat: the water column of uncontaminated neritic areas, and the interstitial water of littoral sediments. Each of these shows variation but these two categories justify separation because of the higher levels of FAA in interstitial water corresponding to the much greater organic content of sediment. Reports based on analysis of inshore surface water samples after desalting and reports based on estimation of total primary amines by fluorometric techniques agree very well; levels fall in the range of $0.2\text{--}2 \times 10^{-6}$ moles l^{-1} approximately 20–200 μg l^{-1} liter (Siegel and Degens, 1966; North, 1975; Crowe et al., 1977). These values are low, but are in the same general range as the availability of organic material via phytoplankton. Interstitial water from inshore sediments collected by several different procedures proves to be approximately two orders of magnitude richer in FAA; levels range roughly from $0.1\text{--}2 \times 10^{-4}$ moles l^{-1} or 1 to 20 μg l^{-1} (Stephens, 1975; Crowe et al., 1977; Stephens et al., 1978; Henrichs and Farrington, 1979).

The presence of these high levels of FAA in interstitial water is of considerable interest and requires further investigation. Marine sediments are the locus of intense microbial activity; under aerobic conditions, organic materials including FAA disappear rapidly (personal observation). However, marine sediments are largely anaerobic, aerobic conditions obtaining for only a very short distance beneath the surface (Fenchel and Riedl, 1970). Stephens (1975) observed a characteristic distribution of FAA in interstitial water of sediments such that FAA concentration decreases with increasing depth beneath the surface. The presence of a polychaet irrigating its burrow increases FAA concentrations in the deeper areas of sediment adjacent to the burrow. This effect can be mimicked by slow artificial irrigation of sediment with aerated sea water. It is known that the presence of burrowing infauna increases oxygen consumption of sediment disproportionately higher than the metabolism of the burrowing organisms (Hargrave, 1970) and stimulates bacterial growth (Fenchel, 1972). It is also known that the transition zone between superficial aerobic and deeper anaerobic conditions is a region of particularly intense microbial activity (Fenchel and Hemmingsen, 1974). Finally, there are well characterised responses of facultative aerobic microorganisms leading to liberation of large quantities of FAA during growth; such organisms can be isolated from marine sediments (Yamada et al., 1972). Taken together, these findings suggest that the irrigating activity of the burrowing infauna stimulates microbial activity and hastens utilisation of detritus of plant origin. This provides the burrowing metazoa with additional food resources, both as increased microbial biomass which is cropped by ingestion, and as increased levels of FAA available for transepidermal uptake and assimilation.

Most reports identifying individual amino acids in samples from either the water column or interstitial water list glycine, alanine, serine, valine, glutamate, aspartate and lysine as major constituents of the FAA pool with other amino acids making more modest contributions to the total. It should be noted that reports based solely on fluorometric determination of total primary amines may include compounds other than amino acids; however, where this has been checked, the majority of primary amines prove to be FAA and, as noted, the general levels obtained by this procedure agree well with those obtained from more detailed analysis (Stephens et al., 1978).

To summarise, organisms exposed to uncontaminated surface waters will rarely encounter FAA concentrations in excess of 1 µM; members of the nearshore infauna will typically encounter levels which are considerably greater.

5 Nutritional significance of FAA: selected examples

There is a great deal of information about the kinetics and distribution of transepidermal transport of amino acids in marine organisms. However, the number of cases where all of the critical elements outlined above (kinetics, metabolic requirements, availability) have been measured is more restricted. If one demands, as now seems proper, that the data should also directly address the relation between influx and net exchanges of FAA between organism and environment, the number of cases is still further restricted.

The following paragraphs summarise three such cases where information is available in all of the required categories and where the relation between influx and net flux has been examined. Each has different features of interest.

'MYTILUS CALIFORNIANUS'

The following account is based principally on the work of Wright and his colleagues. *Mytilus californianus*, the common California mussel, is a widely distributed and important element of the near-shore fauna. Although individuals are sometimes buried beneath sediment for long periods, they are typically attached to coastal rocks. The mussel obtains food and oxygen by ciliary-mucoid filter feeding by means of pumping and filtering organs (the ctenidia) which are also the principal site of amino acid transport. Isolated ctenidia of bivalves, including *Mytilus*, have been attractive and widely used preparations for the study of kinetics of FAA transport. Isolated preparations have the great advantage of ease of handling and results are not dependent on the vagaries of pumping behavior. Such studies showed that J^i_{max} for influx of FAA in such preparations was high compared to that reported for the sediment infauna, but it was disturbing that K_t's were also apparently quite high; for example, Wright and Stephens (1977) report a K_t for glycine of 35 μM. If this is correct, only a very small fraction of the glycine transport capacity (J^i_{max}) will be used under natural conditions. In fact, one can calculate that influx of FAA by such a transport system can make virtually no contribution to the animals despite the very high maximum velocity of influx.

Despite the manipulative attraction of preparations of the isolated gill, this is scarcely a physiologically normal preparation and gill perfusion is grossly disrupted by the isolation procedure. Consequently, Wright and Stephens (1978) devised a technique for studying transport in the intact animal. A small cannula is placed in the excurrent siphon, care being taken not to touch the edge of the mantle. Animals are unrestrained, and continued to pump normally. Small water samples can be drawn from the cannula and compared with samples taken at the incurrent margin of the animal. Chemical changes in the sea water after a single passage across the gill can be measured, given the sensitivity of radiochemical and fluorometric techniques available. In such experiments, the gill functions as a self-perfusing transport epithelium and results obtained can be extrapolated to normal physiology with much more confidence.

The results of this approach are interesting as an illustration of the speed and effectiveness of FAA transport in this animal. A typical mussel of approximately 5 gm soft tissue wet weight removed roughly 30% of a 1 μM solution of glycine during a single passage across its gill while pumping at a rate of 6 to 10 litres h^{-1}. At this pumping rate, water was in contact with the gill for 2 to 4 seconds. This figure of 30% for efficiency of glycine influx is based on fluorometric data and represents net intake of FAA into animals. Net influx at comparable rates could be established at ambient concentrations as low as 0.3 μM. Figure 3 presents typical data.

The efficiency of FAA influx at these low concentrations indicated that K_t of the transport system responsible must be relatively low. Extensions of the cannulation procedure were used to estimate kinetic parameters in two logically independent ways. Values for K_t ranged from 1–5 μM; more recent data (Wright et al., in press) indicated that the lower end of this range is probably a more accurate estimate. The discrepancy between K_t's obtained using isolated gill preparations and those obtained with intact animals is the result of the failure of normal perfusion of the gill upon isolation. The resulting unstirred layer of water separates the transport sites from the bulk phase which leads to substantial overestimates of K_t's in isolated gill preparations (Wright et al., in press, see Winne, 1973, for a theoretical discussion of this effect).

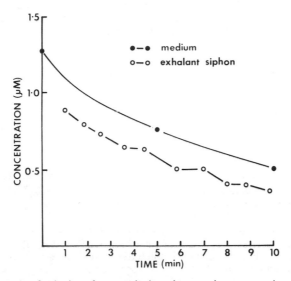

FIG. 3 Removal of glycine from solution by an intact specimen of *Mytilus californianus*. Solid circles represent the level of glycine, measured with fluorescamine, remaining in a 400 ml test solution containing an intact, actively pumping 5 g mussel; the solid line describes medium depletion assuming depletion to be first order. The open circles indicate the levels of glycine measured in samples collected from the exhalent water stream. The average clearance of glycine occurring during a single passage of water through the animal was 25%

Table III compares influx rates in *Mytilus* at an assumed FAA concentration of 1 μM with metabolic needs estimated on the basis of data presented by Bayne *et al.* (1976). This estimate of the contribution of FAA is probably high for several reasons. First it assumes that all naturally occurring FAA can be handled as efficiently as glycine. Second, Hammen (1979) has shown that *Mytilus*' metabolic needs are probably greater than those estimated from oxygen consumption, even in well aerated water. However, it is clear that *Mytilus* possesses an extremely effective transport system adapted to the very low concentrations of FAA in its habitat which is capable of providing a continuous supplement of reduced carbon and amino nitrogen to its food intake from particulate resources.

'DENDRASTER EXCENTRICUS'

The following account is based principally on the work of Stephens *et al.* (1978). *Dendraster excentricus* is the common west coast sand dollar and is a conspicuous and sometimes dominant member of the subtidal inshore fauna. It can feed by filtering water when it is partially immersed in sediment

TABLE III   Nutritional role of the transepidermal accumulation of amino acids in *Mytilus californianus*: examination of critical parameters

Oxygen consumption	$0.54 \, mlO_2h^{-1}$	Bayne *et al.* (1976)
Environmental concentration	0.2–2.0 μM	North (1975)
Kinetics of transport[a] (intact animal)		
V_{max}	$0.85 \, \dfrac{\mu moles}{litre \ pumped}$	Wright and Stephens (1978)
K_t	1.3 μM	
% oxidative requirements accounted for by net uptake of amino acids at environmental level of 1 μM	37	Wright and Stephens (1978)

[a] These values were derived from studies of influx using radio-labelled substrates; estimates of rates of acquisition at environmental levels of substrate were made using measured values of net flux, and hence are lower than influx

(Timko, 1976) but individuals also spend time completely submerged below the sediment surface (Chia, 1969); members of populations occur completely and partially buried in sediment. Feeding of echinoids is of particular interest since metabolic communication between the viscera and the extensive epidermal structures (e.g. spines, pedicellariae) is apparently very poor. Ferguson (1967, 1970) has shown by autoradiography that labelled amino acids introduced into the coelom do not result in labelling of the epidermis of the starfish, *Echinaster*; conversely, supplying labelled amino acids in the medium leads to intense labelling of epidermis but virtually no export to subepidermal tissues. Other workers have reported comparable isolation of epidermis and deep tissues in a variety of echinoids and asteroids. This is reasonable in view of the rather poor morphological substrate for nutrient distribution but it raises the interesting question of how the active and complex epidermis obtains its sustenance.

Stephens *et al.* (1978) examined the transport of seven amino acids into *Dendraster* measuring influx radiochemically and net influx fluorometrically in duplicate samples. Influx of neutral amino acids occurs rapidly and this represents essentially a net influx at concentrations greater than 5 μM. Net influx occurs at lower concentrations but is accompanied by a slow efflux comparable to that shown in Fig. 1 for *Ophionereis*. Thus influx estimated radiochemically is an overestimate of net influx at low initial concentrations. Rates of influx are not modified in animals pretreated with antibiotics though of course these cannot be considered axenic. Nevertheless, it appears that microbial uptake is not a significant factor in this system.

Metabolic requirements were estimated by measuring oxygen consumption of intact animals as well as oxygen consumption of isolated portions of the test (epidermis and underlying skeletal support). Rates were proportional to area and were not significantly different for whole animals versus isolated tests. This suggests that echinoids might be interesting subjects for direct calorimetry. However, metabolism of the epidermis is presumably aerobic so oxygen consumption can be simply converted into a minimal requirement of these structures for food input expressed as amino acid.

Dendraster was shown to accumulate FAA from samples of naturally occurring interstitial water using fluorometric techniques. Chromatography of water samples before and after exposure to the animal established the disappearance of amino acids from the medium. Calculations based on the observed rate of uptake from natural samples and the oxygen consumption of the animals indicated that *Dendraster* can obtain sufficient FAA to meet its minimum metabolic needs provided it has access to interstitial water in which the primary amine concentration exceeds 35 μM. The average concentration of primary amines in samples of interstitial water collected in the immediate vicinity of the population was 115 μM; of this 70–85% was identified as free amino acids.

Thus sand dollars can obtain sufficient reduced carbon from dissolved resources to support their epidermal metabolism provided they spend a reasonable fraction of the time in the submerged orientation.

MARINE PHYTOPLANKTERS
Many phytoplankters have been cultured axenically. Frequently, amino acids can serve as the sole source of nitrogen in such cultures (Danforth (1962)). North and Stephens (1967), working with the chlorophyte, *Platymonas*, showed that this phytoplankter could be cultured axenically on several single sources of nitrogen including glycine. Studies of uptake of glycine using labelled substrate indicated that input by this pathway could provide a modest supplement (a few percent) to inorganic nitrogen sources at the realistic levels of environmental FAA outlined earlier (i.e. 0.2–2.0 μM).

In these experiments, *Platymonas* was harvested from fixed volume culture flasks which had been supplied with nitrogen at levels of 2×10^{-3} g atoms l^{-1}, a typical value selected on the basis of reports of culture media in the preceding literature. When nitrogen levels were restricted using continuous culture techniques (North and Stephens, 1969), there were three major effects: generation time increased, nitrogen content per cell decreased, and rates of amino acid influx increased dramatically. The interaction of these three changes increased the contribution of glycine uptake from an ambient concentration of 0.5×10^{-6} M so that it could provide 380% of the nitrogen required to support the observed growth. The presence of NH_4^+ or NO_3^- 10^{-4} M did not modify these rates of glycine influx (North and Stephens, 1971).

These estimates of the importance of amino acid transport as a nitrogen supply are based on calculation of the nitrogen required to support the observed rate of cell growth. Glutamate uptake by *Platymonas* is an illustration of the utility of this indirect approach and of the necessity of using it quantitatively. Glutamate is a competent nitrogen source for an axenic culture of *Platymonas* when supplied at 2×10^{-3} M as the sole available form of nitrogen. *Platymonas* accumulates ^{14}C-glutamate from an ambient concentration of 0.5×10^{-6} M. However, when one calculates the generation time for *Platymonas* on a sole supply of nitrogen as glutamate at these natural levels, it is found that this particular amino acid must be a trivial source of nitrogen for this organism under natural conditions.

North (1975) subsequently studied influx of glycine in *Platymonas* using labelled glycine to study disappearance of radioactivity and reduction in primary amines in parallel medium samples. Influx measured radiochemically and net influx measured fluorometrically were essentially identical. She also exposed *Platymonas* to samples of water from Newport Bay which had been filtered to remove other organisms. Naturally occurring FAA in samples of Bay water were removed by axenic *Platymonas* at a rate sufficient to support a doubling time of 30 hours, estimated from influx rates and nitrogen content of the cells.

Platymonas was a good test of organism for the work just described since it occurs abundantly throughout the year in Newport Bay. However, it is not a major member of the plankton community in general. Therefore it was of interest to study the distribution among phytoplankters of this ability to take up amino acids at rates which might contribute significantly to nitrogen inputs. Wheeler *et al.* (1974) studied amino acid uptake in 25 species of unicellular marine algae in four divisions. Algae were in axenic culture. Nine amino acids were tested for their ability to support growth when supplied in high concentration in batch culture. Uptake of each amino acid was also studied radiochemically using organisms from nitrogen-rich and nitrogen-restricted cultures. Approximately 75% of the phytoplankters were able to utilise one or more amino acids as a sole nitrogen source in culture. Fourteen of the phytoplankters tested were able to accumulate one or more of the amino acids sufficiently rapidly to provide enough nitrogen for a generation time of 10 days or less. This level was selected as indicating a contribution from this pathway of potential significance though mean natural phytoplankton generation times are significantly lower than this. In general, inshore and epibenthic forms proved more effective in obtaining and utilising FAA than open ocean forms.

Wheeler *et al.* (1977) addressed the question of the ability of phytoplankters to compete for available DOM resources in a mixed plankton population. Uptake of ^{14}C-glycine by various size fractions of natural plankton communities was measured. The simultaneous occurrence of glycine uptake, plant pigments and photosynthetic $^{14}CO_2$ fixation in the same size fractions, as well as autoradiography of plankton samples provided evidence of a substantial role of phytoplankters in utilisation of DOM. Phytoplankters proved to be responsible for 50% or more of the observed glycine uptake. It was important to separate size fractions of the community prior to offering the labelled substrate. If these steps are reversed, the disruption of fragile organisms during the filtration process obscures the results.

In summary, a number of inshore and epibenthic phytoplankters are capable of obtaining significant nitrogen inputs from FAA at ecologically realistic concentrations. Naturally occurring FAA is available and utilisable. Finally, phytoplankters compete effectively with other members of the natural inshore plankton community for available DOM resources.

6 Conclusion

The main conclusion to be advanced here is simply that there is a significant nutritional input to many marine organisms from FAA, which has been selected for discussion since it is the fraction of the DOM pool about which we have most information. Transepidermal transport of glucose has been studied (Ahearn and Gomme, 1975), but information concerning environmental availability of this substrate is sparse. Particular attention should also be directed to the case for urea as a significant nitrogen source for phytoplankton (McCarthy et al., 1977), another relatively small but apparently metabolically important fraction of the total DOM pool.

Unfortunately, the case for a trophic role of FAA which is developed in the preceding discussion is fundamentally an indirect one with one interesting exception. Shick (1975) was able to use the developmental process of strobilisation in the jellyfish, Aurelia, to provide direct evidence for a trophic role of dissolved FAA. Polyps were starved for an extended period and then induced to strobilate. They were less able to carry out this developmental step than were starved polyps which had been provided with low levels (1 μM) of dissolved glycine. This is the single example of a direct nutritional effect of ecologically meaningful levels of FAA.

It would be more satisfactory to be able simply to cultivate invertebrates on sterilised dilute artificial broth, the qualitative and quantitative composition of which matched naturally occurring DOM. However, even on the assumption that axenic suspensions of soft-bodied invertebrates could be obtained and the technical problems surmounted, the probability of success is low. The evidence obtained indirectly suggests that the trophic role of FAA is a supplementary one, though there may be exceptional cases, such as the pogonophorans, where dissolved resources may provide the sole input. However, in general one would not expect DOM to serve as the sole or even primary food resource for many marine organisms. Virtually all of the animals which have been studied have obvious functional methods of obtaining particulate food, and the algae which have been examined are obligate phototrophs.

A comment should be made concerning the nature of the examples used in the preceding discussion and the extent to which they may be representative of marine organisms generally. In each case, the organisms discussed are important members of their respective ecocystems. In the case of marine inshore phytoplankton, we noted that naturally occurring phytoplankters do appear to be able effectively to compete for and make use of dissolved resources. *Mytilus* and *Drendraster* were selected on the ground they have been extensively described so that information in all the categories required to permit an estimate of the significance of FAA inputs was available. There are perhaps a dozen other genera that have been studied using modern techniques for which our information is less complete but broadly consistent with the cases presented (e.g. Stephens, 1975; C. B. Jørgensen, 1976; Crowe et al., 1977; N. L. G. Jørgensen, 1979). Prior to the introduction of fluorometric procedures, a large number of studies were based solely on radiochemical observations. Although final judgment on the extent to which the reported influx of labelled substrates does indeed reflect a net input to the organism must be suspended, it appears very likely that many of these organisms will prove to be capable of obtaining and utilising naturally occurring FAA. It is also likely that the level of such inputs from dissolved resources will vary widely among different organisms. However, the examples put forward here as 'best cases' seem not to be significantly different from other representatives from these ecosystems.

The concept of a trophic role of DOM, though attractive in some ways, is also troubling. As noted, DOM is normally an imperceptible feature of marine environment; its detection and analysis is not part of the accustomed research of most marine biologists. However, the work of the past two decades, and particularly of the past few years, provides strong evidence for the existence of such a role. In selected cases, that role appears not only real but quantitatively important. It is hoped that further work will more firmly establish and clarify that role and will allow it to be assimilated into the interpretation of the trophic structure of marine ecosystems.

Acknowledgements

I am grateful to Torben Davidsen, August Krogh Institute, University of Copenhagen, for introducing me to the protocol illustrated in Fig. 1 and for permitting me to examine unpublished data. The data in Fig. 1 were kindly supplied by Michael Rice, University of California, Irvine. This was written during tenure of NSF Grants OCE 78-09017 and PCM 78-09576 and includes discussion of work so supported.

References

Ahearn, G. A. and Gomme, J. (1975). Transport of exogenous D-glucose by the integument of polychaet worm (*Nereis diversicolor* Muller). *J. Exp. Biol.* **62**, 243–264

Anderson, J. A. and Stephens, G. C. (1969). Uptake of organic material by aquatic invertebrates. VI. Role of epiflora in apparent uptake of glycine by marine crustaceans. *Mar. Biol.* **4**, 243–249

Bayne, B. L., Bayne, C. J., Carefoot, T. C. and Thompson, R. J. (1976). The physiological ecology of *Mytilus californianus*. *Oecologia* **22**, 229–250

Chia, F. A. (1969). Some observations on the locomotion and feeding of the sand dollar, *Dendraster excentricus*. *J. Exp. Mar. Biol. Ecol.* **3**, 162–170

Chien, P. K., Stephens, G. C. and Healey, P. L. (1972). The role of ultrastructure and physiological differentiation of epithelia in amino acid uptake by the bloodworm, *Glycera*. *Biol. Bull.* **142** (2), 219–235

Crowe, J. H., Dickson, K. A., Otto, J. L., Colón, R. D. and Farley, K. K. (1977). Uptake of amino acids by the mussel, *Modiolus demissus*. *J. Exp. Zool.* **202**, 323–332

Danforth, W. F. (1962). Substrate assimilation and heterotrophy. *In* "Physiology and Biochemistry of Algae" (R. A. Lewin, ed.) pp. 99–123. Academic Press, New York

Fenchel, T. (1972). Aspects of decomposer food chains in marine benthos. *Verhandl. der Deutschen Zool. Gesell.* **65**, 14–22

Fenchel, T. and Hemmingsen, B. B. (1974). "Manual of Microbial Ecology". Akademisk Forlag, Copenhagen

Fenchel, T. and Riedl, R. J. (1970). The sulfide system: a new biotic community underneath and oxidized layer of marine sand bottoms. *Mar. Biol.* **7**, 255–268

Ferguson, J. C. (1967). An autoradiographic study of the utilization of free exogenous amino acids by starfishes. *Biol. Bull.* **133**, 317–329

Ferguson, J. C. (1970). An autoradiographic study of the translocation and utilization of amino acids by starfish. *Biol. Bull.* **138**, 14–25

Hammen, C. S. (1979). Metabolic rates of marine bivalve molluscs determined by calorimetry. *Comp. Biochem. Physiol.* **62A**, 955–959

Hargrave, B. T. (1970). The effect of a deposit-feeding amphipod on the metabolism of benthic microflora. *Limnol. Oceanogr.* **15**, 21–30

Henrichs, S. M. and Farrington, J. W. (1979). Amino acids in interstitial waters of marine sediments. *Nature* **279**, 955–959

Johannes, R. E., Coward, S. J. and Webb, K. L. (1969). Are dissolved amino acids an energy source for marine invertebrates? *Comp. Biochem. Physiol.* **29**, 282–288

Jørgensen, C. B. (1976). August Pütter, August Krogh, and modern ideas on the use of dissolved organic matter in aquatic environments. *Biol. Rev.* **51**, 291–328

Jørgensen, N. L. G. (1979). Uptake of L-valine and other amino acids by the polychaete *Nereis virens*. *Mar. Biol.* **52**, 42–52

Krogh, A. (1931). Dissolved substances as food of aquatic organisms. *Biol. Rev.* **6**, 412–442

Lavoisier, A. and LaPlace, P. (1780). Memoire sur la chaleur. *Mem. Acad. Sci., Paris. In* "Great Experiments in Biology (1955)" (M. L. Gabriel and S. Fogel, eds) pp. 85–93. Prentice-Hall, Englewood Cliffs

Lee, C. and Bada, J. L. (1977). Dissolved amino acids in the equatorial Pacific, the Sargasso Sea, and Biscayne Bay. *Limnol. Oceanogr.* **22**, 502–510

McCarthy, J. J., Taylor, W. R. and Taft, J. L. (1977). Nitrogenous nutrition of the plankton in the Chesapeake Bay. 1. Nutrient availability and phytoplankton preferences. *Limnol. Oceanogr.* **22**, 996–1011

North, B. B. (1975). Primary amines in California coastal waters: utilization by phytoplankton. *Limnol. Oceanog.* **20**, 20–26

North, B. B. and Stephens, G. C. (1967). Uptake and assimilation of amino acids by *Platymonas*. *Biol. Bull.* **133**, 391–400

North, B. B. and Stephens, G. C. (1969). Dissolved amino acids and *Platymonas* nutrition. *Proc. 6th Int. Seaweed Symp.*, Madrid, pp. 263–273

North, B. B. and Stephens, G. C. (1971). Uptake and assimilation of amino acids by *Platymonas*. II. Increased uptake in nitrogen-deficient cells. *Biol. Bull.* **140**, 252–254

Pütter, A. (1909). "Die Ernährung der Wassertiere und der Stoffhaushalt der Gewässer". Fischer, Jena

Schultz, S. G. and Curran, P. F. (1970). Coupled transport of sodium and organic solutes. *Physiol. Rev.* **50**, 637–718

Sepers, A. B. J. (1977). The utilization of dissolved organic compounds in aquatic environments. *Hydrobologia* **52**, 39–54

Shick, J. M. (1975). Uptake and utilization of dissolved glycine by *Aurelia aurita* scyphistomae: temperature effects on the uptake process; nutritional role of dissolved amino acids. *Biol. Bull.* **148**, 117–140

Siegel, A. and Degens, E. T. (1966). Concentration of dissolved amino acids for saline waters by ligand-exchange chromatography. *Science* **151**, 1098–1101

Stephens, G. C. (1963). Uptake of organic material by aquatic invertebrates. II. Accumulation of amino acids by the bamboo worm, *Clymenella torquata*. *Comp. Biochem. Physiol.* **10**, 191–202

Stephens, G. C. (1964). Uptake of organic material by aquatic invertebrates. III. Uptake of glycine by brackish water annelids. *Biol. Bull.* **126** (1), 150–162

Stephens, G. C. (1972). Amino acid accumulation and assimilation in marine organisms. *In* "Nitrogen Metabolism and the Environment" (J. W. Campbell and L. Goldstein, eds) pp. 155–184. Academic Press, New York

Stephens, G. C. (1975). Uptake of naturally occurring primary amines by marine annelids. *Biol. Bull.* **149**, 397–407

Stephens, G. C., Volk, M. J., Wright, S. H. and Backlund, P. S. (1978). Transepidermal transport of naturally occurring amino acids in the sand dollar, *Dendraster excentricus*. *Biol. Bull.* **154**, 335–347

Timko, P. L. (1976). Sand dollars as suspension feeders: a new description of feeding in *Dendraster excentricus*. *Biol. Bull.* **151**, 247–259

West, B., deBurgh, M. and Jeal, F. (1977). Dissolved organics in the nutrition of benthic invertebrates. *In* "Biology of Benthic Organisms" (B. F. Keegan, ed.) pp. 587–593. Pergamon Press, London

Wheeler, P. A., North, B. B. and Stephens, G. C. (1974). Amino acid uptake by marine phytoplankters. *Limnol. Oceanog.* **19**, 249–259

Wheeler, P., North, B., Littler, M. and Stephens, G. C. (1977). Uptake of glycine by natural phytoplankton communities. *Limnol. Oceanogr.* **22**, 900–910

Williams, P. M., Oeschger, H. and Kinney, P. (1969). Natural radiocarbon activity in the northeast Pacific Ocean. *Nature* **224**, 256–258

Winne, D. (1973). Unstirred layer, source of biased Michaelis constant in membrane transport. *Biochim. Biophys. Acta* **298**, 27–31

Woodwell, G. M., Whittaker, R. H., Reiners, W. A., Likens, G. E., Delwiche, C. C. and Botkin, D. B. (1978). The biota and the world carbon budget. *Science* **199**, 141–146

Wright, S. H. (1979). Effect of activity of lateral cilia on transport of amino acids in gills of *Mytilus californianus*. *J. Exp. Zool.* **209**, 209–220

Wright, S. H. and Stephens, G. C. (1977). Characteristics of influx and net flux of amino acids in *Mytilus californianus*. *Biol. Bull.* **152**, 295–310

Wright, S. H. and Stephens, G. C. (1978). Removal of amino acid during a single passage of water across the gill of marine mussel. *J. Exp. Zool.* **205**, 337–352

Wright, S. H. and Stephens, G. C. Transepidermal transport of amino acids in the nutrition of marine invertebrates. *In* "Ecosystem Processes in the Deep Oceans" (J. Morin and W. G. Ernst, eds). Prentice-Hall, (in press)

Wright, S. H., Becker, S. A. and Stephens, G. C. Influence of temperature and unstirred layers on the kinetics of glycine transport in isolated gills of *Mytilus californianus*. *J. Exp. Zool.* (in press)

Yamada, K., Kinoshita, S., Tsunoda, T. and Aida, K. (eds) (1972). "The Microbial Production of Amino Acids". Kodansha, Ltd., Tokyo. (John Wiley and Sons, New York)

10 Microheterotrophic Organisms in Marine Ecosystems

YU. I. SOROKIN

1 Introduction

Marine microheterotrophs are microscopic organisms less than 100 to 200 μm in size, which pass through nets used for sampling mesozooplankton. They are missed during routine biological surveys (Sorokin, 1975; Sieburth *et al.*, 1978). They include microflora (bacteria, fungi, yeasts), and also microzooplankton, mostly protoza (ciliates, and zooflagellates) and also the larval stages of planktonic and benthic animals.

Attempts to quantify the biomass, production and metabolism of microheterotrophs demonstrate their extremely important role in the functioning of aquatic ecosystems. They appear to be responsible for 70 to 80% of total heterotrophic metabolism and production (Pomeroy and Johannes, 1968; Sieburth, 1977; Sorokin, 1975, 1978a), even in tropical planktonic ecosystems where direct grazing of phytoplankton is most important. The role of microheterotrophs is greater in temperate regions, where the maxima of phytoplankton blooms and of heterotrophs are separated in time (Sorokin, 1977). These organisms are also very important in lakes, estuaries and coastal waters (Sorokin, 1972a; Eriksson *et al.*, 1977), where allochthonous organic matter augments local phytoplankton production (see Tables III and IV) as a substrate for heterotrophic activity.

Tables of energy balance and schemes of energy flow in various pelagic ecosystems show that most energy passes through the microheterotrophs – through the bacteria and microzooplankton (see Table IV and Figs 4 and 5). They demonstrate also, that microheterotrophs act to incorporate the energy of dead organic matter into the living food web so that they form the basic food sources for the larger zooplankton as well as for the bottom fauna (Rodina, 1963; Parsons and Strickland, 1962; Sorokin, 1975, 1978a).

Modern studies of the functioning of aquatic ecosystems are based upon holistic and mathematical analysis, and cybernetic modelling. Such an approach urgently needs an adequate account of the functions of every component of an ecosystem. Studies of microheterotrophs thus become more and more urgent (Sorokin, 1979b).

To quantify the roles of microheterotrophs in aquatic ecosystems it is necessary to obtain quantitative data upon: (a) the vertical structure of communities of microheterotrophs, and their microdistribution and aggregation, (b) their *in situ* biomass, production and metabolism, (c) their energy sources and efficiency of utilisation, (d) their importance as food for consumers at subsequent trophic levels, and (e) their share in the total production, metabolism and energy flux.

Of great significance for the solution of problems of functioning of ecosystems is also information on the species composition of communities of microheterotrophs, as well as some physiological characteristics of the mass species: metabolic and production rates, efficiency of food utilisation for growth, etc.

Progress in acquiring such information for natural populations of microscopic organisms seems to be one of the most difficult in modern marine ecology as well as being one of the most urgent. Nevertheless, we do now have some suitable methods, though most of them are rather often subjected to criticism. Criticism especially concerns the philosophy of the adequacy of rates measured by the use of bottle techniques compared with *in situ* ones (Banse, 1974), as well as the use of formaldehyde or other fixatives in counts of naked protozoa (Sorokin, 1975, 1977). But in order to progress at all towards satisfactory holistic studies of marine ecosystems we must proceed on the basis of present methodology, while recognising its shortcomings.

2 Microbial populations

COMPOSITION

Marine microbial populations are composed primarily of bacteria, but also include viruses, actinomycetes, yeasts and moulds (Meyers, 1967; Jones, 1976; Simidu, 1974; Fell, 1976; Sieburth *et al.*, 1978). For the study of species composition of natural populations the traditional methods of cultivation of isolates on culture media are rather inadequate. More than 99% of natural populations are represented by small oligocarbophylic bacterial forms which do not grow on nutrient media usually employed (Rasumov, 1962; Watson *et al.*, 1977; Hoppe, 1976; Azam and Hodson, 1977). The exceptions are some specific physiological groups of bacteria such as sulphate reducers or methane oxidisers. Therefore for the study of microflora direct microscopy is mostly used, employing phase contrast, fluorescence, transmission and scanning electron microscopy methods (Perfiljev and Gabe, 1969; Wood, 1965; Nikitin and Kuznetsov, 1967; Hobbie *et al.*, 1972). The use of electron microscopy recently gave good insight into the morphology of the dominant species in some natural microbial populations (Gorlenko *et al.*, 1977; Hirsh and Parkratz, 1979; Lapteva, 1976; Young, 1978; Bowden, 1977; Todd and Kerr, 1972; Weise and Rheinheimer, 1978; Sieburth, 1975; Zimmerman, 1977; Ferguson and Rublee, 1976). These studies demonstrated that the planktonic microflora in oligotrophic and mesotrophic marine basins is composed mostly by a small coccoid, a short rod, and 'horseshoe' forms 0.3 to 0.8 µm size. Many of these bacteria have prostecate or fimbriate ultramicroscopic filaments, which enlarge the surface of their cells and promote the absorption and consumption of organic molecules. In the deep oceanic waters, the microbial population includes large 'olive' cells – unidentified microbial forms (Fournier, 1970). A significant part of the planktonic microflora is represented by forms such as *Caulobacter* which are attached to detrital particles, and to phytoplankton cells with the aid of stalks. In rich eutrophic waters, mobile rods such as *Pseudomonas*, and also mycobacteria and yeast are usually abundant (Schneider, 1977), as well as relatively large rods attached to detritus particles (Ferguson and Rublee, 1970).

Among the periphytonic microbial communities, filamentous forms of bacteria, such as the flexibacteria *Leucothrix* or chain-form bacteria *Crenothrix*, *Cladothrix* and *Leptothrix* are common. Attached large rods are also abundant there (Sieburth, 1975; Sorokin, 1973a). In the surface layers of bottom sediments, mobile and attached rods predominate over filamentous bacteria; parasitic microbial forms like *Bdellvibrio*, *Mycoplasma* (Marbach *et al.*, 1975; Gorlenko *et al.*, 1977), and also actynomycetes and moulds are also present there. Within the mass of sediments most of the microbial population is formed by latent coccoid bacteria and spores (Sorokin, 1970a). In a juvenile eutrophic ecosystem, microbial populations are less diverse and are composed mostly of large freely staining rods which decompose proteinaceous substances; in mature oligotrophic ecosystems, the microbial populations are more diverse. There they are composed of a great variety of small poorly-staining, mostly coccoid microbial forms, which are strictly oligocarbophylic and do not grow in the usual nutrient media. Therefore, the ratio 'direct count: plate count' in oligotrophic water is usually 1 to 2 orders more than that in eutrophic: 1×10^2 to 1×10^3 in eutrophic water, and 5×10^3 to 2×10^4 in oligotrophic water (Hoppe, 1976).

VERTICAL STRUCTURE OF COMMUNITIES AND AGGREGATION

Vertical structure in microbial communities is studied by measuring parameters which characterise the abundance or relative activity of natural microbial populations in a series of samples taken in vertical profiles. The most convenient of these are the relative activity of heterotrophic microflora, measured as the rate of assimilation of ^{14}C-labelled organic compounds (Sorokin, 1970b), and the total microbial number and biomass by direct microscopy or ATP-methods (see below). The main difficulty is the selection of the depths of sampling. They should not be standardised because of variations in the vertical distribution of planktonic microflora. Microbial maxima in the water column appear to occur at boundaries between the water masses or between phases, and so their position in vertical profiles has a certain predictability. Therefore, optimum sampling depths to correspond with the maxima and minima of microbial concentration can be selected using physical or biophysical indicators such as temperature, turbidity or bioluminescence profiles (Sorokin, 1971, 1973b, 1978a).

The main maximum of microbial biomass and activity is situated, as a rule, at the upper boundary of the thermocline. This coincides well with the main maximum of bioluminescence and of water turbidity (Fig. 1). Values for total biomass of relative activity of microbial populations in the main maximum are up to an order more than the euphotic zone background values (Sorokin, 1977). The main maximum of bacterioplankton coincides well with maximum concentrations of other components of plankton communities: phytoplankton, micro- and mesozooplankton (Fig. 1). So in the stratified basins at the upper boundary of the thermocline there constantly exists a kind of layer-biocoenosis in which the components of

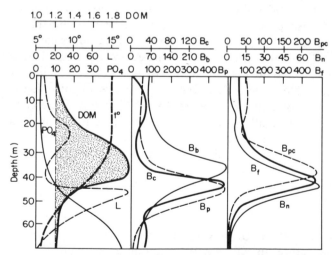

FIG. 1   Vertical structure of microplankton community in the Sea of Japan in June. PO_4 = inorganic phosphorus, 10^{-2} µg at l^{-1}; L, bioluminiscence, % of maximum value; t°, temperature of water; DOM, dissolved organic matter, mgC l^{-1}; B, biomass mg m^{-3}, wet weight; B_b, of bacteria; B_p, of phytoplankton; B_c, of ciliates; B_f, of zooflagellates; B_n, of nauplii; B_{pc}, of *Paracalanus*

planktonic community are closely interconnected (see also chapter 14). The causes of its formation are increased stability within this layer and the equilibrium between light penetration and upward nutrient flux, which here provides ecological stability for a layer of high phytoplankton biomass (Sorokin, 1971, 1973a).

A similar layer-biocoenosis also exists within the surface film, where a specific hyponeustonic community exists (Zaitzev, 1970; Tsiban, 1970; Sieburth, 1976a,b; Deitz *et al.*, 1976; Sorokin, 1978a; McIntyre, 1974; Wangersky, 1976). The concentrations and activity of total microbial plankton (and also the plate counts) within this layer is usually 1 to 3 orders higher than their average levels within the euphotic zone. The formation of this hyponeuston layer is connected with the accumulation within the surface film of phytoplankton and of organic matter, nutrients and detritus (Williams, 1967; Parker and Barsom, 1970; Nishizawa, 1971; Duce *et al.*, 1972; Sieburth, 1976a).

A third permanent microbial maximum exists in the tropical ocean at the upper boundary of the Intermediate Antarctic Water at 400 to 600 m (Fig. 2). Accumulation of detritus, particulate protein and zooplankton in this layer have also been recorded (Sorokin, 1971; Gundersen *et al.*, 1972; Holm

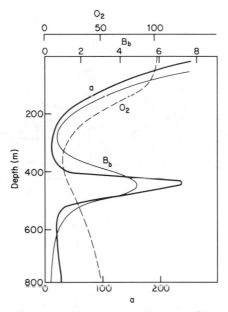

FIG. 2   Distribution of the microbial biomass (B_b, mg m^{-3}) on vertical profile in the equatorial Pacific; O_2, dissolved oxygen, % of saturation, a, microbial activity, relative values

Hansen, 1969; Ichikawa and Nishizawa, 1975). A high rate of microbial activity in this layer is one of the main causes of the formation there of an oxygen minimum (Sorokin et al., 1977).

In meromictic basins such as the Black Sea, with an anaerobic H_2S-zone, the largest maximum of bacterioplankton is situated within the redox zone, where redox potential values (Eh) are approximately -10 to -30 mv (Fig. 3). The microbial maximum is moreover larger within the redox-zone in meromictic basins in cases in which light penetrates to its upper boundary, so that photobacteria can develop. In this case the biomass of bacteria within this layer can reach 20 to $30 \times 10^6 ml^{-1}$, which is more than an order greater than the euphotic zone average (Sorokin, 1964; 1970a, 1972b; Sorokin and Donato, 1975).

Such a layered vertical structure of microbial populations exists constantly in the tropical ocean (Sorokin, 1971, 1973a). In temperate waters it exists only during the summer period of vertical stratification (Sorokin, 1977). The formation of layers of accumulated bacterioplankton has great

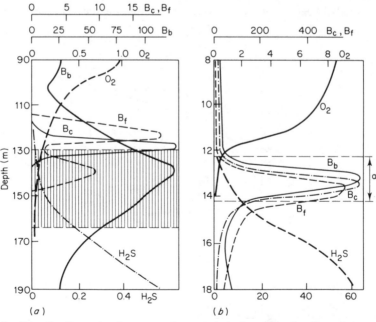

FIG. 3   Distribution microheterotrophs on vertical profiles in the redox zones of meromictic marine basins: the Black Sea (a) and Faro Lagoon (b): O_2, dissolved oxygen (mg l^{-1}); H_2S, mg l^{-1}; B, biomass mg m^{-3}, wet weight: B_b, of bacteria; B_f, of zooflagellates; B_c, of ciliates; a, redox zone. (Adapted from Mamaeva and Moiseev, 1979; Sorokin and Donato, 1975)

ecological importance. It provides for the energetically profitable filter feeding of zooplankton which also accumulates within the same layers (cf. Fig. 1), while the mean concentration of particulate food in the water coluumn, of which bacteria compose a significant part, is usually insufficient for that (Sorokin, 1973a). Another specific feature of the microdistribution of bacterioplankton, namely its aggregation (Sorokin, 1971; Seki, 1971; Ferguson and Rublee, 1976; Paerl, 1975) also has ecological importance. About 20 to 30 % of the total microbial population in sea water exists within aggregates exceeding 5 µm in size. The formation of such aggregates is a specific feature of the growth of microbial population in aquatic habitats (Sorokin, 1970c; Marshall, 1976). Surface film material, membranes and macromolecules excreted by microplankton organisms, as well as the shells of larvae and tininnides were suggested to be important agents in the formation of aggregates (Alldredge, 1972; Wheeler, 1975). The adhesion of bacteria within aggregates proceeds by the excretion of microfilaments of polysaccharides (Fletcher and Floodgate, 1973; Marshall et al., 1971). Aggregation makes accessible a large part of the bacterio-plankton for consumption by coarse filterers such as calanoids, which are unable to ingest single microbial cells (Sorokin, 1971). Microbial aggregates seem also to be an important part of macroaggregates of about 2 to 7 mm in size. These are visible as the flakes of 'marine snow', and seem to contain a significant part of the total microplankton. The concentration factor of bacteria and protozoa within these aggregates can be as much as 100 to 1000 times (Silver et al., 1978).

TOTAL NUMBER, BIOMASS AND PRODUCTION OF PLANKTONIC MICROFLORA

The total number and biomass of bacteria are measured by direct microscopy counts and captured on white membrane filters stained with erythrosin, or on black nucleopore filters stained with acridin orange, and observed with the aid of epifluorescent microscope (Zimmerman, 1977; Dale, 1974; Hobbie et al., 1977; Ferguson and Rublee, 1976).

Several biochemical methods of microbial biomass estimation have been employed. Among these is the ATP-method of Holm Hansen (1969). But this method actually measures the biomass of total microplankton including also the phytoplankton (Sorokin, 1975; Jassby, 1975). Therefore it gives values difficult to interpretate, which usually exceed the real microbial biomass by 1.5 to 4 times (Sorokin and Mikheev, 1979). Nevertheless, empirical correlations have been found for some oceanic habitats which make it possible to use ATP-method to obtain some information on microbial biomass (Sorokin and Lazarev, 1978). The hope that the use of nucleopore filters will assist separation of bacterioplankton from the larger microplankton organisms for ATP estimation of only microbial biomass (Sieburth et al., 1978) appears to be too optimistic, based on our experience (Sorokin, 1978a). Even 5 μm nucleopore filters retain a significant part of the aggregated bacterioplankton, yet at the same time these filters pass a large part of all microflagellates.

Estimation of microbial biomass based upon the measurements of muramic acid as a component of procaryotic cell wall was employed by Morearty (1975) in detritus and marine sediments. The limitation of this method is the presence in shallow sediments of blue-green algae, which, being procaryotes, also contain muramic acid. Watson et al. (1977) estimated microbial biomass in sea water measuring the lipopolysaccharide content which is specific for gram negative bacteria. Among the possible methods of estimation of number of total bacterioplankton, of its biomass and of the percentage of active cells in natural microbial populations is ^{3}H radioautography (Azam and Hobson, 1977; Hoppe, 1977), and transmission or scanning electron microscopy (Watson et al., 1977; Hobbie et al., 1977).

For the estimation of in situ rates of microbial production, of respiration, the ^{14}C-method by Romanenko (1964) has been employed (Sorokin, 1971). This is based upon the ratio between dark ^{14}CO$_2$ assimilation and microbial production which is 3 to 6% (Sorokin, 1965). Campbell and Baker (1968) used the simultaneous measurements of ^{35}S and ^{3}H-glucose uptake for the estimation of microbial production. Microbial production P can also be calculated by physiological methods employing the relationship: $P = 0.8$ R mg m^{-3} of wet biomass day^{-1}, where R = rate of microbial respiration in mgO$_2$m^{-3}day^{-1}. Calculation of microbial production has also been based upon data on rates of microbial multiplication, by using the bottle techniques of Ivanoff or Gak, by observations in chemostats (Jannasch, 1967), in diffusion chambers (Sieburth et al., 1978), or in microcolonies cultivated in natural sea water (Meyer-Reil, 1977).

For measurement of microbial respiration in water and sediments in mesotrophic and eutrophic habitats direct estimations of oxygen consumption by BOD bottle techniques can be used with in situ temperature and short-term (1–2 days) exposures (Tchepurnova, 1976; Dale, 1978; Sorokin and Mamaeva, 1979). In oligotrophic and deep-sea water the ^{14}C-method (Sorokin, 1971) or the electron-transport (ETS) analysis of microbial respiration (Packard et al., 1971; Hobbie et al., 1972) have been used.

Criticism of bottle techniques in situ (Steemann Nielsen, 1972; Banse, 1974) was a sequence of some misunderstandings (Sorokin, 1975, 1978a) and subsequent attempts to evaluate the rate of microbial respiration, avoiding the use of bottles by employing a diffusion chamber, and also by its calculations from diurnal variations of dissolved organic carbon, gave the same range of values as those obtained by bottle techniques (Sieburth, 1977; Sieburth *et al.*, 1977).

The available data are summarised in Table I. They show that in the upper layers of eutrophic habitats such as lagoons, estuaries, and upwelling areas the total number of bacteria varies within the range 1.0 to 5.0 $\times 10^6 ml^{-1}$, and their biomass[1] from 0.5 to 1.5 g m^{-3}. Microbial production in eutrophic water in summer averages 0.3 to 0.7 g m^{-3}day^{-1}, with a generation time of 30 to 40 hours, and P/B ratios of 0.3 to 0.6 per day. Rates of microbial respiration in such water vary within the range 0.1 to 0.4 mgO$_2$l^{-1}day^{-1}. The values of P/B ratios for total microbial populations appear to be regulated mostly by water temperature and by grazing pressure rather than by the level of primary productivity. Extremely high numbers and biomass of bacterioplankton were recorded in redox zone layers of meromictic lakes and lagoons: up to 30 to 40 $\times 10^6$ cells ml^{-1}, and biomass of 10 to 15 g m^{-3}. These dense populations are formed in such basins within the redox zones largely by photosynthetic bacteria. In the Black Sea, the redox zone is situated far below the boundary of euphotic zone, and the maxima of microbial biomass and production are here formed by a chemoautotrophic microflora – mainly of thiobacilli (Sorokin, 1964).

In the upper layers of mesotrophic temperate seas or in tropical divergence areas the average total number of bacteria varies between 0.3 to 0.8 $\times 10^6$ and their biomass between 0.05 to 0.3 g m^{-3}. These numbers are usually several times greater in the maximum layer at the upper boundary of thermocline (see above). Average values of microbial production and respiration in the euphotic zone of these parts of the ocean are: 0.05 to 0.2 g m^{-3}day^{-1}, and 0.03 to 0.15 mgO$_2$l^{-1} correspondingly.

In the oligotrophic tropical ocean microbial numbers are 0.05 to 0.15 $\times 10^6 ml^{-1}$ and biomass 10 to 30 mg m^{-3}, again increasing several times in the maximum layer (Table I). Microbial production in these areas is 20 to 40 mg m^{-3}day^{-1}, and respiration 0.01 to 0.03 mgO$_2$l^{-1}day^{-1}. The P/B coefficients in these waters are rather high (1–2 day^{-1}) because of high water temperature and grazing pressure. In the stratified layers the corresponding parameters are usually several times more than above given average values, except for P/B ratios and generation times, which usually decrease in layers of maximal concentrations.

Below the euphotic zone, the density and the activity of the microbial population usually decreases quickly. Excluding the microbial maximum layer, which in tropical trade wind areas occurs between 450 and 550 m,

---

[1] All values of biomass and production of bacterioplankton are given in units of wet weight having a carbon content of 10%.

TABLE I Total number (N), biomass (B), production (P) and respiration rates (M) of microbial populations in different aquatic habitats (based on Sorokin, 1971, 1973a, 1977, 1978a,b; Sorokin and Fedorov, 1978; Sorokin and Kovalevskaya, 1980; Sieburth et al., 1977; Holm Hansen, 1969; Hobbie et al., 1971; Williams and Carlucci, 1976)

Basin	Habitat	N 10^6ml^{-1}	B mg m^{-3}	P mg m^{-3}day^{-1}	M μgO$_2$l^{-1}day^{-1}
Eutrophic meromictic Faro Lagoon (Sicily)	redox zone	10–30	5000–15 000	500–700	—
Eutrophic temperate coastal waters Tatarsky strait, Japan Sea	euphotic zone	1–5	200–2000	100–700	200–500
Eutrophic waters of Peruvian upwelling	euphotic zone	1–5	500–2000	200–500	140–400
Eutrophic waters of coral reef lagoons	surface layer	0.8–1.5	600–1500	150–500	100–400
Mesotrophic Lake Dalnee (Kamchatka)	euphotic zone	0.5–3	500–1500	200–800	120–600
Mesotrophic temperate waters of Japan Sea	average in euphotic zone	0.15–0.3	50–200	40–100	30–70
	layer of basic maximum	1–2	600–1000	200–400	150–300
Mesotrophic waters of equatorial divergence, Pacific Ocean	euphotic zone	0.3–1	100–400	100–300	60–200
Mesotrophic temperate waters of Black Sea	euphotic zone	0.1–0.2	30–100	20–60	15–50
	redox zone	0.2–0.5	150–300	40–80	—
Mesotrophic waters of Antarctic region	euphotic zone	0.1–0.2	20–40	3–6	2–5
Oligotrophic tropical waters of oceans, trade wind area	average in euphotic zone	0.05–0.15	10–30	20–40	10–30
	layer of basic maximum	0.2–0.5	60–200	50–120	40–100
Intermediate Antarctic waters in Central Pacific, 300–1200 m	average	0.01–0.02	1–2	0.04–0.1	0.02–0.05
	layer of maximum at 500–600 m	0.05–0.15	10–30	5–20	3–10
Deep oceanic waters in Central Pacific, 1200–5000 m	average	0.003–0.006	0.3–1	0.003–0.015	0.002–0.01

total numbers of bacteria between 300 and 1200 m averages 10 to 15 $\times$ 10^3 cells ml^{-1}, and their biomass 2 to 4 mg m^{-3}, that is 10 to 20 times less than the euphotic zone above (Table I and Fig. 2). Microbial production thus decreases in these waters to 0.04 to 0.1 mg m^{-3}day^{-1} and the respiration to 0.02 to 0.05 $\mu gO_2 l^{-3}$day^{-1}. In the deep oceanic waters below 1000 m, biomass and the activity of baterioplankton are quite negligible. Total number of bacteria is only 3 to 6 $\times$ 10^3ml^{-1}, their biomass only 0.5 to 1 mg m^{-3} and production 3 to 15 g m^{-3}day^{-1}. Microbial respiration in deep oceanic waters is only 0.002 to 0.01 mgO$_2 l^{-1}$day^{-1}, or about 0.6 to 3 $\mu gO_2 l^{-1}$yr^{-1}. The cause for such low activity of microbial populations at depth is less likely to be a lack of utilisable organic matter, but rather the cumulative effects of low temperature and high pressure upon a larger part of the microbial population of deep waters, which appears to be non-psychrophylic and non-pressure-tolerant (Sorokin, 1969, 1971, 1973a, 1978a; Jannasch et al., 1976; Jannasch and Wirsen, 1973; Wirsen and Jannasch, 1975).

MICROBIAL POPULATIONS IN DETRITUS

Detritus is an important source of food for planktonic and benthic filter- and sediment-feeders (Darnell, 1967; Nishizawa, 1966; Seki, 1972; Lenz, 1977; Finenko and Zaika, 1970; Petipa, 1978; Sorokin, 1978a). In the open deep sea its standing stock approximates to 300 to 400 g m^{-2} dry weight, which far exceeds annual primary production. Detritus particles are formed from the decaying remains of organisms, of faecal pellets and of terrestrial organic material (Wangersky, 1977). Detritus particles are the foci of microbial and microheterotrophic activity. They harbour not only bacteria but also protozoa which form microcosms around and within the detritus particles (Fenchel, 1970). The biota actively utilise the organic matter and incorporate it into utilisable microbial biomass (Khailov, 1971; Rodina, 1963; Briggs et al., 1979). The organic matter of microheterotrophs forms up to 3 to 5% of its total organic matter (Sorokin, 1978b). Microbial numbers in detritus particles reach 5 to 25 $\times$ 10^9g^{-1}, approaching in this respect the active sludge of sewage plants. Oxygen consumption by mature detritus per unit of dry weight from the metabolism of the microheterotrophs inhabiting it (10 to 20 mgO$_2$g^{-1}day^{-1}), approaches the respiration rate of a planktonic crustacean of the same weight (Johannes and Satomi, 1966). Therefore detritus cannot be thought of as 'non-living organic material' Detritus is actually a microbial microcosm, and should be recognised as an active component of marine ecosystems. Its trophic value derives from the biosynthetic activity of microheterotrophs which comprise the assimilable portion of mature detritus. Energy budgets of marine ecosystems (Figs 4, 5) show that 'non-living' organic material (together with

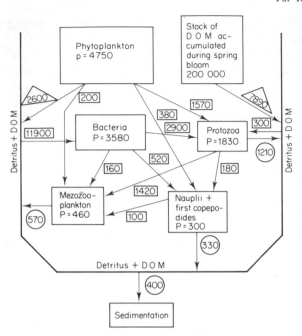

FIG. 4 Scheme of energy flow (kcal m⁻²day⁻¹) in the pelagic ecosystem of the Sea of Japan in the middle of June: P, production; numbers in squares, food rations of consequent components of food chain, numbers in triangles, nonconsumed phytoplankton production used via the stage of the dead organic matter; numbers in circles, nondigested part of consumed food

the dissolved organic matter which is involved in the formation of detritus via the participation of microheterotrophs) may be their principal energy source. This basic stock of energy makes them more stable and less dependent on the fluctuations of the level of primary autotrophic production.

MICROBIAL POPULATIONS IN BOTTOM SEDIMENTS

The most abundant marine microbial populations are those which inhabit coastal, lagoon or estuarine sediments, with constant large inputs of allochthonous organic matter. In these sediments and detritus, the total number of bacteria ranges from 3 to 10×10^9cells g⁻¹, and their biomass from 2 to 5 mg g⁻¹. Microbial biomass in detrital sediments composes 2 to 5% of total organic matter (Table II), and microbial production reaches 0.3 to 0.5 mg g⁻¹day⁻¹. High microbial biomass and production in such a sediment provide much of the food for benthic faunas, especially by species which directly ingest the sediments. In the upper layer of sediments of the

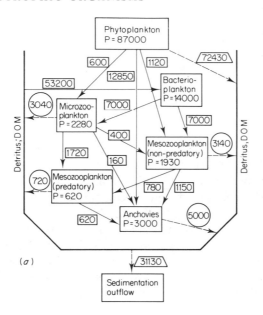

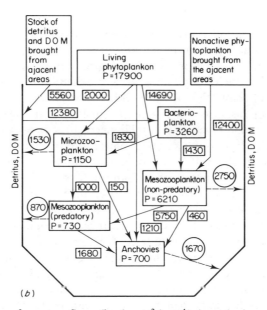

FIG. 5 Scheme of energy flow (kcal m^{-2}day^{-1}) in pelagic ecosystem of the Peruvian upwelling (Chimbote area): (a) in the patch of diatom bloom after intensive local upwelling, (b) area to the west of this patch after the front; for designations, see Fig. 4

inland seas and continental shelves the total number of bacteria varies within 0.1 to 1×10^9cells g^{-1}, their biomass from 0.02 to 0.2 mg g^{-1}, and their production from 5 to 50 μg g^{-1}day^{-1}. In the sediments of the central deep parts of the oceans, the total number of biomass of bacteria are extremely low, being only 0.1 to 1×10^7cells g^{-1} and 2 to 5 μg g^{-1}. Microbial production decreases here to about 2 orders (0.3–0.03 μg g^{-1}day^{-1}) below that of shallow seas.

Microbial biomass is actually a principal nutritional component of marine sediments, and its level may regulate the distribution of the biomass of benthic fauna (Table II). Within the sediments the number and activity of microflora quickly decreases. In shelf sediments below 30 to 50 cm, and in deep sea pelagic sediments below 10 cm, microbial activity is practically absent (Sorokin, 1970a).

The respiration of bottom microflora in shallow coastal areas is 0.3 to 0.8 gO$_2$ m^{-2}day^{-1} (Dale, 1978), in slope sediments 0.1 to 0.16 g, and in deep oceanic sediments 1 to 1.5 mg (Sorokin, 1970a). The integral respiration of bottom microflora in the water column has been calculated to be around 2% of that of microbial plankton (Sorokin, 1978a) in the oceans generally.

EFFICIENCY OF MICROBIAL PRODUCTION AS A SOURCE OF ENERGY

The efficiency of microbial production (the coefficient K_2) can be expressed as a ratio of production P to the amount of assimilated food material A. In natural microbial populations it is measured in prefiltered samples to remove grazers: in such samples microbial production P is measured by the ^{14}C-method, and respiration M by Winkler titration. K_2 is calculated as the ratio $P/(P + M)$, because $A = P + M$. K_2 in Peruvian upwelling water has values of 0.28 to 0.36 (average 0.32). Higher values of K_2 were obtained in water enriched with utilisable organic matter after a 'red tide' phenomenon (Sorokin and Kogelshatz, 1979; Sorokin and Mamaeva, 1979). K_2 obtained for natural microbial populations utilising natural organic matter is somewhat less than estimated in the cultures (0.45–0.55) or in sea water enriched with protein hydrolizate (0.35–0.45) (Sorokin, 1970b).

These data suggest that the process of aerobic microbial decomposition of organic matter in sea water results in the formation of newly produced proteinaceous material of microbial biomass eqaual to about 30% of the organic material decomposed. This is, then, an average efficiency for the microbial production of particulate food in marine environments.

Energy sources for microbial production can be identified only by the study of total energy balance within an ecosystem. In practice, this kind of study requires analysis of biomass, production and metabolism of all of the biotic components of a local ecosystem per unit area per day, or per year.

TABLE II Characteristics of total number (N), biomass (B) and production (P) of microbial populations and the biomass of meiobenthos in the surface layer of bottom sediments of different marine basins (data by Sorokin, 1964, 1970a, 1978a, 1978b; Dale, 1974)

Habitat, type of sediment	Depth	N $10^9 g^{-1}$	B $\mu g\ g^{-1}$	P $\mu g\ g^{-1} day^{-1}$	B (%) of total organic matter	Biomass of meiobenthos $g\ m^{-2}$
Shallow estuaries and lagoons, detrital sediments	1–2	6–10	1200–5000	500–800	4–8	50–200
Coral sands	1–10	2–4	1000–1500	300–700	1–6	100–500
Neritic zone grey sandy alevrites	100–1000	0.2–1	50–300	5–50	0.1–0.3	1–10
Continental slopes, sandy alevrite-carbonate sediments	1000–4000	0.05–0.1	10–30	2–5	0.1	0.03–1
Deep internal sea (Japan Sea), gray alevrite-pelite	2000	1.6	300	10	0.3	5–15
Deep meromictic sea (Black Sea)	2000	0.3	40	2	0.1	0
Deep oceanic areas of equatorial divergence, radiolarian pelites	4000–5000	0.03–0.05	10–20	1–2	0.03	0.1–0.3
North Pacific Ocean, deep red clays	5000	0.015	5	0.3	0.01	0.10–0.15
South Pacific Ocean, deep red clays	5000	0.001	2	0.03	0.005	0.05–0.1

There are two main potential sources of energy (Sorokin, 1972a; Sorokin *et al.*, 1977): the first is the surplus of local primary production over grazing rates together with unassimilated plant remains; the second is allochthonous organic matter reaching the system from the land, or from other sea areas by advection (see Fig. 5). Relative quantification of these two sources is extremely important for evaluation of primary energy input in the analysis of the energy balance of aquatic ecosystems. Primary energy input comprises these two energy sources combined to sustain the local microbial population. Even when analysing short-term energy balance in a local plankton community (which in practice is more frequently the case than long-term observations) we have to include as a primary energy input also the allochthonous organic matter produced by phytoplankton during the previous step in the succession process, and accumulated within the system as a stock of dissolved and suspended organic matter, to be used by the microbial population at the time of the investigations (Sorokin, 1977); Fig. 4 shows an example of this.

The allochthonous organic material incorporated by microbial activity into the food web A may be evaluated by calculation of mirobial production derived from its utilisation P_a. This value is calculated as a difference between total microbial production P_t and microbial production P apparently produced from the nongrazed plus nonassimilated local photosynthesic production D. In this case, $P_1 = 0.32D$, where 0.32 is the efficiency coefficient K_2; $P_a = P_t - P_1$, and $A = (P_a/K_2) = (P_t - 0.32D/0.32)$.

Calculations of external sources of energy incorporated via microbial synthesis into the energy balances of some aquatic ecosystems at different steps in their successions are given in Table III. These data show that during the autotrophic period of succession (the spring bloom in temperate waters or the bloom in upwelling areas) the plankton ecosystem not only does not depend upon external energy sources but produces a surplus load of organic matter not used by local heterotrophes during this period. But during the following heterotrophic period of succession (the summer phytoplankton minimum in temperate waters or the final step in the succession of plankton communities in upwelling areas) the energy from allochthonous organic matter incorporated into the food web by bacteria may comprise 30 to 60% of total energy input (Sorokin, 1972a, 1977, 1978a, 1978c; Sorokin and Mikheev, 1979).

MICROFLORA AS FOOD FOR ANIMALS AT SUBSEQUENT TROPHIC LEVELS

Concepts concerning the importance of microbial populations as a source of particulate food in aquatic habitats have been developed by ZoBell (1946), Rodina (1949), Gorbunov (1946), Parsons and Strickland (1962). Quantitative data on the feeding of marine animals on natural microbial

TABLE III Sources of the formation of primary energy input in aquatic ecosystems at different steps of succession, and the share (PM, %) of microheterotrophes in summary heterotrophic production. (Based on Sorokin, 1972, 1978a; Sorokin and Paveljeva, 1977; Sorokin and Mikheev, 1979)

Habitat	Phase of succession	Time taken for analysis (days)	Formation of primary input of energy, kcal m^{-2} per time taken for analysis			PM (%)
			Local photosynthesis of algae	External energy included via microbial function (+used; −released)	Primary energy input	
Dalnee Lake	autotrophic phase, spring	1	45.5	−34.0	11.5	94
	heterotrophic phase, summer	1	6.7	+26.8	33.5	85
	whole period of vegetation	150	2070	+870	2940	82
Rybinsk reservoir	whole period of vegetation	180	500	+1280	1780	85
Japan Sea	heterotrophic phase, summer	1	4.7	+6.5	11.2	88
Tropical Pacific Ocean, equatorial area	upwelling, autotrophic phase	1	33.0	−17.3	15.7	61
	weak divergence, heterotrophic phase	1	6.3	+7.3	13.6	86
Peruvian upwelling	patch of bloom, autotrophic phase	1	69.3	−37.7	31.6	65
	a week later, heterotrophic phase	1	24.9	+42.5	67.4	86
	quasipermanent upwelling area	1	17.9	+12.4	30.3	57
Tropical Pacific Ocean trade wind area	heterotrophic phase	1	1.4	2.5	3.9	88

populations have been mostly obtained by the use of ^{14}C as a tracer (Sorokin, 1968, 1971, 1978a; Sorokin *et al.*, 1970; Petipa *et al.*, 1971, 1974, 1977). Such studies show that bacteria are a normal source of nutrition for mass occurrence species of many filtering planktonic and benthic biota: ciliates, hydroids, coral polyps, appendicularians, sponges, cladocerans, copepods and their larval stages, euphausiids, bivalves, polychaetes and ascidians. Fine filterers like some cladocerans, appendicularians, sponges, and polychaetes can satisfy their nutritional demands using bacterioplankton at unaggregated concentrations (0.2 to 1.0 g l^{-1}) as a sole dietary source. This is also true of phagotrophic microzooplanktic ciliates and flagellates. For most of these, bacteria are a more important source of nutrition than phytoplankton (see below).

The coarse filterers like most copepods, bivalves, and euphausiids are less able to ingest single microbial cells, and mostly utilise the aggregated part of the bacterioplankton. Even predatory animals like *Euchaeta* in part also ingest clumped bacterioplankton and detrital particles. Optimal concentrations of bacterioplankton for fine filterers was found to be from 0.15 to 0.4 g m^{-3}, and for coarse filterers 1.0 to 1.5 g m^{-3}. The former range of concentrations occurs in the open sea maximum layer, and the latter in coastal

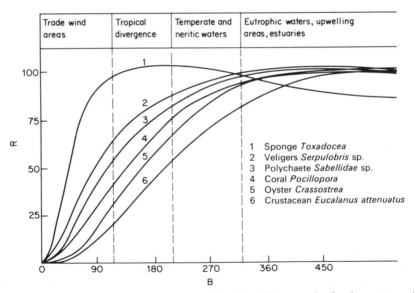

FIG. 6   Dependence of the relative rate of feeding (R) upon the food concentration of bacteria in different planktonic and benthic filter-feeders. B, biomass of bacteria mg m^{-3}, wet weight. Vertical lines show the levels of bacterial biomass, in specified ocean areas

areas and upwellings. So in heavy pelagic marine and benthic habitats fine filterers can satisfy their food demand solely with bacteria, while the coarse filterers can do this only in upwelling areas having an extremely rich microbial plankton (Fig. 6).

Assimilation of bacteria by most animals tested is about the same as that of phytoplankton, from 40 to 60%. Daily rations related to body weight for fine planktonic filterers feeding on natural bacterioplankton at optimal concentrations ranged from 50 to 100% (Table IV). Respiratory losses of these animals are 15 to 20%, so they can cover their requirements with bacterioplankton at concentrations 2 to 3 times less than the optimal. For example, *Penilia avirostris* can do this at bacterial concentrations of 50 mg m^{-3}, which is actually less than usually occurs in the neritic habitats of this species on the Black Sea coast (Pavlova and Sorokin, 1970). An important intermediate trophic link between bacterioplankton and meso-zooplanktonic coarse filterers and predators are the protozoa and other microzooplankton which very actively ingest bacteria (Monakov and Sorokin, 1972).

Experiments show that benthic suspension feeders such as bivalves, sabellids and sponges can normally feed on bacterioplankton at its normal

TABLE IV Daily food rations (R) and assimilability (I) of different aquatic invertebrates by their feeding with bacteria. (From data by Goryacheva, 1976; Kopylov *et al.*, 1979; Chorik and Shubernetzki, 1978; Kopylov, 1977; Sorokin *et al.*, 1980; Sorokin, 1978b)

Animals	R (%) of body carbon	I (%)
Parabodo attenuatus: zooflagellate	300	40
Carchesium pectinatum (ciliate)	700	43
Carchesium polipianum: ciliate	290	41
Gastropoda sp.: veliger	51	59
Penilia avirostris: cladoceran	105	35
Paracalanus parvus: copepod	25	57
Rhincalanus cornutus: copepod	12	45
Clausicalanus mastigophorus: copepod	75	41
Eucalanus attenuatus: copepod	10	43
Oicopleura dioica: appendicularian	75	40
Pennaria tiarella: hydroid	52	54
Pocillopora damicornis: coral	15	64
Serpullidae sp: polychaete	11.8	63
Littorina scabra: gastropod	6.7	22
Ophiodesoma spectabilis: holothurian	10.4	26
Toxoidocea violacea: sponge	3.9	85

concentration in near-bottom layers (Sorokin, 1978b). The microflora of detritus and sediments is assimilated by sediment- and seston-feeders such as gastropods and holothurians (Table IV). The direct correlation between the biomass distribution of bottom fauna and bacterial abundance (Table II) also emphasises the significance of microflora in determining the nutritional quality of sediments. Thus, experimental data confirm that particulate microbial protein is an important food source in most planktonic and benthic marine habitats.

THE CONTRIBUTION OF BACTERIA TO PRODUCTION,
METABOLISM AND ENERGY FLOW

A quantitative evaluation of the role of microbial populations as a component of marine ecosystems can only be done by complete component analysis, which has to include all the above-listed analyses of microbial populations together with similar estimations concerning all other components of the given ecosystem. This type of study has been accomplished in the Rybinsk reservoir, in Dalnee Lake (Sorokin, 1972; Sorokin and Paveljeva, 1978), in the Japan Sea and in the Pacific Ocean (Sorokin, 1977; Sorokin and Mikheev, 1979; Sorokin et al., 1977; Vinogradov and Shushkina, 1979). An example of data on biomass, production and metabolism of all of the components of an ecosystem are given in Table V. Examples of energy flow schemes based on such tables are given in Figs 4 and 5. These tables and schemes enable us to evaluate the role of microbial populations in the total energy balance of the given ecosystem (Table VI). It can be seen that the contribution of bacteria is greater compared with that of the other groups of heterotrophs in most types of ecosystem: their share in the metabolism and energy flow of the heterotrophic population usually being 60 to 90%. These numbers are as high even in mature pelagic communities in trade wind areas where direct grazing of phytoplankton predominates.

A simple calculation proves the reality of these numbers. The maximum possible direct utilisation by grazing of phytoplankton production appears to be around 50% (Sorokin, 1973a). The remaining 50% ungrazed primary production, together with the unassimilated faeces should be respired by microflora.

The contribution of bacteria to total respiration of the heterotrophic part of planktonic communities is especially high (more than 80%) at the late heterotrophic phase of seasonal succession, when the microbial population utilises organic matter accumulated during the preceding phytoplankton bloom. The share of bacteria in total heterotrophic production is of the same order as that of total respiration: usually 60 to 75%.

TABLE V Parameters of heterotrophic activity of different components of plankton ecosystems of some basins at different phases of their succession

Locality, period of observation, phase of succession	Ratio: total heterotrophic respiration : primary production	Ratio: production of micro-zooplankton : production of meso-zooplankton	Part of primary production used via detritus (%)	Share of main groups of pelagic communities in total heterotrophic respiration (%)			
				Bacteria	Micro-zooplankton	Meso-zooplankton	Fish
Dalnee Lake, whole period of vegetation	1.35	1.7	86	71	11	12	6
Japan Sea, summer, heterotrophic phase	2.4	5.0	67	71	19	8	2
Eastern tropical Pacific Ocean, area of upwelling, autotrophic phase	–	0.9	73	81	9	8	2
Central tropical Pacific Ocean, weak divergence, heterotrophic phase	1.9	2.0	73	84	5	10	1
Peruvian coastal upwelling, patch upwelling area, autotrophic phase	0.45	10.7	57	61	14	3	22
Same patch, a week later, heterotrophic phase	2.0	35.0	79	80	11	1	8
Peruvian upwelling, quasipermanent upwelling area, heterotrophic phase	1.4	0.23	32	43	5	44	8
Central Pacific Ocean, trade wind area, heterotrophic phase	3.1	1.5	70	74	12	13	1

TABLE VI Elements of energy balance in the ecosystems of the equatorial Pacific Ocean

Habitats	Components of community	Ration	Production	Non-assimilated food	Respiration, kcal m^{-2}	Respiration, % of Sum.
Equatorial upwelling area, 97°W	Phytoplankton	—	33.0	—	—	—
	Bacteria	15.70	4.70	—	11.00	72
	Microzooplankton	3.76	1.33	1.28	1.22	8
	Mesozooplankton (herbivorous)	2.50	0.62	1.62	1.26	8
	Mesozooplankton (predatory)	3.30	0.82	0.66	1.82	12
Equatorial area in Central Pacific, 154°W	Phytoplankton	—	6.30	—	—	—
	Bacteria	16.30	4.90	—	11.4	85
	Microzooplankton	2.66	1.15	0.63	0.69	5
	Mesozooplankton (herbivorous)	1.27	0.32	0.31	0.64	5
	Mesozooplankton (predatory)	1.13	0.28	0.22	0.63	5

3 Animal microheterotrophs

COMPOSITION OF COMMUNITIES IN MARINE HABITATS

Animal microheterotrophes are microscopic organisms 3 to 200 μm in size. This group of zooplankton includes colourless phagotrophic zooflagellates, ciliates, planktonic amoebae, sarcodines (radiolaria, foraminifera) and also larval or juvenile stages of benthic and planktonic organisms.

The zooflagellates (Zoomastigophorea) are represented in marine habitats by four orders: Kinetoplastida (suborder Bodonina), Protomonadina, Choanoflagellida, Bicosoecida. These small colourless flagellates have usually one or two flagella of which one is directed anteriorly and provides mobility. The second is used to attach these organisms to solid surfaces or loricas, and also for catching food, usually bacteria and small algae. The cytostome of zooflagellates is provided with a system of microtubules which provide for flexibility of the oral aperture which makes possible the swallowing of relatively large food particles (Brooker, 1971; Gorjacheva *et al.*, 1978). Choanoflagellates have silicified loricas (Throndsen, 1974). Systematic examination of zooflagellates on living or lugol fixed specimens can be made using phase contrast or water immersion objectives. Choanoflagellates can be identified by their loricas with electron microscopy (Leadbeater and Manton, 1974).

Systematic studies of marine zooflagellates have so far mostly concerned the order Choanoflagellida. Systematics of these organisms have been worked out by Kent (1882), Ellis (1929), Scuja (1956), Throndsen (1969, 1974) and Leadbeater (1972, 1974). The study of the three other orders is still in its infancy. Systematics of free living Bodonia was developed by Hollande (1952) and later revised by Zukov (1971), though mostly concerning fresh water species. The marine Bicosoecida were studied by Valcanov (1970), Moestrup and Thromsen (1976) and recently reviewed by Zukov (1979). Transparent or coloured loricas are used for taxonomic purposes in this group.

An example of the species composition of a pelagic zooflagellate community was obtained in October in the eastern Black Sea by Moiseev (1980). Among the mass species there were: *Bodo saltans, Parabodo attenuatus, Bodomorpha reniformis* (suborder Bodonina), *Monosiga marina Salpingoeca spinifera* (order Choanoflagellates), *Bicoeca* sp. (order Bicosoecida), *Monas* sp., *Actinomonas* sp., *Oicomonas* sp. (order Protomonadida). The species of Choanoflagellida usually inhabit the mixed layer above the thermocline and the species of Bodonina the deeper anoxic layers.

Systematic studies of aquatic zooflagellates have concentrated on planktonic forms. Less is known about the zooflagellate fauna of detritus, periphyton and sediments.

Marine planktonic ciliates range in size from 15 to 200 μm. They include a variety of species, mostly of the order Oligotrichida. Two main groups can be separated: the loricated, and the naked ciliates. Populations of loricated ciliates or tintinnids in neritic and temperate waters mostly comprise genera of the families Tintinnidae, Codonellidae, Coxliellidae (*Tintinnidium, Tintinnopsis, Codonella, Metacylis, Acanthostomella, Stenosemella, Helicostomella, Favella*). In tropical and subtropical oceanic waters tintinnids of the genera *Dictyocysta, Parundella, Epiplocylis* are also abundant. Most naked ciliates belong to the suborders: Oligotrichina (families Strombidiidae, Strobilidiidae and Halteriidae), Prorodontina (family Colepidae) and Heterotrichina (family Peritromidae). Common genera are: *Strombidium, Strobilidium, Lohmanniella, Coleps, Tontonia* and *Peritromus*, and these usually represent the mass species in planktonic ciliate populations. In coastal areas their populations often include epibiotic planktobenthic and benthic forms, including species of the suborder Holotrichida, such as *Tiarina* and *Prorodon* and Hypotrichida, such as *Diophrys, Euplotes, Urostyla* (Tumantzeva, 1979; Agamaliev, 1977; Sournia et al., 1974).

Loricated cilicates can be detected, counted and identified much more easily than the naked forms because their loricas are more resistant to the concentration and fixation procedures, and also because the form of lorica is a very useful systematic character for this group. Because they are relatively easy to collect, preserve and determine we have come to think that tintinnids are, if not the only, at least the dominant group of ciliates in the sea. Recent attempts to study also the naked forms by the Utermöl sedimentation procedure (Beers and Stewart, 1967, 1969; Beers et al., 1971), and especially as intact samples in the living state (Sorokin, 1977, 1978c; Tumantzeva and Sorokin, 1977; Tumantzeva, 1979; Pavlovskaya, 1976) have demonstrated that tintinnids are more often only a minor component of ciliate populations. Seventy to ninety-five percent of total ciliate biomass in most marine planktonic habitats is formed usually by the naked ciliates of the genus *Strombidium*, sometimes also by *Tiarina fusus*, in anoxic layers by *Pluronema marinum* or by benthoplanktonic forms such as *Urostyla*. In coastal areas, and especially in areas of coastal upwellings a significant part of the biomass of planktonic ciliate population is also composed by benthoplanktonic forms of Hypotrichida (*Euplotes*) and Hymenostomidae (*Colpidium*).

The ciliate populations studied in intact water samples in the equatorial area and in the Peruvian upwelling by Tumantzeva (1979) includes about 200 species belonging to 23 families and 59 genera. Among them 170 species belonged to loricated forms. Nevertheless their biomass usually did not exceed 5 to 15% of the total ciliate biomass. Mass species were: *Strombidium strobilus, S. lagenula, S. conicum, Tontonia gracillima, Lohmanniella spiralis*,

Strobilidium acuminatum, *S. marimum*, *S. pelagicum*. Tintinnids species usually did not contribute to lists of species of mass occurrence.

In the Mediterranean Sea the population of tintinnids includes about 180 species (Treguboff, 1978). In the open Black Sea in autumn Mamaeva (1980) found 17 species of ciliates, including 8 species of tintinnids. There the mass species were: *Tiarina fusus* and *Metacylis angulata*. In a coastal area near Gelendzhik Bay five mass species dominate at this season: the tintinnids *Tintinnopsis baltica*, *Tintinnidium mucicola*, *Coxliella helix* and the naked ciliates *Tiarina fusus* and *Strombidium sulcatum*. The tintinnid population in the Black Sea was thoroughly studied by Morosovskaya (1969). The mass species were *Tintinnopsis beroidea*, *T. tubulosa*, *T. meuneri*, *Stenosemella ventricosa*, *Coxliella helix*.

In the coastal area near Sebastopol the ciliate species of the genera *Strombidium*, and also *Prorodon*, *Coxliella*, *Tintinnopsis* and *Metacylis* were most abundant (Pavlovskaya, 1973a). In a Pacific neritic area near Vladivostok the fauna of tintinnids includes around 30 species, among which species of *Tintinnopsis*, *Favella* and *Leptotintinnus* predominate (Konovalova and Rogachenko, 1974).

In some natural as well as eutrophicated coastal marine habitats a mass bloom of chloroplast-bearing protozoa sometimes occurs such as euglenoids and the ciliate *Mesodinium rubrum*. The former includes an organic chloroplast, while *M. rubrum* harbours a pseudo-symbiotic one. Being facultative autotrophs and also capable of phagotrophic feeding such forms dominate in areas where water rich with nutrients and particulate organic matter lies close below a surface layer impoverished with nutrients. In such areas, they often cause 'red tides', accumulating during the day time near the surface, where their biomass sometimes exceeds 50 g m^{-3}. At night, they migrate below the thermocline to 20 to 30 m and there replenish their cells with inorganic and organic nutrients (Barber *et al.*, 1969; White *et al.*, 1977; Sorokin and Kogelshatz, 1979; Sorokin, 1979c).

Marine planktonic protozoan populations also include naked amoebae (Davis *et al.*, 1978), and sarcodines: radiolarians (such as *Nassellaria*, *Phaeodaria*, *Spumellaria*, *Acantharia*), and planktonic foraminifera, represented mostly by the families Globigerinidae, Globorotaliidae, Cymbaloporidae. The fauna of radiolarians was reviewed by Petrushevskaya (1971), Cachon and Cachon (1970), and foraminifera by Treguboff (1978) and Boltovskoy (1973).

Benthic habitats have a specific ciliate fauna composed mostly by the orders Hypotricha, Holotricha and Spirotricha (families Holophryidae, Spirostomidae, Oxytrichidae, Colepidae, Euplotidae). Especially common there in such habitats are the genera *Trachelostyla*, *Uronema*, *Urostyla*, *Coleps*, *Euplotes*, *Diophrys*, *Remanella*, *Uronychia* (Dragesco, 1954; Vacelet,

1961; Fenchel, 1967). The ciliate fauna in sandy sediments of the Caspian sea includes 130 species (Agamaliev, 1969) and that of the White Sea around 200 species (Burkovsky, 1969). Systematics of ciliates was developed by Kent (1882), Kahl (1935) and reviewed by Kudo (1966), Jankovsky (1967) and Corliss (1979).

Studies of multicellular microzooplankton animals are much easier. They are larger, and are not destroyed by fixatives. They include: (a) small or juvenile forms of zooplankters such as young *Oikopleura* or rotifers, (b) small nauplii of copepods, sometimes are less than 50 μm in length, (c) small larvae of benthic animals, such as veligers and trochophores, abundant in coastal waters (Zaika *et al.*, 1976; Pavlovskaya, 1973a; Tumantzeva and Sorokin, 1977). The composition and the distribution of pelagic larvae of benthic animals in the sea was studied by Thorson (1961) and Mileikovsky (1971, 1972).

METHODS OF QUANTITATIVE ANALYSIS OF THE NUMBERS
AND BIOMASS OF MICROZOOPLANKTON

For a long time microzooplankton was ignored during quantitative biological oceanography. A more careful approach to this problem was recently stimulated by recognition of its importance in the functioning of aquatic ecosystems. Separate methods for each of the components discussed above had to be developed. Zooflagellates can be counted only in the living state in the intact samples taken less than 1 hour before analysis. They are counted in a standard sample (5 ml) in a 7 to 8 cm Petri dish with a phase-contrast ($\times 10$, $\times 20$), and water immersion ($\times 60$) objective for species enumeration and sizing (Zukov, 1973). At sea, zooflagellates are counted in glass chambers 1.5 mm depth and of 2 to 3 ml volume with a movable glass lid and a gridded glass bottom under a phase contrast microscope at $\times 200$ and $\times 300$ (Sorokin, 1977). For this purpose an inverted phase contrast microscope can be employed with phytoplankton counting chambers.

Ciliates should also be counted only in the living state and as soon as possible after sampling. This concern applies equally to naked forms and also tintinnids. To say nothing about the use of pumps for sampling, and the screens for concentration (Beers and Stewart, 1967), even the most gentle possible methods of concentrating water samples to count protozoa with the use of the 7 μm nylon net, of 1 or 5 μm nucleopore filters or of membrane filters were all found to be rather useless (Sorokin, 1977; Tumantzeva and Sorokin, 1977; Pavlovskaya, 1976). During such concentration procedures a large part of the total ciliate population, observable in the intact sample, disappears. This loss averages more than 95 % for both naked and loricated forms. Thus even the most gentle concentration methods can be used only for examination of species composition of tintinnid populations.

The Utermöl sedimentation method cannot be used quantitatively for naked ciliates because they are destroyed (or at least lose their recognisable form) in any known fixatives available for large volumes of sea water. It is also impossible to get quantitative data on tintinnids by this method (Sukhanova and Ratkova, 1977).

Beers and Stewart (1967) counted tintinnids directly and on membrane filters. Counts of ciliates can be done on unconcentrated samples in the chambers described above for zooflagellates. To count small naked forms chambers of 1.5 to 2.0 mm depth are used, at a magnification of $\times 200$. For ciliates of 30 to 70 µm, which are normal in the open sea, chambers 4 mm deep and of 30 ml volume are used, at magnification of $\times 60$, or $\times 100$. For species identification and sizing, specimens of ciliates are removed from the chamber and observed at higher magnification after staining with neutral red (Sorokin, 1977; Tumantzeva and Sorokin, 1977; Tumantzeva, 1979). When taking subsamples from the water bottles for counting protozoa it is necessary to limit pressure gradients and to avoid the use of pipettes to prevent mechanical injury.

The numbers of ciliates in sediments also can be counted in living samples of sediments resuspended in water and placed into the counting chamber (Tchorik, 1968), or ciliates can be induced to leave the sediment by the action of gradients of salinity or temperature (Uhlig, 1964; Fenchel, 1967).

Multicellular microzooplankton can be concentrated for counts with the aid of 10 to 15 µm nylon mesh or with the aid of 1 to 5 nucleopore or millipore filters by direct filtration, or by reverse filtration using apparatus described by Sorokin (1979e). Water samples (3 to 10 l) are concentrated to a volume of 10 to 20 ml and fixed. Microzooplankton organisms are counted by settlement (Tumantzeva and Sorokin, 1977). Alternatively, they may be counted directly on membrane filters after filtering 0.5 to 2.0 l of sea water (Beers and Stewart, 1967; Sorokin, 1977).

VERTICAL POPULATION STRUCTURE

Where the oxygen profile is normal, populations of microzooplankton usually form two main maxima in vertical profiles. One is situated within the surface film, and another within the basic plankton maximum at the upper boundary of the thermocline. The microzooplankton in the surface film layer has a specific composition dominated by several specific forms of microplankton and tintinnids, which are a part of the hyponeuston (Zaitzev, 1970; Morosovskaya, 1969). Their accumulation in this layer is connected with the accumulation there of bacteria and phytoplankton.

A large part of the microzooplankton population usually occurs within the euphotic zone above the thermocline (Fig. 7). In temperate waters in summer, in tropical waters and in upwelling areas the maximum concen-

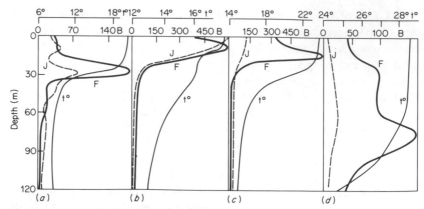

FIG. 7 Vertical distribution of protozoa: ciliates (J) and zooflagellates (F) in different marine planktonic habitats: (a) in the Black Sea; (b) in the Peruvian coastal waters south of intensive upwelling; (c) in the Peruvian upwelling region after the 'red tide'; (d) in the trade wind area of the South Pacific. B, biomass, mg m⁻³, wet weight; t°, temperature. (From Mamaeva and Moisseev, 1970; Sorokin, 1978c, 1979d)

tration can often be observed near the upper boundary of the thermocline (Sorokin, 1977; Sorokin and Mikheev, 1979; Sorokin and Kogelschatz, 1979). Accumulation of microzooplankton in this layer is connected with the accumulation of their food to form a biocoenosis with intimate trophic and metabolic relations. The biomass of protozoa and meroplankton within this layer often exceeds the euphotic zone average by one order of magnitude (Fig. 1), and may greatly exceed the mesozooplankton biomass. Microzooplankton species composition usually alters down vertical profiles (Beers and Stewart, 1969; Tumantzeva, 1979).

A significant proportion of all planktonic protozoa, and especially of zooflagellates, exist within the large organic aggregates known as 'marine snow' (Silver et al., 1978). This phenomenon is very important in the ecology of marine protozoa, especially in oligotrophic waters, providing them with better trophic conditions. Dead zooplankton and faeces of zooplankton are quickly colonised by bacteria and protozoa, and so detritus microcosms are formed. Protozoa often colonise living and also dead phytoplankton cells.

Below the euphotic zone in the open sea, microzooplankton populations are extremely sparse until at a depth of 500–600 m they reappear (mainly zooflagellates). In the meromictic basis, with an anoxic hydrogen sulphide layer, such as the Black Sea or inlets like the Faro Lagoon in Sicily an abundant protozoa fauna exists within the redox zone at the interface between the oxygenated and H₂S layers (Fig. 3). In the Faro Lagoon a dense population of brown bacteria (Chlorobium phaeovibrioides) supports an abundant population of two hypotrichid ciliate species: Trachelostyla sp. and

Urostyla sp. (Sorokin and Donato, 1975). The layer of 'red' water, where the brown bacteria flourish, is also inhabited by a rich population of zooflagellates, mostly of the genus *Bodo* (*Bodo sorokini*: Zukov, 1975). Living zooflagellates are present in water down to depths where the H_2S concentration exceeded 50 mg l^{-1}. The same phenomenon, a layer of high protozoan biomass also occurs within the redox zone of the Black Sea at 120 to 130 m (Mamaeva and Moiseev, 1979), mostly by a single species of ciliate (*Pleuronema marinum*) and by several zooflagellates of the families Bicosoecidae and Bodonidae.

The resistance of some protozoa to hydrogen sulphide and their ability for anaerobiosis enables them to inhabit reduced sediments (Fenchel, 1970; Fenchel and Reidl, 1970; Baas Becking and Wood, 1955). The most abundant protozoan populations inhibit the water-mud interface and the fine layers of the soft sediments, especially in the most superficial layer (Goulder, 1971; Johannes, 1965). The biomass of ciliates per 1 m^2 of the bottom surface of sea coast varies between 0.1 to 2 g m^{-2}, and may even reach 5 g m^{-2} (Burkovsky, 1969; Fenchel, 1967).

NUMBERS AND BIOMASS OF MICROZOOPLANKTON

The maximum numbers and biomass of 3 to 8 μm planktonic zooflagellates in the surface layers of the open sea are in the range 8 to 13 × $10^6 l^{-1}$, and 300 to 700 mg m^{-3} (Table VII). Such values were recorded during the heterotrophic phase of community succession in the temperate Sea of Japan in June, and in upwelling areas during the decay of a bloom of the 'red tide' peridinian *Gymnodinium splendens*, and of diatoms following intensive upwelling (Sorokin, 1977, 1978c, 1979d).

In eutrophic brackish lagoons, and fish culture ponds their numbers sometimes reach 20 to 60 × $10^6 l^{-1}$ and 1 to 3 g m^{-3} (Table VII). We have observed such values in lagoons near Comacchio and Valpisani in Italy.

A dense population of zooflagellates inhabits the redox zones of the meromictic Faro Lagoon (biomass up to 1.5 g m^{-3}), and of the Black Sea (biomass up to 0.4 g m^{-3}). Relatively high biomass was also observed in the trade wind area of the Central Pacific Ocean after a bloom of blue-green algae (*Trichodesmium*). Here zooflagellates were a principal component of plankton biomass (Table VII). Average biomass values of zooflagellates in the euphotic zone varies depending on productivity levels, and during seasonal succession varies between 20 and 100 to 150 mg m^{-3} (Sorokin, 1977, 1978c, 1979d; Sorokin and Fedorov, 1979; Sorokin and Kogelschatz, 1979).

Marine planktonic ciliates are usually 20 to 100 μm long and 3 to 40 × $10^3 μl^3$ in volume. Their numbers in surface waters varies in different habitats between 0.5 to 15 × $10^3 l^{-1}$, and their biomass between 10 and 200 mg m^{-3} (Pavlovskaya, 1976; Beers and Stewart, 1969; Tumantzeva and

TABLE VII Number (N) and biomass (B, mg m^{-3}) of ciliates and zooflagellates (maximum values) total microzooplankton biomass (mg m^{-3}) in different marine habitats (maximum values), and integral biomass of protozoa within the water column

Habitat	Zooflagellates		Ciliates		Total microzooplankton, B, mg m^{-3}	Integral biomass of protozoa in the euphotic zone, g m^{-2} (wet wt)
	N 10^6 l^{-1}	B	N 10^3 l^{-1}	B		
Brackish fish pond, lagoons Valpisani (Italy)	4.8	450	370	11 700	13 000	8.0
Meromictic Faro Lagoon (Sicily), redox zone	23.0	1500	40	840	2400	6.0
Lagoon of Venice (Italy), April	4.0	240	17	430	750	0.8
De Castri Bay, Japan Sea, NE coast, July	2.0	120	19	440	650	3.3
Japan Sea, pelagic part, June	9.5	450	9	190	700	10.2
Peruvian upwelling, patch of diatom bloom, late phase of succession	6.0	320	86	800	1200	14.0
Peruvian upwelling, area of 'red tide'	13.0	720	145	1800	2500	20.0
Peruvian upwelling frontal area	0.7	40	22	810	950	6.8
Equatorial upwelling Eastern Pacific Ocean, 97°W	0.4	16	30	50	250	2.5
Peruvian upwelling area of strong upwelling near San Juan	9.0	500	21	650	1300	12
Equatorial divergence, Central Pacific Ocean, 154°W	0.5	20	7	10	40	1.6
Trade wind area of South Pacific Ocean	3.0	100	1.5	30	150	1.3
Black Sea, coastal area	4.0	200	3.7	140	600	3.5
Black Sea, open area	1.8	100	0.45	20	100	2.8
Black Sea, upper boundary of redox zone, 130 m depth	0.7	40	0.04	8	50	—
Antarctic Ocean, mid-summer	—	—	29	113	—	2, 8

Sorokin, 1977; Sorokin, 1977, 1978c). But in some cases, usually during the early stages of the heterotrophic phase of plankton succession when the decay of phytoplankton production from the previous bloom has started but the grazing pressure of mesozooplankton is not yet high, ciliate biomass may attain very high values of 1.0 to 1.5 g m^{-3}. Similar values occurred in the Peruvian upwelling area after a 'bloom' of *Gymnodinium splendens* due to the swarming of naked ciliates *Strombidium* spp. (Tumantzeva, 1979). Daily observations in the Rybinsk reservoir revealed that a bloom of ciliates occurred in June after the spring diatom bloom during a very short period – only about one week, when their biomass exceeded 7.0 gm^{-3} (Sorokin, 1972a) and similar brief blooms may occur in temperate seas (Sorokin, 1977). But such brief peaks of protozoa (and too often, the peaks of phytoplankton blooms as well) are usually missed by ecologists, who must select their sampling frequency and timing mostly by prearranged ship's schedules or by reference to weather conditions, rather than from a close monitoring of the sequence of plankton succession.

The biomass of sarcodine planktonic protozoa (amoebae, foraminifera, radiolaria) usually comprises a minor part of total microzooplankton biomass (1–5%). But in some areas of the tropical Pacific this percentage increased to 10% (Table VIII). In these areas numbers of radiolaria reached 13.5 × 10^{3}m^{-3}, of foraminifera 9.5 × 10^{3}m^{-3} in the euphotic zone. These data were obtained by reverse filtration (see above). On the other hand, counts of foraminifera in plankton net samples in the Atlantic Ocean give average numbers of 5 to 100 per m^3, and reached no more than 1.0 × 10^{3}m^{-3} (Boltovskoy, 1973; Parker, 1973).

Biomass of multicellular microzooplankton in the euphotic zones of neritic areas and in eutrophic upwelling water often varies within 100 to 300 mg m^{-3} and comprise 10 to 50% of total microzooplankton biomass. In oligotrophic and mesotrophic pelagic areas it is more usually 10 to 100 mg m^{-3} (Zaika *et al.*, 1976; Pavlovskaya, 1973a, 1976; Tumantzeva and Sorokin, 1977; Tumantzeva, 1979; Sorokin, 1977), comprising 10 to 60% of total microzooplankton biomass (Table VII).

In shallow coastal habitats of Vostok Bay, in the Japan Sea, planktonic larvae comprise the bulk of the microzooplankton populations. In spring gastropod veligers are abundant (13 × 10^{3}m^{-3}) and in summer and autumn the larvae of gastropods, decapods, cirripedes and polychaetes are also abundant (Kasjanov *et al.*, 1978). These observations are typical of other neritic and continental shelf regions.

PRODUCTION AND METABOLISM OF MICROZOOPLANKTON

Measurements of the production and metabolism of these microscopic and highly motile animals create serious methodological problems, especially in

TABLE VIII The biomass and composition of microzooplankton in layer 0–150 m at the section along the Equator in the eastern Pacific Ocean. (After Tumantzeva and Sorokin, 1977)

Position of stations along the Equator	Total microzooplankton biomass, g m^{-2}	Ratio of biomasses of microzooplankton and mesozooplankton	Biomass of separate components of microzooplankton % of its total biomass				
			Zooflagellates	Ciliates	Sarcodines	Nauplii
97°W	2.43	0.2	9	20	2	69
122°W	2.76	0.5	50	26	9	15
139°W	1.91	0.6	27	53	6	14
154°W	1.79	0.5	17	54	3	26

the evaluation of these parameters for a natural microzooplankton population. The parameters of production rate of protozoans and rotifers (coefficients P/B or generation time) have been evaluated from the generation times of species in culture (Fenchel, 1968a; Zaika, 1973) or of a natural population in bottle samples (Petrova et al., 1969). The production of microzooplankton organisms (P) may also be calculated by employing physiological methods: $P = (M \times K_2/1 - K_2)$, where M is the respiration rate, and K_2 the efficiency coefficient of production, which for rapidly growing microzooplankton animals can be accepted as being between 0.50 to 0.60 (average 0.55). The respiration rate of microzooplankton can be measured in cultures by direct Winkler oxygen titration, as well by Cartesian respirometry (Klekovsky, 1971; Klekovsky et al., 1977; Zeuthen and Hamburger, 1977). Production rates for nauplii were measured from the growth rates of specimens under laboratory conditions (Edmondson and Winberg, 1971; Grese, 1978). The generation time (T) zooflagellates (Bodo) in cultures was found to be 8 to 24 hours depending upon food quality (Gorjacheva et al., 1978). In natural sea water of the Black Sea it was 6 to 8 hours in summer (Kopylov, personal communication). Data on values for T in ciliates and meroplankton were reviewed by Zaika (1973): at optimum temperatures (around 20° C) T for ciliates varies from 3 to 48 hours, usually within the most common are range 10 to 30 hours (Zaika and Pavlivskaya, 1970). T appears to be size and temperature-dependent (Fenchel, 1968a, 1974; Lee and Fenchel, 1972). The dependence of generation time upon cell volume (V) in ciliates was expressed as: $T = K_2 \times V \times 0.44$.

The coefficients Q_{10} of dependence of T for ciliates on temperature from 4 to 20 °C varied from 2.3 to 3.5 for different species. Temperature adaptation in species inhabiting the Antarctic Ocean, temperate waters and tropical waters has been recorded; temperature tolerance ranges were: $-3 + 10°$, $-1 + 29°$, and $+3 + 39$ °C, respectively.

Measurements of T in natural populations of planktonic ciliates in lakes give lower values (T 20–40 hours). The rate T in ciliates is also dependent upon the food concentration (Pavlovskaya, 1973b).

Production rates (P/B ratios) for rotifiers (Sychaeta baltica) inhabiting coastal areas of the Black Sea is around 0.2 per day. For Oikopleura dioica this values was 0.35, and for nauplii and early copepodites 0.15 to 0.20 (Zaika, 1973; Grese, 1978).

Experimental studies of respiration rates of ciliates were reviewed by Winberg (1949). Average values are 2 to 10 μl h^{-1} per 1 mg of wet weight, depending of the size of animals. Recent studies suggest the following relationships between respiration rate R (h^{-1}) and wet body weight W: $kR = 0.716 \times W^{Q76}$ (Khlebovich, 1974); $R = 0.0676 \times W^{Q53}$ (Choria and Shubernetzki, 1978); $\log R = 0.966 + 0.4171 \times \log W$ (Laybourn and Finlay, 1976).

SOURCES OF NUTRITION AND ROLE IN FOOD CHAINS

A wide spectrum of data relating to this problem are now available. The methods employed are: the direct observations of grazing and analysis of ingested food by cytological methods or electron microscopy (Pavlovskaya, 1973b; Moestrup and Thomsen, 1976; Laval, 1971); observations of grazing rates using batch and continuous cultures, and mathematical analysis of prey-predator relationships (Gause, 1969; Hamilton and Preslan, 1970; Gold, 1971; Legner, 1973; Umorin, 1975; Berk et al., 1976; Villareal et al., 1976); the use of the radiocarbon method (Sorokin, 1968; Sorokin and Wyskwartzev, 1973; Monakov and Sorokin, 1972; Kopylov, 1977; Mamaeva and Kopylov, 1977).

These studies show that the main food resources for zooflagellates are bacteria, and dissolved organic matter (Goryacheva, 1976; Umorin, 1976; Kopylov et al., 1979), while ciliates feed mostly on bacteria and unicellular algae though they also utilise dissolved organic matter, detritus and the corpses of larger animals (Paveljeva and Mamaeva, 1976; Umorin and Klain, 1977; Canter and Lund, 1968). Experiments with benthoplanktonic ciliates demonstrate selectivity of ingestion for different species of algae (Pavlovskaya, 1971; Zaika and Pavlovskaya, 1970). Most multicellular microzooplankton animals are filter feeders, utilising bacteria and other nannoplankton organisms (Sorokin, 1978b). Ciliated forms such as trochopores or *Echinocardium* also utilise dissolved organic matter, up to 3% of body weight per day.

The food intake of microzooplankton varies from 50 to 800% of body weight per day. The dependence of a daily ration R upon body weight of ciliates W can be expressed by the equation $R = 0.65 \, W^{0.8}$. The coefficient of assimilation for growth (K_1) was found to be 0.2 to 0.5 in ciliates, depending on the type of food. Total assimilation of bacteria and algae was estimated to be 30 to 60% (Kopylov, 1977; Pavlovskaya, 1971; Sorokin, 1968, 1978b). The concentration of food required by protozoan phagotrophic forms estimated experimentally is rather high: 1 to 5 g m^{-3} of biomass (Hamilton and Preslan, 1970; Pavlovskaya, 1971). Such a level of microbial and algal biomass can only be found in eutrophic waters, or rarely in mesotrophic waters. But ciliates and zooflagellates also flourish in less productive marine planktonic habitats and even in oligotrophic tropical areas (see Table VII). This paradox could be solved if we take into account that: (a) the most intensive development of protozoa occurs during the heterotrophic phase of succession when the concentration of bacteria is highest (Sorokin, 1977, 1978a; Sorokin and Fedorov, 1977); (b) feeding of microzooplankton largely proceeds within the high biomass layers, and within the microcosms of macroaggregates, where the concentration of food is much larger than the background average which we measure with our present techniques of sampling and filtration, and (c) that the above data on the necessary levels of food concentration were obtained with species of ciliates isolated from eutrophic coastal biotopes, while oligotrophic waters are populated mostly by ciliates of genus *Strombidium* and with tintinnids that are still impossible to cultivate; these can have a lower demand in food concentration levels, and their food is composed not only of bacteria but also of zooflagellates and nannophytoplankton.

In the benthic and periphytonic biotopes where the density of bacteria is high these are a main food source of protozoa inhabiting them while some of the benthic and periphytonic protozoa are predators on other protozoan species (Fenchel, 1968b; Salt, 1967; Orlovskaya, 1973). The hystiophage ciliates which devour the tissues of dead and even of living animals play an important role in the processes of mineralisation (Fenchel, 1970; Czapic, 1974).

Some planktonic ciliates contain chloroplasts (*Mesodinium rubrum*), zoochlorellae (*Paramaecium bursaria*), or zooxanthellae, and therefore utilise the products of their photosynthesis or gradually digest them (Brown and Neilson, 1974). And, on the contrary, some flagellated algae are capable of phagotrophic feeding, and so can participate in heterotrophic processes. Among them there are some Chrysophyta, such as *Ochromonas ovalis* (Balonov and Jagodka, 1977), or Pyrrophyta, such as *Oxyrrhis marina* (Droop, 1957).

The microzooplankton itself is an important food source for larger planktonic animals, especially during its peaks during the heterotrophic phase of plankton succession (Takahashi and Moskins, 1978; Heinbockel, 1978; Berk et al., 1977; Sorokin, 1977). That ciliates are high quality food for calanoids was proved by Petipa et al. (1977) and Pavlovskaya and Pechen (1971). They compose a significant part of the total ration of omnivores such as *Metridia*, *Clausocalanus*, *Rhincalanus*, *Undinula*. During the mass development of tintinnids the stomachs of *Metridia* may be packed with their thecae (Sorokin, 1977). Meroplankton and rotifers are a food source for small marine predators such as fish larvae. Benthic protozoa are an important part of the microbenthos, comprise the main food component of settled detritus, of sediments, and of periphyton, which are consumed by detritus and sediment feeding benthic macro invertebrates and even by some fishes such as the mullets, *Mugil* (Briggs et al., 1979).

ROLE OF MICROSCOPIC ANIMALS IN TOTAL METABOLISM AND
PRODUCTIVITY OF HETEROTROPHS

By analysis of whole ecosystems in different marine and fresh water habitats, it appears that in most of them microzooplankton comprise from 5 to 25% of total heterotrophic production, and from 4 to 10% of the total respiration and energy flow, and is often greater than mesozooplankton metabolism (Table V, Figs 4 and 5). This is especially true in early stages of the heterotrophic phase of succession, when many decaying algae are present, but larger zooplankton is still scarce, being represented mostly by juvenile stages still functionally within the microzooplankton. In such cases microzooplankton respiration can be at least one order of magnitude higher than that of mesozooplankton (Sorokin and Mikheev, 1979). This phenomenon was demonstrated first by Pomeroy and Johannes (1968), who found that the respiration of microplankton concentrates exceeded by several times that of the net mesozooplankton in the same volume of water. In fact, the joint production and respiration of heterotrophic microplankton usually equals to 80 to 95% of total heterotrophic production and metabolism. So microheterotrophs together with phytoplankton comprise the bulk of the utilisable food and of energy flow within at least some plankton ecosystems (Tables III and V).

4 Ecological implications of the role of microheterotrophs

As Table III shows, microbial populations which are basically secondary producers in many cases actually participate in the formation of primary energy input in ecosystems at the first trophic level. This happens whenever allochthonous organic matter is incorporated in the metabolism of an aquatic community. The incorporation of energy from allochthonous organic matter and also from organic matter derived from previous phytoplankton blooms, proceeds mostly by microbial biosynthesis. It results in the production of new particulate microheterotroph protein. The inclusion of new production into a food chain from non-living organic matter is the most important function of microheterotrophs in the marine ecosystems. Indeed this function provides the energy coupling: (a) between terrestrial and marine ecosystems; (b) between separate biotopes (in space and in time) within the same ecosystem; (c) between ecosystems of different regions of the oceans (Sorokin, 1978a). Analysis of different types of marine ecosystems show that in most of them such a 'detritus' type of food chain predominates in which a major part of local phytoplankton production is used not by direct grazing by herbivores but by the utilisation of non-living organic matter, in fact, directly or indirectly, composes the bulk of the diet of all herbivores (Petipa, 1978; Suschenya, 1968; Finenko and Zaika, 1970). This is equally the case in pelagic plankton communities, and in estuarine and coastal ecosystems, including coral reefs (Odum and de la Cruz, 1963; Darnell, 1967; Sorokin and Fedorov, 1976; Fenchel, 1972; Sorokin, 1978b).

These conclusions have been reached by analysis of models of the energy balance of several different types of marine ecosystems, given in Figs 4 and 5 and in Table VI. The first scheme (Fig. 4) demonstrates the participation of microheterotrophs in the energy flow in the pelagic ecosystem of the Japan Sea in summer. After the formation of a thermocline nutrients are quickly utilised (Fig. 1), and the diatom population developed during the spring bloom began decaying. Only a small part of it remained above the upper limit of the thermocline. The plankton population during this phase of the seasonal succession was used by microheterotrophs. They mediated about 90% of the total energy flow. The basic energy source for planktonic community during this period was detritus formed during the spring phytoplankton bloom. The main trophic function of microheterotrophs in ecosystems of temperate waters is thus the transformation of the load of organic matter accumulated during the autotrophic phases of seasonal succession into particulate protein which during the intervening heterotrophic phases of succession of phytoplankton minima provides for the nutrition of larger zooplankton.

Analysis of the energy balance of tropical pelagic ecosystems shows that the nutritional demands of heterotrophs there are not usually satisfied by local primary production. We suppose these ecosystems to be using allochthonous dissolved and particulate organic matter formed in more productive areas and brought there by advection and the global oceanic circulation (Sorokin, 1971, 1973a).

An example of this type of energy coupling by the advection of organic matter and its subsequent use by microheterotrophs is given in Table VI, where the energy balance of the equatorial upwelling areas in the eastern central part of the equatorial Pacific are analysed. The export of excess autotrophic production westwards and its utilisation by the plankton community of the Central Pacific is discussed in detail in chapter 3.

Somewhat similar spatial energy coupling between planktonic ecosystems occurs in the Peruvian coastal upwelling (Sorokin and Mikheev, 1979; Vinogradov and Shushkina, 1978). In an intensive phytoplankton bloom in a patch of upwelled water only about 50% of phytoplankton production is used by the local juvenile heterotrophic community. Surplus organic production is advected to the frontal area at the boundary of the upwelled patch, where it is used by a rich heterotrophic community composed mainly by bacteria, protozoa, and juvenile euphausiids (Fig. 5).

The participation of allochthonous or locally accumulated non-living organic matter in ecosystem metabolism as an energy source occurs wherever total respiration of heterotrophs exceeds local primary production (Table V). The same table shows that microzooplankton production usually greatly exceeds that of mesozooplankton, especially during the early heterotrophic phase of succession.

The extremely important role of microheterotrophs in the metabolism and productivity of estuarine, coastal and benthic communities is less controversial (Corliss, 1973; Sorokin, 1973b; Rheinheimer, 1977; Joris, 1977; Lackey, 1967; Eriksson et al., 1977; Konovalova, 1972). Microbial populations in these habitats transform microbial biomass into the energy of the non-living organic matter of bottom sediments, detritus, remains of macrophytes and periphyton, as well as the reduced products of anaerobic decomposition.

In coral reef ecosystems (chapter 5), microbial periphyton which covers the surface of lime rocks, of dead corals, and of coral debris is one of the most metabolically active components of the reef system. The respiration rate of branches of dead corals covered with periphyton is comparable with that of the similar branches of living corals (Sorokin, 1979a). The periphytonic microflora and microbial populations in coral sand are the main nutritional support of the periphyton and sediment-feeding benthic fishes (Sorokin, 1978b).

In reducing layers, the ecological function of microheterotrophs is to couple the aerobic and anaerobic biotopes. Anaerobic decomposition of organic matter results in the formation of methane, H_2S, and low molecular weight organic acids, which include about 90% of the energy of the initial organic material which is available for microbes. The microbial biomass thus produced is incorporated into the food chain largely by protozoa (Sorokin, 1965, 1972b, 1978a; Mamaeva and Moiseev, 1979; Fenchel and Reidl, 1970).

In polluted marine ecosystems, microheterotrophs play the role of the main agent of self-purification. When proceeding within an intact ecosystem, which is not stressed by an overload of organic matter, nutrients or toxic substances, this process normally results in an improvement of water quality coupled with an increase in its biological productivity (Sorokin, 1978a). However, in some polluted coastal habitats the damage to the animal population inhibits the grazing and mineralisation of a mass of accumulating bacterial biomass and so water quality can continue to deteriorate. Bacteriovorous organisms, and especially the phagotrophic zooflagellates, are most resistant to stress (Gorjacheva et al., 1978; Liepa, 1978). They, therefore accomplish the function of grazers and mineralysers of microbial biomass in polluted habitats wherever the development of larger filterers is inhibited by anthropogenic stress. We have observed this phenomenon in the Venetian lagoon which is subjected to industrial and urban pollution and to hypereutrophication. Here, protozoa absolutely dominate the animal grazers. Their biomass reached as high as 0.5 to 0.8 g m^{-3}, whilst that of herbivorous zooplankton was several times less.

The respiratory and metabolic activities of microheterotrophs is also a major factor in controlling the oxygen regime, redox conditions and nutrient regeneration (Redfield *et al.*, 1963; Sorokin, 1964, 1970c, 1978a). It has long been assumed that mineralisation of organic matter and nutrient regeneration is due mostly to the activity of bacteria. More recently it has been shown that animal microheterotrophs are also an important agent in these processes by respiring 5 to 20% of total organic matter subjected to the heterotrophic decomposition (Table V). The participation of microzooplankton in the nutrient regeneration is important because: (*a*) they are all main grazers and mineralysers of bacterial biomass, which is especially important in the pelagic marine habitats where there are few fine filterers among the mesozooplankton; (*b*) they practically do not form faeces and therefore do not deplete water layers of nutrients, as do larger zooplankters; (*c*) rates of nutrient regeneration in microheterotrophes per unit of biomass is several times higher than that in mesozooplankters (Johannes, 1965).

The present status of studies of aquatic ecosystems based upon a holistic approach (Sorokin, 1979b; Shushkina and Vinogradov, 1979) is characterised by a rapid increase in the attention that is being given to microheterotrophs (Pomeroy, 1974; Rheinheimer, 1974; Zaika *et al.*, 1976; Joris, 1977; Sieburth, 1976b; Sorokin, 1978a; Kuznetzov, 1977; Sieburth *et al.*, 1978). This tendency gives us hope for further progress in the understanding of the structure and function of marine ecosystems, which forms such a necessary theoretical background for their rational management.

References

Agamaliev, F. G. (1969). Ciliates in the mesopsammon of the Caspian Sea. *In* "Comm. III Internat. Congress on Protozoology", pp. 193–194. Nauka, Leningrad

Alldredge, A. L. (1972). Abandoned larvacean houses: a unique food source for the pelagic environment. *Science* **177**, 885–887

Azam, F. and Hobson, R. E. (1977). Size distribution and activity of marine microheterotrophes. *Limnol. Oceanogr.* **22**, 492–501

Baas Becking, L. C. M. and Wood, E. J. F. (1955). Biological processes in the estuarine environment. *Proc. Koninkl. Ned. Akad. Wetenschap., Ser. B.* **60**, 88–102

Balonov, I. M. and Jagodka, S. N. (1977). On the bacterial feeding of chrysophyte algae *Ochromonas ovalis. Inform. Bull. Inst. Biol. Inland Water (Borok)* **33**, 13–18 (in Russian)

Banse, K. (1974). On the role of bacterioplankton in the tropical ocean. *Mar. Biol.* **24**, 1–6

Barber, R. T., White, A. W. and Siegelman, H. W. (1969). Evidence for a cryptomonad symbiont in the Ciliate *Cyclotrichium meuneri. J. Phycol.* **5**, 86–88

Beers, J. R. and Stewart, G. L. (1967). Microzooplankton in the euphotic zone at five locations across the California current. *J. Fish. Res. Board Can.* **24**, 2053–2068

Beers, J. R. and Stewart, G. L. (1969). The vertical distribution of microzooplankton and some ecological observations. *J. Cons. Expl. Mer.* **33**, 67–87

Beers, J. R. and Stewart, G. L. (1970). Numerical abundance and estimated biomass of microzooplankton. *Bull. Scripps. Inst. Oceanogr.* **17**, 67–87

Beers, J. R., Stevenson, M. R., Eppley, R. N. and Brooks, E. (1971). Plankton populations and upwelling off the coast of Peru, June 1969. *Fish. Bull.* **69**, 859–868

Berk, S. G., Colwell, R. R. and Small, E. B. (1976). A study of feeding response to bacterial prey by estuarine ciliates. *Trans. Amer. Microscop. Soc.* **95**, 514–520

Berk, S. G., Brownlee, D. C., Heinle, D. R., Kling, H. J. and Colwell, R. R. (1977). Ciliates as a food source for marine planktonic copepods. *Microb. Ecol.* **4**, 27–40

Bishop, J. K., Edmond, J., Ketten, D., Bacon, M. and Silker, W. (1977). The chemistry, biology and vertical flux of particulate matter in equatorial Atlantic ocean. *Deep-Sea Res.* **24**, 511–548

Boltovskoy, E. (1973). Daily vertical migration and absolute abundance of living planktonic foraminifera. *J. Foraminif. Res.* **3**, 84–94

Bowden, W. B. (1977). Comparison of two direct count techniques for aquatic bacteria. *Appl. Environ. Microbiol.* **33**, 1229–1232

Briggs, K. B., Tenore, K. R. and Hansom, R. B. (1979). The role of microfauna in detrital utilization by the polychaetes. *J. Exp. Mar. Biol. Ecol.* **36**, 225–235

Brooker, B. E. (1971). Fine structure of *Bodo saltans* and *Bodo caudatus* (Zoomastigophora, Protozoa). *Bull. Brit. Mus. Lond.* **22**, 87–102

Brown, J. A. and Nielsen, P. J. (1974). Transfer of photosynthetically produced carbohydrate from endosymbiotic *Chlorellae* to *Paramaecium bursaria*. *J. Protozool.* **21**, 569–570

Burkovsky, I. V. (1969). Quantitative study of the ecology of psammophylic ciliates of the White sea. *In* "Comm. III Internat. Congress of Protozoology", pp. 195–196. Nauka, Leningrad (in Russian)

Cachon, J. and Cachon, M. (1970). La système axopodial des Radiolaries Nasselaires. *Proc. II. Plankton Conference, Rome* 1–23

Campbell, P. G. and Baker, J. H. (1968). Estimation of bacterial production by simultaneous measurements of S^{35} and H^3 glucose uptake in the dark. *J. Fish. Res. Board Can.* **24**, 939–946

Canter, H. M. and Lund, J. W. (1968). Importance of protozoa in controlling abundance of planktonic algae. *Proc. Linn. Soc. Lond.* **179**, 203–209

Choric, F. P. and Shubernetzki, J. W. (1978). Intensity of respiration in some ciliate species. *In* "Problems of Ecology of Protozoa", pp. 66–75. Nauka, Leningrad (in Russian)

Corliss, J. O. (1973). The role of protozoa in ecological problems. *Am. Zool.* **131**, 145–148

Corliss, J. O. (1979). "The Ciliated Protozoa". Pergamon Press, Oxford

Czapic, A. (1974). Hystophagic ciliates. *Prz. Zool.* **18**, 350–354

Dale, D. (1973). Feeding of ciliate *Colpidium*. *Water Res.* **7**, 695–706

Dale, N. (1974). Bacteria in intertidal sediments. *Limnol. Oceanogr.* **19**, 509–518

Dale, T. (1978). Total, chemical and biological oxygen consumption in sediments in Lindaspollene, Western Norway. *Mar. Biol.* **49**, 333–341.

Darnell, R. M. (1967). Organic detritus in relation to estuarine ecosystem. *In* "Estuaries" (G. H. Lauff, ed.). *Publ. Am. Assoc. Adv. Sci.* **83**, 376–382

Davis, P. G., Caron, D. A. and Sieburth, J. M. N. (1978). Oceanic amoebae from the North Atlantic: Culture, Distribution and Taxonomy. *Trans. Amer. Microscop. Soc.* **97**, 73–88

Deitz, A. S., Albright, L. J. and Tuoninen, T. (1976). Heterotrophic activities of bacterioplankton and bacterioneuston. *Can. J. Microbiol.* **22**, 1699–1709

Dragesco, R. (1954). Diagnoses préliminaires des quelques ciliés nouveaux des sables. *Bull. Soc. Zool. Fr.* **79**, 62–70

Droop, M. R. (1957). Auxotrophy and organic compounds in the nutrition of marine phytoplankton. *J. Gen. Microbiol.* **16**, 3–15

Duce, R. A., Quinn, J. and Olney, C. E. (1972). Enrichment of heavy metals and organic compounds of the surface microlayer of Narragansett bay. *Science* **176**, 161–163

Edmondson, W. T. and Winberg, G. G. (1971). "A Manual on Methods for the Assessment of Secondary Production in Fresh Waters". Blackwell Scientific Publications, Oxford and Edinburgh

Ellis, W. M. (1929). Recent researches of the Choanoflagellata, freshwater and marine. *Annales. Soc. R. Zool. Belg.* **60**, 49–88

Eriksson, S., Sellei, C. and Wallström, K. (1977). The structure of the plankton community of the Öregundsgrepen. *Helgol. Wiss. Meeresunters.* **30**, 582–597

Fell, J. W. (1976). "Yeasts in Oceanic Regions". *In* "Recent Advances in Aquatic Mycology" (E. Jones, ed.) pp. 93–123. Wiley, Interscience, London

Fenchel, T. (1967). The quantitative importance of ciliates as compared with metazoans in various types of sediments. *Ophelia* **4**, 121–137

Fenchel, T. (1968a). Reproductive potential of ciliates. *Ophelia* **5**, 123–136

Fenchel, T. (1968b). The food of marine benthic ciliates. *Ophelia* **5**, 73–121

Fenchel, T. (1970). Studies on decomposition of organic detritus derived from turtle grass. *Linmol. Oceanogr.* **15**, 14–20

Fenchel, T. (1972). Aspects of decomposer food chains in marine benthos. *Verhandl. Deutch. Zool. Ges.* **65**, 14–22

Fenchel, T. (1974). Intrinsic rate of natural increase: the relationship with body size. *Oecologia* **14**, 314–326

Fenchel, T. and Reidl, R. (1970). The sulphide system: a new biotic community underneath the oxidised layer of marine sandy sediments. *Mar. Biol.* **7**, 255–268

Ferguson, R. L. and Rublee, P. (1976). Contribution of bacteria to standing crop of coastal plankton. *Limnol. Oceanogr.* **21**, 141–145

Finenko, S. S. and Zaika, V. E. (1970). Particulate organic mattter and its role in the productivity of the sea. *In* "Marine Food Chains" (J. Steele, ed.) pp. 32–44. Oliver and Boyd, Edinburgh

Fletcher, M. and Floodgate, G. D. (1973). An electron microscopic demonstration of an acid polysaccharide involved in the adhesion of solid surfaces. *J. Gen. Microbiol.* **74**, 325–334

Fournier, R. O. (1970). Studies of pigmented micro-organisms from aphotic marine environments. *Limnol. Oceanogr.* **15**, 675–682

Gause, G. F. (1969). "Struggle for Life". Hafner, New York

Gold, K. (1971). Growth characteristics of mass reared tintinnid *Tintinnipsis beroidea*. *Mar. Biol.* **8**, 105–108

Gorbunov, K. V. (1946). Cellulose bacteria as a link in the food chain of freshwater habitats. *Mikrobiologya* (Moscow) **15**, 160–168 (in Russian)

Gorjacheva, N. V. (1976). On the feeding of free living bodonides. *Trans. Inst. Biol. Inland Waters (Borok)* **31** (34), 103–112 (in Russian)

Gorjacheva, N. V., Zukov, B. F. and Mylnikov, A. P. (1978). Biology of free living Bodonides. *In* "Biology and Systematics of Lower Organisms". *Trans. Inst. Biol. Inland Waters (Borok)* **35**, 29–50 (in Russian)

Gorlenko, V. M., Dubinina, G. A. and Kuznetzov, S. I. (1977). "Ecology of Aquatic Micro-organisms". Nauka, Moscow (in Russian)

Goulder, R. (1971). Vertical distribution of some ciliated Protozoa in two freshwater sediments. *Oikos* **22**, 199–202

Grese, V. N. (1978). Production in animal populations. *In* "Marine Ecology" (O. Kinne, ed.) Vol. 4, pp. 89–114. Interscience, London and New York

Gundersen, K., Mountain, C. W., Taylor, D., Ohye, R. and Shen, J. (1972). Some chemical and microbiological observations in the Pacific ocean off the Hawaii Islands. *Limnol. Oceanogr.* **17**, 524–531

Gurds, C. R. and Bazin, M. J. (1977). Protozoan predation in batch cultures. *In* "Advances in Microbiology", Vol. 1, pp. 115–175. Academic Press, New York

Hamilton, R. D. and Presland, J. E. (1970). Observations on the continuous culture of planktonic protozoa. *J. Exp. Mar. Biol. Ecol.* **5**, 94–104

Heinbockel, J. F. (1978). Studies of the functional role of tintinnids in the Southern California Bight. II. Grazing rate of field populations. *Mar. Biol.* **47**, 191–197

Hirsh, P. and Pankratz, S. H. (1970). Study of the bacterial population in natural environments by use of submerged electron microscopy grids. *Zt. Allg. Mickobiol.* **10**, 589–605

Hobbie, J. E., Daley, R. J. and Jasper, S. (1977). Use of Nucleopore filters for counting bacteria by fluorescent microscopy. *Appl. Environ. Microbiol.* **33**, 1225–1228

Hobbie, J. E., Holm, Hansen, O. Packard, T. T., Pomeroy, L. R., Sheldon, R. W. and Thomas, J. P. (1972). A study of the distribution and activity of microorganisms in ocean waters. *Limnol. Oceanogr.* **17**, 544–555

Hollande, A. (1952). Ordre des Bodonides. *Traité de Zoologie* **1**, 669–693

Holm Hansen, O. (1969). Determination of microbial biomass in ocean profiles. *Limnol. Oceanogr.* **14**, 740–747

Hoppe, H. G. (1976). Determination of properties of actively metabolising heterotrophic bacteria by means of radioautography. *Mar. Biol.* **36**, 291–302

Hoppe, H. G. (1977). Analysis of actively metabolising bacterial population with the autoradiographic method. *In* "Microbiological Ecology of Brackish Water Environments". (G. Rheinheimer, ed.) pp. 179–197. Springer-Verlag, Berlin

Ichikawa, T. and Nishizawa, S. (1975). Particulate organic carbon and nitrogen in the eastern Pacific Ocean. *Mar. Biol.* **29**, 129–138

Jankovsky, A. V. (1967). New system of ciliated Protozoa (Ciliophors). *Trans. Zool. Inst. Acad. Sci. USSR* (Leningrad), **43**, 3–52 (in Russian)

Jannasch, H. W. (1967). Growth of marine bacteria at limiting concentration of organic carbon in sea water. *Limnol. Oceanogr.* **12**, 264–271

Jannasch, H. W. and Wirsen, C. O. (1973). Deep sea microorganisms: *in situ* response to nutrient enrichment. *Science* **180**, 641–643

Jannasch, H., Wirsen, C. O. and Taylor, C. D. (1976). Undercompressed microbial population from the deep sea. *Appl. Environ. Microbiol.* **32**, 360–367

Jassby, A. D. (1975). An evaluation of ATP estimation of bacterial biomass in the presence of phytoplankton. *Limnol. Oceanogr.* **20**, 646–648

Johannes, R. E. (1965). Influence of protozoa on nutrient regeneration. *Limnol. Oceanogr.* **10**, 434–442

Johannes, R. E. and Satomi, M. (1966). Composition and nutritive value of fecal pellets of a marine crustacean. *Limnol. Oceanogr.* **11**, 191–197

Jones, E. B. (ed.) (1976). "Recent Advances in Aquatic Mycology". Wiley, London

Joris, C. (1977). On the role of heterotrophic bacteria in marine ecosystems. *Helgol. Wiss. Meeresunters.* **30**, 611–621

Kahl, A. (1930–1935). "Urtiere oder Protozoa". Jena

Kasjanov, V. L., Konovalova, G. V., Krjuchkova, G. A. and Gorkhova, V. N. (1978). Dynamics of planktonic larvae and phytoplankton in Vostok Bay, Sea of Japan. *In* "Regularities of Distribution of Ecology of Coastal Ecosystems", pp. 157–159. U.S.S.R.–U.S.A. Symposium, Leningrad (in Russian)

Kent, W. S. (1882). "A Manual of the Infusoria". London

Khailov, K. M. (1971). "Ecological Metabolism in the Sea". Naukova Dumka, Kiev (in Russian)

Khlebovich, T. V. (1974). Rate of respiration in ciliates of different size. *Cytologya* (Leningrad) **16**, 103–105 (in Russian)

Klekovski, R. Z. (1971). Cartesian microrespirometry for aquatic animals. *Polskie Arch. Hydrobiol.* **18**, 93–114

Klekovski, R. Z., Kukina, N. J. and Tumantzeva, N. I. (1977). Respiration in the microzooplankton of the equatorial region in the eastern Pacific Ocean. *Polskie Arch. Hydrobiol.* **24**, 467–490

Konovalova, G. V. (1972). Seasonal characteristics of phytoplankton in Amursky Bay (Sea of Japan). *Oceanologya* (Moscow), **12**, 123–128 (in Russian)

Konovalova, G. V. and Rogachenko, L. A. (1974). The species composition of planktonic ciliates (*Tintinnina*) and dynamics of their number in the Amursky Bay of the Sea of Japan. *Oceanologya* (Moscow) **14**, 699–703 (in Russian)

Kopylov, A. I. (1977). On the feeding of aquatic ciliates. *Inform. Bull. Inst. Biol. Inland Waters* (*Borok*) **33**, 19–23 (in Russian)

Kopylov, A. I., Mamaeva, T. I. and Batzanin, S. F. (1979). Studies of energy balance in *Parabodo attenuatus*. *Oceanologya* (Moscow), **20**, 1073–1078 (in Russian)

Krsinic, F. (1977). Tintinnids of the eastern coast of the middle Adriatic. *Rapp. et Proc-Verb. reun. Comm. Int. expl. Mer. Mediterr.* **24**, 95–96

Kudo, R. R. (1966). "Protozoology". C. C. Thomas, Springfield, Illinois

Kuznetzov, S. I. (1977). Trends in the development of aquatic microbiology. *In* "Advances in Aquatic Microbiology", Vol. 1, pp. 1–42. Academic Press, New York

Lackey, J. B. (1967). The microflora of estuaries and their roles. *In* "Estuaries" (G. H. Lauff, ed.) Vol. 83. Publ. Am. Assoc. Adv. Sci.

Lapteva, N. A. (1976). Electron microscopy study of microflora of the Rybinsk reservoir. *Mikrobiologya* (Moscow) **45**, 547–550 (in Russian)

Laval, M. (1971). Ultrastructure et mode de nutrition du choanoflagellates. *Protistologica* **7**, 325–336

Laybourn, J. and Finlay, B. (1976). Respiratory energy losses related to weight and temperature in ciliated protozoa. *Oecologya* (Berlin) **24**, 349–355

Leadbeater, B. S. (1972). Identification by means of electron microscopy of flagellate nannoplankton from the coast of Norway. *Sarsia* **49**, 107–211

Leadbeater, B. S. (1974). Ultrastructural observations on nannoplankton collected from the coast of Jugoslavia and the Bay of Algier. *J. Mar. Biol. Assoc. U.K.* **54**, 179–196

Leadbeater, B. S. and Manton, I. (1974). Preliminary observations on the chemistry and biology of the lorica in collared flagellate. *J. Mar. Biol. Assoc. U.K.* **54**, 269–276

Lee, C. C. and Fenchel, T. (1972). Temperature responses in ciliates from Antarctic, temperate and tropical habitats. *Arch. Protistenk.* **114**, 237–244

Legner, M. (1973). Experimental approach to role of Protozoa. *Amer. Zoologist* **13**, 177–192

Lenz, J. (1977). On detritus as a food source for pelagic filter feeders. *Mar. Biol.* **41**, 39–48

Liepa, P. A. (1978). Ecology of free living protozoa in rivers of Latvia. *In* "Problems of Ecology of Protozoa", pp. 58–65. Nauka, Leningrad (in Russian)

Mabrach, A., Varon, M. and Shilo, M. (1975). Properties of marine Bdellovibrios. *Microb. Ecol.* **2**, 284–295

Mamaeva, N. V. (1980). Microzooplankton in the open part of the Black Sea. *In* "Ecosystems of Pelagic Parts of the Black Sea" (M. Vinogradov, ed.). Nauka, Moscow (in Russian)

Mamaeva, N. V. and Kopylov, A. I. (1977). On study of feeding of freshwater ciliates. *Cytologya* (Leningrad) **20**, 472–474 (in Russian)

Mamaeva, N. V. and Moiseev, E. V. (1979). Protozoa near the upper boundary of hydrogen sulphide zone of the Black Sea. *Doklady Acad. Sci. U.S.S.R.* **248**, 506–508 (in Russian)

Marshall, K. C. (1976). "Interfaces in Microbial Ecology". Harvard University Press, Cambridge, Mass.

Marshall, K., Stout, R. and Mitchell, R. (1971). Selective sorption of bacteria from seawater. *Can. J. Microbiol.* **17**, 1413–1416

McCarthy, J. J., Taylor, W. R. and Taft, J. L. (1975). The dynamics of nitrogen and phosphorus cycling in open waters of Chesapeake Bay. *In* "Marine Chemistry in the Coastal Environments" (H. Church, ed.) Vol. 18. Am. Chem. Soc. Symp. Ser.

McIntyre, F. (1974). The top millimeter of the ocean. *Sci. Amer.* **230**, 62–77

Meyer-Reil, L. A. (1977). Bacterial growth rate and biomass production. *In* "Microbial Ecology of a Brackish Water Environment" (G. Rheinheimer, ed.) pp. 223–243. Springer-Verlag, Berlin

Meyers, S. P. (1967). Yeasts from the North Sea. *Mar. Biol.* **1**, 118–123

Mileikovsky, C. A. (1971). Types of larval development in marine bottom invertebrate, their distribution and ecological significance: a re-evaluation. *Mar. Biol.* **10**, 193–213

Mileikovsky, S. A. (1972). The pelagic 'larvation' and its role in the biology of the World Ocean. *Mar. Biol.* **16**, 13–21

Moestrup, O. and Thomsen, H. A. (1976). Fine structural studies of flagellates of the genus *Bicoeca maris*. *Protistologica* **12**, 101–120

Moiseev, E. V. (1980). Zooflagellates of the open Black Sea. *In* "The Ecosystems of the Open Black Sea" (M. Vinogradov, ed.). Nauka, Moscow (in Russian)

Monakov, A. V. and Sorokin, Yu. I. (1972). Some results of investigations on nutrition of aquatic animals. *In* "Productivity Problems of Freshwaters" (Z. Kajak and Hillbricht, eds) pp. 765–774. Warszawa-Krakow

Morearty, D. J. W. (1975). A method for estimating the biomass of bacteria in aquatic sediments and its application to trophic studies. *Oecologia* **20**, 219–230

Morosovskaya, O. I. (1969). Composition and distribution of ciliates of suborder *Tintinnoinea* in the Black Sea. *In* "Biological Problems of the Oceanography of the Black Sea", pp. 110–122. Naukova Dumka, Kiev (in Russian)

Nikitin, D. I. and Kuznetzov, S. I. (1967). The use of electron microscopy for the study of aquatic microflora. *Mikrobiologya* (Moscow) **36**, 938–942 (in Russian)

Nishizawa, S. (1966). Suspended material in the sea: from detritus to symbiotic microcosmos. *Inform. Bull. Planktonology* (Japan) **3**, 1–33

Nishizawa, S. (1971). Concentration of organic and inorganic material in the surface film at the equator, 155°W. *Bull. Plankt. Soc. Japan* **18**, 42–44

Norris, R. E. (1965). Neustonic marine choanoflagellates from Washington and California. *J. Protozool.* **12**, 589–602

Odum, E. P. and Cruz, A. A. (1967). Particulate organic detritus in a Georgia salt marsh estuarine ecosystem. *In* "Estuaries" (G. H. Lauff, ed.) Vol. 83, pp. 333–388. Publ. Am. Assoc. Adv. Sci.

Orlovskaya, E. E. (1973). "Selection of Food by Protozoa". Thesis, Leningrad University (in Russian)

Packard, T. T., Healy, M. L. and Richards, F. A. (1971). Vertical distribution of the activity of the respiratory electron transport system in marine plankton. *Limnol. Oceanogr.* **16**, 60–70

Paerl, H. W. (1975). Microbial attachment of particles in marine and freshwater ecosystems. *Microb. Ecol.* **2**, 70–77

Parker, B. and Barsom, G. (1970). Ecological and chemical significance of surface microlayers in aquatic ecosystems. *Bioscience* **20**, 87–93

Parker, F. L. (1973). Living planktonic foraminifera from the Gulf of California. *J. Foraminif. Res.* **3**, 70–77

Parsons, T. R. and Strickland, J. D. N. (1962). Oceanic detritus. *Science* **136**, 313–314

Paveljeva, E. B. and Mamaeva, N. V. (1976). On the participation of ciliates in the utilization of phytoplankton in the Rybinsk reservoir. *Ecologya* (Moscow) **3**, 76–80 (in Russian)

Pavlova, E. V. and Sorokin, Yu. I. (1970). Bacterial feeding of planktonic crustacean, *Penilia*, from the Black Sea. *In* "Biology of the Sea", pp. 166–182. Naukova Dumka, Kiev (in Russian)

Pavlovskaya, T. V. (1971). "Feeding and Multiplication of Mass Species of Ciliates of the Black Sea". Thesis, Inst. Biol. South Seas, Sebastopol (in Russian)

Pavlovskaya, T. V. (1973a). Microzooplankton of the coastal waters of the Black Sea. *In* "Results of the Symposium on the Investigation of the Black and Mediterranean Seas", pp. 135–137. Naukova Dumka, Kiev (in Russian)

Pavlovskaya, T. V. (1973b). Influence of feeding conditions upon the feeding and reproduction of ciliates. *Zoological J.* (Leningrad) **52**, 1451–1457 (in Russian)

Pavlovskaya, T. V. (1976). Distribution of microzooplankton in the coastal waters of the Black Sea. *In* "Biological Studies in the Black Sea", pp. 75–83. Naukova Dumka, Kiev (in Russian)

Pavlovskaya, T. V. and Pechen, G. A. (1971). Ciliates as a food for some mass species of marine zooplankton. *Zoological J.* **50**, 633–640 (in Russian)

Perfiljev, B. V. and Gabe, D. P. (1969). "Capillary Methods of Investigating of Micro-organisms". Oliver and Boyd, Edinburgh

Petipa, T. S. (1978). Matter accumulation and energy expenditure in planktonic ecosystems at different trophic levels. *Mar. Biol.* **49**, 285–294 (in Russian)

Petipa, T. S., Pavlova, E. V. and Sorokin, Yu. I. (1971). Studies of the feeding of mass species of plankton of the tropical Pacific with the aid of ^{14}C. *In* "Functioning of Pelagic Communities in the Tropical Regions of the Ocean" (M. Vinogradov, ed.) pp. 123–141. Nauka, Moscow (in Russian)

Petipa, T. S., Monakov, A. V., Pavljutin, A. P. and Sorokin, Yu. I. (1974). Feeding and energy balance in marine copepods. *In* "Biological Productivity of the South Seas", pp. 136–152. Naukova, Dumka, Kiev (in Russian)

Petipa, T. S., Monakov, A. V., Sorokin, Yu. I., Voloshina, G. V. and Kukina, I. V. (1977). Balance of substance and energy in copepods of tropical upwellings. *Polskie Arch. Hydrobiol.* **24** (suppl.), 413–430 (in Russian)

Petrova, M. N., Bochkareva, N. A. and Salova, N. V. (1969). Rate of multiplication of planktonic ciliates in a lake. *In* "Second Symposium on Energy and Matter Turnover in Lakes". Listvinichnoe, Baikal (in Russian)

Petrushevskaya, M. G. (1971). Radiolarians *Nassellaria* in the plankton of the

World Ocean. *In* "Radiolarians of the World Ocean". Acad. Sci. U.S.S.R., Leningrad (in Russian)

Pomeroy, L. R. (1974). Oceanic productivity – a changing paradigm. *Bioscience* **24**, 499–501

Pomeroy, L. R. and Johannes, R. E. (1968). Occurrence and respiration of ultraplankton in the upper 500 m of the ocean. *Deep-Sea Res.* **15**, 381–391

Rasumov, A. S. (1962). The microbial plankton of water. *Trans. Hydrobiol. Soc.* (Moscow) **12**, 60–190

Redfield, A. C. and Ketchum, B. H. and Richards, F. A. (1963). The influence of organisms on the composition of sea water. *In* "The Sea" (M. N. Hill, ed.) Vol. 3, pp. 26–77. Interscience, London and New York

Rheinheimer, I. (1974). "Aquatic Microbiology". John Wiley, New York

Rheinheimer, G. (1977). "Microbial Ecology of a Brackish Water Environment". Springer-Verlag, Berlin

Rodina, A. G. (1949). Bacteria as a food for aquatic animals. *Priroda* (Moscow) **10**, 23–26 (in Russian)

Rodina, A. G. (1963). Microbiology of detritus of lakes. *Limnol. Oceanogr.* **8**, 388–393

Romanenko, V. I. (1964). Heterotrophic CO_2 assimilation by aquatic microflora. *Mikrobiologya* (Moscow) **34**, 679–683 (in Russian)

Roth, I. F., Ahearn, D. G. and Fell, J. W. (1962). Ecology and taxonomy of yeasts from various marine substrates. *Limnol. Oceanogr.* **7**, 178–185

Salt, G. W. (1967). Predation in experimental protozoan population. *Ecol. Monogr.* **37**, 113–144

Schneider, J. (1977). Fungi. *In* "Microbial Ecology of a Brackish Water Environment" (G. Rheinheimer, ed.) pp. 90–102. Springer-Verlag, Berlin

Scuja, H. (1956). Taxonomische und Biologische Studien über das Phytoplankton Schwedischer *Binnengewässer, Nova Acta Soc. Sci. Uppsala*, ser. 4, **16**, 3, 1–404

Seki, H. (1971). Microbial clumps in seawater in the Saanich Inlet. *Mar. Biol.* **9**, 4–8

Seki, H. (1972). The role of micro-organisms in the marine food chain with reference to organic aggregates. *Mem. Ital. Ist. Idrobiol.* **29**, 245–259

Sieburth, J. McN. (1975). "Microbial Seascapes". Baltimore, Maryland

Sieburth, J. McN. (1976a). Dissolved organic matter and heterotrophic microneuston in the surface microlayers of the North Atlantic. *Science* **194**, 1415–1418

Sieburth, J. McN. (1976b). Bacterial substrates and productivity in marine ecosystems. *Ann. Rev. Ecol. Syst.* **7**, 259–285

Sieburth, J. McN. (1977). Report on biomass and productivity of microorganisms in planktonic ecosystems. *Helgol. Wiss. Meeresunters.* **30**, 697–704

Sieburth, J. McN., Smetacek, V. and Lenz, J. (1978). Pelagic ecosystem structure: heterotrophic compartments of plankton and their relationship to plankton size fractions. *Limnol. Oceanogr.* **23**, 1256–1261

Sieburth, J. McN., Johnson, K. M., Burnery, C. M. and Lavone, K. M. (1977). Estimation of *in situ* rates of heterotrophy using diurnal changes of organic matter. *Helgol. Wiss. Meeresunters.* **30**, 565–574

Silver, M. W., Shanks, A. L. and Trent, J. (1978). Marine snow: microplankton habitat and source of small scale patchiness in plankton populations. *Science* **201**, 371–373

Simidu, U. (1974). The taxonomy of marine bacteria. *In* "Marine Microbiology" (N. Taga, ed.) pp. 45–65. Tokyo University Press, Tokyo

Sorokin, Yu. I. (1964). Primary production and microbial activity in the Black Sea. *J. du Conseil Expl. Mer.* **29**, 41–65

Sorokin, Yu. I. (1965). On the trophic role of chemosynthesis and bacterial biosynthesis in water bodies. *Mem. Ist. Ital. Idrobiol.* **18**, 187–205

Sorokin, Yu. I. (1968). The use of ^{14}C in the study of nutrition of aquatic animals. *Rep. Intern. Assoc. of Theor. and Appl. Limnol.* **16**, 1–41

Sorokin, Yu. I. (1969). Experimental studies of the influence of pressure and temperature upon the metabolism of deep sea microflora. *Mikrobiologya* (Moscow) **38**, 868–877 (in Russian)

Sorokin, Yu. I. (1970a). Characteristics of the number, activity and production of bacteria in bottom sediments of the central Pacific. *Oceanologya* (Moscow) **10**, 1055–1065 (in Russian)

Sorokin, Yu. I. (1970b). On the estimation of activity of heterotrophic bacteria in the ocean with the use of labelled organic matter. *Mikrobiologya* (Moscow) **39**, 149–156 (in Russian)

Sorokin, Yu. I. (1970c). Aggregation in marine bacterioplankton. *Doklady Acad. Sci. U.S.S.R.* (Moscow) **192**, 905–907 (in Russian)

Sorokin, Yu. I. (1970d). Interrelation between sulphur and carbon turnover in meromictic lakes. *Arch. Hydrobiol.* **66**, 391–420

Sorokin, Yu. I. (1971). On the role of bacteria in the productivity of tropical oceanic waters. *Int. Rev. ges. Hydrobiol.* **56**, 1, 1–48

Sorokin, Yu. I. (1972a). Biological productivity of Rybinsk reservoir. *In* "Productivity Problems of Freshwaters" (Z. Zajak and A. Hillbricht, eds) pp. 293–304. Warszawa-Krakov

Sorokin, Yu. I. (1972b). The bacterial population and the process of hydrogen sulphide oxidation in the Black Sea. *J. Cons. perm. int. Explor. Mer.* **34**, 423–454

Sorokin, Yu. I. (1972c). Microbial activity as a biochemical factor in the ocean. *In* "The Changing Chemistry of the Oceans" (D. Dyrssen and D. Jagner, eds) pp. 189–204. Almquist, Stockholm

Sorokin, Yu. I. (1973a). Data on biological productivity of the western tropical Pacific Ocean. *Mar. Biol.* **20**, 177–196

Sorokin, Yu. I. (1973b). Microbiological aspects of productivity of coral reefs. *In* "Biology and Geology of Coral Reefs" (R. Endean, ed.) pp. 17–45. Academic Press, London and New York

Sorokin, Yu. I. (1975). Heterotrophic microplankton as a component of marine ecosystems. *J. gen. Biol.* (Moscow) **36**, 716–730 (in Russian)

Sorokin, Yu. I. (1977). The heterotrophic phase of plankton succession in the Sea of Japan. *Mar. Biol.* **41**, 107–117

Sorokin, Yu. I. (1978a). Decomposition of organic matter and nutrient regeneration. *In* "Marine Ecology" (O. Kinne, ed.) Vol. 4, pp. 501–616. Interscience, London and New York

Sorokin, Yu. I. (1978b). Microbial production in the coral reef community. *Arch. Hydrobiol.* **83**, 281–323

Sorokin, Yu. I. (1978c). Characteristics of primary production and heterotrophic microplankton in the Peruvian upwelling area. *Oceanologya* (Moscow) **18**, 3–9 (in Russian)

Sorokin, Yu. I. (1979a). On the methodology of studies of lake ecosystems. *Arch Hydrobiol.* in press

Sorokin, Yu. I. (1979b). Microbial populations of bottom sediments and in periphyton of coral reefs. *J. gen. Biol.* **41**, 241–252 (Moscow) (in Russian)

Sorokin, Yu. I. (1979c). Red tide in the Peruvian upwelling area. *Doklady Acad. Sci. U.S.S.R.* **249**, 253–256 (in Russian)

Sorokin, Yu. I. (1979d). Zooflagellates as a component of communities in eutrophic and oliogotrophic waters of the Pacific ocean. *Oceanologya* (Moscow) **20**, 484–488 (in Russian)

Sorokin, Yu. I. (1979e). On the method of concentration of phytoplankton. *Hydrobiological J. Kiev* **15**, 71–76 (in Russian)

Sorokin, Yu. I. and Donato, N. (1975). On the carbon and sulphur metabolism in the meromictic Lake Faro (Sicily). *Hydrobiologya* **47**, 241–252

Sorokin, Yu. I. and Fedorov, V. K. (1976). Productivity of microplankton in the northern part of the Tatarsky Strait. *Biology of the Sea* (*Vladivostok*) **5**, 48–56 (in Russian)

Sorokin, Yu. I. and Kadota, H. (1972). "On the Characteristics of Microbial Production in Freshwaters", pp. 1–112. IBP Handbook 23. Blackwell, London

Sorokin, Yu. I. and Kogelschatz, J. (1979). Analysis of heterotrophic microplankton in upwelling areas. *Hydrobiologya* **66**, 195–208

Sorokin, Yu. I. and Kovalevskaya, R. Z. (1980). Biomass and production of the aerobic zone of the Black Sea. *In* "Ecosystems of Pelagic Parts of the Black Sea" (M. Vinogradov, ed.). Nauka, Moscow (in Russian)

Sorokin, Yu. I. and Lazarev, S. (1978). Comparative evaluation of two methods of estimating the biomass of planktonic microflora. *Oceanologya* (Moscow) **18**, 358–364 (in Russian)

Sorokin, Yu. I. and Mamaeva, T. I. (1979). Microbial production and decomposition of organic matters of Pacific ocean near the Peruvian coast. *In* "Ecosystems of Pelagic areas of Peruvian Upwelling" (M. Vinogradov, ed.). Nauka, Moscow (in Russian)

Sorokin, Yu. I. and Mikheev, V. N. (1979). On the characteristics of the Peruvian upwelling ecosystem. *Hydrobiologya* **62**, 165–189

Sorokin, Yu. I. and Paveljeva, E. D. (1977). On structure and functioning of the ecosystem in a salmon lake. *Hydrobiologya* **57**, 25–48

Sorokin, Yu. I. and Wyshkwartzev, D. I. (1973). Feeding on dissolved organic matter by some marine animals. *Aquaculture* **2**, 141–148

Sorokin, Yu. I., Paveljeva, E. B. and Vasiljeva, M. J. (1977). Productivity and the trophic role of bacterioplankton in the area of equatorial divergence. *Polskie Arch. Hydrobiol.* **24** (suppl.), 241–260

Sorokin, Yu. I., Petipa, T. S. and Pavlova, E. R. (1970). The quantitative evaluation of the food importance of marine bacterioplankton. *Oceanologya* (Moscow) **10**, 332–340 (in Russian)

Steemann Nielsen, E. (1972). The rate of primary production and the size of standing stock of zooplankton in the ocean. *Int. Rev. Ges. Hydrobiol.* **57**, 513–516

Sukhanova, I. N. and Rat'kova, T. N. (1977). Comparison of the number of phytoplankton in samples collected by the technique of double filtration and by standard sedimentation method. *Oceanologya* (Moscow) **17**, 691–693 (in Russian)

Suschenya, L. M. (1968). Detritus and its role in the aquatic productivity. *Hydrobiologya* **4**, 77–84 (in Russian)

Takahashi, M. and Moskins, K. D. (1978). Winter conditions in marine plankton populations on Saanich Inlet II. Microzooplankton. *J. exp. Mar. Biol. Ecol.* **32**, 27–38

Tchepurnova, E. A. (1976). Characteristics of the intensity of metabolism in marine bacterioplankton. *In* "Biology the of Sea", Vol. 39, 12–19. Naukova Dumka, Kiev

Tchorik, F. P. (1968). "Free Living Ciliates in the Water Bodies of Moldavia'. Ac. Sci. Moldavia Republic, Kishenev (in Russian)

Thomsen, H. A. (1973). Studies of the choanoflagellates. I. Silicified choanoflagellates. *Ophelia* 12, 1–26

Thorson, G. (1961). Length of pelagic larval life in marine bottom fauna as related to larval transport by ocean currents. *In* "Oceanography" (M. Sears, ed.). Amer. Assoc. Adv. Sci., Washington

Throndsen, J. (1969). Flagellates of Norwegian coastal waters. *Nytt. Mag. Bot.* 16, 161–216

Throndsen, J. (1974). Plankton choanoflagellates from North Atlantic waters. *Sarsia* 56, 95–122

Todd, R. L. and Kerr, T. J. (1972). Scanning electron microscopy of microbial cells on membrane filters. *Appl. Microbiol.* 23, 6, 1160–1162

Treguboff, G. (1978). Foraminifera and Ciliata. *In* "Manuel de Planktonologie Méditerranéenne" (G. Treguboff and M. Rose, eds) Vol. 1, pp. 129–139, 229–232. Paris

Tsiban, A. V. (1970). "Bacterioplankton and Bacterioneuston in Coastal Waters of the Black Sea". Naukova Dumka, Kiev (in Russian)

Tumantzeva, N. I. (1979). Microzooplankton in pelagic areas of the Peruvian upwelling. *In* "Ecosystems of Pelagic Areas of Peruvian Upwelling" (M. Vinogradov, ed.). Nauka, Moscow

Tumantzeva, N. I. and Sorokin, Yu. I. (1977). Microzooplankton of the equatorial upwelling in the Pacific Ocean. *Polskie Arch. Hydrobiol.* 24 (suppl.), 271–280

Uhlig, G. (1964). Eine einfache Methode zur Extraction der valigen Microfauna aus Marinen Sedimenten. *Helgol. Wiss. Meeresunters.* 11, 178–185

Umorin, P. P. (1975). On the calculation of the rate of grazing of bacteria by ciliates in the experimental conditions. *Inform. Bull. Inst. Biol. Inland water* (*Borok*) 28, 31–34 (in Russian)

Umorin, P. P. (1976). Relationships of bacteria and zooflagellates during the decomposition of organic matter. *J. gen. Biol.* (Moscow) 37, 6 (in Russian)

Umorin, P. P. and Klain, N. P. (1977). The influence of Protozoa upon the decomposition of organic matter by bacteria. *J. gen. Biol.* (Moscow) 38, 4 (in Russian)

Vacelet, E. (1961). La faune infusorienne des sables à amphioxus des environs de Marseille. *Bull. Inst. Oceanogr.* (Monaco) 1202, 1–12

Valkanov, A. (1970). Beitrag zur Kenntnis der Protozoen des Schwarzen Meers. *Zool. Anz.* 184, 241–290

Villarreal, E., Canale, R. and Akcasu, Z. (1976). Transport equations of microbial predator-prey community. *Microb. Ecol.* 3, 31–143

Vinogradov, M. E. and Shushkina, E. A. (1979). Some development patterns of plankton communities in the upwelling areas of the Pacific Ocean. *Mar. Biol.* 48, 357–366

Wangersky, P. J. (1976). The surface film as a physical environment. *Ann. Rev. Ecol. Syst.* 7, 161–176

Wangersky, P. J. (1977). The role of particulate matter in the productivity of surface waters. *Helgol. Wiss. Meeresunters.* 30, 546–564

Watson, S. N., Novitsky, T. J., Quinby, H. L. and Valois, F. W. (1977). Determination of bacterial number and biomass in the marine environments. *Appl. Environ. Microbiol.* 33, 940–946

Weise, W. and Rheinheimer, G. (1978). Scanning electron microscopy and epifluorescent investigation of bacterial colonisation of marine sand sediments. *Microb. Ecol.* **4**, 175–188

Wheeler, J. R. (1975). Formation and collapse of surface films. *Limnol. Oceanogr.* **20**, 338–342

White, A. W., Sheath, R. G. and Hellebust, J. A. (1977). A red tide caused by the ciliate *Mesodinium rubrum*. *J. Fish. Res. Board Can.* **34**, 413–416

Williams, P. M. (1967). Organic carbon nitrogen and phosphorus in surface waters. *Deep-Sea Res.* **14**, 791–800

Williams, P. M. and Carlucci, A. F. (1976). Bacterial utilisation of organic matter in the deep ocean. *Nature (Lond.)* **262**, 810–811

Winberg, G. G. (1949). Intensity of respiration in protozoa. *Adv. Mod. Biol.* (Moscow) **2** (5), 226–245 (in Russian)

Wirsen, C. O. and Jannasch, H. W. (1975). Activity of marine psychrophylic bacteria at elevated hydrostatic pressures and low temperatures. *Mar. Biol.* **31**, 201–208

Wood, E. J. F. (1965). "Marine Microbial Ecology". Chapman and Hall, London

Young, L. I. (1978). Bacterioneuston examined with critical point drying and transmission electron microscopy. *Microb. ecol.* **4**, 267–277

Zaika, V. E. (1973). "Specific Production of Aquatic Invertebrates". Wiley Interscience, London

Zaika, V. E., Averina, T. Yu., Ostrovskaya, N. A. and Zalkina, A. V. (1976). "Distribution of Marine Microzooplankton". Naukova Dumka, Kiev (in Russian)

Zaika, V. E. and Pavlovskaya, T. V. (1970). Feeding of marine protozoa with unicellular algae. *In* "Biology of the Sea", Vol. 19, pp. 80–90. Naukova Dumka, Kiev (in Russian)

Zaitzev, Yu. P. (1970). "Marine Neustonology". Naukova Dumka, Kiev (in Russian)

Zeuthen, E. and Hamburger, K. (1977). Microgasometry with single cells using ampulla divers operated in density gradients. *Tech. Biochem. Biophys. Morphol.*, Copenhagen **3**, 59–79

Zimmermann, R. (1977). Estimation of bacterial number of epifluorescent microscopy and scanning electron microscopy. *In* "Microbial Ecology of a Brackish Water Environment" (G. Rheinheiner, ed.) pp. 90–102. Springer-Verlag, Berlin

ZoBell, C. E. (1946). "Marine Microbiology". Chronica Bot. Co. Waltham, Mass.

Zukov, B. F. (1971). Systematics of free living zooflagellates of suborder bodonina (Hollande). *Trans. Instit. Biol. Inland Waters Acad. Sci. U.S.S.R.* **21–24**, 241–283 (in Russian)

Zukov, B. F. (1973). Colourless flagellates. *Hydrobiol. J.* (Kiev) **9**, 28–31 (in Russian)

Zukov, B. F. (1975). A new flagellate (*Bodonina* Holl.). *Inform. Bull. Inst. Biol. Inland Waters (Borok)* **25**, 23–25 (in Russian)

Zukov, B. F. (1979). Systematics of colourless zooflagellates of the order Bicosoecida (Zoomastigophorea). *In* "Biology and Systematics of the Lowest Organisms" (M. Kamshilov, ed.) pp. 3–28. Nauka, Moscow (in Russian)

11 Autotrophic Production of Particulate Matter

RICHARD W. EPPLEY

1 Introduction

This chapter is concerned primarily with the regulation of primary production in the sea emphasising those areas where the euphotic zone is essentially two-layered with an upper nutrient-limited layer and a lower, light-limited layer (Dugdale and Goering, 1967). Such regions include much of the subtropical and tropical ocean. Upwelling regions within them, where nutrient-rich water reaches the surface, are treated in chapters 2 and 3. My emphasis is on understanding the driving forces for production, the mechanisms, and more especially the rates of nutrient input into the euphotic zone.

I will refer frequently to the concepts of new and regenerated production (Dugdale and Goering. 1967). New production is that resulting from allochthonous nutrient inputs to the euphotic zone, such as from deep waters, river inputs, atmospheric deposition and including the fixation of molecular nitrogen. Regenerated production is that resulting from the recycling of nutrients within the surface waters. In the case of nitrogenous nutrients these processes can, with certain reasonable assumptions, be separately measured; new production as the assimilation of nitrate and N_2 (as acetylene reduction) and regenerated production as the assimilation of ammonium, urea and other forms of organic N using ^{15}N isotopic labels. This assumes, of course, that nitrate and N_2 are not formed to an appreciable extent by recycling processes in the euphotic zone, and that ammonium is not entering as an allochthonous input from the bottom sediments or the atmosphere.

The production dynamics of temperate and boreal seas will be treated only briefly in the context of light and stratification of the water column as necessary conditions for production.

The vertical stratification of the phytoplankton is of current interest and some examples are presented of subsurface chlorophyll maximum layers. The species assemblages forming these and the mechanisms apparently responsible for them appear to vary somewhat over time and from place to place.

I will briefly discuss new production as quantitatively equivalent to the organic matter which can be exported from the production system of the

343

euphotic zone without its running down. New production may be quantitatively equivalent to the sinking flux of particulate matter from the euphotic zone and therefore the input source term for metabolism in the deeper waters (Eppley and Peterson, 1979). Finally, I will propose a possible index for the quantitative evaluation of the extent of grazing control of primary production.

Reid (1962) showed that zooplankton volumes in the Pacific were closely related to phosphate concentration at 100 m depth. Zooplankton volumes were used as surrogate, in the absence of adequate phytoplankton data, as an indication of production. Indeed meridional sections of nutrient concentrations through the major oceans show high nutrient levels in the surface waters of boreal and temperate seas, depression of the nutrient isopleths across the anticyclonic central gyres, and doming of the isopleths in the equatorial divergences (McGowan and Williams, 1973). The depth of nutrient isopleths roughly parallel the production rates shown in large scale maps of ocean production as in Koblenz-Mishke *et al.* (1970). The large scale features of the circulation clearly establish the biological provinces seen in zooplankton species distribution (Reid *et al.*, 1978) and in primary production (Steemann Nielsen, 1955). Typical production rates for inshore and offshore areas of the oceans are shown in Table I.

Recently the relationships between production and the vertical position of nutrient concentration gradients have been examined in several tropical and subtropical sea areas. Some examples will follow.

The physical mechanisms whereby nutrients are injected from deep water into the upper, nutrient-depleted layer in the tropical and subtropical regions have become an important consideration in studies of the production biology of these regions. These mechanisms include Ekman transport (i.e. wind-driven upwelling); upwelling at current divergences, horizontal advection of allochthonous water as in eastern boundary currents, river inputs, Gulf Stream and Kuroshio rings, nitrogen fixation and eddy diffusion (McGowan and Hayward, 1978).

2 Primary production in tropical and subtropical seas

Phytoplankton standing stocks and production are very low in the central, subtropical and tropical oceans with gyre-like anticyclonic circulations, and are correlated inversely with the depth of the euphotic zone (Steemann Nielsen, 1955; Lorenzen, 1976). Most of the incident sunlight is absorbed by the water rather than by the photosynthetic apparatus of the phytoplankton cells.

The changing depth of seasonal wind mixing of the surface layers may cause increased nutrient inputs and seasonal bursts of production at the poleward limits of these waters, as at 35°N near Bermuda in the northern Sargasso Sea

TABLE I Annual phytoplankton production in the world ocean in offshore (depth >200 m) and inshore (depth <200 m) zones. From Platt and Subba Rao (1975). Production in coastal upwelling areas, including the Antarctic ocean, is listed under shelf production

Ocean	Area (10^6 km^2)			Primary production				
	Total	Shelf	Offshore	Shelf offshore (gCm^{-2}day^{-1})		Shelf offshore (10^9tons C yr^{-1})		Total (19^9tons C yr^{-1})
Indian	73.82	2.80	71.02	0.71	0.23	0.725	5.875	6.60
Atlantic	92.57	8.65	83.92	0.41	0.28	1.295	8.461	9.76
Pacific	177.56	10.67	165.89	0.52	0.15	2.037	9.360	11.4
Antarctic[a]	23.8–11.8	4.80	6.99	0.89				3.3
Arctic	13.10	6.11						0.013
Total						4.057	23.692	31.1

[a] A more recent estimate of total production in open waters of the Antarctic is 0.65×10^9 tons C yr^{-1} (Holm-Hansen et al., 1977)

(Menzel and Ryther, 1961). However winter mixing apparently does not extend as deep as the nutrient concentration gradients (~ 150 m) in the central N. Pacific at 28°N (McGowan and Hayward, 1978). The multiple year time series described in the two reports cited above indicate differences in production between years. Greater vertical mixing was proposed to account for the years of higher production: a year of greater wind mixing in 1957 off Bermuda (Menzel and Ryther, 1961) and of shallower maximum frequency of temperature inversions and variance in isotherm depths, perhaps related to shear-induced turbulence or breaking internal waves in the central N. Pacific gyre (McGowan and Hayward, 1978).

PRODUCTION AND VERTICAL NUTRIENT GRADIENTS

Natural variations in the depth of nutrient concentration gradients has allowed the relation between these depths and phytoplankton stocks and activities to be examined. In the Gulf of Guinea in the eastern tropical Atlantic (0°–15°S) Herbland and Voituriez (1977, 1979) found a strong inverse correlation between phytoplankton production and the depth of the nitrate concentration gradient (Table II). This depth, Z_n, was coincident with the oxycline and the chlorophyll maximum. The primary nitrite maximum was 10–12 m below the top of the nitracline. In the Southern California Bight (31–32°N) variations in the depth of the nitracline accounted for about 50% of the variability in phytoplankton concentration and production rate per m^3 (Eppley et al., 1979).

Production under unit area of sea surface was a linear function of Z_n in the eastern Tropical Atlantic between latitudes approximately 15°N–15°S (Table II). In the Southern California Bight, 31°N–33°N, production per m^2 was not significantly correlated with Z_n. Production per unit volume was a non-linear function of Z_n, however (Table II), as would be expected if vertical diffusive transport of nutrients were the driving force for production (Eppley et al., 1979).

Voituriez and Herbland (1979) found also that the nutrient enrichment of the surface layer in the Gulf of Guinea could be related to the high salinity core of the equatorial undercurrent from October to June. They found a linear, negative relationship between the nitrate enrichment of the surface layer and the salinity of the core as if vertical shear between the westward south equatorial current and the eastward equatorial undercurrent resulted in the observed mixing and nutrient transport to the surface layer.

VERTICAL NUTRIENT TRANSPORT AS A DRIVING FORCE FOR PRODUCTION

Ambient concentrations of nutrients such as ammonium and nitrate are essentially undetectable with present methods in the surface waters of tropical and subtropical seas (McCarthy and Carpenter, 1975). Expected concen-

TABLE II Regression equations relating phytoplankton parameters to the depth of the nitrate concentration gradient, Z_n. Z_n was defined as the depth at which nitrate first began to increase. In the eastern tropical Atlantic the depth of the chlorophyll maximum layer, total chlorophyll a integrated over the depth of the euphotic zone, and primary production are linear functions of Z_n when Z_n is between 15 and 100 m depth (from Herbland and Voituriez, 1979). In the Southern California Bight, the phytoplankton standing stock and primary production rate on a volume basis varied in a linear fashion with the reciprocal of Z_n from Eppley et al., 1979). The number of depth profiles (n) and the correlation coefficient for the regression equations (r) are indicated

Equation	r	n
Depth of Ch1 Max $= 0.95 \, Z_n + 3.6$	0.95	126
Integrated ch1$^a = -0.17 \, Z_n + 22.5$ (mg m^{-2})	-0.80	56
Production$^a = -0.87 \, Z_n + 90.2$ (mgCm^{-2}h^{-1})	-0.84	17
Phytoplankton C $= 23 + 2113 \, (1/Z_n)$ (mgCm^{-3})	0.67	112
Production $= -8.9 + 608 \, (1/Z_n)$ (mgm^{-3}day^{-1})	0.72	112

^{a} for Z_n between 15 m and 100 m depth range

trations of ammonium, for example, are of the order 50 nanomolar (Eppley et al., 1977) or somewhat less, based on results with N-limited chemostat cultures (Caperon and Meyer, 1972; Mickelson et al., 1979). Thus, attempts to relate phytoplankton growth to ambient nutrient levels have been unsuccessful. Menzel and Ryther (1961) suggested that ambient concentrations did not serve as an index of nutrient availability and that vertical diffusion may account for the enrichment of the surface waters while the nutrients are consumed as quickly as they become available. Recent comparisons of production and nutrient transport across the nitracline are consistent with this idea.

There is historical precedent for this view as well. For example, Riley (1956) and Sapozhnikov and Galerkin (1971) have estimated production from calculations of vertical transport of nutrients into the euphotic zone. In the recent work, apparent eddy diffusivities of nitrate were calculated from the nitrate concentration gradient at the top of the nitracline and the measured uptake rate of ^{15}N-nitrate above that depth by King and Devol (1979), for the Costa Rica Dome area of the eastern tropical Pacific, and by Eppley et al. (1979) for the Southern California Bight. Values were of the order 0.05– 0.5 cm^{2}s^{-1} and varied, as expected, with the magnitude of the thermal gradient at that depth (Table III).

TABLE III   Apparent vertical eddy diffusivity coefficients (K_{z_n}) for the transport of nitrate into N-depleted surface layers assuming steady state (i.e. nitrate input to the surface layer = nitrate uptake in that layer), negligible vertical advection of nitrate and negligible regeneration of nitrate via nitrification in the surface layer

Equation	Reference
Eastern tropical Pacific $\ln K_Z = -1.34 \ln (10^6 E) + 4.774$ $(cm^2 s^{-1})$	King and Devol (1979)
Southern California Bight $K_Z = -18.0 \dfrac{\Delta T}{\Delta Z} + 3.34$ $(m^2 day^{-1})$	Eppley et al. (1979)

E = stability = 10^{-3} $(\Delta \sigma_t / \Delta z)$ with z in meters. 1 $m^2 day^{-1}$ = 0.116 $cm^2 s^{-1}$

   While the nitracline is below the euphotic zone in the N. Pacific central gyre (28°N, 155°W), McGowan and Hayward (1978) found evidence of increased vertical mixing during times of increased production there also, as noted earlier. Measurements there of the depth variation of microstructure in temperature and salinity suggest vertical eddy diffusion coefficients only a little higher than molecular diffusion in summer (Gregg et al., 1973) and only somewhat higher in winter and spring (Gregg, 1977).

   The input of nitrate into the euphotic zone of the Southern California Bight was inferred from the rate of nitrate assimilation. This new production varied with total production, measured as carbon assimilation and with regenerated production measured as ammonium assimilation (Eppley et al., 1979). Thus nitrate input appears to drive both new and regenerated production. Total production and phytoplankton specific growth growth rates were directly proportional to the ratio new production/total production (Fig. 1) over a range of total production rates up to 200 $gCm^{-2}y^{-1}$ in various subtropical and tropical waters (Eppley and Peterson, 1979). Higher rates of production are largely restricted to the nutrient-rich waters of the temperate and boreal oceans and to upwelling areas.

ROLE OF NITROGEN FIXATION IN SUBTROPICAL AND TROPICAL SEAS
Both Oscillatoria (Trichodesmium) species and Richelia spp. (endosymbionts in Rhizosolenia spp. and certain other oceanic diatoms) are capable of assimilating molecular nitrogen (Mague, 1977). This constitutes another pathway for new production. Its quantitative importance has been examined in several recent studies in the Atlantic and Pacific oceans with emphasis on Oscillatoria spp. This new production is doubtless important in re-charging

oligotrophic waters with deep nitrate gradients. However, the magnitude of N_2 fixation appears quite small in the central N. Pacific (Mague *et al.*, 1977) and in the Sargasso Sea (Carpenter and McCarthy, 1975), even in comparison with new production from nitrate. Molecular nitrogen assimilation can account for the full N-requirement of *Oscillatoria thiebautii*. However, the growth rate of this alga is very low and generation times are 15 days or longer (McCarthy and Carpenter, 1979). Long generation times can be inferred for other oceanic N-fixers as well, judging from ratios of N_2 fixation rate/cellular N content (Mague, 1977, Table 3.2). Yet *Oscillatoria* contributed as much as 60 % of the chlorophyll and 20 % of the primary production in the upper 50 m of the eastern Caribbean Sea (Carpenter and Price, 1977). They attributed this relatively great importance of *Oscillatoria* in the Caribbean to phosphate

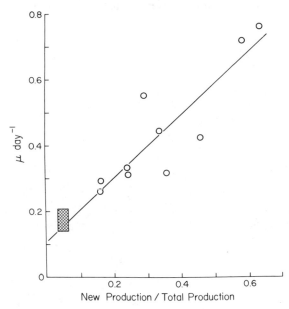

FIG. 1   Estimated specific growth rate (μ) of natural phytoplankton assemblages vs. the ratio new production:total production. μ was calculated approximately as the daily rate of carbon (^{14}C) assimilation ($gCm^{-2}day^{-1}$) divided by the phytoplankton standing stock as carbon (gCm^{-2}). Parameters on the abscissa are based upon ^{15}N assimilation ($gNm^{-2}day^{-1}$): new production as ^{15}N-nitrate uptake and total production as the sum of nitrate + 1.5 ammonium uptake. (The value 1.5 ammonium uptake is used to approximate the uptake of ammonium + urea + amino acid-N, etc.). Open circles represent individual offshore stations in the Southern California Bight. The hatched rectangle represents several stations in the central N. Pacific near 28°N 155°W

enrichment of surface waters during passage of the water over the Lesser
Antilles Banks. *Oscillatoria thiebautii* requires an unusually high phosphate
concentration for phosphate assimilation (McCarthy and Carpenter, 1979).
The Carpenter and Price (1977) result is plotted in Fig. 2 assuming N_2 fixation
was the only pathway of new production. The point falls slightly below the
other points of Fig. 2, perhaps reflecting low predation rate on *Oscillatoria*
observed directly by others.

3   Primary production in temperate and boreal seas

In contrast to the subtropical and tropical oceans, both temporal and spatial
variation in phytoplankton production in temperate and boreal seas is
explained primarily in terms of insolation and stratification of the water
column (Ryther, 1963) with nutrient input rate playing a subordinate role as
nutrients are often present in the surface waters even in the summer growing

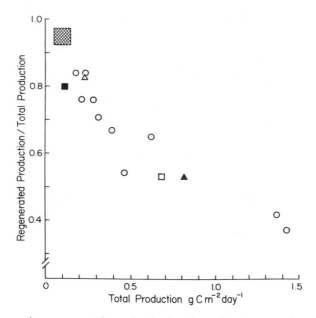

FIG. 2   The ratio regenerated production/total production as a function of total
production. See Fig. 1 for methods of calculation. Open circles are individual
offshore stations in the Southern California Bight, the hatched rectangle represents
several stations in the central N. Pacific, solid square symbol is the Caribbean data
of Carpenter and Price (1977), open square is offshore Monterey Bay (G. Knauer
and R. Eppley, unpubl.). Other data are averages for several stations from MacIsaac
and Dugdale (1972) and Dugdale (1976) as follows: Solid triangle is eastern tropical
Pacific; open triangle is eastern Mediterranean Sea

season. Cushing (1975) recently reviewed the extensive literature on seasonal cycles of production and the historical development of the critical depth concept (Gran and Braarud, 1935). Mathematical models describing seasonal production cycles and their historical development were also reviewed (Cushing, 1975).

Typically in high latitude temperate and boreal waters production is nil in winter due to low insolation and the lack of stratification of the water column. The critical depth is the depth at which integrated photosynthesis in the waters above equals plant respiration, integrated over the same depth range. When the depth of water mixing exceeds the critical depth, assuming the rate of mixing is fast compared with phytoplankton generation times, then there can be no net growth. This is the winter situation. Insolation increases in the spring and the surface waters warm. The resulting stratification reduces the depth of surface water mixing. When the mixing depth becomes shallower then the critical depth net plant growth is possible. Parsons et al. (1966) were able to predict the timing of the spring onset of the growing season in the northeast Pacific from critical depth theory, for example, and provide details of the calculations.

In many areas a classical 'spring bloom' takes place early in the growing season. This bloom may deplete the nutrients that were mixed into the surface waters over the winter. When this occurs crops decline and may remain at low levels over the summer, stratified period. Subsurface chlorophyll maximum layers are typical of the summertime in such locations (Taniguchi and Kawamura, 1972; Holligan, 1979). A second, fall bloom, is typical of some locations. At least three mechanisms have been suggested for the fall pulse; (i) renewed nutrient input to the surface layer by physical mixing, (ii) relaxation of zooplankton grazing intensity, and (iii) mixing of the subsurface chlorophyll maximum layer into the surface waters.

Changes in the species composition of the phytoplankton assemblage accompany the succession of changes in production, insolation, stratification and nutrient input rate (reviewed by Guillard and Kilham, 1977).

Temporal alternations between mixing and stratification take place over the growing season in shallow coastal waters of the continental shelves. This results from strong tidal currents and the mixing associated with them (Pingree, 1978; Chapters 7 and 8). The 'fronts' at the boundary separating the stratified areas from the well-mixed areas are associated with higher phytoplankton crops and higher production rates than are found on either side of the frontal regions (Pingree et al., 1975). Pingree (1978; and earlier references therein) and colleagues have been able to predict the location of the fronts, and hence they developed a tool for predicting spatial and temporal variation in production, from considering the velocity of the tidal currents, the dissipation of tidal energy due to work done against the frictional drag of the bottom, and the water depth. The rather complex physical processes involved in the energy dissipation have been reduced to a simple, useful equation

$$S = \log_{10}\left[\frac{h}{C_D|\bar{u}|^3}\right] = 1.5$$

where S is the Simpson-Hunter stratification parameter, h is water depth, C_D is a drag coefficient ~ 0.0025, and $|\bar{u}|$ is the average value over the tidal cycle of the tidal stream velocity, and 1.5 is a critical value.

Such fronts move somewhat and production rates vary as the tidal current velocities change over the neap-spring tidal cycle and the phytoplankton blooms move with them (Pingree, 1978).

In spite of such apparent complexity the production dynamics of limited areas of the coastal, temperate sea can be relatively simple. Gieskes and Kraay (1977) showed, for example, that essentially all of the variability in phytoplankton production was accounted for by surface irradiance and the chlorophyll concentration in the turbid, nutrient-rich waters off the Dutch Coast. The plume of the Rhine River accounts for the uniqueness of this area.

4 Vertical stratification of phytoplankton and production

Subsurface maxima are often noted in chlorophyll depth profiles in the tropical and subtropical oceans and in temperate and boreal waters in summer. In the Gulf of Guinea (Table II) and the Southern California Bight these maxima are closely associated with Z_n (i.e. the top of the vertical concentration gradient of nitrate). In the central North Pacific, however, the chlorophyll maximum layer is usually well above Z_n (Venrick et al., 1973). There it appears not to represent a biomass peak at depth but rather an increase in the chlorophyll content per cell (Beers et al., 1975). Thin layer chromatography of the pigments of the deep chlorophyll layer off NW Africa showed much allomeric chlorophyll and other chlorophyll derivatives. These are thought to be decomposition products of microzooplankton and bacterial activity (Gieskes et al., 1978).

A bioluminescence probe was used to locate plankton maxima at depth in the western tropical Pacific. Two maxima were found. The lower maximum (60–100 m) corresponded to the depth of a sharp increase in nitrogen and phosphorus concentrations (Vinogradov et al., 1971), i.e. approximately Z_n in the present context.

Sometimes these chlorophyll maxima at depth are coincident with maxima in production or at least they contribute substantially to production in sub-tropical and tropical waters, as well as in temperate oceanic areas (Anderson, 1969; Taniguchi and Kawamura, 1972; Holligan, 1979). Recent estimates of the compensation intensity for photosynthesis of natural phytoplankton assemblages (Platt and Jassby, 1976) indicate values corresponding with a light depth approximately 0.1–0.3 % of surface light at noon. Thus chlorophyll maxima below the 1 % light depth, the traditional base of the euphotic zone, can be expected to represent growing phytoplankton in oceanic situations.

Subsurface chlorophyll maxima are also noted after the spring diatom outburst in temperate coastal waters which are sometimes dominated by dinoflagellates (Holligan and Harbour, 1977; Holligan, 1979). Dinoflagellate dominance of the layer is also common in the subtropical Southern California Bight (Reid *et al.*, 1978) and the Angola Dome (Dandonneau, 1977).

Mathematical models of plankton distribution in relation to light, nutrients and the physical structure of the water column, predict the existence of subsurface phytoplankton maxima and permit calculation of production rates (Jamart *et al.*, 1977). Holligan (1979) estimated 30–80% of summer production on the continental shelf about Britain takes place in the chlorophyll maximum layer, a production equivalent to the spring diatom outburst.

5 Regenerated production and nutrient recycling

Broecker (1974) calculated from a geochemical, two-box model of the global ocean that a nutrient element would be recycled in the surface layer ten times, on average, before sinking out in particulate matter. The ratio regenerated production/new production provides a measure of the extent of nutrient recycling on the surface layers (Table IV). The ratio for the central N. Pacific is

TABLE IV Theoretical relationships between new production, regenerated production and total production assuming steady-state

$\dfrac{\text{New}}{\text{Total}}$	= sinking flux as a fraction of total production (values range from ~ 0.05 to ~ 0.8). Values >0.5 indicate the sinking of intact phytoplankton. Values <0.5 indicate the sinking of fecal pellets and other particulates derived from heterotrophs. Values are high in nutrient-rich waters as during upwelling and lowest in oligotrophic central oceans
$\dfrac{\text{Regenerated}}{\text{Total}}$	= fraction of production being recycled by heterotrophs, i.e. a nutrient recycling index and potentially an index of grazing pressure on phytoplankton. Values have ranged from ~ 0.2 (upwelling) to ~ 0.95 in the central N. Pacific
$\dfrac{\text{Regenerated}}{\text{New}}$	= number of times a nutrient element is recycled in the euphotic zone before sinking out as particulate matter. Experimental values have varied from <1 during upwelling to ~ 20 in the central N. Pacific

about 19 ($=0.95/0.05$ from Figure 2, Eppley and Peterson, 1979). The ratio is less than 1 at least transiently in nutrient-rich waters (MacIsaac and Dugdale, 1972; Dugdale, 1976).

Nutrient recycling in the euphotic zone is intimately related to ocean food webs (Dugdale and Goering, 1967). Phytoplankton eaten by herbivores, such as copepods, which defecate and respire the ingested organic matter constitute one major pathway for N and P recycling. Bacterial-microzooplankton food webs are another major pathway. There is evidence that the importance of this pathway has been underestimated in the past (Pomeroy, 1979; Sorokin, Chapter 10 this volume; Sieburth, 1977). Macrozooplankton may account for about one-half of the N-recycling in the surface waters of the central N. Pacific or approximately 100 nmole $m^{-3}day^{-1}$ (compare Mullin et al., 1975, Figure 2 with Eppley et al., 1977, conservative estimate of Table 3). The microbial-microzooplankton pathway is the more important one nearshore in the Southern California Bight (Harrison, 1978). Entire CEPEX enclosures filled with coastal water were dosed with ^{15}N-labelled nutrients to study the pathways of nitrogen flow (Harrison and Davies, 1977; A. Hattori and I. Koike, personal communication) but unfortunately such experiments have not been designed for oceanic waters.

AN INDEX FOR ASSESSING THE EXTENT OF GRAZING
CONTROL OF PRIMARY PRODUCTION

Grazing may regulate primary production and the standing stocks of phytoplankton in the ocean at various times and places. The simplest evidence offered in support of grazing control of production is that grazing is actively taking place and that the growth of phytoplankton is not sufficient to deplete ambient concentrations of nutrients. Thomas (1979) noted three possible explanations for such observations: (i) upwelling exceeds utilisation of nutrients, (ii) grazing controls the phytoplankton, (iii) minor nutrients may be limiting phytoplankton growth. The discussion of the problem would be facilitated if a quantitative index of the extent of grazing pressure were available. Lorenzen (1967) proposed one such index of grazing pressure upon the phytoplankton based upon the ratio of concentrations of phaeopigments (pheophorbide) to chlorophyll in the water column. I propose that the ratio regenerated production/total production (abbreviated here as ϕ) provides such an index. It takes values between 0 and 1.0. When grazing is high, regenerated production will be a large fraction of the total and ϕ approaches 1.0. When regenerated production/total production is low ϕ approaches zero. Some examples are shown in Fig. 2. In the more oligotrophic regions of the central, anticyclonic gyres ϕ is about 0.95, indicating a high degree of regenerated production and possibly of grazing control of production. During transient spring blooms in temperate and boreal waters, when surface nutrients are high as a result of winter mixing and overwintering zooplankton populations are low, ϕ can be expected to take relatively low values, although actual measurements have not been reported. Low values (<0.5) have been measured during upwelling episodes (MacIsaac and Dugdale, 1972; Dugdale, 1976; Eppley et al., 1979) corresponding to transient periods of high total production. The utility of ϕ as an index of grazing control on phytoplankton growth depends upon whether nutrient recycling can take place directly via phytoplankton mortality, independent of grazing activity, and the subsequent microbial degradation of the phytoplankton or alternatively, whether phytoplankton mortality is directly or indirectly mediated by grazing activity. It can be assumed for near steady-state oceanic systems that phytoplankton are growing exponentially. Nevertheless autoradiographic techniques indicate not all the phytoplankton are photosynthetically active even in productive coastal waters (A. F. Carlucci, personal communication). Whether this results from cell damage in the grazing process, i.e. 'sloppy feeding' (Dagg, 1974; Marshall and Orr, 1962), parasitism, or spontaneous mortality is not clear. In any case, the index may prove to be a useful supplement to the phaeophorbide/chlorophyll index of grazing pressure of Lorenzen (1967).

6 Problems and challenges for future study

Put briefly, the fundamental questions asked by physiologically oriented plankton biologists are: what organisms are present and at what concentrations and biomass, what are their specific growth rates, what regulates their growth rates and how do the organisms interact with one another and with the environment to form the communities or assemblages we encounter in sampling them. (For a different view of the objectives of biological oceanography not concerned with rate processes, see, for example, McGowan, 1974.)

Nearly all of our insight on planktonic rate processes in the oceans comes from experimental incubations of plankton in bottles. Incubation times of samples in the bottles are usually 6, 12, or 24 hours. Considerable mortality of microplankton and changes in relative species composition can take place over these time periods in 250 ml bottles (Venrick et al., 1977) and there is often little difference in measurements of photosynthetic production between 6 and 24 hour incubations of central North Pacific water (Eppley and Sharp, 1975). Phytoplankton specific growth rates based on 24 hour ^{14}C and ^{15}N incubations and biomass estimates of phytoplankton C and N suggested specific growth rates (Fig. 1) of the order of 0.2–0.3 doubling day^{-1} (0.14–0.21 day^{-1}) (Eppley et al., 1973).

On the other hand, incubation of Sargasso Sea plankton suggested modal generation times as short as 3 hours based upon increases in total particle volume (Sheldon et al., 1973) or ATP (Sheldon and Sutcliffe, 1979). Indirect evidence for short generation times and high specific growth rates includes estimates of carbon flux via bacteria (see summaries of Sorokin, 1978; Sieburth, 1977; Pomeroy, 1979) and the observation that Redfield ratios for the elemental composition of particulate matter (and plankton; 106 C: 16 N: 1 P by atoms) require specific growth rates approaching the maximum expected for the species (Goldman et al., 1979).

These conflicting results need further examination and new methods of estimating specific growth rates that require no incubation need to be applied to the problem before final judgment can be made. However, it appears quite possible at present that incubation methods may under-estimate production as a result of grazing within the bottles and unspecified phytoplankton mortality perhaps associated with physical fragility or heavy metal poisoning from the use of metal-contaminated vessels or isotope preparations.

Some potential solutions to estimating specific growth rates without long incubation include assays of enzymes unique to the S-phase of the cell cycle (Sapienza and Mague, 1979), short-term DNA labelling with tritiated thymidine (Polikarpov and Tokoreva, 1970), and enumeration of the frequency of paired cells and cells containing paired nuclei (Weiler, 1978) or reproductive bodies (Swift et al., 1976) in fresh samples. To date these methods have not been applied to the nanoplankton that dominate the biomass of the phytoplankton in oligotrophic central oceans (Beers et al., 1975).

In view of all this uncertainty it is surprising that the data from the central North Pacific are consistent with data from the coastal region off Southern California. An example is shown in Fig. 1 where the specific growth rate (μ) of phytoplankton is plotted with the ratio new production/total production. The coastal waters alternate between periods of oligotrophy and nutrient enrichment (i.e. upwelling), as indicated by the ratio new production/total production and the phytoplankton display a range of values in estimated specific growth rate proportional to the ratio new production/total produc-tion. Estimates of these parameters for several cruises to the central North Pacific are more uniform over time and location and fall on the regression line fit to the coastal data. Bottle incubations were used throughout: μ was calculated from 24 hour ^{14}C uptake and phytoplankton biomass estimated as carbon, new production as ^{15}N-nitrate uptake, and total production as the sum of ^{15}N-nitrate plus ammonium plus urea-N uptake. This graph suggests that if the bottle incubations err, then the error increases with the degree of oligotrophy if either μ (and production) is constant in both regions or is higher in the central North Pacific than in the coastal region.

An example of internal consistency within the central North Pacific data comes from comparing estimates of μ from ^{14}C and ^{15}N data. In several experiments ^{15}N-ammonium, nitrate and urea were added in 24-hour, 4-liter-bottle incubations to measure uptake rate as a function of the concentration of the added ^{15}N substrates (J. Sharp and E. Renger, unpublished). This allowed calculations of the turnover time of these substances (Wright and Hobbie, 1966) without concern for their ambient concentrations in the sea water (which are below the limit of detection with present spectrophotometric methods – see McCarthy and Carpenter, 1975). The flux of ammonium into the phytoplankton can be calculated as the quotient of the ambient concentration (< 50 μM; Eppley et al., 1977) and the turnover time, average 8.7 days, to be $\leqslant 5.7$ nmole day^{-1}. The phytoplankton N content is about one-third of the total particulate organic nitrogen (Eppley et al., 1973) or about 50 nM. The specific growth rate due to ammonia uptake is

$$\mu_{NH_4} = \frac{N \text{ flux}}{\text{Phyto N}} \leqslant 5.7/50 \leqslant 0.11 \text{ day}^{-1}.$$

The flux of urea, nitrate and N_2 into the phytoplankton will increase this about one-third to give $\mu_N \sim 0.15$ day^{-1}, a value similar to that given from the carbon production rate and biomass data.

If the incubation methods are indeed underestimating primary production in the oligotrophic oceans it must be regenerated production rather than new production which is underestimated. The vertical diffusive transport of nitrate into the euphotic zone, estimated from the slope of the nitrate concentration gradient and Kz from microstructure studies (Gregg et al., 1973), plus likely inputs of nitrate from rainfall are an order of magnitude less than observed ^{15}N nitrate uptake rates based upon conservative interpretation of the data (and molecular nitrogen fixation rates are based on only brief incubations). Thus new production rates are probably overestimated to some degree already. Clearly some new approaches are needed to examine regenerated production in the oligotrophic ocean in relation to the flux of organic matter passing through bacteria-microzooplankton portions of the food web.

References

Anderson, G. C. (1969). Subsurface chlorophyll maximum in the northeast Pacific Ocean. *Limnol. Oceanogr.* **18**, 386–391

Anderson, G. C. and Munson, R. E. (1972). Primary production studies using merchant vessels in the North Pacific Ocean. *In* "Biological Oceanography of the Northern North Pacific" (A. Y. Takenouti, ed.) pp. 245–251. Idemitsu Shoten, Tokyo

Beers, J. R., Reid, F. M. H. and Stewart, G. L. (1975). Microplankton of the North Pacific Central Gyre. Population structure and abundance, June 1973. *Int. Rev. ges. Hydrobiol.* **60**, 607–638

Broecker, W. S. (1974). "Chemical Oceanography". Harcourt Brace Jovanovich, New York

Caperon, J. and Meyer, J. (1972). Nitrogen-limited growth of marine phytoplankton. II. Uptake kinetics and their role in nutrient limited growth of phytoplankton. *Deep-Sea Res.* **19**, 619–632

Carpenter, E. J. and McCarthy, J. J. (1975). Nitrogen fixation and uptake of combined nitrogenous nutrients by *Oscillatoria* (*Trichodesmium*) *thiebautii* in the western Sargasso Sea. *Limnol. Oceanogr.* **20**, 389–401

Carpenter, E. J. and Price, C. C. (1977). Nitrogen fixation, distribution, and production of *Oscillatoria* (*Trichodesmium*) spp. in the western Sargasso and Caribbean Seas. *Limnol. Oceanogr.* **22**, 60–72

Cushing, D. H. (1975). "Marine Ecology and Fisheries", Chapters 1–4. Cambridge University Press

Dandonneau, Y. (1977). Variations nycthemerales de la profondeur du maximum de chlorophylle dans le Dôme d'Angola (Ferrier-Mars, 1971). *Cah. O.R.S.T.O.M. Ser. Oceanogr.* **15**, 27–37

Dagg, M. J. (1974). Loss of prey body contents during feeding by an aquatic predator. *Ecology* **55**, 903–906

Dugdale, R. C. (1976). Nutrient cycles. *In* "The Ecology of the Seas" (D. H. Cushing and J. J. Walsh, eds) pp. 141–172. W. B. Saunders Co., Philadelphia

Dugdale, R. C. and Goering, J. J. (1967). Uptake of new and regenerated forms of nitrogen in primary productivity. *Limnol. Oceanogr.* **12**, 196–206

Eppley, R. W. and Peterson, B. J. (1979). The flux of particulate matter to the deep ocean. *Nature* **282**, 677–680

Eppley, R. W. and Sharp, J. H. (1975). Photosynthetic measurements in the central North Pacific: the dark loss of carbon in 24 h incubations. *Limnol. Oceanogr.* **20**, 981–987

Eppley, R. W., Renger, E. H. and Harrison, W. G. (1979). Nitrate and phytoplankton production in Southern California coastal waters. *Limnol. Oceanogr.* **24**, 483–494

Eppley, R. W., Renger, E. H., Venrick, E. L. and Mullin, M. M. (1973). A study of plankton dynamics and nutrient cycling in the central gyre of the North Pacific Ocean. *Limnol. Oceanogr.* **18**, 534–551

Eppley, R. W., Sharp, J. H., Renger, E. H., Perry, M. J. and Harrison, W. G. (1977). Nitrogen assimilation by phytoplankton and other microorganisms in the surface waters of the central North Pacific Ocean. *Mar. Biol.* **39**, 111–120

Gieskes, W. W. C. and Kraay, G. W. (1977). Primary production and consumption of organic matter in the southern North Sea during the spring bloom of 1975. *Netherl. J. Sea Res.* **11**, 146–167

Gieskes, W. W. C., Kraay, G. W. and Tijssen, S. D. (1978). Chlorophylls and their degradative products of the deep pigment maximum layer of the tropical North Atlantic. *Netherl. J. Sea. Res.* **12**, 195–204

Goldman, J. C., McCarthy, J. J. and Peavey, D. G. (1979). Growth rate influence on the chemical composition of phytoplankton in oceanic waters. *Nature* **279**, 210–215

Gran, H. H. and Braarud, T. (1935). A quantitative study of the phytoplankton in the Bay of Fundy and the Gulf of Maine (including observations on hydrography, chemistry and turbidity). *J. Biol. Bd. Can.* **1**, 279–467

Gregg, M. C. (1977). Variations in the intensity of small-scale mixing in the main thermocline. *J. Phys. Oceanogr.* **7**, 436–454

Gregg, M. C., Cox, C. S. and Hacker, P. W. (1973). Vertical microstructure measurements in the central North Pacific. *J. Phys. Oceanogr.* **3**, 458–469

Guillard, R. R. L. and Kilham, P. (1977). The ecology of marine planktonic diatoms. *In* "The Biology of Diatoms" (D. Werner, ed.) pp. 372–469. Blackwell, Oxford

Harrison, W. G. (1978). Experimental measurements of nitrogen remineralization in coastal waters. *Limnol. Oceanogr.* **23**, 684–694

Harrison, W. G. and Davies, J. M. (1977). Nitrogen cycling in a marine planktonic food chain: Nitrogen fluxes through the principal components and the effects of adding copper. *Mar. Biol.* **43**, 299–306

Herbland, A. and Voituriez, B. (1977). Production primaire, nitrate et nitrite dans l'Atlantique tropical. I. Distribution du nitrate et production primaire. *Cah. O.R.S.T.O.M. ser Oceanogr.* **15**, 47–55

Herbland, A. and Voituriez, B. (1979). Hydrological structure analysis for estimating the primary production in the tropical Atlantic Ocean. *J. Mar. Res.* **37**, 87–101

Holligan, P. M. (1979). Patchiness in subsurface phytoplankton populations on the northwest european continental shelf. *In* "Spatial Pattern in Plankton Communities" (J. H. Steele, ed.) pp. 221–238. Plenum Press, New York and London

Holligan, P. M. and Harbour, D. S. (1977). The vertical distribution and succession of phytoplankton in the western English Channel in 1975 and 1976. *J. Mar. Biol. Assoc. U.K.* **57**, 1075–1093

Holm-Hansen, O., El-Sayed, S. Z., Franceschini, G. A. and Cuhel, R. L. (1977). "Primary Production and the Factors Controlling Phytoplankton Growth in the Southern Ocean". Proc. 3rd SCAR Sympt., Antarctic Biology, pp. 11–50, Gulf Publ. Co., Houston

Jamart, B. M., Winter, D. F., Banse, K., Anderson, G. C. and Lam, R. K. (1977). A theoretical study of phytoplankton growth and nutrient distribution in the Pacific Ocean off the northwestern U.S. coast. *Deep-Sea Res.* **24**, 753–773

King, F. D. and Devol, A. H. (1979). Estimates of vertical eddy diffusion through the thermocline from phytoplankton nitrate uptake rates in the mixed layer of the eastern tropical Pacific. *Limnol. Oceanogr.* **24**, 645–651

Koblentz-Mishke, O. J., Volkovinsky, V. V. and Kabanova, J. G. (1970). Plankton primary production of the world ocean. *In* "Scientific Exploration of the South Pacific" (W. Wooster, ed.) pp. 183–193. NAS, Washington, D.C.

Lorenzen, C. J. (1967). Vertical distribution of chlorophyll and phaeo-pigment: Baja California. *Deep-Sea Res.* **14**, 735–745

Lorenzen, C. J. (1976). Primary production in the sea. *In* "The Ecology of the Seas" (J. J. Walsh and D. H. Cushing, eds) pp. 173–185. W. B. Saunders Co., Philadelphia and Toronto

McGowan, J. A. (1974). The nature of oceanic ecosystems. *In* "The Biology of the Oceanic Pacific" (C. B. Miller, ed.) pp. 9–28. Oregon State University Press, Corvallis

McGowan, J. A. and Haywood, T. L. (1978). Mixing and oceanic productivity. *Deep-Sea Res.* **25**, 771–793

McGowan, J. A. and Williams, P. M. (1973). Oceanic habitat differences in the North Pacific. *J. exp. Mar. Biol. Ecol.* **12**, 187–217

MacIsaac, J. J. and Dugdale, R. C. (1972). Interactions of light and inorganic nitrogen in controlling nitrate uptake in the sea. *Deep-Sea Res.* **19**, 209–232

Mague, T. H. (1977). Ecological aspects of dinitrogen fixation by blue-green algae. *In* "A Treatise on Dinitrogen Fixation. IV. Agronomy and Ecology" (R. W. F. Hardy and A. H. Gibson, eds) pp. 85–140. J. Wiley & Sons, New York

Marshall, S. M. and Orr, A. P. (1962). Food and feeding of copepods. *Rapp. Proc.- Verb. Reunions, Cons. Perm. Int. Explor. Mer* **153**, 92–98

Menzel, D. W. and Ryther, J. H. (1960). The annual cycle of primary production in the Sargasso Sea off Bermuda. *Deep-Sea Res.* **6**, 351–367

Menzel, D. W. and Ryther, J. H. (1961). Nutrients limiting the production of phytoplankton in the Sargasso Sea, with special reference to iron. *Deep-Sea Res.* **7**, 276–281

Mickelson, M. J., Maske, H. and Dugdale, R. C. (1979). Nutrient-determined dominance in multispecies chemostat cultures of diatoms. *Limnol. Oceanogr.* **24**, 298–315

Mullin, M. M., Perry, M. J., Renger, E. H. and Evans, P. M. (1975). Nutrient regeneration by oceanic zooplankton: a comparison of methods. *Mar. Sci. Comm.* **1**, 1–13

Parsons, T. R., Giovando, L. F. and LeBrasseur, R. J. (1966). The advent of the spring bloom in the eastern subarctic Pacific Ocean. *J. Fish. Res. Bd. Can.* **23**, 539–546

Pingree, R. D. (1978). Mixing and stabilization of phytoplankton distributions on the northwest European continental shelf. *In* "Spatial Pattern in Plankton Communities" (J. H. Steele, ed.). Plenum Press, New York

Pingree, R. D., Pugh, P. R., Holligan, P. M. and Forster, G. R. (1975). Summer phytoplankton blooms and red tides along tidal fronts in the approaches to the English Channel. *Nature* **258**, 672–677

Platt, T. and Jassby, A. D. (1976). The relationship between photosynthesis and light for natural assemblages of coastal marine phytoplankton. *J. Phycol.* **12**, 421–430

Platt, T. and Subba Rao, D. V. (1975). Primary production of marine microphytes. *In* "Photosynthesis and Productivity of Different Environments", Vol. 3, pp. 249–280. International Biological Programme. Cambridge University Press

Polikarpov, G. G. and Tokareva, A. V. (1970). The cell cycle of the dinoflagellates *Peridinium trochoideum* (Stein) and *Gonyaulax polyedra* (Stein) (microautoradiographic study). *Hydrobiol. J.* **6**, 54–57

Pomeroy, L. R. (1979). Secondary production mechanisms of continental shelf communities. *In* "Conf. on Ecological Processes in Coastal and Marine Systems" (R. J. Livingston, ed.) p. 163. Tallahassee, Fla.

Reid, F. M. H., Stewart, E., Eppley, R. W. and Goodman, D. (1978). Spatial distribution of phytoplankton species in chlorophyll maximum layers off southern California. *Limnol. Oceanogr.* **23**, 219–226

Reid, J. L. (1962). On circulation, phosphate-phosphorus content, and zooplankton volumes in the upper part of the Pacific Ocean. *Limnol. Oceanogr.* **7**, 287–306

Reid, J. L., Brinton, E., Fleminger, A., Venrick, E. L. and McGowan, J. A. (1978). Ocean circulation and marine life. *In* "Adv. Oceanogr." (H. Charnock and G. Deacon, eds) pp. 65–130. Plenum Publ. Corporation

Riley, G. A. (1956). Oceanography of Long Island Sound, 1952–1954. IV. Production and utilization of organic matter. *Bull. Bingham Oceanogr. Coll.* **15**, 324–344

Ryther, J. A. (1963). Geographic variations in production. *In* "The Seas" (M. N. Hill, ed.) Vol. 2, pp. 347–380. Interscience Publ., New York

Sapienza, C. and Mague, T. H. (1979). DNA polymerase activity and growth rates in *Artemia salina* nauplii. *Limnol. Oceanogr.* **24**, 572–576

Sapozhnikov, V. V. and Galerkin, L. I. (1971). On the computation of primary production in the tropical waters based on the vertical flow of nutrient salts. *In* "Functioning of Pelagic Communities in the Tropical Regions of the Ocean" (M. E. Vinogradov, ed.) pp. 65–69. Nauka, Moscow

Sheldon, R. W. and Sutcliffe, W. H. Jr. (1978). Generation times of 3 h for Sargasso Sea microplankton determined by ATP analysis. *Limnol. Oceanogr.* **23**, 1051–1055

Sheldon, R. W., Prakash, A. and Sutcliffe, W. H. Jr. (1973). The production of particles in the surface waters of the ocean with particular reference to the Sargasso Sea. *Limnol. Oceanogr.* **18**, 719–733

Sieburth, J. McN. (1977). International Helgoländ Symposium: Convenor's report on the informal session of biomass and productivity of microorganisms in planktonic ecosystems. *Helgoländ wiss. Meeresunters.* **30**, 697–704

Sorokin, Yu. (1978). Decomposition of organic matter and nutrient regeneration. *In* "Marine Ecology" (O. Kinne, ed.) Vol. 4, pp. 501–616. J. Wiley & Sons, New York

Steemann Nielsen, E. (1955). Production of organic matter in the oceans. *J. Mar. Res.* **14**, 374–386

Swift, E., Stuart, M. and Meunier, V. (1976). The *in situ* growth rates of some deep-living oceanic dinoflagellates: *Pyrocystis fusiformis* and *Pyrocystis noctiluca*. *Limnol. Oceanogr.* **21**, 418–426

Taniguchi, A. and Kawamura, T. (1972). Primary production in the Oyashio region with special reference to the subsurface chlorophyll maximum layer and phytoplankton-zooplankton relationships. *In* "Biological Oceanography of the Northern North Pacific" (A. Y. Takenouti, ed.) Vol. 17, pp. 231–243. Idemitsu Shoten, Tokyo

Thomas, W. H. (1979). Anomalous nutrient-chlorophyll interrelationships in the offshore eastern tropical Pacific Ocean. *J. Mar. Res.* **37**, 327–335

Venrick, E. L., McGowan, J. A. and Mantyla, A. W. (1973). Deep chlorophyll maxima in the oceanic Pacific. *Fish. Bull.* **71**, 41–52

Venrick, E. L., Beers, J. R. and Heinbokel, J. F. (1977). Possible consequences of containing microplankton for physiological rate measurements. *J. exp. mar. Biol. Ecol.* **26**, 55–76

Vinogradov, M. E., Gitelzon, I. I. and Sorokin, Yu. I. (1971). On the spatial structure of the communities in the euphotic zone of the tropical ocean. *In* "Functioning of Pelagic Communities in the Tropical Regions of the Ocean" (M. E. Vinogradov, ed.) pp. 255–264. Nauka, Moscow

Voituriez, B. and Herbland, A. (1979). The use of the salinity maximum of the equatorial undercurrent for estimating nutrient enrichment and primary production in the Gulf of Guinea. *Deep-Sea Res.* **26A**, 77–81

Weiler, C. S. (1979). Phased cell division in the dinoflagellate genus *Ceratium*: temporal pattern, use in determining growth rates, and ecological implications. Ph.D. Thesis, University of California, San Diego

Wright, R. T. and Hobbie, J. E. (1966). Use of glucose and acetate by bacteria and algae in aquatic ecosystems. *Ecology* **47**, 447–464

12 Nutritional Strategies for Feeding on Small Suspended Particles

ROBERT J. CONOVER

1 Introduction

Water has a number of properties which make it unique as an environment. It has relatively high density and viscosity, so that many organic materials are close to neutral buoyancy, while objects moving through it meet with some resistance. It is also quite opaque to light energy, wave lengths being selectively absorbed depending on its color and the amount and kind of suspended particulate matter present. As a consequence, photosynthesis can only occur to a depth of about 100 m in the open ocean and to much shallower depths near shore so that most plants, being unable to grow attached to the bottom, are suspended as single cells or in colonies near the surface. Small size increases the relative amount of cell surface, which, in addition to facilitating trans-membrane molecular movement, also increases the frictional resistance of the particle, retarding its rate of sinking. Not surprising then the primary producers in the sea are mostly small.

When untreated water is passed through a counting and sizing device, the number of particles, including the suspended cells, is found to decrease exponentially in successively larger categories, though larger particles have a proportionately larger volume and may contain more organic matter. To emphasise this important observation, Sheldon and Parsons (1967) devised a scale in which particles are grouped according to the log of their mean spherical diameter so that the geometric mean volume of each successively larger particle category exactly doubles that of the preceding category. A plot of the volume of particles falling into each size category or spectral band often shows peaks representing biomass accumulations available as food for fine particle feeders.

The herbivorous animals, which depend on this suspended particulate matter for nutrition, must optimise the particle capture process to yield a relatively large net return of energy, to be used for growth and reproduction, for calories expended in the search and capture, and they must achieve this in an environment containing only small and frequently widely dispersed particles.

363

2   Mechanisms for feeding on small particles

Human innovators have used filters made of woven fabric, matted fibers, closely-packed sediment or quality-controlled pores in artificial membranes to purify liquids and recover desirable suspensions for a few thousand years. In comparison, the most primitive invertebrates have probably used similar devices since the evolution of the heterotrophic mode of existence, and they can do so with better energetic efficiency than we can. How are they able to do this?

HYDRODYNAMICS AND SMALL SUSPENDED PARTICLES

There is a relationship between the velocity of a fluid v, its density ρ and its viscosity μ such that their ratio, the Reynolds number, describes forces operating during flow around an object. If the object moves through the water or the water moves around it, size L also becomes important so that $Re = \rho\mu L/N$. As density and viscosity of water change little with temperature or salinity, the rate of flow and particle size must influence the magnitude of the Reynolds number. A moving ship or a large fish swimming rapidly involves turbulent forces and Re is large ($> 10^3$). In contrast, a flagellate, being small and slow, would have to contend with a viscous environment ($Re < 1$). The flows around the appendages of small, swimming zooplankters are also viscous (Alcaraz et al., 1980).

There are advantages and disadvantages to being in a viscous environment. If a particle moves from one point to another in space it is difficult to escape the molecules immediately surrounding it. Purcell (1977) has calculated that a small flagellated cell must move at a rate of at least 30 μm s^{-1} before it changes its molecular environment faster than occurs by simple diffusion. To beat a flagellum or wave a tail in a perfectly reciprocal fashion is ineffectual for small swimming organisms, as the viscous molecules are not permanently displaced easily. The movement of suspended particles therefore is tied to that of the viscous medium, which can be demonstrated by dropping a few fragments of breakfast toast into the honey jar and watching their movement as a knife is stirred around. Note that when the knife is stopped so is the motion of the crumbs. The moving cirri of small feeding appendages can cause analogous movement of food particles without actually contacting them. The viscosity of a liquid also causes reduction of speed of flow near the surface of an object and Zaret (1980) believes that small pelagic animals can detect such rate differences, using them to avoid large objects or predators.

For suspension feeding animals, a filter or sieve would seem to be the logical device for removing small particles from a fluid medium, but such a system has its limitations. If the pores are too small, energy requirements to push water through a sieve are high, and flow rate is low. Clogging may also occur. Aperture size probably defines the range of particles which can be removed: to be sure, Rubenstein and Koehl (1977) have described several means by which particles small enough to pass through the pores may be captured, but neither inertial nor gravitational accelerations would seem to be important at viscous flows. Probably direct interception of particles by structural elements of the filter explains most food capture by grazers and generally fits the concept that size selection is largely a passive, mechanical process (Boyd, 1976). Filtering devices must also incorporate some means of transferal of food particles to the mouth, and, conceivably, during this process some qualitative and quantitative selection can take place.

MECHANISMS INVOLVING CILIA

Cilia can cause very small animals to move about, or may move fluid when attached to the relatively immobile surface of larger animals. Along with water, sheets of mucus and any embedded foreign material, can be transported by the coordinated mass action of these organelles. Since these very small structures must function in a viscous medium their action cannot be reciprocal and symmetrical. Instead, cilia beat out of phase in a metachronal wave which generally moves at right angles to the effective stroke.

The ultrastructure of a cilium consists of a circle of nine paired microtubules surrounding two single units, the whole surrounded by a ciliary membrane. The sub-fibers in each pair are of unequal length and the longer one has spaced along it protein projections of ATPase (dynein) which point toward the shorter fiber of the adjacent doublet. ATP, produced by mitochondria in the cell, diffuses out into the cilium and causes it to beat. There is no contraction as in muscle fibers, but the bundles of microtubules slip back and forth past each other in a differentially reciprocating fashion under the influence of ATP (Satir, 1974).

The energetic costs of ciliary activity have been calculated for *Mytilus* and *Sabellaria* cilia by Sleigh and Holwill (1969). On the basis of the ciliary geometry from high speed photographs, they showed that the effective stroke in *Sabellaria* required 2.3×10^{-8} ergs and the recovery stroke 3.3×10^{-8} while for *Mytilus* the energetic costs were only about half as great. On the basis of the number of dynein sites per microtubule pair, one molecule of ATP per dynein molecule for each effective, and one for each recovery stroke would be required to meet these energy requirements.

Ciliary tracks in invertebrate larvae

Invertebrate larvae use bands of cilia for locomotion and to catch small particulate food. In the simplest arrangement, characteristic of echinoderms and hemichordates, there is a single row of cilia and particles are captured on its upstream side by an instantaneous reversal of beat, though the stimulus causing this reversal is not known (Strathmann, 1975). Unwanted particles pass over or through the row of cilia. Once the particles reach the oral region they may be caught up in a string of mucus which may be ingested or rejected by a massive reversal of ciliary beat (Strathmann and Bonar, 1976).

Trochophore larvae of molluscs and polychaetes and some rotifers use an opposed-band mechanism for food gathering. In this arrangement, particles are directed into a ciliated food groove by a preoral row of cilia and retained there by opposing postoral cilia. They are then transferred along the groove toward the mouth in a mucous stream. Particles can be rejected at the mouth or by cessation of activity by preoral cilia and those of the food groove (Strathmann et al., 1972).

A wide range of diatoms and flagellates, as well as some unidentifiable material, are ingested by these larvae. Particles less than 1 or 2 μm do not cause a reversal of beat in the single band system and particles up to 75 μm are taken; opposed band larvae seem capable of ingesting particles larger than about 1 μm. A clearance rate of 1.7 ml day^{-1} per mm of ciliated band has been calculated for a pluteus larva (Strathmann et al., 1972).

Choanocyte-driven filtration in sponges

Virtually the entire sponge is a complex filter. The external surface is covered with dermal pores (ostia) which vary somewhat in size from species to species, but probably have a minimum diameter of about 4 μm and a maximum less than 80 μm. Internally a complicated system of inhalent tubules, initially up to 400 μm in diameter, becomes more and more dendritic until the openings into the flagellated chambers have a diameter of only 1 to 6 μm. During this process the cross-sectional area of the system increases 20 to 30 times, ensuring a decrease in flow rate. The driving force is millions of flagellated collar cells (choanocytes) lining these chambers; apparently their beat is without any form of coordination. For water to enter the chambers it must pass around the bases of the cells and through spaces of 0.11 to 0.12 μm between the microvilli forming the collar. By this time the flow rate has been reduced to 0.2 to 0.4% of that at the opening to the inhalant canals. Excurrent canals collect the effluent, the total cross-sectional area of the tubing being reduced as the anastamosing branches become larger in diameter until the water is ejected at considerable velocity from the osculum, either a single jet in tubular forms or a number of smaller apertures in 'solid' forms. Oscular area is less than 1% of inhalant area (Reiswig, 1975a).

Particles can be phagocytised upon coming in contact with virtually any surface of the sponge. Those too large to enter the dermal pores can be ingested by dermal cells. Particles which enter the inhalent system, but are too large to reach the flagellated chambers may be consumed *en route* by amoebocytes and collenocytes. Choanocytes capture virtually all the remainder of the particulate matter. Bacteria can be captured with greater than 95% efficiency by several species (Reiswig, 1975b). Inorganic or non-nutritious particles may also be ingested by phagocytising cells, but whether any degree of selection can take place is unknown.

Ciliary mucous nets of mollusc gills

The molluscan gill can take various morphological shapes but in higher forms consists of a pair of ctenidia each bearing paired, V-shaped demi-branchs (Fig. 1). Water enters by the incurrent siphon or by the incurrent gape in the mantle, passes through spaces between ctenidial filaments into the infrabranchial space and eventually returns to the outside via the excurrent passages. The right side of the diagram illustrates a feeding mussel and the left, a non-feeding animal with demibranchs collapsed and mantle at least partially restricting flow (position 3). Most particulate matter is removed from the incoming water by lateral-frontal cilia. The cilia, which are compound structures bearing pinnae, form a fine reticulum when fully extended (Moore, 1971), but which can be raised to let greater amounts of particulate matter pass between the filaments if particle concentrations become too great for efficient retention (Dral, 1967).

Once captured, particles are incorporated into mucous strands and transported ventrally by frontal cilia. According to Foster-Smith (1975) they travel by several different pathways determined by their size: the smallest enter the marginal food groove by the most direct route but the biggest must enter the marginal food groove ventrally. Figure 1 also demonstrates how the concentration of particles entering the food grooves can be restricted by closing off certain tracts.

At the oral end of the marginal food groove, mucous streams spread out on to the pallial organs where a final selection or rejection of food takes place. The palps themselves bear rows of parallel ridges with ciliated tracks moving in different directions at different levels. A mucous stream can be directed to the mouth, can be shunted off to a rejection tract or can be resorted. While small mucous strings have a greater probability of being accepted, the sorting process appears to be completely mechanical.

Internal mucous nets of salps and tunicates

Tunicates and salps have similar feeding mechanisms consisting of a supporting pharyngeal basket containing ostia and lined with a mucous sheet which is constantly renewed by a ventral endostyle. Water entering by the oral aperture passes through the mucus, which intercepts the particulate

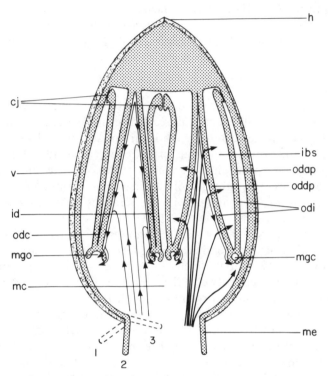

FIG. 1   Diagram of cross-section of mussel, indicating main water currents when the gill transports water both through the interfilamentar spaces and along the frontal surface of the gill filaments (right side of diagram, demibranchs are inflated) and when only the frontal surface of the filaments transports water (left side of diagram, demibranchs collapsed). Heavy lines indicate strong water currents: fine lines, weak currents. Currents are only shown between an outer and an inner demibranch. The diagram also illustrates shapes of the marginal groove from closed to widely open, as well as the varying orientation of the mantle edges in fully open animals. cj, ciliary junctions; h, hinge; id, inner demibranch; ibs, infrabranchial space; mc, mantle cavity; me, mantle edge; mgc, marginal food groove, closed; mgo, marginal groove, open; odc, outer demibranch, collapsed; odi, outer demibranch, inflated; odap, outer demibranch, ascending part; oddp, outer demibranch, descending part; v, valve. 1, mantle edges diverging; 2, mantle edges parallelly oriented; 3, mantle edges converging. (After Jørgensen, 1975)

matter, then passes through the ostia and over the gill bar, eventually leaving by the atriopore. At the dorsal side of the pharynx the mucous web is rolled up into a food string which is transported by cilia along the gill bar to the mouth. The mucous sheet may not always be continuous, and early workers thought that food was strained from the water by the ostia in ascidians (Jørgensen, 1966). However, Madin (1974) observed only continuous mucous nets in undisturbed salps studied *in situ* with SCUBA. Flow through the organism is probably largely under ciliary control in benthic ascidians (Carlisle, 1966) but in salps both locomotion and feeding are accomplished by muscular contraction of circular bands in the body wall in coordination with the opening and closing of muscular valves at the oral and atrial apertures (Madin, 1974).

Feeding efficiency is quite high in tunicates and there is no selection. Madin (1974) observed acceptance and successful ingestion of particles ranging from 1 μm to 1 mm by salps. In three species of ascidians 65 to 90% of the water pumped through the animal was swept clear of particulate matter (Fiala-Médioni, 1978). The specific filtration rate is much higher in salps, being about $48\,l\,h^{-1}$ (g dry wt.)$^{-1}$ in *Pegea confederata* (Harbison and Gilmer, 1976) and between 4 and $9\,l\,h^{-1}$ (g dry wt.)$^{-1}$ for three ascidian species (Fiala-Médioni, 1978).

External trapping devices in larvaceans and pseudothecosomes
Larvaceans, such as *Oikopleura*, have a pharyngeal basket, an endostyle and secrete a mucous net similar to that of tunicates, but they also secrete an elaborate concentrating device, the 'house' within which they live. Beating of the larvaecean tail within the house causes an inhalent current to enter through a coarse filter and forces water into the ventral level of a two-layered system of tubules separated by a net-like membrane with meshes about 0.8 to 0.1 μm (Fjerdingstad, in Jørgensen, 1966). Water presumably passes through this membrane into the dorsal system whence it is discharged to the outside, while particulate matter concentrates in a median groove from which it ultimately reaches the animal's mouth by some still uncertain mechanism.

Different patterns of swimming and feeding within the house were observed by Alldredge (1976) for several species studied *in situ* with SCUBA. New houses, already secreted on the trunk, were expanded in 1.5 to 5 minutes but old ones were not readily discarded. Besides food collection, the house appears to provide considerable protection against invertebrate predators, such as chaetognaths and medusae, and it also seems to provide buoyancy.

Pseudothecomsomes have been observed to secrete mucous webs up to 2 meters in diameter which are used as food collecting devices. After the web has been secreted the animal remains attached to it by its proboscis, while both sink slowly through the water; food particles and mucus are pulled into the proboscis and consolidated into a mucous string by ciliary action prior to ingestion (Gilmer, 1972).

Other systems; tentacles, tube feet, etc.
Many coelenterates and their near relations could probably be regarded as suspension feeders. Seafans, anemones, and corals expose a large surface to prevailing water currents and small planktonic animals touching the nematocyst-bearing tentacles can be dispatched and eaten. Similarly many medusae and cydippid ctenophores increase the probability of contacting suitable prey by exposing a tentacular web, armed with stinging or entangling cells, to a large volume of water.

   Corals can also use cilia and mucus to capture fine particulate matter. In most species examined, cilia beat toward the mouth only in its immediate vicinity; elsewhere they beat away from it out along and between the tentacles so that mucus and retained particles tend to accumulate between polyps. Doubtless this pattern of currents tends to keep the colony surface clean, but such accumulations of mucus can also be captured and redirected to the mouth by inward folding of the tentacles (Lewis and Price, 1978).

   A number of echinoderms are also variously modified for fine particle feeding. In *Antedon bifida*, a crinoid, ten arms bear pinnules about 1 to 1.4 mm apart, which in turn support tube feet of unequal length in groups of three; the longest and most lateral of these, when fully extended, fill in the interpinnular space to form a fine reticulum. There is a food groove down the center of each pinnule and food particles gathered by the extended tube feet are directed toward the mouth in a mucous stream by movements of the smaller tube feet and cilia (Nichols, 1960). Dense aggregations of ophiuroids are frequently found in current-washed areas and, at current speeds less than about 20 cm s^{-1}, arms are extended vertically with the ambulacral side facing the current. Two rows of finely papillate tube feet oriented at about 90° from each other extend laterally on either side of the arm. Adhering particles are gathered into a bolus and transported toward the mouth by a metacronal wave of tube feet (Warner and Woodly, 1975). Some species may spin mucous strands between spines on the arms which are subsequently harvested along with their captured particles by the tube feet (Fontaine, 1965). Among the echinoids the sand dollar *Dendraster excentricus* orients its ventral surface to a current with its anterior end partially buried in the sand and removes small particles largely by means of spines and tube feet (Timko, 1976). While most sea cucumbers are deposit feeders, some such as *Psolus chitinoides* spread a dendritic web of sticky tentacles to capture a range of particle sizes and types. However, stickiness is largely confined to wart-like adhesive papillae on the surface of the tentacles (Fankboner, 1978).

CIRRI

Cirri have no power of movement but must be driven by muscles. Crustacean cirral systems are part of the chitinous exoskeleton and are discarded and regenerated at each ecdysis. For the most part, cirral systems do not rely on mucus or other sticky substance for adhesion. Particles are generally captured by a net-like structure and then grasped, poked, pushed, shredded and crushed on their way to being ingested. While choice would seem to depend largely on the size of the particle, there is usually sufficient opportunity for the grazer to 'examine' what has been captured before consuming it.

'Passive' filters used by copepods and Cladocera

Most copepods can swim with or without feeding. The ultimate filter is usually the second maxilla, shown from the mid-line looking to the right in Fig. 2, in juxtaposition with the maxilliped. Cannon (1928) described a metachronal rhythm initiated posteriorly by the maxillipeds, which generated laterally swimming vortices and medially, coupled and countering feeding vortices. This feeding current supposedly moves in toward the midline under the tips of the swimming feet, forward, and then out through the interstices of the filter formed by the second maxillae, which Cannon thought were largely motionless.

Since that time several workers have emphasised the deviations from this pattern shown by copepods. Lowndes (1935) suggested that the range of limb movements were considerably more varied and that the second maxillae actively open and close the spaces between the cirri. In *Calanus hyperboreus* larger food items can also be captured if they contact the maxillipeds or second maxilla from the front (Conover, 1966). Note in Fig. 2 the effective screen provided by the limbs to particles passing posteriorly along the ventral body surface during normal swimming. Using a modern photomicrographic technique, Alcaraz *et al.* (1980) have shown that the copepod *Eucalanus crassus* can detect a particle at a distance, perhaps by chemoreceptors on the filtering appendages (Friedman, 1977), and can alter movements of the appendages to bring it to the mouth. In *Eucalanus*, feeding movements may occur in short bursts followed by a pause, perhaps to reassess the position of prey particles. There is no loss of efficiency or particles, however, as the remarkable ciné film of Alcaraz *et al.* reveals. Since viscous forces prevail at these scales of motion, when the limbs stop moving so do the proximal particles.

In most Cladocera the carapace forms a large suction chamber similar to

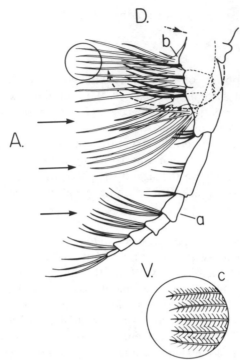

FIG. 2 Semi-diagrammatic view of left second maxilla and left maxilliped of *Calanus* as seen from the mid-line with right structure removed and showing interdigitated setae on both structures. Also shown are generalised current patterns during two modes of feeding. Solid arrows: currents during large particle feeding; Broken arrows: currents during filter feeding; a, maxilliped; b, second maxilla; c, details of setules on setae of second maxilla; D, dorsal; V, ventral; A, anterior

that of the bivalve molluscs. As the five pairs of thoracic limbs are moved forward and laterally, a current enters the anterior portion of the carapace, moving toward the dorsal food groove and through the overlapping filters formed by the endites of the third and fourth trunk limbs. Trunk limbs also combine to comb the filtering setae using tufts of setules on the next anterior limb. Once in the food groove, particles are swept forward to the region under the labrum where they are pushed into the mouth by the maxillule or possibly rejected. Unwanted food is flicked away by the post-abdominal abreptor (Lochhead, 1950).

In general, Cladocera are able to filter finer particles than can copepods. The marine *Penilia avirostris* is a non-selective feeder on particles smaller than 8 μm (Pavlova, 1959). Studies with natural foods using the Coulter Counter[R] suggest that particles less than 2–3 μm are not ingested by neritic copepods (Conover, 1978a).

Active cirral systems in euphausiids and barnacles
As in the copepods, feeding in euphausiids is an integral part of swimming activities but both need not occur simultaneously. Forward motion is caused by rhythmic beating of the abdominal pleopods. When not feeding, the water current passes from anterior to posterior through immobile thoracic limbs and over the gills. When feeding begins, rotatory movements of the exopods of the thoracic limbs divert a portion of the water flowing past, medially and than anteriorly into the food groove between the bases of the thoracic appendages (Mauchline and Fisher, 1969) or through coarse distal cirri between the legs (Berkes, 1975). Regardless of its route, once in the food groove, water moves forward and out through cirral filters on the proximal portion of legs and on the first and second maxillae. Setules on cirri of the mostly herbivorous euphausiids are generally less than 10 μm apart; such filter feeding forms, including *Euphausia superba*, are generally found in areas where large-sized phytoplankton predominate.

In the barnacles, feeding is accomplished by six pairs of biramous cirri morphologically associated with the head and thorax. In the Balanidae three pairs are rhythmically extended and withdrawn like a sweep net and the other three pairs are shorter, remaining largely within the shell. Other species groups have four or more pairs of cirri which they can extend. In areas of high current velocity, cirri may be extended without movement for long periods or until some prey object is captured. The extensible cirri usually have rather coarse inter-setal spacing and the particles consumed are usually small zooplankters. However, fine, hair-like setae are found on several of the mouthparts, especially the outer or second maxilla, and particles less than 2 μm have been found in the gut. Probably the fine particles are removed by these setae from the respiratory current generated by body movements within the shell. Such feeding can take place in the absence of rhythmic behaviour by the extended cirri (Crisp and Southward, 1961).

Gill rakers and other pharyngeal devices
In order to respire, fishes must take in water through the mouth, pass it across some kind of gill, and expel it under the operculum. In bony fishes the gills are supported on up to five gill bars and the spaces between them are frequently bridged by cirrus-like structures called gill-rakers. In some families of fishes the gill-rakers have been variously modified to capture small particles. For example, in the menhaden, *Brevoortia tyrannus*, which consumes phytoplankton readily, the gill rakers are numerous, long and close together on all five gill arches, the gaps between them being further subdivided by closely spaced hooks. Although earlier workers suggested that menhaden could remove particles as small as 1 or 2 μm, Durbin and

Durbin (1975) found that effective filtration began with particles in the range of 13 to 16 µm.

Some filter feeding fish have epibranchial organs, consisting of paired pharyngeal diverticuli, just posterior to the gap between the fourth and fifth gill arches. In threadfin shad, the rakers of the inner series of the lower branch of the fourth gill arch and those of the fifth form a funnel directed to the opening of the pharyngeal sac, which in this species is quite muscular, lined with mucoid epithelium and protected by cartilage. These structures are believed to facilitate the capture of plankton-sized food, perhaps gathering and compacting it, with the aid of mucus, into a bolus which presumably would be easier to ingest (Miller, 1964).

Baleen
Certain whales behave as fine-particle feeders using a baleen fringe to separate zooplankton and small fish from a large volume of water. Nemoto (1970) describes two basic feeding types, skimmers and swallowers, among the baleen whales, although some species such as the sei whale seem to use both methods. Among the skimmers are the right whales, which swim for relatively long periods with their mouths open feeding mostly on copepods. Their baleen cirri may be over two meters long, slender (0.2 mm in diameter), with from 35 to 70 individual elements per cm of jaw. The ratio of filtering area in m² to animal length in m is about 0.8. In contrast, the swallowers, such as the blue and fin whales, have short, stiff baleen fringes and a filtering area ratio around 0.2 or less. The swallowers can only meet their energetic needs in areas of dense plankton concentrations by opening the mouth and ingesting plankton and a large volume of water, which is then discharged through the baleen plates to effect the separation. The volume of water swallowed is enormous, however, up to 60 m³ in the blue whale, which is accomplished by expanding the mouth cavity, aided by pleat-like furrows lying under the throat (Pivorunas, 1979). Prey are generally larger and dominated by euphausiid shrimps.

3   Behavior and performance

As we have seen in this brief review of mechanisms, a wide range of different kinds of animals have evolved the ability to remove fine particles from suspension. The process cannot be specifically related to trophic level, habitat or taxonomic group, but many of these diverse kinds of organisms at times would seem to compete for the same energy resource. To compare performance of different organisms and to determine the capacity of a particular environment to support suspension feeders the concept of filtering rate or volume of water swept clear has been derived. Filtration rate as a function of the rate of oxygen uptake has been employed by Jørgensen (1966) to facilitate evaluation of an organism's potential to survive and grow in a particular environment. Sometimes the volume swept clear is related to some measure of the organism's size. In Table I are assembled some filtration rates for representative pelagic suspension feeders which have been calculated on approximately the same weight-specific basis, either by myself from data presented by individual authors or by the authors themselves. While considerably more examples could have been extracted from the literature on copepod feeding, there is very little data for comparison on suspension feeding in larger animals; indeed these estimates, particularly for the vertebrates, have necessitated a number of questionable assumptions. Still within very wide limits, there appears to be a trend toward reduced specific filtration rates at larger body sizes, although the range of variation would seem to be only three orders of magnitude in contrast with the 14 decade difference in size.

Filtration rate has also been related to size in certain groups of animals. Sushchenya and Khmeleva (1967) showed that ration I in g(animal)$^{-1}$day^{-1}, for a wide range of crustaceans, was related to live weight W in grams by $I = 0.0746\ W^{0.8}$. Conover (1978b) calculated a similar regression from data of Paffenhöfer (1971) and obtained $I = 2.02\ W^{0.74}$ for a system involving herbivorous copepods, but here I and W are in μg carbon. Both exponential constants fall within the range of those obtained by relating respiration and size.

Filtration rates for pelagic animals might be significantly greater than those for benthic filterers as already mentioned (p. 369). Harbison and Gilmer (1976) found that their measurements on salps were 10 to 40 times higher than those in the literature for sedentary tunicates. Similar calculations to those in Table I which I made on a range of filtration rates for bivalves from the literature generally yielded rates less than 0.1×10^4ml day^{-1}mgC^{-1}, but there is no data for a comparable-sized pelagic mollusc. Judging by data summarised in Ali (1970) and Winter (1978) filtration in molluscs is also strongly size dependent.

PHYSICAL FACTORS AFFECTING THE RATE OF SUSPENSION FEEDING

Remarkably little has been published concerning the effect of temperature on filtration or ingestion rates of marine zooplankton. Conover (1956) showed that some form of seasonal temperature acclimation occurred in the filtration rate of two copepods belonging to the genus *Acartia* in Long Island Sound. During fall and early winter the species which was dominant during the cold months, *Acartia clausi* (probably *A. hudsonica* Pinhey), had a higher specific filtration rate at low temperature than *A. tonsa*, the summer dominant, while at high temperatures the reverse was true. However, the

TABLE I Specific filtration rates for selected pelagic suspension feeders

Group and species	Approximate biomass[a] per individual (g wet wt.)	Experimental food source	Food concentrations ($\mu gC\ l^{-1}$)	Experimental temperature (°C)	Specific filtration rate[c] ($ml\ day^{-1}\ mgC^{-1}$)	Reference
PELAGIC ANIMALS						
Tintinnids						
Tintinnopsis cf. *acuminata*	10^{-8}g or 10^{-4} μm³	*Isochrysis galbana* *Monochrysis lutheri*	ca. 150	18	5.6×10^4	Heinbokel (1978)
COPEPODS						
Calanus pacificus[b] NV	10^{-5}	*Lauderia borealis*	49	15	4.9×10^4	Paffenhöfer (1971)
CI	5×10^{-5}	*Lauderia borealis*	49	15	3.0×10^4	Paffenhöfer (1971)
CIII	1.5×10^{-4}	*Lauderia borealis*	49	15	3.1×10^4	Paffenhöfer (1971)
CV	7.3×10^{-4}	*Lauderia borealis*	49	15	1.1×10^4	Paffenhöfer (1971)
C. cristatus CV	17.6×10^{-3}	Natural seston	212	4.9–7.4	0.0319×10^4	Taguchi and Ishii (1972)

Eurytemora N affinis	10^{-5}	Natural seston	633[d]	10	0.17×10^{4}[e]	Allan et al. (1977)
CIV	5×10^{-5}	Natural seston	400[d]	10	0.21×10^{4}[e]	Allan et al. (1977)
SALPS *Pegea confederata*	0.5–30	*Thalassiosira pseudonana*	range	25–29	1.93×10^{4}	Harbison and Gilmer (1978)
FISH Menhaden (*Brevoortia tyrannus*)	ca. 100 (260 mm)	*Acartia tonsa*	ca. 80–400	19.6	ca. 0.9×10^{4}	Durbin and Durbin (1975)
MAMMALS Sei Whale (*Balaenoptera borealis*)	ca. 18×10^{6}	*Calanus tonsus*	1.5×10^{3}– 2.2×10^{3}	ca. 2	ca. 0.004×10^{4}	Kawamura (1974)

[a] Assumed to be equivalent to volume, calculated from linear dimensions in some cases

[b] Called *C. helgolandicus* in original paper

[c] To make calculations, carbon was assumed to be 4% of net weight of suspension feeder unless given by author

[d] Calculated from author's particle volume data, using a regression calculated from similar Bedford Basin data, $gC\ l^{-1} = 87.3\ mm^{3}l^{-1} + 135.0$ (Conover, unpublished)

[e] Calculated from regression for maximum filtration rate against dry weight given by author

curve describing specific grazing rate at different temperatures for *A. clausi* was translated upward during the summer, virtually doubling the rates, while that of *A. tonsa* remained about the same. Anraku (1964) also presented some evidence for a seasonal adjustment in grazing rates for neritic copepods, including *Calanus finmarchicus*, and Fernández (1978) observed similar adjustments in some Mediterranean species. My studies of feeding by communities of neritic zooplankton on their natural food supply do not suggest a strong correlation with temperature (Fig. 3); food supply is apparently more important.

The effect of temperature on feeding and growth is also somewhat equivocal. In laboratory cultures, rates of ingestion and growth were accelerated at higher temperature, but the total amount of food consumed between instars in *Calanus* and *Rhincalanus*, and the ultimate size of the individuals, were little affected (Mullin and Brooks, 1970a). Growth efficiency was not affected by temperature, but decreased with increasing concentration of food. At lower concentrations, efficiency remained high but smaller individuals resulted (Mullin and Brooks, 1970b). The kind of food was probably as important as temperature in determining feeding and growth (Mullin and Brooks, 1970a,b; Conover, 1964).

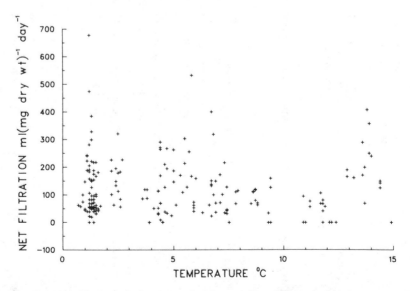

FIG. 3 Relationship between weight-specific, net filtration rate and temperature for the zooplankton community of Bedford Basin, Nova Scotia feeding on natural particulate matter at environmental concentration and at environmental temperature

Little can be said about the effects of other physical factors in the marine environment. Diurnal feeding rhythms and vertical migration patterns are well established and are probably cued by light, but scientists cannot agree on the benefits of this behaviour to the individual organisms involved. The organisms also differ in their responses between stages and at different seasons. Feeding was depressed by high light intensity in some copepods (Fernández, 1977). Oxygen levels vary with depth but their effect on feeding in the sea can as yet only be surmised from observations such as those of Kils (1979) on other changes in physiological behaviour. Decreasing salinity caused reduced feeding in *Calanus* (Anraku, 1964) and in two species of *Acartia* (Lance, 1964), but *Sulcanus conflictus* survived better with abundant food when salinities were raised above those of its normal habitat (Rippingale and Hodgkin, 1977). Presumably the extra food was necessary as an energy source to support osmoregulation.

The responses by filter feeders to physical factors in the environment are often unpredictable and this is to be expected if studies are carried out over only a short time relative to the life-span of the experimental organisms. Genetic adaptation and acclimation of metabolic rates to a changing physical environment, especially to temperature, are well-established for many invertebrates, including zooplankton, and are essential to an organism's survival. Much of the variability we measure in attempting to simulate natural conditions experimentally may be the direct result of our failure to recognise that the true response of an organism comes only after sufficient exposure for internal adjustment of the metabolic machinery.

The concept of temperature acclimation is better understood in the bivalve molluscs. In mussels there is no clear separation between respiration and food-getting functions and acclimation of each proceeds in a similar but not exactly parallel fashion (Fig. 4). Over its normal range of environmental temperature (5° to 20 °C) *Mytilus* regulates to provide a nearly constant filtration rate throughout the year (Bayne *et al.*, 1976). Acclimation of oxygen uptake occurs but is generally not complete. The standard rate does not acclimate at all, routine metabolism seems to be regulated best and active metabolism lies somewhere between. Because the responses to temperature of the different components of metabolism varies in some measure, the excess of energy assimilated over that used for maintenance, acquisition and digestion of food, the 'scope for growth', will have a rather complicated multidimensional response to other variables such as food supply (Widdows, 1978).

FOOD QUANTITY AND FEEDING RATES

The usual mathematical description of filtration rate implies that the amount of water swept clear of particles per unit time should be in-

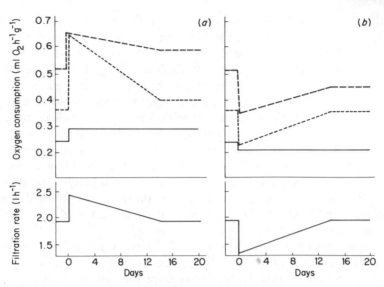

FIG. 4 A diagram to illustrate the change in active (– –), routine (——) and standard (——) rates of oxygen consumption, and in the filtration rate, of *Mytilus edulis* following a maintained change in temperature. (*a*) Increase in temperature from 15 to 20°C; (*b*) Decrease in temperature from 15 to 10°C. (After Bayne *et al.*, 1976)

dependent of particle concentration, at least over rather broad limits. Following the terminology of Frost (1972), the filtration rate $F = Vg/N$ where V is the volume of the experimental container, N is the number of grazers and g is an exponential constant describing the instantaneous rate of removal of particles from the container derived from an expression such as $C_2{}^* = C_1{}^* \, e^{(k-g)(t_2-t_1)}$. Here $C_1{}^*$ and $C_2{}^*$ represent particle concentrations in containers with grazers at time t_1 and t_2 and k is the coefficient of growth for the particle community.

Thresholds and switching
Many organisms, when given unlimited food, feed until they become satiated. Parsons *et al.* (1967) used a modified exponential function to describe zooplankton feeding so that $r = R_{max}(1 - e^{-kp} \times e^{kp_0})$, where r is the ration consumed, R_{max} is the saturation level, or upper limiting threshold, k is an exponential constant operating on particle concentration p and p_0 is the threshold concentration necessary to initiate a feeding response. Several other mathematical formulations of the functional response of a grazer to an unlimited food supply have been derived, but

Mullin *et al.* (1975) were unable to demonstrate statistical superiority among them with the zooplankton grazing data available.

The use of the Coulter Counter[R] has facilitated grazing studies with the natural assemblage of phytoplankton and particulate matter. In some cases, it has been possible to show the existence of a saturation asymptote with natural food concentrations (Parsons *et al.*, 1967, 1969). Over the course of a year, natural populations of neritic zooplankton responded in a linear fashion to changes in ambient concentration in the environment (Mayzaud and Poulet, 1978) and when saturation could be demonstrated, the level of R_{max} varied appreciably (Conover, 1978a). Earlier, Reeve and Walter (1977) found no fixed upper critical concentration in the feeding behavior of *Acartia tonsa*.

The Coulter Counter has provided information about the sizes of particles preferred by grazers. Poulet (1973, 1974) noted that neritic copepods fed on peaks in the particle spectrum, switching from one area of concentration to another as the first was grazed down. Presumably there was a threshold, analogous to p_0, which determined the concentration at which a certain-sized particle was no longer favoured. Hence, the lower threshold is probably some kind of a multivalued function, extending across the particle spectrum but not absolutely fixed, being influenced by the quantity of particles in other parts of the spectrum, and probably also by their quality (Conover, 1978a). Because the lower threshold is not single valued, it is scarcely surprising that previous investigations have given rather equivocal results (Frost, 1975).

The ecological significance of thresholds and switching has proved to be as controversial as determining their magnitude. Poulet (1973) emphasised that switching afforded a 'natural conservation' mechanism for the phytoplankton being grazed and Steele (1974) showed how the presence of a lower threshold could stabilise a pelagic population model. However, Landry (1976), by altering the form of the predation function and making a few other modifications in Steele's (1974) model, obtained a stable simulation without the assumption of thresholds. Certainly, the role of predation in stabilising populations has recently received considerable emphasis (Murdoch and Oaten, 1975), and switching is one option available to an optimal forager, whether herbivore or predator.

Controlling the amount of food ingested and absorbed
If we accept that the functional response curve of a suspension feeder approaches an asymptote in the presence of unlimited food, there must be a means of regulating its ingestion. Copepods can simply stop feeding or, as Esterly (1916) first described, an unwanted bolus of food can be flicked away by the second maxillae. *Daphnia magna* controls its food intake by

reducing the rate of movement of thoracic appendages and of swallowing while increasing its rejection rate (McMahon and Rigler, 1963). In the mussel, *Mytilus edulis*, ingestion can be kept constant over a range of concentration of *Dunaliella marina* from 10×10^6 to 40×10^6 cells 1^{-1}, but pseudofeces are produced at higher concentrations. Curiously, if some sediment is mixed in with the food, filtration, ingestion and growth rates all increase beyond those exhibited over this optimal range of food abundance (Winter, 1978).

When bivalve molluscs are fed natural seston instead of laboratory cultures, the story becomes more complicated. The natural particulate matter may have a relatively low organic content, for example only 5 to 25% in the Lynher estuary, while its total concentration varies several hundred percent. In such circumstances, pseudofecal production, which may be initiated at concentrations of 2 to 5 mg 1^{-1}, becomes a variable to be assessed in order to determine true ingestion rate. Pseudofeces production and seston concentration proved to be approximately proportional up to quite high levels, while the useful organic matter ingested was about the same as in the more typical response curve obtained with cultures (Widdows *et al.*, 1979).

If the amount of food ingested is constant over a range of concentration, the percentage absorbed should also remain constant, and this appears to be the case in *Mytilus* populations. Where the functional response to increasing concentration of food does not show a tendency toward saturation, as in the neritic plankton community studied by Mayzaud and Poulet (1978), reduction in digestive efficiency might be expected, but Conover (1980) found no significant change in percentage carbon absorbed, even when concentrations of natural particulate matter were increased several times higher than ambient. We would argue then that zooplankton populations normally are able to acclimate to changes in food supply about as fast as the changes themselves occur (Conover, 1978a; Mayzaud and Poulet, 1978).

SELECTIVE FEEDING

The use of some kind of net by a suspension feeder to capture fine particles would suggest that size-selective feeding should predominate. If the net has fixed apertures, their dimensions can be measured and the resulting distribution used to predict retention efficiencies for different particle sizes, assuming they have a more or less compact form (Boyd, 1976; Nival and Nival, 1976). On the other hand, some grazers may be able to alter the distances between cirri on their filtration appendages (Lowndes, 1935; Wilson, 1973; Friedman, 1977). A number of studies support the concept that calanoid copepods tend to select large particles preferentially (e.g., Mullin, 1963; Frost, 1977).

Studies with electronic particle counters are not in complete agreement with a wholly size-selective feeding hypothesis. The problem is further complicated by some differences in the definition of 'selective feeding' among authors. Poulet (1974) considers feeding in *Pseudocalanus* to be 'opportunistic' rather than selective, because it is concentrated on peaks in the particle biomass spectrum as they occur rather than on a specific size category. Although certain portions of the particle spectrum yielded positive electivity indices, selective utilisation could not be statistically demonstrated (Poulet and Chanut, 1975). Evidence for selection was graphically demonstrated by Allan *et al.* (1977) who plotted filtration rate against particle size for different stages of juvenile copepods feeding on natural particulate matter. By their definition, a uniform filtration rate at all size categories across the particle spectrum should indicate nonselective feeding, a pattern most nearly duplicated by nauplii of *Eurytemora affinis* and *Acartia* spp. Later developmental stages of these copepods showed a strong tendency for peaks in filtration rates which quite often coincided with biomass peaks in the particle spectrum as found earlier by Poulet (1973, 1974).

When feeding is concentrated on a single biomass peak in the particle spectrum, it is still difficult to rule out a size-selective mechanism such as the 'leaky sieve' proposed by Boyd (1976) or some form of pore size modification (Wilson, 1973). However, when copepods concentrate on two or more peaks of a polymodal particle spectrum, as shown by Richman *et al.* (1977), it is harder to accept a strictly mechanistic model of particle selection on the basis of present knowledge of mouthpart morphology and behaviour. In my own experiments (Conover, 1978a), the insertion into the particle spectrum of a biomass peak between 40 and 64 µm, consisting mainly of dinoflagellates belonging to the genus *Ceratium*, completely inhibited feeding on that size category, where it had occurred prior to the manipulation, even though the zooplankton continued to feed on both larger and smaller sizes. Post-capture rejection was probably involved and also in another experiment in which inert plastic spheres were ignored when fed to *Acartia* together with diatoms that were just smaller and just larger in size, both of which were readily consumed (Donaghay and Small, 1979).

Some kind of chemical cue is probably involved in selective feeding as well. Conover (1966) described how *Calanus hyperboreus* first brought its own fecal pellets to the mouth region as though to taste them, after which they were discarded. Chemoreceptors have been described for copepods by Elofsson (1971) and Friedman and Strickler (1975). Chemosensory selection, regardless of particle size, for microcapsules which contained a phytoplankton homogenate has been demonstrated for two neritic copepod species by Poulet and Marsot (1978).

Chemoreception seems to be important in barnacles (Crisp, 1967; Allison and Dorsett, 1977) suggesting a strong preference for animal tissue.

Very little can be added to the subject of selective feeding for other suspension feeders, aside from the mechanical sorting systems already described. Mussels show a slight positive selection for *Phaeodactylum* cells compared with other particles in the same Coulter Counter[R] size range, which may indicate that the molluscan gill, like the copepod maxilla, can act like a 'leaky sieve' in sorting irregularly shaped objects (Bayne *et al.*, 1977).

PREVIOUS FEEDING HISTORY

Some years ago Harvey (1937) showed that *Calanus* prefed on a smaller diatom, *Lauderia*, selected it in the presence of the larger, normally preferred *Ditylum*. We will now examine in more detail some of the effects of conditioning or the lack of it on the behaviour of suspension feeders.

Starvation

When *Calanus hyperboreus* is brought into the laboratory for physiological studies, over most of the year the metabolism and feeding rate are depressed, but both will increase after several weeks captivity (Conover, 1962). Only during the biological spring were grazing and respiration rates initially high.

At the other end of the scale, a period of starvation may induce a marked deviation from the normal functional response curve for a grazer as demonstrated for *Daphnia magna* by McMahon and Rigler (1965) and for *Calanus pacificus* by Frost (1972). This over-shoot phenomenon is probably a normal characteristic of feeding behavior in predators whose food supply may be sporadic (Conover, 1978b). Reeve and Walter (1977) have pointed out that some flexibility in the response of grazers to a sudden increase in food concentration would be of considerable advantage in a patchy environment.

Enzymes and acclimation

As mentioned in an earlier section, there is a linear correlation between the ingestion rates of the major zooplankton species and their food supply in Bedford Basin, a coastal inlet in Nova Scotia (Mayzaud and Poulet, 1978) but no correlation of filtration rate with environmental temperature (Fig. 3). Even though the total quantity ingested varied widely, the fraction absorbed remained roughly constant at about 60% (Conover, 1980). A partial explanation for this apparent paradox is perhaps associated with flexibility in concentration of at least six digestive enzymes in the zooplankton which are positively correlated with substrate concentration (Mayzaud and Poulet, 1978).

At present, we are uncertain about the period required for the zooplankton to adjust their enzyme complement, but it is probably greater than 24 hours and less than 6 days. It has also been difficult to demonstrate conclusively enzyme induction in small neritic copepods in the laboratory, although some recent unpublished results suggest that high concentrations of cultured algae as food may induce amylase production in *Acartia clausi*. Best support for the enzyme induction hypothesis comes from still unpublished studies in which *Euphausia pacifica*, previously fed *Artemia* nauplii, showed significant laminarinase production in about two days when switched to an algal diet (Cox, 1979).

A somewhat analogous behavior may be associated with digestion in lamellibranchs. Rhythmic patterns of feeding and relative quiescence have been observed which may be endogenous, daily or tidal. Accompanying these changes in behavior are cyclic changes in digestive tissues, such as the dissolution and regeneration of the enzyme-containing crystalline style, formation of fragmentation granules in the stomach, absorption and digestion in the cells of the digestive diverticulum, followed by their disintegration as the nourishment is distributed elsewhere in the body (Morton, 1973). Regrettably, correlative enzyme cycling has not yet been quantitatively demonstrated.

Experience
That previous and prolonged contact with a set of food conditions can influence selectivity has been reported several times since the first observations of Harvey (1937). Conover (1966) observed that prefeeding on large *Coscinodiscus* or small *Thalassiosira* predisposed female *Calanus hyperboreus* to select one or the other in mixtures. The copepods could also acquire a taste for their own eggs and subsequently would select them instead of a large diatom. Skiver (1979) preconditioned three common neritic copepod species to specific bead sizes and then examined their responses to various mixtures. Each of the species behaved differently: Unconditioned *Pseudocalanus* preferred smaller particles regardless of their concentration but shifted their preference with conditioning; *Acartia hudsonica* had a variable response to acutely offered particle mixtures, but could be preconditioned to small particles only in summer, perhaps because its natural food supply had shifted toward small flagellates; *Temora longicornis*, however, could not be preconditioned, showing a strong preference for large particles when available, and feeding on other sizes in proportion to their relative abundance. On the other hand, in the experiments of Donaghay and Small (1979), experienced *Acartia clausi* rejected beads intermediate in size between two diatoms, whether offered singly or in combination, but those conditioned to diatoms seemed to make some mistakes before learning to

reject beads. The larger diatom species was always preferred by the copepods preconditioned to it, but those experienced with the smaller tended to feed on each size of diatom according to its theoretical capture efficiency based on setal spacing of the maxillary filter (Nival and Nival, 1976). As mentioned earlier, postcapture rejection was suggested as a mechanism for avoiding bead ingestion, but the process reduced efficiency compared with a no-bead food supply.

Experience probably has a role in other experiments of Skiver (1980) although just how it operates is still uncertain. In this series, he took natural sea water and diluted it by half before reconstituting the original particle spectrum with appropriately-sized beads. He then compared feeding behavior for the same three species of copepods mentioned above, singly and in various combinations. In two of the three there were seasonal shifts in resource utilisation, but, more interesting, sometimes, but not always, the pattern also changed when two species interacted.

4 Models, budgets and life history strategies

I want now to try to pull these numerous observations together by discussing a few recent models. Over the past few years, beginning with the work of Emlen (1966) and MacArthur and Pianka (1966), a series of models concerned with optimal foraging have been developed. Through the natural selection process, organisms have evolved greater fitness as food gatherers by becoming either 'energy maximisers' or 'time minimisers' (Schoener, 1971). Optimisation models have been concerned with decisions about kind of food, optimal patch type, how long to spend in a patch and the patterns and speed of movement while feeding or getting to the feeding site (Pyke *et al.*, 1977). Two recent papers have applied optimisation concepts to pelagic grazers and will be considered in some detail.

In the first of these, Lam and Frost (1976) assumed a three phase functional response curve with reduced filtration below some threshold concentration above which filtration rate was maximal, while ingestion and growth increased up to a critical concentration above which filtration rate was again reduced. Assimilation was assumed to be constant while respiration and filtration were equated to the size of the grazer. They optimised filtering speed, assuming turbulent drag forces, but argued that the assumption of viscous forces would not greatly affect the form of the function, at least at higher filtering rates. Feeding was assumed to be entirely size-selective with either an adjustable filter to facilitate capture of larger particles or a fixed filter, in each case the filter area being proportional to the length of the grazer. The model more nearly duplicated reality when a filter with variable apertures was assumed.

The second optimal foraging model, that of Lehman (1976), assumes that flow is viscous around the filtering apparatus and that assimilation is a variable related to gut passage time. He first optimised the difference between energy assimilated and that expended in filtration, the major restraint being gut volume. Filtering efficiency was imposed later, firstly assuming that some ingested particles might be rejected, and secondly, considering various types of fixed sieves operating as selective devices. The ingestion model supported the 'leaky sieve' and did not depend on a filter with variable pore sizes. Although the final conclusions are similar to those of Lam and Frost (1976), the way of reaching them required assumptions more relevant to a typical cladoceran than to a copepod.

In the Lam and Frost (1976) model, the option of a variable mesh filter affords greater flexibility to the grazer in dealing with variable food supplies, in agreement with empirical evidence (e.g. Richman et al., 1977; Conover, 1978a; Donaghay and Small, 1979). Cladocera, like the molluscan filter feeders, do not appear to be selective and apparently can only regulate filtering and rejection rates (compare McMahon and Rigler, 1963; Widdows et al., 1979). The Lehman model animal obtains its best performance with a filter having low variance, assuming most of the particles are similar in size and large enough to be efficiently captured. It seems that the Lam and Frost assumptions describe an animal better able to cope with oligotrophy and diversity, while that of Lehman is specialised for eutrophy and particle-size homogeneity. In the sea, Cladocera are largely confined to the more eutrophic coastal zone.

The Lam and Frost (1976) model has been incorporated, with a few modifications, into a model of a boreal planktonic community by Steele and Frost (1977) in which nutrients and light interact to produce phytoplankton of several sizes which are grazed by a large herbivore Calanus finmarchicus and a medium-sized herbivore Pseudocalanus minutus which in turn are subjected to two forms of predation. In this model food capture was accomplished with a fan-like filter proportional to the weight of the grazer, and drag was taken so as to simplify the energy term while yielding a Reynold's number not far from 1. The system is optimised at any food concentration for a feeding rate that maximises growth below some maximum rate.

The simulation makes several good points concerning community structure. With a big grazer only, small cells come to dominate the system and the grazer is eliminated. Smaller grazers lead to a generally larger-sized phytoplankton community and survival of the larger grazer. If aperture sizes of the maxillary filters overlap greatly between the two species, the smaller Pseudocalanus drives Calanus to extinction following the spring bloom. If predation is independent of prey size, Calanus is most affected and can be eliminated if the rate becomes too great. To simulate differing types of predators in the community, predation was also made proportional to the cube root of the mass of prey in which case both herbivores survived ninety days although the phytoplankton community shifted toward smaller sizes. With certain nutrient-growth relations for phytoplankton the system also tended toward very small cells and the extinction of both grazers. Inclusion of microzooplankton in the model, capable of feeding on the smaller particles, might stabilise the simulation, but was not tried. Alternatively, grazers with a capability to feed over a broader spectrum of particle sizes, such as salps (Madin, 1974) or *Calanus plumchrus*, which despite its relatively large size has a rather fine sieve (Steele and Frost, 1977), might be helpful, but could also lead to extinction of the least efficient, probably the larger, competitor.

Certainly these models are far from perfect even though certain processes in the ecosystem can be realistically simulated with the proper set of conditions. That the model fails to deal perfectly with nature is an advantage because it emphasises the weaknesses in our understanding and can also serve as a point of departure for discussion of organisms and behaviour patterns which do not conform to those simulated.

One weakness of these grazing models, emphasised by Lehman (1976), but also alluded to by Lam and Frost (1976), is that they are based on a functional relationship that is best understood when the food concentration is high enough for the filtering rate to be maximal. When there is plenty of suitable food, selection for the biggest cells that can be handled is obviously good energetic policy. However, when the resource is already partially exploited so that growth by the grazer is limited or could become negative if the proper strategy were not followed, the selection of curves and functions by the modeller is guess work.

The choice of *Calanus* and *Pseudocalanus* as model grazers was logical because from our still limited understanding of copepod feeding behavior, these two come closest to following a relatively stereotyped feeding pattern. Their maxillae are relatively unspecialised, they may use more or less continuous high frequency limb motions to generate feeding currents, their feeding may be strongly size selective and they may demonstrate a typical saturation type of functional response to increasing food supply (Frost, 1977).

While this is one pattern of their behavior, it is not the only way that they respond to feeding conditions. As I have mentioned, members of both genera can be conditioned to different foods (Harvey, 1937; Conover, 1966; Skiver, 1979). Over a season *Pseudocalanus* does not have a critical food density above which clearance rate is reduced (Mayzaud and Poulet, 1978) nor can it always be shown to reduce its filtering rate when the food supply is increased substantially beyond its ambient background (Conover, unpublished data). Younger stages of *Calanus plumchrus* sometimes show similar behavior (Figure 3b in Parsons *et al.*, 1969). *Pseudocalanus* probably can alter its enzyme complement with environmental conditions (Mayzaud and Conover, unpublished data) and so apparently can *Calanus finmarchicus* (Hirche, personal communication). *Calanus* spp. do not have to feed on phytoplankton cells, being also quite capable of capturing and ingesting small animals (Corner *et al.*, 1976).

When feeding on large particles, including small animals, filtering is not practiced by *Calanus hyperboreus*. Instead the second maxillae and maxillipeds are used to seize large cells and other prey raptorially when encountered or sensed remotely. Indeed the processes of large particle feeding and of filtration may be mutually exclusive (Conover, 1966), which is sound logic considering that vastly greater volumes of water could be 'searched' by swimming normally at low energy cost (Vlyman, 1970) compared with that which could be pumped through a maxillary filter with a very unfavorable balance between energy expended for that gained. This is clearly one alternative strategy that the grazer may utilise at food concentrations approaching the lower limiting threshold, and it may well explain why there have been many equivocal experimental results from studies of this part of the functional response curve (Frost, 1975). Moreover, carnivores are generally larger than their prey by a considerable amount so that the efficiency advantage probably shifts from the smaller grazer to the bigger, facultative omnivore when filter feeding becomes really unprofitable. Omnivory clearly pays for the majority of zooplankton in the mid- and upper layers of relatively oligotrophic parts of the ocean, judging by the numbers of species with this mode of nutrition found there. It also facilitates efficient entry of the smallest producing particles, the μ-flagellates, into the food webs by consumption of their predators, the microzooplankton. This is part of the particle size optimisation process (Conover, 1978a); while it makes for food chain efficiency, because it reduces the amount of energy that goes to the decomposer side of the energy balance sheet, it does not actually increase the net production of the system (Conover, 1979).

While omnivory-carnivory is one strategy when filter feeding becomes energetically uneconomic, there is also the 'opt-out' option of hybernation or diapause. Many *Calanus* species and near relatives, especially the larger forms such as *Calanus plumchrus* and *C. hyperboreus*, follow this route to survival following the spring increase. By the end of the bloom they have accumulated much high-energy storage products in the form of wax esters (Lee, 1974). By reducing their metabolic rate and sinking to a considerable depth, they avoid the predation pressures of the near-surface zone, save energy by placing themselves in cold storage, and avoid competition with

Pseudocalanus-sized grazers when the Steele-Frost computer simulation says that they will lose the competition. These copepods were said by Heinrich (1962) to reproduce independently of the spring bloom. In fact, the opposite is true, because their life cycles are dependent on reproducing, presumably by some remarkably accurate biological clock, just sufficiently prior to the spring bloom that their developmental stages have a head start on the smaller, more-efficient-in-the-short-run filter feeders which are spawned after a period of parental grazing. The larger size of the hibernating copepods probably also makes them more fecund than their smaller competitors (Hopkins, 1977).

One final point about the feeding physiology of optimal foragers: It makes little sense energetically to approach your only meal of the year with a limit on the amount of energy that you can ingest such as a fixed upper critical concentration for efficient food utilisation. We find no saturation of feeding of small neritic copepods during the spring bloom in Bedford Basin and recently my colleague Mark Huntley (personal communication) has found exactly the same thing for *Calanus finmarchicus*, *C. glacialis* and *C. hyperboreus*, as well as *Metridia longa*, in the Labrador Sea during a spring bloom that peaked around 14 mg Chl$a$ m^{-3}.

References

Alcaraz, M., Paffenhöfer, G.-A. and Strickler, J. R. (1980). Catching the algae: A first account of visual observations on filter-feeding calanoids. *In* "The Evolution and Ecology of Zooplankton Communities" (W. C. Kerfoot, ed.) Special Symposium III, Amer. Soc. Limnol. Oceanogr. pp. 241–248. University Press of New England, Hanover, London

Ali, R. M. (1970). The influence of suspension density and temperature on the filtration rate of *Hiatella arctica*. *Mar. Biol.* **6**, 291–302

Allan, J. D., Richman, S., Heinle, D. R. and Huff, R. (1977). Grazing in juvenile stages of some estuarine copepods. *Mar. Biol.* **43**, 317–331

Alldredge, A. L. (1976). Field behavior and adaptive strategies of appendicularians (Chordata: Tunicata). *Mar. Biol.* **38**, 29–39

Allison, P. and Dorsett, D. A. (1977). Behavioral studies on chemoreception in *Balanus hameri*. *Mar. Behav. Physiol.* **4**, 205–217

Anraku, M. (1964). Influence of the Cape Cod Canal on the hydrography and on the copepods in Buzzards Bay and Cape Cod Bay, Massachusetts. II. Respiration and feeding. *Limnol. Oceanogr.* **9**, 195–206

Bayne, B. L., Thompson, R. J. and Widdows, J. (1976). Physiology: I. *In* "Marine Mussels: Their Ecology and Physiology" (B. L. Bayne, ed.) pp. 121–206. Cambridge University Press, London and New York

Bayne, B. L., Widdows, J. and Newell, R. I. E. (1977). Physiological measurements on estuarine bivalve molluscs in the field. *In* "Biology of Benthic Organisms" (B. F. Keegan, P. O. Ceidigh and P. J. S. Boaden, eds) pp. 57–68. Pergamon Press, Oxford and New York

Berkes, F. (1975). Some aspects of feeding mechanisms of euphausiid crustaceans. *Crustaceana* **29**, 266–270

Boyd, C. M. (1976). Selection of particle sizes by the filter-feeding copepods: a plea for reason. *Limnol. Oceanogr.* **21**, 175–180

Cannon, H. G. (1928). On the feeding mechanism of the copepods *Calanus finmarchicus* and *Diaptomus gracilis. J. exp. Biol.* **6**, 131–144

Carlisle, D. B. (1966). The ciliary current of *Phallusia* [Ascidiacea] and the squirting of sea squirts. *J. mar. biol. Ass. U.K.* **46**, 125–127

Conover, R. J. (1956). Oceanography of Long Island Sound, 1952–1954. VI. Biology of *Acartia clausi* and *A. tonsa. Bull. Bingham Oceanogr. Coll.* **15**, 156–233

Conover, R. J. (1962). Metabolism and growth in *Calanus hyperboreus* in relation to its life cycle. *Rapp. P.-v. Réun. Cons. int. Explor. Mer* **153**, 190–197

Conover, R. J. (1964). Food relations and nutrition of zooplankton. *In* "Proceedings of Symposium on Experimental Marine Ecology", Occ. Publ. No. 2, pp. 81–91. Grad. School Oceanogr., University of Rhode Island

Conover, R. J. (1966). Feeding on large particles by *Calanus hyperboreus* (Kröyer). *In* "Some Contemporary Studies in Marine Science" (H. Barnes, ed.) pp. 187–194. George Allen and Unwin, London

Conover, R. J. (1978a). Feeding interactions in the pelagic zone. *Rapp. P.-v. Réun. Cons. int. Explor. Mer* **173**, 66–76

Conover, R. J. (1978b). Transformation of organic matter. *In* "Marine Ecology" (O. Kinne, ed.) Vol. IV, pp. 221–499. John Wiley and Sons, Chichester and New York

Conover, R. J. (1979). Secondary production as an ecological phenomenon. *In* "Zoogeography and Diversity of Plankton" (S. van der Spoel and A. C. Pierrot-Bults, eds) pp. 50–86. Bunge Scientific Publishers, Utrecht

Conover, R. J. (1980). Zooplankton populations and what is required for their well being. Proc. Marine Science and Ocean Policy Symposium. University of California, Santa Barbara, 17–20 June, 1979, pp. 37–50

Corner, E. D. S., Head, R. N., Kilvington, C. C. and Pennycuick, L. (1976). On the nutrition and metabolism of zooplankton. X. Quantitative aspects of *Calanus helgolandicus* feeding as a carnivore. *J. mar. biol. Ass. U.K.* **56**, 345–358

Cox, J. L. (1979). Enzyme induction in zooplankton. Poster Session, Marine Science and Ocean Policy Symposium. University of California, Santa Barbara. 17–20 June 1979

Crisp, D. J. (1967). Chemoreception in cirripedes. *Biol. Bull., mar. biol. Lab., Woods Hole* **133**, 128–140

Crisp, D. J. and Southward, A. J. (1961). Different types of cirral activity of barnacles. *Phil. Trans. Roy. Soc. (B)* **243**, 271–308

Donaghay, P. L. and Small, L. F. (1979). Food selection capabilities of the estuarine copepod *Acartia clausi. Mar. Biol.* **52**, 137–146

Dral, A. D. G. (1967). The movements of the latero-frontal cilia and the mechanism of particle retention in the mussel (*Mytilus edulis* L.). *Netherlands J. Sea Res.* **3**, 391–422

Durbin, A. G. and Durbin, E. G. (1975). Grazing rates of the Atlantic menhaden *Brevoortia tyrannus* as a function of particle size and concentration. *Mar. Biol.* **33**, 265–277

Elofsson, B. (1971). The ultrastructure of a chemoreceptor organ in the head of copepod crustaceans. *Acta Zool.* **52**, 299–315

Emlen, J. M. (1966). The role of time and energy in food preference. *Am. Nat.* **100**, 611–617

Esterly, C. O. (1916). The feeding habits and food of pelagic copepods and the question of nutrition by organic substances in solution in the water. *Univ. Calif. Publ. Zool.* **16**, 171–184

Fankboner, P. V. (1978). Suspension-feeding mechanisms of the armoured sea cucumber *Psolus chilinoides* Clark. *J. exp. mar. Biol. Ecol.* **31**, 11–25

Fernández, F. (1977). Efecto de la intensidad de luz natural en la actividad metabólica y en la alimentación de varias especies de copépodos planctonicas. *Inv. Pesq.* **41**, 575–602

Fernández, F. (1978). Metabolismo y alimentación en copépodos planctonicos del Mediterráneo: Respuesta a la temperatura. *Inv. Pesq.* **42**, 97–139

Fiala-Médioni, A. (1978). Filter-feeding ethology of benthic invertebrates (ascidians). IV. Pumping rate, filtration rate, filtration efficiency. *Mar. Biol.* **48**, 243–249

Fontaine, A. R. (1965). The feeding mechanisms of the ophiuroid *Ophiocomina nigra*. *J. mar. biol. Ass. U.K.* **45**, 373–385

Foster-Smith, R. L. (1975). The role of mucus in the mechanism of feeding in three filter-feeding bivalves. *Proc. malac. Soc. Lond.* **41**, 571–588

Friedman, M. M. (1977). "Electron Microscopic Studies of the Filter-Feeding of Calanoid Copepods". Ph.D. Thesis, The Johns Hopkins University, Baltimore

Friedman, M. M. and Strickler, J. R. (1975). Chemoreceptors and feeding in calanoid copepods (Arthropoda: Crustacea). *Proc. Nat. Acad. Sci. U.S.A.* **72**, 4185–4188

Frost, B. W. (1972). Effects of size and concentration of food particles on the feeding behavior of the marine planktonic copepod *Calanus pacificus*. *Limnol. Oceanogr.* **17**, 805–815

Frost, B. W. (1975). A threshold feeding behavior in *Calanus pacificus*. *Limnol. Oceanogr.* **20**, 263–266

Frost, B. W. (1977). Feeding behavior of *Calanus pacificus* in mixtures of food particles. *Limnol. Oceanogr.* **22**, 472–491

Gilmer, R. W. (1972). Free-floating mucus webs: a novel feeding adaptation for the open ocean. *Science* **176**, 1239–1240

Harbison, G. R. and Gilmer, R. W. (1976). The feeding rates of the pelagic tunicate *Pegea confederata* and two other salps. *Limnol. Oceanogr.* **21**, 517–528

Harvey, H. W. (1937). Note on selective feeding by *Calanus*. *J. mar. biol. Ass. U.K.* **22**, 97–100

Heinbokel, J. F. (1978). Studies on the functional role of tintinnids in the Southern California Bight. II. Grazing rates of field populations. *Mar. Biol.* **47**, 191–197

Heinrich, A. K. (1962). The life histories of plankton animals and seasonal cycles of plankton communities in the oceans. *J. Cons. int. Explor. Mer* **27**, 15–24

Hopkins, C. C. E. (1977). The relationship between maternal body size and clutch size, development time and egg mortality in *Euchaeta norvegica* (Copepoda: Calanoida) from Loch Etive, Scotland. *J. mar. biol. Ass. U.K.* **57**, 723–733

Jørgensen, G. B. (1966). "Biology of Suspension Feeding". Pergamon Press, Oxford and New York

Jørgensen, C. B. (1975). On gill function in the mussel *Mytilus edulis* L. *Ophelia* **13**, 187–232

Kawamura, A. (1974). Food and feeding ecology in the southern sei whale. *Sci. Repts. Whales Res. Inst. No.* **26**, 25–144

Kils, U. (1979). Performance of Antarctic krill *Euphausia superba*, at different levels of oxygen saturation. *Meeresforsch.* **27**, 35–48

Lam, R. K. and Frost, B. W. (1976). Model of copepod filtering response to changes in size and concentration of food. *Limnol. Oceanogr.* **21**, 490–500

Lance, J. (1964). Feeding of zooplankton in diluted sea-water. *Nature, Lond.* **201**, 100–101

Landry, M. R. (1976). The structure of marine ecosystems: An alternative. *Mar. Biol.* **35**, 1–7

Lee, R. F. (1974). Lipid composition of the copepod *Calanus hyperboreus* from the Arctic Ocean. Changes with depth and season. *Mar. Biol.* **21**, 313–318

Lehman, J. T. (1976). The filter-feeder as an optimal forager, and the predicted shapes of feeding curves. *Limnol. Oceanogr.* **21**, 501–516

Lewis, J. B. and Price, W. S. (1976). Patterns of ciliary currents in Atlantic reef corals and their functional significance. *J. Zool., Lond.* **178**, 77–89

Lochhead, J. H. (1950). *Daphnia magna. In* "Selected Invertebrate Types" (F. A. Brown, Jr., ed.) pp. 399–406. Wiley, New York and London

Lowndes, A. G. (1935). The swimming and feeding of certain calanoid copepods. *Proc. Zool. Soc. Lond.* 1935, 687–715

MacArthur, R. H. and Pianka, E. R. (1966). On optimal use of a patchy environment. *Am. Nat.* **100**, 603–609

Madin, L. P. (1974). Field observations on the feeding behavior of salps (Tunicata: Thaliacea). *Mar. Biol.* **25**, 143–147

Mauchline, J. and Fisher, L. R. (1969). The Biology of Euphausiids. *In* "Advances of Marine Biology" (F. S. Russell and M. Yonge, eds) Vol. 7, pp. 1–454. Academic Press, New York and London

Mayzaud, P. and Poulet, S. A. (1978). The importance of the time factor in the response of zooplankton to varying concentrations of naturally occurring particulate matter. *Limnol. Oceanogr.* **23**, 1144–1154

McMahon, J. W. and Rigler, F. H. (1963). Mechanisms regulating the feeding rate of *Daphnia magna* Straus. *Can. J. Zool.* **41**, 321–332

McMahon, J. W. and Rigler, F. H. (1965). Feeding rate of *Daphnia magna* Straus in different foods labelled with radioactive phosphorus. *Limnol. Oceanogr.* **10**, 105–113

Miller, R. V. (1964). The morphology and function of the pharyngeal organs in the clupeid *Dorosoma petenese* (Gunther). *Chesapeake Sci.* **5**, 194–199

Moore, H. J. (1971). The structure of the latero-frontal cirri on the gills of certain lamellibranch molluscs and their role in suspension feeding. *Mar. Biol.* **11**, 23–27

Morton, B. (1973). A new theory of feeding and digestion in the filter-feeding Lamellibranchia. *Malacologia* **14**, 63–79

Mullin, M. M. (1963). Some factors affecting the feeding of marine copepods of the genus *Calanus. Limnol. Oceanogr.* **8**, 239–250

Mullin, M. M. and Brooks, E. R. (1970a). Growth and metabolism of two planktonic, marine copepods as influenced by temperature and type of food. *In* "Marine Food Chains" (J. H. Steele, ed.) pp. 74–95. Oliver and Boyd, Edinburgh

Mullin, M. M. and Brooks, E. B. (1970b). The effect of concentration of food on body weight, cumulative ingestion and rate of growth of the marine copepod *Calanus helgolandicus. Limnol. Oceanogr.* **15**, 748–755

Mullin, M. M., Stewart, E. F. and Fuglister, F. J. (1975). Ingestion by planktonic grazers as a function of concentration of food. *Limnol. Oceanogr.* **20**, 259–262

Murdoch, W. W. and Oaten, A. (1975). Predation and population stability. *Adv. Ecol. Res.* **9**, 1–131

Nemoto, T. (1970). Feeding patterns of baleen whales in the ocean. *In* "Marine Food Chains" (J. H. Steele, ed.) pp. 241–252. Oliver and Boyd, Edinburgh

Nichols, D. (1960). The histology and activities of the tube-feet of *Antedon bifida. Q. Jl. microsc. Sci.* **101**, 105–117

Nival, P. and Nival, S. (1976). Particle retension efficiencies of an herbivorous copepod, *Acartia clausi* (adult and copepodite stages): Effects of grazing. *Limnol. Oceanogr.* **21**, 24–38

Paffenhöfer, G.-A. (1971). Grazing and ingestion rates of nauplii, copepodids and adults of the marine planktonic copepod *Calanus helgolandicus. Mar. Biol.* **11**, 286–298

Parsons, T. R., LeBrasseur, R. J. and Fulton, J. D. (1967). Some observations on the dependence of zooplankton grazing on cell size and concentration of phytoplankton blooms. *J. Oceanogr. Soc. Japan* **23**, 10–17

Parsons, T. R., LeBrasseur, R. J., Fulton, J. D. and Kennedy, O. D. (1969). Production studies in the Strait of Georgia. Part II. Secondary production under the Fraser River plume, February to May, 1967. *J. exp. mar. Biol. Ecol.* **3**, 39–50

Pavlova, Ye, V. (1959). O pitanii *Penilia avirostris. Tr. sevast. biol. St.* **11**, 63–71

Pivorunas, A. (1979). The feeding mechanisms of baleen whales. *Am. Sci.* **67**, 432–440

Poulet, S. A. (1973). Grazing of *Pseudocalanus minutus* on naturally occurring particulate matter. *Limnol. Oceanogr.* **18**, 564–573

Poulet, S. A. (1974). Seasonal grazing of *Pseudocalanus minutus* on particles. *Mar. Biol.* **25**, 109–123

Poulet, S. A. and Chanut, J. P. (1975). Nonselective feeding of *Pseudocalanus minutus. J. Fish. Res. Board Can.* **32**, 706–713

Poulet, S. A. and Marsot, P. (1978). Chemosensory grazing by marine calanoid copepods (Arthropoda: Crustacea). *Science* **200**, 1403–1405

Purcell, E. M. (1977). Life at low Reynolds number. *Am. J. Phys.* **45**, 3–11

Pyke, G. H., Pulliam, H. B. and Charnov, E. L. (1977). Optimal foraging: a selective review of theory and tests. *Quart. Rev. Biol.* **52**, 137–154

Reeve, M. B. and Walter, M. A. (1977). Observations on the existence of lower threshold and upper critical food concentrations for the copepod *Acartia tonsa* Dana. *J. exp. mar. Biol. Ecol.* **29**, 211–221

Reiswig, H. M. (1975a). The aquiferous systems of three marine demospongiae. *J. Morph.* **145**, 493–502

Reiswig, H. M. (1975b). Bacteria as food for temperate water sponges. *Can. J. Zool.* **53**, 582–589

Richman, S., Heinle, D. R. and Huff, R. (1977). Grazing by adult estuarine calanoid copepods of the Chesapeake Bay. *Mar. Biol.* **42**, 69–84

Rippingale, R. J. and Hodgkin, E. P. (1977). Food availability and salinity tolerance in a brackish water copepod. *Aust. J. mar. Freshw. Rev.* **28**, 1–7

Rubenstein, D. I. and Koehl, M. A. R. (1977). The mechanisms of filter feeding: some theoretical considerations. *Am. Nat.* **111**, 981–994

Satir, P. (1974). How cilia move. *Sci. Am.* 231 (Oct. 1974), 45–52

Schoener, T. W. (1971). Theory of feeding strategies. *Ann. Rev. Ecol. Syst.* **11**, 369–404

Sheldon, R. W. and Parsons, T. R. (1967). A continuous size spectrum for particulate matter in the sea. *J. Fish. Res. Board Can.* **24**, 909–915

Skiver, J. (1979). Active food size selection and the influence of prior feeding history upon the feeding pattern of marine copepods. (Unpublished manuscript)

Skiver, J. (1980). Seasonal resource partitioning patterns of marine calanoid copepods: species interactions. *J. exp. mar. Biol. Ecol.* **44**, 229–245

Sleigh, M. A. and Holwill, M. E. J. (1969). Energetics of ciliary movement in *Sabellaria* and *Mytilus*. *J. exp. Biol.* **50**, 733–743

Steele, J. H. (1974). "The Structure of Marine Ecosystems". Harvard University Press, Cambridge

Steele, J. H. and Frost, B. W. (1977). The structure of planktonic communities. *Trans. Roy. Soc. Lond.* 280B (976), 485–534

Strathmann, R. R. (1975). Larval feeding in echinoderms. *Am. Zool.* **15**, 717–730

Strathmann, R. R. and Bonar, D. (1976). Ciliary feeding of tornaria larvae of *Ptychodera flava* (Hemichordata: Enteropneusta). *Mar. Biol.* **34**, 317–324

Strathmann, R. R., Jahn, T. L. and Fonseca, J. R. C. (1972). Suspension feeding by marine invertebrate larvae: clearance of particles by ciliated bands of a rotifer, pluteus, and trochophore. *Biol. Bull., mar. biol. Lab., Woods Hole* **142**, 505–519

Sushchenya, L. M. and Khmeleva, N. N. (1967). Consumption of food as a function of body weight in crustaceans. *Doklady (Biol. Sci.) Akad. nauk S.S.S.R.* **176**, 559–562

Taguchi, S. and Ishii, H. (1972). Shipboard experiments on respiration, excretion, and grazing of *Calanus cristatus* and *C. plumchrus* (Copepoda) in the northern North Pacific. *In* "Biological Oceanography of the Northern North Pacific Ocean" (A. Y. Takenouti *et al.*, eds) pp. 419–431. Idemitsu Shoten, Tokyo

Timko, P. L. (1976). Sand dollars as suspension feeders: a new description of feeding in *Dendraster excentricus*. *Biol. Bull., mar. biol. Lab., Woods Hole* **151**, 247–259

Vlymen, W. J. (1970). Energy expenditure of swimming copepods. *Limnol. Oceanogr.* **15**, 348–356

Warner, G. F. and Woodley, J. D. (1975). Suspension-feeding in the brittle-star *Ophiothrix fragilis*. *J. mar. biol. Ass. U.K.* **55**, 199–210

Widdows, J. (1978). Physiological indices of stress in *Mytilus edulis*. *J. mar. biol. Ass. U.K.* **58**, 125–142

Widdows, J., Fieth, P. and Worrall, C. M. (1979). Relationships between seston, available food and feeding activity in the common mussel *Mytilus edulis*. *Mar. Biol.* **50**, 195–207

Wilson, D. S. (1973). Food size selection among copepods. *Ecology* **54**, 909–914

Winter, J. E. (1978). A review of the knowledge of suspension-feeding in lamellibranchiate bivalves, with special reference to artificial aquaculture systems. *Aquaculture* **13**, 1–33

Zaret, R. E. (1980). The animal and its viscous environment. *In* "The Evolution and Ecology of Zooplankton Communities" (W. C. Kerfoot, ed.) Special Symposium III, Am. Soc. Limnol. Oceanogr. pp. 3–9. University Press of New England, Hanover, London

13 *The Role of Large Organisms*

G. CARLETON RAY

1 Introduction

All organisms obviously influence the ecosytems of which they are a part, but the question is: to what *extent* is this so? And how can we learn to *detect* and *measure* this relationship? A comparative approach would be to examine coastal, pelagic, or benthic organisms and to compare roles played by diverse groups, such as large fishes, squids, and the like. Or, one could examine the adaptive radiation of one taxon in greater detail. I take the latter approach and choose marine mammals as subjects.

The Pinnipedia (seals, seal lions, and walruses), sea otter, Sirenia (dugong and manatees), and Cetacea (whales, dolphins, and porpoises) form a diverse group; they have undergone considerable adaptive radiation into all seas and most marine habitats, and most are top predators. They are relatively easier to study than most other large marine organisms with the result that knowledge of their ecological relationships and social behavior permits predictive speculation about their ecosystem relationships.

The role of such large organisms as marine mammals within ocean ecosystems has become a major management issue and a challenge to the scientific community. It is inherent in the discussions of 'multi-species' management (May *et al.*, 1979), and it is encodified into law by the US Marine Mammal Protection Act of 1972:

... such species and population stocks should not be permitted to diminish beyond the point at which they cease to be a significant functioning element in the ecosystem of which they are a part.

Holt and Talbot (1978) summarised views expressed at a conference on this subject in a set of principles which included:

The ecosystem should be maintained in a desirable state such that
a. consumptive and nonconsumptive values could be maximized on a continuing basis,
b. present and future options are ensured, and
c. risk of irreversible change or long-term adverse effects as a result of use is minimized.

They did not say, however, how these principles might be applied. Similarly,

397

FAO/ACMRR (1978) asked: '... what are the *roles* of the various groups of marine mammals in the web of ocean life, and what do we do to that web when we catch them or compete with them for food?' This report noted, for example, that removal of marine mammals from an ecosystem may result in an alternate system state, in contrast with the assumption of reversibility of yield-oriented fisheries models. The report also noted that the 'operational biological unit for management should not be the species, but the ecosystem'.

In order to place this matter in perspective, it is well, first, to consider that our understanding of the ecology and behavior of large marine organisms has been hindered because the subject has not generally been of high priority for either biological oceanographers or fisheries biologists; the former have traditionally been concerned with such processes as productivity, the latter with single-species dynamics and yields. Thus, it is still the case that: 'The control of numbers in an animal population is one of the fundamental biological problems and it remains mysterious' (Cushing, 1975). Obviously, the more complex problem of ecosystem management remains mysterious as well. For example, the role of large marine organisms in influencing their ecosystems through predation or by storing and transferring nutrients has barely been examined. The prejudice has been that the role of large organisms of higher trophic levels is not important because they are not large in terms of total biomass.[1]

Similarly, yield-oriented fisheries biology has largely ignored ecological relationships. Wagner (1969) summarised how fisheries biology contrasts with game management, being more quantitative, but less concerned with behavior and ecology. Its models only implicitly involve such subjects. Clearly matters such as reproduction and mortality must be made ecologically explicit; we need to know how density-dependent and, especially, density-independent factors influence populations and community structure. For example, Larkin (1977) has pointed out that fish have been assumed to be

... integrators of their environment by MSY (maximum sustainable yield) models ... [but] from a biological point of view the concept of MSY is simply not sufficient. ... The species of assemblages of fishes that we observe today must be profoundly different in their composition and interrelationships from the assemblages of a century ago, and so are the organisms on which they feed.

---

[1] For example, during the discussion on the subject of 'The significance of vertebrates in the Antarctic marine ecosystem' (Llano, 1977), the question was asked: 'Could any oceanographers comment on the flow of nutrients into and out of the Antarctic in current systems compared with the losses through the movement and harvesting of whales?' The reply, in part, was: 'The carbon mass represented in very large particles (dead krill or whales) is very small compared to the integrated amounts of DOC (dissolved organic carbon) or POC (particulate organic carbon)'. Clearly, the question concerns the ecological role of whale populations. An answer should consider the time–space relationships of feeding, as related to carbon cycling.

I take the view that such large animals as marine mammals exert important effects on their ecosystems. Several examples support this view. Laws *et al.* (1975) showed that African elephants are finely tuned to their habitat. When over-hunted, social structure may be altered, with the result of habitat damage. Some consider elephants to be behaviorly analogous to sperm whales (Laws, personal communication). Estes and Palmisano (1974) elegantly showed how sea otters influence alternate states of nearshore communities. Finally, an extreme example is the case of Amazonian crocodilians which were shown by Fittkau (1973) to be the major factor in controlling the nutrient supply to their environment. Examples such as these yield insight into how we might approach the roles of marine mammals in general.

Steele (1974) pointed out:

We wish to know not only how many species there are, how each individual behaves, or how much energy it exchanges with the rest of the food web; we hope to discover why, in any ecosystem, there is a particular distribution of species and how this distribution persists over periods of time. . . . The answers are unlikely to come from any single method of study.

For the North Sea, his study indicated that the critical point of ecosystem control appears to be the phytoplankton-zooplankton link. I believe it likely, however, that for areas dominated by marine mammals and birds, such as the Southern Ocean and the Bering Sea, control may well be shared by higher trophic levels.

Clearly, the links must be made between population dynamics, population ecology, oceanography, and social behavior in order to perceive the role of large organisms in the sea. But it is important that we address this subject in a pragmatic way. Hence, I will examine feeding, social behavior, *r–K* selection, and ecosystem relationships according to measurable factors subject to relatively straightforward field research. Then I will relate these factors in a 'role model' in the hope that this will provide some further insight for both future research and conservation.

2   Marine mammal adaptations

FEEDING AND SOCIAL BEHAVIOR

Marine mammals, like sea birds and marine turtles, have adapted an essentially terrestrial morphology, physiology, and behavior to the medium of the sea. They have, furthermore, done so widely and successfully; marine mammals are found in all seas and fill many quite different ecological niches. In some seas, particularly in high latitudes, they are clearly an important component, from points of view of both biomass and energy requirements.

Bartholomew (1970) refers to the 'adaptive complex' exhibited by animals in which 'behavior cannot be isolated functionally from other aspects of the biology of a species ... including physiology, morphology, ecology, and distribution'. For the temperate pinnipeds which he mostly considered, terrestrial parturition and marine feeding have led to limited terrestrial mobility, isolated rookeries, male territoriality, female gregariousness, increased fat storage, large size, a capacity for fasting, and a number of other adaptations. Therefore, patterns of feeding and reproduction interact to determine, to a large extent, the ecological impacts, in time and space, of marine mammal populations.

A major consideration is the degree to which certain species of marine mammals are obligate or facultative in their food preferences. Cushing (1975) has stated that in an evolutionary sense, the reproduction of fishes has become adapted to the structure of food webs: 'The process of feeding in the sea is the simplest determinant of population size, mortality, and competition. It follows that food chains have an evolutionary history'. It also follows, I believe, that predators of higher trophic levels are a determinant of that history. Almost nothing is known, however, of the feeding of marine mammals *as an ecological process*, for example, in order to develop a species' time/energy budget or its time/space relationships to food supply. We must settle for some information on food preferences, feeding behavior, and guesses about the quantity of food marine mammals may extract from their ecosystems; quantification alone can be quite misleading, however.

Gaskin (1976), in a comprehensive treatment of Cetacea, noted their great variability nutritionally and identified four major adaptive types: fish-eaters, squid-eaters, flesh-eaters (implying eaters of warm-blooded vertebrates), and plankton-eaters. But he also noted that species may not fall neatly into these niches; some mysticete whales are strictly plankton-eaters in the Southern Ocean, but may be more facultative in northern latitudes. Also, the large whales' distributional ecology involves disjunct feeding and breeding habitats; they feed in high latitudes where oceanic productivity is seasonally high and reproduce in warmer climes for reasons of thermoregulation of the newborn.

The same sorts of observations may be made about Pinnipedia. Walruses are benthic invertebrate feeders and take a wide variety of species. The otariid seals (fur seals and sea lions) are both pelagic and coastal; some are extremely facultative and one appears to be almost exclusively a krill-eater. The phocid seals, similarly, are both coastal and pelagic and contain species that are extremely facultative or highly obligate, an example of the latter being the crabeater seal, *Lobodon carcinophagus*, which presently accounts for most Southern Ocean krill consumption (Laws, 1977; Green, 1977).

Wilson (1975) noted the preeminence of food among the requirements of animals and described two contrasting time/energy extremes: (*i*) '"time minimizers", for which a predictable, reliable amount of energy is available so long as the energy source is protected', and (*ii*) '"energy maximizers", species that consume all of the energy available regardless of the cost of time'. The consequences for both social behavior and food resource partitioning are significant in either case. In general, I would suppose that those marine mammals which are territorial or solitary, or travel in small groups or are socially highly organised at sea, may be time minimisers; those which travel in large groups and which are non-territorial or do not exhibit marked social structure at sea, beyond simple gregariousness, may be energy maximisers.

In addition, Altmann (1974) stated several principles on the relationship between ecology (especially food) and social structure. For example:

If a slowly renewing resource is both sparse and patchy, it can be exploited more effectively by small groups. . . . Large groups will be more effective if a resource has a high density but a very patchy distribution and the patches themselves occur with low density, so that the resources tend to be concentrated in a few places.

Therefore, studies of animal groupings and sociobiology undoubtedly can provide important insights into food supply and, therefore, ecosystem relationships.

We have very limited knowledge of marine mammal social behavior at sea. However, it is sufficient to be useful in defining certain ecological relationships. For example, we know of territoriality of the solitary bearded seal, *Erignathus barbatus* (Ray *et al.*, 1969) in contrast to the 'mobile lek' or 'female defense polygyny' of the walrus, *Odobenus rosmarus* (Fay *et al.*, in press). These two species are, respectively, solitary and gregarious the year long. Both are benthic feeders on similar items in the Bering and Chukchi Seas. Both are numerous and possibly of roughly equivalent biomass. Yet their ecological effects must be markedly different. The effects of the bearded seal's feeding will be 'smoothed out' by virtue of relatively even spacing. The walrus' effects will be more localised by virtue of the marked clumping of individuals. The localised benthic bioturbation of walrus feeding is dramatic, as I have personally observed (Ray, 1973). The conclusion is that the impacts of these two species are not to be seen in quantitative terms only, but also according to time/space relationships and social behavior. In the case of walrus feeding, marine bioturbation may no doubt induce altered benthic community structure, at least on a localised basis.

Similar observations may be made for cetaceans which also exhibit a variety of social patterns. The large *Balaenoptera* species (blue and fin whales and the like), travel in relatively small groups and congregate on rich planktonic pasturages, particularly in the Southern Ocean. The sperm whale, *Physeter catodon*, in contrast, travels in moderately large groups and exhibits marked seasonal segration of sexes. Other species are pelagic and travel in groups numbering in the thousands, or are solitary and inhabit coastal waters, etc. And feeding behavior among cetaceans is as variable as for pinnipeds.

In conclusion, feeding and social behavior interrelate closely and it is difficult to consider one without the other. This is particularly true for such highly socially organised animals as marine mammals. Indicators of their ecological relationships are how obligate or facultative they are in their food preferences, what their time/energy budgets are, and what their social organisation is during feeding and reproduction. Such considerations may well be more important, even, than quantitative assessments of how much they eat.

'r–K' SELECTION STRATEGIES

FAO/ACMRR (1978) stated:

Marine mammals are basically adapted to environments which undergo seasonal changes but are, in general, otherwise relatively unvarying. Such adaptation has been called *K selection* with reference to the parameter *K*, which in simple population models of the logistic family defines the carrying capacity of a particular environment for each species. ... This is in contrast with species of other groups of animals which evolved to take advantage of fluctuating, randomly changing, unpredictable features of environments through selection for high reproductive rates; such species are said to have adopted an *r selection* evolutionary strategy, *r* being the parameter in models which defines the maximum net rate of population increase.

May (1978) has noted that the

... deliberately oversimplified notions of *r* and *K* selection ... evaluation of an organism's general bionomic strategy should always be the first step in any programme of rational conservation or management; methods for regulating duck hunting will not be applicable to sand-hill cranes.

The latter statement seems a truism, yet if it is important to distinguish between birds, it is no less important to distinguish the adaptive strategies of various species of marine mammals. The questions are: how do marine mammals fit into the *r–K* continuum and what insight can this provide for the various species' roles in the ecosystem?

Southwood (1976) notes: 'As large organisms are basically *K*-selected and small ones *r*-selected, it is not unreasonable to consider vertebrates as *K*-selected and insects as *r*-selected'. (However, see Boyce, 1979, for comments on this relationship.) Southwood also makes the important distinction that 'within each group a spectrum has evolved'. Unfortunately, this spectrum was not emphasised by Estes (1979) in his otherwise stimulating discussion –

one of the first of its kind for marine mammals. Estes stresses that marine mammals are probably food limited and that many species 'probably play an important role in the structure and organisation of marine communities'. He also points out that man is managing them as if they were r-selected.

The major distinction between r and K selection lies in reproductive rate. Marine mammals have low reproductive rates, as none can produce and successfully raise more than one young a year. Most species of phocid seals mature at 3–6 years of age and their parental care periods are short – from less than two weeks to about two months – after which pups and adults separate. The otariid seals mature at about the same age or a little later, but parental care lasts several months and the pups commonly associate with the herd following the lactation period. The Pacific walrus effectively matures at age 14–16 (Fay et al., in press), parental care and nursing last about two years, and females produce young every second or third year.

Cetaceans are similar; reproduction may be annual or only once in 2–3 years; maturity occurs at ages of 3–4 to over 10 years. Gaskin (1976) presented a tabulation of reproductive data of large whales. Variation is considerable; for example, gestation ranges from 9 to 19.4 months and lactation from 6 months to two years, implying a large variation in r.

For both pinnipeds and cetaceans, however, it is difficult to be precise about either reproductive rates or natural rates of population increase. This is due to the general difficulty of obtaining data for many species, particularly the unexploited ones. It is also not legitimate to compare species which are recovering from exploitation and which may have 'artificially' high rates of population increase, with species at either very depressed population levels or near K levels. Suffice is to say that the reproductive capacity of marine mammals ranges over much more than an order of magnitude. Some are capable of producing ten young by age 15, whereas others may produce only a single young by that age. Estimates for annual population increments generally vary from about 2–3% for such species as large whales to over 10% for some phocid seals (unpublished papers of FAO/ACMRR, 1978, conference), and to 16.8% for a recovering Arctocephalus gazella population (Payne, 1977). These percentages, furthermore, may indicate physiological maxima, indicating considerable differences in reproductive rate adaptation.

In conclusion, marine mammals may be said to be K-selected only by gross comparison with some other mammalian groups. We are perhaps beyond the stage of lumping group characteristics so crudely as to say that marine mammals are K-selected and should, alternatively, seek to measure species differences. Marine mammals may have high, medium, or low rates of increase – varying over more than an order of magnitude – and this is an important matter when conceptualising their ecological role. In general, we may suppose that those species which have high reproductive rates are 'opportunists' which may have less influence on mature ecosystem structure. Those which have lower rates may be intimately tied to such structure.

ECOSYSTEMS OF MARINE MAMMALS

Marine mammals are important members of the guild of marine predators, especially in certain ecosystems where they are particularly abundant – high latitudes, for example. Does this also indicate that they are shapers of their communities? Some species appear to travel in predictable ways through predictable worlds – the migrations of baleen whales are a notable example. Are such species members of 'stable' ecosystems? Other species appear to wander opportunistically. Are those the 'pioneers'?

Obviously, 'ecosystem' may be defined according to the perspective of its user. But if we recognise a definition based upon functional unity, the role of marine mammals must also be defined on a functional level. Reichle *et al.* (1975) state:

A system is a complex of interacting subsystems which persists through time due to the interaction of its components. The system possesses a definable organization, temporal continuity, and functional properties which can be viewed as distinctive to the system rather than its components.... The ecosystem possesses a definite organization in its trophic structure and, indeed, this structure remains relatively constant in spite of disturbances.

Essential to the function and persistence of ecosystems are: 1. establishment of an energy base, 2. development of an energy reservoir, 3. recycling of elements, and 4. regulation of rates. Reichle *et al.* (1975), note that ecosystems evolve towards 'maximum persistent biomass'. Material cycling, energy transfer, and other ecological processes are thus dependent upon the biota, especially in mature ecosystems, and it is difficult to resist the view that marine mammals and other large organisms of high trophic levels are important from the points of view of transfer of materials and process rate regulation.

Some insight into the degree to which large organisms are ecosystem regulators emerges from whole ecosystem models. Green (1977) conceptualised carbon transfers of the Southern Ocean and divided the whole into 24 compartments, of which marine mammals were one. Hers is a generalised model which treated data only on an annual and whole ocean basis, i.e. geographic and temporal fluctuations were not considered. She concluded, for example, that of a total standing stock of 250×10^6MT of krill, crabeater seals now consume 100×10^6MT and whales only about 30×10^6MT. This must be a vastly different picture than in pre-whaling days. Furthermore, whales are much more efficient than seals, requiring only about 4 × body weight of food per season, whereas crabeater seals might require 23 × body weight. Seals and whales consume krill at different places, at different seasons, and at different rates. Laws (1977) considered the vertebrates within the Southern Ocean specifically. He pointed out that whales were the most important component previously and that their biomass reduction due to whaling has been 43.09×10^6MT, implying an annual krill 'surplus' of 153×10^6MT. He stated: 'There is direct evidence for population increases for fur seals and penguins, and indirect evidence for increased reproductive rates of baleen whales and crabeater seals'.

Thus, one can hardly ignore marine mammals when calculating carbon transfer in the Southern Ocean. The rates of carbon transfer and its distribution must be vastly different now than in pre-whaling days. One is also forced to conclude that: (i) ecological displacement among species groups indicates an altered ecosystem state and (ii) this new state may be 'stable', in which case the whales will not recover to former abundance.

Taking into account the difficulties of validating whole ecosystem models, perhaps a more direct habitat-oriented approach is worthy of pursuit. Gaskin (1976) noted that large whales are known to congregate in three highly productive ocean areas: (i) ocean fronts, (ii) ocean eddys, and (iii) upwelling areas. He stated: 'Investigation of the oceanographic conditions which favour concentrations of prey species will thus perhaps enable us to unravel ... the reasons for the great short-term variations observed in whale distributions, and also analyse any long-term changes which may occur in these distributions'. This approach associates marine mammal feeding functions with ocean function.

I would add that another area needs to be included – the benthos. Hedgpeth (1977) pointed out that benthic organisms may function as residual concentrators of near-surface productivity. Such benthic feeders as grey whales, bearded seals, and walruses take advantage of this. The question of whether such benthic feeders release nutrients back into the water column through feeding or bioturbation appears to me to be a research objective of high priority.

Answers to such questions as I asked at the beginning of this section are possible only through a process ecology approach. The program on Processes and Resources of the Bering Sea Shelf (PROBES 1977–1978) is shedding light on oceanic processes, ecological succession, and the distribution of marine mammals and birds there. Iverson et al. (1979) state: 'Observations support a hypothesis that major food webs leading to large stocks of pelagic fauna and benthic fauna are separated in space and are organized in relation to the fronts which exist in the southeastern Bering Sea'. Essentially, there are three systems there: (i) a 'mature' oceanic system outside the shelfbreak front, (ii) a 'mature' benthic system, and (iii) a 'pioneer' shelf, nearshore system. The distributions of birds and whales correlate with these three systems.

Southwood (1976) characterises habitats according to their 'duration stability', 'temporal stability' and 'spatial heterogeneity'. The ecosystems and habitats of marine mammals exhibit all three. Indeed, the migrations and wanderings of marine mammals allow them to be members of several ecosystems according to where their populations are located and for what purpose. Ray et al. (1978, 1979) have used a method which allows one to conceptualise such matters in graphic terms. Figure 1 presents a model for habitat relationships of the Pacific walrus. It shows the relationships between walrus home range, ice conditions, and food distribution. All three of these factors vary according to climatological, oceanic, and ice conditions. Obviously, there are places and times where and when walruses have maximum impact upon their ecosystem, and such impacts are highly variable according to environmental and population factors. Furthermore, the removal of such an important animal would no doubt result in a different ecosystem state than at present.

In conclusion, in light of what we now know of ecosystems and of the importance of marine mammals in some ecosystems, it is clear that the role of marine mammals may be inferred from their habitat characteristics. A habitat approach which considers the degree of maturity of the ecosystems which marine mammals inhabit may serve to guide our understanding of functional ecological relationships.

3 A role model

The task now is to put these three aspects together. First, we have discussed the adaptive complex of feeding and behavior. It would appear that the simplest indicator in this regard is the degree of obligateness with regard to food. Social behavior and time/energy budgets offer significant clues to exploitation of food supply, but too little is known of these features of marine mammal ecology at sea for purposes of conceptual modeling. Second, reproductive rate is a measure of the degree to which a species is K-selected. K-selected species experience ecosystem feedbacks in which population levels are under both density-dependent and density-independent control; such species influence their ecosystems in ways that r-selected species probably do not. And third, the relative maturity of ecosystems inhabited by marine mammals indicates the possible co-evolution and feedback relationships of species and ecosystems. Mature ecosystems exhibit either steady state or oscillatory stability (Dunbar, 1977) and so do the populations of species within them.

Table I illustrates that marine mammals differ widely in these three

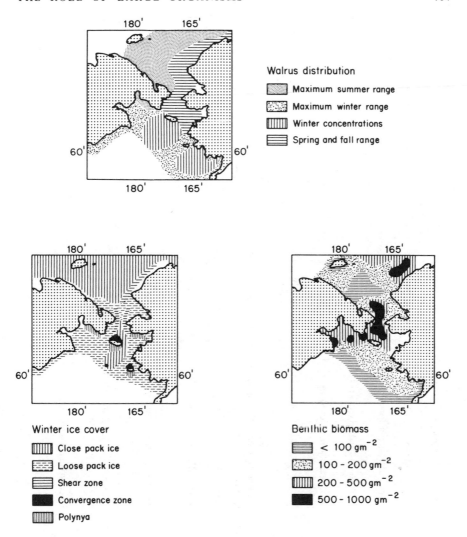

FIG. 1 Some habitat relationships of the Pacific walrus. Walruses concentrate within the area shown on the top figure. However, they tend to avoid close pack ice and prefer thick, but loose, pack ice shown on bottom left. Thus, seasonal variations in ice dynamics partly determine the distribution of walrus winter concentrations. In addition, walrus' food supply is not evenly distributed. Note from the bottom right map that the benthic biomass, which includes walrus food species, is less in areas of their winter concentration than elsewhere. Thus, the walrus' 'critical habitat' is subject to interrelationships among sea ice dynamics, the structure and biomass of benthic communities, and walrus behavior. Further, walrus are 'vulnerable' to decrease of winter food supply through future commercial fishery operations

TABLE I Characteristics of selected marine mammal species

Species	Food obligateness	Reproductive rate	Ecosystem distribution
1 Harbor seal *Phoca vitulina*	Very catholic in food habits	Matures early High rate	Coastal only (Pioneer ecosystem)
2 Crabeater seal *Lobodon carcinophagus*	Almost exclusively a krill eater	Matures early High rate	Oceanic only (Mature ecosystem)
3 Walrus *Odobenus rosmarus*	Moderately restricted in food habits	Matures late Low rate	Benthic feeder (Mature ecosystem)
4 California sea lion *Zalophus californianus*	Very catholic in food habits	Matures fairly early Moderate rate	Feeds along coasts and adjacent ocean (Pioneer to mature ecosystem)
5 Northern fur seal *Callorhinus ursinus*	Moderately catholic in food habits	Matures fairly early Moderate rate	Feeds only in open ocean (Mature ecosystem)
6 Sea otter *Enhydra lutris*	Catholic in food habits	Matures fairly early Moderate rate	Benthic feeder (Mature, but alternate stable state, ecosystem)
7 Belukha whale *Delphinapterus leucas*	Very catholic in food habits	Matures fairly late Moderate rate	Almost entirely coastal, but also in open ocean (Mostly pioneer ecosystem)
8 Gray whale *Eschrichtius robustus*	Mostly an amphipod-eater	Matures fairly late Low rate	Benthic feeder along coasts and open ocean (Pioneer to mature ecosystem)
9 Bowhead whale *Balaena mysticetus*	Restricted in food mostly to small zooplankton	Matures late Low rate	Mostly coastal feeder, but also in open ocean (Mostly pioneer ecosystem)
10 Blue whale *Balaenoptera musculus*	Exclusively a krill-eater	Matures fairly late Low rate	Feeds only in open ocean (Mature ecosystem)

aspects. Further, when these aspects are graphed, as on Fig. 2, we observe quite distinct species' positions in 'ecological space'. Three adaptive peaks seem justified, despite some intuitive judgments which I have made. These are indicated by the closed circles on Fig. 2: 1. Species of obligate food habits, inhabiting mature ecosystems and having high reproductive rates, 2. Species of obligate food habits, feeding in pioneer ecosystems, and having low reproductive rates, and 3. Species which are obligate feeders, inhabiting mature ecosystems, and with low reproductive rates. The first two are

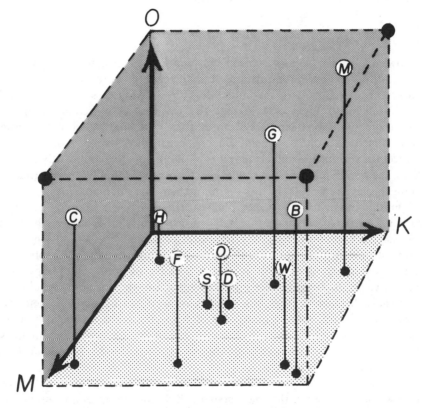

FIG. 2 'Ecological space' differences among marine mammals. Food habits, $r-k$ selection, and states of ecosystem inhabited by each species are related to illustrate adaptive radiation. Their ecological roles differ according to these and other factors See text for further explanation.

O, degree of food obligateness; K, degree to which K-selected; M, degree to which distributed in mature ecosystem; ●, adaptive extremes (theoretical). H, harbor seal; C, crabeater seal; W, walrus; S, California sea lion; F, northern fur seal; O, sea otter; D, Belukha whale; G, gray whale; M, bowhead whale; B, blue whale

theoretically a bit jarring, but appear factual. Apparently absent are species which are obligate feeders, but which maintain high reproductive rates and which inhabit pioneer ecosystems.

This model at least provides a matrix for practical research. Validation of the model is possible through study of food preferences, reproductive rates, social behavior, and integration of oceanographic data with information on the natural history of marine mammals. Although we are forced at present to dwell largely in theoretical constructs, with continued research directed towards these matters, we should be able to emerge from the closets of comparative ignorance in which we now find ourselves.

4 Conclusion

Knowledge of the role of marine mammals in ecosystems has important implications for conservation and management of both species and their ecosystems. According to the model developed in Fig. 2, such species as the blue whale, and to a lesser extent the walrus, are extremely subject to alterations of either their ecosystem or their food supply; whole ecosystem management is vital for them. Such species may also be comparatively subject to ecological displacement if their numbers are significantly reduced. Also, their role in structuring their ecosystems is probably great. However, these species are not likely to overshoot their carrying capacities.

On the other hand, species such as the harbor seal may overreach carrying capacity. They are probably not of much influence on their own ecosystems except as they may temporarily deplete local food supply. In contrast, for such species as crabeater seals, food supply is probably by far the most important controlling factor because of their stenophagy. They may have an important effect on ecosystems, but they are not so subject to ecological displacement as blue whales or walruses because of their more rapid reproductive rates. Finally, for such species as the bowhead, management is, above all, a matter of conservation of the ecological processes which result in continued availability of their food. Their effects on ecosystems may not be great.

Of course, these are but guesses. As Eberhardt (1977) points out: 'There is a great hiatus between the management of living resources and research in ecology'. There are hazards in generalised theory as neither animal populations nor ecosystems fit snugly into niches. However, conclusions about the roles of animals in ecosystems are presently more determined by crude guesswork about the significance of ecological relationships than by our ability to monitor those relationships. Rather than assume relative unimportance of marine mammals to ecosystems, as has been the case in the past, I would judge that we may learn more about ecosystems by looking at the process ecology of higher trophic levels than by traditional productivity and biomass-oriented oceanographic studies. May (1976) and Dunbar (1977) stress, for example, that complex, steady state ecosystems are more dynamically fragile than relatively simple, oscillatory ones. Marine mammals inhabit both types of ecosystems. Comparative studies of large organisms such as marine mammals could yield important insights into ocean ecosystem dynamics. How this may be done depends on development of simple models for testing and new methods of population monitoring (Laws, 1979).

Finally, reflecting on the yield-oriented management studies of the past, I would like to emphasise the words of Bartholomew (1974):

The history of human thought is replete with unsuccessful attempts to apply ideas based on data obtained from one frame of reference to situations in which fundamentally different factors are operating. We can perhaps avoid this pitfall if we use as our point of departure not fishery biology, but the ecology and social behavior of large mammals. . . . However difficult this approach, we need only to point out that the present methods, despite the large amount of information they have yielded, have been depressingly unsuccessful in leading to effective resource management.

And, I might add, such a focus would, I believe, reveal much about the role of large animals in ecosystems.

Acknowledgements

This paper results from research and is partially supported by the Office of Naval Research (Oceanic Biology), contract number NOOO14-75-C-0701, and the National Aeronautics and Space Administration, contract number NAS 2-9300, both to The Johns Hopkins University. I thank M. J. Dunbar, R. J. Hofman, K. S. Norris, J. H. Steele and D. Wartzok for comments on the manuscript.

References

Altmann, S. A. (1974). Baboons, space, time, and energy. *Amer. Zool.* **14**, 221–248
Bartholomew, G. A. (1970). A model for the evolution of pinniped polygyny. *Evol.* **24** (3), 546–559
Bartholomew, G. A. (1974). The relation of the natural history of whales to their management. *In* "The Whale Problem: A Status Report" (W. E. Schevill, G. C. Ray, K. S. Norris, eds) pp. 294–302. Harvard University Press, Cambridge, Mass.
Boyce, M. S. (1979). Seasonality and patterns of natural selection for life histories. *Amer. Nat.* **114** (4), 569–583
Cushing, D. H. (1975). "Marine Ecology and Fisheries". Cambridge University Press, Cambridge, London, New York and Melbourne
Dunbar, M. J. (1977). The evolution of polar ecosystems. *In* "Adaptations within Antarctic Ecosystems" (G. A. Llano, ed.) pp. 1063–1076. Smithsonian Inst., Washington D.C.

Eberhardt, L. L. (1977). Optimal policies for conservation of large mammals, with special reference to marine ecosystems. *Env. Conserv.* **4** (3), 205–212

Estes, J. A. (1979). Exploitation of marine mammals: *r*-Selection of *K*-strategists? *J. Fish. Res. Bd. Can.* **38** (8), 1009–1017

Estes, J. A. and Palmisano, J. F. (1974). Sea otters: their role in structuring nearshore communities. *Science* **185**, 1058–1060

FAO (1978). "Mammals in the Seas", Vol. 1. Report of the FAO Advisory Committe on Marine Resources Research, Working Party on Marine Mammals, with the cooperation of the United Nations Environment Programme. FAO, Rome

Fay, F. H., Ray, G. C., Kibal'chich, A. A. and Gol'tsev, U. N. Time and location of mating and associated behavior of the Pacific walrus, *Odobenus rosmarus divergens* Illiger. *Fish. Bull.* (In press)

Fittkau, E. J. (1973). Crocodiles and the nutrient metabolism of Amazonian waters. *Amazonia* **4**, 102–133

Gaskin, D. E. (1976). The evolution, zoogeography and ecology of Cetacea. *Oceanogr. Mar. Biol. Ann. Rev.* **14**, 247–346

Green, K. A. (1977). Antarctic marine ecosystem modeling: revised Ross Sea model, general Southern Ocean budget, and seal model. Final report for Marine Mammal Commission contract MM6AC032. Nat. Tech. Inf. Serv. pub. PB-270375. Washington D.C.

Hedgpeth, J. W. (1977). The Antarctic marine ecosystem. *In* "Adaptations within Antarctic Ecosystems" (G. A. Llano, ed.) pp. 3–10. Smithsonian Inst., Washington D.C.

Holt, S. J. and Talbot, L. M. (1978). New principles for the conservation of wild living resources. *Wildlife Monogr.* No. 59. Suppl. to *J. Wild. Man.* **43**, 1–33

Iverson, R. L., Coachman, L. K., Cooney, R. T., English, T. S., Goering, J. J., Hunt, G. L. Jr., Macauley, M. C., McRoy, C. P., Reeburg, W. S. and Whitledge, T. E. (1979). Ecological significance of fronts in the Southeastern Bering Sea. *In* "Ecological Processes in Coastal and Marine Systems" (R. J. Livingston, ed.) pp. 437–466. Plenum Publ. Corp., New York

Larkin, P. A. (1977). An epitaph for the concept of maximum sustainable yield. *Trans. Amer. Fish. Soc.* **106**, 1–11

Laws, R. M. (1977). The significance of vertebrates in the Antarctic marine ecosystem. *In* "Adaptations within Antarctic Ecosystems" (G. A. Llano, ed.) pp. 411–438. Smithsonian Inst., Washington D.C.

Laws, R. M. (1979). Monitoring whale and seal populations. *In* "Monitoring the Marine Environment". (D. Nichols, ed.) pp. 115–140. Praeger, New York

Laws, R. M., Parker, I. S. C. and Johnstone, R. C. B. (1975). "Elephants and their Habitats". Clarendon Press, Oxford

Llano, G. A. (ed.) (1977). "Adaptations within Antarctic Ecosystems". Smithsonian Inst., Washington D.C.

May, R. M. (1976). Patterns in multi-species communities. *In* "Theoretical Ecology, Principles and Applications" (R. M. May, ed.) pp. 142–162. Saunders, Philadelphia and Toronto

May, R. M. (1978). Factors controlling the stability and breakdown of ecosystems. *In* "The Breakdown and Restoration of Ecosystems" (M. W. Holdgate and M. J. Woodman, eds) pp. 11–25. Plenum Publ. Corp., New York

May, R. M., Beddington, J. R., Clark, C. W., Holt, S. J. and Laws, R. M. (1979). Management of multi-species fisheries. *Science* **205**, 267–277

Payne, R. M. (1977). Growth of a fur seal population. *Phil. Trans. R. Soc. Lond.*, **B279**, 67–79

Processes and Resources of the Bering Sea Shelf (PROBES) (ms). Report to National Science Foundation, Washington D.C.

Ray, G. C. (1973). Underwater observation increases understanding of marine mammals. *Mar. Tech. Soc. Jour.* **7**, 16–20

Ray, G. C., Dobbin, J. A. and Salm, R. V. (1978). Strategies for protecting mammal habitats. *Oceanus* **21**, 55–67

Ray, G. C., Salm, R. V. and Dobbin, J. A. (1979). Systems analysis mapping: an approach towards identifying critical habitats of marine mammals. Final report for Marine Mammal Commission contract MM6AC011. Nat. Tech. Inf. Serv. pub., 80–111594. Washington, D.C.

Ray, G. C., Watkins, W. A. and Burns, J. J. (1969). The underwater song of *Erignathus* (bearded seal). *Zoologica* **54**, 79–83 plus Plates I–III and phonograph disk

Reichle, D. E., O'Neill, R. V. and Harris, W. F. (1975). Principles of energy and material exchange in ecosystems. *In* "Unifying Concepts in Ecology" (W. H. van Dobben and R. H. Lowe-McConnell, eds) pp. 27–43. Dr W. Junk, The Hague

Schevill, W. E., Ray, G. C. and Norris, K. S. (eds) (1974). "The Whale Problem: A Status Report". Harvard University Press, Cambridge, Mass.

Southwood, T. R. E. (1976). Bionomic strategies and population parameters. *In* "Theoretical Ecology: Principles and Applications" (R. M. May, ed.) pp. 26–48. Saunders, Philadelphia and Toronto

Steele, J. H. (1974). "The Structure of Marine Ecosystems". Harvard University Press, Cambridge, Mass.

Wagner, F. H. (1969). Ecosystem concepts in fish and game management. *In* "The Ecosystem Concept in Natural Resource Management" (G. M. Van Dyne, ed.) pp. 259–307. Academic Press, New York and London

Wilson, E. O. (1975). "Sociobiology, The New Synthesis". Belknap Press of Harvard University Press, Cambridge, Mass.

14 *Significance of Spatial Variability*

ALAN R. LONGHURST

1 Environmental matrix of patchiness

Aggregation is a necessary condition of life in the ocean, and an inevitable consequence of the physical environment, but we have only recently begun to quantify its nature and the advantages accruing to both aggregated and dispersed organisms.

Most recent reviews have emphasised mathematical analysis of patchiness and of the methods used to measure it, principally in the horizontal plane: this chapter, on the other hand, will examine the significance of patchiness for trophic relationships between biota, and the relative significance of variability in the vertical and horizontal planes.

The principal causes of aggregation have been formalised by Stavn (1971) who proposed five kinds: *vectorial*, by the action of regular environmental gradients, as in the vertical plane; *stochastic-vectorial*, by the action of nonregular advection by wind or water movement; *reproductive*, by the action of an aggregated parent stock in generating an aggregation of reproductive products; *social*, by the action of biota in active schooling or swarming; and *coactive*, by the trophic interaction between plants and herbivores or predators and prey. These suggestions form a useful framework for the discussion of patchiness.

A conceptual framework for time-space scales of variability has been proposed by Haury *et al.* (1978), whose three-dimensional distributions of biomass variability demonstrate the positive correlation between size and time, ranging from ephemera on the centimetre scale to variability over tens of thousands of kilometres on the time scale of climatic changes. Patchiness, in the usual sense of the term, occurs across the scales from centimetres to kilometres; the interpretation of horizontal variability on these scales requires study, these authors suggest, of time-scales of the order of hours to months.

What appears to be lacking in many recent studies of patchiness is a proper definition of the environmental matrix in which patchiness has been observed; this requires a clear description of the physical and biological environment at the next larger space and time scale: a larger area and a longer period must be understood than that selected for the definitive study.

The search for mathematically rigorous models to account for observed patchiness has been so successful in recent years that wider extrapolations have been made from them than are perhaps justified by the physical matrix in which the patchiness occurred, and probably than had been originally intended by their authors. Mesoscale differences between regions of the ocean are so great that extrapolations should be made only within similar physical regimes – one should not necessarily expect that what is fundamental to the patchiness observed in, say, the Peruvian upwelling should have relevance to high latitude continental shelves.

The greatest regional difference in small scale patchiness probably occurs between continental shelf waters and the deep ocean, though neither is uniform with respect to patchiness. However, the influence of M_2 tidal streams imposed upon advective flows over continental shelves profoundly affects mixing processes – and hence conservation and dispersion of spatial variability – in a manner that cannot be compared with vertical current shear in the deep ocean. Only recently has a regional description of the consequences of tidal friction and mixing on a major continental shelf become available, from which it is evident that studies of continental shelves must in future be performed purposefully within mixed, stratified, or frontal areas, and extrapolation between such areas carefully supported (e.g. Pingree, 1978). Patch studies must also recognise the occurrence of estuarine upwelling fronts, river discharge plumes and their frontal aspects (Bowman and Iverson, 1978), retrograde fronts and cascading at shelf break regions, prograde fronts at the boundaries of low-latitude coastal upwelling plumes (Mooers et al., 1978), and of ice-edge upwelling phenomena: each of these processes characterise circulation regimes which must be expected to have individual patch dispersion-conservation mechanisms.

2 Spatial variability of phytoplankton

PATCH DIMENSIONS

The development of continuous pumping equipment (Cassie, 1963), of the continuous in vivo fluorescence technique for measuring chlorophyll-a and related plant pigment (Lorenzen, 1966), and latterly continuous profiling equipment (Boyd, 1973; Herman and Denman, 1977), have enabled a comprehensive model of phytoplankton distribution in the ocean to be formulated in the last decade which accounts satisfactorily for the major spatial patterns observed in the ocean. It is now clear that there are three principal states: (i) a subsurface chlorophyll maximum under stratified conditions which lies close to the bottom of the euphotic zone, whatever its depth, and rather often within the pycnocline, (ii) a surface maximum extending to the bottom on continental shelves or to well below the newly developing mixed layer during spring-bloom conditions but which usually includes no layer with peak chlorophyll values, and (*iii*) the situation during high-latitude winters when very sparse chlorophyll is rather unevenly distributed throughout the upper layers.

Early studies with underway continuous *in vivo* fluorescence techniques showed that phytoplankton distribution within the mixed layer can be either extremely uniform over many tens of kilometres, or rather variable. Lorenzen (1971) reviewing such data from about 32 000 km of track in mid-latitudes showed that in mid-ocean areas with low chlorophyll values (0.1 to 0.5 mg m^{-3}) these may be uniformly monotonous over great distances, and that changes in plant pigments are in about 70 % of the cases associated with temperature variation. Lorenzen found much greater horizontal variability in relatively eutrophic (3 to 10 mg m^{-3}) regions off Peru, with changes being associated with plumes of upwelled water; Steele (1978) found similar variability along tracks across the Rockall Trough, and over the Fladen Ground in the North Sea. Similarly, continuous tracks in the southern Bay of Biscay (Fasham and Pugh, 1976) and in the Gulf of St Lawrence (Denman, 1976) show strong mesoscale features, and greater variability between features, than Lorenzen found to be normal in open oceans in mid-latitudes.

Bainbridge (1957) suggested that phytoplankton patches occurred at all sizes, though two distinct families could be recognised: linear aggregations some metres in width and hundreds of metres in length, and much larger patches of less regular form with diameters most commonly about 50 km. There appears to be little evidence in recent data that the larger patches are most often of the size suggested by Bainbridge; rather, there appears to be diversity in patch size probably associated with the diversity of aggregation-dispersion mechanisms, especially in tidally and topographically dominated continental shelf seas. As Steele (1976) points out, dye patch studies do, in fact, confirm the reality of the linear patches postulated by Bainbridge. Such patches, aligned with the surface current and wind direction, commonly result from dye injections and are assumed to be associated with the Langmuir circulation which forms alternating lines of convergence at moderate wind speeds. These accumulate lines of flotsam and neuston at convergences, and of diel migrating organisms at divergences (Langmuir, 1963; Sutcliffe *et al.*, 1963; Owen, 1966; Stavn, 1971).

PATCH INDUCTION

Recent studies of tidally-induced fronts on the European continental shelf during the summer clarify one important process which induces mesoscale variability in the horizontal plane. Fronts are formed (Simpson and Hunter,

1974; Pingree, 1975; Pingree and Griffiths, 1978) between regions where tidal friction maintains a fully mixed water column and those where stability from summer warming and vertical salinity gradients overcome tidal mixing forces and allows the formation of a thermocline. During the early part of the summer the thermocline, and hence the front, advances across the continental shelf from deep water, as the balance between tidal friction and solar warming changes. During the summer, after nutrient depletion within the mixed regions, strong phytoplankton growth continues only near the base of the thermocline in the stratified regions, and also where the thermocline intercepts the surface at the thermal front: here nutrients are constantly replenished if the front occurs over water sufficiently deep (Pingree, 1978). The fronts are unstable, and patchiness of phytoplankton along them appears to be on a scale of about 20 km, due to cyclonic eddies bringing tongues of colder, mixed water across to the warmer, stratified side of the front. Within the spiral arms of such eddies the vertical displacement of the thermocline brings high chlorophyll water to the surface, dominated by the dinoflagellates characteristic of the chlorophyll maximum (Pingree, 1979).

Especially in shelf regions, the effect of internal waves must be considered when analysing patchiness data. Trains of internal waves in shallow seas can be generated by the passage of tidal streams over shallow banks in whose lee the waves originate, having different characteristics at different stages of the tide; Haury et al. (1979) have described the formation of packets of internal waves in the lee of Stellwagen Bank in Massachusetts Bay. These have a period of 6 to 8 minutes and a vertical amplitude of 10 to 15 m and carry with them the passive variables of the water column. Horizontally directed sampling with LHPRs (Longhurst and Williams, 1976) intercepting the peaks of the wave train produce serial samples that could be thought to include indications of horizontal patchiness if the wave train was not known to be present. Coherence occurred between temperature (indicating passage through the peak of a wave) and chlorophyll, salinity, and three groups of zooplankters. These observations in Massachusetts Bay confirm the concern of Denman (1976) who investigated the lack of coherence between chlorophyll and temperature variability in the Gulf of St Lawrence by two simultaneous sets of horizontal tows separated vertically by only 4 m. He suggested that most of the variability in these series was due to the effects of internal waves and vertical mixing. Haury et al. (1979) also showed that as the internal waves steepened, they occasionally overturned with important biological consequences possibly reflected in subsequent horizontal patchiness.

In the open ocean, fronts and other processes also cause aggregation but except at convergence zones between contrary-flowing currents such phenomena do not usually produce such strong spatial variability as occurs on continental shelves, or as is usual in the vertical plane.

Vertical variability

A subsurface chlorophyll maximum (SCM) associated with the thermo-cline occurs remarkably widely (Semina, 1979), and because a high propor-tion of active chlorophyll at any location actually occurs within the feature, its horizontal homogeneity, its stability, and its seasonal occurrence are matters of great importance.

An SCM has been recorded very widely in the Pacific Ocean, and Venrick *et al.* (1973) show that it is possible to follow a continuous subsurface chlorophyll feature from Alaska to New Zealand, from California to Easter Island, and from Peru to New Zealand. It occurs throughout the year from 20°N to 20°S in the eastern Pacific and Shulenberger (1978) has followed it along 28°N from the American coast to the vicinity of Japan. In these observations the SCM lies near a depth of 100 m except near the American coast (where it slopes up towards the land) and in a western Pacific mesoscale eddy field, where it deepens. It also occurs in the Indian Ocean (Yentsch, 1965), and in the tropical Atlantic (e.g. Dandonneau, 1977). Longhurst and Williams (1979) found a summer subsurface maximum, deepening southwards, along a line from Iceland to northern Spain, and one occurred in a section normal to the axis of the Canary current to about mid-Atlantic (Gieskes *et al.*, 1978).

The SCM is obviously a universal feature in the oceans. Where winter overturn occurs it become reestablished each summer; following the seasonal increase in stability a layer of abundant chlorophyll extends down to about 50 m in the California current (Venrick *et al.*, 1973), to the sea bed on the European continental shelf (Holligan and Harbour, 1977), and to at least 150 m at 59°N in the Atlantic (Williams and Robinson, 1973). Nutrients are rapidly depleted, an important fraction of the cells is lost to deep water or to the sea bed, and plant growth comes to be restricted to a layer near the base of the summer thermocline. The initial spring bloom is dominated by diatoms, while the chlorophyll maximum in the thermocline is dominated by dinoflagellates and chlorophyte flagellates (Lasker, 1975; Holligan and Harbour, 1977).

The use of continuous profiling equipment recently has shown that the oceanic and continental SCM may be a remarkably sharp phenomenon, extending over only a few metres or tens of metres, and containing peak values often an order of magnitude higher than background; using a 20-cm spaced array of free-flushing water bottles, Owen (pers. comm.) has shown that even on this scale there is important variability, particle counts varying by as much as a factor of 5 over only 2 m of depth.

The existence of this concentrated layer of phytoplankton a few tens of metres deep, but traceable in the horizontal plane for thousands of kilometres, and having a specialised flora (Gieskes *et al.*, 1978) with a physiology adapted to low light levels (Eppley *et al.*, 1973), represents spatial variability which cannot be matched in the horizontal plane: it must be the most significant 'patch' for other biota occurring in the same water column.

ANALYSIS OF HORIZONTAL PATCHINESS

However, most studies of phytoplankton patchiness in recent decades have concentrated on spatial variability in the horizontal plane in a variety of environments, though mostly in continental shelf seas. These studies have proceeded from relatively simple observations of underway *in vivo* fluorescence, through rigorous evaluations of the covariability of temperature (as an indicator of physical structure rather than for its direct physiological effect) and chlorophyll.

For example, data from towed submersible pumps in the St Lawrence estuary (Platt, 1972; Denman and Platt, 1975; Denman, 1976) were searched for characteristic scales of phytoplankton patchiness in their power and variance spectra by techniques more usually applied to time-series data (Platt and Denman, 1975); it was possible to separate the scales of patchiness due to turbulent diffusion, over which phytoplankton would be expected to act as a passive scalar from those scales of patchiness where biological effects would predominate.

Platt and Denman concluded that at wave lengths from about 100 m to several kilometres there was strong coherence between chlorophyll and temperature; however, over very small distances this coherence broke down for reasons connected with the expected behaviour of energy cascades at low wave numbers. Across environmental discontinuities there also occurred major changes in variance and in mean values of temperature and chlorophyll, which were found to be the cause of large standard errors in some power spectra estimates. Over the range of wave lengths for which coherence occurred, the relationship between temperature and chlorophyll was positively correlated for some runs, negatively correlated for others. Temperature and chlorophyll profiles having slopes of opposite sign would produce a negative correlation, while those of similar sign would produce a positive correlation.

Platt (1978) discussed a large number of chlorophyll power spectra and showed that they characteristically contained a break in slope at reciprocal wave numbers corresponding with a patch dimension of about 1 km; he suggested that this was because when phytoplankton exponential growth rate (r^{-1}) is large in relation to τ (the time scale required for turbulent eddies to transfer their energy to eddies half their size) the chlorophyll spectra will follow temperature spectra (chlorophyll will be a passive scalar), while at wave numbers at which $\tau \gg r^{-1}$ variability of phytoplankton distribution may be strengthened by population growth or grazing before the eddies which cause them are themselves dissipated.

These observations have been confirmed by the horizontal pumped series reported by Fasham and Pugh (1976), though these were taken in an oceanic environment at sampling depths well below the subsurface chlorophyll maximum; in these data, phytoplankton behaves as a passive contaminant of fluid motion between scales of 40 m and 1 km.

However, the situation is probably not as simple as these data might suggest. Horwood and Cushing (1978) showed that a series of horizontal data from the North Sea resulted in power spectra which contained the expected break in slope only prior to the spring bloom. Horwood thought that this was because this region is subjected to much higher tidal friction and turbulence than the micro-tidal Gulf of St Lawrence, while Fasham (1978) and Platt (1878) suggest that the data are in fact in accordance with expectation: the Denman-Platt theory would only be expected to be validated under conditions of relative biological equilibrium, as indeed Horwood and Cushing observed before the spring bloom. The original observations were made during the summer, after biologically stable conditions had been established in the Gulf of St Lawrence.

Two sets of Pacific data have been reported by Star and Mullin (1980). These show that there is a wide range of frequencies over which the chlorophyll variance is small and constant; variability in temperature and chlorophyll is confined to wave lengths greater than 600 m. The major difference between inshore and oceanic data in this series is not in apparent patch size, but rather in the difference between patch and background levels. Such results suggest that though spectral analysis of variance in horizontal series is a powerful tool in the study of patchiness, the kinds of patches actually observed in phytoplankton are very diverse, and a single solution to their description may not yet have been reached.

Despite its attractiveness in solving several trophic paradoxes within the pelagic ecosystem, the Langmuir circulation has not been specifically addressed in studies of the covariability of chlorophyll and physical variables in the ocean; a consistent relationship between the shape of patches and wind direction, so that patchiness was ordered into stripes rather than irregular-shaped forms would be expected to have shown itself strongly in the sets of horizontal data series on which the most rigorous analyses of patchiness have been performed. But this data does not appear to have been the case, and one must ask why this should be so? Whatever the reason, it is clear that this process requires investigation.

MODELS OF AGGREGATION-DISPERSION MECHANISMS

The interaction between physical patch-dispersing and biological patch-reinforcing mechanisms has been studied in the relatively simple two-dimensional situation in which a patch is diffused and strengthened within a single homogeneous water-layer. Such, however, is seldom the actual case; Bowden (1965) suggested that horizontal diffusion could occur solely as a result of vertical current shear, and Kullenberg (1972) suggested that mixing and current shear alone could satisfactorily account for the observed diffusion of injected dye patches; without recourse to horizontal diffusivity, he showed that vertical profiles of dye fluorescence in a mature dye-patch closely resemble chlorophyll fluorescence profiles and he suggested that it is not unreasonable to assume that similar processes influence both dye and chlorophyll profiles.

Evans et al. (1977) and Evans (1978) discuss a simple two-dimensional model (vertical + one horizontal dimension) which is capable of simulating interaction between nutrients, phytoplankton and a herbivore in a stratified sea having a deep reservoir and three layers between which shear can be differentially applied. Evans (1978) shows that in the absence of shear this model generates oscillations having a 50-day period. The application of shear between the upper two layers generates patchiness which appear to be diffusion-induced in the model, though the effects of food on vertical migrations of herbivores cannot be ruled out.

Within homogeneous mixed layers with no vertical current shear, the interaction between physical processes and biological growth is determined by the balance between phytoplankton growth rates and physical diffusion, including both mixing and dispersion; such interactions have been discussed recently by Steele (1976), Okubo (1978), and Wroblewski et al. (1975). From dye-patch studies at relatively small scales, and from salinity distribution fields over larger scales, Okubo (1971) has proposed a relationship between horizontal diffusivity (K_a cm^{2}s^{-1}) over space scales (ρ) from 10^4 to 10^8 cm. The relationship is linear, though there are order of magnitude variations in K_a/ρ at some values of ρ (Okubo, 1978), because of the diverse nature of the physical environment across so large a range of space scales.

A simple model relating phytoplankton growth rate (α) to the dispersion coefficient (D) to identify threshold dimensions for sustaining patches was proposed independently by Okubo (1978), by Kierstead and Slobodkin (1953) and Skellam (1951); in two-dimensional space the critical radius (R_c) is given by

$$R_c = 2.4048(D/\alpha)^{1/2}$$

For phytoplankton growth rates with doubling times of 1 day and 10 days this model predicts critical patch sizes of the order of 2 and 50 km radius. Okubo (1978) tabulates a series of estimates of R_c from related models, those of Joseph and Sender (1958), Ozmidov (1958), and Okubo (1971), and shows that agreement is within relatively close limits, especially at fast rates of growth. As Steele (1976) points out, however, the high degree of variability in the dispersion coefficient K_a means that critical patch size must be correspondingly variable.

Bowman and Esaias (1978) have commented that though such models may predict patch threshold size, they do not account for patch initiation. Although patch initiation is inherent in the covariance of chlorophyll with environmental variables and in the Evans model of vertical shear effects, processes driven by topography may also be important. Bowman and Esaias suggest that in Long Island Sound, within the nearshore boundary of a strong tidally-induced jet current, dinoflagellate chlorophyll concentrations may be higher by a factor of 3 than seawards of it; the coastal jet may inject periodic pulses of high chlorophyll water into the main water mass of the Sound. Similarly, the cyclonic eddies observed by Pingree (1979) at the tidally-generated Ushant front are well within the suggested critical scales reviewed by Okubo (1978), being of the order of 20 km diameter, and may well be effective in initiating subsequent patchiness.

3   Spatial variability of zooplankton

DIMENSIONS OF HORIZONTAL PATCHINESS

Zooplankton patchiness occurs at all scales from centimetres to tens of kilometres in the horizontal plane, and though some of the concepts developed for phytoplankton patch models may be relevant, no general model has yet been evolved, and there has even been a remarkable lack of success in defining characteristic patch dimensions at sea.

It has been obvious since the last century that variance in plankton nets samples reflected over-dispersion of plankton and from the earliest days of plankton surveys the reliance that could be placed on their results has been disputed. During the 1890s, while Hensen was investigating sample variance, and Kofoid the performance of samplers with respect to mesh pressure and extrusion, Haeckel concluded that the signal-to-noise ratio in survey data was always likely to be unacceptably low. These problems are still with us today though there is now wide agreement as to the sources of variance: diel migration, net avoidance, mesh clogging, the existence of linear patches, and so on.

It has also come to be widely accepted that plankton surveys do produce horizontal fields of plankton abundance that can be contoured unambiguously to show variability on the 10 to 100 km scale, often related to topographic or circulation features. This is the basis of the fish egg and larva surveys used as a tool of fishery management, and is capable of very rigorous statistical control. As Haury *et al.* (1978) have reminded us, such patches reflect the smaller units of geographical distributions of species in relation to oceanic fronts, coastal upwelling events, island wakes, and eddies captured over banks; but there are other instances in which a co-active, rather than a vectorial or stochastic-vectorial origin for the patch, must be assumed. Well-described mesoscale patches occur in the North Sea, where Savage and Wimpenny (1936) found patches of *Rhizosolenia* and Cushing (1962) and Cushing and Tungate (1963) investigated *Calanus* patches of the same size (30–40 km diameter) which maintained their integrity during most of the summer, with a slow southerly advection. Remarkably, Cushing found a consistent population difference between two adjacent patches throughout one summer: differences in the size of each of the three earliest copepodite stages were maintained between the two patches. Whether this was a result of genetic identity, or of different temperature or feeding regimes, was not resolved.

A recent study of Georges Bank (Ware, pers. comm.) of the population of *Sagitta* described by Clarke *et al.* (1943) has demonstrated another characteristic of large patches: some are able to maintain their location despite residual tidal currents. A patch (5 × 8 km) of *Sagitta* located in 1978 maintained its position throughout a ten-day period despite the fact that advection by tidal residuals should have carried it about 60 miles downstream. Ware and others have suggested a mechanism to account for this stability, mediated by an interaction between diel migration and the state of the tide.

The early work of Barnes and Marshall (1951) and an elegant series of observation by Cassie (1959, 1963) in the early days of continuous pumping systems showed that patchiness could occur in the ocean down to the metre scale, just as Tonolli (1949) had already shown for lake plankton. Cassie was able to correlate changes in the abundance of small copepods (*Pseudocalanus* spp.) with changes of temperature and salinity of about 0.1 °C and 0.2°/$_{\circ\circ}$ in a nearshore area of New Zealand. Both Tonolli and Cassie found autocorrelation (animals of the same species tended to aggregate together) and interspecific correlation (species having trophic relationships had a tendency to aggregate together). From serial samples taken as close as 10 m apart, Cassie suggested that the variability he was observing was related to patchiness with a characteristic diameter of between 10 and 100 m.

More recently, Smith *et al.* (1976) found similar patches of two species of *Acartia*, one of *Pseudocalanus* and one of *Oithona* lying across the direction of the longshore current off Oregon. The patches were of about the same dimension as those recorded by Cassie, being some tens of metres across, in which maximum abundances for all these species coincided.

Between the scale of Cassie's patches and those described by Cushing there is a very large size range within which patchiness must occur, as demonstrated by the variance of serial or replicate net samples. Although there have been many attempts to discover general solutions to the nature of patchiness within this range of sizes, very little progress has been made since the days of Hardy and Gunther's remarkable investigations almost 50 years ago. Using a prototype of his continuous plankton recorder (CPR) Hardy made 29 tows in the South Atlantic between Cape Town and South Georgia which clearly showed patchiness in the open ocean of many organisms: for example, in the second tow, covering 43 km, there was a patch of *Salpa longicauda*, with its associated *Sapphirina* copepod, of only 8 km in diameter. Patchiness of this dimension, measured with a precision of one or two miles, was found in tropical and Antarctic water, and in all organisms investigated. Two series of ordinary horizontal net tows, each tow covering about 800 m, were taken over distances of up to 50 km, and again demonstrated patchiness; *Beroe*, *Pareuchaeta*, and *Parathemisto* appeared in patches less than 1.0 km across, *Calanus* about 3.2 km, and *Euphausia* and *Salpa* appeared in rather larger patches.

The search for patch dimension continues today with pumping systems and multiple serial plankton recorders but these have added little to the classical observations. As Fasham (1978) notes, dominant zooplankton patch size frequently seems to be smaller than the sampling interval of the gear used; all that appears to have been achieved is to confirm the overdispersion of zooplankton.

Wiebe (1970) studied a layer of *Limacina inflata* at 75–100 m depth in the California current. A square with 500 m sides was covered with a grid of 12 horizontal LHPR tows (Longhurst and Williams, 1976), samples being spaced at about 15 m intervals. Overdispersion was analysed in six taxa, showing 25 m diameter patches, approximately round in plan, and containing aggregations of animals from two to six times above background levels. At greater depths (500 m), over greater distances (about 5000 m), and with greater sampling intervals (about 60 m), Fasham et al. (1974) used the Skellam (1951) model for overdispersion to analyse patch-size for 63 species of ostracods and copepods, and concluded that this was characteristically 208 ± 86 m. Rejecting social or other biological aggregation mechanisms, Fashim et al. examined the correlation between temperature and abundance for all species, with rather inconclusive results.

MECHANISMS FOR ACTIVE AGGREGATIONS

However, it may be very difficult to distinguish between patchiness formed by social, by reproductive or by vectorial mechanisms. On the km scale in the open ocean it is much easier to demonstrate patchiness in zooplankton (Hardy) than for chlorophyll (Lorenzen) and one may reasonably invoke social and reproductive aggregation as the causal factor for animal patches on such scales. It is increasingly obvious that social swarming does occur in a wide range of habitats and taxonomic groups. What is not certain is the extent to which the dense swarms of invertebrates that have recently been described have the same function in reducing the effectiveness of predators as Brock and Riffenberg (1960) have suggested occurs in shoaling fish. Since in at least some cases (euphausiids and whales) the relative scales of predator and prey are such that large proportions of a swarm may be engulfed in a single predation act, it seems unlikely that this is a universal mechanism. To the long-established occurrence of dense swarms of euphausiids that occur in the Antarctic (Hardy and Gunther, 1935) have in recent years been added mysids in nearshore environments (Clutter, 1969) as well as copepods. In the clear waters around coral reefs (where they can be easily observed), copepods form daytime swarms from 1–2 to about 30 m in diameter during the daytime, and disperse again at night (Emery, 1968; Hamner and Carleton, 1979). Similar swarms occur in the tidal channels of seagrass lagoons, and in other locations around Australia and Palau (Hamner and Carleton, 1979); the extent to which comparable social aggregation accounts for the variance of plankton in net samples in the open ocean is presently quite unknown, but Antarctic euphausiid swarms are not restricted to the vicinity of nearshore topographic features and *Calanus* has been described as forming discrete swarms near the surface at sea (Marshall and Orr, 1955) within which Bainbridge (1952) was able to observe social patches of about a dozen individuals. Okubo (1972) has introduced a term for the centripetal attraction of organisms, a process which he calls attraction-dispersion. At present there appears to be little prospect of a practical test of this theoretical proposition, and the term does not yet appear to have been applied in models describing the formation of plankton patches.

VERTICAL PATTERNS OF VARIABILITY

Though there may be no recurrent pattern in horizontal patchiness there is a strong and consistent vertical pattern, the existence of which was evident from the first divided plankton hauls (Murray and Hjort, 1912), and was well explored long before the existence of the subsurface chlorophyll maximum was known. It has long been evident that there is a more consistent pattern in vertical profiles than in the same distance horizontally.

Very many zooplankton profiles from all seas have now been taken with opening-closing nets usually having 5 to 10 depth intervals within the upper

1000 m, though other equipment such as Clarke-Bumpus samplers or Hardy indicators have been used to describe profiles of greater precision in shallower water. The most comprehensive set of data is that from the Soviet *Vityaz* cruises, the first attempt at surface-to-bottom plankton sampling over great depths and covering whole ocean basins (Vinogradov, 1968). From these samples, there are now available descriptions of vertical distributions of particular taxonomic groups, the vertical structure of whole communities in great taxonomic detail, the layering of plankton and micronekton in relation to acoustic records, and the relations between the vertical distribution of organisms and their physical environment.

From these studies has emerged a general understanding that the zooplankton profiles are dominated by a layer of abundant biomass near the surface, with deeper layers containing sparser plankton though in the open ocean often including a subsurface maximum principally revealed by acoustic techniques. Vinogradov (1968) showed that there is, in very general terms, a decrease of biomass of about an order of magnitude for each 1000 m of depth; it must be remembered that although the 9000–10 000 m over ocean trenches seems a very great distance, passes through many distinct life-zones and includes a biomass difference of as many orders of magnitude, it is a linear distance smaller than many of the patches studied in the near-surface waters whose pattern still remains difficult to generalise.

We now also have a good understanding that the general profiles result from the differential distributions of very many individual species with highly diverse responses to depth and very different migration patterns diel, seasonal, or ontogenetic. It is becoming increasingly obvious (Fig. 1) that individual species, and growth stages of species, have specialised requirements that are satisfied by specialised individual depth distributions.

The most obvious depth specialists are the interzonal species having daytime residence depths 300 to 600 m below the surface, where discrete single-species layers occur; individual species may also layer differentially within the upper 50 m at night. In general, larger species lie deeper than smaller but related species, and younger or smaller stages of the same species lie shoaler than older and larger stages. Tseytlin (1976) has suggested that plankters in general inhabit the greatest depth at which they can compensate their energy expenditure for all purposes by their food intake. The extent to which this is a general phenomenon is quite clear from the high-precision zooplankton profiles now becoming available, and it may occur in even quite shallow water bodies, such as the Arctic sea water lake studied by McLaren (1963) where growth stages of *Pseudocalanus* were distributed progressively deeper.

It may also occur in organisms in which the reproduction occurs rather deep in the water column. As Coombs *et al.* (1979) show, the blue whiting

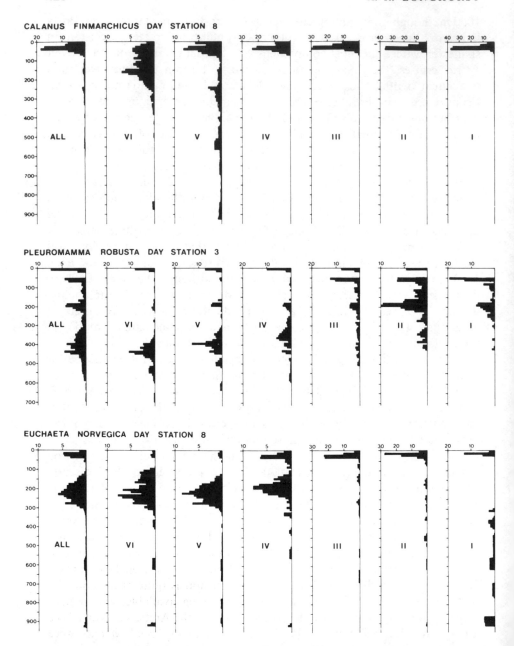

FIG. 1   Depth-differential percentage numerical distributions of the growth stages of three copepods at a series of stations in the north-east Atlantic. Depths in metres, x-axis are percentages. Adults (VI) and five copepodite stages (I–V) are separated

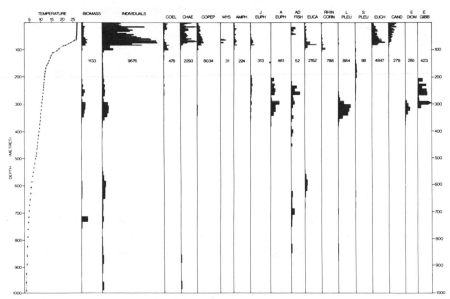

FIG. 2   Zooplankton profile from tropical eastern Pacific Ocean from single Longhurst-Hardy Plankton Recorder haul, to show the relations between epiplankton, plantocline and planktostad and the thermocline. Note also the differential depth distributions between two groups of *Pleuromamma* species (columns 14 and 15) and between species of the genera *Euchaeta* and *Candacia* (16 and 17) within the epiplankton. The horizontal row of numbers represents the absolute total of that category captured in the haul, as a measure of relative abundance

(*Micromesistius potassou*) spawns at 300 400 m, and very young larvae are found mostly at this depth; within about a week, they have migrated to close to the surface, subsequently submerging progressively with growth. The North Atlantic population of the copepod *Euchaeta norvegica* behave very similarly: Williams (1973) found reproducing females in a year-round series of LHPR hauls only at 800–900 m, at which depth substantial numbers of copepodite 1 larvae also occurred; the rapid migration of stage 1 copepodites to the surface, and their subsequent submergence as they pass through later larval stages can be well seen in Williams' profiles.

The high-precision of LHPR profiles of zooplankton abundance which have become available in the last few years from the Atlantic and Pacific have enabled a general model for the detailed form of zooplankton profiles to be proposed (Longhurst, 1976), based on profiles observed in deep water beyond continental shelves.

In this model (Fig. 2), the near-surface layer of abundant plankton (the *epiplankton*, following Fowler, 1898) is bounded by a layer across which plankton abundance decreases very rapidly (the *planktocline*), below which abundance continues to decrease only slowly to the greatest depths as a *planktostad*. In the simplest case, there is a single peak of maximum abundance within the epiplankton and if a secondary peak is present it is very close under the surface forming a *hyponeuston*. Within the planktostad there frequently occur depth-discrete layers of higher abundance. Most of the biomass change within the upper 1000 m occurs across the planktocline within a few tens of metres; planktostad dry weight biomass is about 0.1–1.0 mg m^{-3} in both Atlantic and Pacific above 1000 m, while epiplankton biomass was usually in the range 25–4000 and 1–1000 mg m^{-3} in the survey regions in the Atlantic and Pacific respectively.

BIOTIC-ENVIRONMENTAL INTERACTIONS IN THE VERTICAL PLANE
But, more importantly, the fine depth intervals of the profiles on which this model is based have allowed for the first time comprehensive suggestions to be made about the relations of the plankton profile with the features of the physical water column; the epiplankton maximum usually coincides with

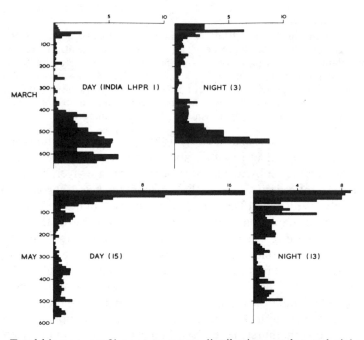

FIG. 3   Total biomass profiles as percentage distributions at day and night station pairs at Ocean Weather Ship INDIA in the north-east Atlantic at the end of March and the beginning of May, 1975, to show the effect on the vertical distribution of zooplankton of the seasonal upward migration of *Calanus finmarchicus* from the depths at which the population overwinters

the bottom of the mixed layer, while the planktocline, which is most strongly developed in stratified water columns, occurs very close to the base of the thermocline as defined by Wyrtki (1964). These relations remain approximately constant in regions with diverse water column characteristics, and anomalies can be explained in terms of special conditions, such as seasonal ontogenetic migrations.

This pattern is most distinct in stable water columns with relatively shallow mixed layers and a single distinct thermocline, as in low latitudes in the eastern part of oceans, and in mid-latitudes towards the end of summer. In the equatorial region, where divergence brings the thermocline much closer to the surface, the epiplankton is crowded surfacewards. At 20°S, where a very deep mixed layer containing several weak thermoclines occur, the epiplankton extends very deep and the planktocline is nonexistent or difficult to demonstrate. In the North Atlantic in early spring, the upper 100 m are effectively isothermal, and there is a downwards increase of biomass to a maximum abundance at 500 to 600 m, caused entirely by the presence there of dense overwintering populations of calanoids; by late spring the well known upwards seasonal migration of this species to exploit the spring bloom has completely re-ordered the pattern to one much more closely resembling the general model (Fig. 3).

Insufficient high-precision profiles have yet been taken over continental shelves to indicate the extent to which the open ocean model is applicable, and apparently none under stratified two-layered conditions in shallow water when the existence of a distinct epiplankton bounded below by a planktocline would be expected to develop.

There are still difficulties in associating features provided by zooplankton sampling systems and the many profiles now available from continuous profiling or pumping systems for temperature, density, illumination, chlorophyll, nutrients, small organic particulate material, bacterioplankton, and microzooplankton; however, the evidence is strong that consistent relations exist between zooplankton and those more easily sampled variables. Indeed, it would be most remarkable if this were not the case, and there is a regular and consistent pattern in the depth relations of these and other variables that indicate functions within the planktonic ecosystem (Vinogradov et al., 1970; Longhurst, 1976; Bishop et al., 1977; Chester, 1978; Vinogradov and Shushkina, 1978). As has been pointed out several times recently, the coincidence of the main gradients of light and temperatures with gravity in the vertical plane in fact assure a consistent pattern, more regularly structured and ordered than any known patterns on similar scales in the horizontal plane. That the major features of the profiles in the upper kilometre are all within the range of the directed locomotion of even very small zooplankters (such as the first copepodite *Euchaeta norvegica* discussed above) determines that it is in this plane, from first principles, that one would seek spatial variability significant to the lives of plankton animals.

4 Vertical–horizontal interactions

EXCLUSION AND ATTRACTION HYPOTHESES

Purposive vertical migration, whether diel or seasonal, in the presence of vertical current shear may cause migrating plankters to be displaced laterally more actively than if they were advected within either of the two water bodies: it is, as Alister Hardy wrote, '. . . as if each tiny creature were given ten-league boots to set it striding through the sea'. Performed seasonally, this may serve to maintain populations within their normal areas of distribution, while performed daily it has been invoked as a cause of, and a reaction to, phytoplankton patchiness.

The 'animal exclusion' hypothesis of Hardy (1936) suggested a mechanism to account for the already well known inverse relation between phytoplankton and zooplankton biomass; more active vertical migration of zooplankters in the presence of strong patches of 'distasteful' phytoplankton would maximise the distance which zooplankters placed between themselves and the phytoplankton. Using an opposite argument, Isaacs *et al.* (1974) suggested that since the presence of abundant phytoplankton was a greater factor in reducing illumination – the principle cue in vertical migration – than either cloudiness or sea-surface roughness, diel migrants would consequently migrate less deeply in the presence of phytoplankton abundance, and hence *reduce* their lateral displacement relative to the plant biomass. It was suggested that this tactic might enable zooplankters to seek, and stay with, patches of phytoplankton satisfactorily dense for their proper nutrition. Evidence that this mechanism exists was taken from records of DSLs in the eastern Pacific which do indeed descend to shallower daytime depths in water of lower transparency, and such differences would reduce lateral displacement from 10 to about one mile daily.

MODELS FOR CONTINENTAL SHELF PATCH FORMATION

Riley (1976) has proposed a model for the generation of patchiness based on similar principles through the interaction of diel migration, tidal ellipses and residual currents of continental shelf seas. The necessary condition is that diel migrants should spend more time at depth than at the surface (Riley assumes 9 h up, 15 h down, basing his argument on a midnight descent). When surface residuals are greater than at depth, and where tidal ellipses differ in the two layers, Riley suggests that differential drift between phytoplankton and zooplankton will be reinforced during some days of the tidal cycle, and reduced during others. Riley shows that during a tidal cycle there will be a period of several days when overlap of more than 50% of the grazed area will occur each night, though during the remainder of the cycle overlap will be insignificant. Thus a patch of zooplankton might graze out a 'window' in its associated phytoplankton population rather like its own dimensions, once every lunar tidal cycle. Converse arguments can be made, for the generation of a patch of high phytoplankton concentration by a discrete area of unusually low zooplankton abundance.

Dubois (1975) and Dubois and Adam (1976) have taken a different approach to explain the patchiness observed on continental shelves, though they ignore vertical terms in their initial model. They suggest that patches in prey-predator systems which begin at the same point in space and time, come to be distributed differentially by trophic interaction, in such a manner that ring-like structures move outward from the origin, herbivores following phytoplankton as successive 'biological waves'. This, they suggest, can only occur under certain conditions of grazing thresholds and constraints which appear to be reasonable; the hypothesis is an extension of the much simpler explanation of the inverse relations between phytoplankton and grazers that was proposed very early on, for example by Harvey (1933): that where there are many grazers, phytoplankton must be of low abundance due to the effect of their feeding. It is also an extension of the observations of the Soviet planktologists (chapter 3) of successive populations of primary producers, herbivores, and predators arranged successively outwards from the linear equatorial divergence.

Evans (1978) introduced vertical shear into a three trophic level (nutrient, phytoplankton, herbivore) model which is a development of Dubois' work; this is capable of generating successive peaks or 'waves' of phytoplankton and herbivores. Herbivore vertical migrations respond to phytoplankton concentrations, and chlorophyll peaks occur in mid-water rather than at the surface, as would be expected. The progressive peaks in the output combine with vertical shear to produce spatial biological waves, in effect resembling those in the simulations from the model of Dubois and Adams.

5 Spatial variability and threshold feeding

CRITICAL NUTRITIONAL THRESHOLD HYPOTHESIS

Dispersion and energy balance

The non-random distribution of plants and animals in the ocean is, obviously, a necessary condition of life for many of them. It is not necessary to quantify the obvious fact that if codfish were uniformly distributed throughout the upper waters of the ocean nobody could make a living catching them. In biological terms, the energy expended during search and capture of randomly dispersed prey would not balance the energy thereby obtained. This has been well understood for many years for baleen whales feeding on euphausiids (chapter 4). Brodie *et al.* (1978) have recently shown that if euphausiids in the Gulf of St Lawrence were distributed uniformly instead of in dense horizontal layers, baleen whales would be unable to make an energy profit from feeding on them.

Though this principle is now understood to operate at all trophic levels and in all food chains, the number of well-documented examples in marine ecosystems is extremely limited, and we presently have to be content with general inferences drawn from disconnected and incomplete data. It is the contention of this review that at the trophic levels of zooplankton and micronekton an organism will have the greatest chance of encountering different levels of food availability by vertical rather than by horizontal migration, and that active horizontal search for food patches is probably confined – for these trophic levels – to the effects of Hardy's 'ten-league boots'.

Experiments with larval fish
Fish larvae are peculiarly suitable for the investigation of patches of water having feeding conditions above or below a critical threshold for growth and survival. They are good subjects for the determination of feeding rates and for critical levels of nutrition, and they are very sensitive to below-threshold levels, so that a point-of-no-return level of starvation can rapidly be determined. It is no surprise, therefore, that the most satisfactory demonstration of a critical feeding threshold in the field should have been based on the larvae of the California anchovy *Engraulis mordax*: this is a small clupeid larva (3.5 mm SL) for which a great deal of background experimental information is available (e.g. Lasker, 1975). As with all fish larvae, available particle size at first feeding is critical and must be 60–80 mm; first feeding larvae have only a 10 % capture efficiency, though both the number of feeding strikes, and their success rate, is dependent on the particle concentration. First feeding larvae have a limited search ability, though this rises from about 200 ml h^{-1} to about $1.0 \, l \, h^{-1}$ by 1 cm SL. Rate of growth and survival depend not only on the number and size of particles ingested, but also their nature. *Gymnodinium splendens*, for example, supports growth in the first days, while other phytoplankters may not. Finally, yolk-sac larvae are able to find and maintain themselves within patches of phytoplankton cells of suitable dimension and higher than background concentrations.

Laboratory experiments indicated that larvae could thrive only above a food particle threshold concentration that seemed unlikely to occur at sea except in the denser layers of particles at the subsurface chlorophyll maximum, and water from several depths at many stations was tested (Lasker, 1975) for its ability to support larval growth. As expected, growth could only be sustained in water from a layer of high chlorophyll, dominated by *Gymnodinium splendens*, that occurred at 20 m depth along at least 100 km of the southern California coastline during March and April 1974: at no depths outside the layer of *Gymnodinium* could growth of the test larvae be demonstrated.

That this ecological situation was unreliable, and the growth of anchovy larvae far from assured, was shown by one windy day which caused sufficient turbulent mixing in the upper 20 m to disperse the dinoflagellate layer, and produce a situation where anchovy larvae were unable to find above-threshold food concentrations anywhere in the water column off southern California. An anchovy larva has a very limited mobility (500–700 m) before it must light upon a patch of food particles of suitable concentration or reach its physiological point of no return, and this suggests that the existence of a layer of dense food particle concentration is essential to the survival of a year-class. California is a site of active, but intermittent, coastal upwelling (Bakun, 1973), and such events disperse or prevent the formation of deep chlorophyll layers. Some indications are already available (Lasker, 1980) from the study of only a few years, that high survival only occurs in years likely to induce dense layers of organisms of the right kind.

This example has been discussed at some length here not just because of its implications (which may be considerable) for the important problem of year-class variation in commercial fish, but because it may serve as a model for suggesting that vertical search, rather than horizontal search, may be critical to the survival of plankters. Although Laskers' study did not specifically address itself to the problem, there is no reason to suppose that horizontal spatial variability could provide the equivalent of a sixfold change in food concentration over 10 m linear distance along a light and gravity gradient perceptible to biota. A horizontal random walk without visual or other cues must be presumed to have a lower probability of a successful outcome than that provided by ontogenetic or diel vertical migration.

OBSERVATIONS ON COPEPODS

Unfortunately, the extent to which we can extrapolate to other planktonic taxa from Laskers' fish larvae is very slight. In general, though the literature on the mechanisms and rate of particle ingestion by planktonic filter feeders is enormous, there have been few studies which have integrated the rate of intake of natural food in the field with calculations about the level of intake required to sustain all metabolic needs, and we remain extraordinarily ignorant on this vital question.

A series of grazing experiments with copepods and natural unconcentrated sea water samples from the same stations (Parsons and Lebrasseur, 1970) suggested that at many stations the measured intake by zooplankton was below what they would require for natural metabolic demands. Because copepods and sea water with contained phytoplankton were not necessarily taken at the same depth in these observations it is not unreasonable to suggest that the copepods might have had a higher rate of intake in nature than in the experiments, if they were actually residing in a depth-discrete layer of suitable food at suitable concentration.

That this possibility exists was shown by a study on the nutrition of *Calanus pacificus* off southern California (Mullin and Brooks, 1976) which showed that energy demands of the natural population could be met at some depth at each station; such depths were frequently below the mixed layer, within which (at these stations) phytoplankton concentrations were below the threshold for sustained growth. At a majority of the stations, when growth and intake were balanced over the whole sampled water column, the population appeared to be in a negative trophic balance: a very significant proportion of the population was apparently located in a nutritional environment insufficiently rich to enable the individuals to meet their metabolic needs.

Though it was not so clear from Mullin and Brooks' study as from the work of Lasker that the significant spatial variability was in the vertical plane, this seems the most tenable hypothesis. As Mullin and Brooks point out, the survival of an important fraction of the population in this situation was dependent on the rapidity with which patchiness became re-ordered – or, they might have added, the copepods responded to unsuitable conditions by seeking higher food concentrations. Dagg (1977), also discussing the importance of vertical spatial variability, has been able to demonstrate a different physiological response to temporal variability of food concentration, over scales resembling the effects of small scale spatial patchiness, by four species of copepods. Dagg suggests that whether or not patchiness is significant to plankters depends on their reaction to periods of low rations while an organism is 'between patches'. He was able to show that *Acartia* and *Centropages* responded to a period of starvation by a sudden and rapid onset of mortality on the third day, with almost no individuals surviving beyond 6 to 10 days. *Calanus* (late copepodites) and adult *Pseudocalanus* showed very little increased mortality for as long as 15 to 25 days under the same experimental conditions. Similarly, *Acartia* and *Centropages* had significantly lower rates of egg production under conditions of intermittent feeding regimes, while under the same conditions of 'patchy' food, *Pseudocalanus* egg production was unchanged.

Dagg reflects the same opinion as Lasker: that 'spatial patchiness of food in the horizontal plane is on too large a scale for individuals to find their way to a patch, while . . . in the vertical it is on a scale that enables adult copepods to find a patch readily'. He suggests that over the continental shelf *Pseudocalanus* is much better adapted than *Acartia* or *Centropages* to survive the frequent breakdown of stability produced by wind events early in the season, with their consequent destruction of the subsurface chlorophyll maximum expected under stable conditions. He suggests that it is no coincidence that in the New York Bight it is *Pseudocalanus* that dominates the copepod plankton in winter-spring, when wind events occur at 3 to 5 day intervals, while during the summer period of stratification *Centropages* rapidly increases to high numbers. Population of both species decline in late summer, due either to predation or insufficient food at any depth in the water column.

References

Bainbridge, R. (1952). Underwater observations on swimming of marine zooplankton. *J. mar. biol. Ass. U.K.* **31**, 107–112

Bainbridge, R. (1957). The size, shape, and density of marine plankton concentrations. *Biol. Rev.* **32**, 91–115

Barnes, H. and Marshall, S. M. (1951). On the variability of replicate plankton samples and some application of contagious series to the statistical distribution of catches over restricted periods. *J. mar. biol. Ass. U.K.* **30**, 233–263

Bishop, J. K. B., Ketten, D. R. and Edmond, J. M. (1977). The chemistry, biology and vertical flux of particulate matter from the upper 400 m of the Cape Basin in the southeast Atlantic Ocean. *Deep-Sea Res.* **25**, 1121–1162

Bowden, K. F. (1965). Horizontal mixing in the sea due to a shearing current. *J. Fluid Mech.* **21**, 83–95

Bowman, M. J. and Esaias, W. E. (1978). Coastal jets, fronts and phytoplankton patchiness. *In* "Oceanic Fronts and Coastal Processes" (M. J. Bowman and W. E. Esaias, eds) pp. 54–63. Springer-Verlag, Berlin

Bowman, M. J. and Iverson, R. L. (1978). Estuarine and plume fronts. *In* "Oceanic Fronts and Coastal Processes" (M. J. Bowman and W. E. Esaias, eds) pp. 87–93. Springer-Verlag, Berlin

Boyd, C. M. (1973). Small-scale spatial patterns of marine zooplankton examined by an electronic *in situ* zooplankton detecting device. *Neth. J. Sea Res.* **7**, 103–111

Brock, V. and Riffenburgh, R. H. (1960). Fish schooling: a possible factor in reducing predation. *J. Cons. int. Explor. Mer* **25**, 307–317

Brodie, P. R., Sameoto, D. D. and Sheldon, R. W. (1978). Population densities of euphausiids off Nova Scotia as indicated by net samples, whale stomach contents, and sonar. *Limnol. Oceanogr.* **23**, 1264–1267

Cassie, R. M. (1959). Micro-distribution of plankton. *N. Z. J. Sci.* **2**, 398–409

Cassie, R. M. (1963). Microdistribution of plankton. *Oceanogr. Mar. Biol. Ann. Rev.* **1**, 223–252

Chester, A. J. (1978). Microzooplankton relative to a subsurface chlorophyll maximum layer. *Mar. Sci. Comm.* **4**, 275–292

Clutter, R. I. (1969). The microdistribution and social behaviour of some pelagic mysid shrimps. *J. Exp. Mar. Biol. Ecol.* **3**, 125–155

Coombs, S. H., Pipe, R. K. and Mitchell, C. E. (1979). The vertical distribution of fish eggs and larvae in the eastern North Atlantic, and North Sea. *In* "Proc. 2nd Int. Symp. Early Life History of Fish". ICES/ELH Symp. DA: 3, 1–17

Cushing, D. H. (1962). Patchiness. *Rapp. P.V. Reun. Cons. Int. Explor. Mer.* **153**, 152–163

Cushing, D. H. and Tungate, D. S. (1963). Studies of *Calanus* patch I. The identification of a *Calanus* patch. *J. mar. biol. Ass. U.K.* **43**, 327–337

Dagg, M. (1977). Some effects of patchy food environments on copepods. *Limnol. Oceanogr.* **22**, 99–107

Dandoneau, Y. (1977). Variations nycthemerales de la profondeur du maximum de chlorophyll dans le Dôme d'Angola. *Cah. ORSTROM Ser. Oceanogr.* **15**, 27–37

Denman, K. L. (1976). Covariability of chlorophyll and temperature in the sea. *Deep-Sea Res.* **23**, 99–107

Denman, K. and Platt, T. (1975). Coherences in the horizontal distributions of phytoplankton and temperatures in the upper ocean. *Mem. Soc. R. Sci. Liege* **7**, 19–30

Dubois, D. M. (1975). Simulation of the spatial structuration of a patch of prey-predator plankton populations in the southern bight of the North Sea. *Mem. Soc. R. Sci. Liege* **7**, 75–82

Dubois, D. M. and Adam, Y. (1976). Spatial structuration of diffusion prey-predator biological populations: simulation of the horizontal distribution of plankton in the North Sea. *In* "System Simulation in Water Resources", pp. 343–356. Elsevier, North-Holland, New York

Emery, A. R. (1968). Preliminary observations on coral reef plankton. *Limnol. Oceanogr.* **13**, 293–303

Eppley, R. W., Renger, E. H., Venrick, E. L. and Mullin, M. M. (1973). A study of the plankton dynamics and nutrient recycling in the central gyre of the North Pacific Ocean. *Limnol. Oceanogr.* **18**, 534–551

Evans, G. T. (1978). Biological effects of vertical-horizontal interactions. *In* "Spatial Patterns in Plankton Communities" (J. H. Steele, ed.) pp. 157–179. Plenum Press, London

Evans, G. T., Steele, J. H. and Kullenberg, G. B. H. (1977). A preliminary model of shear diffusion and plankton populations. *Scott. Fish. Res. Rep.* **9**, 1–20

Fasham, M. J. R. (1978). The statistical and mathematical analysis of plankton patchiness. *Oceanogr. Mar. Biol. Ann. Rev.* **16**, 43–79

Fasham, M. J. R. and Pugh, P. R. (1976). Observations on the horizontal coherence of chlorophyll *a* and temperature. *Deep-Sea Res.* **23**, 527–538

Fasham, M. J. R., Angel, M. V. and Roe, H. S. J. (1974). An investigation of the spatial pattern of zooplankton using the Longhurst-Hardy Plankton Recorder. *J. Exp. Mar. Biol. Ecol.* **16**, 93–112

Fowler, G. H. (1898). Contribution to our knowledge of the plankton of the Faeroe Channel. *Proc. Zool. Soc. London* **4**, 545

Gieskes, W. W. G., Kraay, G. W. and Tijssen, S. B. (1978). Chlorophylls and their degradation products in the deep pigment maximum layer of the tropical north Atlantic. *Neth. J. Sea Res.* **12**, 195–204

Hammer, W. M. and Carleton, J. H. (1979). Copepod swarms: attributes and role in coral reef ecosystems. *Limnol. Oceanogr.* **24**, 1–14

Hardy, A. C. and Gunther, E. R. (1935). The plankton of the South Georgia whaling grounds and adjacent waters, 1926–27. *Discovery Rep.* **11**, 1–456

Hardy, A. C. (1936). Observations on the uneven distribution of oceanic plankton. *Discovery Rep.* **11**, 511–538

Harvey, H. W. (1933). The measurement of phytoplankton population. *J. mar. biol. Ass. U.K.* **19**, 761–773

Haury, L. R., McGowan, J. A. and Wiebe, P. H. (1978). Patterns and processes in the time-space scales of plankton distribution. *In* "Spatial Patterns in Plankton Communities" (J. H. Steele, ed.) pp. 277–328. Plenum Press, London

Haury, L. R., Briscoe, M. G. and Orr, M. H. (1979). Tidally generated internal wave packets in Massachusetts Bay. *Nature (London)* **278**, 312–317

Herman, A. W. and Denman, K. L. (1977). Rapid underway profiling of chlorophyll with an *in situ* fluorometer mounted on a "Batfish" vehicle. *Deep-Sea Res.* **24**, 385–397

Holligan, P. M. and Harbour, D. S. (1977). The vertical distribution and succession of phytoplankton in the western English Channel in 1975 and 1976. *J. mar. biol. Ass. U.K.* **57**, 1075–1093

Horwood, J. W. and Cushing, D. H. (1978). Spatial distributions and ecology of pelagic fish. *In* "Spatial Patterns in Plankton Communities" (J. H. Steele, ed.) pp. 355–383. Plenum Press, London

Isaacs, J. D., Tont, S. A. and Wick, G. L. (1974). Deep scattering layers: vertical migration as a tactic for finding food. *Deep-Sea Res.* **21**, 651–656

Joseph, J. and Sendner, H. (1958). Uber die horizontale diffusion in Meere. *Dtsch. Hydrogr. Z.* **11**, 49–77

Keirstead, H. and Slobodkin, L. B. (1953). The size of water masses containing phytoplankton blooms. *J. Mar. Res.* **12**, 141–147

Kullenberg, G. (1972). Apparent horizontal diffusion in stratified vertical shear. *Tellus* **24**, 17–28

Langmuir, I. (1963). Surface motion of water induced by wind. *Science* **87**, 119–123

Lasker, R. (1975). Field criteria for survival of anchovy larval: the relation between the inshore chlorophyll layers and successful first feeding. *U.S. Fish. Bull.* **71**, 453–462

Lasker, R. (1980). Factors contributing to variable recruitment of the northern anchovy in the California current: contrasting years, 1975 through 1978. *In* "Proc. 2nd Int. Symp. Early Life History of Fish". ICES/ELH Symp./PE: 11

Longhurst, A. R. (1976). Vertical migration. *In* "The Ecology of the Seas" (D. H. Cushing and J. J. Walsh, eds) pp. 116–137. Blackwell, Oxford

Longhurst, A. R. and Williams, R. (1976). Improved filtration systems for multiple-serial plankton samplers and their deployment. *Deep-Sea Res.* **23**, 1067–1073

Longhurst, A. R. and Williams, R. (1979). Materials for plankton modelling: Vertical distribution of Atlantic zooplankton in summer. *J. Plankt. Res.* **1**, 1–28

Lorenzen, C. J. (1966). A method for the continuous measurement of *in vivo* chlorophyll concentration. *Deep-Sea Res.* **13**, 223–227

Lorenzen, C. J. (1971). Continuity in the distribution of surface chlorophyll. *J. Cons. int. Explor. Mer* **1**, 18–23

Marshall, S. M. and Orr, A. P. (1955). "The Biology of a Marine Copepod *Calanus finmarchicus* (Gunnerus)". Oliver and Boyd, Edinburgh

McLaren, I. A. (1963). Population and production ecology of zooplankton in Ogac Lake, a land-locked fjord on Baffin Island. *J. Fish. Res. Board Can.* **26**, 1485–1559

Mooers, C. N. K., Flagg, C. N. and Boicourt, W. C. (1978). Prograde and retrograde fronts. *In* "Oceanic Fronts and Coastal Processes" (M. J. Bowman and W. E. Esaias, eds) pp. 43–58. Springer-Verlag, Berlin

Mullin, M. M. and Brooks, E. R. (1976). Some consequences of distributional heterogeneity of phytoplankton and zooplankton. *Limnol. Oceanogr.* **21**, 784–796

Murray, J. and Hjort, J. (1912). "The Depths of the Ocean". Macmillan, London

Okubo, A. (1971). Oceanic diffusion diagrams. *Deep-Sea Res.* **18**, 789–802

Okubo, A. (1972). A note on small-organism diffusion around an attractive center: a mathematical mode. *J. Oceanograph. Soc. Japan* **28**, 1–7

Okubo, A. (1978). Horizontal dispersion and critical scales for phytoplankton patches. *In* "Spatial Pattern in Plankton Communities" (J. H. Steele, ed.) pp. 21–42. Plenum Press, London

Owen, R. W. (1966). Small-scale, horizontal vortices in the surface layer of the sea. *J. Mar. Res.* **24**, 56–66

Ozmidov, R. V. (1958). On the calculation of horizontal turbulent diffusion of pollutant patches in the sea. *Dokl. Akad. Nauk S.S.S.R.* **120**, 761–763

Parsons, T. R. and LeBrasseur, L. J. (1970). The availability of food to different trophic levels in the marine food chain. *In* "Marine Food Chains" (J. H. Steele, ed.) pp. 325–343. Oliver and Boyd, Edinburgh

Pingree, R. D. (1975). The advance and retreat of the thermocline on the continental shelf. *J. mar. biol. Ass. U.K.* **55**, 965–974

Pingree, R. D. (1978). Mixing and stabilisation of phytoplankton distributions on the northwest European continental shelf. *In* "Spatial Patterns in Plankton Communities" (J. H. Steele, ed.) pp. 181–220. Plenum Press, London

Pingree, R. D. (1979). Baroclinic eddies bordering the Celtic Sea in late summer. *J. mar. biol. Ass. U.K.* **59**, 689–698

Pingree, R. D. and Griffiths, D. K. (1978). Tidal fronts on the shelf seas around the British Isles. *J. Geophys. Res.* (Oceans and Atmospheres), Chapman Conference Special Issue

Platt, T. (1972). Local phytoplankton abundance and turbulence. *Deep-Sea Res.* **19**, 183–187

Platt, T. (1978). Spectral analysis of spatial structure in phytoplankton populations. *In* "Spatial Patterns in Plankton Communities" (J. H. Steele, ed.) pp. 73–84. Plenum Press, London

Platt, T. and Denman, K. L. (1975). Spectral analyses in ecology. *Ann. Rev. Ecol. Syst.* **6**, 189–210

Riley, G. A. (1976). A model of plankton patchiness. *Limnol. Oceanogr.* **21**, 873–879

Savage, R. E. and Wimpenny, R. S. (1936). Phytoplankton and the herring. Part II. 1933 and 1934. *Fish. Invest.* Ser. 2, **15**, 1–88

Shulenberger, E. (1978). Vertical distributions, diurnal migrations and sampling problems of hyperiid amphipods in the North Pacific central gyre. *Deep-Sea Res.* **25**, 605–624

Skellam, A. (1951). Random dispersal in theoretical populations. *Biometrics* **38**, 196–218

Simpson, J. H. and Hunter, J. R. (1974). Fronts in the Irish Sea. *Nature* **250**, 404–406

Smith, L. R., Miller, C. B. and Holton, L. R. (1976). Small scale horizontal distribution of coastal copepods. *J. Exp. Mar. Biol. Ecol.* **23**, 241–253

Star, J. L. and Mullin, M. M. (1980). Horizontal undependability in the planktonic environment. *Marine Science Communications* **5**, 31–46

Stavn, R. H. (1971). The horizontal-vertical distribution hypothesis: Langmuir circulations and *Daphnia* distribution. *Limnol. Oceanogr.* **16**, 453–466

Steele, J. H. (1976). Patchiness. *In* "The Ecology of the Seas" (D. H. Cushing and J. J. Walsh, eds) pp. 98–115. Blackwell, Oxford

Steele, J. H. (1978). Some comments on plankton patches. *In* "Spatial Patterns in Plankton Communities" (J. H. Steele, ed.) pp. 1–20. Plenum Press, London

Tonolli, V. (1949). Struttura spaziale del popolaments mesoplanctonico. Eterogeneita delle densita dei popolameutic orizontale e sua variazione in funzione della quota. *Mem. Inst. Ital. Idrobiol.* **10**, 125–152

Tseytlin, V. G. (1977). Hunting strategy and vertical distribution of pelagic zooplanktophages in the tropical ocean. *Oceanologya* **16**, 507–513

Venrick, E. L., McGowan, J. A. and Mantyla, A. W. (1973). Deep maximum of photosynthetic chlorophyll in the Pacific Ocean. *U.S. Fish. Bull.* **71**, 41–52

Vinogradov, M. E. (1968). Vertical distribution of the oceanic zooplankton. *Nauka Publications*

Vinogradov, M. E., Gitelzon, I. I. and Sorokin, Yu. I. (1970). The vertical structure of a pelagic community in the tropical ocean. *Mar. Biol.* **6**, 187–194

Vinogradov, M. E. and Shushkina, E. A. (1978). Some characteristics of the vertical structure of a planktonic community in the equatorial Pacific upwelling region. *Oceanologya* **17**, 345–350

Wiebe, P. H. (1970). Small-scale spatial distribution in oceanic zooplankton. *Limnol. Oceanogr.* **15**, 205–217

Williams, R. (1973). Vertical distribution and the development of generations of copepods at OWS India. *Proc. Challenger Sci.* **4** (5) (unpaged)

Williams, R. and Robinson, G. A. (1973). Biological sampling at OWS India in 1971. *Annls. Biol. Copenh.* **28**, 57–59

Wroblewski, J. S., O'Brien, J. J. and Platt, T. (1975). On the physical and biological scales of phytoplankton patchiness in the ocean. *Mem. Soc. R. Sci. Liege* **7**, 43–57

Wyrtki, K. (1964). The thermal structure of the eastern Pacific Ocean. *Dtsch. Hydrogr. Z.* **6**, 11–24

Yentsch, C. S. (1965). Distribution of chlorophyll and phaeophytin in the open ocean. *Deep-Sea Res.* **12**, 653–666

15 *Temporal Variability in Production Systems*

DAVID H. CUSHING

1 Introduction

The study of production in the sea must comprise studies of the ecosystem both in terms of trophic levels and of its component populations. The two kinds of study cannot yet be integrated because there is not enough information on the causes of change in the numbers of single populations. The description of temporal variability in numbers is one (but not the only) way of gaining insight into ecological processes. To be useful, time-series should extend over many generations, perhaps somewhere between ten and a hundred. For whales that live for 25 years or more we have information for less than three decades. For fish that live for up to 20 years or so, our best information extends for three to five decades. The most remarkable point is that there are still no satisfactory time-series for the algae and the small herbivores that live for a few days, or a few weeks, respectively.

During the last 20 years or so the reproduction of algae in the sea has been well described so that from a number of environmental variables an average algal reproductive rate can be estimated. Unfortunately, we tend to think in terms of nutrient uptake which is convenient for making a model system budgeted in nutrients. But such an approach is of little value in understanding an ecosystem: only data from a reliable field method of estimating algal reproductive rates is of much use for that purpose since it can be directly used to estimate the parameters of the herbivore populations. In the subtropical ocean such animals reproduce quickly and perhaps continuously and the five to ten generations recorded for some fish populations would appear in a year or less. Considerable efforts have been expended on the small scale spatial variability of chlorophyll and of small animals in an attempt to separate physical and biological processes which might in the future yield solutions to such problems as those posed above (see chapter 14).

In this chapter the scales of temporal variability in the better known parts of the marine ecosystems are reviewed. Simple interpretations are attempted

443

and possible causal mechanisms discussed. These are primarily in terms of the stock and recruitment problem of fisheries biologists, or, more generally, of population stability mechanisms. Until such processes are more fully understood, studies of the whole ecosystem, apart from simple budgets, are hardly worth attempting.

2 Scales of variability

Differences in the quantities of phytoplankton in the sea are caused principally by differences in the reproductive rate of algae and in the grazing activity of zooplankton. In temperate or high latitude seas, the seasonal production of algae may be considered to be a wave of high amplitude lasting six weeks to three months after which it subsides to much lower levels. In tropical and subtropical seas outside upwelling areas (where cycles analogous to those in temperate waters occur), the seasonal differences are obviously much less pronounced, but are sufficient to generate an analogous wave of much lower amplitude. Production of zooplankton may be described similarly, but it endures for longer, three to six months poleward of 40°, where there is no production in winter. Although production can be summed over hours, days, weeks or months, an annual summation is the most useful because this integrates all temporal differences that derive from seasonal variation in the environment.

Because seasonal changes are considerable, the annual changes that integrate them from year to year are best expressed in periods of years, decades, or even centuries. Dickson (1971) observed salinity anomalies in the Baltic outflow and elsewhere with a periodicity of approximately five years. Observations on the recruitment to fish populations extend back for decades: some stocks increase or decline slowly over such periods, whereas others remain steady, only varying somewhat about a mean. Observations on marine ecosystems for long time periods are sparse but some do exist; indeed, perhaps the most important data come from the study of plankton populations at International Station E 1 in the western English Channel (Russell, 1973). For a few fish populations, records extend back over several centuries and for some fish populations off Southern California there are abundance indices extending back for 2000 years in cores taken from anoxic sediments. Thus the scales of information on variability in production systems range from one to two thousand years, with most information concentrated in the few decades before the present day.

PLANKTON
A long series of macroplankton samples has been collected from E 1 near the Eddystone Lighthouse in the western English Channel from 1924 onwards;

it comprised weekly observations until 1939 but subsequently samples were taken much less frequently. The Hardy continuous plankton recorder data (Colebrook, 1965), based on merchant shipping routes, is the most extensive series of records of epiplankton in the North Sea and North Atlantic at monthly intervals; the survey was started by Sir Alister Hardy in 1931, and standardised records extend from 1948 to the present day, the zooplankton caught on 60 mesh silk represent 10 mile sections of track at 10 m and are expressed as deviations about a long-term mean. There are also estimates of 'greenness' (or phytoplankton abundance) from algae trapped on the silk.

In Lake Windermere (in north-west England), Lund (1964) examined the production cycle with weekly observations for 15 years by counting one predominant species *Asterionella formosa* in water samples (Fig. 1a); although the production on average rises from 1 cell/ml to 7000 cells/ml seasonally the year to year variation in production changes by only a factor of three. Recently, Boalch *et al.* (1978) have published records of primary production measurements throughout each year for a decade (Fig. 1b). Such data as these show that primary production in aquatic ecosystems is quite variable and presumably may rise and fall with time.

Another long time series of plankton data has been collected over a very large grid of stations off California and Mexico since 1950; the stations were occupied monthly from 1951 to 1966, and at longer intervals since then. Data for fish larvae, plankton biomass and some plankton species, chlorophyll and physical oceanography are published in the CalCOFI Atlas series. These data are of special interest because they cover a biological unit, the California coastal upwelling, where meteorological, oceanographic, and biological data can be related.

FISH

Data on fish catches, fish stock biomass and recruitment indices exist in fisheries laboratories throughout the world, but mainly from temperate or high latitude stocks. Many fishes live for more than a decade and their stocks comprise many age groups, augmented each year by a new brood, the 'recruiting' year class. Long time-series are not very available, but Fig. 2 (Garrod, personal communication) shows several decadal series of recruitment in numbers, most estimated by cohort analysis. The Karluk salmon stock, *Oncorhynchus nerka*, has declined steadily for the best part of a century, while the two herring populations, *Clupea harengus*, demonstrate both upward and downward trends in recruitment; the three cod stocks have remained more or less steady for a long time. It will be shown below that the abundance of West Greenland cod rose and fell during the recent period of warming as it responded to climatic change in the success of its recruitment. The logarithmic standard deviation of these time series of

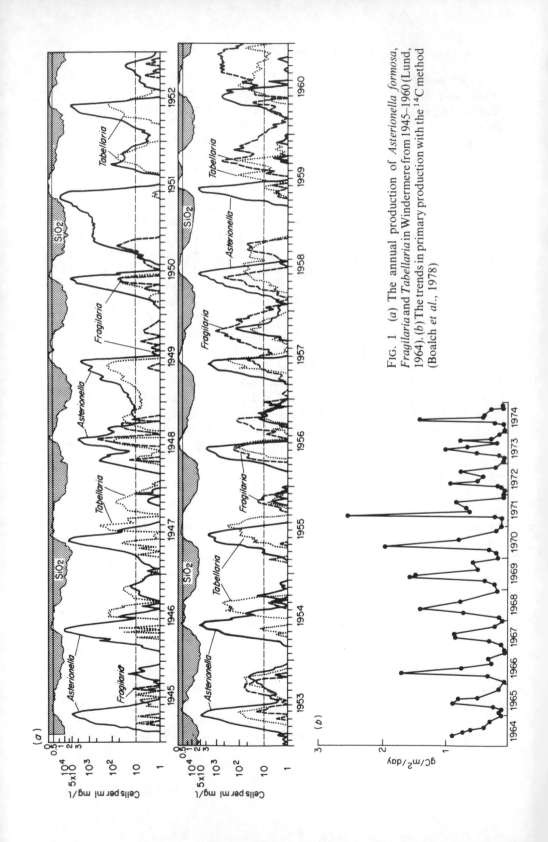

FIG. 1 (a) The annual production of *Asterionella formosa, Fragilaria* and *Tabellaria* in Windermere from 1945–1960 (Lund, 1964). (b) The trends in primary production with the ^{14}C method (Boalch *et al.*, 1978)

recruitment is about a factor of two. The quality of observation is probably about the same for the fish recruitment and for the plankton sampling and the trends and variability observed are real.

3 The recent period of warming

The circumpolar circulation described in the temperature distribution at the 500 mbar level in the atmosphere usually shows five Rossby or planetary waves about which the depressions are driven by the rotation of the earth. Hence it is not surprising that events should be linked across great distances, for example, a negative pressure anomaly over Cape Farewell at the southern tip of Greenland is highly correlated with a positive one over northwestern Europe. Similarly, other events might be expected to be correlated across the same great distance. Between 1880 and 1890 and 1940 and 1945, the mean surface air temperature in the northern hemisphere increased by 0.5 °C after which it declined. The same trend appears in the number of days of westerly wind over the British Isles; they increased from 1890 to about 1925 to 1930, after which they declined to 1970. As the westerlies gave way to meridional winds, evaporation in the north-east Atlantic decreased and the sea surface temperature rose in the later part of the period of atmospheric temperature change (Cushing and Dickson, 1977).

Three spectacular biological events occurred during this period. The first was the general appearance of southern animals in boreal latitudes and the appearance of boreal species in high latitudes. The animals recorded were conspicuous and relatively long lived. The fish *Balistes*, the siphonophore *Velella*, the planktonic mollusc *Ianthina*, the ascidian *Salpa fusiformis*, the albacore *Thunnus alalunga*, and the goose barnacle *Lepas* appeared in the decade 1925 to 1935 between the coast of Brittany and the Isle of Skye. Boreal animals appeared off the Faroe Islands, off Iceland, West Greenland, Jan Mayen and in the Barents Sea: in 1928, cod (*Gadus morhua*) and haddock (*Melanogrammus aeglefinus*) were caught in the middle Barents Sea, off Novaya Zemlaya, and in 1930 and 1932 cod and herring were caught off Jan Mayen. Twaite shad (*Alosa finta*), pollack (*Pollachius pollachius*), and swordfish (*Xiphias gladiens*) were caught off the Faroe Islands, dragonets (*Callionymus maculatus*) and swordfish off Iceland in addition to mackerel (*Scomber scombrus*), bluefin tuna (*Thunnus thynnus*), basking shark (*Cetorhinus maximus*), horse mackerel (*Trachurus trachurus*), and conger (*Conger conger*). Sharks, rays, catfish (*Anarrhichas minor*), Greenland halibut (*Rheinhardtius hippoglossoides*), and redfish (*Sebastes marinus*) were taken off East Greenland and cod were caught on the offshore banks off West Greenland and halibut *Hippoglossus hippoglossus*, coalfish

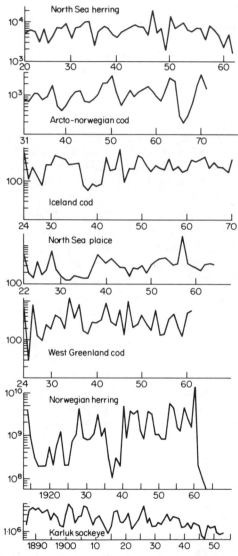

FIG. 2   Recruitment to a number of fish stocks estimated for
decades by cohort analysis (Garrod, pers. comm.)

Pollachius virens, and piked dogfish, *Squalus acanthias* spawned there for the first time. Two arctic species, the sea tadpole (*Careproctus longipinnis*) and the catfish appeared in 1932, 1935, and 1936 in the Bay of Fundy and off Nova Scotia, and southern species, the greater silver smelt (*Leptagonus decagonus*), the sea snail (*Neoliparis atlanticus*), and sucker (*Remora remora*) appeared in the same regions in 1932 and 1935.

All these events occurred in an oceanographically remarkable decade. Certain years were important, 1926, 1931, and 1935, when northerly invasion and colonisation became prominent. Some elements of the oceanic circulation may have intensified: the Gulf Stream, Labrador current, North Atlantic drift, Irminger current, West Spitzbergen current, and the North Cape current.

The second event was the well documented colonisation that occurred during this period by a cod stock of the offshore banks of West Greenland. Although cod populations lived in Greenland fjords, none had been found offshore during the late nineteenth century or the first decade of the twentieth. Some were found between 1906 and 1914 and fishing started between 1912 and 1923. Strong year-classes occurred in 1917, 1922, 1926, 1934, 1936, 1942, and 1945. By the fifties up to 450 000 tons were landed each year and good catches were also being made off East Greenland. The last good year-class appeared in 1963 and the last significant one in 1968 since when the population has declined. The peak abundance in the West Greenland cod fishery occurred just after the peak of the recent period of warming.

We can understand something of these events because of a remarkable tagging experiment (Hansen *et al.*, 1935; Taning, 1937). Between 1924 and 1936 mature or maturing fish were tagged all along the coast of West Greenland and recaptures on the spawning ground at Iceland started in 1930. At the same time fish were tagged on the Icelandic spawning ground and 443 were recaptured there and 17 at West Greenland. After 1945 no such migration back to West Greenland was detected; indeed, the proportion of West Greenland fish recaptured off Iceland was reduced after 1945. It is possible that eggs and larvae were drifted westwards in the Irminger current away from the Icelandic spawning ground and that they reached West Greenland in the mixed East Greenland and Irminger currents. Dickson *et al.* (1975) suggested that when the Greenland high shifted in the middle sixties, westerly winds across the Denmark Strait were replaced by northerlies and so the larvae and juvenile cod no longer reached Greenland. This type of mechanism may also govern the appearance of the West Greenland cod fishery during the periods of warming.

The 'Russell cycle' is the third event characteristic of the recent period of warming between 1925 and 1935. Recruitment to the Plymouth herring stock started to decline with the 1925 year-class and in 1926 pilchard (*Sardina pilchardus*) eggs were recorded in considerable numbers, and by 1935 they were abundant. The winter phosphate values declined by one-third between 1925 and 1935 with the sharpest decrease in 1929 or in 1930. Macroplankton biomass fell by a factor of four in the autumn of 1931, the numbers of summer-spawned fish larvae declined in the summer of 1932, and those of spring-spawned larvae fell in 1935. In the autumn of 1931, the 'indicator species' *Sagitta elegans* was replaced by *Sagitta setosa*.

A reversal of this sequence of events started in 1965 when the numbers of spring-spawned fish larvae increased by about an order of magnitude. In 1970/1, macroplankton increased to the level of the twenties, winter phosphate returned to its pre-1930 level, and summer-spawned fish larvae increased in numbers again. *Sagitta elegans* replaced *Sagitta setosa* in the summer of 1972. From 1966 onwards, summer-spawning pilchards were replaced by autumn spawners (Southward and Demir, 1974) and during the seventies herring were once more caught in the western channel. During the thirties the barnacle (*Balanus balanoides*) retreated westward on the coast of Cornwall, but Southward (1967) reported that it had again started to migrate eastward and to increase in numbers on the Cornish coast. Many of these changes are summarised in Fig. 3 and they reflect thoroughgoing changes within the ecosystem. Each niche sampled in the ecosystem was changed and the implication is that whatever caste of players is on stage at any time there is an alternative set of understudies waiting in the wings. This implies linkages between a matrix of stock and recruitment relationships throughout the ecosystem (Cushing and Dickson, 1976). Thus, the recent period of warming and subsequent cooling has been characterised by the

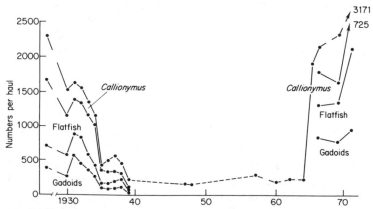

FIG. 3 The Russell cycle; the changes in numbers of flatfish, gadoid and *Callionymus* larvae are shown between 1924 and 1972 (Russell, 1973)

widespread northward movement of conspicuous animals associated with the intensification of many ocean currents. A great cod fishery was established off West Greenland and the stock subsequently collapsed due to a naturally induced recruitment failure. Barnacles migrated back and forth along the western channel coast of England and extensive changes occurred within the whole marine ecosystem in the same region.

4 Long-term trends in fisheries

There are long-term records of catches in a number of pelagic fisheries. A group of such records back to the fifteenth century is summarised in Fig. 4; the blocks represent periods of presence or absence in the two Scandinavian fisheries for herring, and periods of high and low catches in the two Japanese fisheries for herring and sardine (*Sardinops melanosticta*). The most detailed record is that from the Bohuslån fishery in Sweden: the circles above the blocks indicate years of peak catches and the crosses show the 110-year period fitted (somewhat improperly) by Ljungman (1892). The last period ended probably in the nineties of the last century, but some fishing continued, as indicated by the cross-hatched block. It is likely that the recent Norwegian period started with the 1904 year-class, and catches subsequently were not as high as those in the thirties. The years of high catches are shown in the two Japanese fisheries. The records show a rough alternation between the Swedish Bohuslån fishery and the Norwegian spring fishery. The last Swedish period occurred during a period of deteriorating climate whereas the last Norwegian period between the outstanding 1904 year-class and the extinction of the fishery in 1967 occurred during the recent period of warming.

In Fig. 4, the Japanese sardine fishery can be seen to alternate with that for the Swedish herring, while the variations of the Hokkaido herring bear no relation to those of the other three. Zupanovitch (1968) has compared periods of high and low catches for Adriatic and Japanese sardines and has shown that catches tend to vary together; further, the Californian sardine was abundant in the thirties and forties at the peak of the recent warm period. Thus the Norwegian herring and Japanese sardine (and the Californian sardine) all appear to be linked with the Adriatic sardine and all were abundant at the same time.

Soutar and Isaacs (1974) estimated the abundance of scales of five fish species from cores in anaerobic sediments in the Santa Barbara basin off California and the Soledad basin off Baja California. For the limited period during which the stocks of these species were well sampled at sea the rate of sardine scale deposition was well correlated with the stock of two-year-old sardines; scales of anchovies in the cores are similarly well correlated with

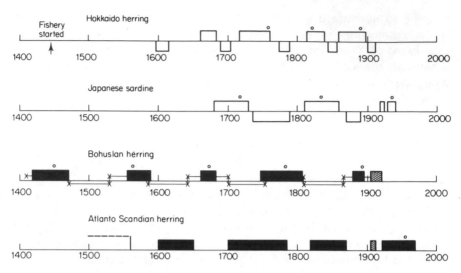

FIG. 4 Long-term records of four pelagic fisheries. The blocks for Hokkaido herring and Japanese sardine represent periods of high and low catches and the circles represent years of peak catches. The blocks for the two Scandinavian fisheries (Bohuslan herring and Atlanto-Scandian or Norwegian herring) represent periods when the fishery existed; in other words, the periods represent presence or absence of fisheries. The line of crosses represents Ljungman's (1882) fitted periods of 110 years in the Swedish Bohuslan herring fishery (Cushing and Dickson, 1976)

total anchovy biomass. Smith (1978) has converted rates of scale deposition to species biomass in five-year periods from 1785 to 1970 for sardine, anchovy, hake, and mackerel.

Figure 5 shows the changes in biomass on a logarithmic scale. For each time series the logarithmic mean is shown as a horizontal line and positive or negative anomalies are indicated.

There are three conclusions to be drawn from this figure. First, there are two sardine periods – 1815 to 1870 and 1896 to 1940 – which correspond to some degree to the Norwegian herring periods (Fig. 4); however, the sardine collapsed in 1949–50, 20 years or more before the Norwegian herring fishery was extinguished, perhaps because the California stock suffered first from recruitment overfishing. Figure 4 shows the decrease in stock under the pressure of fishing from 1930 onwards and its collapse after 1950, as was shown more conventionally by Murphy (1966). The second conclusion from the figure is that both hake and anchovy declined after 1925; indeed, the anchovy may have decreased before then. The anchovy recovered during the sixties and the hake may have started to recover at the end of the period. More detailed study (Ahlstrom, 1966) has shown that the anchovy stock increased in the middle fifties as if it were filling a niche occupied by the sardine albeit with a lag of two or three generations. The third conclusion is that the mackerel and saury reached peaks in abundance between 1925 and 1960, and this may have occurred with the hake about 1800. The decline of hake and anchovy, together with an increase in mackerel and saury, occurred in the period 1925 to 1970, that of the Russell cycle in the English Channel.

The periodicity of the Norwegian herring has been correlated with the amount of ice cover north of Iceland (Beverton and Lee, 1965), so indicating a link with climatic events. Thus, fisheries on a global scale in temperate and

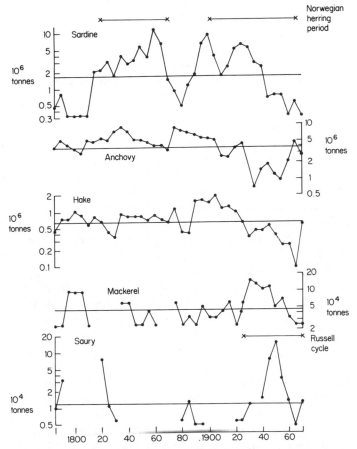

FIG. 5   The records of scale depositon in the anoxic sediments in the Santa Barbara basin off California from 1785 to 1970 (Smith, 1978). Tonnages of stock are estimated from regressions of tonnage on scale deposition given in Soutar and Isaacs (1974)

Mediterranean waters appear to be related rather approximately on a temporal scale, and this may indicate that the planetary pattern exhibited by the Rossby wave system has common effects throughout the northern hemisphere.

5   Analysis of events off California and in the North Sea

In recent years the publication of Bakun's (1973) upwelling indices for the west coast of North America has encouraged simple analyses of biological events there. Considerable changes have occured in the North Sea since 1962, and there has been some discussion whether they were determined by the indirect effects of fishing on the upper levels of the ecosystem or whether they were the result of external events such as variation in climatic factors which might increase the availability of food to larval fish, or indeed which might increase the larval production itself. There is no suggestion that the two regions are physically linked, but the scientific studies in each region have yielded analogous results.

EVENTS OFF CALIFORNIA

Since the collapse of the sardine fishery in the late forties a considerable biological and physical program has been sustained in the waters off California and it can now be shown that the production of pelagic fish and zooplankton are each linked to the intensity of upwelling. Zooplankton was estimated as displacement volume/1000 m^3 by month and quarter between 1951 and 1966 at many stations up to 500 km from the shore on a repetitive grid survey. The data are summarised in Smith (1971), and the quantities used here are the annual averages between Point Concepcion and Punta San Eugenio for the 16-year period.

Upwelling indices were averaged for the same region by years. Such indices are expressed as Ekman transports calculated from the winds estimated from the pressure fields in the north-east Pacific. There is an association between upwelling and the production of zooplankton but a regression of zooplankton on upwelling index is poor ($r^2 = 0.16$). An invasion of the same region by the pelagic crab *Pleuroncodes planipes* from more southerly waters was recorded in 1958 and 1959 (Longhurst, 1967). It is likely that this invasion was associated with the flooding of the eastern tropical Pacific with warm water at the same time as the transequatorial flow from that region towards the coast of Peru which is El Nino: indeed, this point was established by Walsh (1978). Thus the invasion of southern Californian waters in 1958 and 1959 was linked to the El Nino of 1957 and 1958. A regression of zooplankton volumes upon upwelling indices, excluding the years 1958 and 1959, is significant ($r^2 = 0.47$). Upwelling continues during the southern years but the population originates from the south and not from the north and so one should not expect the same quantities to be produced. The conclusion from the regression is that higher production is generated by more intense upwelling.

Peterson and Miller (1975) showed that more plankton is produced in years of high upwelling off Oregon. Peterson (1973) related upwelling indices and the annual catches of the Dungness crab, *Cancer magister*, between Washington State and Northern California; he detected a lag of about one and a half years which suggests the good survival of a year-class. Botsford and Wickham (1975) have confirmed the lag by cross-correlation, but have shown additionally by auto-correlation a lag of 9 or 12 years, probably due to cannibalism. Again, Parrish (1976) linked the recruitment of the Pacific mackerel stock (*Scomber japonicus*) to Bakun's upwelling indices on the spawning ground.

Thus, not only is the production of zooplankton enhanced in years of high upwelling, but also the production of benthic animals and pelagic carnivores is also increased. In other words, elements of the whole ecosystem are related to differences from year to year in the upwelling index. If the upwelling generates a production cycle of the same form as that in temperate waters, then the differences in production of the higher parts of the ecosystem may well be related to differences in primary production from year to year.

EVENTS IN THE NORTH SEA

During the sixties the gadoid stocks in the North Sea increased by about a factor of three. Andersen and Ursin (1977) suggested that the decline of herring and mackerel stocks released food from which the gadoids and sandeels profited. Their evidence is given in Fig. 6a, which shows the changes on a logarithmic scale in 11 stocks in the North Sea during the sixties. The herring stock decreased after 1963 and that of mackerel after 1965; both were reduced to half by 1968 and this decline occurred mainly during the second half of the decade.

The cod, haddock, whiting, and saithe stocks increased from 1962 onwards and the increase was maintained in all four stocks at the same rate until the end of the decade. Thus the gadoid outburst, as it has been called, started before the stocks of herring and mackerel began to decline and each of the four gadoid stocks appeared to increase at the same rate during the decade as shown by the quantities plotted on a logarithmic scale. The stocks of Norway pout and sandeels decreased to 1965 after which both rose sharply; the Norway pout year-classes were not related to this event, which therefore may have been the consequence of the increased exploitation rate. The stocks of flatfish increased during the decade, but not by very much.

Figure 6*b* shows that Norway pout year-classes were larger than average from the beginning of the sixties so that their increase was really part of the gadoid outburst. It is possible that the increase in sandeels was the effect of increased exploitation of the northeastern North Sea.

In the sixties, there were six outstanding cod year-classes, two strong haddock year-classes and two of whiting, but an earlier one (1931) was also

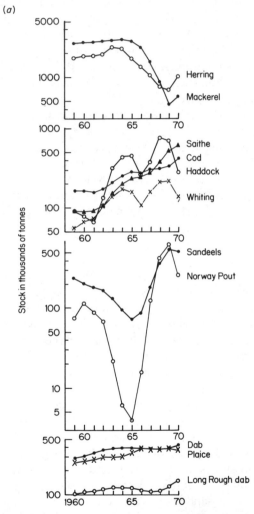

FIG. 6   (*a*) The changes in the biomass of 11 stocks in the North Sea as estimated by Andersen and Ursin (1977); the most prominent change is the gadoid outburst during the whole decade at a rate of increase common to the four species

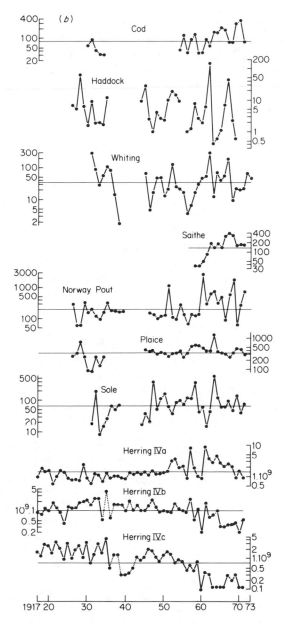

FIG. 6 (b) Time series of recruiting year-classes to a number of stocks in the North Sea from the twenties to the present day. Numbers of recruits are presented in logarithms about a logarithmic mean. The outstanding year-classes of herring, plaice, sole, and gadoids are shown strong. Saithe year-classes were strong from 1962 onwards as compared with the previous four years. Norway pout broods were high in the sixties, although there was also a strong one in 1951. Plaice year-classes tended to be a little stronger after 1957, but those of the sole remained equally variable from 1931 onwards; in both flat fish stocks, the 1963 year-class predominated.

Stocks of herring increased in the central North Sea during the thirties and in the northern North Sea in the fifties and sixties (Burd, 1978). But in the southern North Sea (area IVc) the year-classes declined after 1953. Those of the central North Sea (area IVb) after 1959 and those of the northern North Sea decreased after 1963 due to recruitment overfishing (the collapse of a stock due to failing recruitment under the pressure of heavy fishing). There may have been a decline in recruitment in the southern North Sea during the forties, but there was certainly a partial recovery subsequently. In conclusion, the herring broods declined from south to north between 1953 and the late sixties, and the gadoids increased during the sixties. The flatfish increased slightly from 1957 onwards.

Figure 7 shows the increase in length of four-year-old herring in the Buchan fishery (Cushing and Bridger, 1966), of four-year-old female plaice in Lowestoft catches, in the estimated length for soles, and in length of four-year-old haddock and whiting. Cod did not increase in length at all during this period (Daan, 1974). For flatfish and gadoids, the increase was about 2.5 cm/decade, but herring was only about 1.5 cm/decade. Because of the increase in length many stocks must have increased in weight considerably, perhaps as much as a factor of two over three decades.

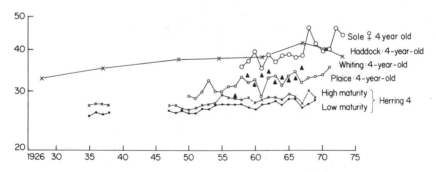

FIG. 7 The increase in length of 4-year-old herring (of high and low maturity stages) in the northern North Sea (Saville, 1978), the increase in length of plaice in Lowestoft catches (Bannister, 1978), the increase in the estimated length for soles (de Veen, 1978), and the increase in length of 4-year-old haddock and whiting (Jones and Hislop, 1978)

Numerical recruitment to the plaice and sole stocks did not increase during the period. Herring recruitment in the North Sea increased between 1953 and 1962. Haddock and whiting year-classes were high in 1962 and 1967. The cod stock increased by a factor of three in weight during the sixties, but the growth of individual fishes remained constant, although length at first maturity fell from about 72 to 56 cm for females (Oosthuizen and Dann, 1976). The increase in weight and possibly earlier maturation must have augmented egg production, but this in itself should not have altered the average rate of recruitment by very much, though it might have altered the dynamics of recruitment.

Iles (1968) and Hubold (1978) have shown that the growth rate of herring in their first year is density dependent. That of adult soles is also density dependent (Houghton, pers. comm.) but in the plaice and cod stocks growth rate increases independently of stock causing an increase in stock weight. The growth rate of whiting increased during the sixties, that of cod remained steady and that of haddock reached a peak in 1967 after which it decreased a little. Because the increase in growth rate antedated the gadoid outburst, we may conclude that food became more available from the thirties onward and that augmentation of egg production dates from then.

The question arises whether it is the markedly different food for adults or for juveniles which is important. The exploitation of adult stocks must release more food for adults and therefore the growth rates of such stocks must increase. If the growth of juvenile fishes is density dependent, food must limit the growth of larger year-classes. The question raised by Andersen and Ursin cannot yet be answered, but there may be two distinct effects of fishing: (i) if an array of stocks is reduced by exploitation to the level at which the maximum sustainable yield is taken, more food becomes available, leading to a greater growth rate and more egg production which might do little more than delay the onset of recruitment overfishing; (ii) if the array of stocks is reduced by exploitation beyond the point at which the maximum sustainable yield is taken, recruitment overfishing may supervene for some stocks, more food is available for the juveniles, which might augment recruitment for the remainder. Then we should not search for the transfer of food amongst adult stocks but rather between juveniles in the form of a multi-species stock and recruitment relationship. This would resemble the Andersen and Ursin model superficially but would be concerned with the transfer of food not between adults but between larvae or juveniles.

Discussion of events in the North Sea has centred on whether the changes were induced by recruitment overfishing of herring or mackerel or whether they were the result of climatic changes in increasing primary production, so making more food available to the larvae and juveniles; Dickson et al.

(1973) suggested that cod year-class strength is inversely related to temperature, so that as the climate deteriorated, the gadoid outburst became more likely. The difference of opinion can only be resolved with a model of the mechanism of recruitment which includes stock, potential switching, and climatic factors at the same time.

A tenuous physical link has been proposed between the two regions, that concomitant shifts of wind strength and direction might have taken place at the same time. The more important link is a conceptual one when analyses of the two ecosystems are compared. From the Californian study it was shown that differences in upwelling and presumably primary production were related to differences in the production of zooplankton, mackerel, and crabs. The system worked as an engine, more primary production leading to more secondary and more tertiary production; such a proposition is inherent in comparative and in model studies, but has not been shown within one system.

In the North Sea the gadoid outburst has generated speculation on the nature of the marine ecosystem. From the Californian experience we might conclude that the gadoid outburst was caused by an increment in primary production. Yet from the study of the Russell cycle, switches from one array of species to an alternative set seems to be possible and stimulated by some form of climatic change. Lastly, fishing may release food for adults (thus stimulating egg production) and the effects of recruitment overfishing may release food for larvae and juveniles and thus stimulate forms of Russell cycle species switching.

6 Possible mechanisms

The basis of our study of possible mechanisms is Ottestad's (1960) correlation between catches of cod in the Vestfjord and the width of pine tree rings in the same area. From the tree ring data he extracted four periods, added them and fitted the resulting complex curve by least squares to the annual catches lagged by seven years. The result is given in Fig. 8. He concluded that if differences in catch, lagged by seven years, indicate differences in recruitment to the stock then those differences are related to the same factors that govern tree ring width, for example solar radiation and wind strength and direction. In the sea, the only way in which such factors can modify the recruitment to the fish stocks is through the production cycle. Colebrook (1965) showed that this cycle varied in amplitude, spread, and time of onset, due to differences in wind strength, wind direction, and

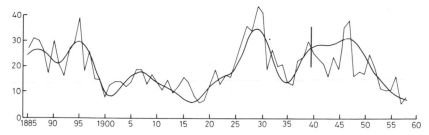

FIG. 8   Ottestad's (1960) correlation between catches of cod in the Vestfjord in northern Norway lagged by 7 years and the sum of four periods extracted from the time series of pine tree widths in the area

solar radiation and such factors are probably linked to those that modify tree ring width.

Another approach emerges from the work of Templeman (1963, 1972) who noticed that outstandingly good or poor year-classes of gadoids and of herring were common to many stocks on both sides of the Atlantic. For the period 1941 to 1971 he tabulated gadoid year-class strengths by scores of 1 to 5, 5 representing a strong year-class. Figure 9 shows the added scores for all gadoid stocks in the north-east and north-west Atlantic for the three decades. The annual scores reached a peak about 1950 and there has been some decline subsequently. The figure includes the gadoid outburst in the North Sea and one might imagine that as the climate deteriorated in the North Atlantic after 1945 the optimal conditions for cod production shifted from Greenland waters towards those of the North Sea. The decline in common recruitment as the climate deteriorated suggests that a determining factor may well be the changes in wind strength and direction associated with the Rossby waves as they shift across the ocean.

The argument is carried a stage further by Garrod and Colebrook (1978). They show that the first principal component of the annual variation of zooplankton has declined steadily in the north-east Atlantic since the early fifties; in the central and northern North Sea the onset of decline may have occurred a little later. A corresponding decline in the number of days of westerly wind and in the first principal component of annual differences in sea temperature is also given. In the second half of the paper the principal components of the recruitment to 18 fish stocks in the North Atlantic were related to periods of predominantly high and low pressure in 20° to 30°W. Figure 10 shows that the first principal component tends to be positive during periods of predominantly high pressure in the central North Atlantic. In other words, the recruitments tend to be positively correlated when northerly winds blow in the north-east Atlantic and when southerly winds blow in the north-west Atlantic.

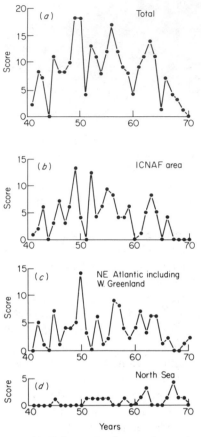

FIG. 9   The summed scores (1–5) for each of annual recruitments to gadoid stocks
in the North Atlantic between 1941 and 1971 (Templeman, 1972). The distributions
are given for the whole ocean, the north-west Atlantic, the north-east Atlantic, and
for the North Sea

The question arises how can the differences in recruitment be affected by
the annual differences in primary production which are themselves modified
directly by physical factors. In temperate waters, fish tend to spawn for
longish periods about a fixed date (Cushing, 1969). Hence the production of
eggs and larvae is modified by the initial stock, by the temperature
coefficients of development, and probably by food density coefficients of
development, which are as yet unknown. But the production of larval food
depends directly on primary production itself varying in amplitude, spread
and time of onset. Hence the distribution in time of larval fish production
and that of their food may overlap, and the degree of overlap may determine

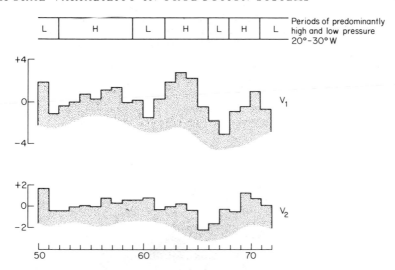

FIG. 10 The first principal component of recruitments to 18 stocks in the North Atlantic for as long a time series as possible and the periods of predominantly high and low atmospheric pressure between 20° and 40°W in the North Atlantic (Garrod and Colebrook, 1978)

the success of recruitment. It is now possible to forecast recruitment from year-old cod, haddock, plaice, and soles in the North Sea and from cod and haddock of the same age in the Barents Sea. Indeed, Rauck and Zijlstra (1978) have shown that the year-class strength of plaice is probably determined before the little fish arrive on the beaches. Hence the magnitude of recruitment is established in the first year of life; if it is governed by those physical factors that modify primary production, then it is reasonable to suppose that the match or mismatch of larval production to that of their food as illustrated in Fig. 11 is a reasonable hypothesis.

There are two additional points to the hypothesis. The first is that as development rate is an inverse power function of temperature, the distribution of larval production is shifted to the right in Fig. 11 in cooler water. Hence the chance of match is improved. The second point is that at high stock more eggs are produced throughout a possibly longer season and again the chance of match is improved, irrespective of whether high stocks generate additional compensatory mechanisms which might modify recruitment as it is being generated.

The match/mismatch hypothesis presupposes that differences in the residual drift from spawning ground to nursery ground are not of great importance. To put it another way, any stock occupies a number of

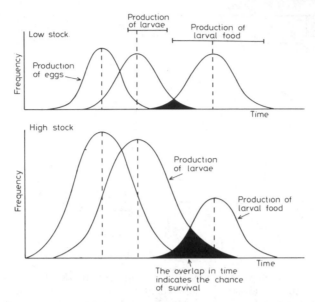

FIG. 11   The match or mismatch of larval production to that of their food (Cushing, 1975). In temperate waters fish spawn at a fixed season and the standard deviation of the peak spawning date is low, perhaps a week or so. The production cycle varies in amplitude, spread, and time of onset and hence we would expect the production of larval food to vary in the same way as shown by the extensive error bar. That for the larval production is asymmetric because it is an inverse power function of temperature; hence there is a great chance of match at lower rates of development in relatively cooler water. At high stock (in the lower diagram), the overlap between the two distributions is greater or the chance of match is improved

spawning and nursery grounds and differences in residual drift can carry larvae from one spawning ground to a nursery ground beyond the usual one. But the recruitment to the stock is in common from all the nursery grounds and although some larvae may be lost to the system, only a very small proportion of the gonad is devoted to random sources of mortality. If this extension to the hypothesis were true we need only examine events along the tracks of the larval drifts.

Other mechanisms are concerned with the relationship between recruitment and parent stock. Figure 12 shows a standard curve of recruitment on parent stock with a bisector; where the curve cuts the bisector the stock stabilises itself. The figure also shows the variability of recruitment at low stock and at high stock. In each case the bar of variation extends below the bisector at low stock or above it at high stock and a single observation at such points is of no consequence. If, however, there is a succession of such

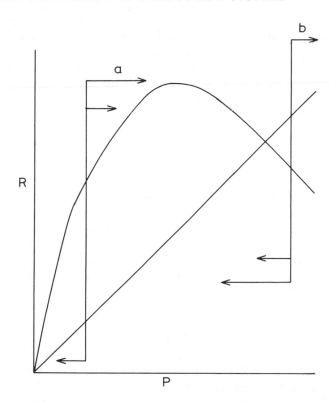

FIG. 12   The dependence of recruitment in numbers on parent stock in numbers with respect to a bisector. At low stock a succession of power year-classes below the bisector will generate a sharp reduction in stock and at high stock the reverse process will generate a sharp increase – in both cases of orders of magnitude (Cushing, 1969)

observations, a sharp decline of stock will take place at low stock and a sharp increase at high stock. In such ways can a stock increase or decrease naturally by orders of magnitude as observed in the secular series of Fig. 2, in the 1904 year-class of herring, or in the Russell cycle.

The simple stock and recruitment curve is:

$$R = \alpha P \exp - \beta P$$

where R is recruitment in numbers,
        P is stock in number of eggs,
        α is density independent survival,
and β is density dependent mortality (Ricker, 1958).

One might expect climatic factors to alter density independent survival and because the number of eggs produced is more or less fixed, the density dependent mortality is also sensitive to climatic factors. Thus an increase in α will cause an increase in β, and vice versa, provided that the climatic change does not alter the carrying capacity of the environment, that is, the quantity of food available at the critical stages of the life history. Generally an increase in α as a consequence of increased β means that the population has become more resilient and can respond more quickly to the annual changes in recruitment.

In the frame of the standard stock and recruitment curve, the match/mismatch hypothesis might account for the variability in recruitment at constant stock. Then an upward or downward trend of recruitment in time might be caused by an increasing degree of match or mismatch so that increments or decrements of recruitment are transferred to the stock. Trends of upward or downward recruitment might also be affected by an alteration in the density independent survival by climatic factors. More drastic changes might be brought about by the succession of abnormally low or high year-classes as shown in Fig. 12.

Such mechanisms are put forward in terms of the simple stock and recruitment curve. A more complex relationship in which recruitment is expressed in terms of food available and predatory mortality might express more fully the three mechanisms given above. Available food and abundance of predators may be determined by climatic factors, match, or mismatch might be expressed only as the quantity of food available, and long-term changes in either food or predators might affect the recruitment directly. Such a relationship has been developed by Shepherd and Cushing (in press) and may be useful in the development of multi-species relationships.

7 Discussion

Temporal changes in marine ecosystems are not often recorded directly as primary production. To some degree, however, such changes can be inferred from recorded changes in the fish stocks or changes in the animal plankton. The production of zooplankton off California responds directly to the degree of upwelling and we must presume that primary production does so as well. Hence it is likely that the annual differences in primary production there are as great as those in the English Channel. Similarly, the decline in phytoplankton during recent decades in the north-east Atlantic includes indication of a progressive delay in the time of onset of primary production each spring.

From studies of the Californian upwelling area, fish stocks and benthic

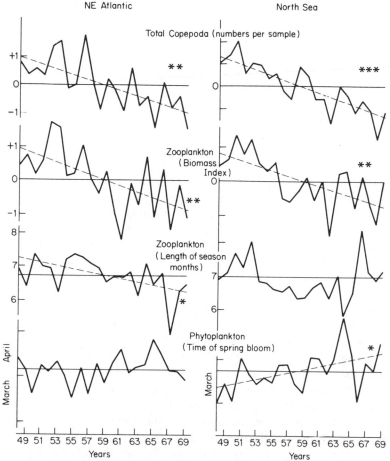

FIG. 13   The decline in zooplankton biomass and in some species in the north-east Atlantic since 1948 as shown form the plankton recorder network (Glover *et al.*, 1972). A delay in the time of onset of the production cycle of nearly a month occurred during the same period

assemblages respond directly to the strength of upwelling and presumably to the magnitude of primary production. The dependence of cod recruitment upon factors that also determine tree ring width implies an almost direct link with the primary production itself. If this is so the remarkable changes in the Russell cycle and in the last two decades in the North Sea might well be responses to changes in the magnitude or timing of the primary production. The direct effect of climatic factors upon primary production has been implicit, if not always stated, since the time of Sverdrup and Riley.

Our concept of marine ecosystems has been governed by the magnitude of transfer coefficients, the nature of trophic levels, and the possibility of species switching mechanisms. For such constructions a fixed value for primary production has often been assumed as a convenience and this convention has quietly become part of our belief. If the initial primary production is variable and responsive to temporal shifts in wind strength and direction, then our preconceptions of how an ecosystem works must change considerably.

I make no apology for an obsession with the problem of the stabilisation of populations, the stock and recruitment problem of fisheries biologists. The essential links between trophic levels are more direct for fishes because they grow through them. But only recently has it been established that the year-class strength of fishes (or more rigorously, one species of flatfish) is determined during larval life or at metamorphosis, and the link between primary production and recruitment has become a little more plausible. Then the primary variation can be directly transmitted, and the trends in time of onset of the spring bloom are manifest in the shift between species – as in the gadoid outburst in the North Sea.

This simplification, if it can be considered one, arises because fishes like all animals in the sea progress through several trophic levels as they grow. Hence in an evolutionary sense the quantities of adults are 'predicted' in the preceding numbers of larvae or juveniles. Therefore there is always enough food (or nearly enough) for the adults, but the competition for it takes place amongst the larvae or juveniles.

The usual ideogram of an ecosystem comprises a few trophic levels, not necessarily comprising adult organisms. The engine works because adequate food for larvae presupposes adequate or even superabundant food for the adults, so that the usual concept of trophic level becomes suspect. To understand the changes in time of any of the animals in the ecosystem, we need to follow those transient inhabitants of a trophic level that determine the magnitudes of the adult populations. This is true not only of fishes, but of fish food, their competitors, and all the inhabitants of the complex web. In this sense the model of the ecosystem tells us only what is uninteresting – that we can't get something for nothing; another model comprising sets of stabilising populations might teach us how an ecosystem really works. The present groping towards an understanding of multi-species interactions represents an attempt to reach such a model without facing the essential problems of stabilisation in populations.

References

Ahlstrom, E. H. (1966). Distribution and abundance of sardine and anchovy larvae in the California current region off California and Baja California, 1951–64: a summary. *Spec. Sci. Rep. U.S. Dept. Int. Fish and Wildlife Service (Fisheries)* **534**

Andersen, K. P. and Ursin, E. (1977). A multi-species extension of the Beverton and Holt theory of fishing, with accounts of phosphorus circulation and primary production. *Medd. Dan. Fisk. Havunders* **7**, 319–435

Bakun, A. (1973). Coastal upwelling indices, west coast of North America, 1946–71. *NOAA Tech. Report NMFS SSRF-671*

Bannister, R. C. A. (1978). Changes in plaice stocks and plaice fisheries in the North Sea. *Rapp. P.V. Int. Explor. Mer* **172**, 86–101

Beverton, R. J. H. and Lee, A. J. (1965). Hydrographic fluctuations in the North Atlantic Ocean and some biological consequences. *In* "The Biological Significance of Biological Change in Britain" (C. G. Johnson and L. P. Smith, eds) pp. 79–107. Institute of Biology, Symposium 14, London and New York

Boalch, G. T., Harbour, D. S. and Butler, E. I. (1978). Seasonal phytoplankton production in the western English Channel 1964–73. *J. mar. biol. Ass. U.K.* **58**, 943–953

Botsford, L. W. and Wickham, D. E. (1975). Correlation of upwelling index and Dungeness crab catch. *Fish. Bull. U.S.* **73**, 901–907

Burd, A. C. (1978). Long term changes in North Sea herring stocks. *Rapp. P.V. Int. Explor. Mer* **172**, 137–153

Colebrook, J. M. (1965). On the analysis of variation in the plankton, the environment and the fisheries. *Spec. Publ. ICNAF* **6**, 291–302

Cushing, D. H. (1969). The regularity of the spawning season of some fishes. *J. Cons. int. Explor. Mer* **33**, 81–92

Cushing, D. H. (1971). Upwelling and the production of fish. *Adv. Mar. Biol.* **9**, 255–335

Cushing, D. H. and Bridger, J. P. (1966). The stock of herring in the North Sea and changes due to fishing. *Fish. Invest. London* **2**, 25 (1)

Cushing, D. H. and Dickson, R. R. (1976). The biological response in the sea to climatic change. *Adv. Mar. Biol.* **14**, 1–123

Daan, N. (1974). Consumption and production in North Sea cod (*Gadus morhua*): an assessment of the ecological status of the stock. *Neth. J. Sea Res.* **9**, 24–55

Dickson, R. R. (1971). A recurrent and persistent pressure anomaly pattern as the principal cause of intermediate scale hydrographic variation in the European Shelf sea. *Dtsch. Hydrogr. Z.* **24**, 97–119

Dickson, R. R., Pope, J. G. and Holden, M. J. (1973). Environmental influences on the survival of North Sea cod. *In* "The Early Life History of Fish" (J. H. S. Blaxter, ed.) pp. 69–80. Springer-Verlag, Heidelberg and New York

Dickson, R. R., Lamb, H. H., Malmberg, S. A. and Colebrook, J. M. (1975). Climatic reversal in the northern Atlantic. *Nature (London)* **256**, 479–481

Garrod, D. J. (personal communication)

Garrod, D. J. and Colebrook, J. M. (1978). Biological effects of variability in the north Atlantic Ocean. *Rapp. P.V. Cons. Int. Explor. Mer* **173**, 128–144

Hansen, P. M., Jensen, A. S. and Taning, A. V. (1935). Cod marking experiments in the waters of Greenland, 1924–33. *Medd. Dan. Fisk. Havunders.* **10**

Houghton, R. G. (Personal communication)

Hubold, G. (1978). Variations in growth rate and maturity of herring in the northern North Sea in the years 1955–1973. *Rapp. P.V. Cons. Int. Explor. Mer* **172**, 154–163

Iles, T. D. (1968). Growth studies on North Sea herring. II. O-group of East Anglian herring. *J. Cons. Int. Explor. Mer* **32**, 98–116

Jones, R. and Hislop, J. R. G. (1978). Changes in North Sea haddock and whiting. *Rapp. P.V. Cons. Int. Explor. Mer* **172**, 58–71

Ljungman, A. (1882). Contribution towards solving the question of the secular periodicity of the great herring fisheries. *U.S. Comm. Fish and Fisheries* **7**, 497–503

Longhurst, A. R. (1967). The pelagic phase of *Pleuroncodes planipes* Stimpson (Crustacea, coalatheidae) in the California current. CALCOFI Report II, 142–154

Lund, J. W. G. (1964). Primary production and periodicity of phytoplankton. *Verh. Int. Ver. Theor. Angew. Limnol.* **15**, 37–56

Murphy, G. I. (1966). Population biology of the Pacific sardine (*Sardinops caerulea*). *Proc. Calif. Acad. Sci.* Ser. 4 **34**, 1–84.

Oosthuizen, E. and Daan, N. (1976). Egg fecundity and maturity of North Sea cod, *Gadus morhua*. *Neth. J. Sea Res.* **8**, 378–397

Parrish, R. (1976). Environmental-dependent recruitment models and exploitation simulations of the California Current stock of Pacific mackerel (*Scomber japonicus*). Ph.D. thesis, Oregon State University, Corvallis, Oregon

Peterson, W. T. (1973). Upwelling indices and annual catches of Dungeness crab *Cancer magister*, along the west coast of the United States. *Fish. Bull. U.S.* **71**, 902–910

Peterson, W. T. and Miller, C. B. (1975). Year to year variations in the planktology of the Oregon upwelling zone. *Fish. Bull. U.S.* **73**, 642–653

Rauck, G. and Zijlstra, J. J. (1978). On the nursery aspects of the Waddensea for some commercial fish species and possible long term changes. *Rapp. P.V. Cons. Int. Explor. Mer* **172**, 266–275

Russell, F. S. (1973). A summary of the observations of the occurrence of planktonic stages of fish off Plymouth, 1924–72. *J. mar. biol. Ass. U.K.* **53**, 347–355

Saville, A. (1978). The growth of herring in the north western North Sea. *Rapp. P.V. Cons. Int. Explor. Mer* **172**, 164–171

Shepherd, D. J. and Cushing, D. H. A mechanism for density-dependent survival of larval fish as the basis of a stock-recruitment relationship. *J. Cons. int. Explor. Mer* (In press)

Smith, P. E. (1971). "Distributional Atlas of Zooplankton Volumes in the Californian Current Region 1951–1956". CALCOFI Atlas Series No. 13

Smith, P. E. (1978). Biological effects of ocean variability: Time and space scales of biological response. *Rapp. P.V. Cons. Int. Explor. Mer* **173**, 117–127

Soutar, A. and Isaacs, J. D. (1974). Abundance of pelagic fish during the 19th and 20th centuries as recorded in anaerobic sediments off California. *Fish. Bull. NOAA*, U.S. Dept. Commerce **72**, 257–275

Southward, A. J. (1967). Recent changes in abundance of intertidal barnacles in southwest England: a possible effect of climatic deterioration. *J. mar. biol. Ass. U.K.* **47**, 81–95

Southward, A. J. and Demir, N. (1974). Seasonal changes in dimensions and viability of the developing eggs of the Cornish pilchard (*Sardina pilchardus* Walbaum) off Plymouth. *In* "The Early Life History of Fish" (J. H. S. Blaxter, ed.). Springer-Verlag, Heidelberg and New York

Steele, J. H. (ed.) (1978). "Spatial Patterns in Plankton Communities". Plenum Press, London

Taning, A. V. (1937). Some features in the migration of cod. *J. Cons. int. Explor. Mer* **12**, 1–35

Templeman, W. (1965). Relation of periods of successful year classes of haddock on the Grand Bank to periods of success of year classes for cod, haddock and herring in areas to the north and east. *Spec./Publ. ICNAF Env. Symp.* **6**, 523–533

Templeman, W. (1972). Year class success in some North atlantic stocks of cod and haddock. *Spec. Publ. ICNAF* **8**, 223–239

Walsh, J. J. (1978). The biological consequences of interaction of the climatic El Niño, and event scales of variability in the eastern tropical Pacific. *Rapp. P.V. Cons. Int. Explor. Mer* **173**, 182–192

Zupanovitch, S. (1968). Causes of fluctuations in sardine catches along the eastern coast of the Adriatic Sea. *Anali Jadramskog Instituta IV*, 401–489

16 Comparative Function and Stability of Macrophyte-based Ecosystems

C. PETER McROY and DENBY S. LLOYD

1 Introduction

Marine macrophytes comprise two fundamentally different groups of plants – the seaweeds, or macroscopic algae, and the vascular plants, including seagrasses, marsh grasses and mangroves. In many coastal regions the shore is characterised by large stands of these benthic plants which can extend from the intertidal zone to depths generally less than 50 meters, ultimately dependent upon the depth at which the lack of light limits their growth. These macrophytes form the structural base for some of the most productive ecosystems in the world (Mann, 1973; McRoy and McMillan, 1977; Teal, 1962), yet the inherent differences in the biological aspects of algae and vascular plants result in major functional differences between their respective systems. Excluding salt marshes and mangrove swamps, since they are most obviously linked to terrestrial habitats (Chapman, 1977), we hope in this chapter to present arguments supporting two conceptual models of function and stability in marine macrophyte systems – an algal-based, or marine, model and a seagrass-based, or terrestrial model.

Although historically the seaweeds and seagrasses have often been lumped together in a single group[1] because of a few morphological similarities and the apparent overlap in habitat, these plants are actually quite different (see Dawson, 1966; den Hartog, 1970). Seaweeds, being algae, are relatively primitive plants; they have no vascular system with which to efficiently translocate nutrients nor a root system to take up nutrients and secure themselves in sediment. Instead, they extract required compounds from the water flowing past them, which is conceptually similar to sessile animals feeding on passing plankton, and they secure themselves to hard substrates by means of a holdfast. For them the bottom is a two-dimensional surface and space is the primary limiting resource (Dayton, 1971, 1975); the substrate serves only as a place of attachment and a

---

[1] On the most recent treatise of seagrass biology (Phillips and McRoy, 1980), the publisher supplied a beautifully illustrated dust cover depicting seaweeds

473

'virtually inexhaustible' supply of inorganic nutrients is available from the surrounding water (Mann, 1973). These truly marine plants contrast sharply with the terrestrially derived vascular plants.

Seagrasses are advanced, flowering plants, complete with both a vascular and root system. These act in concert to absorb nutrients from both the water and sediment and to transport them throughout the plant (McRoy and Barsdate, 1970; McRoy and Goering, 1974). The root-rhizome system also anchors seagrasses into soft sediments, thus making the substrate three-dimensional. Space is not the limiting resource so much as is the essential nutrient regime; inorganic nitrogen in particular can be limiting (McRoy and McMillan, 1977; Patriquin, 1972).

Both groups of macrophytes, and their associated communities, are persistent features of near-shore seas and maintain a dynamic equilibrium in spite of the variable environmental conditions of shallow waters. The processes that maintain the stability of macrophyte-based systems is a primary theme of this paper, yet stability is an ambiguous term ascribed several meanings in the ecological literature. There is general agreement, however, that it refers to the ability of a system to maintain or recover to an equilibrium condition after a disturbance (Horn, 1974; Orians, 1975). Orians (1975) goes further to describe several aspects of stability, and here we are concerned with three of them: persistence, which is the survival time of a system or its components; constancy, which refers to the lack of change in some parameter of a system; and amplitude, which describes the degree of disturbance a system can withstand and still return to its previous state.

May (1975, 1976) develops these latter two definitions into a useful analogy. A system with a large domain of stable values (i.e. poor constancy) has a parameter space resembling a wide and deep valley; all but the most extreme disturbances to the system will eventually allow a return to equilibrium (i.e. high amplitude). A system with a smaller domain of stable values (i.e. high constancy) has a parameter space more like a shallow valley atop a volcano; modest disturbances will cause the system to spill out from the top of the volcano and crash (i.e. low amplitude). The first example, with a large domain of stable values, is called dynamically robust, and the second example, with a small stable domain, is dynamically fragile. What appears to be important, then, given that a system will persist in ecological time (as opposed to longer term, evolutionary time), is how a system functions and what aspects of stability it exhibits.

We emphasise function over structure in describing ecosystems since obviously species change from area to area, but also because it is the set of processes occurring in a system that determine the relationships between components. With this in mind it is more useful to describe the development of an ecosystem in terms of function, or process succession, and to evaluate its current status with functional parameters.

In an important paper Odum (1969) defined ecosystem succession as:

1 An orderly process of community development that is reasonably directional and predictable.

2 Resulting from modification of the physical environment by the community, even though the environment determines the pattern, rate of change, and the limits of development.

3 Culminating in a stabilised ecosystem in which maximum biomass and symbiotic function between organisms are maintained per unit of available energy.

Within the limits set by environmental conditions, systems may progress over time from pioneering, or colonising, to more mature stages. Alternatively, this developmental sequence may be represented spatially, rather than temporally, along a gradient of environmental conditions ranging from severe to optimal (Reichle et al., 1975). Maturity, then, is a relative descriptor of advanced stages of succession, associated with an increase in internal control and less dependence on inputs and events not determined within the confines of the system (Margalef, 1963, 1975). The stability of macrophyte-based ecosystems can best be understood by comparing the processes that lead to their development and maintenance at a certain (equilibrium) level of maturity, or those that return them to equilibrium after a disturbance.

As a hypothesis we propose that two models are appropriate: a truly marine model for the seaweeds, where development and persistence are controlled by grazing, predation, competition, and disturbance; and a terrestrial model for the seagrasses, controlled by organic matter accumulation and decomposition in the sediments and hence nutrient retention or supply. In the former case seaweeds function ecologically much like animals, gaining required nutrients from the passing water column and competing for primary space, while the seagrasses are more typical of plants, developing feedback relationships with the sediment system, thus modifying their environment and ameliorating the stress of the primary limiting factor. Indeed, as we discuss later, seaweeds and their associated fauna might more appropriately be considered a community, since there is little or no influence of the biotic component of the system on the abiotic, as is generally expected in 'ecosystems' (Odum, 1959). Like the deep sea basins and other areas, seaweed-based systems are incomplete as ecosystems, in this case missing the capacity to retain, decompose and remineralise organic material.

2 Seaweed-based systems

The most obvious problem in studying the development of a natural ecosystem is that most of these systems have established a dynamic equilibrium and so have developed to their relative climax (i.e. their appropriate level of maturity). To overcome this problem, several investigators have tried small-scale perturbation experiments (Jones, 1948; Southward, 1964; Dayton, 1971, 1975) in order to observe secondary succession. One difficulty with this approach is that adults of mobile species can migrate from nearby, unaffected areas to preempt a more isolated, successional recolonisation.

The wreck of the M/V *Torrey Canyon* off the coast of Cornwall in the spring of 1967, and the subsequent 'clean up' with toxic dispersants, provided a large-scale experiment in the recolonisation of rocky shores. Southward and Southward (1978) monitored the structural succession of species returning to the denuded areas and discussed the functional interactions they observed during the ten years following the accident.

Appearing first, after a film of diatoms and other unicellular algae, was what they described as a 'green flush' of *Enteromorpha* and *Ulva*. These very simple, annual plants are capable of rapid colonisation due to their frequent broadcasting of gametes and spores. A dense cover of these green algae developed since there was no grazing by the limpet *Patella* or other herbivores. *Patella*, in particular, is a very efficient and effective grazer, and under normal conditions it maintains the algal cover at a low density so that the community of barnacles and limpets dominates.

Later in the first summer, in many places, the 'green flush' was replaced by several species of fucoids, the spores of which may require the protection from dessication afforded by the colonisers *Enteromorpha* and *Ulva* (Neill, 1979). The settlement of *Patella* remained fairly low into the next year so the thick covering of *Fucus* grew and a large upward displacement of more subtidal seaweeds, such as *Laminaria* and *Alaria*, occurred. In some locations this vertical shift was up to 1.5 to 2.0 meters, putting these large kelps in zones where there is usually a community of barnacles and limpets. This upward shift lasted several years, until after a peak in the limpet population had been reached. This startling phenomenon dramatically illustrates that the zonation in intertidal, and probably subtidal, communities is not so strongly controlled by purely physical or physiological forces (Stephenson and Stephenson, 1972; Doty, 1946), but that biological interactions (i.e. grazing and predation) play an important role (Connell, 1972, 1975).

With an increase in grazing pressure by a growing population of *Patella*, the rocky substrate was available for settlement by barnacles, and the limpets soon decreased as their algal food supply diminished. By the ninth year or so, an equilibrium had become established in most areas, resulting in the usual *Patella*-barnacle dominated community.

On the Pacific coast of Washington, Dayton (1975) observed a similar succession of what he called fugitive species. A film of diatoms was succeeded by the ulvoids and *Porphyra*, a red algae. After a few months most of these plants were replaced by longer-lived species of red algae, such as *Halosaccion* and *Iridea*. In this case, however, grazing molluscs did not exert much influence. Further succession proceeded with the fugitive, or colonising, species being displaced by *Hedophyllum sessile* in area of moderate wave exposure, and by *Laminaria setchellii* and *Lessoniopsis littoralis* in the most exposed locations. This last step is an interesting example of community interaction; according to Dayton (1975), *Hedophyllum* loses its dominance in areas of severe wave exposure to *Laminaria* and *Lessoniopsis* even though those sites represent its optimal physiological habitat. These macrophytes are competing for primary space much like the barnacles in Connell's (1961) celebrated experiments.

Although the chitons and limpets are insignificant grazers on the Washington coast, the sea urchin, *Strongylocentrotus purpuratus*, very effectively grazes the lower intertidal and subtidal seaweeds, in many areas setting the lower limits for these macrophytes (Dayton, 1975). A further biological interaction that shapes these communities is predation on the urchins; on the Washington coast the sea star, *Pycnopodia* is a major predator of *S. purpuratus*, while in the Aleutian Islands Estes *et al.* (1978) noted that sea otters perform a similar role. In natural communities, then, a dynamic equilibrium usually exists between predator, grazer, and macrophyte.

In the upper intertidal zone Dayton (1971) suggests that the mussel *Mytilus californicus* and the barnacle *Balanus cariosus* are the potential dominant species since they can escape predation from gastropods, such as limpets and the snail *Thais*, by growing to large size. Predation by the sea star *Pisaster* (Paine, 1966) and physical disturbance by floating logs (Dayton, 1971; Paine, 1979), however, prevent these species from monopolising the primary space. Consequently, more species, such as other barnacles and the algae *Fucus* and *Postelsia*, may compete for this limiting resource.

Postelsia, the sea palm, in particular requires frequent and locally intense disturbance in order to persist, so it is confined to high energy environments (Dayton, 1973; Paine, 1979). Here, patches of primary space are created by disturbance to the dominant mussel beds and patches of barnacles, on which the sea palm may colonise. But soon the mussel bed will encroach and usurp the space again, thus displacing the seaweed. Therefore, a predictable and sufficient, though not catastrophic, level of disturbance is required for this species and its reproductive products to hop-scotch around to suitable habitats. This illustrates the crucial role, not only of disturbance, but its severity and timing, in determining community structure.

The composition of a seaweed-based ecosystem and the nature of its community interactions will vary depending on the physical stresses and the species available to interact (Menge, 1976; Menge and Sutherland, 1976). However, from this discussion of a few examples it is apparent that the seaweeds are functioning much like typical sessile invertebrates. In addition to photosynthesis and reproductive mechanisms, the predominant processes affecting these organisms are physical stress and disturbance, competition for space, predation (grazing), the broadcasting of reproductive products, and the capture of food (nutrients). This macrophyte-based 'ecosystem' might more aptly be considered an open-ended community, as indeed Dayton (1971) does, stressing that much of the primary production and decomposition occur elsewhere.

Chapman (1974) described succession in algal communities, not ecosystems, and suggested that the often divergent paths to different climax communities in the same area indicate that colonising species do not specifically modify the environment. Foster (1975) makes a similar distinction and Connell (1972) found no convincing mechanism through which early colonisers enhanced the settlement of succeeding species. In his experiments Fager (1971) concluded that, 'the observed distributions of the species of large brown algae and of the smaller fish appear . . . to be reasonably explained as the result of chance processes and not by other factors'. Persisting in the face of environmental stress, these communities are dynamically robust, exhibiting relatively low constancy but high amplitude aspects of stability. This is to be expected of a functionally simple, immature system (Horn, 1974; May, 1975; Margalef, 1975).

3 Seagrass-based ecosystems

While the seaweeds require a hard substrate for attachment, all of the seagrasses live rooted in sediments of low energy environments (with the exception of *Phyllospadix*, which uses it roots and rhizomes to keep a purchase on rocks and entrap sediment in areas of strong current or surf). As with terrestrial plants, the seagrasses participate in dynamic interactions with the sediment and its microbial community (Fenchel, 1977; Klug, 1980), resulting in, among other things, a biogeochemical cycling of primary nutrients such as nitrogen and phosphorus (McRoy *et al.*, 1972; Patriquin, 1972). The structural development of a seagrass-based ecosystem relies on a succession of cycling processes much like the development of terrestrial ecosystems. With increasing maturity, these cycles tend to become more internalised within the system (Odum, 1969; Vitousek and Reiners, 1975; Gorham et al., 1979). Internalising these and other processes buffers the system from short-term environmental fluctuations, but reduces its capacity to withstand severe environmental stress.

Den Hartog (1971) summarises several studies describing the structural succession in several types of seagrass systems. The successional sequences he proposed are largely based on descriptions rather than manipulative studies. Yet there are a few studies of structural changes that occurred after natural or human induced disturbances that constitute some of the evidence needed to evaluate seagrass-based systems. Regrettably, there are still too few studies that deal directly with processes in this context.

Along the Mediterranean coast of France *Cymodocea* pioneers in shallow depression on sandy bottoms where some organic matter has accumulated. After the bottom is stabilised by a network of roots and rhizomes, individuals of the genus *Posidonia* take root and eventually out-compete and eliminate the *Cymodocea*. On rocky substrates a small sand-binding algae may accumulate the required sediment for *Cymodocea* colonisation and eventual dominance by *Posidonia*. Sediment continues to accumulate within the *Posidonia* system and presumably an intensifying of internal recycling occurs. Among stable environments in the Mediterranean *Posidonia* represents the climax macrophyte system, with no capacity for further development, while in more stressful habitats succession may be arrested as a persisting *Cymodocea* community.

In seagrass communities in the Caribbean and Gulf of Mexico, den Hartog (1971) presents a sequence that begins with rhizophytic algae and *Halodule* and culminates in *Thalassia*. His conjecture was confirmed by Zicman (1976); the latter followed the recovery of mature *Thalassia* beds in Florida Bay whose rhizome system had been physically disturbed by the propellors of motor boats. The sediments in scars left by boat traffic were severely altered, mostly through oxidation, and even though some scars soon filled with new sediments, in no cases did *Thalassia* recolonise the disturbed areas. However, recolonisation by *Halodule* occurred rapidly, within a few weeks, and only several years later, with accompanying changes in sediment chemistry, did *Thalassia* again establish a mature system. In another area of Florida, a severe stress in the form of sewage effluents led to a decline in *Thalassia* beds, yet the pioneering *Halophila* and *Halodule* persisted in sparse patches near the outfall (McNulty, 1970).

Patriquin (1975) studied the recolonisation of 'blowouts', or grassfree depressions in *Thalassia* meadows in the Caribbean. The floor of the blowout consisted of coarse rubble and sand and was colonised by blue-green and rhizophytic algae; on the lee slope, where the sand layer was thicker, grew *Syringodium* and farthest from the scarp, where sediments were undisturbed, was a mixed *Thalassia-Syringodium* community. By following the advance of *Syringodium* into a blowout and its eventual closing, Patriquin estimated that intial colonisation of the grass-free area occurred about 2 years after blowout formation and that reinvasion by *Thalassia* took 4 to 5 years. Patriquin, like Zieman (1976), noted that the slower recovery of *Thalassia* was probably related to the time required to develop stable, anoxic sediments which are prime *Thalassia* habitat. Where the bottom was continually unstable, pure stands of *Syringodium* persisted.

Apparently the simplest case of structural change associated with succession is for *Zostera marina* (eelgrass), the dominant, as well as the most common and abundant, seagrass in north temperate Atlantic and Pacific coastal waters. *Zostera* is considered a coloniser as well as a climax species – that is, the single species represents the entire range of system development (den Hartog, 1971). Since there are not the usual visual characters of succession (species changes), other indicators must be sought and what appears simple is, in fact, complex.

Initial evidence for understanding the development of the *Zostera* system was provided by a large-scale perturbation that destroyed most of the eelgrass meadows in the North Atlantic in the early 1930s. This was called the 'wasting disease' of eelgrass in the literature; it was thought that the destruction was caused by the slime mold *Labyrinthula* sp. The causal factor has more recently been attributed to abnormally high temperatures (McRoy, 1966; Rasmussen, 1973), and in fact, *Labyrinthula* commonly occurs in healthy *Zostera* meadows (e.g. Kirkman, 1978). Although detailed studies following the demise of eelgrass were not undertaken, it was apparent that simple recolonisation and high productivity alone did not quickly re-establish these systems. Following the disappearance of eelgrass, sediment conditions changed to gravel and rock, and populations of benthic algae appeared (Rasmussen, 1977). In 10 years sparse patches of eelgrass appeared; 20 years later dense stands were noticeable, but it was not until 30 years following its decline that dense, widespread meadows of eelgrass were again apparent.

The evidence of events leading to the recovery of eelgrass suggests that it involved ecosystem processes, primarily the accumulation and decomposition of organic matter in sediments. Both of these processes and the consequent shift in the sediment microbial community are known to be slow, especially when compared to primary production. The refuges, however, where eelgrass persisted during the 'wasting disease' were generally high stress habitats; in these areas succession would be arrested at less mature, or persistently colonising, stages, and these more dynamically robust systems would be better able to withstand severe environmental stress. In recent studies (McRoy and Klug, unpubl. data) several changes in ecological processes during the development of *Zostera* beds have been defined. For example, the process of nitrogen fixation is a significant contributor to the nitrogen economy of the system in colonising beds, but it becomes trivial in mature systems where regenerated nitrogen in the sediments supplies the requirement.

Kirkman (1978) presents another example of disturbance for a different type of *Zostera* system in Australia. In these waters *Z. capricorni* forms a dominant community, but other seagrass species are involved in the succession. For unexplained reasons a flow of sand covered an existing *Z. capricorni* meadow, eliminating it. The new sand habitat was soon colonised by *Halodule unineris*, and Kirkman speculated that this would eventually be replaced by *Z. capricorni*.

The evidence presented by the several studies of the decline and recovery of seagrass-based systems indicates the primary role of severe physical disturbance in determining successional stage. In all cases known a drastic change in sediment type and quality, in particular the destruction of the anoxic, organic rich sediments of mature seagrass systems, led to a renewal of the successional cycle. The colonising seagrass species – *Halodule*, *Halophila*, *Cymodocea*, *Syringodium*, and seedling *Zostera* (at least *Z. marina*) – all have shallow, reduced root-rhizome systems and are able to initiate growth on oxidized, relatively unstable sediments. This is in distinct contrast to the mature system species – *Posidonia*, *Thalassia*, and *Zostera* – that build a thick root-rhizome mat and to a large extent influence the nature and chemistry of associated sediments.

But what about faunal interactions in these seagrass ecosystems? Besides facilitating the remineralisation of organic material, the greatest effect of animals on the system is the cropping of primary production. The major portion, however, of plant material produced in a seagrass meadow, whether temperate or tropical, is not grazed directly, but becomes detritus (McConnaughey and McRoy, 1979; Ogden, 1980). While grazing is not so dominant a process in organisation as in seaweed-based communities, it does come into play. In temperate and higher latitude seagrass systems, species of direct grazers are comparatively few, but an extensive detritus-based food web exists (Kikuchi, 1980; Kikuchi and Pérès, 1977). On the other hand, tropical seagrass systems contain a larger variety of species of grazers that range from invertebrates and fishes to sea turtles and dugongs.

Ogden (1976, 1980) studied faunal relationships of herbivores in Caribbean seagrass beds. These beds are apparently unique in tropical waters because of the large number and biomass of animals that graze directly on seagrasses and their epiphytes. Most of these herbivores, including fishes, invertebrates, and even the green sea turtles (*Chelonia myidas*) consume seagrasses by clipping or biting the leaves (see also Greenway, 1976), but do not disturb the root-rhizome system. This reduces the standing stock of seagrass without changing the basic character of the system. A remarkable feature of coral patch reefs in the Caribbean is the approximately 10-m wide halo of stubble *Thalassia* around the reefs. This halo has been attributed to the grazing of reef fishes (Randall, 1965), but Ogden showed that the sea urchin *Diadema antillarum* is also a major contributor. In a series of exclusion experiments Ogden removed all sea urchins from a patch reef and obtained striking results. After removal of urchins, the halo surrounding the reef disappeared within 6 months; the biomass of *Thalassia* increased, and the sand bottom was no longer visible. This indicates that the mature *Thalassia* systems are not disturbed by this grazing, only the biomass was cropped to a low standing stock. The effect of urchin exclusion on the patch reef itself was more dramatic: the ungrazed reef became dominated by one species of seaweed, *Padina samctae-crucis* indicating that the structure of the seaweed community is forcibly controlled by grazing. This example shows the effect of the same species of grazer on adjacent seagrass and seaweed systems.

The alternative to the example given above where grazing had little or no effect on structure in seagrass systems is that of the dugong that lives in the shallow, tropical waters of the Indo-West Pacific Ocean. This herbivorous mammal is an ardent feeder on seagrasses, primarily species of *Halodule*, *Halophila*, and *Cymodocea* (Heinsohn and Birch, 1972). Dugongs are evidently snouters in that they feed by 'grubbing seagrasses from the bottom' creating distinct troughs in the meadows (Anderson and Birtles, 1978). The suite of seagrasses consumed are all colonisers, and this feeding mechanism that disrupts the sediments insures the continued renewal of early successional stages.

These examples illustrate a fundamental difference between seagrass ecosystems and seaweeds communities: within the former, processes such as nutrient cycling allow a successional development to higher stages of maturity, depending on the severity of environmental stress or disturbance. They can range from pioneering to relatively mature systems, while the seaweed communities have no mechanism to promote further maturity and homeostasis.

4 The models

In a lucid discussion of stability, May (1975) conjectures that,

ecosystems will evolve to be as rich and complex as is compatible with the persistence of most populations. In a predictable environment, the system need only cope with relatively small perturbations, and can therefore achieve [a] fragile complexity, yet persist. Conversely, in an unpredictable environment, there is need for the stable region of parameter space to be extensive, with the implication that the system must be relatively simple.

This is fine in theory, but as he readily admits, 'success in confronting theoretical ideas about stable domains with real world observations will come from the detailed study of simple, low-dimensional systems or subsystems'.

The marine macrophyte-based ecosystems are limited to the shallow water margins of the sea, and so are subject to a high degree of environmental fluctuations. This maintains these systems at a somewhat simple, immature state. Yet, within these limits, there is significant variation in environmental stress and in the accommodation of systems to it. Therefore, between the seaweed- and seagrass-based systems, we may have a reasonable basis from which to study stability and function.

Our model of function in the seaweed-based system is founded on interactions typically associated with populations or communities, such as disturbance, competition, and predation. Since there is almost no modification of the environment, succession is arrested at an immature state and function within the system has a high dependence on outside events. Density-independent factors, such as physical disturbance, obviously play an important role in shaping and maintaining these communities, and the more density-dependent mechanisms of competition and predation act to maintain a dynamic equilibrium between the populations. In this system the macrophytes function ecologically much like animals, being preyed (grazed) upon by them and competing directly with them for space. In both, high reproductive rates indicate their fugitive or colonising status (Horn, 1974; Odum, 1969). Nutrient supply is not a major problem to these plants, so nutrient retention and the establishment of a microbial community to regenerate them is not required. Much like Margalef's (1975) analysis of an upwelling system, these algal communities are prevented from increasing organisation by harsh and unpredictable physical stress, with a flow through the system of nutrients and production. These systems exhibit relatively low constancy but high amplitude while persisting in the face of unameliorated environmental stress.

Contrasting with this marine model, our terrestrial analog for seagrasses involves more conventional notions of development and maturity. Nutrient cycles are gradually internalised (Gorham et al., 1979) and other, less directly measurable aspects of homeostasis develop (Odum, 1969). In colonising conditions these systems function much like the marine model: highly susceptible to, and even dependent on, physical or biological disturbance, dominated by fugitive or colonising, species, and grazing and competition maintain a dynamic equilibrium. However, these seagrass systems can modify the environment, allowing the succession of more constant, yet more dynamically fragile, relatively mature systems.

5 Discussion and conclusions

Specific determinants of maturity, and hence stability, deserve closer scrutiny, and we hope that our discussion here will prompt further inquiry. Biological systems have evolved within certain limits of environmental stress, and along a gradient of this stress they exhibit different characteristics of maturity and stability. As discussed earlier, more severe stress arrests maturity, yet the degree of biological accommodation to this stress is also an important factor, as illustrated in these macrophyte systems. In colonising communities many species require some threshold level of disturbance in order to persist, indicating that they have evolved with similar stress.

One permutation of this effect is the dependence of many species on biological disturbance, such as predation. In several studies (Paine, 1966; Dayton and Hessler, 1972) this has been evoked as a major interaction structuring communities. From our discussion about maturity and stability, however, it would seem that such biological disturbance would be important only in relatively simple, immature systems. The pivotal role of Paine's (1969) concept of a keystone species may only be maintained in simple trophic structures and as maturity and complexity increase, the effect of a single species probably declines.

For example, aboriginal humans usurped the role of the sea otter as a keystone species in certain Aleutian Island seaweed communities (Simenstad et al., 1978), and it could be argued that the Russian conquerors later assumed this role which ultimately led to the seaweed communities in the Aleutian Islands being dependent on the politics of Western Europe.

A similar process is evident in some colonising systems of seagrasses, where dugongs root through the grass beds, creating a biological disturbance that destroys the structure and chemistry of the sediments, thus maintaining the area in a colonising state. In more mature systems there is little such disturbance; turtles, snails, urchins, and fishes only graze the grass blades, leaving the sediments intact.

Physical disturbance, however, can be important to more mature ecosystems as well as colonising ones. In fact, most 'climax' systems are actually a 'blurred mosaic of successional stages' (Horn, 1974) maintained by small, perhaps random disturbances (Loucks, 1970; Pickett, 1976; Bormann and Likens, 1979). Thus, the fugitive species survive and are available to recolonise areas as they open up. The initiation of recolonisation and community development may vary, of course, depending on the meshing of disturbance with the life cycle of the fugitive organisms (Southward and Southward, 1978; Fager, 1971). Extreme disturbances, however, destroy mature systems and secondary succession then proceeds from colonisation to relative climax.

Some other problems, specific to macrophyte ecosystems, are the questions of whether seaweeds and seagrasses are only seral stages in the maturation of other ecosystems. Chapman (1960) suggests that seagrass beds will eventually develop into salt marshes or mangrove swamps; any overlap, however, between these systems appears only to be an edge effect (McRoy and Bridges, 1975). Den Hartog (1971) describes small algae that bind sand prior to invasion of *Cymodocea*. On the other hand, he describes the surfgrass *Phyllospadix* as a precursor to a *Macrocystis* kelp community. Functionally, this last suggestion seems logically reversed since *Phyllospadix* actually entraps sediments around its rhizomes and develops a small-scale sediment community (McRoy, pers. obs.). Phillips (1979) notes that in some cases *Phyllospadix* may accumulate enough sediment to allow invasion by *Zostera*, and he questions whether this might be an example of succession.

The question remains, though – what comes after a seagrass system? What is an appropriate time scale on which to evaluate stability? In what we have referred to as ecological time, systems are capable of succession from colonising to climax conditions, yet this is dependent on the past co-evolutionary history of associated organisms. Over a longer, perhaps evolutionary, time frame the system and its interactions will change. The nature of ecosystem function and stability, however, should not. Stability becomes an even more subjective criterion in this light, but when viewed with respect to immediate environmental conditions and 'non-evolving' species, it retains its immediate usefulness; less mature systems are dynamically robust and more mature systems are dynamically fragile. A practical application of these concepts might prove especially helpful in areas such as the commercial harvest of seaweeds, the large-scale pollution of coastal waters, and the revegetation of disturbed shoreline sediments.

Acknowledgements

This paper is a contribution of the Seagrass Ecosystem Study sponsored by the Office of the International Decade of Ocean Exploration of the National Science Foundation (Grant OCE 77-27050). Contribution No. 399 from the Institute of Marine Science, University of Alaska, Fairbanks.

References

Anderson, P. K. and Birtles, A. (1978). Behavior and ecology of the dugong, *Dugong dugong* (Sirenia): Observations in Shoalwater and Cleveland Bays, Queensland. *Austral. Wildl. Res.* **5**, 1–24

Bormann, F. H. and Likens, G. E. (1979). Catastrophic disturbance and the steady state in northern hardwood forests. *Amer. Scient.* **67**, 660–669

Chapman, A. R. O. (1974). The ecology of macroscopic marine algae. *Ann. Rev. Ecol. Syst.* **5**, 65–80

Chapman, V. J. (1977). Introduction. *In* "Wet Coastal Ecosystems" (V. J. Chapman, ed.) pp. 1–29. Elsevier, Amsterdam

Connell, J. H. (1961). The influence of interspecific competition and other factors on the distribution of the barnacle *Chthamalus stellatus*. *Ecology* **42**, 710–723

Connell, J. H. (1972). Community interactions on marine rocky intertidal species. *Ann. Rev. Ecol. Syst.* **3**, 169–192

Connell, J. H. (1975). Some mechanisms producing structure in natural communities. *In* "Ecology and Evolution of Communities" (M. L. Cody and J. M. Diamond, ed.) pp. 460–490. Belknap-Harvard, Cambridge, Mass.

Dawson, E. Y. (1966). "Marine Botany. An Introduction". Holt, Rinehart and Winston, New York

Dayton, P. K. (1971). Competition, disturbance, and community organization: The provision and subsequent utilization of space in a rocky intertidal community. *Ecol. Monogr.* **41**, 351–389

Dayton, P. K. (1973). Dispersion, dispersal, and persistence of the annual intertidal alga, *Postelsia palmaeformis* Ruprecht. *Ecology* **54**, 433–438

Dayton, P. K. (1975). Environmental evaluation of ecological dominance in a rocky intertidal algal community. *Ecol. Monogr.* **45**, 137–159

Dayton, P. K. and Hessler, R. R. (1972). Role of biological disturbance in maintaining diversity in the deep sea. *Deep-Sea Res.* **19**, 199–208

den Hartog, C. (1970). "The Seagrasses of the World". Elsevier, Amsterdam

den Hartog, C. (1971). The dynamic aspect of the ecology of sea-grass communities. *Thalassia Jugoslavica* **7**, 101–112

Doty, M. S. (1946). Critical tide factors that are correlated with the vertical distribution of marine algae and other organisms along the Pacific coast. *Ecology* **27**, 315–328

Estes, J. A., Smith, N. S. and Palmisano, J. F. (1978). Sea otter predation and community organization in the western Aleutian Islands, Alaska. *Ecology* **59**, 822–833

Fager, E. W. (1971). Pattern in the development of a marine community. *Limnol. Oceanogr.* **16**, 241–253

Fenchel, T. (1977). Aspects of decomposition of seagrasses. *In* "Seagrass Ecosystems: A Scientific Perspective" (C. P. McRoy and C. Helfferich, eds) pp. 123–145. Dekker, New York

Foster, M. S. (1975). Algal succession in a *Macrocystis pyrifera* forest. *Mar. Biol.* **32**, 313–329

Gorham, E., Vitousek, P. M. and Reiners, W. A. (1979). The regulation of chemical budgets over the course of terrestrial ecosystems succession. *Ann. Rev. Ecol. Syst.* **10**, 53–84

Greenway, M. (1976). The grazing of *Thalassia testudinum* (Konig) in Kingston Harbor, Jamaica. *Aquat. Bot.* **2**, 117–139

Heinsohn, G. E. and Birch, W. R. (1972). Foods and feeding habits of the dugong, *Dugong dugong* (Erxleben), in northern Queensland, Australia. *Extrait de Mammalia* **36**, 414–422

Horn, H. S. (1974). The ecology of secondary succession. *Ann. Rev. Ecol. Syst.* **5**, 25–37

Jones, N. S. (1948). Observations and experiments on the biology of *Patella vulgata* at Port St. Mary, Isle of Man. *Proc. Lpool. Biol. Soc.* **56**, 60–77

Kikuchi, T. (1980). Faunal relationships in temperate seagrass beds. *In* "Handbook of Seagrass Biology: An Ecosystem Perspective" (R. C. Phillips and C. P. McRoy, eds) pp. 153–172. Garland STPM, New York

Kikuchi, T. and Pérès, J. M. (1977). Consumer ecology of seagrass beds. *In* "Seagrass Ecosystems: A Scientific Perspective" (C. P. McRoy and C. Helfferich, eds) pp. 147–193. Dekker, New York

Kirkman, H. (1978). Decline of seagrass in northern areas of Moreton Bay, Queensland. *Aquat. Bot.* **5**, 63–76

Klug, M. J. (1980). Detritus-decomposition relationships. *In* "Handbook of Seagrass Biology: An Ecosystem Perspective" (R. C. Phillips and C. P. McRoy, eds) pp. 225–245. Garland STPM, New York

Loucks, O. (1970). Evolution of diversity, efficiency, and community stability. *Amer. Zool.* **10**, 17–25

Mann, K. H. (1973). Seaweeds: Their productivity and strategy for growth. *Science* **182**, 975–981

Margalef, R. (1963). On certain unifying principles in ecology. *Amer. Natur.* **97**, 357–374

Margalef, R. (1975). Diversity, stability, and maturity in natural ecosystems. *In* "Unifying Concepts in Ecology" (W. H. van Dobben and R. H. Lowe-McConnell, eds) pp. 139–150. Junk, The Hague

Margalef, R. (1975). What is an upwelling system. *In* "Upwelling Ecosystems" (R. Boje and M. Tomczak, eds) pp. 12–14. Springer-Verlag, Berlin

May, R. M. (1975). Stability in ecosystems: Some comments. *In* "Unifying Concepts in Ecology" (W. H. van Dobben and R. H. Lowe-McConnel, eds) pp. 161–168. Junk, The Hague

May, R. M. (1976). Patterns in multi-species communities. *In* "Theoretical Ecology" (R. M. May, ed.) pp. 142–162. Saunders, Philadelphia

Menge, B. A. (1976). Organization of the New England rocky intertidal community: Role of predation, competition, and environmental heterogeneity. *Ecol. Monogr.* **46**, 355–393

Menge, B. A. and Sutherland, J. P. (1976). Species diversity gradients: Synthesis of the roles of predation, competition, and temporal heterogeneity. *Amer. Natur.* **110**, 351–369

McConnaughey, T. and McRoy, C. P. (1979). [13]C label identifies eelgrass (*Zostera marina*) carbon in an Alaskan estuarine food web. *Mar. Biol.* **53**, 263–269

McNulty, J. K. (1970). Effects of abatement of domestic sewage pollution on the benthos volumes of zooplankton and the fowling organisms of Biscayne Bay, Florida. *Stud. Trop. Oceanogr.* **9**, 1–107

McRoy, C. P. (1966). The standing stock and ecology of eelgrass (*Zostera marina* L.) in Izembek Lagoon, Alaska. Master's Thesis, University of Washington, Seattle, Washington

McRoy, C. P. and Barsdate, R. J. (1970). Phosphate absorption in eelgrass. *Limnol. Oceanogr.* **15**, 6–13

McRoy, C. P. and Bridges, K. W. (1974). Dynamics of seagrass ecosystems, pp. 374–375. *Proc. First Intl. Congr. Ecol.* The Hague

McRoy, C. P. and Goering, J. J. (1974). Nutrient transfer between the seagrass *Zostera marina* and its epiphytes. *Nature* **248** (5444), 173–174

McRoy, C. P. and McMillan, C. (1977). Production ecology and physiology of seagrass. *In* "Seagrass Ecosystems: A Scientific Perspective" (C. P. McRoy and C. Helfferich, eds) pp. 53–87. Dekker, New York

McRoy, C. P., Barsdate, R. J. and Nebert, M. (1972). Phosphorus cycling in an eelgrass (*Zostera marina* L.) ecosystem. *Limnol. Oceanogr.* **17**, 58–67

Neill, F. X. (1979). Structure and succession in rocky algal communities of a temperate intertidal system. *J. Exp. Mar. Biol. Ecol.* **36**, 185–200

North, W. J. (1971). Introduction and background. *In* "The Biology of Giant Kelps (*Macrocystis*) in California" (W. J. North, ed.) pp. 1–96. Verlag von J. Cramer, Lehre

Odum, E. P. (1959). "Fundamentals of Ecology". Saunders, Philadelphia

Odum, E. P. (1969). The strategy of ecosystem development. *Science* **164**, 262–270

Ogden, J. C. (1976). Some aspects of herbivore-plant relationships on Caribbean reefs and seagrass beds. *Aquat. Bot.* **2**, 103–116

Ogden, J. C. (1980). Faunal relationships in Caribbean seagrass beds. *In* "Handbook of Seagrass Biology: An Ecosystem Perspective" (R. C. Phillips and C. P. McRoy, eds) pp. 173–198. Garland STPM, New York

Orians, G, H, (1975). Diversity, stability and maturity in natural ecosystems. *In* "Unifying Concepts in Ecology" (W. H. van Dobben and R. H. Lowe-McConnell, eds) pp. 139–150. Junk, The Hague

Paine, R. T. (1966). Food web complexity and species diversity. *Amer. Natur.* **100**, 65–75

Paine, R. T. (1969). A note on trophic complexity and community stability. *Amer. Natur.* **103**, 91–93

Paine, R. T. (1979). Disaster, catastrophe, and local persistence of the sea palm, *Postelsia palmaeformis*. *Science* **205**, 685–687

Patriquin, D. G. (1972). The origin of nitrogen and phosphorus for growth of the marine angiosperm *Thalassia testudinum*. *Mar. Biol.* **15**, 35–46

Patriquin, D. G. (1975). "Migration" of blowouts in seagrass beds at Barbados and Carriacou, West Indies, and its ecological and geological implications. *Aquat. Bot.* **1**, 163–189

Phillips, R. C. (1979). Ecological notes on *Phyllospadix* (Potamogetonaceae) in the northeast Pacific. *Aquat. Bot.* **6**, 159–170

Phillips, R. C. and McRoy, C. P. (1980). "Handbook of Seagrass Biology: An Ecosystem Perspective". Garland STPM, New York

Pickett, S. T. A. (1976). Succession: An evolutionary interpretation. *Amer. Natur.* **110**, 107–119

Randall, J. E. (1965). Grazing effect on seagrass by herbivorous reef fishes in the West Indies. *Ecology* **46**, 255–260

Rasmussen, E. (1973). Systematics and ecology of the Isefjord marine fauna. *Ophelia* **11**, 1–495

Rasmussen, E. (1977). The wasting disease of eelgrass (*Zostera marina*) and its effects on environmental factors and fauna. *In* "Seagrass Ecosystems: A Scientific Perspective" (C. P. McRoy and C. Helfferich, eds) pp. 1–51. Dekker, New York

Reichle, D. E., O'Neill, R. V. and Harris, W. F. (1975). Principles of energy and material exchange in ecosystems. *In* "Unifying Concepts in Ecology" (W. H. van Dobben and R. H. Lowe-McConnell, eds) pp. 27–43. Junk, The Hague

Simenstad, C. A., Estes, J. A. and Kenyon, K. W. (1978). Aleuts, sea otters, and alternate stable-state communities. *Science* **200**, 403–411

Southward, A. J. (1964). Limpet grazing and the control of vegetation on rocky shores. *In* "Grazing in Terrestrial and Marine Environments" (D. J. Crisp, ed.) pp. 265–273. Blackwell, Oxford

Southward, A. J. and Southward, E. C. (1978). Recolonization of rocky shores in Cornwall after use of toxic dispersants to clean up the Torrey Canyon spill. *J. Fish. Res. Board Can.* **35**, 682–706

Stephenson, T. A. and Stephenson, A. (1972). "Life Between Tidemarks on Rocky Shores". W. H. Freeman, San Francisco

Teal, J. M. (1962). Energy flow in the salt marsh ecosystem of Georgia. *Ecology* **43**, 614–624

Vitousek, P. M. and Reiners, W. A. (1975). Ecosystem succession and nutrient retention: A hypothesis. *Bioscience* **25**, 376–381

Zieman, J. C. (1976). The ecological effects of physical damage from motor boats on turtle grass beds in southern Florida. *Aquat. Bot* **2**, 127–139

17 Lipids and Hydrocarbons in the Marine Food Web

JOHN R. SARGENT and KEVIN J. WHITTLE

1 Introduction

Lipids are grouped into polar and neutral classes (Fig. 1) with regard to structure and function. Unusual polar lipids exist in numerous marine phyla (Malins and Wekell, 1970) but the major polar lipids (Fig. 1) are akin to those in terrestrial organisms.

Neutral lipids (fats or oils) are reserves both of fatty acids which are catabolised to form ATP and fatty acids in phospholipids which can generate ATP via catabolism but whose major role is in biomembrane structure and function.

Constancy of tissue or organ function is reflected in constancy of (*a*) polar lipid content, usually a few per cent or less of the dry weight; (*b*) polar lipid class composition reflecting specific biomembrane function, e.g. the high

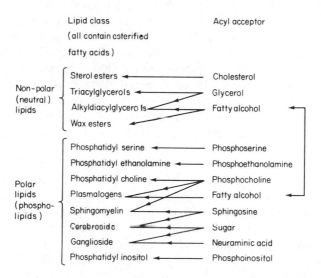

FIG. 1 Reproduced from Sargent (1976)

content of sphingomyelin in nerve; (c) fatty acid composition of the polar lipid; certain polyunsaturated fatty acids (PUFA) are required on position-2 of glycerol and usually 16:0 or 18:1 fatty acid is present on position-1. Neutral lipid content of a tissue can range from a few per cent of its dry weight to more than 90% (for specialised adipose tissues). Fatty acid composition of neutral lipid can vary substantially since unit weights of different fatty acids generate similar amounts of ATP.

The concentration of neutral lipid in an organism is determined by a balance between dietary intake, anabolism and catabolism. The last two processes may occur simultaneously or at different times, and constitute the phenomenon of turnover. A large intake of lipid results in substantial deposition of lipid within an organism, so that the fatty acid composition of tissue neutral lipid reflects that of the diet. However, the biosynthesised fatty acids can be markedly species specific. The level of polar lipid in an organism is relatively constant but it is also subjected to turnover; its fatty acid composition is also dependent on diet since the PUFA in animal polar lipids are essentially dietary in origin and can only be formed *de novo* by plants.

Given the dietary origin of part of the lipids of an organism, lipid analyses can be used to probe predator–prey relationships. The specificity of the approach depends on the extent to which individual lipid components are specific to individual species. Given species-specific 'markers' the transfer efficiency from one trophic level to another can be determined. We need to know the total pools of a given marker at the trophic levels by determining the percentage of the marker in a given organism and the total biomass of that organism. The turnover of the pool must also be known since its size is dictated not only by dietary input but also by internal anabolism and catabolism. Unfortunately, metabolic information is fragmentary if not non-existent in most cases. Study of essential dietary factors, e.g. PUFA, simplifies the problem since their rate of biosynthesis in a predator is zero. They are also tightly conserved, i.e. their rate of catabolism or loss is low. The supposed metabolic inertness of hydrocarbons, both natural and man-made, makes them potentially promising markers. We emphasise however that the notion of metabolic inertness is tenuous – there is scarcely any known lipid compound that is not metabolised to some extent – we are forced to consider turnover as a factor.

Considering lipid analytical data in terms of percentage composition rather than total pool sizes is dangerous. If a relatively inert marker is transferred from one trophic level to another, the total pool in the higher level cannot exceed that in the lower. The percentage of the marker in total lipid in the higher level can be greater than, equal to or less than that in total lipid in the lower level, depending on the extent to which the lipid pool in the higher level is decreased or increased by combined anabolism and catabolism.

This chapter illustrates some of these principles in using lipid analyses to probe the marine food web. We consider only lipids and their constituent fatty acids, and hydrocarbons. Recent reviews on marine lipids and related compounds deal with general lipid biochemistry (Malins and Wekell, 1970; Sargent, 1976), analytical techniques and fatty acid analyses (Morris and Culkin, 1976), sterols (Goad, 1976; Morris and Culken, 1977), marine waxes and wax esters (Nevenzel, 1970; Benson et al., 1972; Sargent et al., 1976; Sargent, 1978) and hydrocarbons (Farrington and Meyer, 1975; Clark and Brown, 1977; Clark and MacLeod, 1977; Hardy et al., 1977; Varanasi and Malins, 1977).

2 Polyunsaturated fatty acids and phospholipids

Fluidity of biomembranes is determined by fluidity of phospholipids which depends on a balance between polyunsaturated fatty acids (PUFA) on position-2 of the glycerol backbone and a saturated or monounsaturated fatty acid on position-1. This role together with their being biosynthesised de novo only by photosynthetic organisms makes PUFA essential dietary factors in animals. PUFA are major constituents of both neutral and phospholipids of photosynthetic organisms (Lee et al., 1971). Animal neutral lipid invariably contains PUFA, but at a much lower concentration than animal phospholipid.

Typical fatty acid analyses of cultures of marine phytoplankton are reproduced in Table I (see also Tables III and IV). The following points emerge (see also Chuecas and Riley, 1969): (i) there are significant differences in the fatty acid compositions of the major phytoplankton groups; (ii) PUFA from C16 to C22 are present; (iii) the abundant diatoms (Bacillariophyceae) and dinoflagellates (Dinophyceae) contain respectively large concentrations of 20:5 (n–3) and 22.6 (n–3) (see also Beach and Holz, 1973; Moreno et al., 1979a); (iv) littoral algae (Rhodophyceae, Porphyridium sp.) contain substantial amounts of (n–6) PUFA, especially 20:4 (n–6), confirmed by analysis of Ascophyllum nodosum, a member of the Phaeophyta (Paradis and Ackman, 1977), and benthic Australian algae (Johns et al., 1979). Thus marine benthic algae contain noticeable concentrations of (n–6) PUFA as do fresh water organisms (Ackman, 1967); planktonic algae contain typically marine high concentrations of (n–3) PUFA.

Marine invertebrates are rich in (n–3) PUFA, especially 20:5 (n–3) and 22:6 (n–3). Benthic and inshore invertebrates including molluscs, coelenterates, echinoderms and crustacea (Morris, 1971; Paradis and Ackman,

TABLE I Fatty acid composition of representative gorups of algae[a]

	Bacillario-phyceae	Chryso-phyceae	Xantho-phyceae	Dino-phyceae	Crypto-phyceae	Chloro-phyceae	Prasino-phyceae	Rhodo-phyceae
14:0	13.9	8.7	7.4	3.3	1.3	—	—	—
16:0	21.2	16.2	21.8	36.0	14.5	15.5	19.7	33.8
16:1 (n–7)	38.8	16.9	14.3	—	1.1	1.0	1.9	—
16:2 (n–7)	1.1	—	—	—	—	—	—	—
16:2 (n–6)	—	—	—	—	—	1.7	—	—
16:2 (n–4)	2.9	2.4	2.0	—	—	—	—	—
16:3 (n–6)	—	—	—	—	—	2.6	—	—
16:3 (n–4)	4.0	—	—	—	—	—	—	—
16:3 (n–3)	—	—	—	—	—	2.7	14.7	—
16:4 (n–3)	—	—	—	—	—	20.5	—	—
16:4 (n–1)	1.8	—	—	—	—	—	—	—
18:0	—	—	—	5.7	1.1	—	—	2.8
18:1 (n–9)	—	2.4	—	7.5	3.0	1.9	8.8	1.9
18:1 (n–7)	—	1.7	—	—	7.2	1.0	3.0	—
18:2 (n–6)	—	3.1	—	—	10.5	5.4	3.3	10.7
18:3 (n–6)	—	—	—	—	—	5.0	—	—
18:3 (n–3)	—	2.2	6.8	—	23.0	32.8	18.9	—
18:4 (n–3)	1.3	9.9	7.9	10.1	16.1	—	7.5	—
20:4 (n–6)	—	—	1.2	—	—	—	1.3	24.9
20:5 (n–3)	9.5	15.1	21.8	7.4	13.8	—	7.5	17.4
22:6 (n–3)	1.4	11.2	3.0	25.4	—	—	1.4	—

[a] Reproduced from Sargent (1976); data compiled from the original work of Ackman et al. (1968). Data represent weight per cent; — signifies absent or less than 1%

1977; Kanazawa *et al.*, 1977; Clarke, 1979; Jones *et al.*, 1979) contain substantial quantities of (n–6) PUFA, especially 20:4 (n–6), in their polar lipids. These (n–6) PUFA must originate in benthic algae. Pelagic invertebrates including copepods, decapods, mysids and euphausiids contain small concentrations of (n–6) PUFA (Ackman *et al.*, 1970; Lee *et al.*, 1971; Bottino, 1974, 1975; Morris and Sargent, 1973) (Tables II, III and IV), reflecting the low concentrations of (n–6) PUFA in offshore phytoplankton. The lipid analyses in Table I show substantial concentrations of C16 and C18 PUFA in phytoplankton. These acids are seldom present in significant amounts in invertebrate polar lipid although C18 can reach sizeable proportions of total fatty acids in e.g. copepod wax esters – see below. Either the shorter chain PUFA in plant material are converted to longer chain PUFA in the invertebrates or they are catabolised. Little precise information is available on the relative importance of these two processes. The major PUFA of *Calanus* phospholipids were 20:5 (n–3) and 22:6 (n–3) even though these were not major PUFA in its algal diet (Lee *et al.*, 1971). *Artemia* forms 20:5 (n–3) on algal diets lacking this fatty acid (Kayama *et al.*, 1963; Hinchcliffe and Riley, 1972). *Paracalanus parvus* converts exogenous 18:3 (n–3) to 22:6 (n–3) albeit at a low rate (Moreno *et al.*, 1979b). We conclude that (*a*) phytoplanktonic PUFA can be deposited unchanged in invertebrate neutral and polar lipid; (*b*) invertebrates can convert shorter chain dietary PUFA to longer chain PUFA through chain elongation and desaturation reactions. Reaction pathways (Fig. 2) have been established for land animals and fresh water species, especially fish (reviewed by Cowey and Sargent, 1972, 1979).

Thus, marine invertebrates have the capacity to elongate shorter chain dietary PUFA. We do not know the extent of these reactions in nature especially where a mixed phytoplankton diet containing diatoms and dinoflagellates results in a large input of 20:5 (n–3) and 22:6 (n–3) into invertebrates. Are the amounts of these acids in natural phytoplanktonic diets sufficient to suppress elongation of shorter chain dietary PUFA? If so, are the latter catabolised to provide metabolic energy? The answers to these questions require quantitative as well as qualitative data.

Marine fish contain 20:5 (n–3) and 22:6 (n–3) as major PUFA in both phospholipid and neutral lipid. Fresh water fish can elongate short chain dietary PUFA, especially 18:3 (n–3) which is a significant constituent of their partly terrestrial diets, to 20:5 (n–3) and 22:6 (n–3) as shown by both radioactive tracers and nutritional balance experiments (reviewed by Cowey and Sargent, 1979); species studied include rainbow trout (see Cowey and Sargent, 1979), eel and ayu (Kanazawa *et al.*, 1979). There is no experimental evidence that marine fish can elongate shorter chain dietary PUFA to 20:5 (n–3) and 22:6 (n–3) at significant rates;

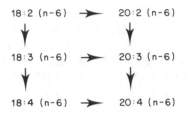

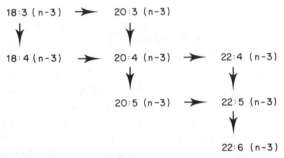

FIG. 2 Pathways of polyunsaturated fatty acid biosynthesis in animals. Desaturation reactions are shown vertically, chain elongation reactions horizontally. The upper scheme is the linoleic acid, 18:2 (n–6), series; the lower is the linolenic acid, 18:3 (n–3), series. Based on Mead and Kayama (1969) and Sinnhuber (1969)

nutritional studies show that 22:6 (n–3) is markedly superior to C18 PUFA in maintaining growth; radioactive tracer studies show that 18:3 (n–3) is converted to 22:6 (n–3) at very low rates, if at all (reviewed by Cowey and Sargent, 1977, 1979); species studied include turbot (Owen et al., 1975), red sea bream (Yone and Fujii, 1975; Fujii and Yone, 1976), rockfish and globefish (Kanazawa et al., 1979). The marked superiority of long chain over short chain PUFA in maintaining growth holds also for the shrimp *Penaeus japonicus* (Kanazawa et al., 1977). It has been proposed (see Cowey and Sargent, 1977) that the carnivorous turbot and red sea bream always have a luxus of C20 and C22 PUFA in their natural diets and seldom if ever form these acids from shorter chain dietary precursors. They have either lost or have never evolved the requisite reactions. A similar situation exists for terrestrial carnivores such as the lion and the cat which have a strict requirement for the preformed long chain terrestrial PUFA, arachidonic acid 20:4 (n–6) (Rivers et al., 1975, 1976).

These considerations apply generally to the marine food web where most species are carnivores. No species beyond the trophic level phytoplankton – invertebrates is likely to experience a dietary deficiency of long chain PUFA and, as seen above, the extent to which short chain dietary PUFA are transformed to higher derivatives by marine herbivores remains undefined.

An example of selective concentration of dietary PUFA occurs in the marine mammals which have high concentrations of (n–6), especially 20:4 (n–6), in their phospholipids (Ackman and Hooper, 1974; Crawford et al., 1976; Bottino, 1977) (Table II). In brain and muscle the ratio (n–3)/(n–6) PUFA is similar in marine and terrestrial mammals (Crawford et al., 1974), despite the low percentage of (n–6) PUFA in non-mammalian marine lipids, especially in euphausiids that are major constituents of the natural diets of whales (Tables II, III and IV). The blubber of whales, largely triacylglycerols, contains large concentrations of (n–3) PUFA reflecting its predominantly dietary origin but the animals' phospholipids are much richer in (n–6) PUFA. They have preferentially incorporated the low percentage of (n–6) PUFA in dietary lipid into their tissue phospholipids. There is a sufficiency of (n–6) PUFA generated in the phytoplankton and transmitted through the euphausiids to maintain a large population of marine mammals. Are not the much larger quantities of C20 and C22 (n–3) PUFA in the phytoplankton sufficient to maintain the non-mammalian carnivorous population including the invertebrates? These considerations emphasise the inadequacy of considering fatty acid analyses in terms only of percentages of individual acids; it is essential to consider total pools of individual acids at given trophic levels.

TABLE II   Polyunsaturated fatty acids in phospholipids of marine animals

	Calanus[a]	Meganycti-phanes[b]	Herring[c]	Cod[d]	Grey seal[e]	Dolphin[f]
18:2 (n–6)	—	2.1	—	—	2.0	1.7
18:3 (n–3)	—	—	—	—	—	—
20:4 (n–6)	—	1.0	1.4	2.9	16.2	13.0
20:5 (n–3)	13.3	14.5	12.2	14.6	3.6	8.2
22:5 (n–3)	—	—	—	1.2	2.4	1.3
22:6 (n–3)	36.5	34.6	32.7	35.4	4.5	11.0

Data are weight per cent of total fatty acids; — Signifies absent or less than 1%.
Data are assembled from [a] Lee et al. (1971) for wild Calanus helgolandicus, total phospholipid; [b] Ackman et al. (1970) for Meganyctiphanes norvegica from the stomach of fin whales, total phospholipid; from [c] Drozdowski and Ackman (1969) for whole herring phospholipid; from [d] Olley and Duncan (1965) for cod flesh phospholipid; from [e] Ackman and Hooper (1974) for liver phospholipid for the harbour seal; [f] from Crawford et al. (1976) for liver phosphatidylethanolamine from Tursiops truncatus

3   Saturated and monounsaturated fatty acids and neutral lipids

NEUTRAL LIPID CLASSES

Many unusual neutral lipids occur in marine animals, often with specific fatty acid compositions.

Triacylglycerols are the major or only neutral lipids of phytoplankton (Lee et al., 1971) although wax esters are prominent in a marine crypto-monad grown photoheterotrophically (Antia et al., 1974). Wax esters consisting of phytol esterified largely to C16, C18, C20 and C22 PUFA are also present in a dinoflagellate (Withers and Nevenzel, 1977). Perhaps major lipid classes other than triacylglycerols occur in natural phytoplank-ton populations under special conditions (see Sargent et al., 1977).

Many marine invertebrates are rich in wax esters (Benson et al., 1972; Sargent et al., 1976), especially calanoid copepods from high latitudes and bathypelagic mysids, decapods and euphausiids. All wax ester-rich animals so far examined can biosynthesise these lipids de novo. At the trophic level zooplankton – clupeoid fish there is virtually complete conversion of wax esters to triacylglycerols (Sargent et al., 1979). Wax ester hydrolysing activity is widely distributed in the guts of fish (Patton and Benson, 1975; Patton et al., 1975).

Wax esters are present in the Antarctic krill, E. crystallophorias (Bottino, 1975), a Thysanoessa sp. (Clarke and Prince, 1976) and E. triacantha (Clarke, personal communication). E. superba contains traces of wax esters (Clarke, personal communication) and has triacylglycerols as its major neutral lipid (Bottino, 1975). Meganyctiphanes and Thysanoessa inernis from the North Atlantic contain only triacylglycerols (Ackman et al., 1970).

Wax esters are absent from benthic organisms, notable exceptions being certain anemones and tropical corals (reviewed by Sargent et al., 1976). Wax esters are produced by algae symbiotic with the coral (Benson and Muscatine, 1974; Patton et al., 1977).

Wax esters are considered to be elaborated in large quantities by species in environments where short periods of food abundance are followed by long periods of scarcity (Lee and Hirota, 1973). This hypothesis has a biochemical corollary that wax ester formation represents a mechanism whereby lipids can be formed in particularly large amounts over short periods (Sargent, 1976). The presence of wax esters in tropical corals is not an exception to these hypotheses since the quantities in corals are much less than in animals in extreme environments (Benson and Muscatine, 1974; Patton et al., 1977; Lee and Hirota, 1973). Wax esters are present in all animal and plant forms at least in trace amounts. It is not their presence in copepods that warrants attention but the very large amounts.

The alkyl (alcohol) moieties of wax esters from calanoid copepods in polar and subpolar surface waters are generally 16:0, 20:1 and 22:1, while the acyl (acid) moieties are most characteristic of fatty acids in the diet, e.g. they are commonly rich in PUFA (see Sargent et al., 1976). The alcohols appear to be biosynthesised de novo from dietary protein and carbohydrate whereas the acids emanate mainly from the diet. The 20:1 and especially the 22:1 alkyl moieties produced in large amounts by copepods are minor constituents of phytoplankton lipid. Wax esters are produced as metabolic energy reserves although secondary roles, e.g. in providing buoyancy, are not excluded. They may be used for maintaining adult or pre-adult copepods during over-wintering; they are rapidly used during the development of fertilised eggs to free-swimming, early copepodites (Lee et al., 1974).

Commercially important marine fish such as clupeoids, gadoids and flatfish have triacylglycerols as their major neutral lipid whether they consume wax ester – rich diets or not. Very abundant pelagic species such as the myctophids and bristlemouths can have wax esters in abundance (Nevenzel, 1970; Sargent et al., 1976). Numerous bathypelagic fish such as Ruvettus and the Moridae (Nevenzel, 1970) as well as Gephyroberyx (Gatten and Sargent, unpublished data) are rich in wax esters.

Unusual lipid classes are common in elasmobranchs, especially squaloid sharks that are rich in alkyldiacylglycerols (Malins and Wekell, 1970; Sargent, 1976). These lipids are biosynthesised from glycerol, fatty acid and fatty alcohol and are intermediate between triacylglycerols and wax esters. The alkyl (alcohol-derived) moieties of alkyldiacylglycerols from mesopelagic squaloids are similar to those of wax esters in crustacea from the same environment (Sargent, 1976). Alkyldiacylglycerols are biosynthesised de novo by Squalus acanthias (Malins and Sargent, 1970) and do not originate in the diets of sharks to any significant extent. Sharks rich in wax esters are known (Malins and Wekell, 1970) but perhaps the most interesting lipoidal compound in these animals is the terpenoid hydrocarbon squalene that can be present in very large quantities in liver and muscle and has been implicated in buoyancy regulation (Corner et al., 1969). Squalene does not originate in significant amounts in the diets of sharks.

The blubber of filter-feeding whales consists largely of triacylglycerols although the animals consume significant amounts of zooplanktonic wax esters in their natural diets. The blubber of the sperm whale consists largely of wax esters as does its spermaceti organ (reviewed by Malins and Varanasi, 1975). Lipids in the head regions of marine cetaceans have been implicated in biosonar (Malins and Varanasi, 1975) and buoyancy regulation (Clark, 1978) and special biosynthetic mechanisms probably exist for their formation, e.g. the formation of isovaleryl units from isoleucine (Malins and Varanasi, 1975). Thus their levels can be maintained independently of possible fluctuations in dietary supply. This stricture need not apply to the blubber of the sperm whale and wax esters similar to those in blubber occur in squid in the animal's stomach contents (Hansen and Cheah, 1969). Deep sea squid are recorded among the wax ester-rich animals listed by Sargent et al. (1976). This is one example of wax esters possibly being transmitted from one animal to another. Such transmission does not preclude hydroloysis and reformation of wax esters during intestinal absorption including even the conversion fatty alcohol → acid → alcohol.

Thus, wax esters and to a lesser extent alkyldiacylglycerols are prominent lipid classes in marine biota with the more usual triacylglycerols. The former lipids contain relatively simple and sometimes unique alkyl units, especially the 22:1 alcohol of copepod wax esters. These lipids are not transmitted intact through the food chain to any significant extent.

SATURATED FATTY ACIDS

The major marine saturated fatty acids are myristic acid, 14:0 and palmitic acid, 16:0. Stearic acid, 18:0, is present in trace amounts in marine lipids except in phospholipids of marine animals (Ackman and Hooper, 1974; Bottino, 1977). The 18:0 acid on position-1 of the phospholipids of marine mammals is replaced by both 16:0 and 18:1 acids in the phospholipids of marine poikilotherms (Menzell and Olcott, 1964; Olley and Duncan, 1965). Palmitic acid is ubiquitous in marine life forms, is extensively biosynthesised de novo, is present in large amounts in all lipid classes, and is of limited use in food chain studies.

Table I shows myristic acid, 14:0, to be a prominent constituent of phytoplankton lipid, especially in the diatoms. It accounts for up to one-fifth of the total fatty acids in mixed Antarctic phytoplankton (Bottino, 1974) (see Table III). Phospholipids of marine animals contain very small amounts of 14:0 acid (Lee et al., 1971; Morris and Sargent, 1973; Lee, 1974; Ackman, 1974; Menzell and Olcott, 1964; Olley and Duncan, 1964). Triacylglycerols contain substantial amounts (Ackman and Burger, 1964; Culkin and Morris, 1969, 1970a,b; Bottino, 1974; Ackman, 1974) with pelagic invertebrates containing more than their benthic counterparts (Paradis and Ackman, 1977). Myristic acid can appear in large quantities in the fatty acid moieties of calanoid wax esters (Sargent and McIntosh, 1974; Ackman et al., 1974).

Since 14:0 is abundant in neutral lipid but absent from phospholipid it is used only for production of metabolic energy. It may, therefore, be a useful marker for lipid metabolic energy transmitted through the trophic levels phytoplankton – zooplankton – fish. Thus the abundance of 14:0 in Antarctic krill is reflected in E. superba lipid that consists of triacylglycerols (Bottino, 1974; Table III). E. crystallorophias does not contain 14:0 fatty

TABLE III   Fatty acid composition of lipids from marine organisms in the southern hemisphere

	Mixed phyto[1]	E. superba[b]	E. cryst-allorophias[c]	Para. gaudi[d]	Sei whale[e]
14:0	16.1	14.0	2.4	3.3	9.2
16:0	18.5	22.3	14.6	14.6	7.5
18:0	2.8	1.0	—	1.6	1.7
16:1	8.3	9.5	7.9	6.8	5.1
18:1	16.6	20.8	46.2	16.7	13.3
20:1	—	—	—	3.8	28.5
22:1	—	—	—	4.6	10.5
18:2 (n–6)	—	—	—	—	—
18:2 (n–3)	5.5	2.4	2.1	—	1.4
18:4 (n–3)	2.9	2.7	1.2	1.2	1.6
20:4 (n–3)	—	—	—	—	3.6
20:5 (n–3)	8.2	13.1	14.4	16.7	2.8
22:5 (n–3)	—	—	—	—	1.7
22:6 (n–3)	6.9	8.1	7.5	17.3	7.0

— Signifies less than 1%.
Data are weight per cent of individual fatty acids and are compiled from Bottino (1974, 1975) for [a] total lipid from mixed Antarctic phytoplankton, [b] total lipid from *Euphausia superba*, [c] total lipid from *E. crystallorophias*, and from Bottino (1977) for [d] total lipid from *Parathemisto gaudi* from the stomachs of sei whales and [e] blubber triacylglycerols from the sei whale

acid (Table III). However the lipids of this animal are abundant in wax esters that consist mainly of 18:1 fatty acid and 14:0 fatty alcohol (Bottino, 1975), the latter constituent could well derive from dietary 14:0 acid. Table III also shows high concentrations of 14:0 acid in the blubber of the sei whale.

Bottino (1974) refers to the large quantities of lipid in phytoplankton generally (2–20% of their dry weight; Parsons *et al.*, 1961) giving a high dietary input of lipid into zooplankton thereby damping the animals' endogenous lipid biosynthesis (but note the continued biosynthesis of lipid in the wax ester-rich animals). The same occurs when the even larger quantities of zooplanktonic lipid rich in 14:0 acid are consumed by vertebrates including the whales. While significant catabolism of 14:0 acid certainly occurs at various trophic levels, this acid may be transmitted through the food chain in very significant quantities with little or no additional biosynthetic input beyond the phytoplankton. Thus 14:0 acid may mark phytoplanktonic lipid used only for energy production by animals higher up the food chain. Evidence similar to that of Bottino (1974) exists for the transmission of 14:0 acid through the northern food chain phytoplankton – *Thysanoessa/Meganyctiphanes* – whales (Table IV and references therein).

MONOUNSATURATED FATTY ACIDS

The 20:1 and 22:1 fatty alcohols abundant in wax esters of calanoid copepods are formed when copepods are reared on algae lacking these alkyl chains (Lee *et al.*, 1971). 20:1 alkyl chains are present in phytoplankton in at most traces and 22:1 chains are absent (Table I). However, these fatty acids are prominent in the oils of clupeoid fishes (Ackman, 1974), presumably deriving from the corresponding alcohols ingested as zooplanktonic wax esters. The oils consist of triacylglycerols. 20:1 and 22:1 fatty acids are essentially absent from the phospholipids of clupeoids (Menzell and Olcott, 1964; Ackman, 1974). Thus 20:1 and 22:1, like 14:0, are not involved in biomembrane function but instead are used for the production of metabolic energy.

The extent to which 20:1 and 22:1 units could be transmitted intact through the food chain is shown in Table IV. Since very large quantities of these alkyl units are present in the zooplankton they are unlikely to be biosynthesised significantly by zooplanktonic predators. The same applies at higher levels in the food chain, e.g. to gadoids preying on clupeoids. Thus 20:1 and 22:1 alkyl units appear to mark metabolic energy reserves produced by certain species of zooplankton, especially the copepods, and passed through the food chain to the highest trophic levels. As noted for 14:0, it is certain that significant catabolism of 20:1 and 22:1 occurs at all the trophic levels.

Relatively few studies have so far been published on the lipids of Antarctic zooplankton (krill apart). Wax esters are present in large amounts in the dominant Antarctic copepods *Rhincalanus gigas*, *Paraeuchaeta antarctica* and *Calanoides acutus* (A. Clarke, personal communication). Wax esters rich in 20:1 and 22:1 fatty alcohols are present in the stomach oils of birds known to consume zooplankton (Clarke and Prince, 1976). 20:1 and 22:1 fatty alcohols present in the blubber oils of the sei whale (see Table II) indicate the involvement of calanoid copepods at some stage in the animal's diet.

Marine monounsaturated fatty acids contain a range of isomers. The 16:1 fatty acid abundant in phytoplankton is the (n–7) isomer palmitoleic acid. This can be formed by a $\Delta 9$ desaturase (numbering from the carboxyl end of the molecule) acting on 16:0. Phytoplanktonic 16:1 must contribute significantly to the quantities of this acid present in invertebrates (Tables III and IV) although the amount biosynthesised by the animals themselves is unknown. Phytoplankton contain lesser but substantial quantities of 18:1 fatty acid including both the (n–9) isomer, oleic acid, and the (n–7) isomer,

TABLE IV Fatty acid compositions of lipids from marine organisms in the northern hemisphere

	Mixed Phyto[a]	Copepod[b] Alc	Copepod[b] Acid	Krill[c]	Capelin[d]	Cod liver[e]	Fin whale[f]	Grey seal[g]
14:0	6.8	2.0	32.2	6.9	6.4	3.5	5.0	3.5
16:0	22.3	12.6	11.3	17.0	9.3	10.4	12.0	9.2
18:0	1.2	—	—	2.3	—	1.2	1.7	—
16:1 (n–7)	16.6	5.5	15.6	3.0	14.4	12.2	12.0	19.1
18:1 total	4.1	5.0	3.9	5.4	7.5	19.6	30.6	31.9
20:1 (n–9)	—	17.2	—	—	16.9	14.6	12.1	4.9
22:1 (n–11)	—	39.0	—	5.7	16.9	13.3	9.7	—
C$_{16}$ PUFA	3.6	—	7.0	—	—	—	—	—
C$_{18}$ (n–6) PUFA	4.2	1.5	—	—	—	—	1.3	1.1
C$_{18}$ (n–3) PUFA	14.3	—	8.1	1.7	1.0	—	1.2	1.0
20:4 (n–6)	2.4	—	—	—	—	1.2	—	—
20:5 (n–3)	10.7	—	8.6	6.8	8.6	5.0	3.7	5.2
22:5 (n–3)	—	—	—	—	—	1.9	1.6	4.9
22:6 (n–3)	4.9	2.0	—	13.8	3.6	10.5	4.1	7.1

— Signifies less than 1 %.

Data are weight per cent of individual fatty acids and are compiled from [a] Ackman et al. (1968), averaged total lipid analyses from 12 species of unicellular algae; [b] Ackman et al. (1974), alcohol (Alc) and acid analyses of wax esters from North Atlantic copepods; [c] Ackman et al. (1970) triacylglycerols in *Meganyctiphanes norvegica* from the stomachs of fin whales; [d] Eaton et al. (1975), triacylglycerols from capelin oil; [e] Ackman and Burgher (1964), cod liver oil; [f] Ackman et al. (1965), blubber oil from the Arctic fin whale; [g] Ackman and Hooper (1974), blubber triacylglycerols from the grey seal

cis-vaccenic acid (Ackman *et al.*, 1968). 18:1 (n–7) is a major fatty acid of benthic algae (Johns *et al.*, 1979). Assuming only Δ9 desaturases to operate in marine organisms (an assumption yet untested) then 18:1 (n–9) can be formed from 18:0 in both plants and animals and 18:1 (n–7) can be formed by the elongation of 16:1 (n–7). These conversions are outlined in Fig. 3. In mammals 18:1 fatty acid is generally assumed to be oleic acid (n–9), and *cis*-vaccenic acid is associated with bacteria such as *E. coli*. Should only Δ9 desaturase operate in animals and should a large dietary input of 16:1 damp down endogenous biosynthesis of monounsaturates in animals, then their (n–7) isomers originate in the diet. Marine fish oils contain both (n–9) and (n–7) isomers of 18:1 with the latter usually accounting for less than one-third of total 18:1 (Ackman, 1980). The extent to which 18:1 (n–7) in fish oils derives from dietary (n–7), either 18:1 (n–7) itself or 16:1 (n–7) after chain elongation, is not known. Ratios of (n–9)/(n–7) monounsaturates could be useful indices of food chain relationships.

The 20:1 unit abundant in zooplankton wax esters is the (n–9) isomer (Pascal and Ackman, 1976). This can be formed from elongation of 18:1 (n–9). The 18:1 acid present in large amounts in the wax esters of *E. crystallorophias* is the (n–9) isomer (Bottino, 1975). The 22:1 unit also abundant in zooplankton wax esters is the (n–11) isomer, cetoleic acid (Pascal and Ackman, 1976). This alkyl chain could be formed by a Δ9 desaturase acting on 20:1 to form 20:1 (n–11) followed by immediate elongation of the latter to 22:1 (n–11) (Fig. 3).

Much remains to be learned about the biosynthetic specificities for monounsaturated fatty acids in marine organisms at all levels of the food chain; the likelihood is that such information will greatly help our understanding of food chain interactions.

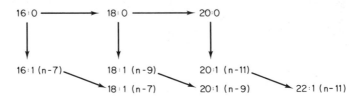

FIG. 3 Pathways of biosynthesis of monounsaturated fatty acids in marine organisms. Desaturation reactions using only a Δ9 desaturase operate vertically. Chain elongation reactions (2C addition) operate horizontally. Based on Pascal and Ackman (1976)

4 Hydrocarbons

Hydrocarbons are ubiquitous in marine organisms but, unlike the lipids considered so far, they are usually minor constituents of tissues, representing a fraction of one per cent of the dry weight. There are exceptions and perhaps the most notable is the major contribution of squalene to the liver oil of squaloid sharks. Others are the unusually high concentrations of pristane in the oil of some calanoid copepods and of heneicosahexaene in some diatoms. An array of hydrocarbons has been identified, normal, branched and cyclic paraffins and olefins as well as aromatic compounds.

In contrast to lipids, these hydrocarbons originate from a variety of sources other than biosynthesis within the marine food web. They enter the marine environment from sources described broadly as biochemical, geochemical and anthropogenic (reviewed by Farrington and Meyer, 1975; Clark and Brown, 1977, and Clark and MacLeod, 1977).

Both marine and land organisms form hydrocarbons *de novo* or from dietary precursors. The contribution from land organisms is unknown but global estimates of the hydrocarbon component of marine primary production range from 1 to 10 million tonnes per year.

Submarine and coastal oil seeps introduce petroleum hydrocarbons at specific points. Global estimates vary widely around 0.6 million tonnes per year. The weathering of soils and sediments, the diagenesis of organic matter in sediments and forest fires contribute small but unknown amounts.

Losses of fossil fuels, mostly petroleum and derived products, contribute hydrocarbons directly and, via combustion processes, indirectly. Again, global estimates vary widely but most fall in the range 1 to 10 million tonnes per year.

In global terms, primary production and petroleum are major sources and may contribute similarly. However, the latter is related particularly to specific incidents such as spills and to defined continuous inputs, so that with time and space, the contribution from each source will vary markedly.

Organisms are exposed to hydrocarbons in their food and in the surrounding aqueous environment. The dispersed or soluble lipophilic compounds are acquired across absorptive surfaces such as gut or respiratory organs. They are adsorbed on the surfaces of living or non-living particulates to augment any biogenic hydrocarbons already present and the particles are then filtered out and ingested by animals. Thus, the compounds identified in marine organisms vary in origin and can show marked variations in the relative contributions of the sources (Whittle *et al.*, 1977a). The analyst, therefore, must sort out which components originate from which sources and which are the products of marine biosynthesis. The differences between the biogenic and petroleum arrays of hydrocarbons

likely to be found in marine organisms were emphasised by Blumer *et al.*
(1972), Farrington and Meyer (1975) and Clark and Brown (1977), and are
often quite well defined. However, most reviewers correctly caution that,
because we are dealing with a minor tissue constituent, and because
hydrocarbons are ubiquitous and present in e.g. dust, soot, aerosols,
solvents, lubricants, plastics and so on, adventitious contamination must be
excluded during analysis.

Awareness of the identity, origin and quantity of hydrocarbons in marine
organisms has increased dramatically in the last 10 to 15 years, encouraged
in part by interest in the presence, behaviour and fate of petroleum-derived
hydrocarbons as pollutants in the marine environment and in the distribution and fate of polynuclear aromatic hydrocarbons, as well as more
widespread use of sophisticated analytical techniques such as capillary gas
chromatography and mass spectrometry. In spite of these advances, the data
base for examining the formation and transfer of hydrocarbons in marine
organisms is rather less complete than that of the lipids just considered.

Below we will discuss the major hydrocarbons of marine origin (other than
squalene and the carotenes) which, as we shall see, are aliphatic compounds.
Biosynthesis of aromatic hydrocarbons by marine organisms has been
suggested (ZoBell, 1971) but the evidence is poor and controversial (Clark
and Brown, 1977), and it is unlikely that algae and higher organisms
produce them. The issue hinges on production by bacteria in sediments but
no evidence of biosynthesis was found in anaerobic bacteria (Hase and
Hites, 1976).

HYDROCARBONS IN ALGAE

Phytoplankton

The composition of cultures from about 40 species has been examined
(Blumer *et al.*, 1970a; Lee *et al.*, 1970; Blumer *et al.*, 1971; Lee and
Loeblich, 1971). All algal classes of the marine phytoplankton were
represented including about 20 species from the Bacillariophyceae,
Dinophyceae and Haptophyceae which are probably the most important in
the sea, although the Cryptophyceae, Chlorophyceae, Cyanophyceae and
Xanthophyceae may be very abundant at times (Parsons *et al.*, 1977). An
omission is the important nanoplankton (Whittle, 1977). Supporting work
on biosynthesis using radiolabelled substrates has been carried out on a few
species including some blue-green algae (Lee *et al.*, 1970; Murray *et al.*,
1977; Murray and Thomson, 1977).

All cultured algae showed the presence of (*a*) trace quantities of *n*-alkanes
covering the range C22 and C30, (*b*) pristane and (*c*) squalene but, most
significant, was the predominance (*d*) of a few olefins in the range C15–C21,
including the polyolefin heneicosahexaene, and (*e*) and *n*-alkanes C15 and

C17 (Table V). The structure 3, 6, 9, 12, 15, 18-heneicosahexaene (*n*-C21:6) was assigned to the polyolefin (Blumer *et al.*, 1970a; Lee *et al.*, 1970). Following the subsequent identification of an isomer in brown algae with a terminal double bond (Halsall and Hills, 1971), structures in all phytoplankters were not confirmed. The *n*-C21:6 predominated in photo-

TABLE V   Hydrocarbon distribution in marine planktonic algae. (Modified after Blumer *et al.*, 1971)

Organism	Predominant hydrocarbons[a]	Trace hydrocarbons
Bacillariophyceae		
Cyclotella nana	*n*–C21:6	Pristane, *n*–p (low)
Rhizoselenia setigera	*n*–C21, *n*–C17, *n*–C21:6	Pristane, *n*–p
Ditylum brightwellii	*n*–C21:6	*n*–p (low)
Lauderia borealis	*n*–C21:6	*n*–p
Skeletonema costatum	*n*–C21:6	Pristane, *n*–p
Thalassiosira fluviatilis	*n*–C21:6	Pristane, *n*–p, *n*–C21:4 (?)
Dinophyceae		
Gonyaulax polyhedra	*n*–C21:6	*n*–p
Gymnodinium splendens	*n*–C21:6	*n*–p (low)
Peridinium trochoideum	*n*–C21:6	*n*–p
Peridinium trochoideum (old culture)	*n*–C21:6, *n*–C17, *n*–C15	*n*–p
Haptophyceae		
Coccolithus huxleyi	*n*–C21:6	Pristane, *n*–p
Isochrysis galbana	*n*–C21:6	Pristane, *n*–p
Phaeocystis poucheti	*n*–C21:6	Pristane, *n*–p
Cryptophyceae *Cryptomonas* sp. (possibly *Chroomonas*)	*n*–C21:6	
Euglenophyceae *Eutrepiella* sp.	*n*–C21:6, *n*–C17, *n*–C15	Pristane, *n*–p
Cyanophyceae		
Oscillatoria woronichinii	*n*–C15	*n*–p
Synechoccus bacillaris	*n*–C17, C–17:1, *n*–C15, *n*–C16	*n*–C16:1, *n*–C15:1, *n*–C18:1, *n*–p
Rhodophyceae *Porphyridium* sp.	*n*–C17, *n*–C15	*n*–C15:1, *n*–p
Chorophyceae		
Dunaliella tertiolecta	*n*–C17:1, *n*–C17:2, *n*–C17	*n*–p
Derbesia tenuissima	*n*–C17:1, *n*–C17, *n*–C19:2, *n*–C15, *n*–C19:3, *n*–C17:2	*n*–C15:1, *n*–p, other olefins

[a] Presumed *n*–C21:6 has the structure all-*cis*-3,6,9,12,15,18-heneicosahexaene; *n*–p *normal paraffins*

TABLE VI Hydrocarbon and polyolefin content of cultured marine phytoplankton. (Modified after Lee and Loeblich, 1971)

Organism	Lipid (% dry wt)	Hydrocarbon[a] (% lipid)	Hydrocarbon[a] (% dry wt)	n–C21:6 polyolefin[b] (% hydrocarbon)	n–C21:6 polyolefin[b] (% dry wt)
Bacillariophyceae					
Chaetoceros curvisetus	9.1	12.5	1.14	95	1.08
Cylindrotheca fusiformis	11.8	3.6	0.43	93	0.40
Ditylum brightwellii	—	2.5	—	99	—
Lauderia borealis	13.2	9.9	1.31	94	1.23
Nitzschia alba	10.6	0.2	0.02	A	A
Skeletonema costatum (3 day) [c]	6.2	4.3	0.27	91	0.25
(8 day)	8.6	15.1	1.30	95	1.24
Dinophyceae					
Cachonina niei	13.6	6.6	0.90	5	0.045
Exuviaella cassubica	5.2	3.6	0.19	20	0.038
Gymnodinium splendens	16.2	0.4	0.06	80	0.048
Noctiluca scintillans	28.0	20.7	5.80	A	A
Oxyrrhis marina	17.3	7.1	1.23	A	A
Peridinium sociale	16.2	0.9	0.15	80	0.12
Haptophyceae					
Cricosphaera carterae	9.7	15.2	1.47	88	1.29
Isochrysis galbana	26.0	4.5	1.17	93	1.09
Prymnesium parvum	4.2	7.4	0.31	82	0.25

Phaeophyceae					
Ectocarpus siliculosus	8.7	7.2	0.63	89	0.56
Streblonema oligosporum	5.6	8.4	0.47	89	0.42
Cryptophyceae					
Rhodomonas lens	16.8	10.3	1.73	51	0.88
Euglenophyceae					
Eutreptia viridis	18.8	3.8	0.71	13	0.09
Cyanophyceae					
Oscillatoria woronichinii	3.4	0.7	0.002	A	A
Rhodophyceae					
Porphyridium cruentum	1.7	0.1	0.002	A	A
Xanthophyceae					
Vaucheria longicaulis	10.1	14.1	1.42	A	A
Chlorophyceae					
Acetabularia calyculus	3.5	2.0	0.07	A	A
Dunaliella tertiolecta	—	1.0	—	A	A
Dunaliella viridis	—	—	—	A	A
Platymonas sp.	6.6	7.5	0.50	85	0.43

[a] Corrected to exclude carotene content for valid comparison with Tables VII, VIII and IX

[b] Presumed all-cis-3,6,9,12,15,18-heneicosahexaene

[c] Assigned n–C21:6 structure above

— Not investigated; A, absent

synthetic organisms from the Bacillariophyceae, Dinophyceae, Hapto-phyceae, Cryptophyceae and Phaeophyceae, contributing up to 80 to 99%
of the hydrocarbons (Table VI). In some cases it accounted for more than
10% of the lipid extracts which themselves constituted about 1.5% of the wet
weight of cultures (Blumer et al., 1970a). The n–21:6 occurred in smaller
amounts in the Euglenophyceae and some Chlorophyceae, but not in Cyano-phyceae, Rhodophyceae, Xanthophyceae and most Chlorophyceae. In these
classes alkanes and alkenes in the C15 to C19 range predominated (Table V).

The diatoms *Skeletonema costatum* (Lee et al., 1970) and *Phaeodactylum
tricornutum* (Murray et al., 1977) biosynthesised n–C21:6 from $^{14}CO_2$, 85%
of the ^{14}C-activity of the hydrocarbon fraction occurring in n–C21:6 in the
latter. In both, the low background of n-alkanes identified in the analyses
was not labelled. The alkane n–C21 was the predominant hydrocarbon in
the diatom, *Rhizosolenia setigera*, presumably formed by hydrogenation of
the hexaene. The n–C21:4 was thought to be present in some diatoms
(Blumer et al., 1971), but n–C21:5 and n–C21:4 were both prominent in the
dinoflagellates and euglenids (Lee and Loeblich, 1971).

The major hydrocarbon in *Porphyridium* species (*Rhodophyceae*) was
n–C17 with smaller amounts of n–C15, confirmed by ^{14}C-labelling studies on
P. aerugineum (Murray et al., 1977) in which 91% of the ^{14}C-activity in
hydrocarbons was n–C17. The remainder was identified tentatively as n–C23
and n–C25 in contrast to the small amount of n–C15 and trace of n–C15:1
observed by Blumer et al. (1971). It is noteworthy that in old cultures of the
dinoflagellate, *Peridinium*, n-heptadecane assumed predominance over
n–C21:6. Such differences, probably related to culture conditions, stage of
growth and physiological state (Patterson, 1967; Brown et al., 1969; Fehler
and Light, 1970), also show up well in analyses of cultures of the fresh water
Xanthophyte, *Tribonema equale*; Lee and Loeblich (1971) identified n–
C22:1 as the major hydrocarbon; Blumer et al. (1971) found n–C15:1,
n–C15, n–C14 and n–C15:2 in decreasing order of importance; Murray et al.
(1977) found at least 97% of the hydrocarbon ^{14}C-activity in n–C15. Similar
differences occur in blue-green algae (Gelpi et al., 1968, 1970; Han et al.,
1968; Winters et al., 1969; Blumer et al., 1971; Lee and Loeblich, 1971;
Murray et al., 1977; Murray and Thomson, 1977). In marine species, the
alkanes and alkenes in the range C15 to C17 seemed most abundant but the
n–C17:1 in the blue–greens was 5-heptadecene compared with 7-heptadecene
in the green algae (Blumer et al., 1971).

Analyses of field samples of microalgae are few but n–C21:6 was the
major hydrocarbon (90%) in a sample of 'red tide' which consisted of 99%
Gonyaulax polyhedra (Lee and Loeblich, 1971). Thompson and Eglington
(1979) identified predominant n–C21:6 and n–C17 above the pollutant
hydrocarbon background in epipelic diatoms from an estuarine sediment.

The diatom population was *Surirella* (60%), *Nitzschia* (30%) and *Gyrosigma* plus *Pleurosigma* (10%). We have identified n–C21:6 in mixed phytoplankton collections from Scottish coastal waters in which *Chaetoceros* was a major component (Mackie and Whittle, unpublished observations).

Benthic algae

In coastal zones the macrophytes (Chlorophyceae, Phaeophyceae and Rhodophyceae) make a major contribution to biomass (Whittaker and Likens, 1975). Their hydrocarbon composition has been examined in some detail (Clark and Blumer, 1967; Youngblood *et al.*, 1971; Lee and Loeblich, 1971; Youngblood and Blumer, 1973; Nishimoto, 1974; Botello and Mandelli, 1978). Alkane and alkene distributions and amounts (Table VII) were reported by Youngblood *et al.* (1971) who carefully excluded epiphytes and diatoms from field collections. Comparison of Tables VI and VII suggests that the hydrocarbons are about 20 times more abundant on a dry weight basis in microalgae (mean 0.95% dry wt) than benthic algae (mean 0.045% dry wt). Comparisons of cultures and field samples must be treated with caution but it is worth noting that the hydrocarbons (mainly n–C21:6) in the 'red tide' sample of *Gonyaulax* (Lee and Loeblich, 1971) were about 10 times more abundant than in benthic algae on a dry weight basis.

Normal alkanes and alkenes, including polyolefins, were the major constituents of the benthic algae together with relatively minor quantities of branched and cyclic compounds. As in the microalgae, a broad range (C20 to C30) of small quantities of n-alkanes, with no marked predominance of odd to even carbon numbers and peaking between C25 and C28, was identified in which a few odd-carbon chain length alkanes or alkenes, in the range C15 to C21, predominated. Pristane was as high as 5% of the n–C17 in the Laminariales but was usually less than 0.1% in other species or not even detected in some red algae (Clark and Blumer, 1967; Youngblood *et al.*, 1971). The hydrocarbons were mostly saturated in red algae and unsaturated in green algae but there were exceptions. Brown algae ranged from saturated to unsaturated and both hydrocarbon content and alkene:alkane ratio varied in morphologically different parts of the plant which may be related to maturity of the tissue (Youngblood and Blumer, 1973). The most prominent alkanes were n–C15 and n–C17. Heptadecane predominated in red algae and green algae but pentadecane was prominent in brown algae (Clark and Blumer, 1967; Youngblood *et al*, 1971; Nishimoto, 1974). As Table VII shows, in some green algae heptadecane was replaced by heptadecene and heptadecanylcyclopropane was tentatively identified also and found to vary seasonally (Youngblood *et al.*, 1971). Several isomeric heptadecenes were identified but other C15 and C17 olefins

TABLE VII   Hydrocarbons in benthic marine algae. (Modified after Youngblood *et al.*, 1971)

Species	% HC	% sat.	% unsat.	C13	C14	C15	C16	C17	C18	C19	C20	C21	C22
Saturated (sat.) and unsaturated (unsat.) hydrocarbon content (HC) of benthic marine algae. (Dry weight basis)				**Normal alkanes in benthic marine algae[a]**									
Chlorophyceae													
Enteromorpha compressa	0.028	3.5	96.5			1.3		2.2	t	t			
Ulva lactuca	0.481	0.5	99.5					0.5					
Spongomorpha arcta	0.024	1.1	98.9			t		t	t				
Codium fragile sp.	0.048	92.7	7.3			1.4	t	89		1.9			
Phaeophyceae													
Ectocarpus fasciculatus	0.023	8.4	91.6			7.0		1.2	t	t			
Pilayella littoralis	0.048	0.9	99.1			0.6		0.4					
Chordaria flagelliformis	0.005	96.5	3.5			96	t	0.8	t		t	t	t
Leathesia difformis	0.001	99.9	0.1			59		3.6		0.4	t	1.2	3.5
Punctaria latifolia	0.077	88.0	12.0		1.6	85	0.2	1.1	0.2	t	t		
Scytosiphon lomentaria	0.003	53.6	46.4			38	0.2	11	t	1.3			0.6
Chorda filum	0.002	40.8	59.2			38		2.3		0.2	0.4	t	t
Chorda tomentosa	0.003	31.5	68.5			31		0.4					
Laminaria agardhii	0.025	93.4	6.6			73		0.7		t		0.7	0.9
Laminaria digitata	0.002	67.1	32.9			64	0.6	2.0	t	0.3	0.6		t
Ascophyllum nodosum	0.024	57.4	42.6			56	0.2	1.2		2.9			
Fucus distichus sp.	0.008	99.9	0.1			98	0.3	0.5	t			0.2	
Fucus spiralis	0.011	96.5	3.5	0.4	1.2	95	2.0	0.3	0.1	t			
Fucus vesiculosus		81.6	18.4			65	0.2	0.2	t				
Rhodophyceae													
Porphyra leucosticta	0.044	16.3	83.7			0.1		17		0.2	0.1		
Dumontia incrassata	0.029	98.9	1.1			1.0	0.4	98					
Chondrus crispus	0.073	99.0	1.0			0.4	0.1	99					
Rhodymenia palmata	0.003	99.7	0.7			1.2	0.1	99	0.1	0.1	0.1	0.1	0.1
Ceramium rubrum	0.051	97.7	2.3			1.3	0.2	96					
Polysiphonia urceolata	0.028	96.8	3.2			0.8	0.1	96	0.1				

[a] Total content of saturated and olefinic hydrocarbons normalised to 100%, individual figures rounded to two significant places
[b] Alkycyclopropane; t < 0.1%

were relatively minor components. The *n*–C21:6 olefin, a major component in a number of microalgal classes, was prominent in some green and brown benthic algae and the pelagic *Sargassum natans*, but was not detected in any red algae examined. There were significant contributions from other poly-olefins, especially *n*–C21:5 and *n*–C19:5 (Lee and Loeblich, 1971; Youngblood *et al.*, 1971; Youngblood and Blumer, 1973; Burns and Teal, 1973). The detailed but limited number of analyses carried out so far suggest that the *n*–C21:6 olefin in green algae is the *cis*-3, 6, 9, 12, 15, 18-isomer (Δ^3) but brown algae contain the *cis*-1, 6, 9, 12, 15, 18-isomer (Δ^1) (Halsall and Hills, 1971; Youngblood and Blumer, 1973).

Origins and functions

Decarboxylation is a possible route for the formation of hydrocarbons in marine algae. However, the difference in hydrocarbon composition of cultured organisms in relation to culture conditions, stage of growth and physiological state, as well as differences in macrophyte tissues, certainly pose problems for the interpretation of simple comparisons of hydrocarbon

Normal alkenes in benthic marine algae[a]

C23	C24	C25	C26	15:1	15:2	17:1	17:2	17:3	Δ17:0[b]	19:1	19:4	19:5	19:6	21:4	21:5	21:6
						96			1.0							
					1.3	88			4.0						2.2	4.0
										3.0		9.4	6.7		22	60
						5.9				1.4						
																91
										0.4			0.8			
t	t			1.9			1.0			0.4			0.8			98
5.7	13	6.0	5.4													
					0.6		1.1						1.3		0.1	5.9
0.7	1.2					0.4	15	0.5	4.2	2.5			19	1.8	37	10
t	t										0.9		2.4		2.5	2.7
2.2	2.4	4.2	5.0												17	47
															2.8	3.8
														4.6	9.7	
				2.1	1.7		1.1	23	4.6	1.0			1.0		5.8	t
				0.1												
				0.2	0.1		0.1								2.2	0.7
				0.4			0.9	0.4	1.0						1.0	15
				15								6.2	62			
				0.3	0.1		0.4									
				0.1			0.1									
0.1	0.1					0.1	0.1	0.1								
							2.2	0.1								
				0.1			0.4				2.7					

and fatty acid compositions, particularly in whole organisms. The predominance of n–C15 and n–C17 alkanes and the occurrence of the n–C16:0 and n–C18:0 fatty acids (Clark and Blumer, 1967) did not correlate well in benthic algae. The structural similarity between n–C21:6 olefin and n–C22:6 fatty acid, a major PUFA in microalgae, suggested (Lee and Loeblich, 1971) that decarboxylation was the likely pathway. The differences in distribution between algal groups were explained by variations in the activity of the proposed specific C22:6 fatty acid decarboxylase. However, Δ^3-isomers of n–C21:6 and n–C21:5 olefins could result from simple decarboxylation of C22:6 (n–3) and C22:5 (n–3) fatty acids respectively, whereas formation of Δ^1-isomers now identified in brown algae would be more complex than simple decarboxylation of the known (n–3) or (n–6) fatty acids (Youngblood and Blumer, 1973).

Detailed analyses of brown algal tissues and of the growth stages of planktonic algal cultures suggest that polyolefins are associated with sites and periods of rapid cell division. Hydrocarbon content, alkene: alkane ratio and polyolefin content are highest in rapidly growing tissue, i.e. in young plants, in growing tissue of older plants and in the logarithmic phase of cell growth (Youngblood and Blumer, 1973). The predominant *n*-alkanes of benthic algae increase from the early stage of growth to the flourishing stage and decrease again before withering (Nishimoto, 1974). These observations and the association of *n*–C21:6 with photosynthetic pigments in microalgae (Lee and Loeblich, 1971) suggest possible physiological and/or structural roles for hydrocarbons in algae and, consequently, may imply rapid turnover rates of hydrocarbons.

HYDROCARBONS IN ANIMALS

Pristane and olefins

The isoprenoid hydrocarbon, pristane, is a relatively minor constituent of algal hydrocarbons but it can be predominant in the saturated fraction (1000 times the background *n*-alkanes) of marine animals, particularly in some species of zooplankton (Blumer *et al.*, 1963 and 1964) and planktonivorous animals (Blumer, 1967). Isoprenoid C19 and C20 olefins are often the major constituents of the unsaturated fraction (Blumer and Thomas, 1965a, 1965b) when pristane is present in large amounts. Pristane was first observed (Tsujimoto, 1917) in basking shark and other shark liver oils and later structurally identified (Sörensen and Sörensen, 1949). It is widely distributed among marine invertebrates and vertebrates both geographically and at all trophic levels (Blumer, 1967; Hashimoto and Mamuro, 1967; Sano, 1968; Lewis, 1969; Ackman, 1971; Lee *et al.*, 1972; Inoue *et al.*, 1973; Linko and Kaitaranta, 1976; Whittle *et al.*, 1977b; Johansen *et al.*, 1977; Mackie *et al.*, 1978). Blumer *et al.* (1964) reviewed its presence in zooplankton (Table VIII) and Table IX shows its distribution in fish and mammals. The calanoid copepods are the richest source of pristane. In *Calanus* pristane is up to 3% of the total lipid extract, 10 to 100 times the concentration in other zooplankton (Blumer *et al.*, 1964). Calanoids are probably the principle source of pristane in mixed zooplankton samples (Blumer *et al.*, 1963) in which the concentration can reach 1.7% of the dry weight (Whittle *et al.*, 1977b).

Although pristane concentration in the lipids varies (Table IX), it is often present in high concentration or predominant in the lipid-rich tissues of planktonivorous animals likely to have pristane-rich zooplankton in their diets (Table IX). Chemically, it is more stable than the polyolefins; it is biodegraded more slowly than the *n*-alkanes; it is likely to be metabolised by fish more slowly than the *n*-alkanes; it is readily absorbed across the gut and becomes associated with the storage lipid in liver or muscle tissues. All these factors contribute to its wide distribution in the marine food web. It is usually a prominent liver hydrocarbon in teleosts which store lipid in the liver but is often in small concentrations in the muscle tissues similar to the background *n*-alkanes (Whittle *et al.*, 1977b).

TABLE VIII Total lipid, and pristane content of zooplankton, (Modified after Blumer et al., 1964)

Organism	Total lipid % of dry wt	Pristane % of dry wt	Pristane, % of total lipid	Feeding type
Chaetognath				
Sagitta elegans	29.6	0.02	0.05	d
Pteropod				
Limacina retroversa	3.7	0.01	0.14	a
Ostracod				
Conchoecia sp.	5.1	≪0.01	0.03	
Amphipod				
Parathemisto gaudichaudii	12.3	<0.01	0.04	d
Euphausids				
Nematocelis megalops	7.9	<0.01	0.09	b
Meganyctiphanes norvegica	5.3	≪0.01	0.02	b
Copepods				
Calanus finmarchicus IV	57.5	0.85	1.47	a
Calanus finmarchicus V		0.73		
Calanus finmarchicus V	52.8	0.77	1.45	
Calanus finmarchicus female	27.0	0.46	1.69	
Calanus finmarchicus male	51.8	0.68	1.31	
Calanus glacialis V		0.45		a
Calanus glacialis V	54.5	0.47	0.86	
Calanus hyperboreus V		0.92		a
Calanus hyperboreus V	51.3	0.84	1.62	
Calanus hyperboreus female	30.8	0.90	2.94	
Rhincalanus nasutus female	57.2	0.01	0.03	a
Rhincalanus nasutus female	42.7	0.01	0.03	
Paraeuchaeta norvegica IV	15.3	0.02	0.14	d
Paraeuchaeta norvegica V	16.2	0.03	0.19	
Paraeuchaeta norvegica female		0.05		
Paraeuchaeta norvegica female	41.8	0.03	0.08	
Metridia longa female		0.01		c
Metridia lucens female	7.9	0.01	0.15	c
Pleuromamma robusta V and adults	11.3	≪0.01	0.01	a
Euchirella rostrata V and female	21.5	0.01	0.04	c
Candacia armata female	7.9	≤0.02	<0.20	d

[a] Predominantly herbivorous, but may also take animal food
[b] Omnivorous, including feeding on detritus
[c] Probably predominantly predatory, but may also take plant food
[d] Predatory

TABLE IX Pristane in marine vertebrates

Species	Tissue/Organ	Pristane % in oil	Pristane % in hydrocarbons[a]	Reference
Cetorhinus maximus	liver	6	—	Warloe (1967)
Basking shark		0.7	99 (est.)	Kayama *et al.* (1969)
		1.1–1.3	97	Blumer (1967)
	cardiac stomach	0.7		
	pyloric stomach	0.4		
	bursa entiana	0.9		
	spiral valve	0.8		
Cetorhinus sp.	liver	4.4	74.5	Gershbein and Singh (1969)
		7.6	98.0	Gelpi and Oró (1968)
Etmopterus spinax	liver	1–2	—	Warloe (1967)
Black dogfish				
Triakis scyllia	liver	0.03	99 (est.)	Kayama *et al.* (1969)
Leopard shark				
Apristurus macrohinus	liver	0.6	99 (est.)	
Centroscyllium ritteri	liver	1.1	99 (est.)	
Centrophorus sp.	liver	0.06	99 (est.)	
Squalus acanthias	liver	0.1	99 (est.)	
Dogfish		—	17.4	Whittle *et al.* (1974b)
	muscle	—	2.9	
Dalatias licha	stomach	0.01–0.43	5.0–48.0	Lewis (1969)
Black shark	liver	0.001–0.23	0.5–57.0	
	intestine	0.03–0.43	7.5–86.0	

Species	Tissue			Reference
Myxine glutinosa Hagfish	whole fish	0.001	—	Johansen et al. (1977)
Clupea harengus Baltic herring	muscle	0.005	—	Linko and Kaitaranta (1976)
Herring	whole fish	0.22	—	Gershbein and Singh (1969) quoted by Ackman (1971)
		<0.035	—	
Winter herring	muscle	0.11	96.9	Whittle et al. (1974b)
	whole fish	0.14	—	Ackman (1971)
Sprattus sprattus Sprat	whole fish	—	76.1	Mackie (unpublished)
	liver	—	67.6	Whittle et al. (1974a)
Scomber scombrus Mackerel	muscle	0.14	82.0	
		0.05	93.8	
Ammodytes americanus Sand launce (summer)	whole fish	—	99 (est.)	Dubravcic and Nawar (1969)
	muscle	—	—	Ackman (1971)
Mallotus villosus Capelin	whole fish	0.05	76 (est.)	Johansen et al. (1977)
Oncorhynchus nerka Salmon	muscle	0.05	>90 (est.)	Inoue et al. (1973)
Sebastes marinus Redfish	liver	0.003	97 (est.)	Johansen et al. (1977)
	muscle	0.001	99 (est.)	
Anarhichas sp. Wolf fish	liver	0.006	—	
	muscle	0.003	55 (est.)	
Gadus ogac Greenland cod	liver	0.004	—	
Gadus morhua Cod	liver	—	60.5–69.1	Whittle et al. (1974b)
	muscle	—	16.2–17.1	

TABLE IX (Cont.)

Species	Tissue/Organ	Pristane % in oil	Pristane % in hydrocarbons[a]	Reference
Merluccius merluccius				
Hake	muscle	—	3.7	Mackie and Platt (unpublished)
Notothenia rossii	liver	—	15.6	
	muscle	—	30.6	
Parachaenichthys georgianus	liver	—	52.5	
Icefish	muscle	—	20.9	
Chaenocephalus aceratus	liver	—	57.5	
Crocodile fish	muscle	—	59.2	
Thaleichthys pacificus	liver	0.03	—	Ackman *et al.* (1968)
Eulachon	stomach	0.02	—	
	head and body	0.02	—	
Reinhardtius hippoglossoides	liver	<0.0001–0.066	—	Johansen *et al.* (1977)
Greenland halibut	muscle	0.006	—	
Hippoglossoides platessoides	liver	0.005	87 (est.)	
American plaice	muscle	0.002	86 (est.)	
Pleuronectes platessa	liver	—	1.8–3.8	Whittle *et al.* (1974b)
Plaice	muscle	—	12.9–18.7	
Balaenoptera borealis	blubber	—	95	Hashimoto and Mamuro (1967)
Sei whale				
Physeter macrocephalus	blubber	0.24	98	Sano (1968)
Sperm whale				

[a] Hydrocarbons excluding squalene; (est.) estimated

The source of pristane is almost certainly the phytyl moiety of chlorophyll in the algal diet of copepods. *Calanus* converted ^{14}C-phytol adsorbed on its algal food to pristane, probably by oxidation to phytanic acid and subsequent decarboxylation, 97% of the activity in the copepod hydrocarbons occurring as labelled pristane (Avigan and Blumer, 1968). After feeding a mixed zooplankton population with *Phaeodactylum* grown on $^{14}CO_2$, the zooplankton hydrocarbon fraction constituted 1.3% of the radioactivity in the lipid and again pristane accounted for 97% of the activity in the hydrocarbons (Murray *et al.*, 1977). The results are similar to the conversion by *Calanus* but the mixed zooplankton was not rich in *Calanus*; it was composed mainly of *Acartia* (17%), *Temora* (16%), *Calanus* (16%), cirriped cyprids (15%), *Centropages* (11%) and other species. In considering conversion rates and efficiencies, 15% of the total radioactivity fed as *Phaeodactylum* was recovered in intact grazing zooplankton over 7 days and labelled pristane represented 0.05% of the radioactivity fed. It remains uncertain whether the copepods or the gut microflora produce pristane; in the latter case other important sources of pristane may occur in the food web.

No evidence was found of significant radioactivity in C19 olefins (zamenes) or C20 olefins (phytadienes) isolated from mixed zooplankton samples (Blumer and Thomas, 1965a,b). In fact, the remaining hydrocarbon radioactivity (ca. 3%) was associated with n–C21:6 and n–C21:5 polyolefins, predominant in the *Phaeodactylum*, and incorporated into the zooplankton lipid. Pristane and n–C21:6, in the ratio 92:1, were predominant in the zooplankton when caught, but, after feeding the n–C21:6-rich algae, the zooplankton became relatively enriched in n–C21:6 (ratio 36:1), further evidence that the chemically less stable polyolefin is quite widespread in the food web (Blumer *et al.*, 1970a).

Field collections show that a few copepods such as *Rhincalanus* and *Euchaeta* can store n–C21:6 as the predominant hydrocarbon (0.1 to 0.5% of lipid) just as *Calanus* sp. store pristane. However, *Calanus helgolandicus* fed n–C21:6-rich algae did not accumulate the polyolefin and were thought to metabolise it instead (Lee *et al.*, 1970). It has also been found in oyster and scallop (Blumer *et al.*, 1970b), in mussel (Lee *et al.*, 1972) and in herring and basking shark livers (Blumer *et al.*, 1970a) but not in the species of a *Sargassum* community (Burns and Teal, 1973) apart from *Sargassum natans* itself. Although clearly not as chemically stable, nor as metabolically inert or widespread as pristane, the polyolefin enjoys some stability and distribution in the food web.

Transfer in the food web
Copepods may or may not store relatively large quantities of pristane

compared with other hydrocarbons. Alternatively, they may or may not selectively metabolise or accumulate the n–C21:6 polyolefin but, their minor components do tend to reflect seasonal and species differences in the phytoplankton of their diet (Blumer et al., 1970a). There is a considerable amount of evidence suggesting that the major hydrocarbon constituents of marine animals can also be related directly to their recent dietary history.

Hydrocarbons in the C15 to C17 range were the main components in algal grazers such as the chiton and the mainly herbivorous sea urchin, the prominent cis-7-heptadecene and n–C17 presumably arising from the green algae in their diets (Yasuda and Fukamiya, 1977a and b). Pristane, minor in benthic algae, was not identified in the herbivorous chiton but was a major component of the more omnivorous sea urchin. The variations observed in the saturated and unsaturated constituents may reflect the changes in alkene:alkane ratio associated with algal growth. The abundance of n–C21:6 in filter-feeding bivalves, oysters and scallops, from the same location in winter closely resembled the composition of a winter diatom in the area, Rhizoselenia setigera (Blumer et al., 1970b). Mussels kept in cages in different locations in the central North Sea for up to 48 hours acquired rather different hydrocarbon patterns, depending on location, from those present initially (Whittle et al., 1978).

Similar relationships between tissue hydrocarbon array and dietary availability occur in higher animals, especially planktonivorous species. Analyses of the digestive tract and liver of a basking shark suggested that it had been feeding on Calanus rich in pristane and the hydrocarbons of the zooplankton had been absorbed and deposited in the liver without fractionation or structural modification (Blumer, 1967). The minor constituents, the C19 and C20 olefins (2% of the pristane concentration), did not decrease in concentration in contrast to the apparent loss of the PUFA and their replacement with monounsaturated acids in the shark liver compared with the copepods. The hydrocarbons of the basking shark liver apparently were indicative of its recent food sources which accounts for the variation in composition of basking shark oils (Blumer, 1967; Gelpi and Oró, 1968; Gershbein and Singh, 1969; Kayama et al., 1969).

Similarly, the hydrocarbons in herring livers were indicative of the principal zooplankton species in the same area (Blumer et al., 1969). Herring, mackerel, sprat and other fatty species store most of their lipid reserves in muscle rather than liver tissue and, as noted in the section on neutral lipids, the fatty acid composition of the reserve lipid resembles the diet. The hydrocarbons of herring and mackerel muscle reflect dietary composition (Mackie et al., 1974; Whittle et al., 1974a) which was also noted at each level in a brief study of a naturally occurring food chain zooplankton, sprat, mackerel (Whittle et al., 1974a). The alkane fraction of the mackerel flesh was similar to whole sprat and plankton suggesting that the hydrocarbons were assimilated virtually intact at each stage. ^{14}C-Hexadecane incorporated into the food of herring was rapidly deposited in the muscle tissue and represented nearly 50% of all the radioactivity recovered from the fish (Whittle *et al.*, 1977c), compared with about 6.5% in the liver. Oils and blubber oils from filter-feeding whales also vary in hydrocarbon composition (Hashimoto and Mamuro, 1967; Sano, 1968), probably indicating dietary differences. Comparisons of hydrocarbon composition are easier in the species described above because particular components such as pristane predominate and compounds of specific origin such as the C19 and C20 olefins are present.

The composition of the diet of planktonivorous species varies with space and time just as the species and hydrocarbon composition of the zooplankton changes. The dietary changes must be quite rapid if animals move in and out of water masses or discrete zooplankton populations. Since hydrocarbon composition apparently varies as the animals' diet changes, it seems that dietary hydrocarbons are not incorporated for long residence times into the relatively hydrocarbon-rich tissues of the predator, but move in and out in response to dietary changes, assuming that metabolic transformation of most marine biogenic hydrocarbons is relatively slow. A small fraction of assimilated ^{14}C-hexadecane was metabolised slowly in herring liver by oxidation to fatty acids (Whittle *et al.*, 1977c) but, in contrast, some copepods were thought to metabolise n–C21:6 (Lee *et al.*, 1970).

This principle, that residence time is related to dietary change seems to hold at least superficially for aliphatic hydrocarbons incorporated directly from the diet into the tissues of animals whose diets normally contain relatively large quantities of lipid and associated hydrocarbon. Its validity needs further testing. It does not hold for the animal's biosynthetic hydrocarbons such as the pristane in calanoid copepods. Pristane is retained during starvation and actually increases in concentration on a dry weight basis, possibly related to maintenance of buoyancy as storage lipid is mobilised (Blumer *et al.*, 1964).

Origin of the background hydrocarbons
So far, we have reviewed the predominant hydrocarbons present in marine algae and animal tissues above a low background of n-alkanes. None may be predominant in some teleost muscle tissues which contain only structural phospholipids and no storage lipids. Their muscle content is low and consists mainly of a background of n-alkanes peaking in the range C24 to C28 (0.1–1.0 µg/g wet wt). These so-called lean fish such as the gadoids store their lipid reserve in the liver where the n-alkane concentration is 5 to 10 times higher and pristane may be predominant.

The ^{14}C-labelling studies (Lee *et al.*, 1970; Murray *et al.*, 1977; Murray and Thomson, 1977) showed that a low background of unlabelled n-alkanes ranging from C20 to C30 was present in the algal cultures maintained on $^{14}CO_2$ and in the zooplankton fed labelled algae. The algae clearly had not synthesised them, even in cultures more than 20 days old (Murray and Sargent, unpublished observations), and they were not recent metabolic products from the zooplankton. This low background of alkanes peaking in the range C24 to C28 and showing no marked odd/even predominance is typical of laboratory cultures and many field collections of marine organisms from areas not chronically or acutely contaminated with petroleum and is not the result of adventitious contamination. It probably reflects a ubiquitous background presence of non-biogenic hydrocarbons and it is interesting to note that apart from conditions of recent petroleum input the alkanes in sea water routinely show the same pattern (Hardy et al., 1977).

The exception to the smooth profile of background n-alkanes is the characteristic predominance of odd-carbon number n-alkanes (particularly C27, C29 and C32), which we noted in the livers or hepatopancreas of many marine vertebrates and invertebrates. They are similar to the array in recent marine sediments uncontaminated with petroleum residues (Farrington and Meyer, 1975). The difference in n-alkane distribution between muscle and liver is found also in pelagic, demersal and benthic animals (Mackie et al., 1974; Whittle et al., 1978). The concentration of n-alkanes in muscle (0.1–1.0 µg/g wet wt) excepting fish which store large quantities of lipid in the muscle tissue, is 5 to 10 times lower than liver (Whittle et al., 1977b), so that the odd-carbon predominance is still noticeable in analyses of whole fish.

Benthos apart, it seems unlikely that recent sediments are the immediate source of these odd-carbon alkanes in animals because of their wide distribution among different types of organisms including the planktonivorous pelagic species. One might speculate that the waxes of land plants rich in the odd-carbon numbered n-alkanes (C25, C27, C29, C31 and C33) which probably contribute the odd-carbon predominance to recent sediments (Farrington and Meyer, 1975), may also be the source of this predominance in the tissues. Particulates derived from land plants, rich in odd-carbon alkanes, enter the sea from the atmosphere (Simoneit, 1977) or land run-off. The few hydrocarbon analyses of particles collected in sediment traps suspended in inshore and offshore waters (Davies et al., 1981; Mackie and Davies, unpublished observations) do show odd-carbon predominance analogous to the terrestrial input. The odd-carbon alkanes on particles could enter the food web by being scavenged by detritus feeders, trapped by filter feeders or caught in the mucus on the respiratory surfaces of fish, again being available to detritus feeders.

This relatively simple explanation does not readily accommodate the apparent specificity of the odd-carbon distribution in liver and hepatopancreas or all the observations of environmental contamination and related experimental uptake studies with petroleum hydrocarbons (reviewed by Varanasi and Malins, 1977). In addition we must postulate that (a) odd-carbon n-alkanes are available to fish mainly via ingestion, (b) liver or hepatopancreas does show some specificity in uptake and deposition in the range C20 to C30 (Hardy et al., 1974; Teal, 1977) and (c) dietary hydrocarbons are preferentially but indiscriminately deposited in the lipid-rich muscle of some species.

There is no evidence of discrimination against even carbon numbers in uptake and deposition in fish tissues. After feeding cod small quantities of crude oil in squid for a number of months (Hardy et al., 1974), n-alkanes in the range C24 to C30 were retained preferentially in the liver masking the odd-carbon predominance but there was no difference in concentration or distribution in the muscle constituents. Discrimination against low carbon numbers was confirmed by the low incorporation into cod liver of ^{14}C-hexadecane fed in the squid diet. In contrast, dietary ^{14}C-hexadecane was incorporated rapidly into herring muscle rather than liver (Whittle et al., 1977c). The evidence available does not exclude biosynthesis of the odd n-alkanes.

The relative importance of the modes of uptake in marine animals, across the respiratory, gut or body surfaces, is a factor in determining the potential contribution of hydrocarbons in the water column (in the 'dissolved' or 'particulate' state) or in food to those ultimately deposited in the tissues (Corner et al., 1975; Teal, 1977). It is also relevant to the availability and uptake of petroleum hydrocarbons, which may be present in much larger local concentrations in the water column, diet or surface sediments than the marine biogenic hydrocarbons. By analogy (Teal, 1977) with studies on organochlorine uptake we might expect that food is a much more important source than water since the critical factor is the relative concentration in water and food, although uptake can be just as efficient by either route. Calculation confirmed that pristane in pristane-rich copepods and fish probably came from the diet, the pristane derived from water being some 1000 times less than the actual concentration of pristane in the copepods and fish. This is of the same order as the difference in concentration between the background n-alkanes and pristane in these animals again suggesting that the former may be derived from the environment.

Detailed consideration of petroleum derived hydrocarbons in the marine food web is beyond the scope of this review. Superimposition on the naturally occurring compounds and the implications for identification, deposition, metabolism and depuration or turnover have been extensively covered elsewhere in recent years (Blumer et al., 1972; Farrington and

Meyer, 1975; NAS, 1975; GESAMP, 1976; Clark and Brown, 1977; Varanasi and Malins, 1977; Lee, 1977; Teal, 1977; Malins, 1977a,b; Whittle *et al.*, 1977a). Neff (1979) has reviewed the origin and fate of the polynuclear aromatic hydrocarbons in the aquatic environment.

5 Perspective

We have attempted to illustrate the value of lipid and hydrocarbon analyses in understanding the structure and function of marine food webs. The major neutral lipid classes have been considered with major polyunsaturated, monounsaturated and saturated fatty acids, space precluding the interesting trace fatty acids; the isoprenoid fatty acids that originate in the phytol-pristane chain (Ackman and Hooper, 1968; Liugh, 1974); the odd-chain fatty acids (Paradis and Ackman, 1976); the non-methylene interrupted dienoic fatty acids that originate in the phytoplankton (Paradis and Ackman, 1977; Johns *et al.*, 1979); and the furan-containing fatty acids (Gunstone *et al.*, 1976). All provide valuable information on predator–prey relationships. The marine biosynthetic hydrocarbons have been emphasised rather than those from anthropogenic sources but it is axiomatic that the latter will generally become entrained into the food web pathways of the biogenic lipids and hydrocarbons.

A cursory glance at the literature on natural marine lipids and hydrocarbons reveals the abundant, detailed, quantitative information, especially on percentage composition of marine oils and lipids which arises from the easy application and great precision of current analytical methods. It is easy to form the impression that the trees are in danger of obliterating the wood completely, and we are concerned that too much attention is paid to application of these methods and too little to integration of the data into the body of marine biology as a whole. For example, few if any serious efforts have been made to exploit the precisely quantitative data gained from lipid analyses in testing current models of marine ecosystems. The flow of a chemically defined energy source, e.g. 22:1 alkyl unit, through a food chain, can surely stand comparison with the flow of 'total energy' as calories or total carbon. We need to know the total pool size of a given constituent at a given trophic level, which demands not only easily obtained (but seldom available in the literature) data on the amount of the constituent in the organism, but also knowledge of the biomass. We also require quantitative data on turnover, bearing in mind that both anabolic and catabolic activities are involved. Turnover may be difficult to determine experimentally especially in natural situations, but it may not be too important in practice. Characteristically, in the marine food web there are many examples of predators consuming such large amounts of lipids that their endogenous lipid biosynthesis is likely to be minimal. Moreover, the large lipid deposits accumulated in these circumstances are commonly used for reproduction so that lipid deposition is separated seasonally and frequently geographically from lipid mobilisation.

Given limiting assumptions, therefore, there is no reason why the modelling of lipids and their constituents as well as hydrocarbons should not be attempted in at least some food chain situations. Artificial enclosures should feature prominently in such studies since, there at least, biomasses can be monitored easily and frequently. We are conscious that a great deal of analytical work remains to be done to establish the lipid and hydrocarbon compositions of groups such as microzooplankton, nanoplankton and bacteria whose role(s) in the food web remain to be evaluated properly. We hope that such future analytical work will be guided by unifying principles developed from the present detailed knowledge of those areas of the food web studied so far.

References

Ackman, R. G. (1967). Characteristics of the fatty acid composition and biochemistry of some freshwater fish oils and lipids in comparison with marine oils and lipids. *Comp. Biochem. Physiol.* **22**, 907–922

Ackman, R. G., Addison, R. F. and Eaton, C. A. (1968). Unusual occurrence of squalene in a fish, the Eulachon (*Thaleichthys pacificus*). *Nature* **220**, 1033–1034

Ackman, R. G. (1971). Pristane and other hydrocarbons in some freshwater and marine fish oils. *Lipids* **6**, 520–522

Ackman, R. G. (1974). Marine lipids and fatty acids in human nutrition. *In* "Fishery Products" (R. Kreuzer, ed.) pp. 112–131. Published for FAO of UN by Fishing News (Books) Ltd., England

Ackman, R. G. (1974). Myocardial alterations resulting from feeding partially hydrogenated marine oils and peanut oil to rats. *Lipids* **9**, 1032–1035

Ackman, R. G. (1980). Fish lipids. *In* "Advances in Fish Science and Technology" (J. J. Connell, ed.) pp. 86–103. Fishing News Books Ltd., Farnham, Surrey, England

Ackman, R. G. and Burgher, R. D. (1964). Cod liver oil: component fatty acids as determined by gas-liquid chromatography. *J. Fish. Res. Bd. Can.* **21**, 319–326

Ackman, R. G. and Eaton, C. A. (1978). Some contemporary applications of open-tubular gas-liquid chromatography in analyses of methylesters of longer-chain fatty acids. *Fette Seifen Anstrichmittel* **80**, 21–37

Ackman, R. G. and Hooper, S. N. (1968). Examination of isoprenoid fatty acids as distinguishing characteristics of specific marine oils with particular reference to whale oils. *Comp. Biochem. Physiol.* **24**, 549–565

Ackman, R. G. and Hooper, S. N. (1974). Long-chain monoethylenic and other fatty acids in heart, liver and blubber lipids of two harbour seals (*Phoca vitulina*) and one grey seal (*Halichoerus grypus*) *J. Fish. Res. Bd. Can.* **31**, 333–341

Ackman, R. G., Eaton, C. A. and Jangaard, P. M. (1965). Lipids of the fin whale (*Balaenoptera physalus*) from North Atlantic waters. I. Fatty acid composition of whole blubber and blubber sections. *Can. J. Biochem.* **43**, 1513–1520

Ackman, R. G., Tocher, C. S. and McLachlan, J. (1968). Marine phytoplankter fatty acids. *J. Fish. Res. Bd. Can.* **25**, 1603–1620

Ackman, R. G., Linke, B. A. and Hingley, J. (1974). Some details of fatty acids and alcohols in the lipids of North Atlantic copepods. *J. Fish. Res. Bd. Can.* **31**, 1812–1818

Ackman, R. G., Eaton, C. A., Sipos, J. C., Hooper, S. N. and Castell, J. D. (1970). Lipids and fatty acids of two species of North Atlantic krill (*Meganyctiphanes norvegica* and *Thysanoessa inermis*) and their role in the aquatic food web. *J. Fish. Res. Bd. Can.* **27**, 513–533

Antia, N. J., Lee, R. F., Nevenzel, J. C. and Cheng, J. Y. (1974). Wax ester production by the marine cryptomonad *Chroomonas salina* grown photoheterotrophically on glycerol. *J. Protozool.* **21**, 768–771

Avigan, J. and Blumer, M. (1968). On the origin of pristane in marine organisms *J. Lipid Res.* **9**, 350–352

Beach, D. H. and Holz, G. G. (1973). Environmental influences on the docosahexaenoate content of the triacylglycerols and phosphatidylcholine of a heterotrophic marine dinoflagellate, *Cryhthecodinium cohnii. Biochem. Biophys. Acta* **316**, 56–65

Benson, A. A. and Muscatine, L. (1974). Wax in coral mucus: energy transfer from corals to reef fishes. *Limnol. Oceanogr.* **19**, 810–814

Benson, A. A., Lee, R. J. and Nevenzel, J. C. (1972). Wax esters: major marine metabolic energy source. *Biochem. Soc. Symp.* **35**, 175–187

Blumer, M. (1967). Hydrocarbons in digestive tract and liver of a basking shark. *Science* **156**, 390–391

Blumer, M. and Thomas, D. W. (1965a). "Zamene", isomeric C19 monoolefins from marine zooplankton, fishes and mammals. *Science* **148**, 370–371

Blumer, M. and Thomas, D. W. (1965b). Phytadienes in zooplankton *Science* **147**, 1148–1149

Blumer, M., Mullin, M. M. and Guillard, R. R. L. (1970a). A polyunsaturated hydrocarbon (3, 6, 9, 12, 15, 18-heneicosahexaene) in the marine food web. *Mar. Biol.* **6**, 226–236

Blumer, M., Souza, G. and Sass, J. (1970b). Hydrocarbon pollution of edible shellfish by an oil spill. *Mar. Biol.* **5**, 195–202

Blumer, M., Mullin, M. M. and Thomas, D. W. (1963). Pristane in zooplankton *Science* **140**, 974

Blumer, M., Mullin, M. M. and Thomas, D. W. (1964). Pristane in the marine environment. *Helgoländer Wiss. Meeresunters.* **10**, 187–201

Blumer, M., Robertson, J. C., Gordon, J. E. and Sass, J. (1969). Phytol-derived C19 di- and triolefinic hydrocarbons in marine zooplankton and fishes. *Biochemistry* **8**, 4067–4074

Blumer, M., Blokker, P. C., Cowell, E. G. and Duckworth, D. F. (1972). Petroleum. *In* "A Guide to Marine Pollution" (E. D. Goldberg, ed.) pp. 19–40. Gordon and Breach, New York

Blumer, M., Guillard, R. R. L. and Chase, T. (1971). Hydrocarbons of marine phytoplankton. *Mar. Biol.* **8**, 183–189

Botello, A. V. and Mandelli, E. F. (1978). Distribution of *n*-paraffins in sea-grasses, benthic algae, oysters and recent sediments from Terminos Lagoon, Campéche, Mexico. *Bull. Environ. Contam. Toxicol.* **19**, 162–170

Bottino, N. R. (1974). The fatty acids of Antarctic phytoplankton and euphausiids. Fatty acid exchange among trophic levels of the Ross Sea. *Mar. Biol.* **27**, 197–204

Bottino, N. R. (1975). Lipid composition of two species of Antarctic krill *Euphausia superba* and *E. crystallorophias*. *Comp. Biochem. Physiol.* **50B**, 479–484

Bottino, N. R. (1978). Lipids of the Antarctic sei whale, *Balaenoptera borealis Lipids* **13**, 18–23

Brown, A. C., Knights, B. A. and Conway, E. (1969). Hydrocarbon content and relationship to physiological state in the green algae (*Botryococcus braunii*). *Phytochemistry* **8**, 543–547

Burns, K. A. and Teal, J. M. (1973). Hydrocarbons in the pelagic *Sargassum* community. *Deep-Sea Res.* **20**, 207–211

Christophersen, B. O. and Bremer, J. (1972). Erucic acid – an inhibitor of fatty acid oxidation in the heart. *Biochem. Biophys. Acta* **280**, 506–514

Chuecas, L. and Riley, J. P. (1969). The component fatty acids of the lipids of some marine phytoplankton species. *J. mar. biol. Ass. U.K.* **49**, 97–116

Clark, M. R. (1978). Buoyancy control as a function of the spermaceti organ in the sperm whale. *J. mar. biol. Ass. U.K.* **58**, 27–71

Clark, R. C. and Blumer, M. (1967). Distribution of *n*-paraffins in marine organisms and sediment. *Limnol. Oceanog.* **12**, 79–87

Clark, R. C. and Brown, D. W. (1977). Petroleum: Properties and analyses in biotic and abiotic systems. *In* "Effects of Petroleum on Arctic and Subarctic Marine Environments and Organisms" (D. C. Malins, ed.) Vol. 1, pp. 1–69. Academic Press, London and New York

Clark, R. C. and MacLeod, W. D. (1977). Inputs, transport mechanisms and observed concentrations of petroleum in the marine environment. *In* "Effects of Petroleum on Arctic and Subarctic Marine Environments and Organisms" (D. C. Malins, ed.) Vol. 1, pp. 91–223. Academic Press, London and New York

Clarke, A. (1979). Lipid content and composition of the pink shrimp *Pandalus montagui* (leach) (Crustacea:Decapoda). *J. Exp. Mar. Biol. Ecol.* **38**, 1–17

Clarke, A. and Prince, P. A. (1976). The origin of stomach oil in marine birds. Analyses of the stomach oil from six species of subantarctic procellariiform birds. *J. Exp. Mar. Biol. Ecol.* **23**, 15–30

Corner, E. D. S., Denton, F. J and Forster, G. R. (1969). On the buoyancy of some deep sea sharks. *Proc. R. Soc. B* **171**, 415–429

Corner, E. D. S., Harris, R. P., Whittle, K. J. and Mackie, P. R. (1976). Hydrocarbons in marine zooplankton and fish. *In* "Effects of Pollutants on Aquatic Organisms" (A. P. M. Lockwood, ed.) Vol. 2, pp. 71–105. Soc. Exp. Biol. Sem. Ser. Cambridge University Press

Cowey, C. B. and Sargent, J. R. (1972). Fish Nutrition. *Adv. Mar. Biol.* **10**, 383 492

Cowey, C. B. and Sargent, J. R. (1977). Minireview. Lipid nutrition in fish. *Comp. Biochem. Physiol.* **57B**, 269–273

Cowey, C. B. and Sargent, J. R. (1979). Nutrition. *In* "Fish Physiology" (W. S. Hoar and D. J. Randal, eds) Vol. 8, pp. 1–69. Academic Press, New York

Cowey, C. B., Owen, J. M., Adron, J. W. and Middleton, C. (1976). Studies on the nutrition of marine flatfish. The effect of different dietary fatty acids on the growth and fatty acid composition of turbot (*Scophthalmus maximus*). *Brit. J. Nutr.* **36**, 479–486

Crawford, M. A., Casperd, N. M. and Sinclair, A. J. (1976). The long chain metabolites of linoleic and linolenic acids in liver and brain in herbivores and carnivores. *Comp. Biochem. Physiol.* **54B**, 395–401

Crawford, M. A., Sinclair, A. J. and Stevens, P. (1974). Specificity of brain grey matter fatty acids in land and marine mammals. *J. Amer. Oil. Chem. Soc.* **51**, Abstr. No. 37 of Fall A.O.C.S. Meeting, 1974

Culkin, F. and Morris, R. J. (1969). The fatty acids of some marine crustaceans. *Deep-Sea Res.* **16**, 109–116

Culkin, F. and Morris, R. J. (1970a). The fatty acids of marine teleosts. *J. Fish. Biol.* **2**, 107–112

Culkin, F. and Morris, R. J. (1970b). The fatty acid composition of two marine filter-feeders in relation to a phytoplankton diet. *Deep-Sea Res.* **17**, 861–865

Davies, J. M., Johnston, R., Whittle, K. J. and Mackie, P. R. (1981). Origin and fate of hydrocarbons in Sullom Voe. *Proc. Roy. Soc. Edin., Sect. B.* **80** (In press)

Drozdowski, B. and Ackman, R. G. (1969). Isopropyl alcohol extraction of oil and lipids in the production of fish protein concentrate from herring. *J. Amer. Oil Chem. Soc.* **46**, 371–376

Dubravcic, M. F. and Nawar, W. W. (1969). Effects of high-energy radiation on the lipids of fish. *J. Agr. Food Chem.* **17**, 639–644

Eaton, C. A., Ackman, R. G., Tocher, C. S. and Spencer, K. D. (1975). Canadian capelin 1970–73. Fat and moisture composition and fatty acids of some oils and lipid extract triglyceride. *J. Fish. Res. Bd. Can.* **32**, 507–513

Farrington, J. W. and Meyer, P. A. (1975). Hydrocarbons in the marine environment. *In* "Environmental Chemistry" (Senior Reporter, G. Eglington) Vol. 1, pp. 109–136. Specialist Periodical Reports, Chemical Society, London

Fehler, S. W. G. and Light, R. J. (1970). Biosynthesis of hydrocarbons in *Anabena variabilis*. Incorporation of [*methyl*-^{14}C] – and [*methyl*-^{2}H$_3$] methionine into 7- and 8-*methyl*-heptadecanes. *Biochemistry* **9**, 418–422

Fujii, M. and Yone, Y. (1976). Studies on the nutrition of red sea bream – XIII. Effect of dietary linolenic acid and ω–3 polyunsaturated fatty acids on growth and feed efficiency. *Bull. Jap. Soc. Sci. Fish.* **42**, 583–588

Gelpi, E. and Oró, J. (1968). Gas chromatographic-mass spectrometric analysis of isoprenoid hydrocarbon and fatty acids in shark liver oil products. *J. Am. Oil Chem. Soc.* **45**, 144–147

Gelpi, E., Oró, J., Schneider, H. J. and Bennet, E. O. (1968). Olefins of high molecular weight in two microscopic algae. *Science* **161**, 700–702

Gelpi, E., Schneider, H., Mann, J. and Oró, J. (1970). Hydrocarbons of geochemical significance in microscopic algae. *Phytochem.* **9**, 603–612

Gershbein, L. L. and Singh, E. J. (1969). Hydrocarbons and alcohols of basking shark and pig liver lipids. *J. Am. Oil Chem. Soc.* **46**, 34–38

GESAMP (1976). The impact of oil on the marine environment. IMCO/FAO/-UNESCO/WHO/WMO/IAEA/UN. Joint group of experts on the scientific aspects of marine pollution (GESAMP). Report of the Working Group, GESAMP, **6**, 1976

Goad, L. J. (1976). The steroids of marine algae and invertebrate animals. *In* "Biochemical and Biophysical Perspectives in Marine Biology" (D. C. Malins and J. R. Sargent, eds) Vol. 3 pp. 213–318. Academic Press, London

Gunstone, F. D., Wijesundera, R. C., Love, R. M. and Ross, D. (1976). Relative enrichment of furan-containing fatty acids in the liver of starving cod. *Chem. Comm.* 630–631

Halsall, T. G. and Hills, I. R. (1971). Isolation of heneicosa-1, 6, 9, 12, 15, 18-hexaene and -1, 6, 9, 12, 15-pentaene from the alga *Fucus vesiculosus*. *Chem. Commun.* 448–449

Han, J., McCarthy, E. D., Caliun, M., Benn, M. H. (1968). Hydrocarbon constituents of the blue-green algae *Nostoc muscorum*, *Anacystis nidulans*, *Phormidium luridum* and *Chorogloea fitschii*. *J. Chem. Soc.* (Sect. C) 2785–2791

Hansen, I. A. and Cheah, C. C. (1969). Related dietary and tissue lipids of the sperm whale. *Comp. Biochem. Physiol.* **31**, 757–761

Hardy, R., Mackie, P. R., Whittle, K. J. and McIntyre (1974). Discrimination in the assimilation of *n*-alkanes in fish. *Nature* **252**, 577–578

Hase, A. and Hites, R. A. (1976). On the origin of polycylic aromatic hydrocarbons in recent sediments: biosynthesis by anaerobic bacteria. *Geochim. Cosmochim. Acta* **40**, 1141–1143

Hashimoto, T. and Mamuro, H. (1967). Unsaponifiable matter of Sei-whale blubber oil. Presence of pristane and squalene. *Yukagaku* **16**, 657–661

Hinchcliffe, P. R. and Riley, J. P. (1972). The effect of diet on the component fatty acid composition of *Artemia salina*. *J. mar. biol. Ass. U.K.* **52**, 203–211

Inoue, N., Hosokawa, Y. and Akiba, M. (1973). Pristane in salmon muscle lipid. *Bull. Fac. Fish. Hokk. Univ.* **23**, 209–214

Johansen, P., Jensen, V. B. and Buchert, A. (1977). "Hydrocarbons in Marine Organisms and Sediments off West Greenland" (R. G. Ackman, ed.). Fisheries and Marine Service Tech. Rep. No. 729

Johns, R. B., Nichols, P. D. and Perry, G. J. (1979). Fatty acid composition of ten marine algae from Australian waters. *Phytochem.* **18**, 799–802

Jones, D. A., Kanazawa, A. and Ono, K. (1979). Studies on the nutritional requirements of the larvae of *Penaeus japonicus* Bate using microencapsulated diets. *Mar. Biol.* **54**, 261–267

Kanazawa, A., Teshima, S. and Tokiwa, S. (1977). Nutritional requirements of prawn – VII. Effect of dietary lipids on growth. *Bull. Jap. Soc. Sci. Fish.* **43**, 849–856

Kanazawa, A., Teshima, S. I. and Ono, K. (1979). Relationship between essential fatty acid requirements of aquatic animals and the capacity for bioconversion of linolenic acid to highly unsaturated fatty acids. *Comp. Biochem. Physiol.* **63B**, 295–298

Kayama, M., Tsuchiya, Y. and Mead, J. F. (1963). A model experiment of aquatic food chain with special significance in fatty acid conversion. *Bull. Jap. Soc. Sci. Fish.* **29**, 452–458

Kayama, M., Tsuchiya, Y., Nevenzel, J. C. (1969). The hydrocarbons of shark liver oils. *Bull. Jap. Soc. Sci. Fish.* **35**, 653–664

Lee, R. F. (1974). Lipid composition of the copepod *Calanus hyperboreus* from the Arctic ocean. Changes with depth and season. *Mar. Biol.* **26**, 313–318

Lee, R. F. (1977). Accumulation and turnover of petroleum hydrocarbons in marine organisms. *In* "Fate and Effects of Petroleum Hydrocarbons in Marine Ecosystems and Organisms" (D. A. Wolfe, ed.) pp. 60–70. Pergamon Press, New York

Lee, R. F. and Hirota, J. (1973). Wax esters in tropical zooplankton and nekton and the geographical distribution of wax esters in marine copepods. *Limnol. Oceanogr.* **18**, 227–239

Lee, R. F. and Loeblich, A. R. (1971). Distribution of 21:6 hydrocarbon and its relationship to 22:6 fatty acid in algae. *Phytochem.* **10**, 593–602

Lee, R. F., Nevenzel, J. C. and Paffenhöfer, G. A. (1971). Importance of wax esters and other lipids in the marine food chain: phytoplankton and copepods *Mar Biol.* **9**, 99–108

Lee, R. F., Nevenzel, J. C. and Lewis, A. G. (1974). Lipid changes during life cycle of marine copepod, *Euchaeta japonica* Marukawa. *Lipids* **9**, 891–898

Lee, R. F., Sauerheber, R. and Benson, A. A. (1972). Petroleum hydrocarbons: Uptake and discharge by the marine mussel, *Mytilus edulis*. *Science* **177**, 344–346

Lee, R. F., Nevenzel, J. C., Paffenhöfer, G. A., Benson, A. A., Patton, S. and

Kavanagh, T. E. (1970). A unique hexaene hydrocarbon from a diatom (*Skeletonema costatum*). *Biochem. Biophys. Acta* **202**, 386–388

Lewis, R. W. (1969). Studies on the stomach oils of marine animals – I. Oils of the black shark *Dalatias licha* (Bennaterre). *Comp. Biochem. Physiol.* **31**, 715–724

Linko, R. R. and Kaitaranta, J. (1976). Hydrocarbons of Baltic herring lipids. La Rivista Italiana delle Sobstanz Grasse. **53**, 36–39

Lough, A. K. (1973). The chemistry and biochemistry of phytanic, pristanic and related acids. *Prog. Chem. Fats Lipids* **XIV**, 5–48

Mackie, P. R., Platt, H. M. and Hardy, R. (1978). Hydrocarbons in the marine environment II. Distribution of *n*-alkanes in the fauna and environment of the sub-antarctic island of South Georgia. *Est. Coastl. Mar. Sci.* **6**, 301–313

Mackie, P. R., Whittle, K. J. and Hardy, R. (1974). Hydrocarbons in the marine environment. I. *n*-Alkanes in the Firth of Clyde. *Est. Coastl. Mar. Sci.* **2**, 359–374

Malins, D. C. (1977a). Metabolism of aromatic hydrocarbons in marine organisms. *In* "Aquatic Pollutants and Biologic Effects with Emphasis on Neoplasia" (H. F. Kraybill, ed.) **298**, 482–496. Ann. N.Y. Acad. Sci.

Malins, D. C. (1977b). Biotransformation of petroleum hydrocarbons in marine organisms indigenous to the Arctic and Subarctic. *In* "Fate and Effects of Petroleum Hydrocarbons in Marine Ecosystems and Organisms" (D. A. Wolfe, ed.) pp. 47–59. Pergamon Press, New York

Malins, D. C. and Sargent, J. R. (1971). Biosynthesis of alkyldiacyglycerols and triacylglycerols in a cell-free system from the liver of dogfish (*Squalus acanthias*). *Biochemistry* **10**, 1107–1110

Malins, D. C. and Varanasi, V. (1975). Cetacean Biosonar II. The biochemistry of lipids in acoustic tissues. *In* "Biochemical and Biophysical Perspectives in Marine Biology" (D. C. Malins and J. R. Sargent, eds) Vol. 2, 237–290. Academic Press, London

Malins, D. C. and Wekell, J. C. (1970). The lipid biochemistry of marine organisms. *Progr. Chem. Fats Lipids* **X**, 339–363

Mead, J. and Kayama, M. (1967). Lipid metabolism in fish. *In* "Fish Oils" (M. E. Stansby, ed.) pp. 289–299. Avi Publishing Co., Westport, Conn.

Menzell, D. B. and Olcott, H. S. (1964). Positional distribution of fatty acids in fish and other animal lecithins. *Biochem. Biophy. Acta* **84**, 133–139

Moreno, V. J., de Moreno, J. E. A. and Brenner, R. R. (1979a). Biosynthesis of unsaturated fatty acids in the diatom *Phaeodactylum tricornutum*. *Lipids* **14**, 15–19

Moreno, V. J., de Moreno, J. E. A. and Brenner, R. R. (1979b). Fatty acid metabolism in the calanoid copepod *Paracalanus parvus*: 1. Polyunsaturated fatty acids. *Lipids* **14**, 313–322

Morris, R. J. (1971). Seasonal and environmental effects on the lipid composition of *Neomysis integer*. *J. mar. biol. Ass. U.K.* **51**, 21–31

Morris, R. J. and Sargent, J. R. (1973). Studies on the lipid metabolism of some oceanic crustaceans. *Mar. Biol.* **22**, 77–83

Morris, R. J. and Culkin, F. (1976). Marine Lipids: Analytical techniques and fatty acid ester analyses. *Ann. Rev. Oceanogr. Mar. Biol.* **14**, 391–433

Morris, R. J. and Culkin, F. (1977). Marine Lipids: Sterols. *Ann. Rev. Oceanogr. Mar. Biol.* **15**, 73–102

Murray, J. and Thomson, A. B. (1977). Hydrocarbon production in *Anacystis montana* and *Botryococcus braunii*. *Phytochem.* **16**, 465–468

Murray, J., Thomson, A. B., Stagg, A., Hardy, R., Whittle, K. J. and Mackie, P. R. (1977). On the origin of hydrocarbons in marine organisms. *Rapp. P.-v. Réun. Cons. int. Explor. Mer.* **171**, 84–90

NAS (1975). National Academy of Sciences Report "Petroleum in the marine environment". Workshop on Inputs, Fates and Effects of Petroleum in the Marine Environment, Airlie, Virginia, U.S.A. NAS, Washington, D.C.

Neff, J. M. (1979). "Polycyclic Aromatic Hydrocarbons in the Aquatic Environment". Applied Science Publishers, Ltd., London

Nevenzel, J. C. (1970). Occurrence, function and biosynthesis of wax esters in marine organisms. *Lipids* **5**, 308–319

Nishimoto, S. (1974). A chemotaxonomic study of *n*-alkanes in aquatic plants *J. Sci. Hiroshima Univ. Ser. A.* **36**, 159–163

Olley, J. and Duncan, W. R. H. (1965). Lipids and protein denaturation in fish muscle. *J. Sci. Fd. Agric.* **16**, 99–104

Owen, J. M., Adron, J. W., Middleton, C. and Cowey, C. B. (1975). Elongation and desaturation of dietary fatty acids in the turbot *Scophthalmus maximus* L. and rainbow trout *Salmo gairdneri. Rich. Lipids* **10**, 528–531

Paradis, M. and Ackman, R. G. (1976). Localisation of a source of marine odd chain length fatty acids. II. Seasonal propagation of odd chain length monoethylenic fatty acids in a marine food chain. *Lipids* **11**, 871–876

Paradis, M. and Ackman, R. G. (1977). Potential for employing the distribution of anomalous non-methylene-interrupted dienoic fatty acids in several marine invertebrates as part of food web studies. *Lipids* **12**, 170–176

Parsons, T. R., Stephens, K. and Strickland, J. D. H. (1961). On the chemical composition of eleven species of marine phytoplankton. *J. Fish. Res. Bd. Can.* **18**, 1001–1016

Parsons, T. R., Takahashi, M. and Hargrave, B. (1977). "Biological Oceanographic Processes", 2nd edn, p. 3. Pergamon Press, Oxford

Pascal, J.-C. and Ackman, R. G. (1976). Long chain monoethylenic alcohol and acid isomers in lipids of copepods and capelin. *Chem. Phys. Lipids* **16**, 219–223

Patterson, G. W. (1967). The effect of culture conditions on the hydrocarbon content of *Chlorella vulgaris. J. Phycol.* **3**, 22–23

Patton, J. S. and Benson, A. A. (1975). A comparative study of wax ester digestion in fish. *Comp. Biochem. Physiol.* **52**, 111–116

Patton, J. S., Abraham, S. and Benson, A. A. (1977). Lipogenesis in coral and isolated zooxanthellae: evidence for a light-stimulated lipid-carbon cycle. *Mar. Biol.* **44**, 235–247

Patton, J. S., Nevenzel, J. C. and Benson, A. A. (1975). Specificity of digestive lipases in hydrolysis of wax esters and triglycerides studied in anchovy and other selected fish. *Lipids* **10**, 575–583

Rahn, C. H., Sand, D. M. and Schlenk, H. (1973). Wax esters in fish. Metabolism of dietary palmityl palmitate in the gourami (*Trichogaster cosby*). *J. Nutr.* **103**, 144–147

Rivers, J. P. W., Sinclair, A. J. and Crawford, M. A. (1975). Inability of the cat to desaturate essential fatty acids. *Nature, Lond.* **258**, 171–173

Rivers, J. P. W., Hassam, A. G., Crawford, M. A. and Brambell, M. R. (1976). The inability of the lion, *Panthera leo* L. to desaturate linoleic acid. *FEBS Lett.* **67**, 269–270

Sano, Y. (1968). Studies on the minor constituents of whale oils – I. Identification of 2,6,10,14-tetramethyl-pentadecane and 2,6,10,14-tetramethyl-2-pentadecene in sperm blubber oil. *Bull. Jap. Soc. Sci. Fish.* **34**, 726–733

Sargent, J. R. (1976). The structure, metabolism and function of lipids in marine organisms. *In* "Biochemical and Biophysical Perspectives in Marine Biology" (D. C. Malins and J. R. Sargent, eds) Vol. 3, pp. 149–212. Academic Press, London

Sargent, J. R. (1978). Marine Wax Esters. *Sci. Prog. Oxf.* **65**, 637–658

Sargent, J. R. and McIntosh, R. (1974). Studies on the mechanism of biosynthesis of wax esters in *Euchaeta norvegica. Mar. Biol.* **25**, 271–277

Sargent, J. R., Gatten, R. R. and McIntosh, R. (1977). Wax esters in the marine environment – their occurrence, formation, transformation and ultimate fates. *Mar. Chem.* **5**, 573–584

Sargent, J. R., Lee, R. J. and Nevenzel, J. C. (1976). Marine Waxes. *In* "Chemistry and Biochemistry of Natural Waxes" (P. E. Kolattukudy, ed.) pp. 50–91. Elsevier, Amsterdam

Sargent, J. R., McIntosh, R., Bauermeister, A. E. N. and Blaxter, J. H. S. (1979). Assimilation of the wax esters of marine zooplankton by herring (*Clupea harengus*) and rainbow trout (*Salmo gairdneri*). *Mar. Biol.* **51**, 203–207

Simoneit, B. R. T. (1977). Organic matter in eolian dusts over the Atlantic Ocean. *Mar. Chem.* **5**, 443–464

Sinnhuber, R. O. (1969). The role of fats. *In* "Fish in Research" (O. Neuhaus and J. E. Halver, eds) pp. 245–259. Academic Press, New York and London

Sörensen, J. S. and Sörensen, N. A. (1949). Studies related to pristane 3. The identity of norphytane and pristane. *Acta Chem. Scand.* **3**, 939–945

Teal, J. M. (1977). Food chain transfer of hydrocarbons. *In* "Fate and Effects of Petroleum Hydrocarbons in Marine Organisms and Ecosystems" (D. A. Wolfe, ed.) pp. 71–77. Pergamon Press, New York

Thompson, S. and Eglington, G. (1979). The presence of pollutant hydrocarbons in estuarine epipelic diatom populations. II. Diatom slimes. *Est. Coastl. Mar. Sci.* **8**, 75–86

Tsujimoto, M. (1917). Saturated hydrocarbons in basking shark liver oil. *J. Ind. Eng. Chem. Japan* **9**, 1098–1099

Varanasi, U. and Malins, D. C. (1977). Metabolism of petroleum hydrocarbons: accumulation and biotransformation in marine organisms. *In* "Effects of Petroleum on Arctic and Subarctic Marine Environments and Marine Organisms" (D. C. Malins, ed.) Vol. II, pp. 175–270. Academic Press, London and New York

Warloe, B. (1967). Hydrocarbon-containing marine oils. *In* "Proc. of 4th Scandinavian Symposium on Fat and Oil Chemistry, 1965" (F. Bramsnaes, ed.) pp. 41–50

Whittaker, R. H. and Likens, G. E. (1975). The biosphere and man. *In* "Primary Productivity of the Biosphere" (H. Keith and R. H. Whittaker, eds) pp. 305–328. Springer-Verlag, New York

Whittle, K. J. (1977). Marine organisms and their contribution to organic matter in the ocean. *Mar. Chem.* **5**, 381–411

Whittle, K. J., Mackie, P. R., Hardy, R. and McIntyre, A. D. (1974a). The fate of *n*-alkanes in marine organisms. ICES CM 1974/E33 Fisheries Improvement Committee, Copenhagen

Whittle, K. J., Mackie, P. R. and Hardy, R. (1974b). Hydrocarbons in the marine ecosystem. *South African Journal of Science* **70**, 141–144

Whittle, K. J., Hardy, R., Holden, A. V., Johnston, R. and Pentreath, R. J. (1977a). Occurrence and fate of organic and inorganic contaminants in marine animals. *In* "Aquatic Pollutants and Biologic Effects with Emphasis on Neoplasia" (H. F. Kraybill, ed.) **298**, pp. 47–49. *Ann. N.Y. Acad. Sci.*

Whittle, K. J., Mackie, P. R., Farmer, J. and Hardy, R. (1978). The effects of the Ekofisk blowout on hydrocarbon residues in fish and shellfish. *In* "Proc. Conf.

Assessment of Ecological Impacts of Oil Spills, June 1978" pp. 541–559. American Institute of Biological Sciences, Keystone, Colorado

Whittle, K. J., Mackie, P. R., Hardy, R., McIntyre, A. D. and Blackman, R. A. A. (1977b). The alkanes of marine organisms from the United Kingdom and surrounding waters. *Rapp. P.-v. Réun Cons. int. Explor. Mer* **171**, 72–78

Whittle, K. J., Murray, J., Mackie, P. R., Hardy, R. and Farmer, J. (1977c). Fate of hydrocarbons in fish. *Rapp. P.-v. Réun. Cons. int. Explor. Mer* **171**, 139–142

Winters, K., Parker, P. L. and Van Baalen, C. (1969). Hydrocarbons of blue-green algae: geochemical significance. *Science* **163**, 467–468

Withers, N. W. and Nevenzel, J. C. (1977). Phytyl esters in a marine dinoflagellate. *Lipids* **12**, 989–993

Yasuda, S. and Fukamiya, N. (1977a). Hydrocarbons of a chiton. *Bull. Jap. Soc. Sci. Fish* **43**, 1249

Yasuda, S. and Fukamiya, N. (1977b). Hydrocarbons of gonads of sea urchin. *Bull. Jap. Soc. Sci. Fish.* **43**, 1175–1180

Yone, Y. and Fujii, M. (1975). Studies on the nutrition of red sea bream – XI. Effect of ω-3 fatty acid supplement in a corn oil diet on growth rate and feed efficiency. *Bull. Jap. Soc. Sci. Fish.* **41**, 73–77

Youngblood, W. W. and Blumer, M. (1973). Alkanes and alkenes in marine benthic algae. *Mar. Biol.* **21**, 163–172

Youngblood, W. W., Blumer, M. Guillard, R. L. and Fiore, F. (1971). Saturated and unsaturated hydrocarbons in marine benthic algae. *Mar. Biol.* **8**, 190–201

ZoBell, C. E. (1971). Sources and biodegradation of carcinogenic hydrocarbons. *In* "Proc. 1971 Joint Conf. on Prevention and Control of Oil Spills", pp. 441–451. American Petroleum Institute, Washington D.C.

18 Elemental Accumulation in Organisms and Food Chains

MICHAEL N. MOORE

1 Introduction

Living organisms accumulate a wide range of elements, including metals, from their environment. This chapter is concerned with the nature of this uptake, the mechanisms of bioaccumulation, as well as the possible consequences for the organism.

In attempting to cover this extensive subject, the chemical nature of elements in the marine and estuarine environments is discussed in relation to bacterial transformations and biotic uptake. This in turn leads to a consideration of biogeochemical cycling of elements and the processes involved in their movement through food chains including cumulative phenomena. These cumulative properties are examined from the viewpoint of the differential affinities of various elements for specific tissues and cells and the important role of compartmentalisation of many elements within membrane-bound intracellular vesicles. The roles of environmental factors such as salinity, temperature and interactive effects of mixtures of metals in uptake are also discussed.

Finally, biochemical and cellular processes involved in the uptake, transformation, toxication and detoxication of metals are reviewed in the light of current thinking, while the physiological responses of organisms to accumulations of elements are discussed in terms of mechanisms of toxic action, impact on ecosystems and the possible development of genetically-linked tolerance.

2 Enrichment of elements in organisms due to biotic processes

ELEMENTAL COMPONENTS OF ORGANISMS

The four main elemental components of protoplasm are oxygen, carbon, hydrogen and nitrogen which together account for more than 95 % of living material. Other elements which are considered as major components of living material are P, Ca, K, S, Na and Cl in order of abundance. Many other

535

elements are required for many of the specific biochemical and physiological functions of the living organism. These include magnesium, iron and elements such as Mn, Cu, I, Co, Zn, Se, Mo, Ba, Sr, Cr, Sn, Pb, Ti, Rb, Li, As and Br (Hoar, 1975) which are normally referred to as trace elements as they are present in only minute quantities in vertebrates, although this is not strictly the case with certain groups of invertebrates as will be discussed later.

This chapter will not be concerned with the first two groups of elements mentioned above with the exception of calcium, as their physiological functions as the major components of living systems have been extensively discussed by many authors. Our main concern will include the roles of calcium and silicon in biomineralisation, but will be particularly focussed on the chemical forms, nature of uptake and bioaccumulation of trace elements as well as their functional relationships, possible toxicity and ecophysiological interactions.

BIOMINERALISATION

The processes of biomineralisation have been intensively studied in the invertebrates during the past decade. Much of this work concerns the calcium deposits which occur as intracellular granules and this has been extensively reviewed by Simkiss (1976). On current evidence it appears that calcareous material frequently occurs in membrane-bound cytoplasmic vesicles in the cells of many types of invertebrate. Berkaloff (1958) described the initial concretions as appearing in small vesicles in insect malpighian tubules while Waku and Simimoto (1974) described them as appearing in the cisternae of the endoplasmic reticulum in the region of the Golgi complex in the epithelial cells of the midgut gland of the silkworm moth *Bombyx mori*. Simkiss (1976) has reviewed the frequent occurrence of these membrane-bound calcareous structures in the cells of a wide range of marine invertebrates including molluscs and crustaceans and discussed the recycling of these granules such as occurs during moulting and mineralisation of new exoskeleton in the Crustacea. It has been suggested that the phosphate which is present in these granules may provide an inorganic store necessary for animals with a high carbohydrate metabolism, while the mineral deposits may also provide a way of protecting an organism with a large permeability to calcium (Desser, 1963) from the problem of mineralisation. In addition, it has been found that the calcareous granules in the barnacles *Balanus balanoides* and *Lepas anatifera* could provide a major route for the removal of contaminants such as zinc ions (Walker *et al.*, 1975).

Silicon is another element which is a major structural component in many bacteria, fungi, protozoa and higher plants. This role of silicon has been reviewed by Allison (1968) in biological systems as a whole, while Comhaire (1967) has discussed its role in plants. In aquatic systems, a major role of silicon is in the silicaceous shells of diatoms and Reimann *et al.* (1966) have shown that the cell wall of the diatom, *Navicula pelliculosa* is composed of a silica shell surrounded by an organic membrane. The growth of the silicaceous shell occurs intracellularly within a vesicle delimited by a lipoprotein membrane, the silicalemma.

TRACE ELEMENTS

The other main source of elements in living organisms concerns the trace elements which are listed above. Many of these elements are now known to be essential for life and are universally distributed. They are generally only required in minute amounts, with the possible exceptions of magnesium and iron, and are all concerned with specific physiological activities (Hoar, 1975). Some elements which may only occur in trace amounts in the vertebrates can be much more abundant in some of the invertebrates such as copper in the blood of crustaceans and some molluscs, where it is associated with haemocyanins (Ghiretti, 1966) which are functional in oxygen transport. Many trace elements form structural components of the prosthetic groups of enzymes such as xanthine oxidase which is a molybdoflavoprotein (Hoar, 1975). Another function can often involve the activation of enzymes such as dipeptidases and aminopeptidases by manganese, magnesium or zinc (Baldwin, 1967).

Many of the data relating to the specific biochemical and physiological functions of trace elements have been derived from mammalian investigations. These have demonstrated, for instance, that enzymes such as cytochrome-*c* oxidase, superoxide dismutase, tyrosinase, uricase, dopamine β-hydroxylase, lysyl oxidase, spermine oxidase, benzylamine oxidase and diamine oxidase are all cuproenzymes containing copper as an essential component (see review by O'Dell, 1976). Zinc has also been shown to be an essential stabiliser of many macromolecules and biological membranes (Chvapil, 1973) as well as being involved in the homeostasis of various inflammatory cells (Chvapil *et al.*, 1976). Chromium is directly involved as an essential cofactor for insulin in the maintenance of normal glucose tolerance and deficiency of this element can result in a reduction of growth and longevity (Mertz *et al.*, 1974; Doisy *et al.*, 1976). Insulin producing cells have been described in the intestines of two fresh water bivalves (Plisetskaya *et al.*, 1978), although the function of this hormone is as yet unclear and there is no information on its possible associations with trace elements; however it is immunoreactive to antibody for mammalian insulin. Manganese is essential for growth, reproduction and skeletal development in birds and mammals (Leach, 1976) and much of this element is localised within the mitochondria where it is associated with the manganese-containing metalloenzyme pyruvate carboxylase (Scrutton *et al.*, 1966). Vanadium, nickel and silicon have also been shown to be essential nutritional requirements for mammals and there are indications that tin and fluorine may also be essential (Nielsen, 1976).

The ascidians have long been known to have rather exotic complements of trace elements since Henze (1911) demonstrated the presence of vanadium in their blood. Webb (1939) showed that the vanadium was present in blood cells termed vanadocytes and it is now known to be a constituent of haemovanadium (vanadin) which is a powerful reducing agent and may be concerned in the synthesis of the cellulose in the tunic (Carlisle, 1968). The more primitive ascidians also concentrate large amounts of related metals including niobium, tantalum, titanium, chromium, manganese and possibly zinc, molybdenum and tungsten and it has been proposed that during the course of their evolution there has been an increase in the selectivity for certain of these metals (see review by Carlisle, 1968).

3   Chemical forms of elements and relative importance of particulate and dissolved forms

SPECIATION OF TRACE ELEMENTS

Definitive knowledge of the speciation of many trace elements is complicated by difficulties encountered with the sensitivity of analytical techniques at concentrations of less than 10^{-6} M. This problem is further compounded by the need to consider complexation of trace elements by organic ligands.

The problem of inorganic speciation of trace elements has been extensively reviewed by Stumm and Brauner (1975) who discussed the use of both chemical models for the equilibrium distribution of chemical species and concentration ratio diagrams for the prediction of the probable main dissolved inorganic species in sea water. A selection of the probable main species of trace elements together with their predicted concentrations are presented in Table I.

Procedures for finding the equilibrium distribution of species are based on the principle that, at equilibrium, the total free energy of the system is at a minimum (Stumm and Brauner, 1975). Proton exchange and electron exchange processes are of primary importance in marine systems. These are expressed as pH and pε (where p$\varepsilon = -\log[e^-]$) respectively and provide convenient devices for facilitating the application of thermodynamic data to environmental problems. Trace element species present at equilibrium are controlled by pH, pε and $\Sigma \, pX_i$ where $X_1, X_2, \ldots$, are the activities of the various reacting species.

DISSOLVED AND PARTICULATE COMPONENTS

The distinction of the chemical form of a trace element into 'dissolved' and 'particulate' components is largely an arbitrary operational one based on fractionation by filtration (Riley, 1975) but this distinction can no longer be

TABLE I Probable main dissolved inorganic species of certain trace elements in sea water and their total concentrations. Data taken from Brewer (1975)

Element	Chemical species	Total concentration $(\mu g\,l^{-1})$
Ca	Ca^{2+}	4.12×10^5
V	$H_2VO_4^-$, HVO_4^{2-}	2.5
Cr	$Cr(OH)_3$, CrO_4^{2-}	0.3
Mn	Mn^{2+}, $MnCl^+$	0.2
Fe	$Fe(OH)_2^+$, $Fe(OH)_4^-$	2
Co	Co^{2+}	0.05
Ni	Ni^{2+}	1.7
Cu	$CuCO_3^0$, $CuOH^+$	0.5
Zn	$ZnOH^+$, Zn^{2+}, $ZnCO_3^0$	4.9
As	$HAsO_4^{2-}$, $H_2AsO_4^-$	3.7
Mo	MoO_4^{2-}	10
Ag	$AgCl_2^-$	0.04
Cd	$CdCl_2^0$	0.1
Sn	$SnO(OH)_3^-$	1×10^{-2}
Au	$AuCl_2^-$	4×10^{-3}
Hg	$HgCl_4^{2-}$, $HgCl_2^0$	3×10^{-2}
Pb	$PbCO_3^0$, $Pb(CO_3)_2^{2-}$	3×10^{-2}
U	$UO_2(CO_3)_2^{4-}$	3.2

considered adequate. Determination of molecular size by filtration through gels (Gjessing, 1973; Means et al., 1974; Sugai and Healy, 1978), ultra-membranes (Gjessing, 1973; Sharp, 1973; Smith, 1976) and dialysis bags (Dawson and Duursma, 1974) are useful methods by which to distinguish macromolecular and colloidal constituents from the truly dissolved species.

This in turn leads us to the highly complex interactions of inorganic species with organic ligands and in particular to chelation of trace elements by humic compounds. Three categories of organic material can be usefully distinguished in sea water in relation to their coordination of trace metals. These are low molecular weight soluble organic substances, polymeric organic substances which contain a sufficient number of hydrophilic functional groups ($-COO^-$, $-NH_2$, $-R_2NH$, $-RS^-$, ROH, RO^-) to allow them to remain in solution despite their molecular size, and thirdly colloidal organic material, either high-molecular weight compounds or organic substances, adsorbed or chemically bound to inorganic colloids (Stumm and Brauner, 1975). Interaction of a given metal ion with a particular ligand group of the macromolecule is a function of all the ionic equilibria in which the various types of ligand groups in the macromolecule may take part (Sillen and

Martell, 1971). Mantoura (1979) has extensively reviewed and discussed the organic complexing capacity of sea water in relation to the main trace metal constituents and has concluded that substantial fractions of the trace elements interact with organic ligands derived either from the ubiquitous humic pool or from specific biological systems.

BIOGEOCHEMICAL PROCESSES IN ESTUARIES

In estuaries and some near-shore coastal waters the patterns of trace element speciation are further complicated by the variations in temperature, geomorphology, ionic concentration (salinity), $p\varepsilon$, pH and the presence of humics of terrestrial and fresh water origins which are encountered in these environments. The work of Morris (1974) and Morris et al. (1978) has indicated that the fresh water–sea water interphase (FSI) is a chemically and biologically interactive zone and that the estuarine chemistries of manganese, zinc and copper are conspicuously different within the $0.1-1.00°/_{oo}$ mixing segment. Cross and Sunda (1978) discussed the relationship between bioavailability of trace metals and geochemical processes in estuaries and there is little doubt that the bioavailability is continually changing in estuaries where the trace metals are exposed to biogeochemical processes such as adsorption, precipitation, dissolution, complexation and biological incorporation. The relative influences that these processes will have on bioavailability will be dependent on the chemical composition of the fresh water inputs, $p\varepsilon$, pH (Morris, 1978) and salinity gradients, physical characteristics of the estuary and the nature of the organic speciation. It is obvious that a better understanding of the FSI could lead to the formulation of more precise estuarine geochemical cycling models as discussed by Millero (1975).

THE ROLE OF SEDIMENTS IN THE CYCLING OF TRACE ELEMENTS

Sediments are of prime importance as traps for trace elements (Chester and Stoner, 1975) as many metals may be lost from the soluble fraction by precipitation and high levels of metals can frequently occur in this environment. Mud-type estuarine sediments contain many substances that may complex or chelate metal ions and these include inorganic ligands, humic compounds, clays and small organic ligands (Rashid, 1971; Hahne and Kroontje, 1973; Steger, 1973; Gardiner, 1974). Clay debris for instance can contribute to the enrichment of trace metals in the sediments due to its large surface area which is suitable for adsorption, and this has been shown to be an important component in the enrichment of Li, B, C, N, F, Zn, Br, I, Pb, As and Sn in the Severn estuary (Hamilton et al., 1979). In addition the humic ligands contribute extensively to metal complexation in the organic rich environments of the sedimentary pore waters (Elderfield and Hepworth, 1975; Nissenbaum and Swaine, 1976; Sugai and Healy, 1978).

The consequences of discharge of trace elements of anthropogenic origin into estuarine environments has been of increasing interest. This type of activity will obviously influence the speciation of the trace metals in both the water and the sediments, and it will be extremely difficult to accurately predict the biogeochemical cycling of such elements in view of the complexity of the estuarine environment as outlined above. The requirement for fundamental chemical research on trace metal speciation is evident in order to be able to understand the processes controlling the bioavailability of anthropogenic elements in estuarine and coastal water as well as retention by sediments and subsequent release by leaching or weathering (Burrows and Hulbert, 1975), or by biologically induced transformations, such as active biological transport or soil oxidation adjacent to plant roots.

BIOAVAILABILITY OF TRACE ELEMENTS

The uptake, accumulation and toxicity of trace elements by marine organisms is dependent on their specification and bioavailability. Complexation of iron and cadmium by both natural and synthetic organic chelating agents has been shown to facilitate their uptake and accumulation by the blue mussel, *Mytilus edulis* (George and Coombs, 1977a,b). Various chelators including natural organic ligands are known to reduce the toxicity of copper to phytoplankton (Sunda and Guillard, 1976; Anderson and Morel, 1978; Sunda and Lewis, 1978) and to the crustacean *Daphnia magna* (Andrew *et al.*, 1977) by reducing the concentration of free cupric ion. There are also some indications that inorganic complexion of free cupric ion reduces the toxicity of copper to several species of fish including rainbow trout and fathead minnows (Pagenkopf *et al.*, 1973). Complexation of cadmium by nitrilotriacetic acid also reduced its toxicity in the grass shrimp, *Palaemonetes pugia* (Sunda *et al.*, 1978).

In contrast, the microbial synthesis of alkylated mercuric compounds (Jensen and Jernelov, 1969) results in organometallic species which are readily accumulated through the gills of bivalves (Mellinger, 1972) and fish (Rucker and Amend, 1969) and are highly toxic (Eyl, 1970).

Considerable interest has been focussed during the past decade on the bioavailability of trace elements present in estuarine sediments although it has generally been difficult to correlate metal concentrations in organisms with those in sediments (Cross *et al.*, 1970; Bryan and Hummerstone, 1973a,b; Halcrow *et al.*, 1973; Stenner and Nickless, 1974, 1975).

Ingestion of sedimentary particles by organisms such as bivalve molluscs and burrowing polychaetes can provide a route of uptake for trace elements. Filter-feeding organisms, such as mussels and oysters which ingest suspended sediment, may mobilise the complexed metals in their digestive tracts due to lowered pH of the gastric fluid (Owen, 1966; Huggett *et al.*, 1975). Deposit feeders such as the clams *Scrobicularia plana*, *Macoma baltica*, *Mya arenaria* and a range of polychaetes are able to accumulate metals such as Cd, Co, Cr, Cu, Ag, Fe, Mn, Ni and Zn from estuarine sediments although there is evidence that bioavailability of ingested sediment-bound metals is dependent on the nature of the complexation or binding (Cross *et al.*, 1970; Bryan and Hummerstone, 1973a,b, 1977, 1978; Bryan and Uysal, 1978; Luoma and Bryan, 1978; Phelps, 1979). Luoma and Bryan (1978) have also demonstrated that the biological availability of lead can be predicted from the lead/iron ratio in 1 N hydrochloric acid extracts of surface sediments indicating that the bioavailability is strongly dependent on the level of readily extractable iron. Hydrous oxides of iron are an important component of oxidised sediments and are probably crucial in the binding of trace metals within such sediments (Jenne, 1968).

4 Trophic and ecological processes by which trace elements pass through ecosystems and are accumulated in biota

PARTITIONING OF TRACE ELEMENTS

As already stated below, most trace elements occur naturally in the estuarine and coastal environments, although certain elements may be enhanced by anthropogenic activities. These elements are partitioned among the three major compartments of estuarine and coastal ecosystems, that is water, sediment and biota (Bryan, 1971; Degrott and Allersma, 1975). The effective concentration of trace elements within these components of the ecosystem is affected by a number of factors including the ambient elemental composition and concentration, the speciation and binding to organic and inorganic ligands, salinity, pH, temperature, bacterial activity, feeding rates, composition of food and physiological processes governing bioavailability, uptake, storage and excretion.

The complexities of chemical speciation of ligand binding have already been dealt with, while the physiological and biochemical processes will be covered in later sections of this chapter. The main concern in this section is to establish the relative importance of the sources of trace elements and to outline the various routes of transference through the ecosystem, including, of course, the question of biomagnification of certain elements within these systems.

UPTAKE OF TRACE ELEMENTS FROM WATER AND SEDIMENT

It is evident from the previous section that the vast majority of the metallic trace elements in estuarine and marine ecosystems are adsorbed and accumulated by sediments which act as sinks (Bryan, 1971; Gardiner, 1974), leaving comparatively small amounts in the water column. This adsorption of trace metallic ions to sediments is dependent on the properties of the chemical species and the organic content (including humics), surface area and cation exchange properties of the sediment (Ramamoorthy and Rust, 1976). Nevertheless, organismal absorption of trace metals from sea water may occur across the general body surface, gut wall or gill epithelia (see review by Bryan, 1971). In examining the uptake of radiolabelled Zn, Mn, Co and Fe from sea water by the blue mussel, *Mytilus edulis*, Pentreath (1973) concluded that uptake of these elements from the water was relatively minor in comparison to uptake from the food, and the importance of the food chain in uptake of trace elements has also been highlighted by a number of recent investigations of metal accumulation at various trophic levels (Wolfe, 1974; Schulz-Baldes, 1974; Cunningham and Tripp, 1975; Cutshall *et al.*, 1977; Young, 1977; Jennings and Rainbow, 1979a,b; Wrench *et al.*, 1979). In a discussion of the seasonal variability of trace elements in two species of scallop, Bryan (1973) concluded that this variation may be linked to phytoplankton productivity where an increase corresponded to decreased metal levels. This was argued on the basis that the greater availability of food probably increased the metabolic rate of the scallops as well as their excretion rate. He also proposed that the greatly increased incorporation of metals by the phytoplankton would reduce the amount present in the water or that released organic compounds would complex the metals thus reducing their bioavailability. With this high algal productivity the levels per phytoplankton cell would be lower than at other times, so that even with increased ingestion by the scallops, the amounts of metals available from the food would be lower than at times of reduced phytoplankton productivity.

Accumulation of trace elements by algae and phytoplankton is obviously an important consideration in gaining an understanding of the transference of these elements within the food web. The use of indicator plants has been proposed for a number of years as a rapid method of determining the patterns of trace elements within aquatic ecosystems (Peterson, 1971). Brown algae have been reported to be useful indicators of a number of trace elements

Accumulation of trace elements by algae and phytoplankton is obviously an important consideration in gaining an understanding of the transference of these elements within the food web. The use of indicator plants has been proposed for a number of years as a rapid method of determining the patterns of trace elements within aquatic ecosystems (Peterson, 1971). Brown algae have been reported to be useful indicators of a number of trace elements (Nickless *et al.*, 1972; Bryan and Hummerstone, 1973; Fuge and James, 1974; Morris and Bale, 1975; Phillips, 1977, 1979), while the use of higher plants such as mangroves has also been suggested as an indication of the metals available in the sediments in which they are rooted (Peterson *et al.*, 1979). Trace element binding within phytoplankton has been shown to be dependent on the algal species composition rather than to total algal biomass or physicochemical parameters (Briand et al., 1978) and there is evidence that the species which are most important in this respect usually constitute a minor fraction of the total algal volume. There is also evidence that trace metals such as Cu, Cd, Zn, Fe, Mn and Pb can be mobilised from sediments by algae (Wolfe, 1974; Laube et al., 1979; Wong et al., 1979) while decomposition products of plant origin, termed 'detritus', are an important component of these ecosystems in terms of trace element binding capacity (Somers, 1978) and microbial transformation (Wolfe, 1974), prior to macrofaunal assimilation.

ABIOTIC AND BIOTIC CYCLING OF TRACE ELEMENTS

The cycling of trace metals is a combination of abiotic and biotic transformations. There is obviously considerable potential for a combination of both abiotic and biotic pathways which could function as a means for trace element transport across sediment-water-organism interfaces (Brinckmann and Iverson, 1975; Jewett et al., 1975). In sediments which support high concentrations of ionic sulphide many trace metals such as iron, copper, nickel and mercury will be sequestered as supposedly 'insoluble' sulphides. However, these metal sulphides can be resolubilised by the action of aerobic bacteria (Duncan and Trussell, 1964; Silverman and Ehrlich, 1964) and this process provides a pathway for substantial release of trace metals with the potential for cycling through the primary trophic level.

The alkylation of certain trace metals including mercury, arsenic, selenium and tellurium can be carried out by a number of bacterial species and these compounds are all highly toxic (see review by Brinckman and Iverson, 1975). A strain of Pseudomonas sp. isolated from Chesapeake Bay (Nelson et al., 1973) is capable of forming elemental mercury or Hg^{2+} from phenylmercuric acetate as well as methylating tin. Abiotic transalkylation between anthropogenic alkyl lead compounds and inorganic mercury has been demonstrated in river sediments (Jernelov et al., 1972), while salinity and photolysis may have important roles in transalkylation of a number of metals including mercury and tin (Jewett et al., 1975). Photoalkylation of mercury may occur in sunlight (Agaki and Takabatake, 1973; Jewett et al., 1975) and there is some evidence that similar chemical pathways may exist for lead, tin and chromium (Janzen and Blackburn, 1969; Ardon et al., 1971). Recent investigations of bacteria isolated from soil, fresh water sediments and sewage effluents have revealed that certain species are also capable of dealkylation of methylated mercury although the ecological significance of this process is not yet evident (Mason et al., 1979; Shariat et al., 1979).

In essence therefore, aquatic bacteria can be involved in processes which may release soluable metallic species from sediments as well as producing alkylated species which may then be involved in abiotic transalkylations. This is indicative of a highly sophisticated system involving complex interactions between biological and non-biological components of the ecosystem.

BIOMAGNIFICATION AND TURNOVER OF TRACE ELEMENTS

The accumulation of certain trace metals such as mercury in marine food chains has been considered to represent a major environmental hazard which has stimulated numerous scientific investigations, although the mechanisms by which many trace elements are accumulated in higher organisms is not completely understood. Evidence for the magnification of mercury concentrations, for instance, through trophic levels has been reported in a number of experimental food chains (Caracciolo et al., 1975; Cunningham and Tripp, 1975; Smith et al., 1975), while other investigations have indicated that magnification does not occur (Jernelov and Lann, 1971; Leatherland et al., 1973; Williams and Weiss, 1973). A recent approach to this problem using a model food chain composed of four trophic levels has indicated that mercury could be concentrated by the two lower trophic levels, namely bacteria and mosquito larvae, but was not magnified in the two higher levels consisting of guppies and cichlids (Hamdy and Prabhu, 1979). This work also indicated that transference of inorganic mercury from water to the bacteria was of greater magnitude than the transference from bacteria to mosquito larvae. These findings contrast sharply with those of Goldwater (1971) who reported that mercury is concentrated at each trophic level.

Biomagnification of other metals such as zinc, copper, arsenic and iron has been demonstrated in polychaetes and barnacles while oysters, clams and crabs showed little tendency to biomagnify (Guthrie et al., 1979). Of these latter organisms the crabs were considered to be the least likely to exhibit this phenomenon. A recent investigation of the metabolism of arsenic in a marine food chain consisting of phytoplankton (*Dunaliella marina*), zooplankton (*Artemia salina*) and shrimp (*Lysmata seticaudata*) has demonstrated the transference of lipid extractable arsenic (presumed to be an organoarsenic) through the experimental food chain (Wrench et al., 1979). There were also indications that much of the conversion of arsenic to an organic form occurred in the phytoplankton and that subsequent uptake of organoarsenic by the higher trophic levels was associated with food and not from the water. These investigations together with those of Young (1975, 1977), who demonstrated no magnification of zinc and iron in the herbivorous winkle, *Littorina littoralis* or in the dog whelk *Nucella lapillus*, indicate that biomagnification is dependent on physiological differences with respect to the metabolism of particular trace elements, and the behaviour of these metabolites in marine food webs.

The turnover of metals within an ecosystem obviously involves highly complex interaction between abiotic and biotic processes. Nevertheless, it has proved possible to make certain assessments of the overall cycling of certain trace elements and Wolfe (1974) using ^{65}Zn in the Newport River concluded that the cycling of zinc in this environment was essentially a closed system. However, if one considers the surface sediments only and not the large sink represented by the deeper sediment compartment, then these would probably be involved in a more rapid dynamic equilibrium with metals present in the overlying water and pore-water due to their constant resuspension, bacterial activity, presence of detritus and biological burrowing activity.

5 Differential affinities of various forms of elements for specific tissues and cells

TISSUES ASSOCIATED WITH BIOMINERALISATION

Biomineralisation involving calcium and silicon have already been mentioned in the first section of this chapter. These processes have been extensively reviewed by Simkiss (1976) and Allison (1968) for calcium and silicon respectively. The role of intracellular granules in calcareous biomineralisation has been briefly discussed above and their widespread occurrence in many tissues of both aquatic and terrestrial invertebrates is well documented (Simkiss, 1976).

In molluscan tissues calcareous granules are present in sites such as the foot, mantle and digestive gland, and these granules have been shown to be released from the calcium cells of the digestive gland in *Helix pomatia* for transport around the body by amoebocytic blood cells (Abolins-Krogis, 1965, 1968, 1970). Calcium-rich granules have been demonstrated in the clam *Mercenaria mercenaria* in association with the intercellular spaces, microvilli and apical cytoplasm of the mantle epithelial cells where they are believed to be involved with calcification of the shell (Neff, 1972). Calcium granules are also stored in the epithelial cells of the posterior caecum of the amphipod *Orchestia cavimana*, as well as in the mid-gut gland of the blue crab *Callinectes sapidus*, at the time of moult, from which they are released or dissolved during mineralisation of the new exoskeleton (Graf, 1971; Becker *et al.*, 1974). During the later stages of opercular regeneration in the serpulid polychaete *Pileolaria granulata*, Golgi vesicles give rise to secretory granules which are subsequently released from the opercular epithelial cells and appear to be responsible for the initiation of calcification in the opercular plate (Bubel *et al.*, 1977).

Associations between these intracellular calcareous granules and the lysosomal-vacuolar system have been demonstrated in several tissues involved in biomineralisation (Chan and Saleuddin, 1974; Bubel et al., 1979), and lysosomes are known to be directly involved in the regulation of certain secretory processes (Farquhar, 1969; Dean, 1977).

Simkiss (1976) has speculated that these granules may provide an alternative site for an intracellular system for regulating cytoplasmic calcium. He also suggested that it could be energetically less expensive for the organism to remove calcium influxes by forming relatively insoluble calcareous deposits in membrane-bound vesicles, than by pumping Ca^{2+} back into the extracellular fluid. It is also possible that certain trace elements such as zinc could become incorporated into these granules and zinc phosphate-rich vesicles have been reported from the mid-gut parenchymal cells of the barnacle Balanus balanoides (Walker et al., 1975). Walker (1977) has also demonstrated the presence of copper-rich granules in the same cell type from barnacles living in an area with high trace metal run-off. These granules are distinct from those containing zinc phosphate and the copper is believed to be in the form of an organic complex. This does, however, indicate that two physiological processes can result in the sequestration of potentially toxic trace elements within membrane-bound cytoplasmic granules.

THE ROLE OF MEMBRANE-BOUND VESICLES

Tissues such as the molluscan mantle epithelium are obviously heavily involved with the metabolism of divalent cations. The uptake and concentration of mercury and iron within cytosomes (morphologically similar to lysosomes) has been demonstrated in the mantle tentacle epithelial cells of the clam Mercenaria mercenaria using energy dispersive X-ray microanalysis and transmission electron microscopy (Fig. 1; Fowler et al., 1975). These lysosomes also had high iron concentrations in comparison to the concentrations of mercury and this may indicate that these organelles are involved in an intracellular mechanism for maintaining iron levels. Kidney epithelial cells in marine bivalves are also noted for their ability to accumulate trace metals within membrane-bound vesicles. Pb, Fe and Zn are all accumulated by endocytosis within these organelles in Mytilus edulis (Hobden, 1969; George et al., 1976; Coombs, 1977; Moore, 1977; Schulz-Baldes, 1978; Lowe and Moore, 1979) and these structures have been shown to be lysosomes (Moore, 1977; Lowe and Moore, 1979). However even closely related species of bivalve such as the scallops Argopecten gibbus and Argopecten irradians can show quite different concentration patterns of metals such as Mn, Zn, Ca, Cd, Cu and Cr in the vesicles of the kidney epithelial cells which may indicate physiological differences (Carmichael et al., 1979). Compartmentalisation of copper and zinc in membrane-bound vesicles has been described in the

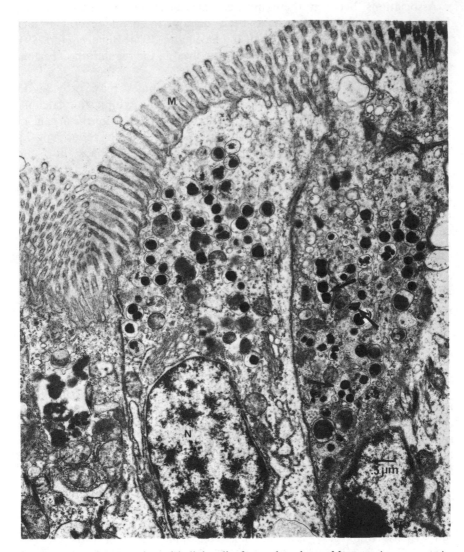

FIG. 1 Mantle tentacle epithelial cells from the clam, *Mercenaria mercenaria*, exposed to 10 p.p.m. Hg^{2+} for 6 days. Large numbers of dense cytosomes which are morphologically similar to lysosomes are present in the apical cytoplasm and aggregates of the cytosomes are prominent near the cellular midline (arrows). These cytosomes have been shown to contain Fe and Hg using energy dispersive X-ray microanalysis. These cells have prominant microvilli (M) and a basally situated nucleus (N). × 10 950. (Reproduced by kind permission of B. A. Fowler, D. A. Wolfe and W. F. Hettler)

amoebocytic blood cells of the oysters *Ostrea edulis, Crassostrea virginica* and *Crassostrea gigas* (Ruddell and Rains, 1975; George *et al.*, 1978). In *Ostrea edulis* there is evidence to show that copper and zinc are sequestered independently in two distinct types of blood cell, the granular acidophils and granular basophils respectively (George *et al.*, 1978). The blood cells in the blue mussel, *Mytilus edulis* do not appear to accumulate metals, although George *et al.* (1976) showed that they are involved in the transport of iron to the kidney cells in mussels exposed to labelled particulate ferric hydroxide. However, a type of cell known as a pigment or brown cell which is believed to be related to the blood cells, does accumulate iron in *M. edulis* (Moore and Lowe, 1977).

Iron associated with the digestive mucus is taken up by endocytosis (pinocytosis) in the tubule epithelial cells of the digestive gland of *M. edulis* and thus enters the lysosomal-vacuolar system of these cells (George *et al.*, 1976). Lead and zinc have also been demonstrated in the lysosomes of these cells (Coombs, 1977; Lowe and Moore, 1979). High levels of metals such as Ag, Cr, Co, Cu, Fe, Mn, Ni, Pb and Zn have been demonstrated in the digestive gland of the clam *Scrobicularia plana* (Bryan and Hummerstone, 1978; Bryan and Uysal, 1978) and it is believed that most of these are derived from ingested sediment. Copper has also been demonstrated in lysosomes of the gastrodermal cells of the coelenterate *Laomedea flexuosa* following treatment with low levels of Cu^{2+}, although it is not known whether the metal is associated with ingestion of food (Moore and Stebbing, 1976).

In the ascidians, vanadium is accumulated within specialised blood cells termed vanadocytes. The metal is apparently scavenged from sea water by the sulphate groups associated with mucopolysaccharides, passed to the pharyngeal cells and hence to the blood in the form of tetravanadate (Carlisle, 1968). Kalk (1963) has shown that the tetravanadate is taken up from the plasma by vanadocytes and is accumulated within membrane-bound vesicles which also contain sulphuric acid. The metal is then complexed within these vesicles to form haemovanadium.

Accumulation of metals within the membrane-bound vesicles of plant cells has been observed in the ulotrichalean alga *Stigeoclonium tenue* (Silverberg, 1975) and in the protozoan ciliate *Tetrahymena pyriformis* (Nilsson, 1978) where lead was sequestered in the digestive vacuole (lysosome). The accumulation of a wide range of metals within vertebrate lysosomes has been extensively reviewed by Sternlieb and Goldfischer (1976).

This accumulation of many trace metals seems, therefore, to be related to the capacity of specific cells or tissues to compartmentalise these elements within membrane-limited vesicles or vacuoles. Many of the cell types mentioned above are actively involved in endocytotic or exocytotic processes and frequently have well developed lysosomal-vacuolar systems for the degradation, transformation or storage of a wide variety of natural products. Obvious sites of endocytotic uptake of metals include the digestive tract and the surface epithelia such as mantle and gills, and pinocytotic uptake of lead and hydrated ferric hydroxide has been demonstrated in the gill epithelial cells of *Mytilus edulis* (George *et al.*, 1976; Coombs, 1977). Accumulation of lead in the kidney epithelial cells of *M. edulis* also appears to occur by endocytosis from the urine in the kidney lumen (Schulz-Baldes, 1978).

6   Effects of environmental factors on cumulative processes

The uptake and accumulation of trace metals by marine organisms and food-chains is known to be subject to a number of environmental variables. These include season, age, feeding mode, salinity, temperature, position in the water column and interactive effects resulting from the presence of several metals, and the relative contributions of these variables has been reviewed by Phillips (1977). The effects of season have already been mentioned in relation to phytoplankton productivity and metal levels in scallops (Bryan, 1973), while seasonal fluctuations in body weight due to gametogenic activity is linked to body burdens of metals in the barnacle, *Balanus balanoides* (Ireland, 1974) and the mussel, *Mytilus edulis* (Simpson, 1979). Variables such as species, age, weight, feeding habits and salinity are of particular importance in metal accumulation although the interrelationships are poorly understood (Phillips, 1977).

The role of environmental variables in the accumulation of trace elements has been extensively investigated in bivalve molluscs although the mechanisms involved are at best unclear. Decreased salinity induces increased accumulation of cadmium in the blue mussel, *Mytilus edulis*, and the influx is dependent on the osmolarity and the relationship between cation uptake and cellular volume (George *et al.*, 1977). Low salinity does not affect the uptake of zinc by *M. edulis*, but decreases the uptake of lead. Similarly, high salinity coupled with low temperature does not affect the uptake of cadmium; however, when both low salinity and low temperature are combined the uptake is reduced (Phillips, 1976). Reduced salinity also increases the uptake of arsenic in the Mediterranean mussel, *Mytilus galloprovincialis*, while increased temperature enhances both arsenic uptake and excretion (Unlu and Fowler, 1979). Increased salinity decreases the uptake ^{137}Cs in the clam *Rangia cuneata*, but uptake is increased with increasing temperature (Wolfe and Coburn, 1970).

The presence of a mixture of trace metals has been shown to influence uptake in certain bivalves. In the oyster *Ostrea edulis*, iron and cobalt depress the uptake of zinc (Romeril, 1971), while in *M. edulis* the uptake of copper is affected by the presence and change in concentrations of zinc, cadmium and lead (Phillips, 1976).

All of these results emphasise the difficulties involved in attempting to make any generalisations concerning the influence of environmental variables on trace metal uptake. The differences in patterns of metal uptake presumably reflect the complexities of chemical speciation which are encountered at various salinities as well as the different physiological mechanisms involved in handling various metals.

7 Biotic processes by which transformation, toxication and detoxication of chemical forms occurs

CELLULAR INTERNALISATION AND BINDING OF TRACE ELEMENTS

The initial step in any trace metal metabolic pathway is transport of the metal ion across a cellular or organelle membrane. Unfortunately this is an area where our knowledge is considerably lacking and most of the current ideas on ion transport are based on the transport of Na^+ and K^+ in association with nerve transmission and osmoregulation. Current ideas on the transport of trace metals across biological lipoprotein membranes have been summarised by Coombs and George (1978) who distinguished three main pathways. The first pathway is the 'Pore' theory, where ions are transported down a potential gradient through a membrane pore, the geometry of which may confer specificity. The second pathway involves a carrier, which can complex with some degree of specificity with a metal ion, thus neutralising the ionic charge and making the complex hydrophobic which allows penetration of the lipoprotein membrane. In some instances the complexing ligand may be attached to the membrane and complexation induces a flip-over mechanism into the cytosol. The third pathway involves certain organic pollutants which may preferentially complex a metal ion and alter its normal pathway resulting in either an excess or a deficiency within the cell. In addition to these pathways, metals may be internalised within cells by the endocytotic mechanisms described in a previous section, although in these instances they will remain separated from the cystosol by a lipoprotein membrane.

In both mammals and lower organisms the intracellular metabolic homeostasis of many trace metals appears to be dependent on binding to low molecular weight proteins such as metallothioneins or by partitioning into the intracellular lysosomal-vacuolar system as discussed in a previous section. Metallothioneins are rich in sulphydryl residues and can bind metals such as Cd, Cu, Hg, Sn, Ag and Zn (Sabbioni and Marafante, 1975; Winge *et al.*, 1975) thus detoxifying these metals and protecting metal-loenzymes from displacement by non-essential or interfering elements

(Goldman, 1970; Brown *et al.*, 1977; Brown and Parsons, 1978). If however, the rate of influx of metals into the cells exceeds the rate at which metallothionein can be synthesised, then there may be a spillover of metals from metallothionein into the metalloenzyme pool (Brown *et al.*, 1977; Brown and Parsons, 1978; Engel and Fowler, 1979; Pruell and Englehardt, 1980; Roesijadi, 1979). This spillover can then result in the manifestation of toxic effects due to the displacement of essential metals from metalloenzymes culminating in conformational changes and dysfunction (Friedberg, 1974). Synthesis of metallothionein is known to be induced by exposure to a range of trace metals (Sabbioni and Marafante, 1975; Winge *et al.*, 1975) and this induction may be at the translational level (m RNA) for low concentrations, or at the transcriptional level (DNA) for higher concentrations (Webb, 1972; Squibb and Cousins, 1974).

THE LYSOSOMAL-VACUOLAR SYSTEM AND METAL SEQUESTRATION

Metallothioneins are generally considered to be cytosolic proteins (Winge *et al.*, 1975) although there is evidence that they may also occur within lysosomes (Porter, 1974). This leads us to consideration of the physiological significance of metal sequestration within the lysosomal-vacuolar system (see review by Sternlieb and Goldfischer, 1976). The occurrence of this phenomenon both in mammals and invertebrates is widespread, which tends to support the role of this system as an important component of the cellular mechanism for handling trace elements. Trace metals may in many instances have a common metabolic pathway with calcium which evolved for the regulation of intracellular calcium associated with mitochondrial function, nerve stimulation, exocrine secretion and muscle contraction (Baker, 1976; Coombs and George, 1978). The recent demonstration of calcium release from dense lysosome-like organelles in neurons in the Japanese land snail, *Euhadra peliomphala*, during neuronal bursting activity (Sugaya and Onozuka, 1978), as well as the examples cited in the previous section tend to confirm the functional relationship between calcium metabolism and the lysosomal-vacuolar system. Certain elements such as gold are compartmentalised within lysosomes in mammalian hepatocytes (Chvapil *et al.*, 1972), and yet, in the proximal renal tubule cells of the kidney this metal is localised within the mitochondria (Galle, 1974). The complexity of this situation is increased by the fact that the proximal renal tubule cells concentrate methylmercury, uranium and chromium in their lysosomal system (Fowler *et al.*, 1974; Galle, 1974; Berry *et al.*, 1978).

Sequestration of trace elements within the lysosomal-vacuolar system can be considered as a detoxication process which is available in many cell types. Nonetheless, caution must be exercised in the physiological interpretation of this process as it is first necessary to determine the chemical nature of the metal species and the binding properties of organic ligands within the lysosomes, as well as the lysosomal interrelationships with carrier proteins such as metallothionein in the cytosol. In many instances however, this system does provide an excretory route for the removal of compartmentalised trace metals, either by exocytosis or apocrine secretion as has been described in the kidney cells of the blue mussel, *Mytilus edulis*, and the cells of the digestive gland in the isopod, *Asellus meridianus* (George *et al.*, 1976; Moore, 1977; Brown, 1978; Schulz-Baldes, 1978; Lowe and Moore, 1979).

Many trace metals are also known to exert pathological effects on lysosomes in mammalian cells (see reviews by Allison, 1969; Sternlieb and Goldfischer, 1976) and excessive concentrations of many metals present either in the cytosol or the lysosomes can lead to alterations in lysosomal structure, membrane stability and enzymic content (see Table II). These changes can result in the release of hydrolytic enzymes from the lysosomes which will damage cytoplasmic components and lead to cytotoxicity. Pathological responses of this type have been described in the coelenterate, *Laomedea flexuosa*, exposed to mercury, cadmium and copper and the blue mussel, *Mytilus edulis*, from an environment with high zinc levels (Moore and Stebbing, 1976; Lowe and Moore, 1979). This suggests that the lysosomal-vacuolar system, like the metallothionein detoxication system, has limited protective capability probably differing among cell types, and can be overloaded. However, it should be noted that these types of lysosomal response are not specific for toxic metals and can be induced in *M. edulis* by other types of stressor (Bayne *et al.*, 1976, 1978, 1979; Moore, 1976, 1979, 1980; Moore *et al.*, 1978a,b, 1979).

TABLE II   Alterations induced in lysosomes by trace elements

Element	Organism	Effect on lysosomes	Reference
Cu	Rat	Lysosomal membranes broken, hydrolases released	Lindquist (1968)
Hg	Mouse	Lysosomal membranes destabilised	Verity and Brown (1968)
Hg	Rat	Lysosomal hydrolases released in urine	Robinson *et al.* (1967)
Co	Rat	Lysosomal hydrolases released	Smith *et al.* (1974)
Cu, Hg, Cd	Hydroid	Lysosomal membrane destabilised	Moore and Stebbing (1976)
Zn	Mussel	Lysosomal membrane destabilised	Lowe and Moore (1979)

8   Biotic effects of accumulation of elements and the responses of organisms to such accumulations

MECHANISMS OF TOXIC ACTION

Nutritional investigations in mammals have indicated that there are many interactions between trace elements and diet as well as among trace elements themselves (see reviews by de Bruin, 1976; Hill, 1976). This work has revealed that there exist marked interrelationships of such a magnitude that the toxic level of an element may be rendered harmless, while a marginal level of an essential element may be rendered deficient. The mechanisms by which lower organisms absorb, transform, store and excrete trace elements have been discussed in detail in the preceding sections, and it is evident that by combining and regulating these physiological processes, the more sophisticated organisms should be able to maintain a homeostatic condition (Fig. 2). However, this must be considered within the real environmental situations where an organism is not only exposed to single trace elements but to complex mixtures. Superimposed on this situation we have complex fluctuations in salinity, temperature, nutrient levels and anthropogenic organic compounds.

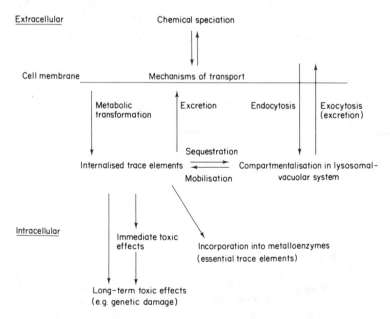

FIG. 2   Diagramatic representation of cellular uptake and metabolic interactions of trace elements

Physiological investigations of the effects of toxic trace metals have revealed three generalised effects of sublethal levels. The first is the induction or depression of enzymes. This includes metalloenzymes which are directly attacked (Friedberg, 1974), enzymes which are involved in energy mobilisation (i.e. glycolysis) or in the production of metabolites for biosynthesis (Calabrese et al., 1977) and lysosomal hydrolases which are involved in normal turnover of cellular components (Sternlieb and Goldfischer, 1976; Dean, 1977). The second effect is the loss of ligand sensitivity by which enzyme reaction rates are regulated (Calabrese et al., 1977). The third effect is the alteration of membrane fluidity or permeability in cells and organelles and this may have far-reaching consequences for marine organisms such as the regulation of cell volume (Riisgård, 1979), uptake of amino acids (Sieberg and Ehlers, 1979) or dysfunction of lysosomes (Moore, 1976, 1980; Moore and Stebbing, 1976; Lowe and Moore, 1979). Such effects will inevitably reduce the metabolic and physiological flexibility of an organism, thus constituting a stress syndrome (Selye, 1950; Bayne, 1975) which will render it more susceptible to further environmental change.

INTERACTIVE EFFECTS

A number of investigations have looked at the toxic effects of mixtures of metals using factorial experiments. The combination of copper and cadmium results in a lower rate of oxygen consumption in the mud snail, Nassarius obsoletus, than either metal alone (MacInnes and Thurberg, 1973). Synergistic actions of copper and zinc have been reported in teleosts (Doudoroff and Katz, 1953; Sprague, 1970) and these metals cause increased mortality in the eggs of the zebra cichlid, Cichlasoma nigrofasciatum (Ozoh and Jacobson, 1979). The mechanisms whereby these metals interact to induce increased toxicity are not clear.

This situation also applies to the effects of interactions of trace elements with salinity and temperature. Complex toxic interactions in oyster larvae have been described for mixtures of mercury and silver, and copper and zinc over a range of temperature (MacInnes and Calabrese, 1978). These results indicated that the metals either singly or in combination were less toxic at 25 °C than at either 20 °C or 30 °C. Interactions between zinc and temperature and zinc and salinity have been reported for larval development of the grass shrimp, Palaemonetes pugio, using response-surface methodology (McKenney and Neff, 1979). Uptake of cadmium is increased by reduced salinity in the blue mussel, Mytilus edulis, and it has been suggested that this effect may be due to increased ion fluxes or the unmasking of a carrier or trans-membrane pore due to physical alteration of the cell membrane caused by swelling (George et al., 1978). Toxicity of both cadmium and chromium decreases with increasing salinity in the blue crab,

Callinectes sapidus (Frank and Robertson, 1979) and this may reflect alterations in speciation and bioavailability as well as physiological changes of the type suggested by George *et al.* (1978).

In addition to these types of interaction, many trace elements inhibit mixed function oxygenases which are involved in the metabolism of steroids and organic xenobiotics (Congiu and Pani, 1978). These microsomal enzymes are responsible for the detoxication of many organic pollutants, and if inhibition occurs then organisms could be rendered more susceptible to their toxic action.

ECOLOGICAL CONSEQUENCES AND DEVELOPMENT OF TOLERANCE

The ecological significance of pollution by trace elements is therefore dependent not only on the relative abundance of individual elements but also upon their toxicity, interactions, differential mobility and bioaccumulation potentials.

Investigations of the impact of toxic trace elements on enclosed planktonic ecosystems have indicated that the changes which occur in population structures appear to be universal irrespective of the nature of the pollutant and seem to be similar to those induced by physical stressors (Gamble *et al.*, 1977; Menzel, 1977). This is indicative of the problems of attempting to extrapolate from toxicity experiments on single organisms or trophic levels to whole ecosystems. Despite these problems, progress has been made in understanding the different tolerances to toxic trace elements which are displayed by different developmental stages of marine organisms. Calabrese *et al.* (1977) has also drawn attention to the fact that the early life history stages tend to be less sensitive to cadmium than to other trace metals with a reversal of this toxicity in the adults. Further investigations in this area could certainly be helpful in any attempt to predict the effects of pollutants on recruitment.

The development of genetically-linked tolerance in organisms exosed to high levels of toxic metals has been described in a number of terrestrial plant species by Bradshaw (1970). There is also evidence of this type of tolerance in copper-resistant populations of the brown alga, *Ectocarpus siliculosus* (Russell and Morris, 1970). Metal tolerance can be acquired during the life of an organism as has been demonstrated in fish (Lloyd, 1960; Pickering and Vigor, 1965; Edwards and Brown, 1967) and Beattie and Pascoe (1978) have shown that pretreatment of the eggs of rainbow trout, *Salmo gairdneri*, with cadmium serves to protect the juvenile fish from this metal. However, this type of tolerance is unlikely to be genetically linked. Bryan and Hummerstone (1971) described a copper-resistant population of the polychaete *Nereis diversicolor* and there is evidence which suggests that this tolerance is genetically linked (Bryan, 1974). Certainly these polychaetes are physiologically different from non-resistant populations and are also resistant to zinc and possibly arsenic. Brown (1978) has also demonstrated that a copper-tolerant population of the isopod *Asellus meridianus* can preferentially detoxify lead by cytoplasmic compartmentalisation and it has been suggested that this may have a genetic basis.

This area of research is obviously of considerable interest in terms of trace metals resulting from anthropogenic activities acting as genetic selection pressures on marine and estuarine organisms. Bryan (1974) has suggested that these selection pressures may be greater in contaminated estuaries due to the increased toxicity of many metals at low salinity.

9 Conclusions

In summary we can say that the overall picture of trace element accumulation will only emerge from investigations of chemical speciation, bioavailability, mechanisms of uptake, cellular metabolic transformation, physiological responses and ecophysiological interactions. One aspect which emerges strongly from this review is the requirement for more detailed knowledge of the chemical and biological interrelationships between essential and non-essential elements. Much of this will involve fundamental investigations of biogeochemical cycling as well as the determination of the exact bases of biochemical interactions such as elemental binding and displacement in metalloenzymes, mechanisms of toxic action, compartmentalisation processes and membrane transport phenomena.

References

Abolins-Krogis, A. (1965). Electron microscope observations on calcium cells in the hepatopancreas of the snail. *Helix pomatia. Ark. Zool.* **18**, 85–92

Abolins-Krogis, A. (1968). Shell regeneration in *Helix pomatia* with special reference to the elementary calcifying particles. *Symp. zool. Soc. Lond.* **22**, 75–92

Abolins-Krogis, A. (1970). Electron microscope studies of the intracellular origin and formation of calcifying granules and calcium spherites in the hepatopancreas of the snail *Helix pomatia. Z. Zellforsch. mikrosk. Anat.* **108**, 501–515

Agaki, H. and Takabatake, E. (1973). Photochemical formation of methylmercuric compounds from mercuric acetate. *Chemosphere* **2**, 131–133

Allison, A. C. (1968). Silicon compounds in biological systems. *Proc. R. Soc. Lond.* B **171**, 19–30

Allison, A. C. (1969). Lysosomes and cancer. *In* "Lysosomes in Biology and Pathology" (J. T. Dingle and H. B. Fell, eds) Vol. 2, pp 178–204. North Holland/American Elsevier, Amsterdam, Oxford and New York

Anderson, D. M. and Morel, F. M. (1978). Copper sensitivity of *Gonyaulax tamarensis. Limnol. Oceanogr.* **23**, 283–295

Andrew, R. W., Biesinger, K. E. and Glass, G. E. (1977). Effects of inorganic complexing on the toxicity of copper to *Daphnia magna. Water Res.* **11**, 309–315

Ardon, M., Woolmington, K. and Pernick, A. (1971). The methylpentaaqua-chromium (III) ion. *Inorg. Chem.* **10**, 2812

Baker, P. F. (1976). The regulation of intracellular calcium. *In* "Calcium in Biological Systems". Symposia of the Society for Experimental Biology No. XXX, pp. 67–88. Cambridge University Press, Cambridge, London, New York and Melbourne

Baldwin, E. (1967). "Dynamic Aspects of Biochemistry", 5th edn. Cambridge University Press, Cambridge

Bayne, B. L. (1975). Reproduction in bivalve molluscs under environmental stress. *In* "Physiological Ecology of Estuarine Organisms" (F. J. Vernberg, ed.) pp. 259–277. University of South Carolina Press, Columbia, South Carolina

Bayne, B. L., Holland, D. L., Moore, M. N., Lowe, D. M. and Widdows, J. (1978). Further studies on the effects of stress in the adult on the eggs of *Mytilus edulis*. *J. mar. biol. Ass. U.K.* **58**, 825–841

Bayne, B. L., Livingstone, D. R., Moore, M. N. and Widdows, J. (1976). A cytochemical and a biochemical index of stress in *Mytilus edulis* L. *Mar. Pollut. Bull.* **7**, 221–224

Bayne, B. L., Moore, M. N., Widdows, J., Livingstone, D. R. and Salkeld, P. (1979). Measurement of the responses of individuals to environmental stress and pollution: studies with bivalve molluscs. *Phil. Trans. R. Soc. Lond. B.* **286**, 563–581

Beattie, J. H. and Pascoe, D. (1978). Cadmium uptake by rainbow trout, *Salmo gairdneri* eggs and alevins. *J. Fish Biol.* **13**, 631–637

Becker, G. L., Chen, C.-H., Greenawalt, J. W. and Lehninger, A. L. (1974). Calcium phosphate granules in the hepatopancreas of the Blue Crab, *Callinectes sapidus*. *J. Cell Biol.* **61**, 316–326

Berkaloff, A. (1958). Les grains de sécrétion des tubes de Malpighi de *Gryllus domesticus* (Orthoptère). *C. r. hebd. Seanc. Acad. Sci. Paris* **246**, 2807–2809

Berry, J. P., Fourdry, J., Galle, P. and Lagrue, G. (1978). Chromium concentration by proximal renal tubule cells: an ultrastructural, microanalytical and cyto-chemical study. *J. Histochem. Cytochem.* **26**, 651–657

Bowen, J. M. (1966). "Trace Elements in Biochemistry". Academic Press, New York and London

Bradshaw, A. (1970). Pollution and plant evolution. *New Scientist* **48**, 497–500

Brewer, P. G. (1975). Minor elements in sea water. *In* "Chemical Oceanography" (J. P. Riley and G. Skirrow, eds) Vol. 1, pp. 415–496. Academic Press, New York and London

Briand, F., Trucco, R. and Ramamoorthy, S. (1978). Correlations between specific algae and heavy metal binding in lakes. *J. Fish. Res. Bd Can.* **35**, 1482–1485

Brinckman, F. E. and Iverson, W. P. (1975). Chemical and bacterial cycling of heavy metals in the estuarine system. *In* "Marine Chemistry in the Coastal Environment" (T. M. Church, ed.) pp. 319–342. American Chemical Society, Washington D.C.

Brown, B. E. (1978). Lead detoxication by a copper-tolerant isopod. *Nature* **276**, 388–390

Brown, D. A. and Parsons, T. R. (1978). Relationship between cytoplasmic distribution of mercury and toxic effects to zooplankton and chum salmon (*Oncorhynchus keta*) exposed to mercury in a controlled ecosystem. *J. Fish. Res. Bd Can.* **35**, 880–884

Brown, D. A., Bawden, C. A., Chatel, K. W. and Parsons, T. R. (1977). The wild-life community of Iona Island jetty, Vancouver, B.C. and heavy-metal pollution effects. *Environ. Conserv.* **4**, 213–216

Bryan, G. W. (1971). The effects of heavy metals (other than mercury) on marine and estuarine organisms. *Proc. R. Soc. Lond. B* **117**, 389–410

Bryan, G. W. (1973). The occurrence and seasonal variation of trace metals in the scallops *Pecten maximus* (L.) and *Chlamys opercularis* (L.). *J. mar. biol. Ass. U.K.* **53**, 145–166

Bryan, G. W. (1974). Adaptation of an estuarine polychaete to sediments containing high concentrations of heavy metals. *In* "Pollution and Physiology of Marine Organisms" (F. J. Vernberg and W. B. Vernberg, eds) pp. 123–135. Academic Press, New York and London

Bryan, G. W. and Hummerstone, L. G. (1971). Adaptation of the polychaete *Nereis diversicolor* to estuarine sediments containing high concentrations of heavy metals. I. General observations and adaptation to copper. *J. mar. biol. Ass. U.K.* **51**, 845–863

Bryan, G. W. and Hummerstone, L. G. (1973). Brown seaweed as an indicator of heavy metals in estuaries in south-west England. *J. mar. biol. Ass. U.K.* **53**, 705–720

Bryan, G. W. and Hummerstone, L. G. (1977). Indicators of heavy-metal contamination in the Looe Estuary (Cornwall) with particular regard to silver and lead. *J. mar. biol. Ass. U.K.* **57**, 75–92

Bryan, G. W. and Hummerstone, L. G. (1978). Heavy metals in burrowing bivalve *Scrobicularia plana* from contaminated and uncontaminated estuaries. *J. mar. biol. Ass. U.K.* **58**, 401–419

Bryan, G. W. and Uysal, H. (1978). Heavy metals in the burrowing bivalve *Scrobicularia plana* from the Tamar estuary in relation to environmental levels. *J. mar. biol. Ass. U.K.* **58**, 89–108

Bubel, A., Thorp, C. H. and Fitzsimons, C. (1977). A histological and electron microscope study of opercular regeneration in the serpulid *Pileolaria granulata* with particular reference to the formation of the calcareous opercular plate. *In* "Proceedings of the 4th International Congress of Marine Corrosion and Fouling", pp. 85–95. Centre des Recherches d'études Oceanographiques, Boulogne

Bubel, A., Thorp, C. H. and Moore, M. N. (1979). A histological, histochemical and ultrastructural study of the operculum of the serpulid *Pomatoceros triqueter* L. with particular reference to the formation of the calcareous opercular plate during opercular regeneration. *In* "Proceedings of the Fourth International Biodeterioration Symposium (Berlin)" (T. A. Oxley, D. Allsopp and G. Becker, eds) pp. 275–290. Pitman, London

Burrows, K. C. and Hulbert, M. H. (1975). Release of heavy metals from sediments: preliminary comparison of laboratory and field studies. *In* "Marine Chemistry in the Coastal Environment" (T. M. Church, ed.). ACS Symposium Series 18. American Chemical Society, Washington, D.C.

Calabrese, A., Thurberg, F. P. and Gould, E. (1977). Effects of cadmium, mercury and silver on marine animals. *Mar. Fish. Rev.* **39**, 5–11

Carlisle, D. B. (1968). Vanadium and other metals in ascidians. *Proc. R. Soc. Lond. B* **171**, 31–42

Carmichael, N. G., Squibb, K. S. and Fowler, B. A. (1979). Metals in the molluscan kidney: A comparison of two closely related bivalve species (*Argopecten*), using X-ray microanalysis and atomic absorption spectroscopy. *J. Fish. Res. Bd Can.* **36**, 1149–1155

Carracciolo, S., Disilvestro, C. and Trevisan, A. (1975). Il contenuto in mercurio totaledi trote iridee di allevamento alimentate con pesci e crostacei di zone di pesca del medio Adriatico prosticienta la costa Abruzzese. *Vet. Ital.* **26**, 13–19

Chan, J. F. Y. and Saleuddin, A. S. M. (1974). Acid phosphatase in the mantle of the shell-regenerating snail, *Helisoma duryi duryi*. *Calc. Tiss. Res.* **15**, 213–220

Chester, R. and Stoner, J. H. (1975). Trace elements in sediments from the lower Severn Estuary and Bristol Channel. *Mar. Pollut. Bull.* **6**, 92–96

Chvapil, M. (1973). New aspects in the biological role of zinc: a stabiliser of macromolecules and biological membranes. *Life Sci.* **13**, 1041–1049

Chvapil, M., Ryan, J. N. and Zukoski, C. F. (1972). The effect of zinc and other metals on the stability of lysosomes. *Proc. Soc. Exp. Biol. Med.* **140**, 642–646

Chvapil, M., Zukoski, C. F., Hattler, B. G., Stankova, L., Montgomery, D., Carlson, E. C. and Ludwig, J. C. (1976). Zinc and cells. *In* "Trace Elements in Human Health and Disease" (A. S. Prasad and D. Oberleas, eds) Vol. I, pp. 269–281. Academic Press, New York, San Francisco and London

Comhaire, M. (1967). The role of silica for plants (Fr). *Agron. Trop. Paris* **22**, 1235–1240

Congiu, L. and Pani, P. (1978). Heavy metal induced liver injury. *In* "Biochemical Mechanisms of Liver Injury" (T. F. Slater, ed.) pp. 591–622. Academic Press, London, New York and San Francisco

Coombs, T. L. (1977). Uptake and storage mechanisms of heavy metals in marine organisms. *Proc. analyt. Div. chem. Soc.* **14**, 219–222

Coombs, T. L. and George, S. G. (1978). Mechanisms of immobilisation and detoxication of metals in marine organisms. *In* "Proceedings of the 12th European Symposium on Marine Biology" (D. S. McLusky and A. J. Berry, eds) pp. 179–187. Pergamon Press, Oxford and New York

Cross, F. A. and Sunda, W. G. (1978). Relationship between bioavailability of trace metals and geochemical processes in estuaries. *In* "Estuarine Interactions" (M. L. Wiley, ed.) pp. 429–442. Academic Press, New York and London

Cross, F. A., Duke, T. W. and Willis, J. N. (1970). Biogeochemistry of trace elements in a coastal plain estuary: distribution of manganese, iron and zinc in sediments, water and polychaetous worms. *Chesapeake Sci.* **11**, 221–234

Cunningham, P. A. and Tripp, M. R. (1975). Accumulation, tissue distribution and elimination of ^{203}Hg Cl$_2$ and CH$_3$ ^{203}Hg Cl in the tissues of the American oyster *Crassostrea virginica*. *Mar. Biol.* **31**, 321–334

Cutshall, N. H., Naidu, J. R. and Pearcy, W. G. (1977). Zinc and cadmium in the Pacific bake *Merluccius productus* off the western U.S. coast. *Mar. Biol.* **44**, 195–201

Dawson, R. and Duursma, E. K. (1974). Distribution of radio-isotopes between phytoplankton, sediment and sea water in a dialysis compartment system. *Neth. J. Sea. Res.* **8**, 339–353

Dean, R. T. (1977). "Lysosomes". Edward Arnold, London

De Bruin, A. (1976). "Biochemical Toxicology of Environmental Agents". North-Holland/American Elsevier, Amsterdam, Oxford, New York

Degroot, A. J. and Allersma, E. (1975). Field observations on the transport of heavy metals in sediments. *In* "Heavy Metals in the Aquatic Environment" (P. A. Krenkel, ed.) pp. 85–104. Pergamon Press, Toronto

Desser, S. S. (1963). Calcium accumulation in larval *Echinococcus mulvilocularis*. *Can. J. Zool.* **41**, 1055–1059

Doisy, R. J., Streeten, D. H. P., Freiberg, J. M. and Schneider, A. J. (1976). Chromium metabolism in man and biochemical effects. *In* "Trace Elements in

Human Health and Disease" (A. S. Prasad and D. Oberleas, eds) Vol. II, pp. 79–109. Academic Press, New York, San Francisco and London

Doudoroff, P. and Katz, M. (1953). Critical review of literature on the toxicity of industrial wastes and their components to fish. Part 2. The metal as salt. *Sewage Ind. Wastes* **25**, 802–839

Duncan, D. W. and Trussell, P. C. (1964). Advances in the microbiological leaching of sulphide ores. *Can. Metall. Qtrly.* **3**, 43–55

Edwards, R. W. and Brown, V. M. (1967). Pollution and fisheries: a progress report. *Wat. Pollut. Control* **66**, 63–78

Engel, D. W. and Fowler, B. A. (1979). Factors influencing the accumulation and toxicity of cadmium to marine organisms. *Environ. Health Perspect.* **28**, 81–88

Elderfield, H. and Hepworth, A. (1975). Diagenesis, metals and pollution in estuaries. *Mar. Poll. Bull.* **6**, 85–87

Eyl, T. B. (1970). Methyl mercury poisoning in fish and human beings. *Mod. Med.* **38**, 135–141

Farquhar, M. G. (1969). Lysosome function in regulating secretion: disposal of secretory granules in cells of the anterior pituitary gland. *In* "Lysosomes in Biology and Pathology" (J. T. Dingle and H. B. Fell, eds) Vol. 2, pp. 462–482. North Holland/American Elsevier, Amsterdam, Oxford and New York

Fowler, B., Brown, H. V., Lucier, G. W. and Beard, M. E. (1974). Mercury uptake by renal lysosomes of rats ingesting methylmercury hydroxide. *Arch. Path.* **98**, 297–301

Fowler, B. A., Wolfe, D. A. and Hettler, W. F. (1975). Mercury and iron uptake in mantle epithelial cells of Quahog clams (*Mercenaria mercenaria*) exposed to mercury. *J. Fish. Res. Bd Can.* **32**, 1767–1775

Frank, P. M. and Robertson, P. B. (1979). The influence of salinity on toxicity of cadmium and chromium to the blue crab, *Callinectes sapidus*. *Bull. Environm. Contam. Toxicol.* **21**, 74–78

Friedberg, F. (1974). Effects of metal binding on protein structure. *Quart. Rev. Biophys.* **7**, 1–33

Fuge, R. and James, K. H. (1974). Trace metal concentrations in *Fucus* from the Bristol Channel. *Mar. Pollut. Bull.* **5**, 9–12

Galle, P. (1974). Rôle des lysosomes et des mitochondries dans les phénomènes de concentration et d'élimination d'éléments minéraux (uranium et or) par le rein. *J. Microscopie* **19**, 17–24

Gamble, J. C., Davies, J. M. and Steele, J. H. (1977). Loch Ewe bag experiment, 1974. *Bull. Mar. Sci.* **27**, 146–175

Gardiner, J. (1974). The chemistry in natural waters: a study of cadmium complex formation using the cadmium specific ion electrode. *Water Res.* **8**, 23–30

George, S. G. and Coombs, T. L. (1977a). Effects of high stability iron-complexes on the kinetics of iron accumulation and excretion in *Mytilus edulis* (L.). *J. Exp. Mar. Biol. Ecol.* **28**, 133–140

George, S. G. and Coombs, T. L. (1977b). The effects of chelating agents on the uptake and accumulation of cadmium by *Mytilus edulis*. *Mar. Biol.* **39**, 261–268

George, S. G., Carpene, E. and Coombs, T. L. (1977). The effect of salinity on the uptake of cadmium by the common mussel, *Mytilus edulis* (L.). *In* "Proceedings of the 12th European Marine Biology Symposium" (D. S. McLusky and A. J. Berry, eds) pp. 189–193. Pergamon Press, Oxford

George, S. G., Pirie, B. J. S. and Coombs, T. L. (1976). The kinetics of accumulation of and excretion of ferric hydroxide in *Mytilus edulis* (L.) and its distribution in the tissues. *J. Exp. Mar. Biol. Ecol.* **23**, 71–84

George, S. G., Pirie, B. J. S., Cheyne, A. R., Coombs, T. L. and Grant, P. T. (1978). Detoxication of metals by marine bivalves: an ultrastructural study of the compartmentation of copper and zinc in the oyster *Ostrea edulis*. *Mar. Biol.* **45**, 147–156

Ghiretti, F. (1966). Molluscan haemocyanins. *In* "Physiology of Molluscs" (K. M. Wilbur and C. M. Yonge, eds) Vol. II, pp. 233–248. Academic Press, New York and London

Gjessing, E. T. (1973). Gel and ultra-membrane filtration of aquatic humus: a comparison of two methods. *Schweiz. Z. Hydrol.* **35**, 286–294

Goldman, D. E. (1970). The role of certain metals in axon excitability processes. *In* "Effects of Metals on Cells, Subcellular Elements and Macro-molecules" (J. Maniloff, J. R. Coleman and M. W. Miller, eds) pp. 275–282. Charles C. Thomas, Springfield, Illinois

Goldwater, L. J. (1971). Mercury in the environment. *Sci. Amer.* **224** (5), 15–21

Graf, F. (1971). Dynamique du calcium dans l'épithelium des caecums posterieurs d'*Orchestia cavimana* Heller (Crustacé Amphiode). Rôle de l'espace intercellulaire. *C. r. Acad. Sci. Paris* **273**, 1828–1831

Guthrie, R. K., Davis, E. M., Cherry, D. S. and Murray, H. E. (1979). Biomagnification of heavy metals by organisms in a marine microcosm. *Bull. Environm. Contam. Toxicol.* **21**, 53–61

Hahne, H. C. H. and Kroontje, W. (1973). Significance of pH and chloride concentration on behaviour of heavy metal pollutants. Mercury (II), cadmium (II), zinc (II) and lead (II). *J. Environ. Qual.* **2**, 444–450

Halcrow, W., MacKay, D. W. and Thornton, I. (1973). Distribution of trace metals and fauna in the Firth of Clyde in relation to the disposal of sewage sludge. *J. mar. biol. Ass. U.K.* **53**, 721–739

Hamdy, M. K. and Prabhu, N. V. (1979). Behaviour of mercury in biosystems. III. Biotransference of mercury through food chains. *Bull. Environm. Contam. Toxicol.* **21**, 170–178

Hamilton, E. I., Watson, P. G., Cleary, J. J. and Clifton, R. J. (1979). The geochemistry of recent sediments of the Bristol Channel-Severn Estuary system. *Mar. Geol.* **31**, 139–182

Henze, M. (1911). Untersuchungen über das Blut der Ascidian. I. Mitteilung. Die Vanadiumverbindung der Blutkörperchen. *Hoppe-Seyler's Z. physiol. Chem.* **72**, 494–501

Hill, C. H. (1976). Mineral interrelationships. *In* "Trace Elements in Human Health and Disease" (A. S. Prasad and D. Oberleas, eds) Vol. II, pp. 281–300. Academic Press, New York, San Francisco and London

Hoar, W. S. (1975). "General and Comparative Physiology". Prentice-Hall, Inc., New Jersey

Hobden, D. J. (1969). Iron metabolism in *Mytilus edulis*. II. Uptake and distribution of radioactive iron. *J. mar. biol. Ass. U.K.* **49**, 661–668

Huggett, R. J., Cross, F. A. and Bender, M. E. (1975). Distribution of copper and zinc in oysters and sediments from three coastal-plain estuaries. *In* "Proceedings of a Symposium on Mineral Cycling in South-eastern Ecosystems, Augusta, Georgia", 1974, pp. 224–238. U.S. Energy Research and Development Administration, Washington, D.C.

Ireland, M. P. (1974). Variations in the zinc, copper, manganese and lead content of *Balanus balanoides* in Cardigan Bay, Wales. *Environ. Pollut.* **4**, 27–35

Janzen, E. G. and Blackburn, B. J. (1969). Detection and identification of short-lived free radicals by electron spin resonance trapping techniques (spin trapping). Photolysis of organolead, -tin, and -mercury compounds. *J. Amer. Chem. Soc.* **91**, 4481–4490

Jenne, E. A. (1968). Controls on Mn, Fe, Co, Ni, Cu and Zn concentration in soils and water: the significant role of hydrous Mn and Fe oxides. *In* "Trace Inorganics in Water" (R. F. Fould, ed.) pp. 337–379. American Chemical Society, Washington, D.C.

Jennings, J. R. and Rainbow, P. S. (1979a). Studies on the uptake of cadmium by the crab *Carcinus maenas* in the laboratory. I. Accumulation from seawater and a food source. *Mar. Biol.* **50**, 131–139

Jennings, J. R. and Rainbow, P. S. (1979b). Accumulation of cadmium by *Artemia salina*. *Mar. Biol.* **51**, 47–53

Jensen, S. and Jernelov, A. (1969). Biological methylation of mercury in aquatic organisms. *Nature* **223**, 753–754

Jernelöv, A. and Lann, H. (1971). Mercury accumulation in food chains. *Oikos* **22**, 403–406

Jernelöv, A., Lann, H., Wennergren, G., Fagerström, T., Asell, P. and Andersson, R. (1972). Analyses of methylmercury concentrations in sediment from the St. Clair system. Unpublished report of the Swedish Water and Air Pollution Research Laboratory, Stockholm (in English)

Jewett, K. L. and Brinckman, F. E. and Bellama, J. M. (1975). Chemical factors influencing metal alkylation in water. *In* "Marine Chemistry" (T. M. Church, ed.) pp. 304–318. American Chemical Society, Washington, D. C.

Kalk, M. (1963). Intracellular sites of activity in the histogenesis of tunicate vanadocytes. *Quart. J. micr. Sci.* **104**, 483–493

Laube, V., Ramamoorthy, S. and Kushner, D. J. (1979). Mobilization and accumulation of sediment bound heavy metals by algae. *Bull. Environm. Contam. Toxicol.* **21**, 763–770

Leach, R. M. Jr. (1976). Metabolism and function of manganese. *In* "Trace Elements in Human Health and Disease" (A. S. Prasad and D. Oberleas, eds) Vol. II, pp. 235–247. Academic Press, New York, San Francisco and London

Leatherland, T. M., Burton, J. D., Culkin, F., McCartney, M. J. and Morris, R. J. (1973). Concentrations of some trace metals in pelagic organisms and of mercury in North-East Atlantic Ocean water. *Deep-Sea Res.* **20**, 679–685

Lindquist, R. R. (1968). Studies on the pathogenesis of hepatolenticular degeneration. 3. Effect of copper on rat-liver lysosomes. *Am. J. Pathol.* **53**, 903–927

Lloyd, R. (1960). The toxicity of zinc sulphate to rainbow trout. *Ann. Appl. Biol.* **48**, 84–94

Lowe, D. M. and Moore, M. N. (1979). The cytochemical distributions of zinc (Zn II) and iron (Fe III) in the common mussel, *Mytilus edulis*, and their relationship with lysosomes. *J. mar. biol. Ass. U.K.* **59**, 851–858

Luoma, S. N. and Bryan, G. W. (1978). Factors controlling the availability of sediment-bound lead to the estuarine bivalve *Scrobicularia plana*. *J. mar. biol. Ass. U.K.* **58**, 793–802

MacInnes, J. R. and Calabrese, A. (1978). Response of embryos of the American oyster, *Crassostrea virginica*, to heavy metals at different temperatures. *In* "Proceedings of the 12th European Marine Biology Symposium" (D. S. McLusky and A. J. Berry, eds) pp. 195–202. Pergamon Press, Oxford and Toronto

MacInnes, J. R. and Thurberg, F. P. (1973). Effects of metals on the behaviour and oxygen consumption of the mud snail. *Mar. Pollut. Bull.* **4**, 185–186

McKenny, C. L. and Neff, J. M. (1979). Individual effects and interactions of salinity, temperature, and zinc on larval development of the grass shrimp *Palaemonetes pugio*. I. Survival and developmental duration through metamorphosis. *Mar. Biol.* **52**, 177–188

Mantoura, R. F. C. (1979). Organo-metallic interactions in natural waters: a review. *In* "Organic Chemistry of Sea Water" (E. K. Duursma and R. Dawson, eds). pp. 179–223. Elsevier, Amsterdam

Martin, J. H. and Flegal, A. R. (1975). High copper concentrations in squid livers in association with elevated levels of silver, cadmium and zinc. *Mar. Biol.* **30**, 51–55

Mason, J. W., Anderson, A. C. and Shariat, M. (1979). Rate of demethylation of methylmercuric chloride by *Enterobacter aerogenes* and *Serratia marcescens*. *Bull. Environm. Contam. Toxicol.* **21**, 262–268

Means, J. J., Crerar, D. A. and Amster, R. L. (1974). Application of gel filtration chromatography to evaluation of organometallic interactions in natural waters. *Limnol. Oceanogr.* **22**, 957–965

Mellinger, P. J. (1972). The comparative metabolism of two mercury compounds as environmental contaminants in the freshwater mussel (*Margaritifera margaritifera*) *Proc. 6th Annu. Conf. on Trace Substances in Environmental Health*, June 13–15, Columbia, Mo., pp. 173–180

Menzel, D. W. (1977). Summary of experimental results: Controlled ecosystem pollution experiment. *Bull. Mar. Sci.* **27**, 142–145

Mertz, W., Toepfer, E. W., Roginski, E. E. and Polansky, M. M. (1974). Present knowledge of the role of chromium. *Federat. Proc.* **33**, 2275–2280

Millero, F. J. (1975). The physical chemistry of estuaries. *In* "Marine Chemistry in the Coastal Environment" (T. M. Church, ed.) pp. 25–55. American Chemical Society, Washington, D.C.

Moore, M. N. (1976). Cytochemical demonstration of latency of lysosomal hydrolases in digestive cells of the common mussel, *Mytilus edulis*, and changes induced by thermal stress. *Cell Tiss. Res.* **175**, 279–287

Moore, M. N. (1977). Lysosomal responses to environmental chemicals in some marine invertebrates. *In* "Pollutant Effects on Marine Organisms" (C. S. Giam, ed.) pp. 143–154. D. C. Heath, Lexington and Toronto

Moore, M. N. (1979). Cellular responses to polycyclic aromatic hydrocarbons and phenobarbital in *Mytilus edulis*. *Mar. Environ. Res.* **2**, 255–263

Moore, M. N. (1980). Cytochemical determination of cellular responses to environmental stressors in marine organisms. *Rapp. P.-v. Reun. Cons. perm. int. Explor. Mer.* **179**, 7–15

Moore, M. N. and Lowe, D. M. (1977). The cytology and cytochemistry of the hemocytes of *Mytilus edulis* and their responses to experimentally injected carbon particles. *J. Invertebr. Pathol.* **29**, 18–30

Moore, M. N. and Stebbing, A. R. D. (1976). The quantitative cytochemical effects of three metal ions on a lysosomal hydrolase of a hydroid. *J. mar. biol. Ass. U.K.* **56**, 995–1005

Moore, M. N., Lowe, D. M. and Fieth, P. E. M. (1978a). Responses of lysosomes in the digestive cells of the common mussel, *Mytilus edulis*, to sex steroids and cortisol. *Cell Tiss. Res.* **188**, 1–9

Moore, M. N., Lowe, D. M. and Fieth, P. E. M. (1978b). Lysosomal responses to experimentally injected anthracene in the digestive cells of *Mytilus edulis*. *Mar. Biol.* **48**, 297–302

Moore, M. N., Lowe, D. M. and Moore, S. L. (1979). Introduction of lysosomal destabilisation in marine bivalve molluscs exposed to air. *Mar. Biol. Letters* **1**, 47–57

Morris, A. W. (1974). Seasonal variation of dissolved metals in inshore waters of the Menai Straits. *Mar. Pollut. Bull.* **5**, 54–59

Morris, A. W. (1978). Chemical processes in estuaries: the importance of pH and its variability. *In* "Environmental Biogeochemistry and Geomicrobiology" (W. E. Krumbein, ed.) Vol. 1, pp. 179–187. Ann Arbor Science, Ann Arbor

Morris, A. W. and Bale, A. J. (1975). The accumulation of cadmium, copper, manganese and zinc by *Fucus vesiculosus* in the Bristol Channel. *Estuar. Coastal. mar. Sci.* **3**, 153–163

Morris, A. W., Mantoura, R. F. C., Bale, A. J. and Howland, R. J. M. (1978). Very low salinity regions of estuaries: important sites for chemical and biological reactions. *Nature* **274**, 678–680

Neff, J. M. (1972). Ultrastructure of the outer epithelium of the mantle in the clam *Mercenaria mercenaria* in relation to calcification of the shell. *Tissue & Cell* **4**, 591–600

Nelson, J. D., Blair, W., Brinckman, F. E., Colwell, R. R. and Iversen, W. P. (1973). Biodegradation of phenylmercuric acetate by mercury-resistant bacteria. *Appl. Microbiol.* **26**, 321–326

Nickless, G., Stenner, R. and Terrille, N. (1972). Distribution of cadmium, lead and zinc in the Bristol Channel. *Mar. Pollut. Bull.* **3**, 188–190

Nielsen, F. H. (1926). Newer trace elements and possible application in man. *In* "Trace Elements in Human Health and Disease" (A. S. Prasad and D. Oberleas, eds) Vol. II, pp. 379–399. Academic Press, New York, San Francisco and London

Nilsson, J. R. (1978). Retention of lead within digestive vacuole in *Tetrahymena*. *Protoplasma* **95**, 163–173

Nissenbaum, A. and Swaine, D. J. (1976). Organic matter-metal interactions in recent sediments: the rôle of humic substances. *Geochim. Cosmichim. Acta* **40**, 809–816

O'Dell, B. L. (1976). Biochemistry and physiology of copper in vertebrates. *In* "Trace Elements in Human Health and Disease" (A. S. Prasad and D. Oberleas, eds) Vol. II, pp. 391–413. Academic Press, New York, San Francisco and London

Owen, G. (1966). Digestion. *In* "Physiology of Mollusca" (K. M. Wilbur and C. M. Yonge, eds) Vol. II, pp. 53–96. Academic Press, New York and London

Ozoh, P. T. E. and Jacobson, C.-O. (1979). Embryotoxicity and hatchability in *Cichlasoma nigrofasciatum* (Guenther) eggs and larvae briefly exposed to low concentrations of zinc and copper ions. *Bull. Environm. Contam. Toxicol.* **21**, 782–786

Pagenkopf, G. K., Russo, R. C. and Thurston, R. V. (1974). Effect of complexation on toxicity of copper to fishes. *J. Fish. Res. Bd Can.* **31**, 462–465

Pentreath, R. J. (1973). The accumulation from water of ^{65}Zn, ^{54}Mn, ^{58}Co and ^{59}Fe by the mussel, *Mytilus edulis*. *J. mar. biol. Ass. U.K.* **53**, 127–143

Peterson, P. J. (1971). Unusual accumulations of elements by plants and animals. *Sci. Prog.* **59**, 505–526

Peterson, P. J., Burton, M. A. S., Gregson, M., Nye, S. M. and Porter, E. K. (1979). Accumulation of tin by mangrove species in West Malaysia. *Sci. Tot. Environ.* **11**, 213–221

Phelps, H. L. (1979). Cadmium sorption in estuarine mudtype sediment and the accumulation of cadmium in the soft-shell clam, *Mya arenaria*. *Estuaries* **2**, 40–44

Phillips, D. J. H. (1976). The common mussel *Mytilus edulis* as an indicator of pollution by zinc, cadmium, lead and copper. I. Effects of environmental variables on uptake of metals. *Mar. Biol.* **38**, 59–69

Phillips, D. J. H. (1977). The use of biological indicator organisms to monitor trace metal pollution in marine and estuarine environments – A review. *Environ. Pollut.* **13**, 281–317

Phillips, D. J. H. (1979). Trace metals in the common mussel, *Mytilus edulis* (L.), and in the alga *Fucus vesiculosus* (L.) from the region of the Sound (Öresund). *Environ. Pollut.* **18**, 31–43

Pickering, Q. H. and Vigor, W. N. (1965). The acute toxicity of zinc to eggs and fry of the fathead minnow. *Prog. Fish Cult.* **27**, 153–157

Plisetskaya, E., Kazakov, V. K., Soltitskaya, L. and Leibson, L. C. (1978). Insulin-producing cells in the gut of freshwater bivalve molluscs *Anodonta cygnea* and *Unio pictorum* and the role of insulin in the regulation of their carbohydrate metabolism. *Gen. Comp. Endocrinol.* **35**, 133–145

Porter, H. (1974). The particulate half-cystine-rich copper protein of new-born liver. Relationship to metallothionein and subcellular localisation in non-mitochondrial particles possibly representing heavy lysosomes. *Bioch. Biophys. Res. Commun.* **56**, 661–668

Pruell, R. J. and Engelhardt, F. R. (1980). Liver cadmium uptake, catalase inhibition, and cadmium thionein production in the killifish (*Fundulus heteroclitus*) induced by experimental cadmium exposure. *Mar. Environ. Res.* **3**, 101–111

Ramamoorthy, S. and Rust, B. R. (1976). Mercury sorption and desorption characteristics of some Ottawa river sediments. *Can. J. Earth Sci.* **13**, 530–536

Rashid, M. A. (1971). The role of humic acids of marine origin and their different molecular weight fractions in complexing di- and trivalent metals. *Soil Sci.* **111**, 298–306

Reimann, B. E. F., Lewin, J. C. and Volcani, B. E. (1966). Studies on the biochemistry and fine structure of silica shell formation in diatoms. II. The structure of the cell wall of *Navicula pelliculosa* (Breb.) Hilse (Chrysophyta). *J. Phycol.* **2**, 74–84

Riisgård, H. U. (1979). Effect of copper on volume regulation in the marine flagellate *Dunaliella marina*. *Mar. Biol.* **50**, 189–193

Riley, J. P. (1975). Analytical chemistry of sea water. *In* "Chemical Oceanography" (J. P. Riley and G. Skirrow, eds) Vol. 3, pp. 193–514. Academic Press, New York and London

Robinson, D., Price, R. G. and Dance, N. (1967). Rat-urine glycosidases and kidney damage. *Biochem. J.* **102**, 533–538

Roesijadi, G. (1979). Influence of copper on the clam *Protothaca staminea*: effects on gills and occurrence of copper-binding proteins. *Biol. Bull.* **158**, 233–247

Romeril, M. G. (1971). The uptake and distribution of ^{65}Zn in oysters. *Mar. Biol.* **9**, 347–354

Ruddell, C. L. and Raims, D. W. (1975). The relationship between zinc, copper and the basophils of two crassostreid oysters, *Crassostrea gigas* and *Crassostrea virginica*. *Comp. Biochem. Physiol.* **51A**, 565–591

Russell, G. and Morris, O. P. (1970). Copper tolerance in the marine fouling alga *Ectocarpus siliculosus*. *Nature* **228**, 288–289

Rucker, R. R. and Amend, D. F. (1969). Absorption and retention of organic mercurials by rainbow trout and chinook and sockeye salmon. *Prog. Fish Cult.* **31**, 197–201

Sabbioni, E. and Marafante, E. (1975). Heavy metals in rat liver cadmium protein. *Environ. Physiol. Biochem.* **5**, 132–141

Schulz-Baldes, M. (1974). Lead uptake from sea water and food, and lead loss in the common mussel *Mytilus edulis. Mar. Biol.* **25**, 177–193

Schulz-Baldes, M. (1978). Lead transport in the common mussel *Mytilus edulis. In* "Proceedings of the 12th European Symposium on Marine Biology" (D. S. McLusky and A. J. Berry, eds) pp. 211–218. Pergamon Press, Oxford and New York

Scrutton, M. C., Utter, M. F. and Mildvan, A. S. (1966). Pyruvate carboxylase. VI. The presence of tightly bound manganese. *J. Biol. Chem.* **241**, 3480–3487

Selye, H. (1950). "The Physiology and Pathology of Exposure to Stress". Acta, Montreal

Shariat, M., Anderson, A. C. and Mason, J. W. (1979). Screening of common bacteria capable of demethylation of methylmercuric chloride. *Bull. Environm. Contam. Toxicol.* **21**, 255–261

Sharp, J. H. (1973). Size classes of organic carbon in sea water. *Limnol. Oceanogr.* **18**, 441–447

Siebers, D. and Ehlers, U. (1979). Heavy metal action on transintegumentary absorption of glycerine in two annelid species. *Mar. Biol.* **50**, 175–179

Sillen, L. G. and Martell, A. E. (1971). Stability constants of metal ion complexes. Supplement No. 1. Special Publication No. 25, The Chemical Society, London

Silverberg, B. A. (1975). Ultrastructural localisation of lead in *Stigeoclonium tenue. Phycologia* **14**, 265–274

Silverman, M. P. and Ehrlich, H. L. (1964). Microbial formation and degradation of minerals. *Adv. Appl. Microbiol.* **6**, 153–206

Simkiss, K. (1976). Intracellular and extracellular routes in biomineralisation. *In* "Calcium in Biological Systems" (J. C. Duncan, ed) pp. 423–444. Symposia of the Society for Experimental Biology, Vol. 30

Simpson, R. D. (1979). Uptake and loss of zinc and lead by mussels (*Mytilus edulis*) and relationships with body weight and reproductive cycle. *Mar. Pollut. Bull.* **10**, 74–78

Smith, A. L., Green, R. H. and Lutz, A. (1975). Uptake of mercury by freshwater clams (Family Unionidae). *J. Fish Res. Bd Can.* **32**, 1297–1303

Smith, R. G. Jr. (1976). Evaluation of combined applications of ultrafiltration and complexation capacity techniques to natural waters. *Anal. Chem.* **48**, 74–76

Smith, R. J., Ignarro, L. J. and Fisher, J. W. (1974). Lysosomal enzyme release: a possible mechanism of action of cobalt as an erythropoetic stimulant. *Proc. Soc. Exp. Biol. Med.* **146**, 781–785

Somers, G. F. (1978). The role of plant residues in the retention of cadmium in ecosystem. *Environ. Pollut.* **17**, 287–295

Sprague, J. B. (1970). Measurement of pollutant toxicity to fish. II. Utilising and applying bioassay results. *Water Res.* **4**, 3–32

Squibb, K. S. and Cousins, R. J. (1974). Control of cadmium-binding protein synthesis in rat liver. *Environ. Physiol. Biochem.* **4**, 24–30

Steger, H. F. (1973). Mechanism of adsorption of trace copper by bentonite. *Clays Clay Miner.* **21**, 429–436

Stenner, R. D. and Nickless, G. (1974). Distribution of some heavy metals in organisms in Hardangerfjord and Skjerstadfjord, Norway. *Water, Air, Soil Pollut.* **3**, 279–291

Stenner, R. D. and Nickless, G. (1975). Heavy metals in organisms of the Atlantic Coast of southwest Spain and Portugal. *Mar. Pollut. Bull.* **6**, 89–92

Sternlieb, I. and Goldfischer, S. (1976). Heavy metals and lysosomes. *In* "Lysosomes in Biology and Pathology" (J. T. Dingle and R. T. Dean, eds) Vol. 5, pp. 185–202. North Holland/American Elsevier, Amsterdam, Oxford and New York

Stumm, W. and Brauner, P. A. (1975). Chemical specification. *In* "Chemical Oceanography (J. P. Riley and G. Skirrow, eds) Vol. 1, pp. 173–239. Academic Press, New York and London

Sugai, S. F. and Healy, M. L. (1978). Voltammetric studies of the organic association of copper and lead in two Canadian inlets. *Mar. Chem.* **6**, 291–308

Sugaya, E. and Onozuka, M. (1978). Intracellular calcium: its release from granules during bursting activity in snail neurons. *Science* **202**, 1195–1197

Sunda, W. G. and Guillard, R. R. (1976). The relationship between cupric ion activity and the toxicity of copper to phytoplankton. *J. Mar. Res.* **34**, 511–529

Sunda, W. G. and Lewis, J. A. M. (1978). Effect of complexation by natural organic ligands on the toxicity of copper to a unicellular alga, *Monochrysis lutheri*. *Limnol. Oceanogr.* **23**, 870–876

Sunda, W. G., Engel, D. W. and Thuotte, R. M. (1978). Effect of chemical speciation on toxicity of cadmium to grass-shrimp, *Palaemonetes pugio*: importance of free cadmium ion. *Environ. Sci. Technol.* **12**, 409–413

Ünlü, M. Y. and Fowler, S. W. (1979). Factors affecting the flux of arsenic through the mussel *Mytilus galloprovincialis*. *Mar. Biol.* **51**, 209–219

Verity, M. A. and Brown, W. J. (1968). Activation and stability of lysosomal acid phosphohydrolase. *Biochim. Biophys. Acta* **151**, 284–287

Waku, Y. and Simimoto, K. (1974). Metamorphosis of midgut epithelial cells in the silkworm (*Bombyx mari*) with special regard to the calcium salt deposits in the cytoplasm. II. Electron microscopy. *Tissue & Cell* **6**, 127–136

Walker, G. (1977). "Copper" granules in the barnacle *Balanus balanoides*. *Mar. Biol.* **39**, 343–349

Walker, G., Rainbow, P. S., Foster, P. and Crisp, D. J. (1975). Barnacles: possible indicators of zinc pollution? *Mar. Biol.* **30**, 57–65

Webb, D. A. (1939). Observations on the blood of certain ascidians, with special reference to the biochemistry of vanadium. *J. Exp. Biol.* **16**, 499–523

Webb, M. (1972). Binding of cadmium ions by rat liver and kidney. *Biochem. Pharmac.* **21**, 2751–2765

Williams, P. M. and Weiss, H. V. (1973). Mercury in the marine environment: concentration in sea water and in a pelagic food chain. *J. Fish. Res. Bd Can.* **30**, 293–295

Winge, D. R., Premakumar, R. and Rajagopalan, K. V. (1975). Metal-induced formation of metallothionein in rat liver. *Arch. Biochem. Biophys.* **170**, 242–252

Wolfe, D. A. (1974). The cycling of zinc in the Newport River estuary, North Carolina. *In* "Pollution and Physiology of Marine Organisms" (F. J. Vernberg and W. B. Vernberg, eds) pp. 79–99. Academic Press, New York, San Francisco and London

Wolfe, D. A. and Coburn, C. B. (1970). Influence of salinity and temperature on the accumulation of cesium[137] by an estuarine clam under laboratory conditions. *Health Physics* **18**, 449–465

Wong, M. H., Chan, K. Y., Kwan, S. H. and Mo, C. F. (1979). Metal contents of the two marine algae found on iron ore tailings. *Mar. Pollut. Bull.* **10**, 56–59

Wrench, J., Fowler, S. W. and Ünlü, M. Y. (1979). Arsenic metabolism in a marine food chain. *Mar. Pollut. Bull.* **10**, 18–20

Young, M. L. (1975). The transfer of ^{65}Zn and ^{59}Fe along a *Fucus serratus* (L.) → *Littorina obtusata* (L.) food chain. *J. mar. biol. Ass. U.K.* **55**, 583–610

Young, M. L. (1977). The roles of food and direct uptake from water in the accumulation of zinc and iron in the tissues of the dogwhelk, *Nucella lapillus* (L.). *J. Exp. Mar. Biol. Ecol.* **30**, 315–325

Part 3

Simulation and Experimental Studies of Marine Ecosystems

These chapters are devoted to studies of marine ecosystems by the use of numerical simulation, by the use of microcosms, and by the manipulation of natural ecosystems.

19 Theory and Observation: Benthic Predator-Prey Relationships

BRIAN L. BAYNE

1 Introduction

To consider predation in the analysis of any ecosystem is to treat with a wide range of biological study, from laboratory experiments on a single predator feeding within a single population of prey, to experiments and observations in nature of the structure and dynamics of communities, where the effects of predation are expressed in the temporal and spatial complexity typical of natural systems. The task of achieving some synthesis of this varied information is eased, however, by a comprehensive body of theory on such topics as optimal diet, foraging behaviour and the results of interaction between predator and prey populations.

Some of the available theory is based on the concept of evolutionary optimisation (Maynard Smith, 1978), with its attendant mathematical formulations, and has been applied to some of the predator-prey interactions that exist amongst benthic macrofauna (Emlen, 1973). On the other hand much of the theory on population interaction derives from arthropod predator-prey (and parasitoid-prey) systems (Hassell, 1976), but with properties of more general application. Studies of some benthic communities have, in turn, provided a rich source of empirical data for understanding some effects of predation in nature (Paine, 1974; Connell, 1975). This chapter attempts a selective view of some of this material, with an emphasis on the species and the communities of the marine benthos. Predation is here interpreted in the strict sense of an animal feeding on other living animals, although some of the theory considered may apply to the feeding process in general.

2 The functional response

Solomon (1949) recognised two aspects of predation; the functional response defines the relationship between the number of prey consumed per predator at different prey densities, whereas the numerical response defines the relationship between predator numbers and prey density (Holling, 1959; Hassell *et al.*, 1976).

Holling (1959) considered the following expression:

$$N_a = a'T_sN_t \tag{1}$$

where N_a is the number of prey eaten per predator, N_t is the prey density, T_s is the time available for search for prey, and a' describes the rate of successful attack by the predator on the prey. However, a finite amount of time must be spent 'handling' the prey, so:

$$T_s = T - T_hN_a \tag{2}$$

where T is the total time the prey are exposed to the predator, and T_h is the handling time. Substituting for T_s in eqn (1) gives:

$$N_a = \frac{a'N_tT}{1 + a'T_hN_t} \tag{3}$$

This is Holling's (1959) 'disc equation', describing the so-called type II functional response (Hassell, 1976) in which the number of prey eaten per predator increases with increasing prey density, but at a decreasing rate as it approaches a maximum.

Royama (1971) discusses two features of the disc equation as applied to experimental data; predator density should be kept to unity since both a' and T_h change with predator density, and depletion of the prey should not occur, since a' and T_h are also decreasing functions of prey density. The disc equation therefore gives only an instantaneous measure of the number of prey attacked. In applying this analysis to situations where exploitation of the prey is not negligible, Rogers (1972) derived the 'random predator equation':

$$N_e = N_t[1 - \exp(-a'(T - T_hN_a))] \tag{4}$$

where N_e is the number of prey eaten. For one predator only, $N_e = N_a$; where predator density (p) is greater than one,

$$N_e = N_t[1 - \exp(-a'p(T - T_hN_a))] \tag{5}$$

Rogers (1972) describes how the parameters a' and T_h in eqn (5) may be derived from experimental data.

In many cases the instantaneous form of the disc equation provides a reasonable fit to data. In Fig. 1a some results from Doi's (1976) experiments with *Astropecten latespinosus* feeding on the gastropod *Ubonium monoliferum* are plotted as fitted disc equations for two prey sizes, 6.0 and 11.0 mm shell diameter.

Royama (1970) modified the disc equation to consider the biomass rather

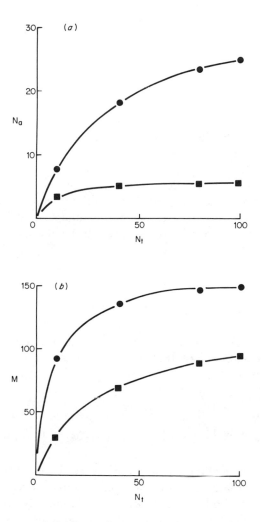

FIG. 1 (*a*) The rate of feeding of *Astropectan latespinosus* on the gastropod *Umbonium monoliferum*. Data from Doi (1976), fitted according to Holling's (1959) disc equation, where N_a is the number of prey eaten per predator and N_t is the initial density of prey. ●, prey shell diam. 6.0 mm; $a' = 1.29$, $T_h = 0.024$. ■, prey shell diam. 11.0 mm, $a' = 0.94$, $T_h = 0.144$. (*b*) As for (*a*), but with N_a converted to the weight of prey eaten (M). ●, weight of individual prey 26.2 mg; ■, weight of individual prey 3.8 mg

than the numbers of prey eaten. If W is the average weight of the prey, then the total biomass (M) of N_a individuals is $W.N_a$, and $N_a = M/W$. Substituting in eqn (3) and solving for M:

$$M = \frac{W_a N_t T}{1 + a' T_h N_t} \tag{6}$$

In Fig. 1b Doi's (1976) data have been recalculated for M to illustrate the effects of altered weight of prey and concomitant change in handling time. The smaller prey individuals must occur at a considerably higher density than the larger in order for the biomass gain to the predator (M) to exceed the value for the larger prey.

Lawton *et al.* (1974) extended the Holling disc equation to a two-prey situation by assuming that the probability of the predator encountering and capturing either prey per unit search time is independent for the two prey types and equal to the probability of capture if the alternative prey were not present. Thus,

$$Na_1 = a'_1 N_{t_1} T / (1 + a'_1 T_{h_1} N_{t_1} + a'_2 T_{h_2} N_{t_2})$$
$$Na_2 = a'_2 N_{t_2} T / (1 + a'_2 T_{h_2} N_{t_2} + a'_1 T_{h_1} N_{t_1}) \tag{7}$$

Lawton *et al.* (1974) integrate these equations, in a manner comparable to Rogers' (1972) treatment of the disc equation, in order to include the reduction of prey density:

$$N_{a_1} = N_{t_1}[1 - \exp\{-a'_1(T - T_{h_1} N_{a_1} - T_{h_2} N_{a_2})\}]$$
$$N_{a_2} = N_{t_2}[1 - \exp\{-a'_2(T - T_{h_2} N_{a_2} - T_{h_1} N_{a_1})\}] \tag{8}$$

These equations allow the predators preference between two prey types to be analysed in situations where prey exploitation occurs. Cock (1978) reviews various methods used to assess preference between prey types. Where prey exploitation is not a factor, the index proposed by Murdoch (1969; see Murdoch, Avery and Smythe, 1975) is convenient:

$$N_1/N_2 = c(H_1/H_2) \tag{9}$$

where N_1 and N_2 are the numbers of prey items 1 and 2 eaten over a fixed period of time, H_1 and H_2 are the densities of 1 and 2, and c is a constant measuring the preference. If c = 1, the predator shows no preference; if c > 1 preference is for prey 1, if c < 1 preference is for prey 2 (see also Murdoch and Oaten, 1975).

However, where prey exploitation is not negligible, Cock (1978) uses eqns (8) to derive a term for preference:

$$\frac{N_{a_1}}{N_{a_2}} = \frac{1 - \exp(-a_1' T_s)N_{t_1}}{1 - \exp(-a_2' T_s)N_{t_2}} \tag{10}$$

where T_s, the time available for search, is given by $(T - T_{h_1}N_{a_1} - T_{h_2}N_{a_2})$. Prey preference is seen to be a complex function of the numbers of each prey type, the handling times and the total time available for search. However, eqns (8) can be used directly to predict the ratio of each prey type eaten (and hence preference) when both prey are present, in a manner discussed by Cock (1978).

If preference for a particular prey is not a simple function of its relative abundance in the environment, but shifts above or below proportionality as the prey increases or decreases, respectively, the predator is said to have 'switched' its preference (Murdoch, 1969). This can be demonstrated from data on the relative abundances of two prey types as a change in the value of c in eqn (9). If switching is to be considered in the context of absolute abundances of alternate prey types, however, eqns (8) can be applied (Lawton et al., 1974). If these equations provide a good fit to data relating the occurrence of a prey in the diet to its occurrence in the environment, the predator has not switched; if the fit is poor, however, switching has possibly occurred.

Only the type II functional response has been discussed so far, but two other forms of response were described by Holling (1959). In type I, the number of prey eaten per predator increases linearly with an increase in prey density; although this response may apply to filter-feeders, it is not feasible for most predators since it requires that the time available for search is constant at all prey densities. In type III, the feeding response is sigmoidal, in which T_s, or perhaps a' (eqn 1), are themselves functions of prey density (Hassell, et al., 1976). Real (1977) considered both the type II and the type III functional responses by analogy with enzyme kinetics, and derived an equation which accounts for both responses, depending on the value of a term for the number of encounters a predator must have with the prey before becoming maximally efficient at utilising that item as food. Real's (1977) treatment of the functional response considers the various ways in which, by learning and other behavioural means, a predator may shift the functional response from a type II to a type III (see also Krebs, 1973; Murdoch and Oaten, 1975). This can be very important in predator-prey population models (May, 1975) as briefly discussed later.

3 The optimal diet

Theoretical work on the foraging behaviour of the predator when faced with more than one type of prey assumes that the predator optimises the food intake by maximising the term E/T, where E is the net gain from the diet during a foraging period of duration T (Emlen, 1966, 1968; Rapport, 1971; Schoener, 1971; Pulliam, 1974). In most studies E is considered in terms of energy. Models to elucidate the rules underlying the optimal choice of prey have been developed independently by many authors (see review by Pyke *et al.*, 1977).

Estabrook and Dunham (1975) make four simplifying assumptions: (*i*) The kinds of prey are distributed independently of one another and individual prey items are encountered singly. (*ii*) The times elapsed between successive encounters by the predator of prey individuals are all equal. (*iii*) The handling time includes pursuit, capture and eating, and also the time taken to re-initiate search. (*iv*) The time taken in search competes with time spent on other essential activities, or is a time of increased risk for the predator, so that increased search or foraging time is a penalty. The model is then formulated that assesses the predators net gain per unit time as a function of the value of the prey item ($V_{(i)}$), its probability distribution ($P_{(i)}$), the abundance of prey (N) and the foraging strategy (Q').

Estabrook and Dunham (1975) discuss the relative importance of N, $V_{(i)}$ and $P_{(i)}$ in setting optimal diet. The most important parameter is the absolute abundance of food, N. For a fixed total abundance of prey, the relative value ($V_{(i)}$) of the potential prey types is more important than their relative abundance in determining the optimal diet, except when the more valuable prey types are rare. The authors also suggest that of these three parameters, $P_{(i)}$ is the most difficult for the predator to evaluate and is also likely to change most rapidly in nature, so its minimal effect on determining the optimal diet is intuitively satisfying.

Hughes (1979) describes a general form of optimal diet model for predators that hunt by chemical or tactile cues. The rate of encounter of a randomly searching predator with prey type i is λ_i. The time the predator is involved with the prey (I_i) is divided into a recognition time (R_i), during which the prey item is evaluated, and a handling time (H_i) which is the time elapsed between the decision to attack and the resumption of search after eating the prey. When a prey is eaten, $I_i = R_i + H_i$, but when it is not eaten $I_i = R_i$. The prey item i yields E_i energy units to the predator, and its value is E_i/I_i. The predator is assumed to be able to rank all potential prey types in decreasing order of energy yield per unit handling time, or E_i/H_i. The energy gain per unit foraging time is:

$$\frac{E}{T} = \frac{\Sigma \lambda_i E_i}{1 + \Sigma \lambda_i I_i} \tag{11}$$

Hughes (1979) discusses a form of this model which includes a finite recognition time, $R_i > 0$. Consider that a predator has a choice of two prey types, where type 1 is more valuable than type 2. Energy yield is optimised by specialising on type 1, rather than by generalising on both types, when

$$\frac{\lambda_1 E_1}{1 + \lambda_1(H_1 + R_1) + \lambda_2 R_2} > \frac{\lambda_1 E_1 + \lambda_2 E_2}{1 + \lambda_1(H_1 + R_1) + \lambda_2(H_2 + R_2)}$$

or

$$\frac{1}{\lambda_1} < \frac{E_1}{E_2} H_2 - (H_1 + R_1) - \frac{\lambda_2}{\lambda_1} R_2 \qquad (12)$$

As the value of type 2 prey decreases relative to type 1, the rarer type 1 must be before it is worth the predator also feeding on type 2. However, unless R_2 is very small relative to H_2, even when type 1 is not scarce the predator should also take type 2 prey if the encounter ratio of type 2 to type 1 is high, in order to compensate for the time spent recognising type 2 items.

A special case of this model considers the situation where recognition time $R_i = 0$. The inequality (12) now becomes:

$$\frac{1}{\lambda_1} < \frac{E_1}{E_2} \cdot H_2 - H_1 \qquad (13)$$

Under these circumstances the conditions for the optimal diet no longer involve λ_2, and prey type 2 should not be eaten, even at high ratios of λ_2 to λ_1.

These conclusions were graphically demonstrated by Krebs (1978) (Fig. 2). The predators net energy intake is E/T; the value of three prey items x, y and z, calculated as E/H are plotted. With a very short recognition time, prey y or z, even if very common, cannot increase the slope of E/T and are not included in the optimal diet; prey item x would increase E/T and should be eaten. The effect of adding a finite recognition time (line c, Fig. 2) is to reduce the slope of E/T and to include item y in the optimal diet set.

The inequality (12) demonstrates that the optimal diet of the predator must always include the type 1 prey, and should also include type 2 when the encounter rate for this prey is high. Since $E_1/H_1 > E_2/H_2$, it can never pay for the predator to specialise on type 2 prey. Under certain circumstances, however, the value of prey types can be transposed, leading to a switch from one preferred prey type to another; those circumstances include the predator learning to recognise a prey type more quickly, to attack it more effectively or to handle it more efficiently. Hughes (1979) considered the effects of learning by assuming that the magnitude of I_i at any given time is the resultant of learning and forgetting, which are respectively increasing and decreasing functions of the encounter rate λ_i. Hughes concludes that

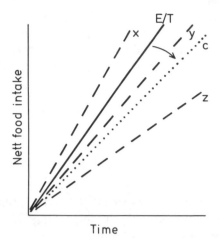

FIG. 2   The net food intake of a predator per unit time (E/T), with the food value (E/H) of three prey items, x, y, z. Item y is not included in the optimal diet set unless E/T is reduced to c, for example by adding a finite recognition time. From Krebs (1978)

any model of the optimal diet where handling time is a decreasing function of encounter rate will demonstrate ranges of the encounter rate ratio λ_1/λ_2 where switching should occur (see also Murdoch and Oaten, 1975). Predators that sense their prey at a distance (by visual or olfactory cues) and for which handling time is significantly long might be expected to switch more readily than predators hunting by a tactile sense, or those for which handling time is short (Cornell, 1976).

4   Some experiments

Elner and Hughes (1978) experimented with the shore crab *Carcinus maenus* feeding on mussels, *Mytilus edulis*, in order to test predictions of the optimal diet theory, in particular the influence of a finite recognition time (crabs manipulated the mussels for 1–2 s before accepting or rejecting them). With unlimited prey available, crabs chose mussels close to the optimum prey value ($=\max. E/T_h$; Fig. 3a). However, suboptimal mussels were incorporated into the crabs diet in proportion to their relative abundance, even when optimally sized mussels were present in excess. Elner and Hughes (1978) showed that the optimally sized mussels were always attacked when encountered, whereas suboptimal mussels were rejected at the first encounter following an encounter with an optimal mussel, but were finally taken by the crabs after subsequent encounters. This is in accordance with the general form of the optimal diet model (Hughes, 1979).

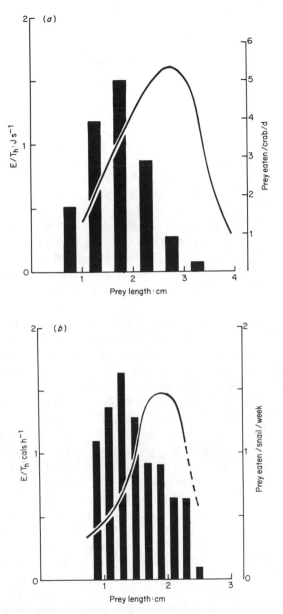

Fig. 3 (a) Histogram of numbers of mussels (*Mytilus edulis*) eaten per day by *Carcinus maenas* of 6.0–6.5 cm carapace width, together with the curve of energy gain per unit prey handling time. Data from Elner and Hughes (1978). (b) Histogram of numbers of mussels (*M. edulis*) eaten per week by *Thais lapillus* of 2.0 cm shell height, together with the curve of energy gain per unit prey handling time

When prosobranch molluscs of the genera *Thais* or *Acanthina* are preying on other molluscs or on barnacles (Murdoch, 1969; J. L. Menge, 1974; pers. obs.) a number of encounters with potential prey often occurs prior to an attack on a chosen individual. In unpublished experiments (Bayne and Scullard) with *Thais lapillus* feeding on *Mytilus edulis*, 1.82 ± 0.92 (SD) encounters were made before feeding, and the mean shell length of rejected prey was 1.90 ± 0.22 cm as compared with 1.72 ± 0.39 cm for the mussels that were attacked and eaten (total of 70 rejection and feeding sequences for 15 snails over 22 days). The time spent in contact with subsequently rejected prey (R_i) was 2.2 ± 1.07 h; handling time ($R_i + H_i$) was 34.7 ± 8.6 h. The recognition time was therefore 6% of handling time, as compared with <0.5% for *Carcinus* feeding on the same prey (Elner and Hughes, 1978). When presented with an abundance of prey of varied size, *Thais* fed mainly on smaller individuals than those of optimum value (Fig. 3*b*).

Hughes and Elner (1978) observed *Carcinus* feeding on *Thais*. With an abundance of prey in the laboratory the crabs did not specialise on the optimal size as predicted by the optimal diet theory; all the encountered prey were attacked but rejected if unbroken after 0.25 to 2.75 min. The handling time for medium sized prey was between 1 and 20 min. Hughes and Elner (1978) also showed that the probability of misidentifying the prey value of an individual *Thais* was high. In these circumstances the predator acts as a number maximiser rather than energy maximiser (Griffiths, 1975).

Landenberger (1968) examined selectivity between preferred and non-preferred (optimal and suboptimal?) prey for the starfish (*Pisaster*) feeding on mussels (the preferred item) or snails (*Tegula funebralis*). When mussels predominated over snails 1:3, selectivity for mussels (% in the diet) was consistently high, irrespective of their absolute density; when snails predominated, however, selectivity for mussels was reduced and density dependent. In these experiments the predator located its preferred prey and fed mostly upon them, whilst taking non-preferred snails in proportion to their encounter frequency. If only non-preferred prey types were available (*Tegula* and *Acanthina*) they were taken strictly in proportion to frequency of abundance. A similar result is postulated for *Carcinus* feeding on mussels (preferred) and *Thais* in the natural habitat (Hughes and Elner, 1978).

In Landenbergers' experiments (1968) *Pisaster* showed strong preference for mussels over other species and there was no evidence of switching from one prey species to another when their relative abundance was altered. Murdoch (1969) distinguished between a 'strong preference' case (*Thais emarginata* feeding on *Mytilus edulis* and *M. californianus*) and a 'weak preference' case (*Acanthina spirata* feeding on *M. edulis* and *Balanus glandula*). In the former case Murdoch could not induce switching by varying the proportions of prey species offered; preference for *M. edulis* remained constant. However, in the weak preference case switching did occur when the predator was trained to a particular prey.

Murdoch and Oaten (1975) reviewed these and other experiments and concluded that predators that show weak average preferences when the densities of alternative prey are similar, but whose preferences vary greatly among individuals, switch when presented with unequal prey ratios. This situation is possibly widespread amongst predators, based, at least in part, on variable rejection rates and a capacity to evaluate varying reward rates in different parts of the habitat. Indeed, patchiness in the distribution of prey probably increases the likelihood of switching by effecting the 'training' of the predator to a particular prey.

Wood (1968) considered one process that could lead to training in some marine invertebrate predators viz. 'ingestive conditioning', which can be defined as a modification of the predators chemotactic response to prey induced by maintenance upon single species diets. Wood demonstrated that the carnivorous gastropod *Urosalpinx cinerea* responded preferentially to water carrying effluent from the barnacle *Balanus*, when given a choice with effluent from the oyster *Crassostrea virginica*, except when the predator had previously been maintained on a diet of *Crassostrea*. Ingestive conditioning may lead to an improvement in the efficiency of the predators attack on a prey item, by concentrating the feeding process on a single species, and so reducing 'involvement time', I (see Lawton et al., 1974).

5 Optimal foraging

The optimal diet (and related considerations of prey choice) is one aspect only of optimal foraging strategy as discussed by Schoener (1971) and Pyke et al. (1977). Other aspects include the effects of patchiness in the distribution of prey, and the currency in which the optimal diet is to be considered.

In the natural environment predators normally forage for food which is patchy in its distribution, with areas of unproductive foraging space between areas of high prey density; patchiness may also occur over time. Royama (1970) supposed that different prey types occupied different patches, so that the predator has to allocate its foraging time between patches. For prey of similar body size the profitability (P) is measured as N_a/T (see eqn (3)) and increases with increased prey density, N_t, but at a decreasing rate towards the asymptote. If P_1 (the profitability in patch 1) is greater than P_2, clearly the predator should spend more time in patch 1. However, as prey densities increase, the difference between profitabilities for the same ratio of densities decreases; it is the profitability, not the density of prey that is important. If the profitability of a patch is high, the predators might be expected to aggregate within the patch (Hassel and May, 1974). If the foraging of the predator then depletes the profitability of the patch, optimal foraging demands that the predator leaves the patch as its food intake declines below its maximised value (Charnov, 1976).

Charnov (1976) and Parker and Stuart (1976) describe the marginal value model to identify some of the rules governing optimal foraging in a patchy environment (Fig. 4). In the graphical solution, cumulative net food gain is plotted against time spent in the patch (G_1); the time spent in travelling to the patch (t) is included. The average intake of food per unit time for the habitat as a whole is represented by the line AB. The optimal time to be spent within the patch (T_{opt}) is then given by the tangent of AB with the cumulative feeding curve, since, with any further time spent within the patch the rate of cumulative net food gain would fall below the value to be expected from the habitat as a whole. In a habitat comprising patches of variable profitability, the optimal foraging strategy is to remain in any one patch only until the rate of food intake falls to the average value for the habitat (the marginal value or mean E/T) (for further discussion of Fig. 4, see p. 591). These ideas have not been tested empirically for macrobenthic predators, but see Krebs *et al.* (1974; birds) and Hubbard and Cook (1978; parasitoid wasps).

Implicit in many theoretical considerations of foraging is the concept of random encounter between predator and prey. However, as Krebs (1978) points out, the assumption of random encounters does not necessarily imply

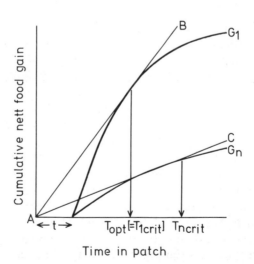

FIG. 4 A graphical representation of the marginal value model, from Parker and Stuart (1976). Discussion in text

random search; if the prey are distributed at random, non-random search could lead to random encounter. The sensory cues by which carnivorous crustacea (Mackie, 1973; Grant and Mackie, 1974; McLeese, 1974; Pearson and Olla, 1977; Pearson et al., 1979) and prosobranchs (Wood, 1968; Carriker and Van Zandt, 1972; Pratt, 1974, 1977) detect and hunt their prey have received much attenton. Pratt (1974) distinguished between the stimuli leading to location of the prey by the oyster drill *Urosalpinx cinerea* and the stimulus to initiate drilling of the prey. *Urosalpinx* is attracted to the effluents of preferred prey species, but the stimulus to attack includes both a tactile and a chemical component. Chemical attraction, coupled with shared ecological requirements and with the results of ingestive conditioning (Wood, 1968) may lead to the concentration of predatory gastropods in a particular area and to the confinement of search behaviour in patches of prey. The means by which such predators may be able to 'sample' the habitat and to be aware of the average net rate of food intake is much less clear although different degrees of satiation or hunger (see later) may play a part.

Hassell and Southward (1978) provide a convenient summary of the factors to be considered in the description of a predator's allocation of time between habitats and between patches of prey within habitats.

[1]
$$T = \sum_{j=1}^{n} T_{Hj} + t_m$$

where $T =$ potential foraging time, $n =$ number of habitats visited, $T_{Hj} =$ time spent in habitat j, and $t_m =$ total migration time between all habitats visited.

[2]
$$T_{Hj} = \sum_{i=1}^{x} T_{Pi} + t_p + t_r$$

where $x =$ number of patches visited, $T_{Pi} =$ time spent in patch i, $t_p =$ total transit time between patches and $t_r =$ total 'resting time' (due to satiation, or to endogenous feeding rhythmicity) spent within the habitat but outside the patches.

[3]
$$T_{Pi} = t_s + t_h Na + t_r$$

where $t_s =$ time spent searching within a patch, $t_h =$ handling time for each food item eaten, $Na =$ number of food items within the patch and $t_r =$ 'resting time' (caused possibly by interference between predators) within the patch (Hassell and Southward, 1978, p. 91).

The currency in which the optimal diet is measured is often energy, for good reasons (Pyke et al., 1977). However, other nutritional factors may be of importance also, and the diet which optimises fitness may not be the diet that maximises the rate of energy intake (Pulliam, 1974). In a later paper, Pulliam (1975) considered the effects of nutritional constraints on the optimal diet and concluded that partial preferences (i.e. prey items included in the diet on some occasions but not on others) may result. A graphical model relating optimal foraging to multiple food resources is offered by Covich (1976) (Fig. 5). The axes represent the quantities of two resources (=prey items), X and Y, which provide the predator with different concentrations of, say, energy and an essential nutrient. The combinations of X and Y that provide identical amounts of energy and of nutrient are indicated by lines Y3–X1 and Y1–X3, respectively. The combinations of the prey X and Y that can be acquired by the predator during the available foraging time are indicated as dashed lines. At the foraging isocline Y2–X2 the optimal combination of prey items (Y0 and X0) is indicated by the intersection E. If both prey were more abundant and the foraging isocline Y4–X4 applied, any combination of prey items along this line would be optimal. A foraging isocline below Y2–X2 would be inadequate for providing sufficient energy and nutrient from these two prey items only.

When discussing optimal foraging behaviour, it is important to consider the time-scale of the response and the need for the predator to 'sample' its food resource (Pyke *et al.*, 1977; Maynard Smith, 1978). In meeting the need to gain experience of its environment, including the abundance of the

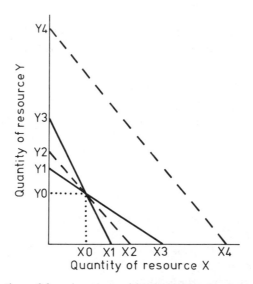

FIG. 5 Optimisation of foraging time with alternative food resources (X and Y) which provide various concentrations of calories and an essential nutrient, from Covich (1976). Discussion in text

different prey items and their ranking in terms of gain, the predator may need to indulge in suboptimal behaviour in the short term. Field studies of optimal foraging on the shore (see below) demonstrate the complexities that arise from the predators need to sample the prey populations, particularly in the context of complex predator-avoidance adaptations by the prey.

6 The effects of hunger

The internal nutritional state of the predator is likely to be a major controlling factor in its feeding behaviour. Hunger provides some measure of nutritional state but must be given an operational definition for use in the analysis of foraging. Holling (1966) did so by equating hunger with the weight of food required to satiate the predator since this weight is a measure of the emptiness of the digestive tract. Holling derived the expression

$$H = HK(1 - e^{-\alpha F})$$ (14)

where H is the hunger level, HK is the maximum amount of food that the gut can hold, F is the time of food deprivation and α is a constant representing the rate of disappearance of food from the gut. In a detailed analysis of the relationship between hunger and various other components of predation, Holling (1966) showed that if hunger was less than a hunger threshold (HT) the predator (a mantid preying upon houseflies) did not attack; the duration of the digestive pause after feeding ($H < HT$), and of the search, pursuit and eating times, could all be expressed in terms of hunger level. A summary and discussion of Holling's comprehensive treatment of attack is given by Watt (1968).

Harris (1974) expanded an equation similar to the disc equation to describe the situation of a predator exposed to any number of prey types; in this model the hunger threshold is considered to be higher for the less preferred prey and the predator returns to a hunger level after feeding that depends both upon the type of prey that has just been consumed and upon the level of hunger prior to this meal. In exploring the behaviour of this model at various prey densities of more than one prey type, and with prey preference included, Harris (1974) demonstrates switching (in the sense of Murdoch, 1969) and points to situations where, at low prey densities, the predator may be expected to attack less preferred, but more easily captured, prey at a higher rate than the normally more preferred type.

In Holling's (1966) operational definition of hunger, the amount of food in the gut is a measure of hunger level (at least over the short term; Beukema, 1968). When the feeding rate of the predator is at equilibrium, the hunger level may provide a measure of E/T, needed in the analysis of optimal foraging (Charnov, 1976). Under some circumstances, therefore, increased hunger level may increase the range of acceptable prey items in the optimal diet (Schoener, 1971). As the encounter rate with preferred prey increases, and the predator becomes satiated, it may be expected to become more selective in its feeding. However, as Pulliam (1974, 1975) points out, these considerations all derive from treating the optimal diet in terms of energy; a partially satiated predator may switch from a diet that maximises energy intake to one with a higher intake of a particular nutrient. Nevertheless, in most discussions of optimal foraging a high energy intake is assumed also to result in a high intake of other essential nutrients.

7 Energy costs

In any full assessment of energy yield to a predator, consideration must be given to the energy costs associated with search, pursuit and feeding, as well as to the energy gains from the meal (Fig. 6). The basic energy equation (Winberg, 1956) may be stated as:

$$C = F + U + \Delta B + R \tag{15}$$

where C is the energy content of the food consumed, F and U are the energy values of the faeces and excretory products, respectively, and ΔB is the change in the energy value of the body, including growth and the production of gametes. R is the total energy of metabolism and is often subdivided into R_s (energy equivalent of standard metabolism), R_a (active metabolism) and R_d (the energy costs of digestion and disposition of food, including the specific dynamic action) (Elliot, 1976).

It follows that

$$\Delta B = C - (F + U + R) \tag{16}$$

where ΔB represents the value to be optimised as the result of the foraging strategy.

The term $\Delta B/C$ represents the gross growth efficiency, K_1; when K_1 is maximal, energy intake C is at an optimum identifying the ration which produces the greatest energy gain for the least energy intake. An analysis of this kind (e.g. the detailed studies by Elliot, 1975, 1976, on trout) therefore allows a precise statement of the optimal diet under a particular set of environmental conditions. Elliot (1976) uses such data to establish a relationship between efficiency, ration size, temperature and body size of the predator.

In such studies the metabolic costs may be estimated indirectly from food intake and growth measurements (Paloheimo and Dickie, 1966), or directly by measurement of the rates of oxygen consumption (Winberg, 1956). The work of Glass (1971; bass, *Micropterus* feeding on guppies, *Lebistes*) illustrates the latter approach. Glass (1968) fitted an equation of the type

$$Y_i = a + be^{cX_i} \qquad (17)$$

to measurements of the rate of oxygen consumption (Y_i) over a period of food deprivation (X_i); a, b and c, are fitted parameters. The asymptotic term a represents the minimal value for metabolic expenditure; the parameter b is related to the maximum gut capacity, whereas c is a function, amongst other things, of the amount of food in the gut at a particular time. Equation (17) therefore expresses a metabolic index of hunger level and provides a useful integration of some of the metabolic costs associated with feeding, to which must be added values for active metabolism (R_a) (Glass, 1971).

Norberg (1977) derives two expressions, for the total number of prey that a predator must ingest per day (N_p) and for the time (T_p) to be spent in pursuit, capture and eating per day:

$$N_p = \frac{24E_B N}{N(e - \varepsilon_p) - \varepsilon_s K} \qquad (18)$$

$$T_p = \frac{24E_B N t_p}{N(e - \varepsilon_p) - \varepsilon_s K} \qquad (19)$$

where E_B is the energy cost per hour for resting metabolism, N is the density of prey, t_p is the pursuit and capture and eating time for one prey, e is the

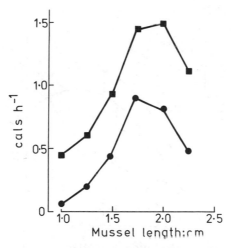

FIG. 6　Energy gain per unit handling time by *Thais lapillus* feeding on *Mytilus edulis*. ■, gross energy gain with no account taken of energy losses during feeding; ●, net energy gain incorporating energy losses due to metabolism during feeding. Data from Bayne and Scullard (1978a,b)

net energy assimilable from one average prey, ε_p is the energy cost for pursuit and capture and eating of one prey, ε_s is the energy cost per hour of search (equivalent to the costs of active metabolism) and K is a constant. Norberg (1977) uses this model to explore the relationships between food intake and costs of search in relation to prey densities, for different methods of foraging. Although derived with birds and terrestrial mammals in mind, the model is applicable also to some invertebrate predators.

8 Mortality risks to the predator, and interference between predators

In most cases, the predator will itself be vulnerable to predation and other potential mortality factors whilst foraging. The optimal behaviour must therefore balance energy gain against the risks of predation, and these risks will apply both in space and time. Covich (1976) related risk (R) to the probability of the forager being attacked at a distance (D) from its refuge. The yield to the forager (Y_1) and the risk are related to distance travelled (D) in a graphical model (Fig. 7). Covich (1976) discusses the slope of the yield and risk curves used in the figure. The optimal distance for the predator to travel in foraging minimises risk whilst it maximises yield and is given by the maximum difference (D_1) between the two curves. If a second accessory resource is considered (Y_2, which does not substitute for Y_1), optimal foraging distance will increase to D_2 to maximise the total yield at minimum risk.

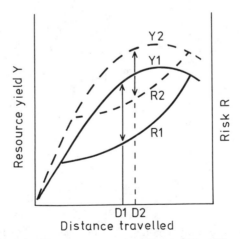

FIG. 7 The effects of increased resource abundances ($Y1$, $Y2$) and/or increased risk ($R1$, $R2$) on the expected optimal distance of travel from a refuge. From Covich, 1976

A related problem concerns mutual interference between individual predators when foraging within the same patch of prey. This might take the form of reduced time available for search following encounters between predators, with resulting decline in searching efficiency (Royama, 1971; Beddington, 1975). Hassell and Varley (1969) discuss a simple version of interference where the term for search (or attack) efficiency in the functional response, a', is related to predator density (P_t) as

$$a' = QP_t^{-m} \tag{20}$$

where m is the mutual interference constant and is the slope (Q the intercept) in the log–log regression of a' and P_t.

In an ecological context the importance of interference probably depends on the extent to which the predators are likely to aggregate in areas of high prey density. In these circumstances interference may condition the dispersal of predators from an area where the prey are being heavily exploited; the dispersing predators then have a greater chance of locating other, less exploited, areas of prey (Hassell, 1976). The optimum number of predators in a particular area of prey should then result from a balance between aggregation and interference. Edwards and Huebner (1977) concluded that interactions amongst snails (*Polinices*) whilst feeding on the bivalve *Mya* reduced growth efficiency through vigorous avoidance reactions; unspecific avoidance reactions may be common in macrobenthic predators.

Parker and Stuart (1976) have used the marginal value theorem to assess the optimal strategy for dispersion (or emigration) of predators sharing a common resource (Fig. 4). The cumulative net food gain G_1 and optimal time in the patch T_{opt} for a single predator are as discussed previously. The cumulative gain per individual for n predators is, of course, less (G_n). Parker and Stuart (1976) deduce that all predators should initially remain in the resource patch, but as time increases, the value to any one predator of leaving becomes greater than the value of staying (T_{crit}, which is equivalent to the optimal time (T_{opt}) in the non-competitive situation), at which time one predator should emigrate. There should then follow a phased emigration until the last predator leaves the patch at T_{ncrit}. Parker and Stuart (1976) further analyse this model to establish preferred rates of emigration under conditions of variable times of arrival of predators in the patch.

The spatial structure of the environment significantly affects many aspects of foraging strategy, not only in the challenge it poses to the predator for allocation of time so as to maximise energy gain, but also in the context of avoiding risks of predation and of minimising the effects of interference (Emlen, 1968, 1973). To the extent that patches of prey may also act as refuges for the predator, risk avoidance should prolong the time spent in the patch beyond the prediction of simple optimal foraging theory. This also applies to risks from physical factors (wave action, dessication).

9 Anti-predator defence

Animals defend themselves against predators in a variety of ways, and many of these will have co-evolved with the predators foraging strategy (Curio, 1976). Defense mechanisms include aposematic and protective colouration, mimicry and colour polymorphism (Edmunds, 1974). Predation that concentrates on the more common morphs of a polymorphic prey could help to maintain the polymorphism by resulting in a selective advantage to the rarer morphs. Clarke (1969) called this apostatic selection; the relationship between switching and apostatic selection is discussed by Murdoch and Oaten (1975), who also review the 'specific search image' concept of Tinbergen (1960; see also Royama, 1970). Smith (1975) discusses one example of polymorphism in a bivalve prey and selection by a predatory crab.

Many benthic invertebrates exhibit escape responses to predators (see, for example, Feder and Christensen, 1966). Mauzey et al. (1968) report field observations (by diving) on the escape responses of a variety of potential prey to asteroids. Dayton et al. (1977) report very effective running escape responses by a variety of prey of the asteroid *Meyenaster gelatinosus*, and one example of a 'counter attack' by an echinoid, using its globiferous pedicillariae (see also Rosenthal and Chess, 1970, 1972). Dayton et al. (1977) also observed that many of the prey species were able to discriminate, in the natural environment, between foraging and non-foraging predators. Stenzler and Atema (1977) discuss the anti-predatory alarm response of *Nassarius obsoletus* (see also Atema and Burd, 1975). The general point to emerge from these studies is the extreme difficulty of obtaining realistic measures of prey availability in the field, with which to evaluate optimal diet models.

Other refuges available to potential prey include camouflage by means of ornamentation (Portman, 1959; Dayton et al., 1974, 1977; Brenchley, 1976), refuge through growth beyond a maximum size capable of successful attack by the predator (Dayton, 1971; Connell, 1972; Paine, 1976; Vince et al., 1976; Seed and Brown, 1978) and by morphological traits consistent with reducing the efficiency of attack by the predator (Vermeij, 1974, 1976, 1977; Reynolds and Reynolds, 1977; Vermeij and Veil, 1978). The relationship between the shell morphology of muriciid prosobranch molluscs and their vulnerability to predation has been the subject of much debate (Ebling et al., 1964; Kitching et al., 1966; Crothers, 1973; Bertness, 1977; Hughes and Elner, 1978). Whelks (*Thais* sp.) from habitats in which they are normally exposed to predation by crabs are less vulnerable than whelks from crab-free habitats.

Some of the effects of patchiness in the prey population on the effectiveness of predation and on the optimal foraging strategy have already been mentioned and are discussed by Huffaker (1958), MacArthur and Pianka (1966), Schoener (1971), Murdoch and Oaten (1975) and Pyke et al. (1977). The topic is central to discussions of predator and prey population stability (p. 596). Spatial heterogeneity may provide prey with structural refuges that reduce predator efficiency, either by lowering the probability of predator-prey encounters, or by decreasing the energetic efficiency of foraging. Glass (1971) examined the energetics of predation by Micropterus on female Lebistes in four habitats of increasing cover for the prey. By multiplying the rate of feeding attempts by the probability of successful capture he estimated the number and hence the energy value of prey captured per hour. This was compared with energy expended, calculated from a detailed assessment of metabolic costs of feeding (p. 589). By these simple means Glass (1971) was able to quantify a decrease in energy available for growth by the predator with increasing environmental complexity, and to suggest a degree of cover or prey refuge that was optimal for survival of both the predator and the prey. Laboratory studies aimed at simulating foraging behaviour in the field must endeavour to reproduce realistic habitat complexity and prey refuge.

The complexity of attributes that comprise potential prey value to the predator including size, energy and nutritional content, speed of movement, presence or absence of defence behaviour, and likelihood of effective refuges, all render the ranking of prey items by the predator very difficult. It seems probable that, in the natural situation, prey value will form a graded series rather than a nicely stepped rank (Hughes, 1979). In addition, the enormous variety of potential prey encountered in the field will also likely lead to a continuous rather than discontinuous ranking of preference. Most conceptions of optimal diet nevertheless assume that the predator is able to rank prey items unambiguously (but see Stenseth and Hansson, 1979). It seems unlikely that the predictions of the optimal foraging models will be borne out in any simple manner by field observation until more of these complicating factors can be formalised.

There have been many studies of foraging behaviour by macrobenthic predators in the field and some have tended both to confirm the broad outlines of theoretical prediction whilst at the same time emphasising the complexities that exist in natural communities. The reader is referred particularly to the following: B. A. Menge (1972a,b; starfish), J. L. Menge (1974; the prosobranch Acanthina), Leviten (1976; the prosobranch Conus sp.), Vadas (1977; sea urchins). Some of the most pertinent tests of foraging theory derive from studies of shore birds feeding on macro-benthic invertebrates (Goss-Custard, 1970, 1977a,b; Krebs, 1973, 1978; Evans, 1976; Horwood and Goss-Custard, 1977; O'Connor and Brown, 1977).

10   Resource depression, the developmental response, and aspects of prey stability

Field studies of the kind described by B. A. Menge (1972a) and J. L. Menge (1974) emphasise the importance of prey availability, in addition to prey abundance, in assessing the predator-prey interaction. Charnov et al. (1976) discuss prey availability in the context of resource depression, i.e. the lowering of predator capture rates, and they distinguish (a) exploitation depression in which prey abundance is reduced, (b) behavioural depression in which the behaviour of the prey leads to reduced capture rates, and (c) microhabitat depression, usually effected by a behavioural change, in which the microhabitat of the prey is altered in order to make capture more difficult. In all of these situations the predator is faced with the decision when to alter or to enlarge its foraging in response to resource depression; the marginal value theorem (p. 584) suggests certain rules for the relevant optimal behaviour. Charnov et al. (1976) discuss some implications of depression as a result of the foraging behaviour of one predator on the success of a second predator employing a similar or different foraging method.

Intuitively, features of resource depression, mutual interference between predators, and of spatial heterogeneity and predator switching may all contribute to the maintenance of stable prey populations (Murdoch and Oaten, 1975). The subject has received considerable theoretical treatment. Of the three types of functional response, only a sigmoidal (type III) response yields density dependent mortality in the prey population. Murdoch (1971, 1973) described a 'developmental response' in which predation rate is a function of predator size. The number of prey killed per predator at a particular prey density may also increase as predator efficiency increases (either as an increase in a' or a decrease in handling time). The result (illustrated by Murdoch for *Thais* feeding on *Balanus*) is a sigmoidal relationship between prey eaten per predator and prey density (and therefore density dependent prey mortality) when considered over a longer time period than is normally considered in evaluating the functional response alone.

Hassell (1976) presents a convenient summary of predator mediated stability in prey populations from research on arthropod predator-prey systems. The theoretical treatment follows from the population model of Nicholson (1933) and Nicholson and Bailey (1935). The inclusion of handling time (T_h) is seen as inevitably destabilising in a population model, but the extent of the instability depends on the value of T_h relative to the time available (T), rather than to the absolute value of T_h. In a realistic population model of the predator-prey interaction incorporating the effects of age, the ways in which both T_h and a' change with predator and prey size need to be considered. Thompson (1975) used Rogers' (1972) random predator equation to estimate a' and T_h for the damselfly (*Ischnura elegans*) feeding on *Daphnia*, over a range of predator and prey size; both increased nearly monotonically, a' with increased predator size and decreased prey size, T_h with decreased predator size and increased prey size. Thompson (1978) has also assessed the effects of temperature on a' and T_h for this predator-prey system.

In Hassell's discussions (1976; Hassell and May, 1973; Hassell *et al.*, 1976) terms for interference between predators (the interference coefficient m in eqn 20), and for the aggregation of predators in areas of high prey density (an aggregative index n) are incorporated into the population model and assessed for effects on stability of the prey population (Hassell and May, 1974) and for possible interrelationships between the various parameters of predation. The stability properties of these models, in which a single predator feeds upon a single prey without resource limitation, are shown to be very sensitive to interference or to the introduction of a non-random search behaviour in which the predators aggregate in areas of high prey density. If two or more prey species are now introduced, and the predator is allowed to switch so that the most abundant prey species at any time suffers the greatest percent predation (Comins and Hassell, 1976), predator-mediated stability is greatly enhanced even in situations where there is considerable interspecific competition between prey (see also Roughgarden and Feldman, 1975).

Tanner (1975) explored the effects of intrinsic growth rates of both predator and prey populations on their stability. He adopted the logistic equation of population growth for both prey and predator, but included a type II functional response to describe mortality from predation in the prey population:

$$dH/dt = rH(1 - HK^{-1}) - wHP(D + H)^{-1} \qquad (21)$$

where H is prey numbers, r the prey intrinsic growth rate, K is the carrying capacity of the resource for numbers of prey, w is the maximum predation rate and D is the constant describing the rate at which the functional response increases with increased prey density. May (1975, 1976) lists other forms of this functional response term, including the type III response. The growth of the predator population (Tanner, 1975) is given by

$$dP/dt = sP(1 - PJH^{-1}) \qquad (22)$$

where P is the predator numbers, s is the predator intrinsic growth rate, and J is the number of prey required to support one predator at equilibrium, when P equals H/J. This expression describes the 'numerical response' of the predator population to prey density (Solomon, 1949; Hollings, 1959; May, 1975), and this also can take different forms as discussed by May (1975) and by Beddington *et al.* (1976).

Tanner (1975) explored the stability properties of a system obeying eqns (21) and (22) (see also May, 1975). When the intrinsic growth rate of the prey population exceeds that of its predators (r/s is large), particularly in an environment of relatively large carrying capacity (K), the predator and prey populations tend towards a stable cycle. Indeed the properties with greatest effect on stability are prey self-limitation and predator searching time, represented in the equations by K and D, respectively. When K/D is small, the predator-prey interaction tends towards a stable, or equilibrium point (May, 1975).

Recently Caswell (1978) has drawn attention to the utility of non-equilibrium concepts in understanding predator-mediated coexistence between competing prey species in 'open' (=spatially heterogeneous) systems, in which the predator is able to migrate between patches (= 'cells') of prey. He points out that in such systems non-equilibrium behaviour of the system components may persist for long periods of time, a conclusion supported by Caswell's model of coexistence of competitors under the impact of predation. The model predicts that the observed effect of predation on the coexistence of prey species will vary with the scale on which such effects are studied. On the scale of a single cell, predation will have a negative effect on coexistence, but on a larger scale predation should have a positive effect, with transient predation pressures opening up new cells for colonisation by prey and then releasing the predation pressure to allow non-equilibrium population growth (see also Hassell and Southward, 1978). Caswell (1978) invokes studies by Paine (1966, 1971), Dayton (1971) and Paine and Vadas (1969) on rocky intertidal shores to support these predictions.

Before considering briefly these and other studies on intertidal environments, it is helpful to consider other effects of predation on prey resource utilisation. Glasser (1978, 1979) points out that there are qualitative as well as quantitative aspects to resource utilisation; by reducing the abundance of prey species (exploitation depression), predators may encourage selective (or discriminate) utilisation of their resources by prey species, so facilitating the coexistence of potential competitors and allowing a greater number of species to coexist than could otherwise do so. Predation may therefore increase both the evenness component of species diversity, as well as species richness, by maintaining prey populations below the level at which they are limited by a particular resource. As with many aspects of the role of predation in structuring communities, empirical evidence is provided by observation and experiment on shore faunas (Paine, 1971, 1974).

11 Predation and community structure

As indicated above, studies on rocky inter-tidal faunas, especially those by Connell (1970, 1972), Paine (1966, 1971, 1974), Dayton (1971) and B. A. Menge (1976), provide the data base for many recent attempts at a theoretical synthesis of the role of predation and other processes in community structure. Connell (1975) considers three methods that have been used to assess interactions between organisms as effecting community structure; statistical description and analysis, a reliance on 'natural experiments', and the use of controlled field manipulations (Connel, 1974) coupled with careful observation in the natural habitat. This last method has been used with great success on rocky inter-tidal shores in temperate regions.

In these rocky shore communities the provision of primary space (Dayton, 1971) is a major potentially limiting resource, except possibly at the highest shore levels where physical stressors may be more significant (Connell, 1972; Paine, 1974). In many of these communities a major competitive dominant tends to exclude other species from occupying primary space. Predation on these dominants lowers their abundance, reduces the severity of competition and so prevents competitive exclusion and increases species diversity (Paine, 1966, 1971, 1974). Other (physical) disturbance factors may enhance the effects of predation (Dayton, 1971; Grant, 1977). Predation may therefore be the main factor limiting the distribution of major prey species on the lower reaches of the shore, but occasional refuges for the prey, both physical (Connell, 1961a,b, 1970; Dayton, 1971) and through growth in size (Paine, 1976) provide a means of escape which contribute towards spatial patchiness.

More recent studies have confirmed these basic processes as affecting community structure in the temperate rocky-intertidal, and Menge and Sutherland (1976) review and discuss the roles of predation, competition and temporal heterogeneity in these habitats. Their hypothesis, relating these processes in communities that differ in complexity and trophic structure, remains to be tested more widely, but provides a convincing synthesis of many of the studies to date. As pointed out by Petersen (1979), however, the communities discussed are mostly dominated by mussels or barnacles, which in turn are the preferred prey for the major predator species, either starfish or prosobranch gastropods (Kitching and Ebling, 1967; Connell, 1970; Paine, 1966, 1971; Dayton, 1971; B. A. Menge, 1976; Louda, 1979). Where predation has been shown to be important, the existence of one or a few potential dominants may be a necessary, although not sufficient, condition (Keough and Butler, 1979).

The importance of predation in other epibenthic communities at Beaufort, North Carolina (Sutherland, 1974; Sutherland and Karlson, 1977; Karlson, 1978) and pier pilings in South Australia (Keough and Butler, 1979) remains to be resolved. In these communities many of the sessile species show considerable temporal variability in recruitment, growth and survival, and no single species exists as a competitive dominant for long periods of time; trophic relationships are often complex. Keough and Butler (1979) concluded that neither competitive nor predatory interactions were important in their study. Rather, seasonal factors, coupled with varying strategies of 'invasion' and 'defence' (Jackson, 1977a,b) lead to a continuously changing community structure, where predation may have little significance. In making 'between community' comparisons (Menge and Sutherland, 1976), the effects of trophic complexity must be given further consideration, as should the effects of spatial scale and persistence (Jackson, 1977b; Karlson, 1978).

In recent years manipulation experiments (cageing to exclude or to enclose predators; Connell, 1974) have been carried out also on sediment shores and infauna communities (Woodin, 1974). Many of these studies emphasise the importance of predation in affecting the structure and the dynamics of both the macrofauna (Brunswig et al., 1976; Virnstein, 1977, 1979; Riese, 1977, 1978; Arntz, 1977) and the meiofauna (Bell and Coull, 1978; Warwick and Gee, pers. comm.) components.

Field studies in epifaunal communities have also elucidated some of the mechanisms whereby predators partition a common prey resource (Birkeland, 1974; see also Fenchel, 1975, 1978 for studies on sediment shores). B. L. Menge (1972b) considered competition between two intertidal starfish, Pisaster and Leptasterias. Leptasterias, the smaller of the two species, has a higher rate of energy intake than Pisaster, but this potential advantage is countered by the latter species' aggression towards its competitor. In a similar habitat, three species of Thais coexist and share a food resource (Connell, 1970; Spight, 1972). Resource allocation between these species is a complex combination of behaviour patterns (Bertness, 1977), distribution on the shore (Emlen, 1966; Connell, 1970), size-related prey preferences (Spight, 1972) and more stochastic events associated with spatial heterogeneity of both predator and prey, seasonality factors, and differing relationships between food availability, reproduction and fecundity (Spight, 1975; Spight and Emlen, 1976).

12 Conclusion

The theories of optimal diet and optimal foraging provide a cogent framework for interpreting predator-prey interactions in natural communities. The gap is still wide, however, between theory and observation. In particular, the implications for optimal behaviour of heterogeneity of spatial structure of the environment, the difficulties of scaling prey availability in a realistic way, and the capabilities of different types of predator for sampling and ranking the value of different prey items all require elucidation by experiment and incorporation into models.

There is a need also to bridge the gap between optimal foraging models and predator-prey population models where predator behaviour is not necessarily assumed to be optimal (Hassell and May, 1973, 1974). A start has recently been made in this by Comins and Hassell (1979), who concluded that although optimal foraging was important to reproductive fitness, the resulting population dynamics are qualitatively similar to predictions from models in which the predator (or insect parasitoid) has a fixed foraging (=aggregation) strategy and is not made to feed optimally. When this gap is bridged, and the effects of competition between predators on population stability of both predator and prey are better understood, ecology will be closer to achieving the goal of a unified theory of community structure (MacArthur, 1972).

References

Arntz, W. E. (1977). Results and problems of an "unsuccessful" benthos cage predation experiment (Western Baltic). *In* "Biology of Benthic Organisms" (B. F. Keegan, P. O. Ceidigh and P. J. S. Boaden, eds) pp. 31–44. Pergamon Press, Oxford

Atema, J. and Burd, G. D. (1975). A field study of chemotactic responses of the marine mud snail, *Nassarius obsoletus*. *J. Chem. Ecol.* **1**, 243–251

Bayne, B. L. and Scullard, C. (1978a). Rates of oxygen consumption by *Thais* (*Nucella*) *lapillus* (L.). *J. Exp. Mar. Biol. Ecol.* **32**, 97–111

Bayne, B. L. and Scullard, C. (1978b). Rates of feeding by *Thais* (*Nucella*) *lapillus* (L.). *J. Exp. Mar. Biol. Ecol.* **32**, 113–129

Beddington, J. R. (1975). Mutual interference between parasites or predators and its effect on searching efficiency. *J. Anim. Ecol.* **44**, 331–340

Beddington, J. R., Hassell, M. P. and Lawton, J. H. (1976). The components of arthropod predation. II. The predator rate of increase. *J. Anim. Ecol.* **45**, 165–185

Bell, S. S. and Coull, B. C. (1978). Field evidence that shrimp predation regulates meiofauna *Oecologia* **35**, 141–148

Bertness, M. D. (1977). Behavioural and ecological aspects of shore-level size gradients in *Thais lamellosa* and *Thais emarginata*. *Ecology* **58**, 86–97

Beukema, J. J. (1968). Predation by the three-spined stickle-back (*Gasterosteus aculeatus* L.): the influence of hunger and experience. *Behaviour* **31**, 1–126

Birkeland, C. (1974). Interactions between a sea pen and several of its predators. *Ecol. Monogr.* **44**, 211–232

Brenchley, G. A. (1976). Predator detection and avoidance: ornamentation of tube-caps of *Diopatra* spp. (Polychaeta: Onuphidae). *Mar. Biol.* **38**, 179–188

Brunswig, D., Arntz, W. E. and Rumohr, A. (1976). A tentative field experiment in population dynamics of macrobenthos in the Western Baltic. *Kiel. Meeresforsch.* **3**, 49–59

Carriker, M. R. and Van Zandt, D. (1972). Predatory behaviour of a shell-boring muricid gastropod. *In* "Behaviour of Marine Animals. I. Invertebrates" (H. E. Winn and B. L. Olla, eds) pp. 157–244. Plenum Press, New York

Caswell, H. C. (1978). Predator-mediated co-existence: a non-equilibrium model. *Amer. Natur.* **112**, 127–154

Charnov, E. L. (1976). Optimal foraging: the marginal value theorem. *Theor. Popul. Biol.* **9**, 129–136

Charnov, E. L., Orians, G. H. and Hyatt, K. (1976). The ecological implications of resource depression. *Amer. Natur.* **110**, 247–259

Clarke, B. C. (1969). The evidence for apostatic selection. *Heredity* **24**, 347–352

Cock, M. J. W. (1978). The assessment of preference. *J. Anim. Ecol.* **47**, 805–816

Comins, H. and Hassell, M. P. (1976). Predation in multi-prey communities. *J. Theor. Biol.* **62**, 93–114

Comins, H. and Hassell, M. P. (1979). The dynamics of optimally foraging predators and parasitoids. *J. Anim. Ecol.* **48**, 335–351

Connell, J. H. (1961a). Effects of competition, predation by *Thais lapillus*, and other factors on natural populations of the barnacle *Balanus balanoides*. *Ecol. Monogr.* **31**, 61–104

Connell, J. H. (1961b). The influence of interspecific competition and other factors on the distribution of the barnacle *Cthamalus stellatus*. *Ecology* **42**, 710–723

Connell, J. H. (1970). On the role of natural enemies in preventing competitive exclusion in some marine animals and in rain forest trees. *Proc. Advance Stud. Inst. Dynamics Numbers Pop. 1970*, 298–312

Connell, J. H. (1972). Community interactions on marine rocky intertidal shores. *Ann. Rev. Ecol. Syst.* **3**, 169–192

Connell, J. H. (1974). Field experiments in marine ecology. *In* "Experimental Marine Biology" (R. Mariscal, ed.) pp. 21–54. Academic Press, New York

Connell, J. H. (1975). Some mechanisms producing structure in natural communities: a model and evidence from field experiments. *In* "Ecology and Evolution of Communities" (M. L. Cody and J. M. Diamond, eds) pp. 460–490. Harvard University Press, Cambridge

Cornell, H. (1976). Search strategies and the adaptive significance of switching in some general predators. *Amer. Natur.* **110**, 317–320

Covich, A. P. (1976). Analysing shapes of foraging areas: some ecological and economic theories. *Ann. Rev. Ecol. Syst.* **7**, 235–257

Crothers, J. H. (1973). On variation in *Nucella lapillus*: shell shape in populations from Pembrokeshire, South Wales. *Proc. malac. Soc. Lond.* **40**, 319–327

Curio, E. (1976). "The Ethology of Predation". Springer-Verlag, Berlin

Dayton, P. K. (1971). Competition, disturbance and community organisation: the provision and subsequent utilization of space in a rocky intertidal community. *Ecol. Monogr.* **41**, 351–389

Dayton, P. K., Robilliard, G. A., Paine, R. T. and Dayton, L. B. (1974). Biological accommodation in the benthic community at McMurdo Sound, Antarctica. *Ecol. Monogr.* **44**, 105–128

Dayton, P. K., Rosenthal, R. J., Mahen, L. C. and Antezana, T. (1977). Population structure and foraging biology of the predaceous Chilean asteroid *Meyenaster gelatinosus* and the escape biology of its prey. *Mar. Biol.* **39**, 361–370

Doi, T. (1976). Some aspects of feeding ecology of the sea stars, genus *Astropecten*. *Publ. Amakusa Mar. Biol. Lab.* **4**, 1–19

Ebling, F. J., Kitching, J. A., Muntz, L. and Taylor, C. M. (1964). The ecology of Lough Ine XIII. Experimental observations of the destruction of *Mytilus edulis* and *Nucella lapillus* by crabs. *J. Anim. Ecol.* **33**, 73–82

Edmunds, M. (1974). "Defences in Animals: A Survey of Anti-predator Defences". Longman, Essex

Edwards, D. C. and Huebner, J. D. (1977). Feeding and growth rates of *Polinices duplicatus* preying on *Mya arenaria* at Barnstable Harbor, Massachusetts. *Ecology* **58**, 1218–1236

Elliot, J. M. (1975). The growth rate of brown trout (*Salmo trutta* L.) fed on reduced rations. *J. Anim. Ecol.* **44**, 823–842

Elliot, J. M. (1976). The energetics of feeding, metabolism and growth of brown trout (*Salmo trutta* L.) in relation to body weight, water temperature and ration size. *J. Anim. Ecol.* **45**, 923–948

Elner, R. W. and Hughes, R. N. (1978). Energy maximization in the diet of the shore crab, *Carcinus maenas*. *J. Anim. Ecol.* **47**, 103–116

Emlen, J. M. (1966). The role of time and energy in food preference. *Amer. Natur.* **100**, 611–617

Emlen, J. M. (1968). Optimal choice in animals. *Amer. Natur.* **102**, 385–390

Emlen, J. M. (1973). "Ecology: An Evolutionary Approach". Addison-Wesley, Massachusetts

Estabrook, G. F. and Dunham, A. E. (1975). Optimal diet as a function of absolute abundance, relative abundance, and relative value of available prey. *Amer. Natur.* **110**, 401–413

Evans, P. R. (1976). Energy balance and optimal foraging strategies in shore birds: some implications for their distributions and movements in the non-breeding season. *Ardea* **64**, 117–139

Feder, H. and Christensen, A. M. (1966). Aspects of asteroid biology. *In* "Physiology of Echinoderms" (R. A. Boolootian, ed.) pp. 87–127. Interscience, New York

Fenchel, T. (1975). Character displacement and coexistence in mud snails (Hydrobiidae). *Oecologia* **20**, 317–326

Fenchel, T. (1978). The ecology of micro- and meiobenthos. *Ann. Rev. Ecol. Syst.* **9**, 99–121

Glass, N. R. (1968). The effect of time of food deprivation on the routine oxygen consumption of large mouth black bass (*Micropterus salmoides*). *Ecology* **49**, 340–343

Glass, N. R. (1971). Computer analysis of predation energetics in the large mouth bass. *In* "Systems Analysis and Simulation in Ecology" (B. L. Patten, ed.) pp. 325–363. Academic Press, New York

Glasser, J. W. (1978). The effect of predation on prey resource utilization. *Ecology* **59**, 724–732

Glasser, J. W. (1979). The role of predation in shaping and maintaining the structure of communities. *Amer. Natur.* **113**, 631–641

Goss-Custard, J. D. (1970). Feeding dispersion in some overwintering wading birds. *In* "Social Behaviour of Birds and Mammals" (J. H. Crook, ed.) pp. 3–34. Academic Press, London

Goss-Custard, J. D. (1977a). Optimal foraging and the size-selection of worms by redshank *Tringa totanus. Anim. Behav.* **25**, 10–29

Goss-Custard, J. D. (1977b). Predator responses and prey mortality in the redshank *Tringa totanus* (L.) and a preferred prey *Corophium volutator* (Pallas). *J. Anim. Ecol.* **46**, 21–36

Grant, P. T. and Mackie, A. M. (1974). "Chemoreception in Marine Organisms". Academic Press, London

Grant, W. S. (1977). High inter-tidal community organization on a rocky headland in Maine, U.S.A. *Mar. Biol.* **44**, 15–25

Griffiths, D. (1975). Prey availability and the food of predators. *Ecology* **56**, 1209–1214

Harris, J. R. W. (1974). The kinetics of polyphagy. *In* "Ecological Stability" (M. B. Usher, ed.) pp. 123–139. Chapman and Hall, London

Hassell, M. P. (1976). Arthropod predator-prey systems. *In* "Theoretical Ecology: Principles and Applications" (R. M. May, ed.) pp. 71–93. Blackwell, Oxford

Hassell, M. P. and May, R. M. (1973). Stability in insect host-parasite models. *J. Anim. Ecol.* **42**, 693–726

Hassell, M. P. and May, R. M. (1974). Aggregation in predators and insect parasites and its affect on stability. *J. Anim. Ecol.* **43**, 567–594

Hassell, M. P. and Southwood, T. R. E. (1978). Foraging strategy in insects. *Ann. Rev. Ecol. Syst.* **9**, 75–98

Hassell, M. P. and Varley, G. L. (1969). New inductive population model for insect parasites and its bearing on biological control. *Nature, Lond.* **223**, 1133–1136

Hassell, M. P., Lawton, J. H. and Beddington, J. R. (1976). The components of arthropod predation. I. The prey death-rate. *J. Anim. Ecol.* **45**, 135–164

Holling, C. S. (1959). Some characteristics of simple types of predation and parasitism. *Can. Ent.* **91**, 385–398

Holling, C. S. (1966). The functional response of invertebrate predators to prey density. *Mem. ent. Soc. Can.* **48**, 1–86

Horwood, J. W. and Goss-Custard, J. D. (1977). Predation by the oystercatcher, *Haematopus ostralegus* (L.) in relation to the cockle, *Cerastoderma edule* (L.) fishery in the Burry inlet, South Wales. *J. appl. Ecol.* **14**, 139–158

Hubbard, S. F. and Cook, R. M. (1978). Optimal foraging by parasitoid wasps. *J. Anim. Ecol.* **47**, 593–604

Huffaker, C. B. (1958). Experimental studies on predation: dispersion factors and predator-prey oscillations. *Hilgardia*, **27**, 343–383

Hughes, R. N. (1979). Optimal diets under the energy maximization premise: the effects of recognition time and learning. *Amer. Natur.* **113**, 209–221

Hughes, R. N. and Elner, R. W. (1978). Tactics of a predator, *Carcinus maenas*, and morphological responses of the prey, *Nucella lapillus. J. Anim. Ecol.* **48**, 65–78

Jackson, J. B. C. (1977a). Competition on marine hard substrata: the adaptive significance of solitary and colonial strategies. *Amer. Natur.* **111**, 743–767

Jackson, J. B. C. (1977b). Habitat area, colonization and development of epibenthic community structure. *In* "Biology of Benthic Organisms" (B. F. Keegan, P. O. Ceidigh and P. J. S. Boaden, eds) pp. 349–358. Pergamon Press, Oxford

Karlson, R. (1978). Predation and space utilisation patterns in a marine epifaunal community. *J. Exp. Mar. Biol. Ecol.* **31**, 225–239

Keough, M. J. and Butler, A. J. (1979). The role of asteroid predators in the organization of a sessile community on pier pilings. *Mar. Biol.* **51**, 167–177

Kitching, J. A. and Ebling, F. J. (1967). Ecological studies at Lough Ine. *Adv. Ecol. Res.* **4**, 197–291

Kitching, J. A., Muntz, L. and Ebling, F. J. (1966). The ecology of Lough Ine. XV. The ecological significance of shell and body form in *Nucella*. *J. Anim. Ecol.* **35**, 113–126

Krebs, J. R. (1973). Behavioural aspects of predation. *In* "Perspectives in Ethology" (P. P. G. Bateson and P. H. Klopfer, eds) pp. 73–111. Plenum Press, New York

Krebs, J. R. (1978). Optimal foraging: Decision rules for predators. *In* "Behavioural Ecology: An Evolutionary Approach" (J. R. Krebs and N. B. Davies, eds) pp. 23–63. Blackwell, Oxford

Krebs, J. R., Ryan, J. C. and Charnov, E. L. (1974). Hunting by expectation or optimal foraging? A study of patch use by chickadees. *Anim. Behav.* **22**, 953–964

Landenburger, D. E. (1968). Studies on selective feeding in the Pacific starfish *Pisaster* in southern California. *Ecology* **49**, 1062–1075

Lawton, J. H., Beddington, J. R. and Bonser, R. (1974). Switching in invertebrate predators. *In* "Ecological Stability" (M. B. Usher, ed.) pp. 141–158. Chapman and Hall, London

Leviten, P. J. (1976). The foraging strategy of vermivorous conid gastropods. *Ecol. Monogr.* **46**, 157–178

Louda, S. M. (1979). Distribution, movement and diet of the snail *Searlesia dira* in the intertidal community of San Juan Island, Puget Sound, Washington. *Mar. Biol.* **51**, 119–131

MacArthur, R. H. (1972). "Geographical Ecology". Harper and Row, New York

MacArthur, R. H. and Pianka, E. R. (1966). On optimal use of a patchy environment. *Amer. Natur.* **100**, 603–609

Mackie, A. M. (1973). The chemical basis of food detection in the lobster *Homarus gammarus*. *Mar. Biol.* **14**, 217–221

Mauzey, K., Birkeland, C. and Dayton, P. K. (1968). Feeding behaviour of asteroids and escape responses of their prey in the Puget Sound region. *Ecology* **49**, 603–619

May, R. M. (1975). "Stability and Complexity in Model Ecosystems", 2nd edn. Princeton University Press, Princeton

May, R. M. (1976). Models for two interacting populations. *In* "Theoretical Ecology: Principles and Applications" (R. M. May, ed.) pp. 49–70. Blackwell, Oxford

Maynard Smith, J. (1978). Optimization theory in evolution. *Ann. Rev. Ecol. Syst.* **9**, 31–56

McLeese, D. W. (1974). Olfactory responses of lobsters (*Homarus americanus*) to solutions from prey species and to seawater extracts and chemical fractions of fish muscle and effects of antennule ablation. *Mar. Behav. Physiol.* **2**, 237–249

Menge, B. A. (1972a). Feeding strategy of a starfish in relation to actual prey availability and environmental predictability. *Ecol. Monogr.* **42**, 25–50

Menge, B. A. (1972b). Competition for food between two intertidal starfish and its effect on body size and feeding. *Ecology* **53**, 635–644

Menge, B. A. (1976). Organization of the New England rocky intertidal community: role of predation, competition and environmental heterogeneity. *Ecol. Monogr.* **46**, 355–393

Menge, B. A. and Sutherland, J. P. (1976). Species diversity gradients: synthesis of the roles of predation, competition and temporal heterogeneity. *Amer. Natur.* **110**, 351–369

604   B. L. BAYNE

Menge, J. L. (1974). Prey selection and foraging period of the predaceous rocky intertidal snail, *Acanthina punctulata. Oecologia* **17**, 293–316

Murdoch, W. W. (1969). Switching in general predators: experiments on predator specificity and stability of prey populations. *Ecol. Monogr.* **39**, 335–354

Murdoch, W. W. (1971). The developmental response of predators to changes in prey density. *Ecology* **52**, 132–137

Murdoch, W. W. (1973). The functional response of ·predators. *J. Appl. Ecol.* **14**, 335–341

Murdoch, W. W. and Oaten, A. (1975). Predation and population stability. *Adv. Ecol. Res.* **9**, 1–132

Murdoch, W. W., Avery, S. L. and Smythe, M. E. B. (1975). Switching in predatory fish. *Ecology* **56**, 1094–2005

Nicholson, A. J. (1933). The balance of animal populations. *J. Anim. Ecol.* **2**, 131–178

Nicholson, A. J. and Bailey, V. A. (1935). The balance of animal populations. Part I. *Proc. zool. soc. Lond.* 1935, 551–598

Norberg, R. A. (1977). An ecological theory on foraging time and energetics and choice of optimal food-searching method. *J. Anim. Ecol.* **46**, 511–529

O'Connor, R. J. and Brown, R. A. (1977). Prey depletion and foraging strategy in the oyster catcher *Haematopus ostralegus. Oecologia* **27**, 75–92

Paine, R. T. (1966). Food web complexity and species diversity. *Amer. Natur.* **100**, 65–75

Paine, R. T. (1971). A short-term experimental investigation of resource partitioning in a New Zealand rocky inter-tidal habitat. *Ecology* **52**, 1096–1106

Paine, R. T. (1974). Intertidal community structure: experimental studies on the relationship between a dominant competitor and its principal predator. *Oecologia* **15**, 93–120

Paine, R. T. (1976). Size-limited predation: An observational and experimental approach with the *Mytilus-Pisaster* interaction. *Ecology* **57**, 858–873

Paine, R. T. and Vadas, R. L. (1969). The effects of grazing by sea urchins, *Strongylocentrotus* spp. on benthic algal populations. *Limnol. Oceanogr.* **14**, 710–719

Paloheimo, J. E. and Dickie, L. M. (1966). Food and growth of fishes. III. Relations among food, body size and growth efficiency. *J. Fish. Res. Bd Can.* **30**, 409–434

Parker, G. A. and Stuart, R. A. (1976). Animal behaviour as a strategy optimizer: evolution of resource assessment strategies and optimal emigration thresholds. *Amer. Natur.* **110**, 1055–1076

Pearson, W. H. and Olla, B. L. (1977). Chemoreception in the blue crab, *Callinectes sapidus. Biol. Bull. mar. biol. Lab., Woods Hole* **153**, 346–354

Pearson, W. H., Sugarman, P. C., Woodruff, D. L. and Olla, B. L. (1979). Thresholds for detection and feeding behaviour in the Dungeness crab, *Cancer magister* (Dana). *J. Exp. Mar. Biol. Ecol.* **39**, 65–78

Petersen, C. H. (1979). The importance of predation and competition in organising the intertidal epifaunal communities of Barnegat Inlet, New Jersey. *Oecologia* **39**, 1–24

Portmann, A. (1959). "Animal Camouflage". University of Michigan Press, Ann Arbor

Pratt, D. M. (1974). Attraction to prey and stimulus to attack in the predatory gastropod *Urosalpinx cinerea. Mar. Biol.* **27**, 37–45

Pratt, D. M. (1977). Homing in *Urosalpinx cinerea* in response to prey effluent and tidal periodicity. *Veliger* **20**, 30–33

Pulliam, H. R. (1974). On the theory of optimal diets. *Amer. Natur.* **108**, 59–75

Pulliam, H. R. (1975). Diet optimization with nutrient constraints. *Amer. Natur.* **109**, 765–768

Pyke, G. H., Pulliam, H. R. and Charnov, E. L. (1977). Optimal foraging: a selective review of theory and tests. *Quart. Rev. Biol.* **52**, 137–154

Rapport, D. J. (1971). An optimization model of food selection. *Amer. Natur.* **108**, 59–79

Real, L. A. (1977). The kinetics of functional response. *Amer. Natur.* **111**, 289–300

Reynolds, W. W. and Reynolds, L. J. (1977). Zoogeography and the predator-prey 'armsrace': a comparison of *Eriphia* and *Nerita* species from three faunal regions. *Hydrobiol.* **56**, 63–67

Riese, K. (1977). Predator exclusion experiments in an intertidal mudflat. *Helgolander wiss. Meeresunters.* **30**, 283–271

Riese, K. (1978). Predation pressure and community structure of an intertidal soft-bottom fauna. *In* "Biology of Benthic Organisms" (B. F. Keegan, P. O. Ceidigh and P. J. S. Boaden, eds) pp. 513–519. Pergamon Press, Oxford

Rogers, D. J. (1972). Random search and insect population models. *J. Anim. Ecol.* **41**, 369–383

Rosenthal, R. J. and Chess, J. R. (1970). Predation on the purple urchin by the leather star. *Calif. Fish and Game* **56**, 203–204

Rosenthal, R. J. and Chess, J. R. (1972). A predator-prey relationship between the leather star, *Dermasterias imbricata*, and the purple urchin, *Strongylocentrotus purpuratus*. *Fish. Bull. U.S.* **70**, 205–216

Roughgarden, J. and Feldman, M. (1975). Species packing and predation pressure. *Ecology* **56**, 489–492

Royama, T. (1970). Factors governing the hunting behaviour and selection of food by the Great Tit (*Parus major* L.). *J. Anim. Ecol.* **39**, 619–668

Royama, T. (1971). A comparative study of models for predation and parasitism. *Res. Popul. Ecol. Suppl.* **1**, 1–91

Schoener, T. W. (1971). Theory of feeding strategies. *Ann. Rev. Ecol. Syst.* **2**, 369–404

Seed, R. and Brown, R. A. (1978). Growth as a strategy for survival in two marine bivalves, *Cerastoderma edule* and *Modiolus modiolus*. *J. Anim. Ecol.* **47**, 283–292

Smith, D. A. S. (1975). Polymorphism and selective predation in *Donax faba* Gmelin (Bivalvia: Tellinacea). *J. Exp. Mar. Biol. Ecol.* **17**, 205–219

Solomon, M. E. (1949). The natural control of animal populations. *J. Anim. Ecol.* **18**, 1–35

Spight, T. M. (1972). Patterns of change in adjacent populations of an intertidal snail, *Thais lamellosa*. Ph.D. thesis, University of Washington

Spight, T. M. (1975). Factors extending gastropod embryonic development and their selective cost. *Oecologia* **21**, 1–16

Spight, T. M. and Emlen, J. (1976). Clutch sizes of two marine snails with a changing food supply. *Ecology* **57**, 1162–1178

Stenseth, N. C. and Hansson, L. (1979). Optimal food selection: A graphic model. *Amer. Natur.* **113**, 373–389

Stenzler, D. and Atema, J. (1977). Alarm response of the marine mud snail, *Nassarius obsoletus*: Specificity and behavioural priority. *J. Chem. Ecol.* **3**, 159–171

Sutherland, J. P. (1974). Multiple stable points in natural communities. *Amer. Natur.* **108**, 859–873

Sutherland, J. P. and Karlson, R. H. (1977). Development and stability of the fouling community at Beaufort, North Carolina. *Ecol. Monogr.* **47**, 425–446

Tanner, J. T. (1975). The stability and the intrinsic growth rates of prey and predator populations. *Ecology* **56**, 855–867

Thompson, D. J. (1975). Towards a predator-prey model incorporating age structure: the effects of predator and prey size on the predation of *Daphnia magna* by *Ischnura elegans*. *J. Anim. Ecol.* **44**, 907–916

Thompson, D. J. (1978). Towards a realistic predator-prey model: the effect of temperature on the functional response and life history of the larvae of the damselfly, *Ischnura elegans* (Odonata). *J. Anim. Ecol.* **47**, 757–768

Tinbergen, L. (1960). The natural control of insects in pine woods. I. Factors influencing the intensity of predation by songbirds. *Archs. Néerl. Zool.* **13**, 265–343

Vadas, R. C. (1977). Preferential feeding: an optimization strategy in sea urchins. *Ecol. Monogr.* **47**, 337–371

Vermeij, G. J. (1974). Marine faunal dominance and molluscan shell-form. *Evolution* **28**, 656–664

Vermeij, G. J. (1976). Interoceanic differences in vulnerability of shelled prey to crab predation. *Nature, Lond.* **260**, 135–136

Vermeij, G. J. (1977). Patterns in crab claw size: the geography of crushing. *Syst. Zool.* **26**, 138–151

Vermeij, G. J. and Veil, J. A. (1978). A latitudinal pattern in bivalve shell gaping. *Malacologia* **17**, 57–61

Vince, S., Valiela, I., Backus, N. and Teal, J. M. (1976). Predation by the salt-marsh killifish *Fundulus heteroclitus* (L.) in relation to prey size and habitat structure: consequences for prey distribution and abundance. *J. Exp. Mar. Biol. Ecol.* **23**, 255–266

Virnstein, R. W. (1977). The importance of predation by crabs and fishes on benthic infauna in Chesapeake Bay. *Ecology* **58**, 1199–1217

Virnstein, R. W. (1979). Predation on estuarine infauna: Response patterns of component species. *Estuaries* **2**, 69–86

Watt, K. E. F. (1968). "Ecology and Resource Management". McGraw-Hill, New York

Winberg, G. C. (1956). "Rate of metabolism and food requirements of fishes". *Fish. Res. Bd Can. Transl. Ser.* No. 194, 1960

Wood, L. (1968). Physiological and ecological aspects of prey selection by the marine gastropod *Urosalpinx cinerea* (Prosobranchia: Muricidae). *Malacologia* **6**, 267–320

Woodin, S. A. (1974). Polychaete abundance patterns in a marine soft-sediment environment: the importance of biological interactions. *Ecol. Monogr.* **44**, 171–187

20 Field Experiments on Benthic Ecosystems

BERNT ZEITZSCHEL

1 Methodology of field experiments

FIELD EXPERIMENTS VERSUS LABORATORY MEASUREMENTS

To understand benthic life in the oceans, especially with regard to its role in marine ecosystems, it is important to obtain detailed knowledge of the structure and functioning of these communities. The structural aspects of benthic communities in shallow water systems down to about 200 m are quite well known, although the causes of variability of species abundance are poorly understood. The knowledge of the standing stock of macro-, meio- and microfauna in deep water is, however, very scarce. This is not surprising because of the vast area to be considered. The functional aspects of benthic plants and animals are even more difficult to investigate. There are two different approaches to this objective:

1 through population dynamics, and
2 through physiology.

For the population dynamics approach, we use data on rates of reproduction, growth and death, and the structure of populations in terms of age and size to calculate secondary production. These data are mostly obtained by intensive field-work, and by exclusion experiments with cages *in situ*. For the physiological approach, experiments are conducted to measure primary production, the rates of feeding, assimilation, respiration, excretion and other metabolic processes. Experiments to obtain data on these processes may be carried out in the laboratory (Lasserre, 1976) or *in situ* (Pamatmat, 1977; Zeitzschel and Davies, 1978). Laboratory experiments in beakers or aquaria yield useful insight into the biology of individual species, or the effect of perturbations on a group of plants or animals. Some methods, such as the measurement of the rate of hydrogen or electron transport activity or the measurement of metabolic heat production have yielded valuable results. They are, however, up to now limited to laboratory analysis (Pamatmat, 1975). The extrapolation of all these autecological experiments to the marine ecosystem is, however, difficult if not impossible.

Further, problems associated with this approach are compounded in that open ocean species especially of deep sea benthos cannot be maintained in the laboratory even for short periods of time.

In situ experiments on the other hand tend to be rather expensive especially when sophisticated gear has to be employed in open waters. The results obtained tend, however, to be more realistic.

According to Menzel and Steele (1978) the main advantage in *in situ* enclosure experiments is the ability to study changes over time of the same population, the ability to study natural species assemblages or parts of the community by excluding predators, for example the ability to manipulate the environment and to incorporate duplicate and experimental replicates containing basically identical populations into research design, and the ability to reduce physical forces (e.g. turbulence) or to reintroduce them by mechanical means to study the effect of such processes on the included biota. The general problem with *in situ* measurements is that by introducing an experimental enclosure on the sea bottom the system to be measured undergoes unavoidable perturbations. For instance, a bell jar placed over a patch of sediment to measure oxygen consumption and/or nutrient flux must be closed to flow from surrounding sea water; thus flow conditions over the substrate and chemical reactions sensitive to flowrate are necessarily altered (Davies, 1975; Berner, 1976). The simulation of flow with a stirrer or a flow through pumping system may overcome this problem as a compromise.

In situ experiments with enclosures are, however, at least from the viewpoint of an ecologist, the only realistic method to carry out controlled environment experiments. With enclosures of an appropriate size it is possible to design experiments with natural populations under near-natural conditions over a relevant time period (Zeitzschel, 1978).

OBJECTIVES OF FIELD EXPERIMENTS ON BENTHIC ECOSYSTEMS

One of the most complex problems in ecology is to explain the natural pattern of species abundance. Predation has been demonstrated to be an important structuring force in many communities. This essential component of biological interactions may be evaluated in field experiments by three relatively simple experimental procedures: removal and exclusion of predators; introduction of predators; and transplantation of prey organisms into areas of varying predator abundance (Arntz, 1977; Reise, 1977; Virnstein, 1979). Besides these structural aspects there are a variety of processes which can be measured adequately in *in situ* experiments to obtain data on the functioning of marine benthic ecosystems. Concerning the plant component, measurement of primary productivitity both for macro- and microalgae is of utmost importance. Examples of this kind of measurement have been published by Grøntvedt (1962), Mann (1972), Nielsen-Aertebjerg (1976) and Schramm (1979).

As to the faunal component, energy transfer in the benthic food chain can be quantified, if measurements of community metabolism can be partitioned for macro-, meio- and microbenthos organisms. Changes of oxygen concentration with time in enclosed bodies of water are used as the main indicator.

The rate of total oxygen uptake by the sediment surface under-estimates community metabolism by the magnitude of anaerobic metabolism in the deeper layers (Pamatmat, 1975). To study anthropogenic influences on marine benthic ecosystems, controlled perturbation experiments may be carried out. These experiments are, up to now, the only way to obtain reliable data on the long-term effect of low level pollutants. The decomposition of organic matter by microorganisms and the release of nutrients are of great importance for pelagic life, especially in shallow water ecosystems (chapter 6).

DESCRIPTION OF RELEVANT INSTRUMENTATION

There are three different types of instruments which have been used successfully in field experiments on benthic ecosystems:

1 Sediment containers and cages (round or rectangular tanks, open on the top with openings at the side, cages of various shapes and with different meshes) on the bottom, or suspended in the water column,

2 bell jars (boxes, cylinders, bags, modified bottom-grabs) which enclose bottom areas of some tens of square centimeters up to a few square meters together with some supernatant water,

3 flow-through systems (annular closed tunnels, flow-through cuvettes, flow-through tunnels) to enclose sediments and to simulate natural water flow caused by currents and tides.

Examples of these different equipments are depicted in Fig. 1, and structural details are summarised in Table I.

Arntz (1980) has summarised the details of predator exclusion experiments with cages of various sizes, enclosing bottom areas of 0.1–11.5 m². A mesh size < 5 mm is necessary to exclude all predators.

Two alternatives to in situ measurements on benthic ecosystems are the use of sediment corers or land-based enclosures (McIntyre et al., 1970; Pamatmat, 1977, Pilson, 1979). With corers a small portion of sediment is cut out together with some overlying water and is used aboard a ship or in the laboratory to measure biological or chemical processes. Land-based enclosures operate in a flow through mode. More or less natural sediments

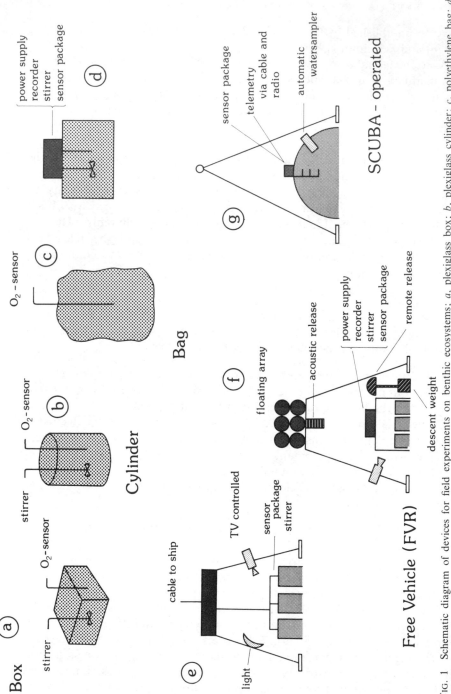

Box

(a)

O₂-sensor

stirrer

(b)

O₂-sensor

stirrer

Cylinder

(c)

O₂-sensor

Bag

(d)

power supply
recorder
stirrer
sensor package

(e)

cable to ship

TV controlled

sensor package
stirrer

light

Free Vehicle (FVR)

(f)

floating array

acoustic release

power supply
recorder
stirrer
sensor package

remote release

descent weight

(g)

sensor package

telemetry
via cable and
radio

automatic
watersampler

SCUBA - operated

FIG. 1 Schematic diagram of devices for field experiments on benthic ecosystems; *a*, plexiglass box; *b*, plexiglass cylinder; *c*, polyethylene bag; *d*, selfcontained bell jar; *e*, TV-controlled bell jar (Pamatmat and Fenton, 1968); *f*, Free Vehicle Respirometer (Smith *et al.*, 1976); *g*, SCUBA operated bell jar (Balzer, 1978). (After Zeitzschel and Davies, 1978)

TABLE I Instrumentation for field experiments on marine benthic ecosystems. (After Zeitzschel and Davies, 1978, supplement and updated)

Apparatus	Material	Dimensions		Sensors	Location	Depth (m)	Remarks	Reference
Continuous flow	Plexiglass	100 × 2 × 5 cm,	200 cm^2	T C, O$_2$, I$_0$	False Bay, San Juan Is.	Shallow	Pumping system	Pamatmat (1965)
Bell jar	Glass	600 ml,	215 cm^2	T C	False Bay, San Juan Is.	Shallow	Stirrer manual	Pamatmat (1966, 1968) sampling
Bell jar	Plexiglass	27 cm; 2–41,	570 cm^2	T C, O$_2$	Puget Sound (Washington)	11–185	TV-controlled telemetry	Pamatmat and Fenton (1968)
Bell jar (box)	Plexiglass	40 × 40 cm,	1600 cm^2	T C, O$_2$, E$_h$, E$_s$	Baltic (Askö, Sweden)	max. 10	SCUBA-operated	Hallberg et al. (1972)
Bell jar (cylinder)	Opaque plexiglass	30.5 × 15 cm	730 cm^2	O$_2$	Castle Habour (Bermuda)	1.5	Recorder	Smith et al. (1972)
Bell jar (cylinder)	Plexiglass		48 cm^2	O$_2$	South of New England	1850	Operated by DSRV Alvin	Smith and Teal (1973)
Continuous flow system	Plexiglass cuvettes	50–5000 ml		T C, O$_2$, S /, I$_0$	Kiel Fjord, Askö, Sweden	1	SCUBA-operated	Schramm (1973)
Bell jar (cylinder)	Plexiglass	30 × 22 cm.	707 cm^2	O$_2$	Baltic (swed. waters)	Shallow	Magnetic stirrer SCUBA-operated	Edberg and v. Hofsten (1973)
Bell jar (cylinder)	Transparent plexiglass	30 × 5 × 10 cm,	730 cm^2	O$_2$	San Diego trough	1230	RUM-operated TV-controlled	Smith (1974)
Bell jar (box)		61 × 30 × 30 cm.	551 cm^2	O$_2$	14 km off San Diego	1230	RUM-operated TV-controlled	Smith and Hessler (1974)
Bell jar (cylinder	Opaque plexiglass		730 cm^2	O$_2$	off New York, off Baja California	23–32	SCUBA-operated, recorder	Smith et al. (1974)
Annular closed tunnel	Translucent plastic	111		O$_2$	Loch Ewe (Scotland)	~30	Sampling port variable waterflow	Davies (1975)
Bell jar (free vehicle respir.)	Acrylic	30.5 × 10 cm.	730 cm^2	O$_2$	northwest of Bermuda	5200	Acoustic contr. time lapse, camera recorder, water-sampler syringe for formalin	Smith et al. (1976)
Bell jar	Plexiglass	2 1001.	3.1 m^2	T C, O$_2$, E$_h$, pH, S /	Baltic (Kiel Bight)	20	SCUBA-operated, automatic water-sampler, telemetry via radio	Balzer (1978)
Flow through tunnel	Polyethylene	5001		T C, O$_2$, I$_0$	Baltic (Kiel Bight)	~1	SCUBA-operated	Schramm and Martens (1976)
Flow through, self flushing	Plexiglass		~0.2 m^2	T C, O$_2$, pH	Flax Pond (Long Island)	Shallow	Circulating water system, periodic flushing	Hall and Tempel (manuscript)
Experimental chamber incl. small bell jar	Plexiglass	1.5 × 0.75 m, 30 × 10 cm	1.8 m^2	O$_2$	Loch Thurnaig (Scotland)	20–30	Fertilisation experiments	Zeitzschel and Davies (1978)
Grab respirometer	Stainless steel	21 × 21 × 30 cm.	420 cm^2	O$_2$	NW Atlantic	2750	Injection-withdrawal system, stirrer	Smith et al. (1978)
Bell jar (cylinder)	Aluminium	30 × 12 cm.	707 cm^2	O$_2$	Eastern Passage N.S.	Shallow	Automatic sampling device of supernatant water	Hargrave and Connolly (1978)

containing organisms may thus be studied under controlled environmental conditions.

MODE OF OPERATION

Field experiments on benthic ecosystems usually follow the following sequence: the exposure of the apparatus at an appropriate location, the measurement of relevant parameters and the control of sampling procedures, the registration and storage of obtained data; and the retrieval of the whole system.

In shallow water, down to approximately 30 m, experiments of this kind are mostly SCUBA-operated. In deeper water they may be controlled by TV-cameras. In recent years DSRV *Alvin* of the Woods Hole Oceanographic Institution, and the remote underwater manipulator 'RUM' of the Scripps Institution of Oceanography, were successfully used to manipulate different types of apparatus to conduct specific experiments in depths down to 2750 m. Smith *et al.* (1976) describe a free vehicle respirometer (FVR, Fig. 1*f*) which measures the oxygen consumption of benthic communities *in situ* to abyssal depth.

It consists of an aluminium tripod supporting a two-command acoustic release-transponder, an oxygen monitoring unit, a glass sphere floating array, and a time-lapse camera system. An acoustic signal actuates the first command mode, to release the oxygen monitoring unit. Settlement is monitored by the camera system. After one to five days, a second acoustic command releases the descent weight, and the free vehicle is brought to the surface by the floating array and recovered.

Recently Smith *et al.* (1978) described a grab-respirometer which was manipulated from *Alvin*. With this device it is possible to measure oxygen consumption and nutrient exchange across the sediment–water interface. The sediment containing the organisms can be retrieved after the experiment. Balzer (1978) and Hargrave and Connolly (1978) used bell jars with automatic sampling devices. Samples of the supernatant water for further chemical analysis are taken at preset fixed or variable time intervals.

Flow-through systems are designed usually to measure oxygen or other variables at the intake and the outlet of a flow-through tunnel. Constant water flow or periodic flushing of the tunnel is maintained by pumps.

Devices for field experiments on benthic ecosystems may be equipped with a variety of sensors e.g. for O_2, $T°$ C, $S°/_{oo}$, E_h, pH. The data thus obtained may be recorded on battery-operated strip-chart recorders or may be transmitted via telemetry to a ship or platform, or by radio to a shore-based laboratory as in the Kiel Bight project of Kiel University (Petersohn and Diekmann, 1978). Measurements may be conducted over a time span of a few hours to some days, or even a few weeks, depending on nature of the problem under investigation.

2 Major results obtained from field experiments

In this section, some typical results are presented from field experiments to study structural and functional aspects of benthic ecosystems. These are results from cage experiments, measurements of the metabolic activity of benthos, nutrient regeneration at the water–sediment interface and the effect of toxic substances on benthic organisms.

Cage experiments on benthic ecosystems have been performed in a variety of environments and on different types of sediments. Wire or nylon mesh cages were used either to exclude predators from, or to confine certain predators to, a small patch of sediment. Blegvad (1928), Naqui (1968), Reise (1977, 1978) and Virnstein (1977, 1979) have shown that the density and diversity of infaunal macrobenthos increased when predators were excluded by cages with small meshes. These experiments revealed that predators of infaunal macrobenthos are important in determining community structure and population densities. In a community studied in Chesapeake Bay infaunal population sizes are limited by predation, not by food or space (Virnstein, 1977). Virnstein (1979) reported results of manipulative field experiments of a subtidal sandy bottom community. The effect of predation by blue crabs (*Callinectes*) and fishes on all species of infaunal macrobenthos was studied. Those benthos species which had tough tubes, which lived deep in the sediment, or which could quickly retract deep into the sediment were shown experimentally not to change much in abundance regardless of whether predators were excluded or included. These species were generally the numerical dominants in the natural community. Other species which lived near the surface, or exposed on the surface, responded to experimentally altered predation intensity with large changes in abundance. These species were either uncommon or only sporadically abundant in the natural community. According to Virnstein (1979) this evidence indicates that the abundant species in the natural community are abundant because they avoid predators.

Virnstein (1978) states that one must not assume that the only effect of caging is predator exclusion or inclusion. Cages may alter the physical environment or attract large predators: cage experiment must therefore be planned carefully, and cautiously interpreted. Reise (1977, 1978) investigated the intensity of predation pressure exerted on the macrofauna in intertidal mud flats in the German Bight. No protection effect was achieved with cages constructed of 20 mm mesh nylon net; however, cages with a screen of 5 mm mesh or smaller resulted in a considerable increase of the infauna. Arntz (1980) gives a detailed review of cage experiments on benthic ecosystems.

The rate of total oxygen uptake by the sediment surface has been assumed to be an integrated measure of metabolic activity in the sediment column

(Teal and Kanwisher, 1961). However, in shallow water down to approximately 200 m benthic algae are able to produce oxygen by
photosynthesis.

An example of an *in situ* record of oxygen production from a *Fucus*-
community is depicted in Fig. 2. This experiment was carried out in June,
1976, near Kämpinge in the Baltic Sea at 1.3 m depth. It is apparent from
this experiment that there is a striking diurnal variation of the rate of
oxygen production of phytobenthos (Schramm, 1979). On the other hand,
oxygen is consumed by biological and chemical processes, and total oxygen
uptake of the sediment can be measured directly by oxygen sensors. To
obtain the chemical oxygen demand, poison (e.g. formaldehyde) is injected
into a replicate enclosure to eliminate biological activity. Benthic community respiration can be calculated as the difference between total and
chemical oxygen consumption. Respiration by bacteria is measured by

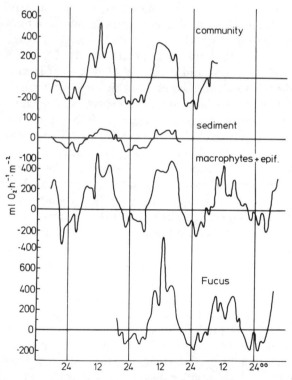

FIG. 2   Diel changes of O_2 production and consumption from a *Fucus* community
at 1.3 m depth in the Baltic. Oxygen values in $mlO_2m^{-2}h^{-1}$ are given for the
entire community, the sediment, macrophytes + epifauna and for *Fucus*. (After
Schramm, 1979)

treating the sample with antibiotics (e.g. streptomycin-SO_4 and neomycin). Estimates of bacterial respiration can be determined by subsequent subtraction of oxygen uptake rates after antibiotic treatment from total community respiration values. The community respiration may be compartmentalised in respiration of macro-, meio- and microfauna, microflora and bacteria.

Data of total oxygen uptake by sediments have been summarised by Pamatmat (1968) and Zeitzschel and Davies (1978). Absolute values in shallow water down to 200 m depth range from 2.6 to 70.3 $mlO_2m^{-2}h^{-1}$. The average value is 18.3 $mlO_2m^{-2}h^{-1}$ ($n = 123$). Values in deep water (>200 m) range from 0.02 to 7.2 $mlO_2m^{-2}h^{-1}$ yielding an average of 2.0 ($n = 83$). Total oxygen uptake can be divided into biological and chemical oxygen consumption. In shallow water the biological portion accounts for 70 to 85% of the total whereas the chemical demand varies from 15 to 30% of the total oxygen consumption. In deep water the biological consumption is around 85% the chemical less than 15%. If total community respiration is compartmentalised the macrofauna accounts for about 25% meio- and microfauna for about 30% whereas bacterial respiration is on average as high as 45% of total community respiration|(chapter 10). In Fig. 3, 206

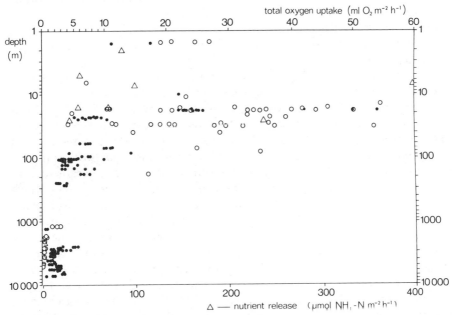

FIG. 3   Rate of total oxygen consumption of the sediment versus depth ($n = 206$) and flux of ammonia at the sediment water interface ($n = 10$). $\bigcirc$, O_2 values from *in situ* experiments; $\bullet$, O_2 values from grab- and core samples; $\triangle$, flux of ammonia. (After Zeitzschel, 1980)

measurements of total oxygen uptake from *in situ* experiments and from core samples are depicted in relation to depth. Not included in this graph are the data by Smith (1973) from measurements off Sapelo Island (Georgia) which range between 56.7 and 110.2 $mlO_2m^{-2}h^{-1}$. Also omitted are the data from Jansson (1969, 140 $mlO_2m^{-2}h^{-1}$ from the Baltic) and high values from specific environments like coral-reef communities. Salt marsh mud in summer has shown the highest rates of oxygen uptake (190 $mlO_2m^{-2}h^{-1}$) for bare soft sediments (Duff and Teal, 1965; Pamatmat, 1977).

It is apparent from Fig. 3 that there is a marked decrease of total oxygen consumption with depth. The data suggest a significant relationship in depths exceeding 100 m. Values from samples from less than 100 m have a much wider range, depending on the communities studied. It should be noted that at the time of writing there are data from only 17 *in situ* measurements recorded from depths exceeding 200 m. Smith (1978) published data from *in situ* measurements of benthic community respiration at 9 stations along the Gray Head-Bermuda transect from depths of 40 to 5200 m. Benthic community respiration rates spanned three orders of magnitude, decreasing from 21.5 $mlO_2m^{-2}h^{-1}$ at 40 m to 0.02 $mlO_2m^{-2}h^{-1}$ at 5200 m. Rates decreased two orders of magnitude between 40 and 1800 m and then significantly declined again between the continental rise (3650 m) and the abyssal plain. Predictive equations for benthic community respiration along the transect reflect a strong correlation with depth of water.

Pamatmat (1971a, 1973) has shown that there is a close agreement between *in situ* and core measurements of community metabolism from samples obtained in Puget Sound (Washington) down to 185 m depth. For deep sea cores, however, the effect of decompression and temperature change is noticeable. Smith and Pamatmat (Smith, 1978) compared data for deep sea samples measured with a grab respirometer *in situ*, and core samples. The data obtained indicate that for samples at comparable depths and environmental conditions the shipboard method yielded values which were generally an order of magnitude higher than the *in situ* measurement. This is apparent also from Fig. 3. Pamatmat (1971b) investigated the seasonal cycle of *in situ* oxygen consumption, chemical oxidation, respiration, oxygen tension, salinity and temperature at a station in Puget Sound, Washington. He comes to the conclusion that these parameters vary with the seasons. Similar findings have been recorded by Pamatmat and Banse (1969), Smith (1973), and Smith *et al.* (1972, 1974). Pamatmat and Banse (1969) suggest that the seasonal changes of rates, and possibly the differences between stations in Puget Sound, are caused primarily by changes of activity of small organisms as governed by the rate of supply of organic matter from the plankton.

According to Pamatmat (1977) it should be pointed out that present estimates of benthic oxygen consumption are based on measurements from undisturbed sediments, while in nature sediments are disturbed periodically by tidal currents (Pamatmat, 1971a), aperiodically by burrowing organisms (Rhoads, 1974; Rhoads et al., 1975), by gas ebullition, or by wind- and wave-generated turbulence, which may be aperiodic as well as seasonal. Even if annual oxygen uptake of disturbed sediments could be estimated, it would still underestimate benthic community metabolism for various reasons. Pamatmat (1975) lists some of these: (i) some by-products of anaerobic metabolism such as N_2 gas and CH_4 are not oxidised by oxygen; (ii) lactate and other by-products of fermentation are of the same oxidation state as the original substrate; (iii) chemical oxidation of sulfides may result in formation of sulfite, thiosulfates etc. of lower oxidation state than sulfate; (iv) furthermore, sulfide from sulfate reduction could be tied up in the formation of humus, which is extremely refractory.

The underestimates caused by all these processes could be negligibly small, but the expectation of a simple relation between oxygen consumption and benthic community metabolism appears too simplistic in view of the complicated dynamic sequence of oxidation-reduction that takes place in anaerobic sediments (Pamatmat, 1977).

Pamatmat (1977) has summarised all available information on the various factors that influence oxygen consumption at the sea bed. Among the most important parameters are the supply of organic matter to the sediment, temperature and oxygen concentration of the bottom water as well as the water flow above the sediment.

The relation between oxygen consumption and primary production in the overlying water column and the content of organic matter in the sediment has been discussed by various authors: Davies (1975), Hargrave (1969, 1973), Pamatmat (1971b; 1973), Rowe (1971), Smith and Teal (1973), Smith et al. (1974).

Primary production in shallow water, both by phytoplankton and phytobenthos, is responsible for most of the energy required by benthic animals. There are diverse arguments, however, whether or not there is a direct input of organic matter produced in the euphotic zone down to the deep sea.

Hargrave (1973) studied the coupling of carbon flow through some pelagic and benthic communities. He concludes that sediment core oxygen uptake is correlated positively with primary production in various aquatic ecosystems for which the input of organic matter is predominantly autochthonous.

Pamatmat (1971b) published data on rates of total oxygen consumption in January and July as related to organic carbon content of the sediment.

From these data from Puget Sound (Washington) no simple correlation can be calculated. Pamatmat concludes that there is good evidence that one important factor causing variability in attempted correlations between rates and organic matter concentration is the flux of organic matter to the sediment. Smith and Teal (1973) suggest from measurements at 1850 m depth that the accumulation of biological oxidisable organic matter is negligible.

According to Smith (1978) the rate of sedimentation of organic matter to the bottom is a very important parameter influencing benthic community respiration, but few measurements of particle flux have been made to date in the open ocean (Zeitzschel, 1980). Preliminary measurements of the sedimentation rate of particulate organic matter have recently been made at three stations along the Gray Head-Bermuda transact which corresponds closely to the location studied by Smith (1978). Sedimentation rates were measured by Rowe and Gardner (1978) with cylindrical PVC traps set for periods of 5 to 15 days from May to September 1976. According to Smith (1978), annual benthic community respiration at three stations in the NW Atlantic could account for the utilisation of from 1 to 2% of the surface productivity, but of the 2 to 7% of surface produced organic carbon which reaches the bottom, 15 to 29% is utilised by the benthic community.

Smith *et al.* (1974) showed that there is a linear relationship of the analysis of total oxygen uptake and chemical oxygen demand versus temperature. There is good evidence that there is a statistically significant correlation between bottom-water temperature and total oxygen uptake as well as chemical and biological oxygen demand of sediments. Similar findings have been described by Smith *et al.* (1972), Smith (1973) and Davies (1975).

There are two fundamentally different approaches to obtain data on the nutrient flux at the sediment–water interface: (*i*) calculation using concentration gradients of constituents of the interstitial water or the water column and (*ii*) measurement of flux directly by *in situ* or laboratory experiments.

For the second approach, the direct measurement of the flux of dissolved constituents, a variety of experimental *in situ* and laboratory apparatus has been designed and described in the literature. Most of them are, however, identical with those devices which are in use for oxygen consumption measurements (Fig. 1).

According to Hargrave and Connolly (1978), the flux of dissolved material into or out of undisturbed sediments can be calculated (m^{-2}) as:

$$\frac{V(C_0 - C_T)}{A} \times \frac{10^4}{T}$$

where V is the volume (liters) of water over the sediment, C_0 and C_T are the dissolved concentrations ($liter^{-1}$) before and after time T, and A is the sediment area (cm^2) enclosed. The calculation requires that the water is homogeneously mixed, that changes in concentration are known or assumed to be linear over time, and that dissolved material is only exchanged at the sediment–water interface.

Actual nutrient flux data from the marine environment are relatively scarce. A selection of published data has been summarised by Zeitzschel (1980).

In Fig. 3 nutrient flux data for ammonia are plotted against depth as was done previously for total oxygen consumption.

It is apparent from these data that they follow the trend already discussed for oxygen consumption data; relatively high nutrient fluxes occur in shallow water, whereas in deep water the values are several orders of magnitude less.

The release of different inorganic nutrients is by no means uniform. According to Nixon et al. (1976) almost all of the inorganic nitrogen released from the bottom was in the form of ammonia. Nitrite fluxes were, in that study, virtually always below the level of detection. All the nitrate in the water was derived from pelagic processes, and the benthic community did not participate directly in the seasonal nitrate cycle in the water.

According to Smith et al. (1978), the basic pattern of nutrient exchange at two deep water stations (2200 and 2750 m) are regeneration of ammonia and uptake of nitrate and phosphate by the benthic community. The authors pointed out that the evolution of ammonia suggests that denitrification processes occur, but not reducing conditions, were observed visually to the maximum depth of grab penetration. The ammonia measured might be a normal excretory product of the fauna, but the uptake of nitrate and nitrite was not easily explained since dissolved oxygen was present as a hydrogen acceptor (Smith et al., 1978).

Chemical reactions at the water–sediment interface fundamentally influence, and are influenced by, biological and physical processes in and above the sediments. Field experiments have shown that four major factors effect the nutrient release across the water-bottom interface. These are: the influence of the input of organic matter; bioturbation by benthos; and the effects of temperature and water flow at the bottom. There is good evidence from experiments that the decomposition of organic matter and the regeneration of nutrients is directly correlated with the input of organic matter to the bottom (Halberg et al., 1973).

The input of organic matter to the sediment is not uniform but takes place as irregular events, depending on physical conditions in the water column and the physiological state of the phytoplankton cells. Time scales of settling events are in the order of a few days (Hargrave, 1975; Smetacek et al., 1978).

High rates of decomposition of organic matter are favoured by high primary production in the overlying water and quick settling and burial to avoid intensive decomposition in the water column. Decomposition in this zone is a function of water depth, turbulence at the sediment–water interface, dissolved oxygen content and rate of burial of enclosing sediment particles (Berner, 1976; Zeitzschel, 1980).

Benthic organisms influence chemical processes at the sediment–water interface. Especially in shallow-water ecosystems, where macrofauna is abundant, intensive biogenic mixing and irrigation of the bottom takes place in the upper few centimeters of the sediment. Mixing activities of benthic organisms, or bioturbation, are important in accelerating vertical diffusion and transport of ions or compounds adsorbed on particles or in solution in pore water (Rhoads, 1974). Nixon et al. (1976) studied the seasonal aspects of nutrient flux of coastal marine bottom communities at three stations in the West Passage of Narragansett Bay, Rhode Island. This study revealed a significant relationship between temperature and release of three nutrients.

The fourth factor, the effect of natural water movement on oxygen consumption and nutrient release was studied by Davies (1975), Rhoads et al. (1975) and Vanderborght et al. (1977). Davies concluded that high flow rates (10–20 cm s^{-1}) cause resuspension of the flocculent material, resulting in notable increases in the rate of oxygen consumption. This seems to be related to the amount of reduced products being formed in the sediment, itself dependent on available food supply and the diffusion into the sediment of aerated water. Ammonia release from sediment at water-flow rates of 3 and 11 cm s^{-1} was rather similar. Davies (1975) suspected that there are probably two processes of nutrient release, a slow steady-state diffusion of nutrients into the water column, and a more sporadic release of nutrients occurring when wind driven turbulence causes resuspension of the bottom sediment.

Enclosures of various size have been used in situ or in land-based facilities to obtain reliable data on the influence of perturbations on marine organisms. The main goal of these experiments is to study the long-term effect of low-level perturbations on natural populations.

Nearly all in situ experiments were conducted on pelagic organisms. For benthic organisms landbased flow-through containers have been used successfully (Zeitzschel, 1978). Saward et al. (1975) used 7 m^3 containers to conduct experiments of 100 days in a flow-through mode. The influence of copper on the bivalve Tellina tenuis in these experimental tanks at Loch Ewe, Scotland show that there is a steady increase of copper accumulation with time. Copper accumulation in the soft tissue of this species was greater than in the shell. After 100 days of exposure to dose rates of 10, 30 and 100 μgCu l^{-1} tissue concentrations of 270, 470 and 1100 μgCu g^{-1} dry weight were recorded (Saward et al., 1975).

Preliminary results of the response of the benthos organisms to oil from the land based experimental facility near Narragansett Bay, Rhode Island (Pilson, 1979) were reported by Grassle (1977). Following the addition of oil in February there was a steady divergence in number of individuals and number of species in the oiled tanks versus the control tanks. The difference between oil tanks and controls was significant in June, July and August samples. A similar trend occurred in each of the three common species *Nucula annulata*, *Yoldia limatula* and *Mediomastus ambiseta*. According to Pilson (1979) who summarised the MERL microcosm experiments for the years 1976 through 1978, the effect of oil on the benthos was rather complex, causing a decrease in total benthic macrofauna and meiofauna but allowing considerable increase in benthic diatoms and ciliates. The microcosm ecosystems continued to function in the presence of oil, but pathways by which the energy, and presumably many of the trace substances, moved through the systems must have been considerably altered (Pilson, 1979).

3 Implications of the field experiments on benthic organisms for the understanding of marine ecosystems

Field experiments on benthic organisms are one tool to verify specific rate terms in simulation models on marine ecosystems.

The ecological significance of field experiments may be summarised as follows: there is good evidence that total oxygen consumption and nutrient release follow the same trend. The measured rates are high in shallow water areas, whereas deep water fluxes are 2–4 orders of magnitude less (Fig. 3). This difference is caused (*i*) by the different input of organic matter to the bottom (25–60% of primary production in shallow water ecosystems compared to 1–10% in open ocean systems, Zeitzschel, 1980), (*ii*) by the relatively low metabolic rates and/or activity of deep water organisms and (*iii*) by relatively slow chemical processes due to low temperature (about 1 °C) in deep water. Released nutrient salts from the bottom are only of direct ecological importance if they are supplied at a time when they are needed in the euphotic zone. For instance, a high nutrient flux in boreal areas in winter will not effect the primary productivity in the euphotic zone because light is the major limiting factor during this time. There is, however, good evidence, at least for some areas like Kiel Bight, Bedford Basin, Narragansett Bay and the mid-Atlantic Bight, that nutrients released from the sediment are mixed into the euphotic zone in summer, to play an important role in inducing and maintaining phytoplankton blooms. Rowe *et al.* (1975), Davies (1975), Rowe and Smith (1977) and Hargrave and Connolly (1978) calculated that the release of nutrients from the sediments may account for between 30 and 100% of nutrient requirements of the phytoplankton. This implies that in shallow water ecosystems bottom regeneration keeps pace with primary production and that primary producers in these areas depend very little on pelagic regenerative processes (Rowe and Smith, 1977).

References

Arntz, W. (1977). Results and problems of an "unsuccessful" benthos cage predation experiment. In "Biology of Benthic Organisms" (B. F. Keegan, P. O. Céidigh and P. J. S. Boaden, eds) pp. 31–44. Pergamon Press, New York

Arntz, W. (1980). Predator exclusion-inclusion experiments – a review. Ber. dt. wiss. Kommn. Meeresforsch. (in press)

Balzer, W. (1978). Untersuchungen über Abbau organischer Materie und Nährstoff-Freisetzung am Boden der Kieler Bucht beim Übergang vom oxischen zum anoxischen Milieu. Rep. Sonderforschungsbereich 95 – Wechselwirkung Meer-Meeresboden, Kiel University 36, 1–137

Berner, R. A. (1976). The benthic boundary layer from the viewpoint of a geochemist. In "The Benthic Boundary Layer" (I. N. McCave, ed.) pp. 35–55. Plenum, New York and London

Blegvad, H. (1928). Quantitative investigations of bottom invertebrates in the Limfjord, 1910–1927, with special reference to the plaice-food. Rep. Danish Biol. Sta. 34, 33–52

Davies, J. M. (1975). Energy flow through the benthos in a Scottish sea loch. Mar. Biol. 31, 353–362

Duff, S. and Teal, J. M. (1965). Temperature change and gas exchange in Nova Scotia and Georgia salt-marsh muds. Limnol. Oceanogr. 10, 67–73

Edberg, N. and v. Hofsten, B. (1973). Oxygen uptake of bottom sediments studied in situ and in the laboratory. Water Res. 7, 1285–1294

Grassle, F. (1977). MERL Benthos report. MERL Newsletter 5, 4–5. Graduate School of Oceanography. Narragansett Marine Laboratory, University of Rhode Island, Kingston, R.I.

Grøntvedt, J. (1962). Preliminary report on the productivity of microbenthos and phytoplankton in the Danish Wadden Sea. Meddr. Danm. Fisk.-og Havunders. N.S. 3, 347–378

Hall, Ch. A. S. and Tempel, N. A benthic chamber for intensely metabolic lotic systems. (Manuscript)

Hallberg, R. O., Bågander, L. E., Engvall, A.-G. and Schippel, F. A. (1972). Method for studying geochemistry of sediment–water interface. Ambio 1, 71–72

Hallberg, R. O., Bågander, L. E., Engvall, A.-G., Lindstrøm, M., Oden, S. and Schippel, F. A. (1973). The chemical microbiological dynamics of the sediment–water interface. Contr. from ASKØ Lab., Univ. Stockholm 2, 1, 1–117

Hargrave, B. T. (1969). Epibenthic algal production and community respiration in the sediments of Marion Lake. J. Fish. Res. Bd Can. 26, 2003–2026

Hargrave, B. T. (1973). Coupling carbon flow through some pelagic and benthic communities. J. Fish. Res. Bd Can. 30, 1317–1326

Hargrave, B. T. (1975). The importance of total and mixed-layer depths in the supply of organic matter to bottom communities. Symp. Biol. Hung. 15, 157–165

Hargrave, B. T. and Connolly, G. F. (1978). A device to collect supernatant water for measurement of the flux of dissolved compounds across sediment surfaces. *Limnol. Oceanogr.* **23**, 1005–1010

Jansson, B.-O. (1969). Factors and fauna of a baltic mud bottom. *Limnologica* **7**, 47–52

Lasserre, P. (1976). Metabolic activities of benthic microfauna and meiofauna: Recent advantage and review of suitable methods of analysis. *In* "The Benthic Boundary Layer" (I. N. McCave, ed.) pp. 95–142. Plenum Press, New York and London

Mann, K. H. (1972). Ecological energetics of the seaweed zone in a marine bag on the Atlantic coast of Canada. II. Productivity of seaweeds. *Mar. Biol.* **14**, 199–209

McIntyre, A. D., Munro, A. L. S. and Steele, J. H. (1970). Energy flow in a sand ecosystem. *In* "Marine Food Chains" (J. H. Steele, ed.) pp. 19–31. Oliver and Boyd, Edinburgh

Menzel, D. W. and Steele, J. H. (1978). The application of plastic enclosures to the study of pelagic marine biota. *Rapp. P.-v. Réun. Cons. int. Explor. Mer* **173**, 7–12

Naqui, S. M. Z. (1968). Effects of predation on infaunal invertebrates of Alligator Harbor, Florida. *Gulf Res. Rept.* **2**, 313–321

Nielsen-Aertebjerg, A. (1976). Undersøgelse over microbenthos algernes primaer produktion, samt pigment inholdet i sedimentet i det Sydfynske Øhav. Thesis University of Copenhagen

Nixon, S. W., Oviatt, C. A. and Hale, S. S. (1976). Nitrogen regeneration and the metabolism of coastal marine bottom communities. *In* "The Role of Terrestrial and Aquatic Organisms in Decomposition Processes" (J. M. Anderson and A. MacFadyen, eds) pp. 269–283. Blackwell, Oxford

Pamatmat, M. M. (1965). A continuous-flow apparatus for measuring metabolism of benthic communities. *Limnol. Oceanogr.* **10**, 486–489

Pamatmat, M. M. (1966). The ecology and metabolism of a benthic community on an intertidal sandflat. (False Bay, San Juan Island, Washington). Ph.D. Thesis, University of Washington, Seattle

Pamatmat, M. M. (1968). Ecology and metabolism of a benthic community on an intertidal sandflat. *Int. Revue ges. Hydrobiol. Hydrogr.* **53**, 211 298

Pamatmat, M. M. (1971a). Oxygen consumption by the seabed. IV. Shipboard and laboratory experiments. *Limnol. Oceanogr.* **16**, 536–550

Pamatmat, M. M. (1971b). Oxygen consumption by the seabed. VI. Seasonal cycle of chemical oxidation and respiration in Puget Sound. *Int. Revue ges. Hydrobiol. Hydrogr.* **56**, 769–793

Pamatmat, M. M. (1973). Benthic community metabolism on the continental terrace and in the deep sea in the north Pacific. *Int. Revue ges. Hydrobiol. Hydrogr.* **58**, 345–368

Pamatmat, M. M. (1975). *In situ* metabolism of benthic communities. *Cah. Biol. Mar.* **16**, 613–633

Pamatmat, M. M. (1977). Benthic community metabolism: a review and assessment of present status and outlook. *In* "Ecology of Marine Benthos" (B. C. Coull, ed.) pp. 89–111. University of South Carolina Press, Columbia

Pamatmat, M. M. and Fenton, D. (1968). An instrument for measuring subtidal benthic metabolism in situ. *Limnol. Oceanogr.* **13**, 537–540

Pamatmat, M. M. and Banse, K. (1969). Oxygen consumption by the seabed. II. *In situ* measurements to a depth of 180 m. *Limnol. Oceanogr.* **14**, 250–259

Petersohn, U. and Diekmann, P. (1978). An adaptable radio-controlled measuring system for oceanographic research in shallow-water. *IEEE J. of Oceanic Engineering* OE^{-3}, 60–64

Pilson, M. E. Q. (1979). The MERL microcosms. A summary of results from 1976 through 1978. *The Marine Ecosystem Research Laboratory*, Univ. Rhode Island, Kingston, R.I.

Reise, K. (1977). Predator exclusion experiments in an inter-tidal mud flat. *Helgoländer wiss. Meeresunters.* **30**, 263–271

Reise, K. (1978). Experiments on epibenthic predation in the Wadden Sea. *Helgoländer wiss. Meeresunters.* **31**, 55–101

Rhoads, D. C. (1974). Organism-sediment relations on the muddy sea floor, *Oceanogr. Mar. Biol. Ann. Rev.* **12**, 263–300

Rhoads, D. C., Tenore, K. and Browne, M. (1975). The role of resuspended bottom mud in nutrient cycles of shallow embayments. *Estuar. Res.* **1**, 563–579

Rowe, G. T. (1971). Benthic biomass and surface productivity. *In* "Fertility of the Sea" (J. D. Costlow Jr., ed.) Vol. 2, pp. 441–454. Gordon and Breach, New York

Rowe, G. T. and Smith, K. L. Jr. (1977). Benthopelagic coupling in the Mid-Atlantic Bight. *In* "Ecology of Marine Benthos" (B. Coull, ed.) pp. 55–65. University of South Carolina Press, Columbia

Rowe, G. T. and Gardner, W. D. (1978). Sedimentation rates in the slope water of the Northwest Atlantic Ocean measured directly with sediment traps. *Contr. NO 4087 from the Woods Hole Oceanographic Institution*

Rowe, G. T., Clifford, C. H., Smith, K. L. Jr. and Hamilton, P. L. (1975). Benthic nutrient regeneration and its coupling to primary productivity in coastal waters. *Nature, Lond.* **255**, 215–217

Saward, D., Stirling, A. and Topping, G. (1975). Experimental studies on the effect of copper on a marine food chain. *Mar. Biol.* **29**, 351–361

Schramm, W. (1973). Langfristige *in situ* – Messungen zum Sauerstoffgaswechsel benthischer Meerespflanzen in einem kontinuierlich registrierenden Durchflußsystem. *Mar. Biol.* **22**, 335–339

Schramm, W. (1979). Phytobenthos Tätigkeitsbericht 1977–79 Sonderforschungsbereich 95. *Wechselwirkung Meer-Meeresboden, Kiel University*

Schramm, W. und Martens, V. (1976). Ein Meßsystem für *in situ* Untersuchungen zum Stoff- und Energieumsatz in Benthosgemeinschaften. *Kieler Meeresforsch. Sonderheft* **3**, 1–6

Smetacek, V., v. Bröckel, K., Zeitzschel, B. and Zenk, W. (1978). Sedimentation of particulate matter during a phytoplankton spring bloom in relation to the hydrographical regime. *Mar. Biol.* **47**, 211–226

Smith, K. L. Jr. (1973). Respiration of a sublittoral community *Ecology* **54**, 1065–1075

Smith, K. L. Jr. (1974). Oxygen demands of San Diego trough sediments: An *in situ* study. *Limnol. Oceanogr.* **19**, 939–944

Smith, K. L. Jr. (1978). Benthic community respiration in the N.W. Atlantic Ocean: *in situ* measurements from 40–5200 m. *Mar. Biol.* **47**, 337–347

Smith, K. L. Jr. and Teal, J. M. (1973). Deep-sea benthic community respiration: An *in situ* study at 1850 meters. *Science* **179**, 282–283

Smith, K. L. Jr. and Hessler, R. R. (1974). Respiration of benthopelagic fishes: *In situ* measurements at 1230 m depth. *Science* **18**, 72–73

Smith, K. L. Jr., Burns, K. A. and Teal, J. M. (1972). *In situ* respiration of benthic communities in Castle Harbor, Bermuda. *Mar. Biol.* **12**, 196–199

Smith, K. L. Jr., Rowe, G. T. and Clifford, C. H. (1974). Sediment oxygen demand in an outwelling and upwelling area. *Tethys* **6**, 223–230

Smith, K. L. Jr., White, G. A., Laver, M. B. and Haugsness, J. A. (1978). Nutrient exchange and oxygen consumption by deep-sea benthic communities: Preliminary *in situ* measurements. *Limnol. Oceanogr.* **23**, 997–1005

Smith, K. L. Jr., Clifford, C. H., Eliason, A. H., Walden, B., Rowe, G. T. and Teal, J. M. (1976). A free vehicle for measuring benthic community metabolism. *Limnol. Oceanogr.* **21**, 164–170

Teal, J. M. and Kanvisher, J. (1961). Gas exchange in a Georgia salt marsh. *Limnol. Oceanogr.* **6**, 388–399

Vanderborght, J.-P., Wollast, R. and Billen, G. (1977). Kinetic models of diagenesis in disturbed sediments. 1. Mass transfer properties and silica diagenesis. *Limnol. Oceanogr.* **22**, 787–793

Virnstein, R. W. (1977). The importance of predation by crabs and fishes on benthic infauna in Chesapeake Bay. *Ecology* **58**, 1199–1217

Virnstein, R. W. (1978). Predator caging experiments in soft sediments: Caution advised. *In* "Estuarine Interactions" (M. L. Wiley, ed.) pp. 261–273, Academic Press, New York

Virnstein, R. W. (1979). Predation on estuarine infauna: response patterns of component species. *Estuaries* **2**, 69–86

Zeitzschel, B. (1978). Controlled environment experiments in pollution studies. *Ocean Management* **4**, 319–344

Zeitzschel, B. (1980). Sediment-water interactions in nutrient dynamics. *In* "Marine Benthic Dynamics" (K. R. Tenore and B. C. Coull, eds) pp. 195–218. University of South Carolina Press, Columbia

Zeitzschel, B. and Davies, J. M. (1978). Benthic growth chambers. *Rapp. P.V. Réun Cons. int. Explor. Mer* **173**, 31–42

21 Microcosms and Experimental Planktonic Food Chains

CARL M. BOYD

Oceanographers frequently feel themselves handicapped by the vastness of their chosen environment and its fluidity, and bemoan that it is impossible to obtain replicated samples of nutrients or populations of plankton. The ocean has moved, the ship has drifted, or the animals have migrated to cause differences between repeated samples that reflect, for example, the spatial distribution of phytoplankton and not changes in the original population of phytoplankton. A fundamental dichotomy in the approach of oceanographers is apparent at this point. One group maintains that the ocean must be studied as a whole, that the interactions between currents, algal growth and zooplankton migrations cannot be separated in the laboratory, and that the ocean is a system that must be studied as an ensemble of processes. The success of this school is impressive, and it is obvious, for example from the work of H. U. Sverdrup and G. A. Riley, that sea-going oceanographers can learn a great deal about their medium.

A second group despairs that the sea is too complicated for direct scrutiny, and argues that it can be better studied as components in the laboratory. Here again, difficulties of scaling arise when it is found that the restricted size of laboratory vessels has introduced uninterpretable artifacts.

Obvious solutions to the dilemma would seem to be to enclose volumes of the sea and to enlarge the laboratory vessel, and it is these solutions that have led to the concept of the 'microcosm', a container of sufficiently large volume to enclose several elements of the biota within confines that allow the elements to be observed and manipulated. The requirement for a microcosm has come both from marine ecologists, who wish to observe natural processes, and from environmentalists, who wish to test the effects of polluting compounds in order to establish guidelines of acceptable concentrations in advance of a catastrophic release of pesticides, hydrocarbons, heavy metals, or radioisotopes.

1 Description of systems

TOWER TANKS

The most logical form of a large experimental vessel is simply an extrapolation in size of the conventional aquarium. The first of these was a 'vertical plankton shaft' 2 m in diameter and 12 m deep ($\sim$ 38 m^3 volume) built prior to World War II at the Oceanographic Institute in Gothenburg, Sweden (Pettersson et al., 1939). Even though the ability to sterilise, filter, cool and illuminate water in the tank gave the experimenter a high degree of control, the tank did not result in extensive publication, a characteristic also shown by the second tower tank, built at the Laboratory for Experimental Limnology, Maple, Ontario, in 1948. This tank, which was 3 m in diameter and 6 m high (42 m^3), was made of steel wrapped in a thermal insulation, and had catwalks at several depths that gave the observer access to viewing ports. A unique feature of this tank was a dome-shaped cover that could be secured to the top, thereby allowing the water in the tank to be pressurised to a few atmospheres to simulate additional depth or to provide pressure change as an experimental variable. The tank was used primarily to study acclimation of fresh water fishes as they responded to temperature changes induced by a cooling fluid that flowed through pipes wound helically on the interior of the tank. This tank as well as the plankton shaft at Gothenburg surely gave the experimenter and observer a better understanding of the problems being investigated, but neither facility resulted in extensive publications.

The two tower tanks presently in use were built at Scripps Institution of Oceanography, La Jolla, California in 1965 and at the Institute of Oceanography, Dalhousie University, Halifax, Nova Scotia, Canada in 1974, and are described by Balch et al., 1978. As typical of the genre, the tower tank at Dalhousie is a cylinder 10.46 m deep and 3.66 m in diameter (volume 108 m^3) made of reinforced concrete with an inner surface of epoxy cement covered with a sprayed-on layer of polyvinyl chloride (Fig. 1). The tank has 26 glass viewing ports 37 cm in diameter placed at eye level along a spiral stairway around the tank (Fig. 1). The plumbing system (Fig. 1a) of the tank is triplicated; each third consists of a sand bed filter, a pump, heat exchangers to heat or cool the water, and a series of inlet-outlet ports through which the water may be withdrawn and re-introduced (Fig. 1b). The ports are arranged to effectively divide the tank into three vertically stacked layers. Flexibility of design allows the experimenter to select components or water source (sea water or fresh water or a mixture) by adjusting the appropriate valves. The tank is structurally isolated from the oceanography building into which it is incorporated, and rests on neoprene pads that de-couple it from sources of high frequency noise associated with

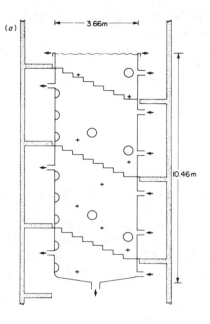

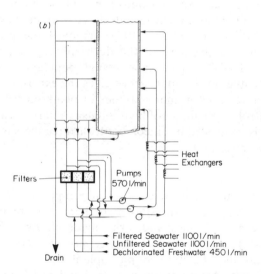

Fɪɢ 1 (a) Sectional view of Dalhousie Tower Tank showing sections of circular stairwell with landings. Circles and semi-circles indicate viewing ports, crosses denote sampling ports, arrows direction of water flow. Depth, 10.46 m; diameter, 3.66 m; volume, 108 m³. (b) Water delivery and conditioning system for Dalhousie tank. (From Balch et al., 1978)

any large building; this was done so that the behaviour of organisms would be affected as little as possible.

It has proven impossible to illuminate a tower tank by natural daylight: the experimenter may (and usually does) wish to alter the daylight cycle, the tank may be shadowed by adjacent buildings, and invariably the low angle of the sun places the lower section of the tank in a shadow. The solution to these problems has been to enclose the tower tank in a light-tight structure and to illuminate it with an artificial light source located directly above the tank.

The tower tank at Dalhousie is illuminated by four 1000 watt phosphor-coated metal halide lamps plus two 500 watt mercury vapour lamps; the phosphor-coated lamps were selected according to their spectral qualities relative to the quantum efficiency of chlorophyll. The lights may be individually switched on or off by clock-controlled switches in order to simulate dawn and dusk, and at 'high noon' provide an intensity at the water surface of 10–40% of that of a summer day – the variance being associated with uneven levels of illumination that are inherent in point-source lights.

The problem of illuminating the Scripps tower tank has been solved by three approaches; (a) the light from a 152 cm diameter searchlight located outside the building was projected onto a mirror positioned over the tank which reflected the light as a vertical shaft down into the tank (Strickland et al., 1969). (b) Mullin and Evans (1974) removed the problem of light intensity by operating the tank in the dark; phytoplankton cells grown elsewhere were added to maintain copepod populations for periods of several weeks. Neither of the approaches gives the experimenter the control over light that is required in many studies of plankton behaviour, and to achieve this control a system, (c) of high intensity quartz incandescent lights whose intensity is controlled by silicon controlled rectifiers is being installed over the Scripps tower tank.

'IN SITU' ENCLOSURES

The concept of isolating large volumes of water in plastic films is deceptively simple, but its use in aquatic studies perhaps was retarded until the expansion of the petrochemical industries made plastics readily available. Edmondson (1942) and Edmondson and Edmondson (1955) had followed the growth of natural populations of phytoplankton in large concrete tanks at Woods Hole, but McAllister, et al. (1961) and Strickland and Terhune (1961) were the first to enclose a large quantity of water in a plastic balloon, to tether this in an arm of the sea in British Columbia, and to monitor the progression of the isolated phytoplankton population (Fig. 2). At last the oceanographer had a quantity of sea water that was subject (more or less) to the same processes as the water which surrounded it but which would not drift away and could therefore be sampled repeatedly. The technique, which must be credited to Strickland and Terhune, has received considerable inspection, adulation, and criticism that ranges from the comment (Parsons, 1978) that Strickland and Terhune had established a *cause célèbre*, to that of Verduin who 'blamed' the invention of the 'big bag' on McAllister *et al.* (Verduin, 1969). By the time the technique was employed again by personnel from the same laboratory (Antia *et al.*, 1963), a proliferation of shapes and designs of plastic enclosures had commenced. Goldman (1962) modified the enclosures to become a simple cylinder of plastic film which, suspended at the surface from a buoyed frame, draped down in the water column to be closed at the bottom by a plastic drumhead, a plankton net screen or a conical sediment trap. Goldman's modification was used in that decade by Geen and Hargrave (1966) in a Nova Scotian bay, and by Kemmerer (1968); in the next decade the scientific literature based on big bags or other 'microcosms' had expanded to approximately 100 publications. Sizes, shapes, and construction materials have been considered by Menzel and Steele (1978) and by Case (1978) who gave detailed engineering information concerning the zenith of the design in the decade of the 1970s, the Controlled Ecosystem Enclosure used in the Controlled Ecosystem Pollution Experiment (CEPEX).

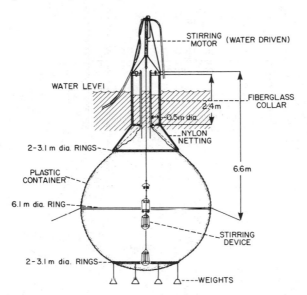

FIG. 2   Submerged spherical container used by Strickland and Terhune (1961) in Departure Bay, B.C., Canada. Diameter 6.1 m; volume 125 m³. (From Strickland and Terhune, 1961)

The final design of Controlled Ecosystem Enclosure used by CEPEX was a cylinder 9.5 m diameter and 29 m deep containing 1700 m³ of water that was effectively suspended from a doughnut-shaped float (Fig. 3) (Menzel and Case, 1977). The lower end of the flexible polyethylene enclosure was terminated in a conical section that allowed sedimenting material to collect at the apex. The enclosures were deployed by collapsing sides of the cylinder to the depth of the conical bottom, and then quickly hoisting the curtain-like walls with ropes up to the surface float. The water that had been trapped in the process had retained its continuity of stratification along with its associated nutrients, other chemical parameters and planktonic populations. The microcosm could be manipulated by addition of toxic or

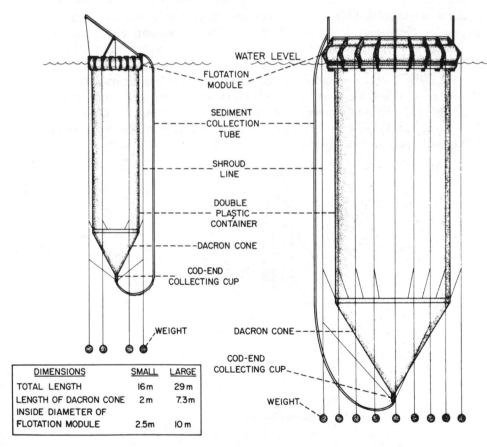

DIMENSIONS	SMALL	LARGE
TOTAL LENGTH	16 m	29 m
LENGTH OF DACRON CONE	2 m	7.3 m
INSIDE DIAMETER OF FLOTATION MODULE	2.5 m	10 m

FIG. 3   Two Controlled Ecosystem Enclosures employed in the CEPEX project in Saanich Inlet, B.C., Canada. The volume of the small enclosure was 64 m³, the large contained 1700 m³. (From Menzel and Steele, 1978)

beneficial compounds, by mixing, etc., and could be sampled repeatedly to give the experimenter a seemingly ideal world in which to test concepts that had been previously elusive (it appeared) when the experimenter was limited to laboratory flasks or the high seas, something 'between beakers and bays', as Strickland wrote (1967). The CEPEX project employed enclosures of 64 m³, 1300 m³, and 1700 m³ volume, and as many as six of the small units were used simultaneously to permit permutations of experimental design and controls. Variations on the theme have been used in several other regions, notably in a Scottish loch (Gamble *et al.*, 1977; Fig. 4) in the

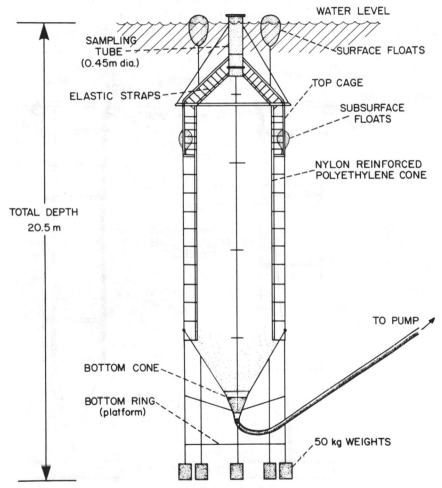

FIG. 4   Experimental container used by Davies and Gamble (1979) in Loch Ewe, Scotland. Volume, 95 m³. (From Davies and Gamble, 1979)

Mediterranean (Berland *et al.*, 1975; Fig. 5), and in Kiel Bay in Germany (von Bodungen *et al.*, 1976; Fig. 6). This latter chamber extended down to the mud-water interface and was oriented toward the investigation of chemical phenomona across this boundary; the study is in contrast to most big bag studies that have attempted to detach themselves from the benthic environment. The enclosures used by Gamble were made of polyvinyl chloride plastic, a substance avoided by Case (*loc. cit.*) in the design of the CEPEX bags; Case used polyethylene rather than PVC on the suspicion that the plasticisers in PVC would be toxic. In chosing polyethylene Case was forced also to accept its structural weakness; polyethylene has about half the strength of PVC. Gamble rinsed the PVC bags employed in Loch Ewe, and felt that possible toxicity of the materials – which he could not

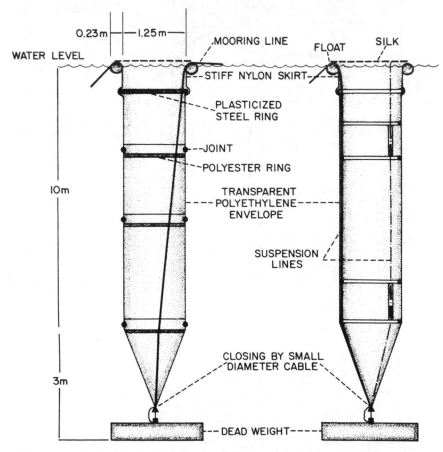

FIG. 5 Experimental enclosure used by Berland *et al.*, 1975, in the Calanque d'En-Vau, near Marseille, France. Volume, 10 m³. (From Berland *et al.*, 1975)

detect – was soon masked by fouling organisms on the plastic that reduced diffusion of toxic materials.

SHORE-BASED MICROCOSMS

The Marine Ecosystems Research Laboratory (MERL) on Narragansett Bay, University of Rhode Island, constitutes a marriage of the tower tanks with the serial perturbation experiments that were the original goal of the CEPEX project. The approach at Rhode Island has been to confine water in rigid cylindrical tanks 2 m diameter and 5.5 m deep made of fiberglass reinforced polyester (Fig. 7) and to locate several (presently 12) of these seaside silos within a laboratory complex adjacent to Naragansett Bay. Since the universe being emulated was the relatively shallow, well mixed

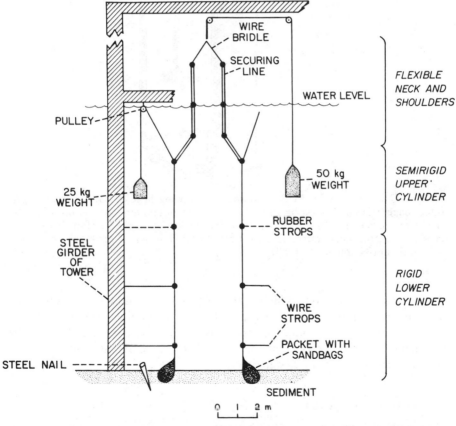

FIG. 6 Experimental enclosure used by von Bodungen et al., 1976 near Kiel, West Germany. The enclosure incorporates a mud-water interface. Volume, 30 m³. (From von Bodungen et al., 1976)

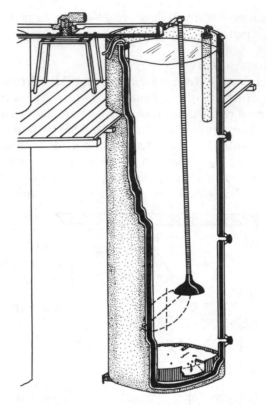

FIG. 7 Cutaway view of MERL microcosm showing heat exchangers hung from upper edge, sampling ports on the side, the sediment container at the bottom, and the mixer device. The mixer motor is located between tanks. The plunger near the centerline of the tank moves in a vertical ellipse shown in dotted lines. Volume, 13 m^3. (From Nixon *et al.*, 1980)

Narragansett Bay, from which water was pumped to fill the tanks, the experiments incorporated a substrate of mud that interacted with the water column to form an analog of the Bay. The system is therefore to be compared with the experiments at Kiel, where the mud-water interface was also a component of the design.

2 Optimal volumes

Although the literature to support the concept is almost non-existent, experience has indicated a relationship between the trophic level of the organism of interest and the amount of water required to allow that organism to assume the essential elements of its behavior. It is this experience, coupled with the pragmatism of finance, that has determined the size of microcosms. Menzel and Steele (1978) describe this relationship as logarithmic, and suggest that herbivorous copepods would require an enclosure of a volume of order of $10^2 m^3$, that planktivorous fishes, ctenophores or chaetognaths would require a volume of $10^3 m^3$, and piscivorous fishes would demand an enclosure of $10^4 m^3$. Obviously ctenophores will live in a few liters of water, but a considerably larger volume is required if a population of ctenophores is to be self-sustaining, and it is perhaps the criterion of reproduction that must be considered as all important.

The relationship, however, has not been rigorously assessed and is not universally accepted. Perez et al. (1977), for example, have presented arguments that a volume of $0.150\,m^3$ is sufficient for simulation of ecosystems that include herbivorous copepods. The use of larger volumes such as those of CEPEX ($1700\,m^3$) is in part based on experience and in part on the intuitive feeling that a larger system will yield results more applicable to the real world. In all these systems a decision must be made as to the highest trophic level to be investigated, and the experimenter must be able to explain the magnitude of difference of phenomena in the microcosm vs those in the real world as they are influenced by exclusion of higher trophic levels. A system including only bacteria and phytoplankton, such as described in the early work of McAllister et al. (1961) will produce anomalously high crops of phytoplankton because no grazers are present, and the experimenter must be able to incorporate this anomaly into the understanding of the system or the experiment will be futile.

Tower tanks were not designed to simulate ecosystems but to study components of a system in isolation; the consideration of size or scale in these installations is based on the requirement that the behavioral pattern under scrutiny is not impeded by the container. Pteropods, for example, apparently swim naturally in a tower tank but in an aquarium of typical size they settle lifelessly to the bottom (Conover and Paranjape, 1977). Copepods (Bohrer, 1980) and squid (Palmer and O'Dor, 1978) will assume predictable patterns of vertical migration in a tower tank, but in an aquarium often respond in a manner impossible to interpret; squid will spawn in a pool of $300\,m^3$ volume, but attempts to study spawning in volumes of a few cubic meters have invariably failed (O'Dor, 1978).

SURFACE TO VOLUME RATIOS

A complication results when organisms grow on the inner walls of microcosms. These organisms block light, metabolise nutrients, are eaten, etc. according to processes that are competitive with those of planktonic organisms in the microcosm, and can be expected to compound problems in the interpretation of nutrient budgets and sunlight penetration (Eppley et al., 1978). Takahashi and Whitney (1977) noted that the fouling community, together with the shadowing effect of the supporting ring of the CEPEX enclosures, reduced the intensity of light inside the bag to about 50% of the outside intensity. The exact relationship, however, varied according to depth, proximity to the supporting ring, and the duration of the experiment. The reduction in light was countered by Gamble et al. (1977), who sent down a team of divers once a week to clean the outer walls of the suspended enclosures ($\sim 110 \text{ m}^3$) moored in Loch Ewe (Fig. 4). Scrubbing the walls has also been the solution at MERL where a mechanised rotary brush is employed every few days to remove algae and other fouling organisms (e.g. tunicates) from the inside walls. Scrubbing the walls of a tank would seem to be a harsh treatment, but no adverse effects on the plankton communities have been noted (Pilson; pers. com.), and the consequences to a microcosm from a build-up of algae, tunicates and mussels over a period of a long experiment would be considerable. Eppley et al. (1978), for example, noted that certain harpacticoid copepods, which are seldom taken as planktonic organism but are associated with benthic communities, became a significant component of the plankton samples taken in a 68 m³ CEPEX enclosure.

The behavior of planktonic organisms inside enclosures is known to alter because the enclosures isolate the organisms from the influences of the turbulent motions associated with wind, waves, and advective processes. Surface to volume ratios suggest that an enclosure of small diameter would pass the thermal changes associated with advection through its walls more rapidly than would occur in an enclosure of large diameter. Such an argument suggests that studies of parameters influenced by turbulence would be better carried out in enclosures of very small diameter (Steele et al., 1977), otherwise turbulence inside the enclosure will be considerably damped relative to the turbulence outside the bag. This solution, however, would maximise the influence of organisms growing on the walls ('wall effects'), and illustrates that an optimum balance exists between diameter and volume, and that – unless mixing is introduced mechanically – enclosures of large volume are of limited use.

REPLICABILITY AND CONTROLS

Perhaps more than in any other phase of experimental marine biological research, the practice of employing replicates and controls of in situ enclosures and tower tanks is impeded by considerations of costs. For example, a study of patterns of vertical migration of copepods when animals have abundant food must be contrasted to results obtained when similar animals are deprived of food. As several authors note, this situation can be extremely frustrating when one has access to only one tower tank (Conover and Paranjape, 1977; Bohrer, 1980; Mullin and Evans, 1974). When working with a single tower tank, two solutions are available: (a) divide the tank with a septum extending down the center, or (b) repeat the control experiment at a later date. Neither of these practices is quite satisfactory because a divided tank has only half the volume, and an experiment duplicated sequentially requires twice as much time (consider, for example, the single experiment of Mullin and Evans of 170 days). Furthermore, the experimenter may not have access to experimental organisms similar to those previously studied. The solution at MERL was to sacrifice certain hardware features of the tower tanks (e.g. windows) and to build 12 tower tanks rather than one. However, the high financial costs of this solution may limit its general applicability. The simplest solution to the equation of costs versus replicability was that of Perez et al. (1977) who restricted their (nevertheless valuable) questions to those that could be answered in 0.150 m³ trash containers; however, as many as 20 tanks could be used in a single experiment.

Two tests are relevant in assessing the utility of microcosms in the simulation of selected components of an ecosystem: (i) does the behavior of the microcosm correspond to the system being simulated, and (ii) do several microcosms, each receiving similar treatments, follow a similar course of events with time? Pilson et al. (1979) in their tests of replicability of tanks at MERL believed that this similarity would be verified if the microcosms maintained 'similar species composition ..., if the metabolic rates in the systems and the major chemical fluxes and transformations were within the range and variability of natural systems, and if the statistical behaviour of the systems was similar to that of the natural systems'. By the time that Pilson had written this, the earlier view, that by working in microcosms one would see less variance and understand more, had disappeared.

The number of tabulated species of phytoplankton growing in any one of the nine replicated MERL tanks was between three and twenty; replicated samples of sea water taken from Narragansett Bay had a similar number of species, but the number of species in the tanks ranged more than in replicated samples from the field. It was as if the various tanks had become 'patchy' between themselves while the number of species in the Bay was rather uniform. Nevertheless the nine tanks still showed close agreement among themselves and with their source water, and could be contrasted with water samples taken from other areas of the Bay which were influenced to a greater extent by terrigenous sources.

Part of the coherence is attributable to experimental design: the MERL tanks received an inflow of about 4% of their volume per day straight from the Bay. This exchange rate, which caused the water in each tank to be completely replaced by Bay water every 30 days, tied the biota of the tanks inextricably to the Bay but still permitted subtle variations.

Comparison, however, begs the question of what variables should be measured as the best expressions of biological processes. Names and concentrations of species of phytoplankton and zooplankton are probably the most sensitive and informative indices in any one system, but are the most difficult to obtain and the most difficult to compare in a global context. Pilson *et al.* (*loc. cit.*) performed a correspondence analysis based on phytoplankton species in the nine tanks and found that a late summer population of diatoms and flagellates occurred jointly in the tanks and in the Bay, and that this mixed population became predominantly a diatom population as the tanks and the Bay both bloomed in November. The nine tanks, however, became less similar as the bloom progressed, and the divergence perhaps detracted from the idea that the tanks were a complete analog of the Bay. However, no consistent pattern existed between the tanks – only that their variance had increased.

A study of replicability of the CEPEX enclosures (Takahashi *et al.*, 1975) also reported a high degree of consistency, but the study points out the necessity of examining species composition when considering responses associated with experimental treatment. Several gross indices (NO_3, NO_2, NH_3, PO_4) progressed similarly inside and outside the bags during the period of a phytoplankton bloom in the first 14 days of the experiment; the concentration of phytoplankton carbon increased and nutrient concentrations dropped correspondingly in all systems. By the end of the 30 day experiment the diatoms in the bags, as represented by *Thalassiosira*, had virtually died out, to be replaced by microflagellates. This important divergence – diatoms persisted in high numbers in the water outside the bags – was not reflected in measurements of the chlorophyll concentration that tended to progress similarly in and out of the bags. The transition inside the bags from a diatom to a microflagellate population is probably attributable to reduced turbulence, and is an important consequence of the experimental concept of big bags. A somewhat similar divergence of algal types occurred in the large plastic bags employed in Loch Ewe (Davis and Gamble, 1979) where large diatoms in the bags declined in number and species relative to populations outside the bags.

Takahashi *et al.* (1975) argue that a variance of even several hundred percent observed from samples from replicated bags could be discounted because the progression of means of samples in a time series would 'show a great similarity in a form of biological production'. A major *raison d'être* in requesting grants for the use of big bags and tower tanks in studying ecological systems was the expectation that such systems would show less unexplained variance than the outside world and that the reduced variance would lead to greater understanding of oceanographic processes. It was without doubt disappointing to these experimentalists to find that parameters inside the bags – be they measurements of biological populations, chemical variables, turbidity, etc. – varied by about the same degree or more as the universe left outside the containers. The high variances that have been seen to date are probably associated with the newness of the techniques and with the problems of confining large volumes of water. Some difficulties that have come from leaky bags, destruction through storms, and sampling problems would be resolved through the experience that comes with trial and error.

In some cases, divergence of parameters can be attributed to an unequal number of upper trophic level predators in replicated containers. An enhanced population of euphausiids, ctenophores, or juvenile fish in an enclosure may exert influences that extend through the population of zooplankton down to phytoplankton (Steele, 1979). The importance of a few higher predators in controlling densities and species composition of their prey has been emphasised by Connel (1972) and is known to have influenced populations in enclosures (Davies and Gamble, 1979). The manner in which enclosures have been filled has made them susceptible to the unequal capture of top predators simply through random sampling of scarce organisms typically distributed in patches. The consequence has been that several papers dealing with microcosms contain vague generalisations that are indicative of the frustration experienced by the authors in dealing with experiments that have produced confusing, inexplicable results. The frustration must be all the more intense in that, by leaving the oceanographic vessels and going to microcosms, the world was to have become more easily manipulated, controlled and understood.

A reasonable approach to interpreting results from microcosms is to employ multivariate techniques (Oviatt et al., 1977; Pilson et al., 1980) that express trends through statistical analyses of many variables of the system. This approach is anathema to the reductionalist, however, who may wish to consider the response of one component to the changes in another component (e.g. copepod development vs algal cell size), an approach that cannot incorporate an influence of a third or nth component.

TURBULENCE

The most surprising result of several experiments on microcosms and probably the most important contribution of these studies to marine ecology has been the inadvertent re-awareness that turbulence and mixing are extremely important to biological processes in the sea. Serious artifacts will result when water is confined in enclosures that isolate the biota from turbulent processes of wind, waves, and advection. The history of this awareness is not clear, for the first marine microcosm experiments (McAllister *et al.*, 1961) incorporated an elaborate mechanism to mix the water within the enclosure; Riley had discussed the importance of turbulence to phytoplankton in several papers (cf. 1942), and Patten (1963) had incorporated energy bonuses into his phytoplankton models that would accrue to those organisms living in a turbulent medium. Verduin, in a prophetic note in 1968, criticised the use of big bags because of the reduced eddy diffusivity within them, and commented that this variable was 'one of the most important physical parameters in the aquatic environment, and one of the most neglected'. It seems unfortunate that his plea was ignored.

The greatest difficulties have arisen in the CEPEX microcosms and in the enclosures in Loch Ewe, where mixing was not employed in the early years of study. In a typical sequence of events, an algal population that was initially a mixture of diatoms and microflagellates progressed in 2–3 weeks to become an assemblage of microflagellates and small diatoms (Loch Ewe; Davies and Gamble, 1979) or almost exclusively microflagellates (CEPEX; Takahashi *et al.*, 1975; Thomas *et al.*, 1977) as the larger diatoms precipitated from the water column. The change of balance of these two components (microflagellates vs diatoms) has ramifications in higher trophic levels that may be compounded if the experiment is designed to, for example, test the toxicity of heavy metals or hydrocarbons on the growth of herbivores. The toxic influence of the pollutant is difficult to separate from the alteration of food supply of the herbivore.

Nor is it obvious how mixing should be introduced once the realisation occurred that it was necessary. Steele *et al.* (1977) found that the coefficient of eddy diffusivity in enclosures at Saanich Inlet and Loch Ewe varied from 0.05 to $0.26\,\mathrm{cm^2 s^{-1}}$; these values are at least an order of magnitude less than typical but highly variable values found in the open sea. Nixon *et al.* (1980), employed a variety of empirical techniques to estimate turbulent mixing in Narragansett Bay with the goal of introducing a similar mixing in the microcosms at MERL, and on the basis of these measurements designed a vertically operated dasher that is raised and lowered in the water column of the MERL tanks. The dasher was operated on a regime of two hours on (causing an eddy diffusivity of $16–20\,\mathrm{cm^2 s^{-1}}$) followed by four hours off during which time the diffusivity decayed to near zero (Nixon *et al.*, *loc. cit.*). The slow, continuous mixing employed at MERL is apparently preferable to the mixing procedures used at CEPEX in later experiments and reported on by Sonntag and Parsons (1979) and by Eppley *et al.* (1978). In these studies air was introduced at depth twice daily for 20 minutes, and then later at intervals of every two or three days (Sonntag and Parsons, 1979); Eppley *et al.* employed a similarly episodic regime of bubbling daily for about a minute. Sonntag and Parsons suggest that such a mixing regime did not simulate the characteristics of upwelling they were attempting to achieve but rather caused intense and episodic vertical mixing which soon decayed. The mixing that was achieved, however, resulted in phytoplankton crops having a larger cell diameter than in unstirred enclosures, and thus broke the pattern of anomalous disappearance of large cells that characterised earlier work.

The importance of mixing to the growth of plankton was well demonstrated (Perez et al., 1977) in a study of scaling of variables necessary for maintenance of microcosms. The studies were carried out in small (0.150 m³) but easily manipulated microcosms, and supported the observation that enhanced algal growth is associated with the increase of turbulence. Statistical analysis allowed segregation of the influences of increased mixing on algal growth and copepod growth, and permitted the conclusion that copepods grew better when turbulence was weak. This observation, remarkable as it is, is in accord with several observations from CEPEX enclosures, where zooplankton populations often increased in the poorly mixed enclosures though populations of large cells decreased. The observation suggests that zooplankton and phytoplankton populations in the sea are perhaps more weakly coupled through trophic interactions than generally held, but are more strongly coupled secondarily through the process of mixing.

REPRESENTATIVE STUDIES

The pioneering work on large in situ enclosures of sea water (McAllister et al., 1961; Anita et al., 1963) dealt only with phytoplankton and bacterial assemblages. The intent was clearly not to mimic the ocean but was rather to isolate selected populations in a somewhat more natural manner than could be achieved in the laboratory. Complications stemming from the presence of herbivores (excepting perhaps some contaminating protozoa) were avoided by filling the bag with filtered water. The design of the submerged sphere-like bag (Fig. 2) prevented atmospheric exchange and permitted estimation of primary production from changes in the concentrations of CO_2 and O_2; these estimates could then be compared with direct counts of algal numbers and with estimates from the ^{14}C technique. Results supported the use of the ^{14}C technique as a measurement of net productivity and pointed out the importance of the loss of dissolved organic matter by phytoplankton when considering total production, dissolved organic matter and the substrate of bacterial populations in sea water.

Nutrient depletion, the phenomenon responsible for the cessation of algal blooms in the bags, was examined in more detail (Strickland et al., 1969) in the tower tank at Scripps Institution of Oceanography. The three algal species (*Ditylum brightwellii, Cachonina niei*, and *Gonyaulax polyedra*) individually grown in a series of three 2-week long experiments assimilated ammonia in preference to nitrate and displayed biochemical patterns associated with the form of nitrogen. *Ditylum* began to accumulate lipids and *Cachonina* carbohydrates after the algae had exhausted the ammonia; the culture of slower growing *Gonyaulax* became contaminated with *Phaeocystis* which masked changes of cellular composition of the dinoflagellate. *Ditylum* cells in the tower tanks were larger, had more silicon and phosphorous but less carbohydrate per gram of carbon than cells grown under similar nutrient conditions in the laboratory, and interestingly, cells in the tower tank were neutrally buoyant while the same cells grown in the laboratory could be kept in suspension only by agitation. Clearly, the diatoms added to the tower tank from a laboratory inoculum altered their metabolic processes, thereby raising the specter that 'there is no assurance that plants grown in the laboratory will be identical to those grown in the sea' (Strickland *et al.*, 1969). The chemical composition of *Gonyaulax* grown in the tank, however, was similar to laboratory grown cells, detracting from any generalisations and directing thought to causes of differing growth patterns of *Ditylum* in the two environments.

The most attractive use of microcosms is perhaps in the examination of trophic interactions of defined populations of phytoplankton and zooplankton. Parsons *et al.* (1977), after fertilising CEPEX enclosures with varying levels of nutrients, followed the production and changes in biomass of phytoplankton, copepods, and ctenophores and added to our understanding of the inverse relationship between primary production and the efficiency of production at the tertiary level. The ability to study a cohort of copepods (Paffenhöfer, 1976) or a set of introduced cohorts of copepods and ctenophores (Mullin and Evans, 1974; Reeve and Walter, 1976) is an extremely attractive feature of microcosms, and establishes them as important experimental vehicles for the exploration of certain ecological concepts.

These studies perhaps represent the best use of microcosms to date. Interestingly, they were all carried out in volumes of less than 70 m^3, and as indicated by Reeve and Walter (1976) could perhaps have been done in much smaller volumes. This is encouraging for it frees the scientist from the need to acquire larger, more expensive enclosures, and holds to him the promise – now that the importance of turbulence in microcosms is appreciated – of carrying out more revealing experiments. One may anticipate, for example, that improved techniques for estimating secondary production will be developed by studying manipulated cohorts of zooplankton as they feed on phytoplankton in the observable confines of a microcosm.

Fishes have only occasionally been added to microcosms, perhaps because it has been realised that the more fundamental mechanisms of nutrient uptake and the growth of phytoplankton and zooplankton must be better understood before yet another trophic level can be added to already complicated systems. Koeller and Parsons (1977), who made preliminary efforts, noted that populations of young salmonids grew at unequal rates in replicated CEPEX enclosures, and ascribed the lower growth rate of one group to a preponderance of smaller copepods (*Corycaeus*) in some enclosures. Young fish exposed to a similar biomass of larger copepods (*Calanus*) in other enclosures had a more normal growth rate. This observation supports the concept that size of the food particle influences the efficiency of growth, a concept developed earlier by Parsons and LeBrasseur (1970) and Paloheimo and Dickie (1966) based on studies of fishes in conventional aquaria.

The complexities inherent in a microcosm containing four trophic levels (phytoplankton, copepods, ctenophores, and fish) (Sonntag and Parsons, 1979) defy interpretation, pointing out that the difficulties associated with a multilevel ecosystem in a 1300 m^3 enclosure are not much less than those encountered when working outside the enclosure in the open sea.

3 Conclusions

When considering microcosms it is inevitable that one think of them as attempts to mimic nature on a scale that can be easily observed. This goal has led to misadventure in many cases. Several of the papers concerning microcosms express the author's unhidden consternation at the inability to explain observations, a consternation that manifests itself as vague generalisations, statements of 'on the one hand', and a chronic inability of the authors to draw conclusions. Oviatt *et al.* (1977) point out, however, that microcosms should not be viewed as attempts to achieve 'an exact duplicate of a particular system'; these authors reflect a philosophy that shows considerable promise for the future of microcosms as an ecological tool.

The relative value of several studies employing microcosms seems to be almost inversely related to the volume chosen for manipulation. The approach at CEPEX was to duplicate nature in an enclosure and to follow the succession of variables as they were influenced by pollutants. The attempt encountered difficulties because the simple act of confining the water imposed a stress on the biota that was difficult to identify and to separate from the responses resulting from experimental perturbation. It is unfortunate that contingencies of funding the large research project initially required that the experiments be directed toward documenting the polluting effects of heavy metals; some scientists in the project noted that the unperturbed enclosures (the controls) were often the most interesting and the most informative. This realisation caused a shift away from a study of pollution to a study of natural processes in the later months of the project.

The other approach, primarily at the University of Rhode Island, was based on the realisation that ecosystems could not be placed under experimental control, but that they could be dissected in smaller volumes to examine the influence of discrete environmental components by altering turbulence, light, or advection, for instance, and by monitoring the consequences. The success of these efforts is extremely valuable, for it indicates that massive budgets are not required, but that the individual scientist, working on chambers of only a few hundred liters volume, may make significant contributions. To the contrary, the inflexibility of manipulation imposed by the constraints of acquiring and handling large volumes of water limits the scope of the individual experimenter. These constraints are compounded by the difficulties of establishing a sufficient range of experimental variables and replicates, and by the need to collectively satisfy the individual requirements of component projects. Since all the properties of the ocean obviously cannot be duplicated in a confined environment, the rationale to opt for small, easily manipulated and duplicated enclosures is strong. The constraints of volume will be greatest in studies of organisms of higher trophic levels, but several studies suggest that these may be performed in smaller volumes than has been assumed (e.g. Baker and Reeve, 1974).

In spite of our attempts to achieve a simulation of nature, processes within microcosms seem to differ in rate and magnitude from those occurring in nature. Nevertheless the results express responses of living organisms to phenomena, and an understanding of the system may be assembled as a composite picture of responses to several perturbations. By this approach, one views microcosms as large *in vitro* experiments designed to test selected components of the ecosystem, and one does not view them as attempts to enclose a multitude of interacting processes. This approach, since it is divested of attempts to contain an ecosystem, can be employed in small volumes, and facilitates relatively inexpensive flexible programs that nevertheless offer a great deal of promise.

The CEPEX project in Saanich Inlet, Canada, brought together a group of scientists who focussed their skills and interests on marine phenomena in controlled enclosures. The contributions of those workers has been important, but their success in sorting out marine problems was, it can be argued, impeded rather than augmented by their intent to study processes in the enclosures.

It is worthwhile noting that J. D. H. Strickland, who with Terhune and others conceived the use of enclosures in marine studies, moved away from them to resume a combination of laboratory and seagoing studies in his latter days, expressing to friends (G. A. Riley; pers. com.) the realisation that, while big bags could be useful in testing some concepts, they were of limited value, and that the sea was a much more revealing medium for studying marine problems.

References

Antia, N. J., McAllister, C. D., Parsons, T. R., Stephens, K. and Strickland, J. D. H. (1963). Further measurements of primary production using a large-volume plastic sphere. *Limnol. Oceanogr.* **8** (2), 166–183

Balch, N., Boyd, C. M. and Mullin, M. (1978). Large-scale tower tank systems. *Rapp. P.V. Réun. Cons. int. Explor. Mer* **173**, 13–21

Berland, B. R., Bonin, D. J. and Maestrini, S. Y. (1975). Isolement *in situ* d'eau de mer naturelle dans des enceintes de grand volume. Rapports scientifiques et techniques, No. 21. Centre National pour L'Exploitation des Ocean

Bodungen, B. von, Bröckel, K. von, Smetacek, V. and Zeitzschel, B. (1976). The plankton tower. I. A structure to study water/sediment interactions in enclosed water columns. *Mar. Biol.* **34**, 369–372

Bohrer, R. N. (1980). Experimental studies on diel vertical migration. *In* "The Evolution and Ecology of Zooplankton Communities". (W. C. Kerfoot, ed.) Special Symposium III. A.S.L.O. University Press, New England. pp. 111–121.

Case, J. N. (1978). The engineering aspects of capturing a marine environment, CEPEX and others. *Rapp. P.V. Réun. Cons. int. Explor. Mer* **173**, 49–58

Connell, J. H. (1972). Community interactions on marine rocky intertidal shores. *Ann. Rev. Ecol. Systematics* **3**, 169–192

Conover, R. J. and Paranjape, M. A. (1977). Comments on the use of a deep tank in planktological research. *Helgoländer wiss. Meeresunters.* **30**, 105–117

Davies, J. M. and Gamble, J. C. (1979). Experiments with large enclosed ecosystems. *Phil. Trans. R. Soc. Lond. B.* **286**, 523–544

Edmondson, W. T. (1955). Factors affecting productivity in fertilized salt water. Papers in Marine Biology and Oceanography; *Deep-Sea Res.* **3** (Suppl.), 451–464

Edmondson, W. T. and Edmondson, Y. H. (1947). Measurements of production in fertilized salt-water. *J. Mar. Res.* **6** (3), 228–246

Eppley, R. W., Koeller, P. and Wallace, G. T. Jr. (1978). Stirring influences the phytoplankton species composition within enclosed columns of coastal sea water. *J. Exp. Mar. Biol. Ecol.* **32**, 219–239

Gamble, J. C., Davies, J. M. and Steele, J. H. (1977). Loch Ewe bag experiment, 1974. *Bull. Mar. Sci.* **27** (1), 146–175

Geen, G. H. and Hargrave, B. T. (1966). Primary and secondary production in Bras d'Or Lake. *Verh. Internat. Verein. Limnol.* **16**, 333–340

Goldman, C. R. (1962). A method of studying nutrient limiting factors *in situ* in water columns isolated by polyethylene films. *Limnol. Oceanogr.* **7**, 99–101

Kemmerer, A. J. (1968). A method to determine fertilization requirements of a small sport fishing lake. *Trans. Amer. Fish. Soc.* **97**, 425–428

McAllister, C. D., Parsons, T. R., Stephens, K. and Strickland, J. D. H. (1961). Measurements of primary production in coastal sea water using a large-volume plastic sphere. *Limnol. Oceanogr.* **6** (3), 237–258

Menzel, D. W. and Steele, J. H. (1978). The application of plastic enclosures to the study of pelagic marine biota. *Rapp. P.V. Réun. Cons. int. Explor. Mer* **173**, 7–12

Mullin, M. M. and Evans, P. M. (1974). The use of a deep tank in plankton ecology. 2. Efficiency of a planktonic food chain. *Limnol. Oceanogr.* **19** (6), 902–911

Nixon, S. W., Alonso, D., Pilson, M. E. Q. and Buckley, B. A. (1980). Turbulent mixing in aquatic microcosms. *In* "Microcoscus in Ecological Research" (J. Giesy, ed.) pp. 819–849. DOE Symposium Series, Augusta, Georgia

O'Dor, R. K. (1978). Editor's note on *I. illecebrosus* larvae hatched in captivity. *In* "Proceedings of the Workshop on the Squid *Illex illecebrosus*". (N. Balch, T. Amaratunga and R. K. O'Dor, eds). Fisheries and Marine Service Technical Report No. 833. Fisheries and Environment Canada, 15.17–15.18

Oviatt, C. A., Perez, K. T. and Nixon, S. W. (1977). Multivariate analysis of experimental marine ecosystems. *Helgölander wiss. Meeresunters.* **30**, 30–46

Paffenhöfer, G.-A. (1976). Continuous and nocturnal feeding of the marine plank-tonic copepod *Calanus helgolandicus*. *Bull. Mar. Sci.* **26** (1), 49–58

Palmer, B. W. and O'Dor, R. K. (1978). Changes in vertical migration patterns of captive *Illex illecebrosus* in varying light regimes and salinity gradients. *In* "Proceedings of the Workshop on the Squid *Illex illecebrosus*" (N. Balch, T. Amaratunga and R. K. O'Dor, eds). Fisheries and Marine Service Technical Report No. 833, Fisheries and Environment Canada: 23.1–23.12

Paloheimo, J. E. and Dickie, L. M. (1966). Food and growth of fishes. II. Effects of food and temperature on the relation between metabolism and body weight. *J. Fish. Res. Bd Canada* **23**, 869–908

Parsons, T. R. (1978). Controlled ecosystem experiments: Introduction. *Rapp. P.V. Réun. Cons. int. Explor. Mer* **173**, 5–6

Parsons, T. R. and LeBrasseur, R. J. (1970). The availability of food to different trophic levels in the marine food chain. *In* "Marine Food Chains" (J. H. Steele, ed.) pp. 325–343. Oliver and Boyd, Edinburgh

Parsons, T. R., Bröckel, K. von, Koeller, P. and Takahashi, M. (1977). The distribution of organic carbon in a marine planktonic food web following nutrient enrichment. *J. Exp. Mar. Biol. Ecol.* **26**, 235–247

Patten, B. C. (1963). The information concept in ecology: Some aspects of information-gathering behaviour in plankton. *In* "Information Storage and Neural Control" (W. S. Fields and W. Abbot, eds) pp. 140–172. Thomas, Springfield, Ill.

Perez, K. T., Morrison, G. M., Lackie, N. F., Oviatt, C. A., Nixon, S. W., Buckley, B. A. and Heltshe, J. F. (1977). The importance of physical and biotic scaling to the experimental simulation of a coastal marine ecosystem. *Helgoländer wiss. Meeresunters.* **30**, 144–162

Pettersson, H., Gross, F. and Koczy, F. (1939). Large-scale plankton culture. *Nature, Lond.* **144**, 332–333

Pilson, M. E. Q., Oviatt, C. A., Vargo, G. A. and Vargo, S. L. (1979). Replicability of MERL microcoms: initial observations. *In* "Advances in Marine Environmental Research" (F. S. Jacoff, ed.) pp. 359–381. EPA-600/9-79-035 Environmental Protection Agency, Narragansett, Rhode Island

Reeve, M. R. and Walter, M. A. (1976). A large-scale experiment on the growth and predation potential of ctenophore populations. *In* "Coelenterate Ecology and Behaviour" (G. O. Mackie, ed.) pp. 187–199. Plenum Press, New York

Riley, G. A. (1942). The relationship of vertical turbulence and spring diatom flowerings. *J. Mar. Res.* **5** (1), 67–87

Sonntag, N. C. and Parsons, T. R. (1979). Mixing an enclosed, 1300 m^3 water column: effects on the planktonic food web. *J. Plankton Res.* **1** (1), 85–102

Steele, J. H. (1979). The uses of experimental ecosystems. *Phil. Trans. R. Soc. Lond. B.* **286**, 583–595

Steele, J. H., Farmer, D. M. and Henderson, E. W. (1977). Circulation and temperature structure in large marine enclosures. *J. Fish. Res. Bd Can.* **34** (8), 1095–1104

Strickland, J. D. H. (1967). Between beakers and bays. *New Scientist* **33** (532), 276–278

Strickland, J. D. H., Holm-Hansen, O., Eppley, R. W. and Linn, R. J. (1969). The use of a deep tank in plankton ecology. I. Studies of the growth and composition of phytoplankton crops at low nutrient levels. *Limnol. Oceanogr.* **14** (1), 23–34

Strickland, J. D. H. and Terhune, L. D. B. (1961). The study of *in situ* marine photosynthesis using a large plastic bag. *Limnol. Oceanogr.* **6** (1), 93–96

Takahashi, M., Thomas, W. H., Seibert, D. L. R., Beers, J., Koeller, P. and Parsons, T. R. (1975). The replication of biological events in enclosed water columns. *Arch. Hydrobiol.* **76** (1), 5–23

Takahashi, M., Wallace, G. T., Whitney, F. A. and Menzel, D. A. (1977). Controlled Ecosystem Pollution Experiment: effect of mercury on enclosed water columns. I. Manipulation of experimental enclosures. *Mar. Sci. Comm.* **3** (4), 291–312

Takahashi, M. and Whitney, F. A. (1977). Temperature, salinity, and light penetration structures: controlled ecosystem pollution experiment. *Bull. Mar. Sci.* **27** (1), 8–16

Verduin, J. (1969). Critique of research methods involving plastic bags in aquatic environments. *Trans. Amer. Fish. Soc.* **98**, 335–336

22 *Principles of Ecosystem Modelling*

WILLIAM L. SILVERT

1 What is a model?

Any construct that we use to simplify and conceptualise nature is a model, and the formal literature of the field has consistently construed the term in the broadest possible fashion.

A model need neither involve computers, nor even be recognisably mathematical. It does not have to involve feedback, hierarchical structure, or any other elaborations and the only really important characteristic of a model is that it should be useful. Models may involve physical representations such as electrical analogue circuits or actual scale models of large systems.

But if the physical manipulation of scale models is viewed as part of modelling, where does one draw the line between modelling and experimentation? In fact, no such line exists. A laboratory experiment is an abstraction of part of nature into a controlled environment, which simplifies it and makes it possible to quantify features of interest: a chemostat can be a model of part of the ocean. Thus an experiment is as much a model as is a computer simulation.

Theoretical modelling should be viewed as an extension of the experimental method, and one can learn a great deal about good modelling technique by following the accepted standards of experimental science. In particular, the accessibility of a model to peer review and critical analysis should be governed by the same standards as apply to evaluation and replication of experimental work. The roles of experimental and theoretical models are complementary, although in a different way from that suggested by Maynard Smith (1974), who wrote that 'mathematical and biological models complement one another': the implication that experimentation is real biology while theoretical work belongs to another discipline is one that fewer and fewer theorists would accept. Both types of modelling involve simplification, but while the experimenter usually simplifies nature by greatly reducing scale and the richness of interactions, the theorist simplifies most drastically in the characterisation of individual living organisms.

This chapter is chiefly concerned with theoretical models, most of which are mathematical in nature. It assumes that a person who constructs models of marine ecosystems must be a marine ecologist, not purely a mathematician or computer programmer; useful models can be constructed by ecologists with limited mathematical ability, but no model can be trusted that has been designed without ecological insight and understanding.

2 Three basic categories of models

Most modelling in marine ecology is concerned with complete ecosystems, and therefore the methodology is characteristically involved with complex systems and vast amounts of data, at least potentially, since much may never actually be measured. It is not practicable to attempt to design a model of an ecosystem which includes all of the available information, and therefore the form of models that are actually feasible depends on external circumstances and is not uniquely determined by the properties of the ecosystem alone. Furthermore, even when a model has been constructed it may not be possible to ascertain its behaviour fully without further simplification, so that often a model of a model needs to be constructed.

A reasonable classification scheme and guide for most modelling applications can be based on three general categories of models. Firstly, *conceptual models* represent complete pictures of a system and encompass all information on which other models can be based. To the individual modeller they represent unique all-inclusive bodies of disorganised knowledge; individual ecologists may disagree about the validity of some of the information, and they may hold different conceptual models of the same system. The second category of models consists of *realisations* derived from a conceptual model. A realisation involves specific identifiable assumptions about the importance of various aspects of the conceptual model and achieves a degree of simplicity or abstraction by assuming that certain effects can be ignored. Because these assumptions are of crucial importance in determining the extent to which the outcome of a modelling program corresponds to reality, the transition from conceptual model to realisation occupies most of this chapter. The third category of model includes actual working models which we will call *implementations*. They may simply be realisations in concrete form or may involve further simplifications and approximations.

Most scientists use a number of different realisations of the same ecosystem, even when working on a single project. One reason for using multiple realisations is to test hypotheses, which can be used as the working assumptions from which the realisation is derived. Another reason is to resolve questions at different hierarchical levels; it is generally not feasible, for instance, to construct a single model that satisfactorily describes both the growth rates of component species and the total productivity of an ecosystem. Often the reason for choosing a particular realisation is to facilitate the move towards implementation; a realisation of feeding based on the assumption that feeding rates are independent of system size makes it possible to build an experimental model in an enclosed tank, while a linear realisation may lead to an easily solved theoretical model.

Just as a single conceptual model may have different realisations, a single realisation may have several implementations; in fact, this is normal. Replication of an experiment is simply duplication of an implementation, a check which is equally appropriate to computer simulations (Mohn, 1979). More fundamentally, a single realisation may be implemented both experimentally and theoretically. It is also not uncommon for several theoretical implementations to be constructed, some of which can be solved by analytical mathematical methods, while others require the use of computers.

The above distinctions are not unvarying, but they define a framework that has proved useful in the organisation of modelling programs. Zeigler (1976) defines a base model, similar to what is here called a conceptual model, and he refers to realisations as lumped models, although his definitions are entirely concerned with mathematical formulations. The construction of alternate realisations is very well treated by Aris (1978). It is important to remember that models are not unique, and that the modelling process generally involves the construction of several different models.

3 Model architecture

The overall form of a model depends chiefly on three considerations, ranked as follows:

1 The purpose of the model
2 The observable behaviour of the system
3 The structure of the system

It should be noted that the structure of the ecosystem does not head the list and does not in general define the structure of the model.

The purpose of a model is defined by its intended output and by the objectives of the programme of which it is a part, although other factors, such as problems of implementation, may also be important. Models which are designed to resolve specific management questions must be designed around the required outputs, even though these may not be conformable with the organisation of the data on which the model is based. The importance of defining clear objectives at the start of any modelling project cannot be overstated (O'Neill, 1975).

The observable behaviour of the ecosystem being studied affects the modelling process in two ways. First, since the model must ultimately be based on real data, the availability of these data must be taken into account in designing the model. If a particular model cannot be implemented without measurements which are impossible to obtain, then the model cannot fulfil its purpose and a different approach must be tried. Second, but much more fundamental, is the fact that what we know about a system is the integrated accumulation of past observations. In interpreting these observations we always risk overlooking relationships which may be significant in ecosystem function. If we base a model on interpretations or theories without direct reference to the accumulated observations we not only exclude these hidden relationships from the model, but we also preclude the possibility of discovering them in a reliable and organised fashion.

This should not be taken to mean that the structure of the model has no relationship to the structure of the ecosystem. In most cases the observed behaviour of an ecosystem is what we expect on the basis of what we know about its structure, and thus a model based on this structure is a good model of the system. We must, however, ensure that the structure is complete enough to account for the observed behaviour, and this requires concentration on behaviour before structure. An additional problem is that some structure may not lead to observable consequences, and incorporation of this structure in the model may lead to unnecessary complexity.

A model should have a coherent architectural framework which reflects these considerations. In particular, it should be clear what the model is supposed to do (its purpose). The relationship between the model and observations, both input (e.g. parameter estimates and external driving forces) and output (quantitative results and other testable predictions) should be transparent. The structure of the model should be clearly specified, ideally in a way that makes the operation of the model clear to the user.

The modelling process is determined by these architectural considerations. Because the model has a well-defined purpose, the information it contains is only part of what is available in the complete conceptual model, and the realisation that is used should utilise that information as effectively as possible.

4 System identification

Ecological modellers constitute a subset of systems theorists, and it is a tautology that the theory of ecosystems is just one of the areas covered by systems theory. The basis of systems theory is that 'the whole is greater than the sum of its parts', or, as Ashby (1956) puts it, 'We must therefore be on our guard against expecting the properties of the whole to reproduce the properties of the parts, and vice versa'. This viewpoint has been echoed as a basis for ecological modelling by Watt (1966), who asserts that:

... a principle attribute of a system is that we can only understand it by viewing it as a whole. ... Clearly, if we are to study systems as whole systems and not just collections of fragments of systems, we must use a strategy of research which at every step is designed in terms of the problem of fitting all the fragments together correctly at the end of the research program.

This viewpoint has been widely echoed, but Mann (1975) pointedly observes that 'Although Odum defines systems ecology as the formal approach to holism, a great deal of the literature on ecosystem models is basically reductionist in its approach'. And, as Patten and Finn (1979) put it, 'the mechanistic approach to systems ecology, that of building wholes from parts, has so far failed to develop realistic and practical ecosystem models'.

The synthesis of models from components may lead to models which are poorly designed, incomplete, and possibly useless, although it must be emphasised that not all models constructed by the synthetic approach succumb to these dangers. The most common design error is a lack of balance between the component submodels, which leads to the complete model being very complex but no better than the weakest component. This danger is stressed in a recent report by the Scientific Committee on Problems of the Environment, which observes that:

Over-dependence on past data and information can easily lead to development of those parts of the model for which relevant data are available, even when these parts are of low priority. Identification of goals and priorities will frequently and drastically reduce the total data set which needs to be incorporated into the model and will indicate those areas for which new data will have to be collected if the model output is to match the requirements of the problem. While most scientists like to emphasize the lack of data within their special field, experience in the modelling of environmental systems suggests that the converse problem is often more inhibiting (SCOPE, 1978).

Even worse is the danger that a critical feature of ecosystem behaviour will be ignored in the process of synthesis. Mann (1979) observes that:

An ecological simulation model can be regarded as a way of summarizing its author's knowledge about the complex set of interactions taking place in an ecosystem. If the author has an incomplete knowledge of important interactions, it is inevitable that the model itself and the predictions of the model will be incomplete and inaccurate.

Mann is clearly using the phrase 'ecological simulation model' to refer to synthetically constructed models since, as will be shown in the next section, the effectiveness of a model does not necessarily depend on how much the modeller knows about the interactions within an ecosystem. In fact, 'interactions' represent couplings between parts of a model which we use to interpret the behaviour of an ecosystem, and thus the interactions may properly be said to arise from the model, as opposed to saying that the model comes in some way from the interactions.

A fundamental problem with models synthesised from independently defined components is that of assuring that the output is relevant to the objectives of the modelling program. It may at times be necessary and appropriate for a modeller to respond to a question with the statement that 'the model doesn't answer that', but if the question is a legitimate one in terms of the objectives of the project then such a reply must be taken as an admission of failure. Unfortunately, when a model is built 'from the ground up', it is generally not possible to assess the precision or possibly even the form of the output until late in the modelling process, at which time it may be discovered that the necessary outputs cannot be obtained (Miller et al., 1973).

An alternative approach to the modelling of complex systems is to start with conceptual framework of the complete model and to incorporate the necessary detail as the model develops, rather than define the detailed structure and if necessary simplify the model by aggregation; this is referred to as 'top-down' modelling, and it is predicated on the principles of model architecture advanced in the preceding section.

The essential feature of top-down modelling is that one starts with the observed behaviour of the system and bases the model description on these observations. The process of building a model which behaves like the system is an inverse problem. Unfortunately, however.

Inverse problems, or system identification problems, are fundamental problems of science. ... Mathematical treatment of physical problems has been devoted almost exclusively to the direct problem. A complete picture of the system is assumed to be given, and equations are derived which describe the output as a function of the system parameters. The inverse problem is to determine the parameters and structure of the system as a function of the observed output (Kagiwada, 1974).

One of the reasons for the preponderance of models based on this approach is that:

While a well-defined and properly posed direct problem generally has a unique solution, if it has any solution at all, the same is not true with an inverse problem. An inverse problem often leads to a multitude of mathematically acceptable solutions out of which one physically acceptable solution, if any such solution exists, has to be selected. Criteria for this selection often depend on the nature of the inverse problem. ... the identification or modelling problem is perhaps the most difficult of all inverse problems. Given a set of inputs and corresponding outputs from a system, find a mathematical description (or a model) of the system (Vemuri, 1978).

One might take issue with Vemuri's contention that one acceptable solution must be selected. 'We are concerned only with the best model that fits the data, when, very often, the conclusion should be that there is inherent ambiguity in the results – no best model is determined, only a class of models, selection among which requires further data' (Gaines, 1978). Each model is really an hypothesis about the operation of the system, and Gaines suggests that modelling in which distinct alternative structures can be derived to describe the behaviour of a single system generates far more interesting experimental questions than simply the precise specification of some of the parameters in a given model structure.

The non-uniqueness of solutions of the system identification problem is illustrated by linear time-invariant models, which are a large subclass of useful and widely applicable models (Beck, 1979). The general input–output structure (which is what we observe) can always be written in the form

$$y(t) = \int_0^\infty f(\tau)x(t - \tau)\,d\tau$$

where $x(t)$ is the time series of input data, $y(t)$ the output data series, and $f(\tau)$ is a function defined for all $\tau > 0$ which completely characterises the behaviour of the system. Since τ can take an infinite number of values, $f(\tau)$ represents an infinite set of parameters. The general time-series model is thus an infinite-parameter model, although filtering techniques are always used to reduce the number of $f(\tau)$ values to a finite and reasonable level. The advantages of a structural model are twofold: it elucidates the basic mechanisms of the system, and it reduces a large number of purely statistical parameters to a small number of realistic and meaningful ones. For example, the single compartment model described by the differential equation

$$dy/dt = x - ky$$

is a one-parameter model which is exactly equivalent to the infinite-parameter model described by

$$f(\tau) = e^{-k\tau}.$$

The difficulties in system identification arise largely because real ecological data usually have high variances, and exact equivalences such as that described above are impossible to verify. No finite set of measurements on a true single-compartment system can establish clearly that it is not actually a two-compartment system governed by the equations

$$dw/dt = x - Kw$$
$$dy/dt = Kw - ky$$

where $K \gg k$. The choice of the simpler model requires scientific judgement and cannot be made on purely mathematical grounds. Often this judgement depends on the nature of the modelling exercise, since treatment of a true two-compartment system as a single-compartment one may be quite adequate in terms of prediction, but the fast compartment (w) may represent an important component of the system which merits further study. As Kerr and Neal (1976) wrote: 'The challenge is to reduce the list of variables without compromising the adequacy of the resulting system representation, or model, that we use to investigate a given problem'.

There are no standard algorithms for solving system identification problems. While there are classes of problems that can be attacked in a well-defined way, these seldom include the ones that ecologists are familiar with. Therefore a common approach is to build a set of different models and see which one works best, about which Kagiwada (1974) writes:

It is possible to solve a given inverse problem by solving a series of direct problems: by assuming different sets of parameters, determining the corresponding outputs from the theoretical equations, and comparing theoretical versus experimental results. By trial and error, one may find a solution which approximately agrees with the experimental data. This is not a very efficient procedure.

The basic problem then is to find a procedure which is more efficient than trial-and-error and leads to satisfactory results in cases where no method for arriving at an exact solution is known.

One approach which has increasingly found favour with ecologists is to identify species in terms of their role in ecosystem functioning rather than through taxonomic concepts, and to group together species which occupy neighboring niches rather than those which are taxonomically closely related. Botkin (1975) refers to aggregations based on resource utilisation as 'functional groups', and the nature of top-down modelling is such that the modeller may identify such functional groups before he has determined which species belong in them. It may even happen that the behaviour of the system provides more detailed information about functional groups than can be obtained with more direct methods of measurement, as with recent evidence that the role of nanoplankton in the oceans may be greater than that determined by direct measurement (Sheldon et al., 1972; Pomeroy, 1974). This approach has also been advocated by the Oak Ridge Systems Ecology Group (1975), which discusses some of the mathematical aspects in greater detail and provides a useful bibliography of relevant literature.

A similar approach has been followed by Lane et al. (1975) in their Gull Lake program. The modelling part of this study is based on the following three objectives:

1   identification of macroscopic parameters of natural ecosystems;
2   elucidation of invariant relationships in these parameters for given states of the system;
3   demonstration of predictable changes in the macroscopic parameters as the system becomes perturbed.

As Levins points out in the discussion of their study (Levin, 1975, pp. 294–295) 'Instead of lumping state variables, we might try to find new variables which are closer to our concern', to which O'Neill (Levin, 1975, p. 295) responds, 'This is an excellent point. I would go so far as to say that proper specification of state variables is the most important single challenge to ecosystem analysis. The state variables must be observable and must capture the essential features of the system's behaviour. I would look in this direction for significant breakthroughs in the science'.

Although Botkin indicates that functional groups can be defined by ecological function at the organismic level, Rosen (1978) points out that the fundamental units of model structure (equivalence classes in his formal terminology) are uniquely specified by the observational behaviour of the system. These units may not in every case correspond to those which might appear most reasonable on biological grounds. In addition, models of the same system with different purposes generally use different sets of observations, and these can identify quite different model structures.

Because this type of approach frequently leads to system descriptions which are quite different from those which synthetic or species-oriented reasoning suggests, some ecologists may feel that the resulting models are unrealistic. For example, Slobodkin (1975) expresses a strong desire that mathematicians

... would not build ecological models in terms of extensive variables. An organism lives with the properties of individual points in space and time impinging upon him. Therefore, if models are presented in terms of quantities which can only be determined by an overview, while they may make perfect sense to the modeller, they make very little sense to the organisms.

Slobodkin uses the term 'extensive variable' in the thermodynamic sense to denote a quantity which is a global constant, such as energy or nutrients (Slobodkin, 1972). However, although measures of the total quantity of some material in a system may not be significant to individual organisms, which respond to intensive variables like densities and concentrations, global constants are of great value to the modeller when they exist and can be defined and measured. Individual organisms can concentrate and transport energy and nutrients, but no matter what the local dynamics may be, they cannot affect the total amount within the system. In principal, of course, one can calculate the global value of an extensive variable by integrating its density (an intensive variable) over the entire system, but this requires measuring the entire ecosystem with a precision that is totally impractical. This means that the system as a whole exhibits 'emergent' properties (e.g. global constants) which cannot be deduced from a detailed description of the system dynamics with any degree of certainty; such properties are characteristic of hierarchical structure, which Rosen (1969) defines as 'one which is (a) engaged simultaneously in a variety of distinguishable activities, for which we wish to account, and (b) such that different kinds of system specification or description are appropriate to the study of these several activities'. Ecosystems certainly fit this description, and it has been noted that the predicted properties of ecosystems depend strongly on how they are described (Kerr, 1976). Perhaps the greatest advantage of top-down modelling is that possible emergent properties cannot lead to failure of the model, since the modelling process starts with an empirical description of the behaviour of the system and this behaviour must be ascribed to whatever components and interactions are defined within the original definition of the model.

Formal techniques of system identification which are especially useful for linear compartmental models or systems of lagged differential equations can be found in the technical literature (Kagiwada, 1974) and will not be further discussed here. Ecologists are more likely to find statistical methods of value, and multiple regression analysis can often be used to suggest an initial plan for model structure (Mott, 1966). Sometimes, however, a system exhibits correlations which are difficult to relate to the causal relationships that make up a meaningful biological model, and in any large system there are bound to be spurious correlations (Greig, 1979). Path analysis, a multiple regression technique which incorporates causal inference, is a valuable technique developed by Sewall Wright and described in a recent text by Li (1975). The user constructs a path diagram of possible causal interactions within the system and uses correlation coefficients to ascertain whether the diagram can correctly model the system, and to determine which of the hypothesised pathways are significant, as well as assigning numerical values (or path coefficients) to the causal links.

If two variables are highly correlated one can generally combine them, even if it is difficult to understand the reason for the correlation. Frequently one arrives at a model in which quite distinct types of organisms are lumped together because of such correlations, while closely related species have to be considered separately. It may even be that different life stages of the same species occupy different functional groups, especially in light of the remarkable life histories of many marine species. There are several statistical techniques for arriving at optimal representations, such as principal component or factor analysis, but these are generally developed only for linear systems and must be used with caution in ecological analysis. It is best to use them only for a guide, but to make sure that the resulting model is adjusted so as to make reasonable biological sense (Fager and Longhurst, 1968; Hughes et al., 1972).

From the foregoing discussion it might appear that the ecosystems analyst is caught in an awkward position, in that trial-and-error modelling is extremely inefficient but formal analytic methods require stringent conditions which cannot be fulfilled. The practical solution is usually a mixture of the two approaches, using statistical and other formal techniques to suggest model structures, but modifying and testing these structures freely to arrive at a reasonable model or set of models. Several ecologists have done this in an organised fashion to see how the results of a modelling program depend on the mix of techniques used to define the model. Weigert (1977) compared five models of a complex food chain representing different patterns of aggregation and found significant differences in model performance. Several interesting examples of how apparently similar models generate unexpectedly disparate output are described in a recent book by Holling (1978, Chapter 7). A somewhat more contentious approach to different ways of modelling a given system is Commoner's (1973) comparison of '... a computer model that encompasses certain assumed interactions ... the cause-and-effect relationships are not derived from the data' with one which '... derives from these data the relationships that appear to govern the interactions among the various parameters, leading to generalizations about the mechanisms'. While none of these studies is so conclusive that it will sway the convinced reductionist, they do show that different approaches to modelling lead to significantly different models, and that the products of top-down modelling programs using the techniques of systems identification generally lead to simpler but better models than those derived by a synthetic approach.

5 Complexity

Because top-down modelling so often leads to models which appear to be less detailed in apparently important areas then seems appropriate, there is a tendency to try to make the models more complicated than they need to be. It is widely felt that because ecosystems are very complex, good models of them also need be complex. This is a misleading line of reasoning, because the complexity of a model is determined chiefly by its function and the observations it is supposed to describe, not by the internal structure. For example, when a seagull drops a clam on a stone to break it open, the trajectory of the clam and its time of fall are describable by a very simple but accurate model that completely ignores the internal structure of the clam. This is an extreme case of a decomposable model (Overton, 1977), but it is not unusual to find that the behaviour of a model does not depend on factors that at first seem important (Holling, 1978, Chapter 7).

There is no general theory that tells us how complex a model needs to be, but information theory provides a useful guide. A model can be viewed as a type of communication channel which converts input data to output data, and thus by a fundamental theorem of information theory (Shannon, 1949) the model must have sufficient capacity to process this information flow. The crucial point is that the information contained in a measurement is proportional to the logarithm of the signal-to-noise ratio, so that most ecological data, which have large relative variances, do not carry a great deal of information. If we are willing to postulate some degree of correspondence between the information-carrying capacity of a model and its complexity, then there is no reason to expect a model which encompasses a large number of very noisy observations to be necessarily more complex than one which handles a small number of very precise measurements. Although this cannot be considered a rigorous justification of the use of simple models in ecology, there is little evidence that very complex ecosystem models have performed significantly better than simple ones, and the most successful complicated models are massive computer simulations of spacecraft orbits, despite the fact that the theory of how an object moves through a vacuum is extremely simple; the complexity of astronomical models of this type is clearly a consequence of the precision with which the results are required to be calculated, not of the fundamental complexity of the system being modelled.

Part of the reason for building complicated models even when simple ones seem to work is a belief that as the amount of detail in a model increases, the performance of the model converges towards an exact representation of the real system. This is not generally true. The classical example is in statistical mechanics, where the behaviour of a gas is believed to be completely governed by Newton's laws of motion. However, no matter how detailed the calculations of forces and molecular collisions becomes, the model never behaves realistically unless a fundamental statistical assumption is made (Nijboer, 1962; the reason is that Newton's laws are unchanged if time is reversed, and therefore they can never give rise to irreversible behaviour). Thus it is possible for a sequence of models having increasingly detailed structure to converge on incorrect behaviour. One can, of course, argue that once the necessary fundamental principles have been incorporated the design of a very complex model is justified, and in fact Rosen (1970) characterises statistical mechanics as 'one of the rare examples of a successful reductionist theory in physics'. However, Rosen does not tell us how to know when all the necessary principles have been incorporated in the model. Kerr (1976) suggests the art of Escher as a model of the dangers inherent in ecosystem analysis, with each work providing a detailed and accurate description of three-dimensional reality but for one subtle and obscure intentional error of perspective which has catastrophic consequences.

Even among very simple models there can be subtle but essential differences which are more significant than much of the omitted detail. In the comparative study of several lumped models by Weigert (1977) mentioned above, he found that the simplest of the lumped models consistently worked as well or better than the more detailed versions. Weigert's conclusion, which is supported by more detailed calculations by Cale and Odell (1979), and O'Neill and Rust (1979), is that aggregation of compartments with comparable characteristic time scales (e.g. generation times or specific growth rates) generally has less effect on model performance than aggregation of compartments with different time scales, even though the latter aggregation may make more sense on functional grounds. For example, attempts to model the size distributions of particles in the sea (Sheldon et al., 1972) led to a dynamic model based entirely on aggregation of particles with equal turnover times without regard to trophic level or other functional properties (Silvert and Platt, 1978).

Care should also be taken to ensure that a simple model is not constructed in a way which artificially distorts valid information. When mathematical forms are fit to scatter plots it is usually advisable to use functions which act reasonably over their entire range and which capture the major qualitative features of the data. In particular, curves which must pass through the origin should only be fit with curves that pass through the origin; plots that show curvature should not be fit with straight lines; asymmetric data should not be fit with symmetric curves; and curves that do not show specific features should not be fit with curves that exhibit these features without clear justification (e.g. the use of the Ricker recruitment function when the recruitment data do not show the existence of a dome). It is almost always possible to find a mathematical function with the appropriate qualitative behaviour which does not require any more parameters to fit when the most popular model appears inadequate.

One way to counteract the tendency to build overly elaborate models is to build a sequence of models in a fashion similar to step-wise regression. The incorporation of additional details at any step can be viewed as a set of hypotheses to be tested by trying them to see what the effects may be. By doing this in a structured manner one can always go back to the appropriate descriptive level. This procedure is basically the same as a structural sensitivity analysis, which will be discussed in the next section.

There are two very common ways in which excess detail tends to creep into models. One is the tendency mentioned previously to use all available data, rather than to accept the fact that some of it is irrelevant. This is especially true when some parts of a system are well understood but others are poorly measured, yet all must be combined. There is an ecological version of Gresham's law which states that no data are better than the worst data, so it is simplest to represent all the input data by a model no more complicated than is justified by the worst data. One can test the effect of adding more detail by sensitivity analysis, but if the added detail makes a significant difference it indicates that the model cannot be of much use until the worst data are improved.

A second pitfall is to build unnecessarily detailed submodels for stylistic consistency. In modelling the response of living organisms to environmental changes it may be desirable to construct an elaborate biological submodel, but there is no particular reason to include equally elaborate submodels of physical processes if these can be equally well predicted by much simpler statistical descriptive models (Silvert, 1979a).

The best existing summary of system identification and complexity is contained in guidelines proposed by Lee (1973). *First*, 'Start with a particular problem that needs solving, not a methodology that needs applying. Work backward from the problem, matching specific methods with specific purposes, and obtaining just enough information to be able to provide adequate guidance'. *Second*, 'Build only very simple models. Complicated models do not work very well if at all, they do not fit reality very well, and they should not be used in any case because they will not be understood. The skill and discipline of the modeller is in figuring out what to disregard in building his model'. This second guideline is prone to misinterpretation, and it does not mean that a careful program of system identification cannot lead to large and complex models; Kagiwada's (1974) book contains many examples to the contrary. However, when complexity is introduced only for the sake of completeness, it generally does more harm than good; Lee goes on to observe that 'there are some researchers who are convinced that it has been the hardware limitations [of computers] that have obstructed progress and that advances in modelling are now possible because of larger computer capacity. There is no basis for this belief; bigger computers simply permit bigger mistakes'.

6 Sensitivity analysis and parameter estimation

There are in general two types of sensitivity analysis: structural and parameter. *Structural sensitivity analysis* refers to how the output of a model is affected by changes in the form of the realisation, while *parameter sensitivity analysis* covers changes in the outputs due to uncertainties in the input parameters. Structural sensitivity analysis typically deals with questions of whether the conceptual model can be realised in different ways, as when a continuous realisation is compared with a discrete model, but in some cases this can be interpreted as a special case of parameter sensitivity analysis. For example, comparison of the differential equation model

$$\mathrm{d}x/\mathrm{d}t = f[x(t)]$$

with one based on the difference equation (often called a delay-differential equation)

$$dx/dt = f[x(t - \tau)]$$

would generally be considered an example of structural sensitivity analysis, even though the differential equation can be considered a special case of the difference equation with zero delay.

Structural sensitivity analysis is strongly subjective, since different realisations reflect different aspects of the basic conceptual model. This is less true of statistical realisations, although statistical models with unreasonable functional transformations offer little promise of representing the true nature of the system. It is generally agreed that in comparing models with equal numbers of parameters, the model with the greatest maximum likelihood is preferable. If all data are normally distributed with equal variance (either by assumption, or by examining replicates) this means that the model with the smallest sum of squared residuals is the best realisation. This criterion should be used with caution, however. If a model with K parameters fits N pairs of data with a minimum sum of squares S_0, then the $1 - \alpha$ confidence limits for the parameters include all sets of parameter values for which the sum of squares is less than

$$S_\alpha = [1 + (K/(N - K))F(K, N - K, 1 - \alpha)]S_0$$

(Silvert, 1979b); if an alternate K-parameter model has a minimum sum of squares less than S_α, it cannot be rejected at this confidence level.

The proper comparison of statistical models with different numbers of parameters is a problem on which some work has been done, but no concensus among statisticians has emerged. Atkinson (1978) has extended the maximum likelihood argument and arrived at the criterion that the preferred model should maximise $LL - (K/2) \log N$, where LL is the log-likelihood function. However, based on a simulation study Atkinson concludes that this criterion unduly favours models with a small number of parameters. A better criterion is that of Akaike (1974) who advocates maximisation of $LL - K$ on information-theoretic grounds.

In the remainder of this section I shall discuss parameter sensitivity analysis, which is of great importance in both the design and evaluation of models. This is a large field with many specific techniques and it is possible only to touch on a few central issues here. The object of sensitivity analysis is to answer the question of how the outputs of the model, y_i, depend on changes in the parameters, p_j. For small changes it is possible to express these dependences through the sensitivity matrix

$$M_{ij} = \partial y_i / \partial p_j,$$

which represents the linear dependence of the outputs on the parameters. However, in most ecosystems there are large uncertainties in at least some of the parameters, so the linear sensitivity matrix should generally be used only as an indicator of the total sensitivity of the system.

Large ecosystem models are likely to have a large number of parameters, and the outputs are often even more numerous. The number of terms in the sensitivity matrix can easily become too large to handle without sophisticated information-processing techniques (e.g. Thornton *et al.*, 1979). For this reason it is frequently desirable to define a measure of model performance which is some function of the model outputs y_i and see how this depends on the parameters p_j. Miller (1974, 1976) uses the summed variances of the model predictions, which depend on the uncertainties of the parameter estimates, as an indicator of model performance, but this implies that all outputs from the model are of equal concern. A more general approach is to define a performance index for the model which is a weighted measure of how well the model predicts various important outputs. For example, the output of a fisheries management model might be yield of adult fish, and a model which predicts this accurately should be preferable to one which is much more accurate in predicting larval biomass but somewhat less good in predicting yield. This can be extended by recognising that different uses correspond to different performance indices, so that a model which performs well from one point of view might be inferior to another when the objective of modelling is changed.

The performance index is a function of the form $\pi(Y, y)$ where Y is the vector of experimental observations and y is the vector of predicted values and depends on the parameter set p. Thus π is an implicit function of the parameters. In particular, the likelihood function

$$L = \exp\left\{ -\sum_i [(Y_i - y_i)/\sigma_{Y_i}]^2 \right\}$$

is a performance index, and the maximum likelihood point estimates of the parameters are obtained by maximising L with respect to the parameters p_j. If the performance index used for model evaluation is different from that used for fitting, or if different sets of data are used for evaluation and for fitting, then the parameter values used will not generally maximise the performance index, and the variance of the performance index as a function of the variances in the parameter estimates is

$$\sigma_\pi^2 = \sum_j (\partial\pi/\partial p_j)^2 \sigma_{p_j}^2$$

$$= \sum \left(\sum \frac{\partial\pi}{\partial y_i} \frac{\partial y_i}{\partial p_j} \sigma_{p_j} \right)^2$$

However, the use of this linear approach is not justified if the variances are large, and it is better to calculate the distribution of the performance index by a Monte Carlo simulation, calculating the distribution of π by evaluating it for a number of sets of p drawn at random from the joint probability distribution for the parameter values (Silvert, 1979b).

From the distribution of π it is possible to decide at an early stage of the modelling process whether the output from the model will be precise enough to be of value, given the anticipated uncertainties in the parameter values. Parameter estimates which make the largest contribution to the variance of π should receive prompt attention to see whether these estimates can be narrowed by further measurement, and this approach to sensitivity analysis is especially valuable in identifying productive avenues for further experimental work. If there is no good prospect of narrowing the variance of these estimates sufficiently to bring the performance index within acceptable limits, it may be necessary to conclude that the model cannot be relied upon for the designated purpose (Miller *et al.*, 1973).

It is important to interpret the result of sensitivity analysis properly, although it may be discouraging to do so. Models are frequently 'validated' by examining their performance with a single set of parameter values. When good results are obtained it is important to recognise that they may be fortuitous, and for this reason the variation of the performance index should be used to establish confidence limits on its value. This is especially important because of the way in which many models evolve, since the evolutionary process often ceases when a good set of results is obtained. Proper use of sensitivity analysis would undoubtedly place many modellers in the discomfiting position of having a model that tracks the data well but of being forced to conclude that the fit may not be significant (Greig, 1979).

Sensitivity analysis plays an especially important role in management modelling, since it provides the kind of information that is required for proper risk assessment. The basic principles of decision analysis will not be discussed here (cf. Raiffa, 1968, and Holling, 1978, for a detailed exposition), but its central theme is that each possible outcome is assigned a utility, and the manager seeks to maximise the probable utility in light of the calculated probabilities of the various outcomes. Sensitivity analysis gives the probability distribution of the various output variables y_i in terms of the distributions of the parameters p_j, and this distribution (which may also be affected by other random effects in a stochastic model) provides the information which a decision-maker needs.

Many of the techniques discussed in this section are not as yet widely used in modelling studies, but there is a growing awareness of their existence: 'Modellers often pay only lip service to the concept of sensitivity analysis and there is need for more research and development in this important area'. (SCOPE, 1978).

7 Structural stability

Closely related to the concept of sensitivity analysis is the question of how the behaviour of a system responds to changes in its structure. The concept is borrowed from dynamical systems theory, where a system is defined as structurally stable if its trajectories do not change drastically when small terms are added to the equations which govern the system (Rosen, 1970).

Structural instability is a problem in some types of conceptually simple models. Conservative systems in particular are always structurally unstable (Rosen, 1970), but the only common model of this type in ecology is the undamped Lotka-Volterra system. Dissipative systems appear to be generally stable in this sense, although no general guidelines are known to exist.

Structural stability poses an especially serious problem for the ecosystem modeller because of the wide range of perturbations that could arise. Examples of such perturbations that normally are not present but may arise during the period modelled are epizootics, invasions by new species, climatic shock, and new forms of anthropogenic mortality such as a new fishery or the release of toxic chemicals. Often a model fails because of some new effect that was not included. There is no general way to avoid problems of this type other than inventing a set of possible scenarios and determining what effect they have on the model. However, when the terms that cause instability are those that have been identified but omitted, this type of error is avoidable.

A known interaction is sometimes omitted from a model on grounds that the interaction coefficient is not well known. Since the coefficient is a parameter of the model, this is equivalent to saying that 'if the confidence range of a parameter is very large, we set its value to precisely zero'. The correct procedure is to incorporate the term and test its importance by sensitivity analysis. If the variation is taken to be about the zero value, this is equivalent to structural sensitivity analysis, and the structural stability of the system with regard to this term is of considerable importance.

8 Validation and verification

Validation involves three completely different aspects of the accuracy of the predictions of a model, and confounding these aspects can lead to considerable confusion. We can summarise validation by stating three questions:

1   Is the model correct?
2   Is the model complete?
3   Is the model accurate?

The independence of these questions can best be seen by considering a number of examples.

Models based on fundamental laws of nature are almost certainly correct, although they may not be complete or accurate. Statistical models, particularly linear models of nonlinear processes, are generally incorrect, even though they may be both complete and accurate. The correctness of a model becomes an interesting subject for validation when the model incorporates unproven hypotheses to be tested through application of the model; however, failure of the model to predict the result of experimentation can disprove the underlying hypotheses only if the model is both complete and accurate.

For a model to be complete requires that it include all significant effects, which is often a matter of subjective judgement. Single-species population models are usually incomplete, but their use is generally defended by the argument that little is known about multi-species interactions. Failure of the model may or may not shed doubt on its basic correctness, since the observed discrepancies may be ascribed to these unknown interactions. Similarly, success should not be taken to mean that the model is complete or correct, since changes in the model environment may lead to unexpected changes in its dynamics.

The accuracy of a model depends on whether all parameters have correct values and thus properly represent the magnitudes of the effects within the real system. Since parameter values can rarely be measured exactly, this aspect requires careful attention to both parameter estimation and sensitivity analysis. When the parameters of the model are evaluated correctly the user can identify confidence intervals for these parameters (Smith, 1979) and can combine this information with sensitivity analysis to obtain probability distributions for the predicted behaviour of the system, rather than simply point estimates. From the width of these distributions it is possible to determine the precision of these point estimates, and this tells the user how accurately the model is likely to describe the real system (Miller *et al.*, 1973).

An important component of the validation process is (or should be) verification. This is the determination of whether the model does what is required of it, and is often a long and difficult procedure. Failure of a model to give good results may be due to errors at many levels: there can be programming or truncation errors in a computer program (failure of implementation); untenable approximations may have been made (failure of realisation); or there can be important missing links in our understanding of the system (failure of conceptual model). If the realisation has been verified by showing that it is a good mathematical representation of the ecosystem, and if the implementation has been verified by showing that, e.g. computer output represents an accurate solution of the equations, then failure of the model to produce reasonable results indicates that the fault lies in our understanding of the ecosystem, and thus invalidation of the model provides useful information; however, unless the realisation and the implementation have been verified, the failure of the model may be due to no more than trivial mathematical or programming errors.

There is no standard procedure for verification, just as there is none for doing good experiments. A mathematical realisation can be verified only by carefully checking all steps, a process which is best carried out by a critical and interactive group; it also helps to have the derivation presented in written report with all assumptions and approximations clearly described. Verification of a computer program or analog simulation is most easily done when there are special cases for which exact solutions are known, as these cases can be used as tests.

There is a natural but unfortunate tendency to build and validate a model, and then if the model works, to skip over the verification process. Unfortunately, it is not uncommon for an incorrect model to give very reasonable output. Caswell (1976) describes a classic example of this which is far from unique; his model of world population growth gives an almost perfect fit to the data, but it also predicts that the population will become infinite in about 50 years. In some cases models have been shown to give fortuitously correct solutions even when the internal structure is fundamentally changed, while in other cases the fitting of the model leads to unrealistic values for some of the parameters involved. It is important to establish that the model is consistent with other information on the system, which requires that the realisation (including parameter values) be verified against the conceptual model.

In most modelling projects there exists a baseline model, which, while seldom written down explicitly, represents a consensus about the dominant factors affecting the system. For example, depth is widely believed to be a principal determinant of benthic productivity. Consequently it should be anticipated that a scientist evaluating the output of a model of benthic productivity along a transect would like to see values for both productivity and depth. It also helps to have quantitative relative measures of the goodness of a model; in this case the correlation between experimental and simulated productivity could instructively be compared with the correlation between experimental productivity and depth. The precise approach depends on the details of the modelling program, but the point is that the user or analyst should be able to determine how much of the agreement between the model output and reality can be ascribed to the assumptions and hypotheses inherent in the model, and how much is simply the result of including established information in the model formulation.

The process of validation therefore requires much more than simply checking the output of a model against experimental data. Since the modelling process involves a series of models (implementations of realisations of conceptual models), it is necessary to verify each stage in order to establish that it does what is intended and does not introduce any errors, approximations, or unspecified assumptions into the line of reasoning. The causes of any discrepancies between model performance and expectation should be carefully analysed to identify weaknesses; often the failure of a model is more informative than its success. In addition, the comparison between model behaviour and system behaviour should be made as objectively as possible. Just as the experimental data on the system can be represented by confidence intervals determined by measurement errors, the predictions of a model can be represented by confidence intervals determined by parameter estimation and sensitivity analysis, and one can use statistical tests to tell whether these two sets of distributions are distinct.

Above all, the user of models should avoid the error of considering a validated model as being categorically correct, any more than an invalidated model should necessarily be considered wrong. A model which is useful for one purpose may be of no value for another, and the basic criterion must be whether a model adequately provides required information; a marine food web model may be quite useful for studying zooplankton dynamics but may give incorrect estimates, if any, of fish production. Validity is part of a functional description, not an absolute attribute of models (O'Neill, 1975; Greig, 1979).

9 Computer simulation

I have tried to stress the general nature of the modelling process, and the computer is by no means an essential component. It is, however, a very useful tool and most theoretical ecologists use it extensively. Because the computer is a specialised tool with which many researchers lack intimate familiarity, its application to ecological modelling merits special consideration.

Not all computer applications can properly be called simulations, but the same guidelines are generally applicable. In addition to simulation, which can be taken to mean a complete implementation on a computer, it is also common to use a computer to process (as well as to log) data, to perform statistical analyses, and to calculate eigenvalues and do other lengthy computations. Some of these applications, notably statistical analysis, frequently use packaged programs which may include subtle assumptions about the model being analysed. For example, nonlinear fitting packages usually assume normal error distributions, while in many biological models the errors are lognormally distributed; this can lead to significant errors in parameter estimation (Silvert, 1979b).

In cases where the modeller writes his own computer program, it must, above all, be fully documented as it is written; adding documentation to an existing program is like filling in a laboratory notebook by memory. It is extremely difficult to verify a computer program, namely to establish that the computations are correct, unless it is possible to reconstruct the precise mathematical assumptions on which it is based. Internal documentation (comments within the actual program) is essential, but it should be augmented by external documentation, i.e. a manual or other description should be provided.

Modular design is important. The main program should be designed to work as a controller with three main tasks:

1   initialise the model and read in any starting information;
2   operate a system 'clock' and call subroutines as appropriate;
3   activate output and termination routines, along with monitoring any interactive modes.

Generally the main program can and should be very short. The main calculations are done within subroutines activated by the system clock in the main program; for example, growth and predation routines would generally be called every cycle, but reproductive calculations would be initiated only on a seasonal basis for some species. The overwhelming advantage of doing this is the ability to test and modify the model in fundamental structural ways. It is possible to change any of the subroutines without recompilation of the main program, and subprogram units incorporating different hypotheses can easily be substituted at execution time. In order to do this effectively the variables should be stored in common blocks, so that early assumptions about the variables affecting a given process do not constrain further modelling developments. This also makes it possible to store all the initial values in a block common program unit, so that initial conditions can be changed by substitution of one unit; this can greatly reduce the time spent reading in massive data sets when only a small set of different initial conditions are needed during the model-building process. The only real objection to modular program structure is that subroutine calls take a small amount of extra time, but this is a minor factor and can be minimised by using common blocks so that the calls do not require the passing of argument lists.

Similar considerations apply to output format, since the type of output generated by a model may be different at various stages of its development and use. The importance of well thought-out graphic output for the presentation of model results to different groups of users has been forcefully presented by Holling and his co-workers (Holling, 1978, Chapter 9; Clark *et al.*, 1978), but developmental work usually requires other types of graphs or numerical output, and often a model is called upon to generate numerical data in a form that can be fed to another program. One solution is to have the program prepare an extensive numerical output file and to construct a number of different programs (post-processors) to convert this information to different types of output formats (for interactive output this can be done by subprogram units to handle output routines).

A final consideration should be transportability, the ability of a program to run on different kinds of computer. A program that can only be run by its author resembles an experiment that cannot be replicated and should be viewed with equal suspicion. Programs in computer languages like BASIC, FORTRAN, and ALGOL, should conform to established standards and should not contain machine-dependent features. In rare cases when machine-dependent optimisation is required, it should be confined to subroutines and less efficient but computationally equivalent standard subroutines should be provided with the documentation. Modellers should encourage others to run their models and should provide tapes or card decks at cost rather than merely sending listings of lengthy programs. When a modeller uses a non-standard compiler it may not be possible to meet these conditions, and this makes the burden of providing complete and lucid documentation of the program even more essential.

Good computer style requires a lot of attention to matters of housekeeping, and there are many parallels to good experimental technique. There is no excuse for using the computer as a screen for muddled modelling.

References

In addition to the specific references cited earlier in this chapter, there is a wide range of monographs and books of direct interest to the ecosystem modeller. These range from disjointed collections of case studies through proceedings of conferences on aspects of modelling to well-organised treatises on the modelling process. One of the best known and certainly the most massive treatment is the four-volume set edited by Patten (1971, 1972, 1975, 1976), which follows the path blazed by Watt (1966). Both of these compendia are dated in parts, and the most up-to-date comprehensive treatment of the subject I have seen is the one edited by Halfon (1979). There is also a forthcoming SCOR report on mathematical models in biological oceanography which promises to be a valuable contribution to the field (Platt *et al.*, 1981). A number of idiosyncratic collections also deserve attention, and of these the proceedings of the 1974 SIAM Conference on Ecosystems which have already been extensively cited in this chapter certainly merit careful study (Levin, 1975). Other collections of this type include those edited by Innis (1975), Hall and Day (1977), and Holling (1978). There are several single-author volumes on modelling which should be consulted, including (roughly in order of increasing sophistication) Jeffers (1978), Zeigler (1976), Vemuri (1978), Aris (1978), and Rosen (1978).

Akaike, H. (1974). A new look at the statistical model identification. *IEEE Trans. Auto. Control AC*-**19**, 716–723

Aris, R. (1978). "Mathematical Modeling Techniques". Pitman, London

Ashby, W. R. (1956). "An Introduction to Cybernetics". Chapman and Hall, London

Atkinson, A. C. (1978). Posterior probabilities for choosing a regression model. *Biometrika* **65**, 39–48

Beck, M. B. (1978). Model structure identification from experimental data. *In* "Theoretical Systems Ecology" (E. Halfon, ed.) pp. 260–289. Academic Press, New York and London

Botkin, D. B. (1975). Functional groups of organisms in model ecosystems. *In* "Ecosystem Analysis and Prediction" (S. A. Levin, ed.) pp. 98–102. SIAM, Philadelphia

Cale, W. G. Jr. and Odell, P. L. (1979). Concerning aggregation in ecosystem modeling. *In* "Theoretical Systems Ecology" (E. Halfon, ed.) pp. 55–77. Academic Press, New York and London

Caswell, H. (1976). The validation problem. *In* "Systems Analysis and Simulation in Ecology" (B. C. Patten, ed.) Vol. IV, pp. 313–325. Academic Press, New York and London

Clark, W. C., Jones, D. D. and Holling, C. S. (1978). Patches, movements, and population dynamics in ecological systems, a terrestrial perspective. *In* "Spatial Pattern in Plankton Communities" (J. H. Steele, ed.) pp. 385–432. Plenum Press, New York and London

Commoner, B. (1973). Alternative approaches to the environmental crisis. *J. Am. Inst. Planners* **39**, 147–162

Fager, E. W. and Longhurst, A. R. (1968). Recurrent group analysis of species assemblages of demersal fish in the Gulf of Guinea. *J. Fish. Res. Bd Can.* **25**, 1405–1421

Gaines, B. R. (1978). General system identification – fundamentals and results. *In* "Applied General Systems Research" (G. J. Klir, ed.) pp. 91–104. Plenum Press, New York and London

Greig, I. D. (1979). Validation, statistical testing, and the decision to model. *Simulation* **33**, 55–60

Halfon, E. (ed.). (1979). "Theoretical Systems Ecology". Academic Press, New York and London

Hall, C. A. S. and Day, J. W. (eds) (1977). "Ecosystem Modeling in Theory and Practice: an Introduction with Case Histories". John Wiley and Sons, New York

Holling, C. S. (ed.) (1978). "Adaptive Environmental Assessment and Management". John Wiley and Sons, New York

Hughes, R. N., Peer, D. L. and Mann, K. H. (1972). Use of multivariate analysis to identify functional components of the benthos in St. Margaret's Bay, Nova Scotia. *Limnol. Oceanogr.* **17**, 111–121

Innis, G. S. (ed.) (1975). New directions in the analysis of ecological systems. *Simulations Councils Proceedings* Vol. 5, Numbers 1 and 2

Jeffers, J. N. R. (1978). "An Introduction to Systems Analysis: With Ecological Applications". Edward Arnold, London

Kagiwada, H. H. (1974). "System Identification: Methods and Applications". Addison-Wesley Publishing Company, Reading, Mass.

Kerr, S. R. (1976). Ecological analysis and the Fry paradigm. *J. Fish. Res. Bd Can.* **33**, 2083–2089

Kerr, S. R. and Neal, M. W. (1976). Analysis of large-scale ecological systems. *J. Fish. Res. Bd Can.* **33**, 2083–2089

Lane, P. A., Lauff, G. H. and Levins, R. (1975). The feasibility of using a holistic approach in ecosystem analysis. *In* "Ecosystem Analysis and Prediction" (S. A. Levin, ed.) pp. 111–128. SIAM, Philadelphia

Lee, D. B. Jr. (1973). Requiem for large-scale models. *J. Am. Inst. Planners* **39**, 163–178

Levin, S. A. (ed.) (1975). "Ecosystem Analysis and Prediction". SIAM, Philadelphia

Li, C. C. (1975). "Path Analysis – A Primer". Boxwood Press, Pacific Grove, California

Mann, K. H. (1975). Relationship between morphometry and biological functioning in three coastal inlets of Nova Scotia. *In* "Estuarine Research" (L. E. Cronin, ed.) Vol. 1, pp. 634–644. Academic Press, New York and London

Mann, K. H. (1979). Qualitative aspects of estuarine modeling. *In* "Marsh-Estuarine Systems Simulation" (R. F. Dame, ed.) pp. 207–220. University of South Carolina Press, Columbia

Maynard Smith, J. (1974). "Models in Ecology". Cambridge University Press, Cambridge

Miller, D. R. (1974). Sensitivity analysis and validation of simulation models. *J. Theor. Biol.* **48**, 345–360

Miller, D. R., Butler, G. and Bramall, L. (1976). Validation of ecological system models. *J. Environ. Man.* **4**, 383–401

Miller, R. D., Weidhaas, D. E. and Hall, R. C. (1973). Parameter sensitivity in insect population modeling. *J. Theor. Biol.* **42**, 263–274

Mohn, R. K. (1979). Sensitivity analysis of two harp seal (*Pagophilus groenlandicus*) population models. *J. Fish. Res. Bd Can.* **36**, 404–410

Mott, D. G. (1966). The analysis of determination in population systems. *In* "Systems Analysis in Ecology" (K. E. F. Watt, ed.) pp. 179–194. Academic Press, New York and London

Nijboer, B. R. A. (1962). General introduction. *In* "Fundamental Problems in Statistical Mechanics" (E. G. D. Cohen, ed.) pp. 1–32. North-Holland, Amsterdam

Oak Ridge Systems Ecology Group. (1975). Dynamic ecosystem models, progress and challenges. *In* "Ecosystem Analysis and Prediction" (S. A. Levin, ed.) pp. 280–296. SIAM, Philadelphia

O'Neill, R. V. (1975). Management of large-scale environmental modeling projects. *In* "Ecological modeling in a Resource Management Framework" (C. S. Russell, ed.) pp. 251–282. Resources for the Future, Washington

O'Neill, R. V. and Rust, B. (1979). Aggregation error in ecological models. *Ecol. Modelling* **7**, 91–105

Overton, W. S. (1977). A strategy of model construction. *In* "Ecosystem Modeling in Theory and Practice: An Introduction with Case Histories" (C. A. S. Hall and J. W. Day, eds) pp. 50–73. John Wiley and Sons, New York

Patten, B. C. (ed.) (1971, 1972, 1975 and 1976). "Systems Analysis and Simulation in Ecology". Vols I–IV. Academic Press, New York and London

Patten, B. C. and Finn, J. T. (1979). Systems approach to continental shelf ecosystems. *In* "Theoretical Systems Ecology" (E. Halfon, ed.) pp. 183–212. Academic Press, New York and London

Platt, T., Mann, K. H. and Ulanowicz, R. E. (1981). "Mathematical Models in Biological Oceanography". UNESCO Monographs in Oceanographic Methodology. UNESCO Paris

Pomeroy, L. R. (1974). The ocean's food web, a changing paradigm. *Bioscience* **24**, 499–504

Raiffa, H. (1968). "Decision Analysis". Addison-Wesley Publishing Company, Reading, Mass.

Rosen, R. (1969). Hierarchical organization in automata theoretic models of biological systems. *In* "Hierarchical Structures" (L. L. Whyte, A. G. Wilson and D. Wilson, eds) pp. 179–199. American Elsevier Publishing Company, New York

Rosen, R. (1970). "Dynamical System Theory in Biology". Vol. I. John Wiley and Sons, New York

Rosen, R. (1978). "Fundamentals of Measurement and Representation of Natural Systems". North-Holland, New York

SCOPE (Scientific Committee on Problems of the Environment). (1978). Simulation Modelling of Environmental Problems (F. N. Frenkiel and D. W. Goodall, eds) SCOPE Report 9. John Wiley and Sons, New York

Shannon, C. E. and Weaver, W. (1949). "The Mathematical Theory of Communication". University of Illinois Press, Urbana

Sheldon, R. W., Prakash, A. and Sutcliffe, W. H. Jr. (1972). The size distribution of particles in the ocean. *Limnol. Oceanogr.* **17**, 327–340

Silvert, W. (1979a). Book review of "A Coastal Marine Ecosystem" by J. N. Kremer and S. W. Nixon. *J. Fish. Res. Bd Can.* **36**, 597–598

Silvert, W. (1979b). Practical Curve Fitting. *Limnol. Oceanogr.* **24**, 767–773

Silvert, W. and Platt, T. (1978). Energy flux in the pelagic ecosystem: a time-dependent equation. *Limnol. Oceanogr.* **23**, 813–816

Slobodkin, L. B. (1972). On the inconstancy of ecological efficiency and the form of ecological theories. *Trans. Conn. Acad. Arts and Sciences* **44**, 293–305

Slobodkin, L. B. (1975). Comments from a biologist to a mathematician. *In* "Ecosystem Analysis and Prediction". (S. A. Levin, ed.) pp. 318–329. SIAM, Philadelphia

Smith, W. R. (1979). Parameter estimation in nonlinear models of biological systems. *Fish. Res. Bd Can.* Tech. Report 889

Thornton, K. W., Lessem, A. S., Ford, D. E. and Stirgus, C. A. (1979). Improving ecological simulation through sensitivity analysis. *Simulation* **32**, 155–166

Vemuri, V. (1978). "Modeling of Complex Systems". Academic Press, London and New York

Watt, K. E. F. (ed.) (1966). "Systems Analysis in Ecology". Academic Press, New York and London

Weigert, R. G. (1977). A model of a thermal spring food chain. *In* "Ecosystem Modeling in Theory and Practice: An Introduction with Case Histories", (C. A. S. Hall and J. W. Day, eds) pp. 289–315. John Wiley and Sons, New York

Zeigler, B. P. (1976). "Theory of Modelling and Simulation". John Wiley and Sons, New York

23 *Simulation Models of Individual Production Processes*

PHILIP J. RADFORD, IAN R. JOINT and ALEX R. HIBY

1 Introduction

Simulation modelling has been applied to whole ecosystems including tundra, coniferous and deciduous forest, grassland, desert, lake and estuarine environments (Patten, 1975; Scavia and Robertson, 1979; Park, 1978; Laevastu and Favorite, 1977). In most cases the study has involved several years of work by multidisciplinary teams of ecologists working in close collaboration with mathematical modelling specialists. The sophistication and specialisation required for such work may have tended to discourage the individual ecologist from considering the application of systems simulation to individual studies of components of an ecosystem. This is to be regretted because a submodel of an individual process may be used to great effect (Radford, 1971), and with a precision and usefulness which can often be superior to whole ecosystem models. This chapter is designed to encourage those with little or no experience of simulation modelling to apply the technique to their own sphere of ecological interest. The various examples given were chosen to illustrate the usefulness of different types of submodels, constructed at different levels of abstraction relevant to their individual objectives.

2 Techniques

There are many techniques which are used to aid both the conceptualisation and implementation of simulation models and in general their use is more a reflection of personal preference than of any inherent superiority of method.

One system of process flow diagrams has been proposed by Levins (1975) which represents state variables and the network of processes by which they interact as a series of circles linked by vectors. From this simple concept he develops the technique known as 'Loop Theory' which enables qualitative predictions to be made of the response of a system when perturbed slightly from equilibrium. No assumptions need to be made regarding the specific

mathematical relationships which govern the rate equations, only a knowledge of the direction of material flow through each pathway. This enables
broad generalisations to be made regarding the behaviour of whole families
of systems, but the technique cannot be used for systems which are not in
equilibrium nor for those affected by major perturbations.

Both qualitative and quantitative measures of system response are
obtainable if the energy circuit language of Odum (1972) is used as the first
step towards complete simulation modelling. The four basic flow diagram
symbols are easy to understand, representing sources, sinks, stocks and
processes respectively, but the further refinement of the language into ten
composite symbols (e.g. cycling receptor, work gate, constant gain amplifier
and active impedence) although enlightening to an electrical engineer tend
only to confuse those not familiar with the theory of analogue computers.
For those willing to invest considerable time and effort into understanding
these more complicated aspects of the energy circuit language many
advantages are promised (Odum, 1976), but it is debatable if the increased
rigour of the technique compensates for increased loss of communication
with the experimental ecologist.

The Industrial Dynamics techniques of Forrester (1961) can cope with a
most comprehensive range of problems, with or without discontinuities,
linear or non-linear, near or away from equilibria, using a disarmingly
simple methodology. The effectiveness of this approach for communicating
the principles of systems simulation to the public at large has been clearly
demonstrated by Meadows *et al.* (1972) and the techniques have been
applied to a variety of disciplines as diverse as urban dynamics (Forrester,
1969) and agricultural research (Jones, 1969). The associated system of
process flow-diagrams uses three basic symbols; 'rectangles' for state
variables, 'valves' for rates and 'clouds' for sources and sinks (Fig. 1), linked
by continuous and discontinuous lines which represent flow of energy and
information respectively. The main advantage of this type of representation
is that it clearly distinguishes between the essentially different types of
variables and their interrelationships without clouding the issue with
detailed specification of the precise form of these relationships. In this way
one flow diagram represents a whole family of models, members of which
can be individually studied when translated into a specific mathematical
model. The translation from a Forrester type flow diagram to a working
computer program using a special purpose simulation language such as
DYNAMO II (Pugh, 1973) or CSMP III (IBM, 1972) is largely a
mechanical exercise (de Wit and Goudriaan, 1974) with the added advantage of the automatic provision of a self-documented listing (Annino and
Russell, 1979). Figure 1 is an example of a flow diagram translation of the
planktonic ecosystem described by Steele (1974, Chapter 5). The details not

specified in this figure are included in the CSMP (IBM, 1967) computer program (Table I) whose broad structure conforms to that of any continuous system simulation model in that it includes INITIAL, DYNAMIC and TERMINAL sections.

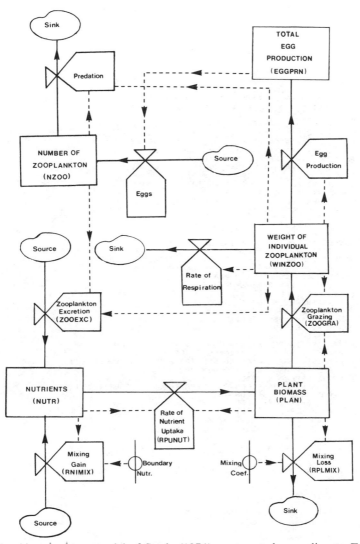

FIG. 1 The plankton model of Steele (1974) represented according to Forrester's (1961) industrial dynamics techniques. The state variables (rectangles) are linked by processes (continuous lines) which are controlled by rate variables (valves). The broken lines indicate information links, and the clouds, sources and sinks

TABLE I   A complete listing of the CSMP/360 computer program which simulates
the planktonic ecosystem described by Steele (1974, Chapter 5). The lines which start
with an asterisk describe the variable defined in the following line

```
                    ****CONTINUOUS SYSTEM MODELING PROGRAM****

               ***PROBLEM INPUT STATEMENTS***

TITLE SIMULATION OF A PLANKTON ECOSYSTEM - JOHN H STEELE -
* CHAPTER 5 FROM THE STRUCTURE OF MARINE ECOSYSTEMS
INITIAL
INCON   INUTR=10.,INPLN=100.,INWZOO=1.,INNZOO=100.,COHORT=1.
PARAM   MRPUNT=0.2,NCONHR=8.0,COEFMX=0.01,NOXEXR=0.4
PARAM   MAXGRA=1.6,PLANTH=75.,PLCONH=4.0,MAXRR=0.5,STRMBS=0.
PARAM   MDENPR=0.02,TNZOPR=0.1,TWZOPR=5.,WZOHPR=50.,PRZOPR=0.
PARAM   MATWTF=75.,LFLIFE=10.,PPREGG=0.3,BLNUTR=10.
DYNAMIC
        NUTR=INTGRL(INUTR,-RPUNUT+RNIMIX+ZOOEXC)
        PLAN=INTGRL(INPLN,RUNUT-RPLMIX-ZOOGRA)
        WINZOO=INTGRL(INWZOO,ASSIMF-STRESP-RRESP)
        NZOO=INTGRL(INNZOO,-DENSPR-PROPPR)
* RATE OF PLANT UPTAKE OF NUTRIENTS
        RPUNUT=MRPUNT*NUTR*PLAN/(NCONHR+NUTR)
* RATE OF NUTRIENT INCREASE BY MIXING
        RNIMIX=COEFMX*(BLNUTR-NUTR)
* ZOOPLANKTON EXCRETION
        ZOOEXC=NOXEXR*(STRESP+RRESP)*NZOO
* METABOLIC BODY SIZE
        MBS=WINZOO**0.7
* RATE OF PLANT LOSS THROUGH MIXING
        RPLMIX=COEFMX*PLAN
* TOTAL ZOOPLANKTON GRAZING OF PLANTS
        ZOOGRA=IZOGRA*NZOO
* PROPORTION OF MAX. PLANT INTAKE DEPENDENT UPON PLANT AVAILABILITY
        PMIDPA=(PLAN-PLANTH)/(PLCONH+PLAN)
* INDIVIDUAL ZOOPLANKTON GRAZING
        IZOGRA=MAXGRA*MBS*PMIDPA
* ASSIMILATED FOOD
        ASSIMF=0.7*IZOGRA
* ROUTINE RESPIRATION
        RRESP=MAXRR*MBS*PMIDPA
* STANDARD RESPIRATION
        STRESP=STRMBS*MBS
* EFFECTIVE ZOOPLANKTON WEIGHT AVAILABLE FOR PREDATION
        EFFWPR=AMAX1(0.,(WINZOO-TWZOPR))
* EFFECTIVE ZOOPLANKTON NUMBER AVAILABLE FOR PREDATION
        EFFNPR=AMAX1(0.,(NZOO-TNZOPR))
* DENSITY DEPENDENT PREDATION
        DENSPR=MDENPR*EFFNPR*EFFWPR/(WZOHPR+NZOO*WINZOO)
* PROPORTIONAL PREDATION
        PROPPR=PRZOPR*NZOO
* INDICATOR OF EGG PRODUCTION
        IEGGPR=COMPAR(WINZOO,MATWTF)
* DURATION OF EGG PRODUCTION
        DUREPN=INTGRL(0.,IEGGPR)
* EGG PRODUCTION BODY WEIGHT
        EGGPRN=INTGRL(0.,IEGGPR*(ASSIMF-STRESP-RRESP))
FINISH DUREPN=LFLIFE
TIMER DELT=1.,FINTIM=360.,PRDEL=1.,OUTDEL=1.
TERMINAL
        COHORT=COHORT+1.
        IF(COHORT.GE.6.)GOTO 1
        IF(TIME.GE.360.)GOTO 1
        INNZOO=PPREGG*EGGPRN/INWZOO
        INUTR=NUTR
        INPLN=PLAN
        CALL RERUN
  1   CONTINUE
END
```

In the INITIAL section the initial conditions (INCON) and parameters (PARAM) are set to their required values. The DYNAMIC section contains all the structural statements of the model including an integral (INTGRL) equation for each of the state variables and a rate equation for each of the processes. In this case these equations describe the development of one cohort of zooplankton from eggs to fertile adults; their ordering is irrelevant because the CSMP compiler sorts the whole of the DYNAMIC section into the required computational sequence before executing the program. The TERMINAL section enables the fate of five consecutive cohorts to be simulated before the program ends. The whole CSMP program is a self documenting, fully explicit and unambiguous statement of the system it attempts to describe. This example illustrates how readily a well-documented model may be transferred between ecologists, thus facilitating a deeper exchange of ideas than could normally be achieved through the scientific press alone. As Steele states in his introduction, his main aim in using this model is to demonstrate how theory, observation and experiment could be combined and how closely each depends upon the other. Such a demonstration is far more strongly reinforced if the model is actually used by those wishing to deepen their understanding of the system it describes.

3 Subsystem models

Individual scientists working at any level of abstraction, from cell to macrocosm, should benefit from the use of subsystem models which can form the structure of the hypothesis stage of the overall iterative scientific method (observation, experimentation, hypothesis etc.). Alternative models may be tested as alternative hypotheses against experimental results especially designed to differentiate between them, established models giving place to more appropriate, though not necessarily more complicated ones. The next four sections of this chapter focus upon specific modelling activities which illustrate some different types of submodels and their usefulness.

1   The section on phytoplankton subsystem models traces the important events in the evolution of models of primary production over the last forty years, showing how each new model is the result of the cumulative knowledge and understanding of the processes involved. The implication is that the rapid rate of progress in this sphere of marine science has been enhanced, if not determined, by the fruitful conjunctive use of models and experimentation; the simple models of the 1930s providing the initial impetus which has led to the present day state of the art.

2   The *Nephtys* subsystem model illustrates how information obtained from the scientific literature, linked with field measurements of key state variables, can aid theoretical studies of individual organisms without

resorting to a full ecosystem model. The results of such theoretical simulations cannot be taken as proven fact but they do provide insight into possible theories, of which the most plausible can become subjects of future experimentation.

3 A model of the carbon and nitrogen balance of *Mytilus edulis* shows how the relevant physiological processes, measured in the laboratory, can be integrated into a subsystem model capable of producing output comparable to measured field performance of the organism. In this case the modelling exercise highlighted our lack of knowledge of the nutritional value of naturally occurring suspended particulate matter and suggested values much lower than expected, which were subsequently verified by field measurements.

4 A further use of modelling is demonstrated in the description of a zooplankton submodel, where the possible interactions between three planktonic groups were examined by postulating purely theoretical feeding patterns dependent only upon their relative sizes. It is only rarely that the validity of such models can be usefully tested against field data, because of the number of unknown parameters and sometimes the vague understanding of the form of the internal relationships inherent in the system. They are valuable in so far as they enable the effects of a variety of theoretical relationships to be investigated, giving insight into the full dynamic implication of the underlying assumptions.

4 Phytoplankton subsystem models

Every subsystem model reflects the evolution of our understanding of the processes occurring in an ecosystem. Perhaps the most modelled aspect of the marine ecosystem is phytoplankton production and Platt *et al.* (1977) have reviewed this subject in detail. Expressions have been used for many years to describe the relationship between the rate of photosynthesis and light intensity, but it is only comparatively recently that models have been developed which realistically simulate the development of phytoplankton in response to a variety of environmental forcing functions.

The most fundamental process in the growth of phytoplankton is photosynthesis. Early workers (Baly, 1935) drew an analogy with the relationship between the rate of photosynthesis and light intensity and the Michaelis-Menten expression for enzyme kinetics; however, this was not very realistic since the initial slope of the photosynthesis vs light curve tended to be much steeper than that predicted by Michaelis kinetics. Smith (1936) suggested an expression which has served as the basis of many subsequent models of primary production; his expression gives a linear relationship between photosynthesis and light at low light intensities, with photosynthesis approaching P*m*, the maximum rate of photosynthesis, at high light intensities. Talling (1957) applied Smith's equation to the estimation of depth integrated photosynthesis of a unit area of water surface; he did not solve the integration analytically but, using planimetric methods, he suggested that the depth integrated photosynthesis was equal to a rectangle given by P*m* and the depth in the water column at which the light intensity was half the surface value. As an approximation of photosynthesis, Talling's solution works reasonably well, although it is imprecise at both high and low light levels. There have been many subsequent expressions to describe the photosynthesis vs light curve up to the point of light saturation and some of these were compared by Jassby and Platt (1976).

A major drawback to these relationships is their failure to attempt to model photoinhibition, the phenomenon of decreasing rate of photosynthesis at high light intensities, and all the early expressions assumed that photosynthesis reached saturation at high light intensities. Steele (1962) proposed an expression which defined the phenomenon of photoinhibition, and which did not break down at low light intensities. The major criticism of Steele's equation is that it places too much emphasis on photoinhibition, and that it defines the linear part of the curve at low light intensity and the photoinhibited part of the curve with the same constant. However, the expression is quite accurate; Parsons and Anderson (1970) found Steele's expression the most applicable of several theoretical models to their data for the North Pacific, and Radford and Joint (1980) demonstrated a close agreement between production measured by ^{14}C incubation in the Bristol Channel, England (a highly turbid estuary) and that predicted by Steele's expression. Because of its simplicity and accuracy, Steele's expression has been used in several models of phytoplankton production (Lehman *et al.*, 1975; Takahashi *et al.*, 1973) and in some large ecosystem models (e.g. Kremer and Nixon, 1978; Scavia and Park, 1976; Radford, 1979).

All the previous expressions have just two parameters and, although remarkably successful in simulating photosynthesis they can only give approximate estimates of production. Vollenweider (1965) has proposed a modification of Smith's expression which is probably the best expression for fitting experimental data. It requires four parameters and is capable of fitting a large variety of photosynthesis-light curves. However, the expression is difficult to integrate with depth and requires elaborate fitting procedures (Fee, 1969); for these reasons it has not been generally adopted by modellers of phytoplankton production.

The process of photosynthesis is only one aspect of the growth of phytoplankton which has been modelled. Takahashi *et al.* (1973) modelled the phytoplankton population production of the Fraser River estuary, assuming that only one of the environmental factors of their model limited phytoplankton growth at any one time; i.e. they subscribed to Leibig's 'law of the minimum'. They used Steele's model for photosynthesis, choosing parameter values which gave a good fit of the linear part of the photosynthesis-light curve to their data, but a poor fit to the photo-inhibition part of the curve. They argued that in turbid, estuarine waters, photoinhibition was restricted to the shallow surface layers and so it was more important to realistically simulate photosynthesis in the rest of the water column. The other environmental factors included in the model were temperature and the concentrations of nitrate and phosphate, the latter represented by a Michaelis-Menten expression. Using the model, Takahashi *et al.* (1973) were able to show the different environmental factors limited photosynthesis at different times of the year; light was usually limiting but temperature limited production in the spring and the lack of nutrients resulted in photoinhibition in August.

A more complex model by Lehman *et al.* (1975) computed growth as an increase in the concentration of cells and depended on light, temperature and nutrient concentration. They included in the model the concept of internal nutrient concentration of the cells and the luxury consumption of nutrients; the uptake of nutrients was a function of the internal, cellular and the external concentration of nutrients but the rate of cell division depended entirely on the internal nutrient concentration.

Iverson *et al.* (1974) developed a model to simulate the effects on phytoplankton dynamics of short-term wind mixing of the water column in a deep estuary. They argued that as the complexity of a model increased so the number of parameters which had to be derived also increased. Since it was difficult to derive values for all the parameters from the field, the choice lay between using values derived from the literature (which are often difficult to apply directly to another system) and ignoring some processes which, although important in an absolute sense, are not directly relevant to the process being modelled.

The processes which Iverson *et al.* (1974) were modelling were the short-term events associated with wind mixing; they were able to omit temperature effects and the grazing by zooplankton but included effects of light, nutrient concentration and mixing coefficients related to wind speed.

Perhaps the most highly developed model of phytoplankton dynamics is that of Winter *et al.* (1975) which simulated the dynamics of spring bloom of phytoplankton in Puget Sound. This model included photosynthesis-light relationships, respiration (including an attempt to account for photorespiration), time-lagged Michaelis-Menten expression for nutrient uptake which allowed for the observation that instantaneous growth rate is a function of the past external nutrient concentration, the effect of grazing and the effect of run-off, winds and tides. Winter *et al.* (1975) used the model as an experimental system to attempt to evaluate the relative importance of the processes governing primary production; they found the expected close relationship between the level of phytoplankton production and circulation, light and the chemical properties of the water but the model also supported the suggestion that mixing introduces algal seed stock into the euphotic zone from depth. Winter *et al.* (1975) conceded that the present knowledge of the response of phytoplankton to environmental variables is too poor to construct predictive models of phytoplankton growth which would be applicable to all fjords. However, they stressed that the modelling process itself was invaluable in concentrating the mind; new relationships between components of ecosystem can be perceived, traditional descriptions of processes are re-evaluated and new experiments are often suggested when one attempts to construct a model of an ecological process.

5 *Nephtys* subsystem model

Even when it is impracticable to model a complete ecosystem there is much to be gained by modelling part of the system in detail and replacing other parts by time-series measurements taken in the field. Warwick *et al.* (1979) produced a model of the secondary production of an estuarine mudflat and used it to investigate a number of trophic relationships, including the nutritional status of the polychaete, *Nephtys*. The model simulates the production of phytobenthos, meiofauna, deposit feeders, suspension feeders and *Nephtys*. The difficulties of modelling such phenomena as the temporal inhomogeneity in phytoplankton biomass and the deposition or resuspension of detritus meant that some components of the ecosystem had to be included as forcing functions. Linear interpolation between mean monthly values from field data were used for daily phytoplankton biomass, organic detritus, heterotrophic microbes, birds and fish. The basic assumptions made in constructing the model were as follows; (*i*) that each food source was consumed in proportion to its abundance and so ingestion rate was limited by the total food available and not by a single component of the diet, (*ii*) that the meiofauna ingestion rate was dependent on temperature with bacterial and phytobenthos as food source, (*iii*) that the gut of deposit feeders is always full but that the rate of passage of material through the gut was temperature dependent (*iv*) that the filtration rate of suspension feeders and their production of pseudofaeces is dependent on the concentration of particulate matter in the water column. A full CSMP listing of the model is given in the original publication (Warwick *et al.*, 1979).

Realistic simulations were obtained for phytobenthos, meiofauna, deposit feeders, suspension feeders and *Nephtys*. Since the nutritional status of

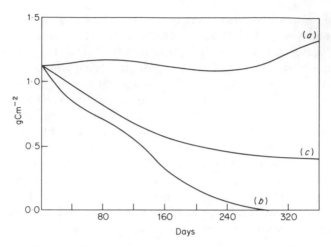

FIG. 2 Simulation results tracing the population biomass of *Nephtys* over an annual cycle assuming a diet of (*a*) phytobenthos, meiofauna, deposit feeders and suspension feeders; (*b*) meiofauna, deposit feeders and suspension feeders only; (*c*) phytobenthos only

Nephtys was in doubt, the model was used to investigate the performance of *Nephtys* on different diets. When the simulated diet of *Nephtys* (Fig. 2) was restricted to meiofauna and small deposit and suspension feeders, the model indicated a rapid population decline, reaching zero biomass within eight months. When considered as a herbivore, feeding exclusively on phytobenthos, the *Nephtys* population stabilised at about half the population density recorded in the field; the standing stock of deposit feeders and suspension feeders in this simulation increased because there was no longer any grazing by *Nephtys*. It was only when phytobenthos, meiofauna, deposit feeders and suspension feeders were all included as components of the diet of *Nephtys* that a realistic simulation of *Nephtys* standing stock could be achieved. *Nephtys* has traditionally been considered as a carnivore because of the structure of its feeding apparatus but it is also the major producer on the mudflat (Warwick and Price, 1975) a fact which is not compatible with a carnivorous mode of nutrition. This simulation exercise was additional evidence that *Nephtys* was an omnivore and stimulated further research on the ecophysiology of *Nephtys* which confirmed the conclusions drawn from the stimulation model.

6 *Mytilus* subsystem model

Models built around a single state variable might be considered by some as trivial, but the complexities of the processes contributing to the dynamics of that variable can justify the use of simulation as a means of integrating the information obtained from a large number of individual laboratory experiments with a view to testing results against field measurements. One such model is *MYTILUS* (Bayne *et al.*, 1976) which synthesises the processes governing the energy balance of *Mytilus edulis* over the three year development period of an individual organism. The model (Table II), which is written in CSMP (IBM, 1967) essentially computes the algebraic sum of the rate of intake of assimilated dry matter, the standard and the routine rate of oxygen consumption, and the rate of egg production. Any reserve or deficit from this calculation necessarily contributes to a change in the dry weight of the animal. The complexity of the model is in the specification of each of these rates because they depend upon a large number of factors including seasonal changes in the metabolism of *Mytilus*, temperature, ration level, salinity and tidal exposure. For example, although the filtration rate of *Mytilus* exhibits a simple log-linear relationship with animal weight the actual dry matter filtered is modified by food concentration and tidal exposure and even then only a proportion is ingested by the animal, some being rejected by the production of pseudofaeces which itself is a function of both metabolic body size and particle concentration. To simulate field conditions the model was driven by seasonal salinity, temperature and food concentration cycles typical of those measured in small estuaries in the south-west of England and the various rate relationships computed according to these data. Results from two such simulations are given in Fig. 3. Two feeding regimes were used; in the first (high ration), food levels were set similar to values recorded in a local estuary (high concentrations of particulate matter with little seasonal change and a low proportion of organic matter) whereas in the second (low ration), the food levels were set similar to offshore conditions (lower concentrations of particulate matter with marked seasonal changes and a high proportion of organic matter). Early runs of this model produced growth rates far in excess of those experienced in the field and this anomaly was traced to an over-estimate of the nutritional value of the suspended particulate matter caused by the assumption that natural particulates would be comparable to those of a laboratory algal culture. Use of the model thus led to a critical study of the composition of naturally occurring suspended particulates which confirmed their relatively low nutritional value. When these realistic values were used in the model the results closely followed the weight gains measured in the field (Fig. 3) both for animals growing in a relatively poor environment and those living in an intermediate food regime.

One of the advantages of using CSMP for this model was the availability of the FUNCTION facility which enables the relationships between variables to be represented by tabular data, and linear or quadratic

```
TITLE MODEL OF THE ENERGY AND NITROGEN BALANCE OF MYTILUS
INITIAL
PARAMETERS INITDW=1.0,INITN=8.,INDAY=151.,NPROPN=.083,...
        FRMBS=3.36,MAE=0.02,COMLGA=89.0,MNEMBS=0.0033
PARAMETERS CFMGAD=.011,CNDMGA=71.43,PRONEG=.084,PPNOM=.28
PARAMETERS PDEXFA=1.,MAXREJ=.8,KINTRJ=-.602
FUNCTION TFA=0.,1.25,15.5,1.5,45.,2.5,74.5,3.5,105.,4.5,...
        135.5,6.0,166.,8.0,196.5,4.5,227.5,4.5,258.,4.5,...
        288.5,4.5,319.,2.5,349.5,1.0,364.,1.25
FUNCTION TCAT=0.,6.1,15.5,4.8,45.,4.5,74.5,5.1,105.,7.0,...
        135.5,11.8,166.,15.7,196.5,17.2,227.5,17.3,258.,15.1,...
        288.5,12.6,319.,9.1,349.5,7.4,364.,6.1
FUNCTION TCSATW=0.,0.617,5.,0.689,10.,0.803,15.,1.0,20.,1.284,25.,1.
8
FUNCTION TCSATS=0.,0.466,5.,0.626,10.,0.803,15.,1.0,20.,1.162,25.,1.
2
FUNCTION TCRATW=0.,0.791,5.,0.791,10.,0.791,15.,1.0,20.,1.265,25.,1.
9
FUNCTION TCRATS=0.,0.575,5.,0.668,10.,0.953,15.,1.0,20.,1.048,25.,1.
3
FUNCTION TRRATW=0.,0.,2.,0.001,3.,0.015,5.,0.030,...
        7.,0.039,9.,0.048,11.,0.054,...
        13.,0.058,15.,0.060,50.,0.060
FUNCTION TRRATS=0.,0.,2.,0.,3.,0.008,5.,0.018,7.,0.025,...
9.,0.031,11.,0.035,...
        13.,0.038,15.,0.039,50.,0.039
FUNCTION TB=0.,0.35,10.0,0.52,20.,0.69,30.,0.86,40.,1.03,50.,1.20
FUNCTION TNEMBW=0.,,106,10.,,027,20.,,007,30.,,0018,...
        40.,,000469,50.,,00012
FUNCTION TSLNTY=0.,25.0,15.5,25.6,45.,22.9,74.5,21.9,105.,28.2,...
        135.5,30.1,166.,31.7,196.5,30.1,227.5,28.2,258.,30.4,...
        288.5,29.3,319.,27.7,349.5,28.2,364.,26.0
FUNCTION TCFATW=0.,0.106,5.,0.305,10.,0.816,15.,1.,20.,1.265,25.,1.7
FUNCTION TCFATS=0.,0.136,5.,0.374,10.,0.877,15.,1.,20.,1.140,25.,1.3
FUNCTION TPNEPN=363.,,2,728.,,3,1093.,,4,1458.,,4
FUNCTION TCOFDD=100.,,0288,250.,,0197,500.,,0144,...
1000.,,012,1500.,,0115,2000.,,011
DYNAMIC
        WMG=DWMYT/1000.
        MBS5=WMG**2.80
        MBS6=WMG*.4
        REJCON=EXP(-MBS6*KINTRJ)
*       DRY WEIGHT OF MYTILUS
        DWMYT=INTGRL(INITDW,RIADM-SR-RR-REP)
*       FOOD AVAILABLE
        FA=AFGEN(TFA,SEASON)
*       SEASON   JAN FIRST=0      DEC 31=365
        SEAS=TIME+INDAY
*       SEASON DITTO FOR SUBSEQUENT YEARS
        SEASON=AMOD(SEAS,365.)
*       SEASON INDICATOR -   0-WINTER     1-SUMMER
        SEASIN=.5*(1.+SIN(((SEASON-151.)*2*3.1416)/364.))
*       DRY MATTER FILTERED PER DAY
        DMFPD=FA*FRMBS*MBS2*PDEXFA
        PREJFI=MAXREJ*(1-EXP(REJCON*(MBS6-FA)))
        OMIPD=DMFPD*PPNOM*(1.-PREJFI)
        COFDD=AFGEN(TCOFDD,DWMYT)
*       METABOLIC BODY SIZE   I.E  W POWER  0.37
        MBS2=DWMYT**0.40
*       OPERATIVE ASSIMILATION EFFICIENCY
        OAE=AMAX1(MAE,0.916-COFDD*OMIPD-0.007*CAT)
*       RATE OF INTAKE ASSIMILATED     DM/DAY
        RIADM=OMIPD*OAE
*       CURRENT AMBIENT TEMPERATURE
        CAT=AFGEN(TCAT,SEASON)
*       METABOLIC BODY SIZE   I.E   W POWER 0.70
        MBS1=DWMYT**0.70
*       STANDARD RESPIRATION OXYGEN     ML O2 /DAY
        SOCMLD=0.055*CSATW*MBS1*(1.-SEASIN)+0.041*CSATS*MBS1*SEASIN
*       ROUTINE RESPIRATION OXYGEN     ML O2 /DAY
        ROCMLD=RRATW*CRATW*MBS1*(1.-SEASIN)+RRATS*CRATS*MBS1*SEASIN
*       STANDARD RESPIRATION OXYGEN     MICRO GM ATOMS
        SRCMGA=SOCMLD*COMLGA
```

The FUNCTION commands define graphical data expressed in 'x, y' pairs which are linearly interpolated when used in an AFGEN function

```
*      CORRECTION FACTOR FOR STANDARD RESPIRATION TO CURRENT AMBIENT T
       CSATW=AFGEN(TCSATW,CAT)
*      CORRECTION FACTOR FOR ROUTINE RESPIRATION TO CURRENT AMBIENT T
       CSATS=AFGEN(TCSATS,CAT)
*      ROUTINE RESPIRATION OXYGEN  MICRO GM ATOMS
       RRCMGA=ROCMLD*COMLGA
*      CORRECTION FACTOR TO AMBIENT TEMPERATURE
       CRATW=AFGEN(TCRATW,CAT)
       CRATS=AFGEN(TCRATS,CAT)
*      ROUTINE RESPIRATION ACCORDING TO RATION  WINTER
       RRATW=AFGEN(TRRATW,DMFPD)
       RRATS=AFGEN(TRRATS,DMFPD)
*      TOTAL OXYGEN CONSUMED IN MICRO GM ATOMS
       TOXMGA=SRCMGA+RRCMGA
*      STANDARD RESPIRATION      DRY MATTER
       SR=SRCMGA*CFMGAD
*      ROUTINE RESPIRATION       DRY MATTER
       RR=RRCMGA*CFMGAD
       PULSE=IMPULS(363.,365.)
       PRODN=INTGRL(0.,GRCWRT-ANPRCD)
       ANPROD=(PRODN*PULSE)/DELT
       TEP=INTGRL(0.,REP)
*      RATE OF EGG PRODUCTION
       REP=73.5*MBS5*PULSE/DELT
*      NITROGEN SECTION      TOTAL NITROGEN MG
       TOTN=INTGRL(INITN,RINN-RNE-RNLEP)
*      RATE OF INTAKE OF NITROGEN
       RINN=OAE*NPROPN*OMIPD
*      NITROGEN EXCRETION MG WINTER
       NEXMGW=NEMBSW*MBS3*SALINE*CFNATW
       NEXMGS=NEMBSS*MBS4*SALINE*CFNATS
*      CURRENT NITROGEN EXCRETION  MG
       NEXMGC=NEXMGW*(1.-SEASIN)+NEXMGS*SEASIN
*      METABOLIC BODY SIZE   I.E VARIABLE ACCORDING TO RATION
       MBS3=DWMYT**AFGEN(TB,OMIPD)
*      NITROGEN EXCRETION PER MBS  WINTER
       NEMBSW=AFGEN(TNEMBW,OMIPD)
       NEMBSS=PDEXFA*MNEMBS
*      METABOLIC BODY SIZE  I.E   W POWER 0.72
       MBS4=DWMYT**0.72
*      EFFECT OF SALINITY ON NITROGEN EXCRETION
       SALINE=2.19-0.036*SLINTY
*      AMBIENT SALINITY
       SLINTY=AFGEN(TSLNTY,SEASON)
*      CORRECTION FACTOR TO AMBIENT TEMPERATURE
       CFNATW=AFGEN(TCFATW,CAT)
*      SUMMER
       CFNATS=AFGEN(TCFATS,CAT)
*      RATE OF NITROGEN EXCRETION
       RNE=NEXMGC
*      TOTAL NITROGEN EXCRETED SINCE START
       TNE=INTGRL(0.,RNE)
*      RATE OF NITROGEN LOSS IN EGG PRODUCTION
       RNLEP=REP*PRONEG
*      OXYGEN TO NITROGEN RATIO FROM MICRO GM ATOMS
       OXNITR=TOXMGA/(NEXMGC*CNDMGA)
*    .. GROWTH RATE DRY WEIGHT
       GROWRT=DERIV(0.,DWMYT)
*    . GROSS EFFICIENCY   GROWTH RATE/DRY MATTER FILTERED
       GROSEF=GROWRT/OMIPD
       GRESRA=GROWRT/((SRCMGA+RRCMGA)*CFMGAD)
*      NET EFFICIENCY
       NETEFF=GROWRT/(OMIPD*OAE)
       PSGREF=AMAX1(AMIN1(GROSEF,0.2),-0.1)
       PSNTEF=AMAX1(AMIN1(NETEFF,0.5),-0.1)
METHOD RECT
TIMER DELT=1.,FINTIM=2190.,PRDEL=10.,OUTDEL=10.
PREPAR DWMYT,RIADM,SR,RR,REP,TEP,FA,SEASON,DMFPD,OAE,CONDD,...
       SOCMLD,ROCMLD,TOXMGA,TOTN,RINN,RNE,TNE,PREJFI,GRESRA,...
       OXNITR,GROWRT,GROSEF,NETEFF,OMIPD,....
       CAT,RRATW,RRATS,PPNOM,...
       SLINTY,PRODN,ANPROD,PSGREF,PSNTEF
END
```

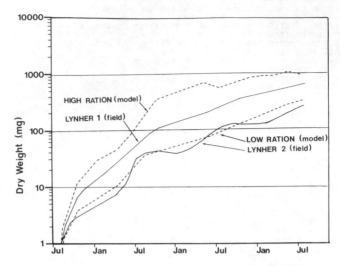

FIG. 3   Simulation results of the change in dry weight of an individual *Mytilus* over a period of three years. The discontinuous lines represent theoretical growth for high and low rations and may be compared to the continuous lines which represent field measurements made in the Lynher (Cornwall) estuary at sites where natural conditions provide low and intermediate rations

interpolation between tabulated points, rather than by an explicit mathematical expression. This rather simple concept allowed the modelling phase of the research to proceed in parallel with the scientific experiments which will eventually provide a more rigorous theoretical base for future studies.

7   Zooplankton subsystem model

The Continuous Plankton Recorder survey (CPR) has been described by Glover (1967) and the analysis of samples and methods of data processing used have been described by Rae (1952) and Colebrook (1975). The results of the survey have been used to describe and monitor the distribution and abundance of the plankton and to establish empirical relationships between plankton and the physical environment (e.g. Edinburgh Oceanographic Laboratory, 1973; Lindley, 1978; Colebrook, 1978).

The plankton communities sampled by the CPR survey may be considered as dynamic systems consisting of a number of interrelated state variables. In addition to describing the state of the system at a given time it would be advantageous to describe its behaviour in order to identify any unusual events, and if possible to make predictions of the state of the system over various lead times.

Where the system under consideration is strongly influenced by the influx of different water masses, relationships between plankton and physical variables associated with the same water mass may be apparent. In general, however, the response of each state variable to the perturbation of a given system input has a transient as well as a steady-state component. As the system inputs are in a state of continuous change, the existence of a transient phase rules out any simple relationship between concomitant values of the input and response. One way of trying to overcome this problem is to model the system as a set of coupled differential or difference equations and solve these numerically for various settings of the system inputs. That is, it is hoped that a simulation model, even if not sufficiently realistic to predict the state of the system, will determine the general form of the response.

The simulation model described below was designed to investigate trophic interactions between major groups of zooplankton in the central North Sea. It was adapted from the model produced by Andersen *et al.* (1973) to study changes in North Sea fish stocks. It applies to the C1/C2 area of the CPR survey (Fig. 4) insofar as CPR long-term monthly mean data from those areas were used to select the major components of the zooplankton system,

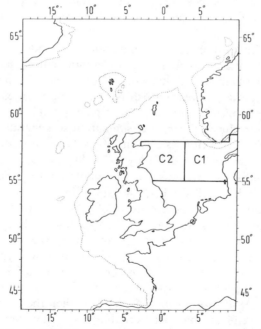

FIG. 4   Chart of the British Isles indicating the areas of the Continuous Plankton Recorder Survey known as C1 and C2

to specify initial conditions, to help to provide production figures for diatom and dinoflagellate groups, and for comparison with model output. The C1/C2 area was chosen as it is subject to a relatively high degree of CPR sampling (Edinburgh Oceanographic Laboratory, 1973) and a relatively low level of hydrographic activity (Hill, 1971). It was assumed that the horizontal components of the velocity field and the concentration gradients for each of the state variables in the area are low enough to allow the area to be regarded as a closed system with respect to exchange with surrounding areas (i.e. the Neumann boundary conditions were set to zero). The model is driven by the introduction of diatom and dinoflagellate production levels for each month throughout the simulation.

Three zooplankton groups are recognised: active predators, mainly chaetognaths and hyperiids; 'large' filter feeders, mainly euphausiids, and 'small' filter feeders, mainly copepods *Calanus* spp., *Pseudocalnus* sp., and *Paracalanus* sp., and thecosomes. Each group is divided into potentially reproductive and post-reproductive animals. Phytoplankton is divided into diatoms and dinoflagellates. For each group, numbers per cubic metre and an average individual wet weight, in grams, are simulated, except that diatoms and dinoflagellates are given fixed weights. Thus the model comprises 14 state variables; wet weight and numbers per cubic metre for six zooplankton groups, and numbers per cubic metre for two phytoplankton groups.

No spatial dimensions are recognised in the model, i.e. the model is 'lumped' and involves only first-order differential equations.

Data from the Danish Institute for Fisheries and Marine Research were used to estimate the numbers of fish per cubic metre in each of 10 weight classes in every month. As in the case of phytoplankton production, these estimates are introduced during the simulation so generating the mortality suffered by the zooplankton as a result of predation by fish.

For convenience the model may be split into two parts – (*i*) the simulation of individual weights, (*ii*) the simulation of population density.

To simulate individual weights, the rate of ingestion is first determined. This is a function of the concentration of potential food, and the feeding behaviour of the groups in question. In this model, the suitability of one individual as food for another is determined by the wet weight ratio of predator to potential prey. Predators are split into two types, filter-feeders and active predators. The distinction is best summarised by Fig. 5 and the functions used are given in Table IV.

The total concentration of food available is then the sum of the biomass of all groups weighted by their suitability as prey for the predator in question. The rate of ingestion is related to the food concentration by a Michaelis-Menten curve, defined by two parameters, a maximum ingestion rate and a half-saturation concentration. Note that no lower feeding threshold is used in this scheme.

The rate of weight change is determined using the von Bertalannfy growth equation:

$$\frac{dw}{dt} = h \cdot w^{2/3} - \kappa \cdot w$$

where w is wet weight in grams. Here h is the product of ingestion rate and assimilation efficiency. Two values for the catabolic rate constant κ are used, a low value for the 'winter' months December, January and February, and a high value for the rest of the year.

The simulation of population density involves the computation of mortality and reproduction. All mortality is the result of predation by other groups and by larger individuals in the same group. The mortality rate for each group is derived by summing the mortalities due to predation by each of the other groups. Prey selection by each group is in proportion to the product of abundance and suitability as food, defined as described above, of each of the other groups. The resulting mortality rate function is given in Table IV.

Energy conservation in the model is ensured by converting wet weight biomass to energy units in all predator-prey interactions. The only energy losses from the system are those implicit in the von Bertalannfy equation, i.e. losses incurred during digestion, assimilation and respiration. No natural mortality is included; for example, phytoplankton loss from the pelagic system due to vertical mixing and sinking is ignored. On the other hand, zooplankton faecal pellets and excreta are considered to be completely lost from the system; for example, feeding on non-living organic particulates is not included. Thus in each case the simplest alternative has been used initially, to be replaced by a more realistic scheme later if this is considered necessary.

Spawning takes place either at a fixed time each year or, for animals spawning more than once per year, when the average individual weight exceeds a given 'critical' value. A fixed proportion of the biomass of the potentially reproductive group is then converted to eggs of a fixed weight, and the remaining biomass is added to that of the post-reproductive group. The state variables and rate determining functions used in this model are given in Tables III and IV respectively.

The transient response of the system is of little interest when explicit phytoplankton production values are used. However, even at this stage, certain insights may be gained by running the model to steady-state. An example of steady-state output from the model is given in Fig. 6. It is taken as axiomatic that the simulation should converge to steady-state when

TABLE III   Key to symbols used to represent the state variables and the parameters of the zooplankton submodel

```
For group i  we define state variables

        W'ᵢ  -  individual wet weight in grams

        Nᵢ   -  nos per cubic meter

and parameters

    α, β         -  digestion and assimilation efficiency constants

  κ, κsᵢ, κωᵢ   -  catabolic rate constants, 'summer' and 'winter'

    hᵢ'          -  maximal ingestion rate constant

    fᵢ'          -  feeding level, function of food availability

    Qᵢ           -  ½ saturation constant for determing fᵢ

    φᵢ           -  biomass available as food to individuals of group i

    Gᵢⱼ          -  proportion of biomass of group j available as food to group i

    ηᵢ, σᵢ       -  parameters for determination of Gᵢⱼ

    Eᵢ           -  wet weight to energy conversion ratio

    Wᵢₒ          -  egg weight

    πᵢ           -  proportion of biomass of potentially reproductive group

                    converted to eggs on spawning.
```

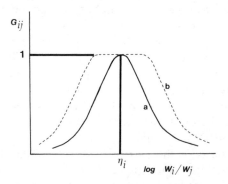

FIG. 5 Distribution of prey preference by predator i as a function of 'predator:prey' wet weight ratio W_i/W_j for, (a) active predator, and (b) filter-feeding predator

For group i we determine the rate of individual weight change

$$\frac{dW_i}{dt} = (1 - \alpha)\, \beta f_i h_i E_i' W_i^{2/3} - K W_i$$

where $\quad f_i = \dfrac{\varphi_i}{\varphi_i + Q_i} \qquad 0 \le f_i \le 1 \qquad$ and $\qquad \varphi_i = \sum_j G_{ij} W_j N_j$

E_i' which is used to allow for different 'energy:wet weight' ratios in the different groups is defined as

$$E_i' = \sum_j \frac{G_{ij} N_j W_j E_j}{\varphi_i E_i}$$

The term $\quad G_{ij} N_j W_j / \varphi_i \quad$ is the fraction of the intake of group i consisting of group j and is again used in determining mortality.

For active predators, G_{ij} has a maximum value of unity and declines symmetrically with the deviation of log ('predator: prey' weight ratio) from the logarithm of the preferred value, η_i (Fig. 5).

$$G_{ij} = \exp\left\{-\left(\log W_i/W_j - \eta_i\right)^2 \Big/ 2\sigma_i^2\right\}$$

For filter-feeders G_{ij} equals one for a range of 'predator:prey' weight ratios and tails off symmetrically for 'predator:prey' ratios outside this range (Fig. 5).

$$G_{ij} = 2\sigma_i \Big/ \left\{\left|\log W_i/W_j - \left(\sigma_i - \eta_i\right)\right| + \left|\log W_i/W_j - \left(\sigma_i + \eta_i\right)\right|\right\}$$

The rate of change of N_i is a function of predation by all groups:

$$\frac{dN_I}{dt} = -\left|\frac{\text{predation rate on group } i}{\text{biomass of group } i}\right| N_i$$

$$= -\left|\frac{1}{N_i W_i}\sum_j f_j h_j W_j^{2/3} N_j \frac{G_{ji} W_i N_i}{\varphi_j}\right| N_i$$

$$= -\left|\frac{\sum_j G_{ji} h_j W_j^{2/3} N_j}{\varphi_j + Q_j}\right|$$

The number of eggs produced by the potentially reproductive group i is given by:-

$$N_{i,o} = \frac{\pi_i N_i W_i}{W_{i,o}}$$

constant forcing functions are applied ('steady-state' and 'constant' being used, in this context, to mean 'having the same seasonal cycle in successive years'). Thus formulations of the model which fail to converge may be rejected.

Originally the time of spawning was determined by weight gain for all groups. However, this scheme is unsatisfactory for animals, such as euphausiids, which spawn once per year (Einarsson, 1945), as it leads to unrealistic multi-year cycles. That is, a sequence of years occurs in which spawning occurs successively later (or earlier) each year. This is followed by a year in which no spawning occurs (or spawning occurs twice), then the sequence is repeated. The problem is due to the fact that the model converges to steady state by adjusting birth rate rather than mortality, which is largely density-independent (except in the case of 'cannibalism' as

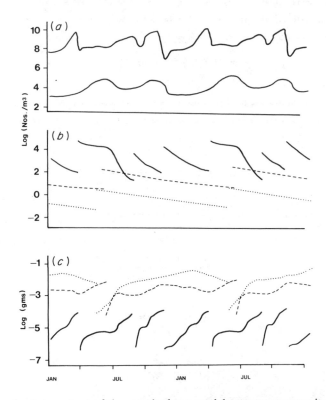

FIG. 6   Steady-state output of the zooplankton model over two successive years. (*a*) No. of diatoms (upper curve) and dinoflagellates; (*b*) No. of 'small' filter-feeders (———), 'large' filter-feeders (– – –), and active predators (· · · · ·); (*c*) Weights of 'small' filter-feeders (———), 'large' filter-feeders (– – –) and active predators (· · · · ·)

described below). When spawning is triggered by weight gain, fecundity is constant and the birth rate is adjusted by changing spawning frequency, which is therefore not generally maintained at one spawning per year. The problem persists even when a delay is introduced between the time the critical weight is exceeded and the time of spawning, and fecundity is related to weight gain during the delay, as suggested by Steele (1974). By setting a fixed spawning time in each year, birth rate is adjusted by changing fecundity. If fecundity could be shown to be density-dependent in field animals, this would suggest that spawning is triggered directly by seasonal changes in the physical system, such as day length or sea-surface temperature, rather than by a purely biological mechanism such as weight gain.

As the suitability of one individual as food for another is assumed to be simply a function of their relative sizes, cannibalism is possible in this model and becomes an important source of mortality following spawning. This provides a stabilising mechanism which, in the absence of threshold feeding levels or other forms of density-dependent mortality, will assume importance on replacement of the explicit phytoplankton production levels by a phytoplankton submodel.

The most difficult problem encountered in working with this form of simulation is that of validation. There is clearly no difficulty in producing any required output given the number of parameters and the degree of uncertainty involved in their estimation. Improved parameter estimation and further information on the processes involved in the simulation helps to overcome this problem, but the final test of the usefulness of such models is their ability to achieve the objective of suggesting useful functional relationships between components of the plankton ecosystem.

8 Conclusions

Simulation modelling can be readily applied to a very wide range of biological systems for a variety of purposes. The few examples given in this chapter are models of subsystems of varying complexity ranging from a simple formula to complex arrays of first order differential equations. They represent systems at the cellular, physiological and crop levels of abstraction and each was designed in order to achieve a different objective. The chief criterion for judging the value of these models must be by an assessment of how well each meets that objective, but a model need not be classified as useless just because it fails that test. The rigour of applying systems analysis and simulation modelling techniques to any problem invariably leads one to ask questions of the system which generate new ideas, new problems, new issues, new experiments and new hypotheses which add creativity to the research programmes.

Acknowledgements

The work reported in this chapter was developed as part of the research programme of the Institute for Marine Environmental Research (IMER) a component body of the United Kingdom Natural Environment Research Council (NERC). The work of the Institute is funded in part by the Development Commission and the NERC, the U.S. Department of the Navy, Office of Naval Research (Contracts N62558-2834/3612 and F61052-67C-0091), the Ministry of Agriculture, Fisheries and Food, the Department of the Environment (Contracts DGR 480/47 and DGR 480/288), and the European Economic Community (Contract 279-77-4 ENV UK).

References

Anderson, K. P., Lassen, H. and Ursin, E. (1973). A multispecies extension to the Beverton and Holt assessment model with an account of primary production. *ICES. C.M. H:* 20

Annino, J. S. and Russell, E. C. (1979). The ten most frequent causes of simulation analysis failure – and how to avoid them! *Simulation* **32**, 137–140

Baly, E. C. C. (1935). The kinetics of photosynthesis. *Proc. Roy. Soc., London* **117B**, 218–239

Bayne, B. L., Widdows, J. and Thompson, R. J. (1976). Physiological integrations. *In* "Marine Mussels" (B. L. Bayne, ed.). International Biological Programme 10, pp. 261–291. Cambridge University Press, Cambridge

Colebrook, J. M. (1975). The Continuous Plankton Recorder Survey: Automatic data processing methods. *Bull. Mar. Ecol.* **8**, 124–142

Colebrook, J. M. (1978). Continuous plankton records: zooplankton and environment, North-east Atlantic and North Sea. *Oceanologica Acta* **1**, 1–23

Edinburgh Oceanographic Laboratory (1973). Continuous plankton records: a plankton atlas of the North Atlantic and North Sea. *Bull. Mar. Ecol.* **7**, 1–174

Einarsson (1945). Euphausiacea. 1. North Atlantic species. *Dana Rep.* **27**, 1–185

Fee, E. J. (1969). A numerical model for the estimation of photosynthetic production, integrated over time and depth in natural waters. *Limnol. Oceanogr.* **14**, 906–911

Forrester, J. W. (1961). "Industrial Dynamics". M.I.T. Press, Cambridge, Massachusetts

Forrester, J. W. (1969). "Urban Dynamics". M.I.T. Press, Cambridge, Massachusetts

Glover, R. S. (1967). The continuous plankton recorder survey of the North Atlantic. *Symp. Zool. Soc. Lond.* **19**, 189–210

Hill, H. W. (1971). Currents and water masses. *In* "North Sea Sciences" (E. D. Goldberg, ed.) pp. 17–42. M.I.T. Press, Cambridge, Massachusetts

IBM (1967). "Continuous System Modelling Program (360A-CX-16X) Users Manual, GH20-0367-3". Techn. Publ. Dept. White Plains, New York

IBM (1972). Continuous System Modelling Program III (CSMP III) (5734-X59), Program Reference Manual, SH 19-7001-0. Techn. Publ. Dept. White Plains, New York

Iverson, R. L., Curl, H. C. and Saugen, J. L. (1974). Simulation model for wind-driven summer phytoplankton dynamics in Auke Bay, Alaska. *Mar. Biol.* **28**, 167–177

Jassby, A. D. and Platt, T. (1976). Mathematical formulation of the relationship between photosynthesis and light for phytoplankton. *Limnol. Oceanogr.* **21**, 540–547

Jones, J. G. W. (1969). "The Use of Models in Agricultural and Biological Research". Grassland Research Institute, Hurley

Kremer, J. M. and Nixon, S. W. (1978). "A Coastal Marine Ecosystem", Ecological Studies, Vol. 24. Springer-Verlag, Berlin, New York

Laevastu, T. and Favorite, F. (1977). Preliminary report on dynamical numerical marine ecosystem model (DYNUMES 11) for eastern Bering Sea, Northwest and Alaska Fisheries Center. Processed Report, U.S. Department of Commerce, Washington

Lehman, J. T., Botkin, D. B. and Likens, G. E. (1975). The assumption and rationales of a computer model of the phytoplankton population dynamics. *Limnol. Oceanogr.* **20**, 343–364

Levins, R. (1975). Evolution in communities near equilibrium. *In* "Ecology and Evolution of Communities" (M. L. Cody and J. M. Diamond, eds). Harvard University Press, Cambridge, Massachusetts and London

Lindley, A. (1978). Population dynamics and production of euphausiids. *Mar. Biol.* **76**, 121–130

Meadows, D. H., Meadows, D. L., Randers, J. and Behrens, W. W. (1972). "The Limits to Growth. A Report for the Club of Rome Project on the Predicament of Mankind. Potomac Associates Inc., Washington D.C.

Odum, H. T. (1972). An energy circuit language for ecological and social systems. Its physical basis. *In* "Systems Analysis and Simulation in Ecology" (B. C. Patten, ed.) Vol. II, pp. 139–211. Academic Press, New York and London

Odum, H. T. (1976). Macroscopic minimodels of man and nature. *In* "Systems Analysis and Simulation in Ecology" (B. C. Patten, ed.) Vol. IV, pp. 249–282. Academic Press, New York and London

Park, R. A. (1978). "A Model for Simulation Lake Ecosystems". Center for Ecological Modelling, Rensselaer Polytechnic Institute, Troy, New York

Parsons, T. R. and Anderson, G. C. (1970). Large scale studies of primary production in the North Pacific Ocean. *Deep-Sea Res.* **17**, 756–776

Patten, C. P. (1975). "Systems Analysis and Simulation in Ecology", Vol. II. Academic Press, New York and London

Platt, T., Denman, K. L. and Jassby, A. D. (1977). Modelling the productivity of phytoplankton. *In* "The Sea: Ideas and Observations on Progress in the Study of the Seas" (E. D. Goldberg, ed.) Vol. VI, pp. 807–856. John Wiley, New York

Pugh, A. L. (1973). "DYNAMO II. Users Manual". Massachusetts Institute of Technology Press, Cambridge, Massachusetts

Radford, P. J. (1971). The simulation language as an aid to ecological modelling. *In* "Mathematical Models in Ecology" (J. N. R. Jeffers, ed.) pp. 277–296. Proc. Symp. British Ecological Society. Blackwell Scientific Publications, London

Radford, P. J. (1979). The rôle of a general ecosystem model of the Bristol Channel and Severn Estuary (GEMBASE). *In* "Tidal Power and Estuary Management" (R. T. Severn, D. Dineley and L. E. Hawker, eds) pp. 40–46. The thirtieth Symposium of the Colston Research Society, Scientechnica

Radford, P. J. and Joint, I. R. (1980). The application of an ecosystem model to the Bristol Channel and Severn Estuary. Institute of Water Pollution Control. Annual Conference. *Water Pollution Control* **79**, 244–254

Rae, K. M. (1952). Continuous Plankton Records: explanation and methods, 1946–1949. *Hull Bull. mar. Ecol.* **3**, 135–155

Scavia, D. and Park, R. A. (1976). Documentation of selected constructs and parameter values in the aquatic model CLEANER. *Ecological Modelling* **2**, 33–58

Scavia, D. and Robertson, A. (1979). "Perspectives on Lake Ecosystem Modelling". Ann Arbor Science, Michigan

Smith, E. L. (1936). Photosynthesis in relation to light and carbon dioxide. *Proc. Nat. Acad. Science, Wash.* **22**, 504

Steele, J. H. (1962). Environmental control of photosynthesis in the sea. *Limnol. Oceanogr.* **7**, 137–150

Steele, J. H. (1974). "The Structure of Marine Ecosystems". Blackwell Scientific Publications, London

Talling, J. F. (1957). The phytoplankton population as a compound photosynthetic system. *New Phytol.* **56**, 133–149

Takahashi, M., Fujii, K. and Parsons, T. R. (1973). Simulation study of phytoplankton photosynthesis and growth in the Fraser River Estuary. *Mar. Biol.* **19**, 102–116

Vollenweider, R. A. (1965). Calculation models of photosynthesis-depth curves and some implications regarding day rate estimates in primary production measurements. *Mem. Inst. Ital. Idrobiol.* **18**, Suppl. 427–457

Warwick, R. M. and Price, R. (1975). Macrofauna production in an estuarine mud-flat. *J. mar. biol. Ass. U.K.* **55**, 1–18

Warwick, R. M., Joint, I. R. and Radford, P. J. (1979). Secondary production of the benthos in an estuarine environment. *In* "Ecological Processes in Coastal Environments" (R. L. Jefferies and A. J. Davy, eds) pp. 429–450. Proceedings of the 1st European Ecological Symposium and the 19th Symposium of British Ecological Society, Blackwell Scientific Publications, Oxford

Winter, D. F., Banse, D. and Anderson, G. C. (1975). The dynamics of phytoplankton blooms in Puget Sound, a fjord in the north-western United States. *Mar. Biol.* **29**, 139–176

Wit, C. T. de. and Goudriaan, J. (1974). "Simulation of Ecological Processes". Centre for Agricultural Publishing and Documentation, Wageningen. pp. 160

24 Holistic Simulation Models of Shelf-Seas Ecosystems

TAIVO LAEVASTU and FELIX FAVORITE

1 Objectives and principles of numerical ecosystem simulation

Man's scientific curiosity as well as his desire to exploit the food resources of the sea drive him to extend his knowledge of marine biology and processes within the marine ecosystem, to assess the abundance of marine resources, their behavior and distribution, and to ascertain the response of these resources to the fishery and possible environmental changes.

In order to obtain a coherent picture of the marine ecosystem we need to summarise the available information quantitatively and in a systematic manner. This task can be accomplished to a considerable extent by large ecosystem simulations on large computers.

Numerical ecosystem simulation is defined here as numerical repro- duction of conditions and processes in the marine ecosystem, based on all available data and knowledge. It might be useful to differentiate between ecosystem models and ecosystem simulations although these terms have been used as synonyms. We usually consider a model to be an abstraction and simplification of a given condition and/or process, whereas, a simu- lation is a reproduction of a system of conditions and processes based on available empirical data and may contain many tested models.

This chapter describes a holistic simulation model in which the emphasis is given to the role of fish in the ecosystem. Ecosystem models which consider primarily plankton are not described here as several description of them are available in existing books (e.g. Kremer and Nixon, 1977). Rather than giving a general review of holistic ecosystem modeling, this chapter describes and provides basic formulas of the Dynamical Numerical Marine Ecosystem (DYNUMES) simulation.

The objectives of numerical ecosystem simulations can be grouped into two main categories:

1 Investigative and digestive (analytical) objectives, including basic ecological research, that permit quantitative determination of the state of

701

the ecosystem, determination of the effects of environmental changes and interspecies interactions in space and time, and the establishment of research priorities.

2 General management guidance, the assessment of fisheries resources, and the effects of exploitation.

The following basic principles are normally followed in ecosystem simulation:

(a) The ecosystem simulation must include all of the essential biological and environmental interactive components of the system.

(b) The ecosystem simulation should have proper space and time resolution, i.e. be three- to four-dimensional (two to three space and one time dimension), and must have a diagnostic and a prognostic phase.

(c) Theoretical conceptualisations should be avoided, unless they have been tested with expirical data and proven to be valid.

(d) Explicit approaches, free from mathematical artifacts, should be preferred, i.e. the mathematical formulas used in the model must reproduce known processes rather than assuming that a mathematical formula presents the behavior of a system.

(e) Biomass balance and trophodynamic computations should start with apex predators (including man); these can be treated as 'forcing functions' of the system.

It has been amply demonstrated in the past that the start of trophodynamic computations from the lower end – i.e. basic organic production, does not lead to reliable quantitative results because the pathways of basic organic production to secondary and tertiary production are very variable in space and time and not fully known quantitatively. Many separate plankton production models exist, which try to alleviate these shortcomings.

2 Modeling developments leading to holistic ecosystem simulation

First attempts to develop mathematical models for fish and other animal populations were made between 1910 and 1925 by Ross, Kevdin, Baranov and Alm. The models proposed and used by these early pioneers comprised predator–prey relations in some form.

Ecosystem modeling concepts originated in the 1940s when relatively simple quantitative explanations of plankton production were attempted by connecting different trophic levels in the ecosystem via food requirements. The development of single-species population dynamics models for commercial fish was also intensified in the mid-50s when also some basis for multispecies theory of fishing was initiated (Beverton and Holt, 1957). More complex ecosystem approaches would have been impossible as large computers were not available at this time.

Several numerical two- and three-dimensional ecosystem models have been developed in the recent past, which deal essentially with planktonic organisms as the basis for marine productivity, e.g. Kremer and Nixon, 1977. The nutrient-plankton-fish energy pathways are, however, greatly variable in space and time, with great lateral losses (e.g. losses into deep water, remineralisation, etc.) that are not yet fully accounted for quantitatively. The large-scale, numerical analysis-forecasting models in meteorology and oceanography developed in the 1960s, have provided methods and approaches which are suitable for, but were until now not applied, to ecosystem modeling.

New needs have recently arisen for ecosystem models. It has become clear that successful fisheries management requires the consideration of the total marine ecosystem because, for example, the fishery on one species affects the abundance and distribution of other species through interspecies interactions, such as predation.

A comprehensive quantitative marine ecosystem simulation model with emphasis on fish components of the ecosystem, has been developed recently in Denmark (Andersen and Ursin, 1977), that includes also primary production and phosphorus circulation. This model, as any extensive simulation model, seeks numerical solutions for established formulations. Various forms of their simulation use 14 to 81 entities of plants, animals, and nutritive matter, calling for simultaneous solution of from 42 to 308 differential equations.

The model emphasises trophodynamics as does the simulation described in this chapter. Growth rates of all species in their model are also a function of the season and availability of food. They have also partitioned natural mortality into various components such as predation mortality, spawning strain, starvation and disease mortalities.

The Andersen and Ursin model is number-based model and has no spatial resolution (i.e. a 'box' model). The DYNUMES model described in this chapter is a biomass-based model with spatial resolution (i.e. a 'gridded' model).

The majority of available water quality management models ignore ecological interactions. The first multipurpose ecosystem model for larger estuaries for water quality management is the GEMBASE model (General Ecosystem Model for Bristol Channel and Severn Estuary, Longhurst 1978; Longhurst and Radford, 1978), which was also designed as a tool for a variety of ecological studies. This model simulates the carbon and nitrogen flow between ecological state variables and seven geographical regions. The whole process requires about 150 equations with 225 parameter values. It uses hydrodynamical models for transfer of materials between adjacent geographical regions. Limited examples of this model are given in this chapter.

Ecosystem simulation is a continuing process, requiring continuous updating and expansion. Many of the processes and conditions described in the previous chapters of this book have not yet been properly included in any simulation. Thus our grandchildren will still have plenty to do.

3   Basic components and processes of a marine ecosystem and their simulation

COMPONENTS AND PROCESSES IN A BASIC MARINE ECOSYSTEM
SIMULATION

There are numerous and varied processes at work in the marine ecosystem, affecting its biological components in a variety of ways. Thus the quantitative computation of changes in the ecosystem requires the use of numerous explicit equations, each adapted to reproduce quantitatively a given process according to available empirical knowledge.

Generalised flow diagrams of ecosystem simulations illustrate some of the emphasis in and peculiarities of the simulations; the essential details are, however, given in computer programs and their documentations. The GEMBASE (Longhurst and Radford, 1978) flow diagram of the estuarine ecosystem is given in Fig. 1. Figure 2 presents the principle processes which are emphasised in DYNUMES simulation. These two diagrams are obviously oversimplified presentations of the full models.

The numerical simulation of ecosystem in three-dimension spaces requires two-dimensional grids (see example of DYNUMES grid on Fig. 3). All computations are carried out at each grid point and time step with prevailing conditions. The advection and migrations occur from grid point to grid point in u and v components. This grid can be repeated for several depth levels (e.g. near-surface layer and bottom). All space-dependent input data are digitised at each grid point, where all computations are carried out in each time step and outputs given in numerical form.

Initial inputs of the DYNUMES model are: depth, sea-land table, surface and bottom temperature, and nature of the bottom. The initial distribution and abundance of species/ecological groups are also given at each grid point as first guess fields.

Numerous species specific coefficients (such as growth coefficients, food requirement, and fishing mortality coefficients, etc.) are introduced in the species computation subroutines. Most of the coefficients in the model, the majority of which present rates of changes, are influenced by a number of factors at each grid point and time step and are thus correspondingly

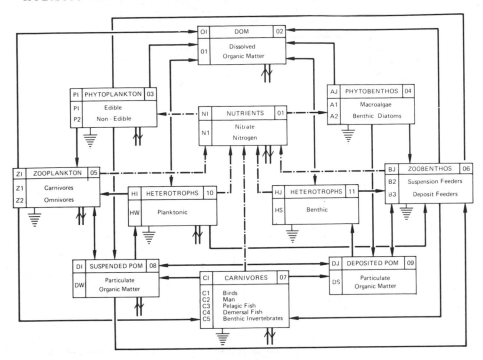

FIG. 1 Process flow diagram of a carbon-based holistic ecological model (GEMBASE) designed as a tool for understanding estuarine ecosystem function (Longhurst and Radford, 1978)

recomputed on the bases of empirical knowledge of their behavior. Thus the model becomes a processs oriented model with rate variables which determine the state variables.

Growth of biomasses, fishery yields, mortalities, consumptions (predation), and migrations are also computed in each time step. The monthly consumption of the given species, which is necessary for the computation of actual month biomass balance, is taken from the summation of predation of this species in the previous month. Great flexibility is allowed in the selection of model outputs, such as monthly distributions of individual species, results of processes, predation, species source and sink areas, etc.

SOME BASIC MATHEMATICAL FORMULATIONS USED IN THE SIMULATIONS
There is usually a lack of reliable quantitative data on marine mammals, birds, and other apex predators present in any given region. Therefore, it is not possible to compute their growth and mortality, but only predation by

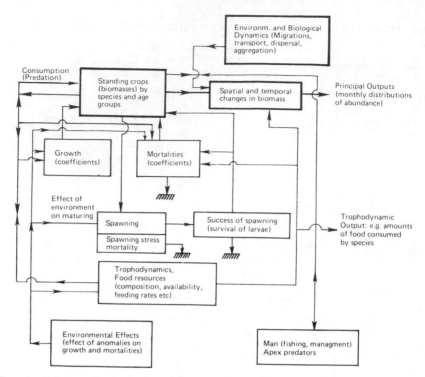

FIG. 2 Scheme of principal processes and interactions in the marine ecosystem

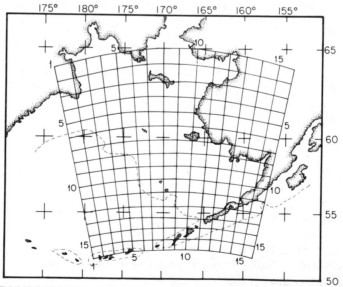

FIG. 3 DYNUMES III grid for eastern Bering Sea, drawn on Mercator projection

them, which becomes one of the forcing functions of the model. (For symbols and abbreviations see section 8):

$$F_{a(t,n,m)} = B_{a(t,n,m)} q_a t_d \qquad (1)$$

The consumption of species i by apex predator a is:

$$C_{i,a(t,n,m)} = F_{a(t,n,m)} p_{i,a} \qquad (2)$$

Consumption of species i by all apex predators is the sum:

$$C_{i,A(t,n,m)} = \sum_a C_{i,a(t,n,m)} \qquad (3)$$

The migrations of individual biomasses are computed with predetermined migration speed components (u, v) either on whole or portion of the biomass, using an 'upcurrent interpolation and direct advection' formulation which is mass conserving. This computation is done in two steps; first the linear gradient of biomass in the 'upcurrent' direction is determined:

$$U \text{ positive: } UT_{(n,m)} = (B_{n,m} - B_{n,m-1})/l \qquad (4)$$

$$U \text{ negative: } UT_{(n,m)} = (B_{n,m} - B_{n,m+1})/l \qquad (5)$$

(The V gradient (VT) is computed in an analog manner.)
Thereafter the gradient is advected to the grid point under consideration:

$$B_{(t,n,m)} = B_{(t-1,n,m)} - (t_d|U_{(t,n,m)}|UT_{(n,m)}) - (t_d|V_{(t,n,m)}|VT_{(n,m)}). \qquad (6)$$

After migration computations, a smoothing (diffusion) operation is performed (which can be considered a random movement of fish).

$$B_{(n,m)} = \alpha B_{(n,m)} + \beta(B_{n-1,m} + B_{n+1,m} + B_{n,m-1} + B_{n,m+1}) \qquad (7)$$

The migrations due to unfavorable environmental conditions or due to scarcity of food are computed by testing the gradients of surrounding points in respect to predescribed criterion and a portion of the biomass at unfavorable grid points is moved towards more favorable conditions:

$$B_{i(n,m)} = B_{i(n,m)} - k_0 B_{i(n,m)} \qquad (8)$$

$$B_{i(n\pm1,m\pm1)} = B_{i(n\pm1,m\pm1)} + k_s B_{i(n,m)} \qquad (9)$$

The coefficient k_0 depends on the amount of grid point values of environmental parameters or food availability exceed the prescribed criterion and on the number of computational passes (ca. 0.03 in two-pass operation). Coefficient k_s depends in addition on the number of favorable surrounding points (two-pass value is ca. 0.008 to 0.03).

The biomass growth is computed with a formula similar to compound interest computation:

$$B_{i(t,n,m)} = B_{i(t-1,n,m)}(2 - e^{-g}) \qquad (10)$$

The time step in the model is sufficiently short so that second order terms can be neglected. In some species the growth is made a function of time (i.e. seasonal variations in the growth coefficient):

$$g = g_0 + A_g \cos (\alpha_i t - \kappa_g) \tag{11}$$

In other, 'temperature sensitive' species the growth is made a function of either surface or bottom temperatures.

$$g = k_t g_0 e^{-1/[T(n,m)]} \tag{12}$$

Temperature (T) in this formulation is restrained between 1 and 18 °C and the coefficient k_t is slightly larger than 1 and is species specific. Furthermore, the growth is dependent on the degree of starvation:

$$g = g_0 - \frac{S_{i(n,m)}}{F_{i(n,m)}} g_0 \tag{13}$$

The food needed by the biomass of species i in a given location $(F_{i(n,m)})$ and the shortage of food to satisfy the food requirement of the species $(S_{i(n,m)})$ are computed in the model at each grid point and time step.

The biomass change formula (growth, mortality, and predation) excluding fishery is:

$$B_{i(n,nm)} = B_{i(t-1,n,m)}(2 - e^{-g})e^{-m} - C_{i(t-1,n,m)} \tag{14}$$

The amount of food required by species i is composed of food requirement for growth and food requirement for maintenance:

$$F_{i(t,n,m)} = B_{i(t,n,m)}(1 - e^{-g})r_i + q_i B_{i(t,n,m)} \tag{15}$$

The consumption of a species (C_i) is the sum of the consumption of this species by all other components of the ecosystem. With 'fixed' food composition (i.e. no spatial and temporal variations) and consumption would be:

$$C_{i(t,n,m)} = F_{i(t,n,m)}p_{i,i} + F_{j(t,n,m)}p_{i,j} + \cdots F_{n(t,n,m)}p_{i,n} \tag{16}$$

The feeding is food density (availability) dependent. Thus, the fractional composition of food can vary from grid point to grid point and from one time step to another. Furthermore, partial starvation can occur. The food requirement of species i with respect to species j as a food item consists of three terms:

$$F_{i,j(t,n,m)} = F_{i(t,n,m)}p_{i,j}a_{j(t,n,m)} + F_{i(t,n,m)}p_{i,j}b_{m(t,n,m)}$$
$$+ F_{i(t,n,m)}p_{i,j}c_{j(t,n,m)} \tag{17}$$

The first term at the right-hand side is the amount of species j taken as food.

This is controlled by availability factor of species j at given grid point. The factor a_j is a function of the fraction of the biomass of species j consumed at this location in previous time step. The second term presents the amount of species j consumed on the bases of starvation array requirements (i.e. substitution of more abundant food item).

The last term on the right hand side presents the requirement of species j as food for the species i, which cannot be satisfied from available biomass at given location and time, and is added to the starvation array. Thus, the computed biomasses can be food availability limited via growth and are also greatly affected by predation.

The biomass mortality from diseases and 'old age' is computed with the conventional formula by multiplying biomass with e^{-m}, eqn (14). Furthermore, a spawning stress mortality is computed on some species during the months of spawning. Consequently the resultant state variables are all determined by rate variables.

The fish catches (yields) are computed using a time and space variable fishing mortality coefficient:

$$P_{u,i(t,n,m)} = B_{i(t,n,m)} - B_{i(t,n,m)}e^{-fi(t,n,m)} \qquad (18)$$

The standing crops of phyto- and zooplankton are simulated with a harmonic formula, which is tuned to available empirical data.

$$P_{s(t,n,m)} = P_{t,0} + A_{1,r}\cos(\alpha_1 t - \kappa_1) + A_{2,r}\cos(\alpha_2 t - \kappa_2) \qquad (19)$$

The annual mean standing stocks for given subregions ($P_{r,0}$), the half-range of primary and secondary annual 'peaks' ($A_{1,r}$ and $A_{2,r}$) and their phase lags (κ_1 and κ_2) must be obtained from available quantitative empirical data, or computed with a plankton submodel. The other two holistic ecosystem models (GEMBASE, Longhurst and Radford, 1978; Andersen and Ursin, 1977) include detailed phytoplankton simulation approaches.

The initial standing stock of benthic fish food is prescribed (digitised) at each grid point. It is assumed to be a function of depth, bottom type, distance from the coast, and the prevailing bottom temperature. The growth, mortality, and consumption of the benthos biomass is computed in each time step with eqn (14).

INPUT DATA ON THE BIOTA AND BIOLOGICAL PROCESSES

The availability, quality, and accuracy of basic input data for the simulation can vary considerably from one region to another. Therefore, only some generalities of input data are considered below.

Estimates of the numbers of marine mammals and birds present in any area are associated with some uncertainties. As marine mammals and birds are consumers only, and their amounts are not large in most areas, the errors introduced by their inaccurate estimates affect the computation of the final biomasses relatively little.

The initial estimates of biomasses of fish species/groups of species are introduced into the program as inputs. A grouping into ecologically and trophically similar groups has been found necessary due to computer limitations – e.g. semidemersal fish (cod, hake), dermersal flatfish (e.g. turbot, halibut), etc. However, some single, dominant species can be treated as single species (e.g. herring, pollock, etc.) and, if desired, divided into age groups.

The recruitment in the biomass-based model such as the DYNUMES simulation model is largely regulated by changes in growth rate and predation pressure, which simulate the regulatory mechanisms which are considered to occur in unexploited natural populations (Ware, 1975). Obviously changes in the recruitment can also be caused by other factors which can be introduced into the model if so desired. The biomass growth changes with age. Thus the computation of the growth rate of a given species biomass requires the knowledge of the distribution of biomass with age within the species, which is computed in an auxiliary model (Laevastu and Favorite, 1978). The growth coefficient must be adjusted to computational time step (e.g. per cent per month). Environmental variables, such as temperature, modify growth. One of the effects of climatic variability can thus be introduced via temperature variability.

The migration speeds of species/groups of species are deduced from empirical knowledge about the seasonal occurrence and migrations (including feeding and spawning migrations) and are prescribed in the model. The migrations due to unfavorable temperature and/or scarcity of food are simulated within the model.

Seasonal or annual mean composition of food (in per cent) must be prescribed for each species/ecological group, using available data, and by considering size dependent feeding and change of food composition with the age (size) of the species. This food composition is changed in computations, considering the availability and suitability of food items.

The food requirement coefficient is divided into two parts in the present model: food requirement for growth, and food requirement for maintenance. The values of these coefficients vary from species to species, depending on activity, growth rates, and normal environmental temperature (re. metabolism). Data for food requirements for growth, maintenance, and reproduction are scarce in the literature, although for some species (gadids) excellent information is available (e.g. Jones, 1978; Daan, 1973), which can be generalised to other species. If an overall food intake coefficient is used (without separation of growth and maintenance requirements), the food coefficient is usually between 0.7 and 1.8% of body weight daily. The food coefficient has seasonal change in higher latitudes, which is approximated by a harmonic curve in the present model.

The uncertainties and errors introduced by trophic coefficients normally do not induce an error in excess of 30% in the final biomass computation according to preliminary experimental determination with the model.

The true natural mortality from old age (senescent mortality) and diseases and possible spawning stress and 'starvation' mortalities are small in exploited populations compared to fishing and predation mortalities. Fishing mortality is computed outside the model from fisheries statistics; and predation mortality, which is the largest component of 'natural mortality', is directly computed within the model. In unexploited and in short-lived species, such as squids, the senescent mortality can, however, be large. The possible errors made in the estimates of fishing and senescent mortality coefficients do not normally cause errors in biomass estimates in excess of 10% in careful work according to preliminary test results with model in which coefficients were changed in reasonable limits.

ENVIRONMENT AND BIOTA INTERACTIONS AND COUPLING OF SIMULATIONS

Most of the pronounced environment-biota interactions must be included in the ecosystem simulations in order to reproduce the ecosystem in a realistic manner. Some fixed environmental data, such as depth, are used directly as a criterion for seasonal migrations of flatfish, abundance of benthos, and in other distribution determinations.

Current as transport mechanism and migrations affect the distribution of most species. The distributional changes in turn affect the predator–prey relations and availability of proper food, thus affecting largely the inter-species interactions. Trophodynamics (feeding relations) and growth variations have been recognised as probably the most important aspects of interspecies interactions in the marine ecosystem (e.g. Andersen and Ursin, 1977). The interactions between growth and predation determine largely the source and sink areas of a given species. In the source area, growth exceeds predation and mortality; and, in sink areas, predation and mortality exceed growth. Examples of these effects are shown in section 5. The temperature affects concurrently the growth, uptake of food, and activity, including migration. Recent work of Jones and Hislop (1978) has provided new data on the effects of food availability and intake on growth and the effect of temperature on metabolic rate.

The ecosystem simulation thus provides a long-sought means of evaluating the environment-biota interactions and the effects of environmental anomalies in all space and time scales, including the study of the effects of climatic changes.

Complete environmental models such as hydrodynamical-numerical models require large computer core and considerable computer time. Thus, it is difficult to run these environmental models simultaneously with ecosystem models. The environmental data fields must either be prescribed (forced) in digital form from preanalysed data (e.g. monthly means) or environmental models must be run separately, storing their outputs on tapes or on discs from where they are read into ecosystem models in desired time steps.

Coupling of different simulations can also be done in biological subjects, such as coupling marine mammal and bird models and/or separate plankton and benthos dynamics simulation models with holistic ecosystem simulation which emphasise nekton ecosystems. The main coupling in this case is via predation. Properly constructed dynamical, time-dependent ecosystem models can use existing single-process and/or small species models or parts of them as adapted, integral parts of holistic simulation models. Exceptions from the above role occur in these holistic models which do not have a diagnostic phase (initial analysis), such as GEMBASE. In the latter models the results from single-process models are used as initial inputs for definition of the initial state and must thus be 'harmonised' (dynamically balanced) with the main model.

4 Simulation of equilibrium biomasses and their long-term fluctuations in marine ecosystems

Several methods are in use for assessment of marine fishery resources, such as direct surveys, virtual population analyses using catch and age composition data, etc. None of the past available methods are fully sufficient *per se* for resource evaluation and none of them give the biomasses of all species and/or ecological groups present, nor the productivity of their biomasses per unit time and/or area. The evaluation of total production, starting with primary production, has not been successful either as the pathways of organic matter transfer are greatly variable in space and time and not known quantitatively, and the concept of distinct trophic levels has been abandoned as unrealistic over-simplification for quantitative resource assessment.

If we, however, apply available empirical knowledge of food requirements of all species, composition of food, and computed growth as affected by a multitude of factors, we can apply an iterative method to find solution to the total utilisation of available food resources in the marine ecosystem. The main objectives of such simplified, essentially trophodynamic, bulk biomass models, are:

1   To determine the abundance of species and/or ecological groups in a given region with available diverse food resources (i.e. determine the carrying capacities and 'equilibrium biomasses' – see definition below – in respect to given species and regions.

2   To determine quantitatively the trophic couplings between different species or groups of species, and to evaluate the marine ecosystem stability.

The equilibrium biomass is defined as the level of the biomass of a given species, or an ecological group of species, which with a given plausible growth rate and plausible ecosystem internal consumption (i.e. lowest plausible food requirements), does neither decline nor increase within the course of a year; seasonal fluctuations are, however, allowed. Mathematically this means that we find a unique solution of a set of equations (eqns 14 and 15) if one species is predetermined.

Equilibrium biomasses in a given region are computed with the approaches and formulas given above (eqns 14 and 15). The basic differences are that only one region rather than each grid point is computed (i.e. no space resolution and no migrations), and that equilibrium conditions are assumed (i.e. growth equals removal by predation, fishery, and other mortalities). Among other limitations of this method are that food composition cannot vary in space and time, and that the obtained equilibrium biomasses are also to some extent dependent on the error in the initial estimate of one or more 'ascertained biomasses'.

Relaxation methods (Shaw, 1953) can be used to solve the equilibrium of ecosystem equation complex; however, a logical (for the particular problem adapted) iteration procedure for the adjustment of biomasses in each January that makes use of the two following criteria, is employed:

$$B_{cg} = B_g - [(B_{1,1} - B_{2,1})/k_c] \tag{20}$$
$$B_{cf} = [(C_1 + B_g)/B_{gg}]B_g \tag{21}$$

where k_c is an iteration constant (3.5 to 10) and

$$B_{gg} = B_g(2 - \exp^{-g}) \exp^{-m} \tag{22}$$

A mean of the above two is formed as the new adjusted biomass:

$$B_c = (B_{cg} + B_{cf})/2 \tag{23}$$

In most cases, 50 years and more computations in real time are required for convergence to a unique solution.

The biomass of one or more species which have been empirically ascertained (e.g. by extensive surveys of spawning biomass, sonar surveys, extensive exploratory fishery, etc.) ('ascertained biomasses') must be kept constant (i.e. are not passed through the iterative adjustment procedure); all other biomasses which were introduced as first guesses are changed in the iteration for a unique solution.

After achieving satisfactory convergence, the model can be run in a predictive mode for various investigations – e.g. long-term and cyclic changes in the ecosystem caused by the fishery and other factors, such as climatic changes. In the predictive mode, density (food availability) dependent feeding must also be used.

Some of the advantages of the bulk biomass model, as compared to some other models, such as virtual population analyses (see Ulltang, 1977), are:

1   It is possible to determine the equilibrium biomasses in little exploited, unexploited (virgin), and extensively exploited stocks with a known fishing mortality.

2   The total ecosystem in a given region is considered, with large-scale quantitative interspecies interactions. Thus it is possible to examine the effects of changing fishing intensity on target species as well as the indirect effects of such fisheries on species that are unfished or little fished, but are trophically related to target species.

3   A detailed, direct computation of predation mortality (i.e. direct determination of largest component of natural mortality) is made.

4   The time variable growth computation allows the simulation of large-scale effects of environmental anomalies via effects on growth.

In contrast to gridded models such as DYNUMES, the bulk biomass models do not allow any spatial resolution. The model is heavily dependent on good, reliable estimates of the quantitative composition of food of species and/or ecological groups.

5   Examples of results from some holistic ecosystem simulations

The marine ecosystem simulations and results from such simulations are location-dependent. Therefore, only a few examples of the results are given below.

An example of outputs from a bulk biomass ecosystem model simulation of equilibrium biomasses (carrying capacity) in the Kodiak Island area in the Gulf of Alaska is given in Table I. The fish biomasses in this table include also prefishery juveniles (the exploitable portion of biomass varies from species to species, e.g. herring 30%, yellowfin sole 46%, pollock 71% of the total biomass). The fish biomass decreases from coastal areas to offshore as expected. The turnover rates (consumption plus other mortality within a year divided by mean standing stock) varies from species to species (0.4 to 1.2), and is in average 0.82 for fish in this region.

TABLE I   Mean biomasses of ecological groups, their consumption and turnover rates in Kodiak area in the Gulf of Alaska as determinated within an ecosystem simulation (all values in tons km^{-2}, except turnover rate)

Species/ecological group and/or other subjects	Coastal areas and continental shelf	Continental slope (150–500 m)	Off continental slope (> 500 m)
Mean biomasses			
Herring	8.92	3.29	1.60
Other pelagic fish	15.56	13.39	6.97
Squids	3.11	2.53	1.35
Salmon	0.44	0.40	0.37
Rockfish	2.68	1.74	0.49
Gadids	7.55	5.10	1.32
Flatfish	3.96	1.99	0.43
Other demersal fish	4.91	3.55	0.66
Crustaceans (of commercial value)	9.03	3.82	1.32
Benthos ('fish food' benthos)	46.91	20.57	3.50
Total finfish	44.02	29.46	11.84
Turnover rates			
Finfish	0.87	0.83	0.75
Crustaceans (of commercial value)	1.17	1.12	1.12
Benthos ('fish food' benthos)	0.85	0.87	0.75
Annual phytoplankton production	1500	1350	1000
Mean phytoplankton standing crop	200	180	135
Annual zooplankton production	200	175	190
Mean zooplankton standing stock	42	35	40
Zooplankton consumption by nekton	165	125	53

The basic organic production in the Kodiak area has been conservatively assumed moderate (100 to 150 gCm^{-2}yr^{-1}). The zooplankton production on the continental shelf in the Kodiak area is about 13% of the phytoplankton production. This is in general agreement with Polyakova and Fedorov (1975) who found that zooplankton production in the White Sea was 5.6 to 15% of primary production.

The requirements of zooplankton as food by nekton communities are high in coastal and continental shelf regions. Two additional factors contribute to this apparently high demand of zooplankton. First, the zooplankton production estimates in the present model were conservative and may be too low. Secondly, the high zooplankton consumption and availability in coastal and slope areas relative to local production may be caused by shoreward transport of zooplankton in deeper layers from the open ocean by the upwelling type circulation that occurs in summer in this area. In addition meroplankton on the continental shelf that obtain part of their food from the surface of the sediment (Gammarids, Mysids and Harpacticoid Copepods), has been included in the benthos. Considering the above it seems plausible that plankton production can sustain the equilibrium biomasses of other marine ecological groups as computed in this model. Furthermore, it is apparent from the lack and uncertainty of plankton data that basic organic and plankton production cannot be used as a reliable sole basis for fish and other biomass production estimates and modeling.

Some additional observations can be made on the bases of data in Table I. First, the nektonic biomasses are greatly dependent upon each other (i.e. feeding upon each other) whereby the younger, juvenile stages provide the greatest contribution. The benthos on the continental shelf is another important food source for fish ecosystem. Advection of zooplankton from deep ocean by upwelling type circulation might contribute to the standing stock of zooplankton on many shelf areas. On the other hand, many pelagic fish and juveniles of semidemersal fish (e.g. pollock, hake), who depend on euphausids as food, spend part of their life feeding in offshore locations where euphausids are plentiful.

The four-dimensional DYNUMES simulation can produce a great variety of outputs. A somewhat smoothed distribution of Pacific herring in the eastern Bering Sea during February, computed with DYNUMES simulation, is given in Fig. 4a. The model estimates of equilibrium biomass of herring in the eastern Bering Sea is 2.75 million tons; the magnitude of annual fluctuation of this biomass is about 0.3 million tons. Shaboneev (1965) found the biomass of wintering herring north and northwest of the

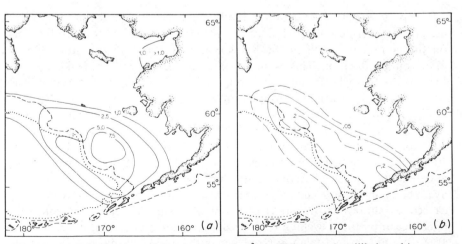

Fig. 4 (a) Distribution of herring (tons/km²) in February (equilibrium biomass, EB, 2.75 million tons), (b) Consumption of herring (tons/km²) in February (EB, 2.75 million tons)

Pribiloffs to be 2.16 million tons, which compares favorably with our model results. For comparison, the biomass of the herring in the North Sea has been estimated by Andersen and Ursin (1977) to be 1.8 million tons at the end of 1959.

The ecosystem internal consumption of herring, as computed within the simulation, is shown on Fig. 4b. Comparison of Figs a and b indicates that predation intensity is not necessarily a function of the density of prey, thus the source and sink areas of herring would be different than its distribution.

The spatial and temporal source-sink mapping provides useful information on many scientific as well as practical fisheries management considerations. Sources and sinks of herring in the eastern Bering Sea in February are presented on Fig. 5a. During the winter months losses of herring biomass exceed increases, except in a small source (increase) area in the southern part of the Bering Sea near the continental slope.

The effects of temperature anomalies in the eastern Bering Sea on the changes of biomass of pelagic fish were investigated with the DYNUMES simulation, using the effect of temperatures on growth. In one of the model runs, a $+1.5\,°C$ temperature anomaly in the surface layers during three winter months was prescribed. The sources and sinks of the herring biomass during February with $+1.5\,°C$ positive temperature anomaly is shown on Fig. 5b which depicts rather pronounced changes from normal conditions (see Fig. 5a). The magnitude of the effect of the temperature anomaly on the biomass changes (via physiological processes) appeared to be considerably

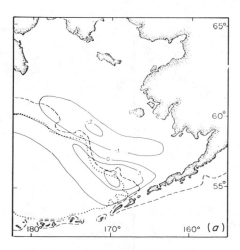 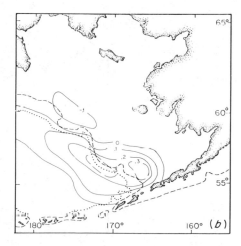

FIG. 5   (a) Herring sources and sinks (tons/km²) in February, (b) herring sources and sinks (tons/km²) in February (1.5°C positive temperature anomaly)

more significant than expected before computations and might explain a great part of the relatively large fluctuations of fish stocks observed in other high latitude areas. These effects can be studied with ecosystem simulation models.

The monthly mean zooplankton biomass (standing stock) rather than production is simulated in the DYNUMES model. Figure 6 shows the percentage of mean zooplankton standing stock consumed as food by fish, mammals, and birds in the eastern Bering Sea ecosystem in February. The utilisation of zooplankton in the northern part of the Bering Sea as well as over deep water, is relatively low, whereas in some parts over the shelf the utilisation exceeds 40% of the mean monthly standing stock.

DYNUMES model runs over several year-spans show that marine ecosystems are unstable and sensitive to changes in growth rates, relative distribution, and abundance of predators/prey, and changes of composition of food. Due to the multiple interactions in the ecosystem, the abundance and distribution of most species show quasi-cyclic variations. An example of changes of biomasses of three different size groups of pollock in the eastern Bering Sea with 1.2 million tons of 'present' fishery and half of this fishery over four and one-half years, is shown on Fig. 7. Medium-size and large pollock (>45 cm length) are cannibalistic (ca 50% food consists of younger pollock). When the fishery removes the older, cannibalistic fish, predation pressure on juveniles is relieved. As the growth in juveniles is high, a higher pollock biomass results.

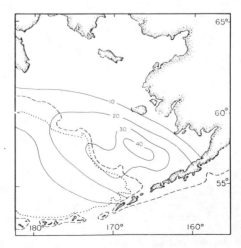

FIG. 6   Percentage of mean zooplankton standing stock consumed in February

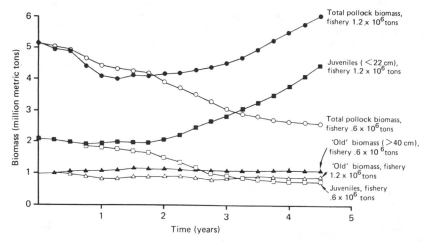

FIG. 7   Change of biomass with time of juvenile, old, and total walleye pollock in the eastern Bering Sea with simulated 1.2 million and 0.6 million tons annual fishery

An example of outputs from Andersen and Ursin (1977) model for the North Sea is shown in Table II, where the North Sea was also run in a virgin state (i.e. no fishery, column C). Comparing the virgin state with the states of ecosystem in 1959 and 1970 (columns A and B) shows that after the cessation of fish, control passed from man to cod, which, with a biomass of 4 million tons, exerted a predation pressure upon most other species with approximately the same effect as the fishing mortality in 1970. The biomass of plaice alone was too large to be effectively controlled by cod.

According to Longhurst (1978), the GEMBASE has produced conceptually realistic simulations of the ecology of the estuary. It has not been systematically exploited as a research tool as yet. However, during model development it has become apparent that it can be used to investigate a variety of ecological questions including theories on ecological relationships. An example of GEMBASE output is given in Fig. 8.

6   Verification and validation

It is necessary to differentiate between verification and validation in large ecosystem simulations. Verification refers to checking of logic and the correctness of individual models and formulas used in the simulation. The models and formulas are verified with available empirical data. Verification includes also the testing of the simulation at large, using various impulses as input, whereby the expected response of the ecosystem to the impulse must

TABLE II   Estimated biomass at the end of the year in 1959, in 1970 and for the almost virgin population (i.e. after 11 years without fishing).
Unit: 1 million tons. Zero means less than 50 000 tons

	Realistic A 1959	fishing effort B 1970	No fishing C
Plaice	0.3	0.4	2.2
Dab	0.3	0.4	0.6
Long rough dab	0.1	0.1	0.2
Saithe	0.1	0.6	0.4
Cod	0.2	0.4	4.2
Haddock	0.1	0.3	0.5
Whiting	0.1	0.1	0.3
Norway pout	0.1	0.3	1.0
Mackerel	2.6	0.6	0.2
Herring	1.8	1.0	0.9
Sandeels	0.2	0.5	1.1
Benthos A	5.7	4.8	4.4
Benthos B	6.6	9.7	8.6
Benthos C	1.6	2.0	1.4
Zooplankton A	0.0	0.2	0.0
Zooplankton B	2.4	1.5	2.1
Zooplankton C	2.0	1.9	2.1
Algae, pelagic	0.5	0.7	0.6
Algae, demersal	0.2	0.2	0.2
Detritus (demersal)	0.1	0.1	0.1
Carcass (demersal)	0.1	0.1	0.1
Fish in all	5.8	4.8	11.6
Benthos in all	14.0	16.5	14.3
Zooplankton in all	4.5	3.6	4.2
Animals in all	24.2	24.9	30.2
Total	25.1	26.0	31.2

be at least qualitatively known. The effect of water temperature on growth in the DYNUMES model was formulated on bases of some earlier available knowledge on the subject, notably Krogh's metabolic curve. When an excellent paper by Jones and Hislop (1978) appeared later, dealing partly with the subject, verification and additional tuning was provided. Further empirical evidence on the effect of temperature on herring abundance was provided by Grainger (1978).

An important part of the verification is the sensitivity analyses. Sensitivity analyses indicates where the influence of possible flaws in the available knowledge has major consequences, thus sensitivity analysis acts also as guidance for further research.

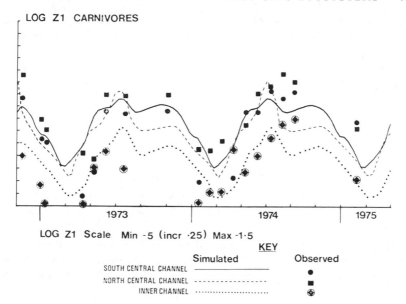

FIG. 8   Example of validation of GEMBASE model (Longhurst and Radford, 1978)

In the past sensitivity analysis in simple, few-parameter models was done by changing one constant (*vice* parameter) at a time. This method is not applicable to large ecosystem models as the number of simulations required is prohibitively large. Behrens (1978) has outlined an analytical method for sensitivity analyses for models consisting of ordinary first order differential equations.

In these ecosystem models which rely heavily on various input parameters for determination of the initial state and for satisfying the parameterised equations in the model, such as GEMBASE, the verification would include the testing of accuracy and reliability of such inputs which are obtained from a multitude of sources.

As both the bulk biomass and the DYNUMES models are deterministic, conventional sensitivity analyses are neither necessary nor possible. In deterministic models, sensitivity analyses become studies of specific responses of the systems to expected changes of parameters (see example in previous section and Fig. 7). The variations in results of deterministic models depend also on accuracy and reliability of the input data, but to a lesser degree than the 'parameterised' models.

Validation of simulation refers to comparison of principal results from simulation with direct observations in the field. Usually these results present either abundance and/or distribution changes of given species if and when a causative factor for these changes has been introduced in the simulation model. Special research projects usually provide validation of the various rate parameters.

One of the basic validation procedures is to compare observed data, e.g. from exploratory surveys with simulation results. Table III presents example of such validation in respect to simulation results in Table I and survey results from the same area by Ronholt *et al.* (1978). Besides errors and uncertainties in interpretation of survey results themselves, additional difficulties arise especially in the use of a catchability coefficient, which is variable in space and time and not well known. Despite these difficulties, the mean equilibrium biomasses computed with the bulk biomass model are in reasonable agreement with mean biomasses of survey results, as adjusted by catchability coefficient and total biomass conversion factors. The largest discrepancy occurs in flatfishes.

Figure 8 presents partial results of GEMBASE model validation with observed data (Longhurst and Radford, 1978). Some indirect validation of simulation output at large in respect to zooplankton utilisation and carrying capacity evaluation, were described in section 5.

7 Use of ecosystem simulations in research guidance and in resource management

The compilation (designing) of the ecosystem simulation serves already as general research guidance, indicating missing data and other shortcomings in data and knowledge. The simulation models serve also as indicators of priorities of research, by suggesting processes of greater importance in terms of greater 'sensitivity' to resultants of larger processes and those of lesser concern and influence. Some of the guidances, in respect to need for new emphasis and direction of research, are local, but many are universal and promote diversity of research as well. Among the examples of new thrust in fisheries research caused by the ecosystem approach is the realisation of shortcomings in single species approaches which lack trophodynamic interactions between the species. However, one of the most important links in interspecies interactions in ecosystem models is the food relations affecting growth and abundance; another example is the realisation of the necessity to deal with age/size dependent mortalities, to ascertain the predation mortalities as well as spawning stress mortalities in ecosystem models.

Quantitative numerical ecosystem simulations bring out processes and resulting changes in the ecosystem which have not been easily observed in the past and permit ecological experiments which would be impossible to conduct in nature.

TABLE III Comparison of total biomass of some groups of species in Kodiak area as computed with Bulk Biomass (BB) model (Table I) and survey data (Ronholt et al., 1978) as adjusted with catchability coefficient and converted to total biomass

Group of species	Exploitable biomass, catchability coefficient = 1, Ronholt et al. (1978)			Adjusted catchability coefficient	Exploitable biomass of total biomass (%)	Adjusted mean biomass (surveys)	Mean biomass from BB model (Table I)
	1960 decade	1970 decade	Mean				
Flatfish	3.4	3.6	3.5	0.75	60	7.78	4.0
Roundfish	1.9	5.3	3.6	0.55	70	9.35	7.6
Rockfish	0.7	0.2	0.45	0.40	50	2.25	2.7
Invertebrates (mainly crustaceans)	2.7	1.6	2.15	0.50	35	12.29	9.0

Among the initial scientific uses of the holistic simulations have been the study of the quantitative effects of environmental anomalies, pollutants, and the fishery on the biotic components of the ecosystem. Furthermore, the determination of the 'carrying capacities' of given regions and the study of fluctuations of abundance of species as caused by various ecosystem internal factors (e.g. cannibalism) using ecosystem simulations has been successfully demonstrated.

The marine ecosystem simulations with emphasis on fish ecosystems provide new powerful tools for fisheries management. These simulations not only allow the determination of magnitudes of the resources and their distributions, but also the simulation of variable space and time responses to any desired and/or prescribed fishery, on target species, as well as indirectly on other species as well, via interspecies interactions. The ecosystem simulations have shown the importance of the determination of the magnitudes and periods of large-scale 'natural fluctuations' in the marine ecosystem which can occur without the influence of fishery but can be caused by fishery as well. Without proper evaluation of these fluctuations, the effects of the fishery on the abundance and distribution of the species cannot be evaluated either (see example of the fluctuations of pollock biomass in section 5).

The applications of ecosystem simulations are indeed numerous and far from being fully explored and utilised. We might visualise these seemingly unlimited possibilities if we consider that in essence we attempt to simulate nature (i.e. the ecosystem) and its functioning quantitatively in computers and can review the whole system of the nature on our desk.

8 List of symbols

A_g	half of the annual range of growth coefficient change
$A_{1,r}$	half-range of annual main (spring) plankton maximum
$A_{2,r}$	half-range of annual secondary (fall) plankton maximum
a_j	fraction of species j requirement as food satisfied (taken from available biomass)
$B, (B_i)$	biomass (of species i)
B_a	biomass of apex predator a
B_c	adjusted biomass in January
B_{cg}	adjusted biomass due to biomass change within a year
B_{cf}	adjusted biomass due to consumption (predation)
B_g	adjusted biomass from previous year
B_{gg}	resultant biomass from growth and mortality
$B_{1,1}$	biomass in January, previous year
$B_{2,1}$	biomass in January, actual year

b_m	fraction of food, substituted
C_i	consumption of species i
$C_{i,a}$	consumption of species i by apex predator a
$C_{i,A}$	consumption of species i by all apex predators
c_j	fraction of the requirement for species j as food, which cannot be satisfied due to low food concentration in given location, and is added to starvation array
C_1	predation (consumption)
$C_{i(t-1)}$	predation (consumption of species i in previous time step $(t-1)$)
e	base of natural logarithms
F_a	amount of food consumed by an apex predator a
F_i	amount of food required by species i
$F_{i,j}$	amount of species j (required) in the food of species i
f_i	time and space dependent fishing mortality coefficient
g	growth coefficient
g_0	basic mean growth coefficient
k_c	iteration constant
k_0	fraction of biomass leaving 'unfavorable' grid point
k_s	fraction of biomass arriving at given grid point from neighboring 'unfavorable' grid point
k_t	coefficient
l	grid length (km)
m	mortality coefficient (from old age and diseases); also space coordinate
n	space coordinate
$p_{i,a}$	decimal fraction of species i in the food of apex species a
$p_{i,j}$	fraction of species j in the food of species i
$P_{r,0}$	annual mean of plankton standing stock in a subregion
P_s	plankton standing stock
P_u	catch (fishery)
q_a	food requirement of apex predator a (in % of body weight daily)
q_i	food requirement of species i (also food requirement for maintenance in % of body weight daily)
$q_{i,j}$	fraction (decimal) of species i in food of species j
r_i	ratio of growth to food required for growth
S_i	shortage of food of species i ('starvation')
$t_1(t_d)$	time, time step
T	temperature
U	u component of migration speed
UT	'upcurrent' (upmigration) gradient of biomass (u component)
V	v component of migration speed

VT 'upcurrent' (upmigration) gradient of biomass (v component)
α smoothing coefficient (horizontal diffusion coefficient)
$\alpha_1(\alpha_2)$ phase speed (e.g. degrees per month)
β $(1 - \alpha)/4$ (secondary smoothing coefficient)
κ_g phase lag of annual growth coefficient change
κ_1, κ_2 phase lag (month of annual maximum)

References

Andersen, K. P. and Ursin, E. (1977). A multispecies extension to the Beverton and Holt theory of fishing with accounts of phosphorus circulation and primary production. *Meddr. Danm. Fisk.-og Havunders.* N.S. **7**, 319–435

Behrens, J. Chr. (1978). A semi-analytical sensitivity analysis of non-linear systems. Inst. of Mathem. Stat. and Oper. Res., Techn. Univ. Denmark, Res. Rpt. 4/1978

Beverton, R. J. H. and Holt, S. J. (1957). On the dynamics of exploited fish populations. *Min. Agr. Fish. and Food, Fish. Invest. London.* Ser. 2 **19**, 1–533

Daan, N. (1973). A quantitative analysis of the food intake of North Sea cod, *Gadus morhua. Netherlands J. of Sea Res.* **6** (4), 479–517

Grainger, R. J. R. (1978). Herring abundance off the west of Ireland in relation to oceanographic variation. *J. Cons. int. Explor. Mer.* **38** (2), 180–188

Jones, R. (1978). Estimates of the food consumption of haddock (*Melanogrammus aeglefinus*) and cod (*Gadus morhua*). *J. Cons. int. Explor. Mer* **38** (1), 18–27

Jones, R. and Hislop, J. R. G. (1978). Further observations on the relation between food intake and growth of gadoids in captivity. *J. Cons. int. Explor. Mer* **38** (2), 244–251

Kremer, T. N. and Nixon, S. W. (1977). "A Coastal Marine Ecosystem. Simulation and Analysis". Springer-Verlag, New York

Laevastu, T. and Favorite, F. (1978). Fish biomass parameter estimations. NWAFC Proc. Rpt.

Longhurst, A. R. (1978). Ecological models in estuarine management. *Ocean Management* **4**, 287–302

Longhurst, A. R. and Radford, P. J. (1978). GEMBASE I. Internal documents, Inst. for Mar. Env. Res., Plymouth

Polyakova, T. P. and Fedorov, V. D. (1975). Production of individual links in the food chain in the White Sea. *Okeanologia* **15** (5), 881–885

Ronholt, L. L., Shippen, H. H. and Brown, E. S. (1978). Demersal fish and shellfish resources of the Gulf of Alaska from Cape Spencer to Unimak Pass 1948–1976 (A historical review). Northwest and Alaska Fisheries Center, Processed Report, August 1978

Shaboneev, I. E. (1965). O biologii i promysle sel'di vostochnoi chasti Beringova morya (Biology and fishing of herring in the eastern part of the Bering Sea). *Tr. Vses. Nauchno-issled. Inst. Morsk. Rybn. Khoz. Okeanogr.* **58** (*Izv. Tikhookean. Nauchno-issled. Inst. Morsk. Rybn. Khoz. Okeanogr.* **53**) pp. 139–154. In Russian. (Transl. by Israel Prog. Sci. Transl., 1968, pp. 130–146 *in* P. A. Moiseev, ed., Soviet fisheries investigations in the northeast Pacific, Pt. 4, avail. Natl. Tech. Inf. Serv., Springfield, Va. as TT 67-51206)

Shaw, F. S. (1953). "An Introduction to Relaxation Methods". Dover, New York

Ulltang, Ø. (1977). Sources of errors in and limitations of Virtual Population Analysis (Cohort Analysis). *J. Cons. int. Explor. Mer* **37** (3), 249–260

Ware, D. M. (1975). Relation between egg size, growth, and natural mortality of larval fish. *J. Fish. Res. Bd Can.* **32**, 2503–2512

Subject Index